T0213743

# Lecture Notes in Computer Science 8583

Commenced Publication in 1973
Founding and Former Series Editors:
Gerhard Goos, Juris Hartmanis, and Jan van Leeuwen

Beniamino Murgante   Sanjay Misra
Ana Maria A.C. Rocha   Carmelo Torre
Jorge Gustavo Rocha   Maria Irene Falcão
David Taniar   Bernady O. Apduhan
Osvaldo Gervasi (Eds.)

# Computational Science and Its Applications – ICCSA 2014

14th International Conference
Guimarães, Portugal, June 30 – July 3, 2014
Proceedings, Part V

 Springer

Volume Editors

*Beniamino Murgante,* University of Basilicata, Potenza, Italy
E-mail: beniamino.murgante@unibas.it

*Sanjay Misra,* Covenant University, Ota, Nigeria
E-mail: sanjay.misra@covenantuniversity.edu.ng

*Ana Maria A.C. Rocha,* University of Minho, Braga, Portugal
E-mail: arocha@dps.uminho.pt

*Carmelo Torre,* Politecnico di Bari, Bari, Italy
E-mail: torre@poliba.it

*Jorge Gustavo Rocha,* University of Minho, Braga, Portugal
E-mail: jgr@di.uminho.pt

*Maria Irene Falcão,* University of Minho, Braga, Portugal
E-mail: mif@math.uminho.pt

*David Taniar,* Monash University, Clayton, VIC, Australia
E-mail: david.taniar@infotech.monash.edu.au

*Bernady O. Apduhan,* Kyushu Sangyo University, Fukuoka, Japan
E-mail: bob@is.kyusan-u.ac.jp

*Osvaldo Gervasi,* University of Perugia, Perugia, Italy
E-mail: osvaldo.gervasi@unipg.it

ISSN 0302-9743                    e-ISSN 1611-3349
ISBN 978-3-319-09155-6           e-ISBN 978-3-319-09156-3
DOI 10.1007/978-3-319-09156-3
Springer Cham Heidelberg New York Dordrecht London

Library of Congress Control Number: 2014942987

LNCS Sublibrary: SL 1 – Theoretical Computer Science and General Issues

*Typesetting:* Camera-ready by author, data conversion by Scientific Publishing Services, Chennai, India

Printed on acid-free paper

Springer is part of Springer Science+Business Media (www.springer.com)

# Welcome Message

On behalf of the Local Organizing Committee of ICCSA 2014, it is a pleasure to welcome you to the 14th International Conference on Computational Science and Its Applications, held during June 30 – July 3, 2014. We are very proud and grateful to the ICCSA general chairs for having entrusted us with the task of organizing another event of this series of very successful conferences.

ICCSA will take place in the School of Engineering of University of Minho, which is located in close vicinity to the medieval city centre of Guimarães, a UNESCO World Heritage Site, in Northern Portugal. The historical city of Guimarães is recognized for its beauty and historical monuments. The dynamic and colorful Minho Region is famous for its landscape, gastronomy and vineyards where the unique *Vinho Verde* wine is produced.

The University of Minho is currently among the most prestigious institutions of higher education in Portugal and offers an excellent setting for the conference. Founded in 1973, the University has two major poles: the campus of Gualtar in Braga, and the campus of Azurém in Guimarães.

Plenary lectures by leading scientists and several workshops will provide a real opportunity to discuss new issues and find advanced solutions able to shape new trends in computational science.

Apart from the scientific program, a stimulant and diverse social program will be available. There will be a welcome drink at Instituto de Design, located in an old Tannery, that is an open knowledge centre and a privileged communication platform between industry and academia. Guided visits to the city of Guimarães and Porto are planned, both with beautiful and historical monuments. A guided tour and tasting in Porto wine cellars, is also planned. There will be a gala dinner at the Pousada de Santa Marinha, which is an old Augustinian convent of the 12th century refurbished, where ICCSA participants can enjoy delicious dishes and enjoy a wonderful view over the city of Guimarães.

The conference could not have happened without the dedicated work of many volunteers, recognized by the coloured shirts. We would like to thank all the collaborators, who worked hard to produce a successful ICCSA 2014, namely Irene Falcão and Maribel Santos above all, our fellow members of the local organization.

On behalf of the Local Organizing Committee of ICCSA 2014, it is our honor to cordially welcome all of you to the beautiful city of Guimarães for this unique event. Your participation and contribution to this conference will make it much more productive and successful.

We are looking forward to see you in Guimarães.

Sincerely yours,

Ana Maria A.C. Rocha
Jorge Gustavo Rocha

# Preface

These 6 volumes (LNCS volumes 8579-8584) consist of the peer-reviewed papers from the 2014 International Conference on Computational Science and Its Applications (ICCSA 2014) held in Guimarães, Portugal during 30 June – 3 July 2014.

ICCSA 2014 was a successful event in the International Conferences on Computational Science and Its Applications (ICCSA) conference series, previously held in Ho Chi Minh City, Vietnam (2013), Salvador da Bahia, Brazil (2012), Santander, Spain (2011), Fukuoka, Japan (2010), Suwon, South Korea (2009), Perugia, Italy (2008), Kuala Lumpur, Malaysia (2007), Glasgow, UK (2006), Singapore (2005), Assisi, Italy (2004), Montreal, Canada (2003), and (as ICCS) Amsterdam, The Netherlands (2002) and San Francisco, USA (2001).

Computational science is a main pillar of most of the present research, industrial and commercial activities and plays a unique role in exploiting ICT innovative technologies, and the ICCSA conference series has been providing a venue for researchers and industry practitioners to discuss new ideas, to share complex problems and their solutions, and to shape new trends in computational science.

Apart from the general track, ICCSA 2014 also included 30 workshops, in various areas of computational sciences, ranging from computational science technologies, to specific areas of computational sciences, such as computational geometry and security. We accepted 58 papers for the general track, and 289 in workshops. We would like to show our appreciation to the workshops chairs and co-chairs.

The success of the ICCSA conference series, in general, and ICCSA 2014, in particular, was due to the support of many people: authors, presenters, participants, keynote speakers, workshop chairs, Organizing Committee members, student volunteers, Program Committee members, Advisory Committee members, international liaison chairs, and people in other various roles. We would like to thank them all.

We also thank our publisher, Springer–Verlag, for their acceptance to publish the proceedings and for their kind assistance and cooperation during the editing process.

We cordially invite you to visit the ICCSA website http://www.iccsa.org where you can find all relevant information about this interesting and exciting event.

June 2014

Osvaldo Gervasi
Jorge Gustavo Rocha
Bernady O. Apduhan

# Organization

ICCSA 2014 was organized by University of Minho, (Portugal) University of Perugia (Italy), University of Basilicata (Italy), Monash University (Australia), Kyushu Sangyo University (Japan).

## Honorary General Chairs

| | |
|---|---|
| Antonio M. Cunha | Rector of the University of Minho, Portugal |
| Antonio Laganà | University of Perugia, Italy |
| Norio Shiratori | Tohoku University, Japan |
| Kenneth C. J. Tan | Qontix, UK |

## General Chairs

| | |
|---|---|
| Beniamino Murgante | University of Basilicata, Italy |
| Ana Maria A.C. Rocha | University of Minho, Portugal |
| David Taniar | Monash University, Australia |

## Program Committee Chairs

| | |
|---|---|
| Osvaldo Gervasi | University of Perugia, Italy |
| Bernady O. Apduhan | Kyushu Sangyo University, Japan |
| Jorge Gustavo Rocha | University of Minho, Portugal |

## International Advisory Committee

| | |
|---|---|
| Jemal Abawajy | Daekin University, Australia |
| Dharma P. Agrawal | University of Cincinnati, USA |
| Claudia Bauzer Medeiros | University of Campinas, Brazil |
| Manfred M. Fisher | Vienna University of Economics and Business, Austria |
| Yee Leung | Chinese University of Hong Kong, China |

## International Liaison Chairs

| | |
|---|---|
| Ana Carla P. Bitencourt | Universidade Federal do Reconcavo da Bahia, Brazil |
| Claudia Bauzer Medeiros | University of Campinas, Brazil |
| Alfredo Cuzzocrea | ICAR-CNR and University of Calabria, Italy |

| Marina L. Gavrilova | University of Calgary, Canada |
| Robert C. H. Hsu | Chung Hua University, Taiwan |
| Andrés Iglesias | University of Cantabria, Spain |
| Tai-Hoon Kim | Hannam University, Korea |
| Sanjay Misra | University of Minna, Nigeria |
| Takashi Naka | Kyushu Sangyo University, Japan |
| Rafael D.C. Santos | National Institute for Space Research, Brazil |

## Workshop and Session Organizing Chairs

| Beniamino Murgante | University of Basilicata, Italy |

## Local Organizing Committee

| Ana Maria A.C. Rocha | University of Minho, Portugal (Chair) |
| Jorge Gustavo Rocha | University of Minho, Portugal |
| Maria Irene Falcão | University of Minho, Portugal |
| Maribel Yasmina Santos | University of Minho, Portugal |

## Workshop Organizers

## Advances in Complex Systems: Modeling and Parallel Implementation (ACSModPar 2014)

| Georgius Sirakoulis | Democritus University of Thrace, Greece |
| Wiliam Spataro | University of Calabria, Italy |
| Giuseppe A. Trunfio | University of Sassari, Italy |

## Agricultural and Environment Information and Decision Support Systems (AEIDSS 2014)

| Sandro Bimonte | IRSTEA France |
| Florence Le Ber | ENGES, France |
| André Miralles | IRSTEA France |
| François Pinet | IRSTEA France |

## Advances in Web Based Learning (AWBL 2014)

| Mustafa Murat Inceoglu | Ege University, Turkey |

# Bio-inspired Computing and Applications (BIOCA 2014)

Nadia Nedjah                    State University of Rio de Janeiro, Brazil
Luiza de Macedo Mourell         State University of Rio de Janeiro, Brazil

# Computational and Applied Mathematics (CAM 2014)

Maria Irene Falcao              University of Minho, Portugal
Fernando Miranda                University of Minho, Portugal

# Computer Aided Modeling, Simulation, and Analysis (CAMSA 2014)

Jie Shen                        University of Michigan, USA

# Computational and Applied Statistics (CAS 2014)

Ana Cristina Braga              University of Minho, Portugal
Ana Paula Costa Conceicao
  Amorim                        University of Minho, Portugal

# Computational Geometry and Security Applications (CGSA 2014)

Marina L. Gavrilova             University of Calgary, Canada
Han Ming Huang                  Guangxi Normal University, China

# Computational Algorithms and Sustainable Assessment (CLASS 2014)

Antonino Marvuglia              Public Research Centre Henri Tudor,
                                  Luxembourg
Beniamino Murgante              University of Basilicata, Italy

# Chemistry and Materials Sciences and Technologies (CMST 2014)

Antonio Laganà                  University of Perugia, Italy

# Computational Optimization and Applications (COA 2014)

Ana Maria A.C. Rocha          University of Minho, Portugal
Humberto Rocha               University of Coimbra, Portugal

# Cities, Technologies and Planning (CTP 2014)

Giuseppe Borruso             University of Trieste, Italy
Beniamino Murgante           University of Basilicata, Italy

# Computational Tools and Techniques for Citizen Science and Scientific Outreach (CTTCS 2014)

Rafael Santos                National Institute for Space Research, Brazil
Jordan Raddickand            Johns Hopkins University, USA
Ani Thakar                   Johns Hopkins University, USA

# Econometrics and Multidimensional Evaluation in the Urban Environment (EMEUE 2014)

Carmelo M. Torre             Polytechnic of Bari, Italy
Maria Cerreta               University of Naples Federico II, Italy
Paola Perchinunno            University of Bari, Italy
Simona Panaro               University of Naples Federico II, Italy
Raffaele Attardi             University of Naples Federico II, Italy

# Future Computing Systems, Technologies, and Applications (FISTA 2014)

Bernady O. Apduhan           Kyushu Sangyo University, Japan
Rafael Santos                National Institute for Space Research, Brazil
Jianhua Ma                   Hosei University, Japan
Qun Jin                      Waseda University, Japan

# Formal Methods, Computational Intelligence and Constraint Programming for Software Assurance (FMCICA 2014)

Valdivino Santiago Junior    National Institute for Space Research
                             (INPE), Brazil

## Geographical Analysis, Urban Modeling, Spatial Statistics (GEOG-AN-MOD 2014)

| | |
|---|---|
| Giuseppe Borruso | University of Trieste, Italy |
| Beniamino Murgante | University of Basilicata, Italy |
| Hartmut Asche | University of Potsdam, Germany |

## High Performance Computing in Engineering and Science (HPCES 2014)

| | |
|---|---|
| Alberto Proenca | University of Minho, Portugal |
| Pedro Alberto | University of Coimbra, Portugal |

## Mobile Communications (MC 2014)

| | |
|---|---|
| Hyunseung Choo | Sungkyunkwan University, Korea |

## Mobile Computing, Sensing, and Actuation for Cyber Physical Systems (MSA4CPS 2014)

| | |
|---|---|
| Saad Qaisar | NUST School of Electrical Engineering and Computer Science, Pakistan |
| Moonseong Kim | Korean Intellectual Property Office, Korea |

## New Trends on Trust Computational Models (NTTCM 2014)

| | |
|---|---|
| Rui Costa Cardoso | Universidade da Beira Interior, Portugal |
| Abel Gomez | Universidade da Beira Interior, Portugal |

## Quantum Mechanics: Computational Strategies and Applications (QMCSA 2014)

| | |
|---|---|
| Mirco Ragni | Universidad Federal de Bahia, Brazil |
| Vincenzo Aquilanti | University of Perugia, Italy |
| Ana Carla Peixoto Bitencourt | Universidade Estadual de Feira de Santana Brazil |
| Roger Anderson | University of California, USA |
| Frederico Vasconcellos Prudente | Universidad Federal de Bahia, Brazil |

## Remote Sensing Data Analysis, Modeling, Interpretation and Applications: From a Global View to a Local Analysis (RS2014)

Rosa Lasaponara                    Institute of Methodologies for Environmental
                                   Analysis National Research Council, Italy
Nicola Masini                      Archaeological and Monumental Heritage
                                   Institute, National Research Council, Italy

## Software Engineering Processes and Applications (SEPA 2014)

Sanjay Misra                       Covenant University, Nigeria

## Software Quality (SQ 2014)

Sanjay Misra                       Covenant University, Nigeria

## Advances in Spatio-Temporal Analytics (ST-Analytics 2014)

Joao Moura Pires                   New University of Lisbon, Portugal
Maribel Yasmina Santos             New University of Lisbon, Portugal

## Tools and Techniques in Software Development Processes (TTSDP 2014)

Sanjay Misra                       Covenant University, Nigeria

## Virtual Reality and its Applications (VRA 2014)

Osvaldo Gervasi                    University of Perugia, Italy
Lucio Depaolis                     University of Salento, Italy

## Workshop of Agile Software Development Techniques (WAGILE 2014)

Eduardo Guerra                     National Institute for Space Research, Brazil

## Big Data:, Analytics and Management (WBDAM 2014)

Wenny Rahayu                       La Trobe University, Australia

# Program Committee

| | |
|---|---|
| Jemal Abawajy | Daekin University, Australia |
| Kenny Adamson | University of Ulster, UK |
| Filipe Alvelos | University of Minho, Portugal |
| Paula Amaral | Universidade Nova de Lisboa, Portugal |
| Hartmut Asche | University of Potsdam, Germany |
| Md. Abul Kalam Azad | University of Minho, Portugal |
| Michela Bertolotto | University College Dublin, Ireland |
| Sandro Bimonte | CEMAGREF, TSCF, France |
| Rod Blais | University of Calgary, Canada |
| Ivan Blecic | University of Sassari, Italy |
| Giuseppe Borruso | University of Trieste, Italy |
| Yves Caniou | Lyon University, France |
| José A. Cardoso e Cunha | Universidade Nova de Lisboa, Portugal |
| Leocadio G. Casado | University of Almeria, Spain |
| Carlo Cattani | University of Salerno, Italy |
| Mete Celik | Erciyes University, Turkey |
| Alexander Chemeris | National Technical University of Ukraine "KPI", Ukraine |
| Min Young Chung | Sungkyunkwan University, Korea |
| Gilberto Corso Pereira | Federal University of Bahia, Brazil |
| M. Fernanda Costa | University of Minho, Portugal |
| Gaspar Cunha | University of Minho, Portugal |
| Alfredo Cuzzocrea | ICAR-CNR and University of Calabria, Italy |
| Carla Dal Sasso Freitas | Universidade Federal do Rio Grande do Sul, Brazil |
| Pradesh Debba | The Council for Scientific and Industrial Research (CSIR), South Africa |
| Hendrik Decker | Instituto Tecnológico de Informática, Spain |
| Frank Devai | London South Bank University, UK |
| Rodolphe Devillers | Memorial University of Newfoundland, Canada |
| Prabu Dorairaj | NetApp, India/USA |
| M. Irene Falcao | University of Minho, Portugal |
| Cherry Liu Fang | U.S. DOE Ames Laboratory, USA |
| Edite M.G.P. Fernandes | University of Minho, Portugal |
| Jose-Jesus Fernandez | National Centre for Biotechnology, CSIS, Spain |
| Maria Antonia Forjaz | University of Minho, Portugal |
| Maria Celia Furtado Rocha | PRODEB and Universidade Federal da Bahia, Brazil |
| Akemi Galvez | University of Cantabria, Spain |
| Paulino Jose Garcia Nieto | University of Oviedo, Spain |
| Marina Gavrilova | University of Calgary, Canada |
| Jerome Gensel | LSR-IMAG, France |

| | |
|---|---|
| Maria Giaoutzi | National Technical University, Athens, Greece |
| Andrzej M. Goscinski | Deakin University, Australia |
| Alex Hagen-Zanker | University of Cambridge, UK |
| Malgorzata Hanzl | Technical University of Lodz, Poland |
| Shanmugasundaram Hariharan | B.S. Abdur Rahman University, India |
| Eligius M.T. Hendrix | University of Malaga/Wageningen University, Spain/Netherlands |
| Hisamoto Hiyoshi | Gunma University, Japan |
| Fermin Huarte | University of Barcelona, Spain |
| Andres Iglesias | University of Cantabria, Spain |
| Mustafa Inceoglu | EGE University, Turkey |
| Peter Jimack | University of Leeds, UK |
| Qun Jin | Waseda University, Japan |
| Farid Karimipour | Vienna University of Technology, Austria |
| Baris Kazar | Oracle Corp., USA |
| DongSeong Kim | University of Canterbury, New Zealand |
| Taihoon Kim | Hannam University, Korea |
| Ivana Kolingerova | University of West Bohemia, Czech Republic |
| Dieter Kranzlmueller | LMU and LRZ Munich, Germany |
| Antonio Laganà | University of Perugia, Italy |
| Rosa Lasaponara | National Research Council, Italy |
| Maurizio Lazzari | National Research Council, Italy |
| Cheng Siong Lee | Monash University, Australia |
| Sangyoun Lee | Yonsei University, Korea |
| Jongchan Lee | Kunsan National University, Korea |
| Clement Leung | Hong Kong Baptist University, Hong Kong |
| Chendong Li | University of Connecticut, USA |
| Gang Li | Deakin University, Australia |
| Ming Li | East China Normal University, China |
| Fang Liu | AMES Laboratories, USA |
| Xin Liu | University of Calgary, Canada |
| Savino Longo | University of Bari, Italy |
| Tinghuai Ma | NanJing University of Information Science and Technology, China |
| Sergio Maffioletti | University of Zurich, Switzerland |
| Ernesto Marcheggiani | Katholieke Universiteit Leuven, Belgium |
| Antonino Marvuglia | Research Centre Henri Tudor, Luxembourg |
| Nicola Masini | National Research Council, Italy |
| Nirvana Meratnia | University of Twente, The Netherlands |
| Alfredo Milani | University of Perugia, Italy |
| Sanjay Misra | Federal University of Technology Minna, Nigeria |
| Giuseppe Modica | University of Reggio Calabria, Italy |

| | |
|---|---|
| José Luis Montaña | University of Cantabria, Spain |
| Beniamino Murgante | University of Basilicata, Italy |
| Jiri Nedoma | Academy of Sciences of the Czech Republic, Czech Republic |
| Laszlo Neumann | University of Girona, Spain |
| Kok-Leong Ong | Daekin University, Australia |
| Belen Palop | Universidad de Valladolid, Spain |
| Marcin Paprzycki | Polish Academy of Sciences, Poland |
| Eric Pardede | La Trobe University, Australia |
| Kwangjin Park | Wonkwang University, Korea |
| Ana Isabel Pereira | Polytechnic Institute of Braganca, Portugal |
| Maurizio Pollino | Italian National Agency for New Technologies, Energy and Sustainable Economic Development, Italy |
| Alenka Poplin | University of Hamburg, Germany |
| Vidyasagar Potdar | Curtin University of Technology, Australia |
| David C. Prosperi | Florida Atlantic University, USA |
| Wenny Rahayu | La Trobe University, Australia |
| Jerzy Respondek | Silesian University of Technology, Poland |
| Ana Maria A.C. Rocha | University of Minho, Portugal |
| Humberto Rocha | INESC Coimbra, Portugal |
| Alexey Rodionov | Institute of Computational Mathematics and Mathematical Geophysics, Russia |
| Cristina S. Rodrigues | University of Minho, Portugal |
| Octavio Roncero | CSIC, Spain |
| Maytham Safar | Kuwait University, Kuwait |
| Chiara Saracino | A.O. Ospedale Niguarda Ca' Granda - Milano, Italy |
| Haiduke Sarafian | The Pennsylvania State University, USA |
| Jie Shen | University of Michigan, USA |
| Qi Shi | Liverpool John Moores University, UK |
| Dale Shires | U.S. Army Research Laboratory, USA |
| Takuo Suganuma | Tohoku University, Japan |
| Sergio Tasso | University of Perugia, Italy |
| Ana Paula Teixeira | University of Tras-os-Montes and Alto Douro, Portugal |
| Senhorinha Teixeira | University of Minho, Portugal |
| Parimala Thulasiraman | University of Manitoba, Canada |
| Carmelo Torre | Polytechnic of Bari, Italy |
| Javier Martinez Torres | Centro Universitario de la Defensa Zaragoza, Spain |
| Giuseppe A. Trunfio | University of Sassari, Italy |
| Unal Ufuktepe | Izmir University of Economics, Turkey |
| Toshihiro Uchibayashi | Kyushu Sangyo University, Japan |

| | |
|---|---|
| Mario Valle | Swiss National Supercomputing Centre, Switzerland |
| Pablo Vanegas | University of Cuenca, Equador |
| Piero Giorgio Verdini | INFN Pisa and CERN, Italy |
| Marco Vizzari | University of Perugia, Italy |
| Koichi Wada | University of Tsukuba, Japan |
| Krzysztof Walkowiak | Wroclaw University of Technology, Poland |
| Robert Weibel | University of Zurich, Switzerland |
| Roland Wismüller | Universität Siegen, Germany |
| Mudasser Wyne | SOET National University, USA |
| Chung-Huang Yang | National Kaohsiung Normal University, Taiwan |
| Xin-She Yang | National Physical Laboratory, UK |
| Salim Zabir | France Telecom Japan Co., Japan |
| Haifeng Zhao | University of California at Davis, USA |
| Kewen Zhao | University of Qiongzhou, China |
| Albert Y. Zomaya | University of Sydney, Australia |

# Reviewers

| | |
|---|---|
| Abdi Samane | University College Cork, Ireland |
| Aceto Lidia | University of Pisa, Italy |
| Afonso Ana Paula | University of Lisbon, Portugal |
| Afreixo Vera | University of Aveiro, Portugal |
| Aguilar Antonio | University of Barcelona, Spain |
| Aguilar José Alfonso | Universidad Autónoma de Sinaloa, Mexico |
| Ahmad Waseem | Federal University of Technology Minna, Nigeria |
| Aktas Mehmet | Yildiz Technical University, Turkey |
| Alarcon Vladimir | Universidad Diego Portales, Chile |
| Alberti Margarita | University of Barcelona, Spain |
| Ali Salman | NUST, Pakistan |
| Alvanides Seraphim | Northumbria University, UK |
| Álvarez Jacobo de Uña | University of Vigo, Spain |
| Alvelos Filipe | University of Minho, Portugal |
| Alves Cláudio | University of Minho, Portugal |
| Alves José Luis | University of Minho, Portugal |
| Amorim Ana Paula | University of Minho, Portugal |
| Amorim Paulo | Federal University of Rio de Janeiro, Brazil |
| Anderson Roger | University of California, USA |
| Andrade Wilkerson | Federal University of Campina Grande, Brazil |
| Andrienko Gennady | Fraunhofer Institute for Intelligent Analysis and Informations Systems, Germany |
| Apduhan Bernady | Kyushu Sangyo University, Japan |
| Aquilanti Vincenzo | University of Perugia, Italy |
| Argiolas Michele | University of Cagliari, Italy |

Athayde Maria Emília
  Feijão Queiroz                  University of Minho, Portugal
Attardi Raffaele                  University of Napoli Federico II, Italy
Azad Md Abdul                     Indian Institute of Technology Kanpur, India
Badard Thierry                    Laval University, Canada
Bae Ihn-Han                       Catholic University of Daegu, South Korea
Baioletti Marco                   University of Perugia, Italy
Balena Pasquale                   Polytechnic of Bari, Italy
Balucani Nadia                    University of Perugia, Italy
Barbosa Jorge                     University of Porto, Portugal
Barrientos Pablo Andres           Universidad Nacional de La Plata, Australia
Bartoli Daniele                   University of Perugia, Italy
Bação Fernando                    New University of Lisbon, Portugal
Belanzoni Paola                   University of Perugia, Italy
Bencardino Massimiliano           University of Salerno, Italy
Benigni Gladys                    University of Oriente, Venezuela
Bertolotto Michela                University College Dublin, Ireland
Bimonte Sandro                    IRSTEA, France
Blanquer Ignacio                  Universitat Politècnica de València, Spain
Bollini Letizia                   University of Milano, Italy
Bonifazi Alessandro               Polytechnic of Bari, Italy
Borruso Giuseppe                  University of Trieste, Italy
Bostenaru Maria                   "Ion Mincu" University of Architecture and
                                    Urbanism, Romania
Boucelma Omar                     University Marseille, France
Braga Ana Cristina                University of Minho, Portugal
Brás Carmo                        Universidade Nova de Lisboa, Portugal
Cacao Isabel                      University of Aveiro, Portugal
Cadarso-Suárez Carmen             University of Santiago de Compostela, Spain
Caiaffa Emanuela                  ENEA, Italy
Calamita Giuseppe                 National Research Council, Italy
Campagna Michele                  University of Cagliari, Italy
Campobasso Francesco              University of Bari, Italy
Campos José                       University of Minho, Portugal
Cannatella Daniele                University of Napoli Federico II, Italy
Canora Filomena                   University of Basilicata, Italy
Cardoso Rui                       Institute of Telecommunications, Portugal
Caschili Simone                   University College London, UK
Ceppi Claudia                     Polytechnic of Bari, Italy
Cerreta Maria                     University Federico II of Naples, Italy
Chanet Jean-Pierre                IRSTEA, France
Chao Wang                         University of Science and Technology of China,
                                    China
Choi Joonsoo                      Kookmin University, South Korea

| | |
|---|---|
| Choo Hyunseung | Sungkyunkwan University, South Korea |
| Chung Min Young | Sungkyunkwan University, South Korea |
| Chung Myoungbeom | Sungkyunkwan University, South Korea |
| Clementini Eliseo | University of L'Aquila, Italy |
| Coelho Leandro dos Santos | PUC-PR, Brazil |
| Colado Anibal Zaldivar | Universidad Autónoma de Sinaloa, Mexico |
| Coletti Cecilia | University of Chieti, Italy |
| Condori Nelly | VU University Amsterdam, The Netherlands |
| Correia Elisete | University of Trás-Os-Montes e Alto Douro, Portugal |
| Correia Filipe | FEUP, Portugal |
| Correia Florbela Maria da Cruz Domingues | Instituto Politécnico de Viana do Castelo, Portugal |
| Correia Ramos Carlos | University of Evora, Portugal |
| Corso Pereira Gilberto | UFPA, Brazil |
| Cortés Ana | Universitat Autònoma de Barcelona, Spain |
| Costa Fernanda | University of Minho, Portugal |
| Costantini Alessandro | INFN, Italy |
| Crasso Marco | National Scientific and Technical Research Council, Argentina |
| Crawford Broderick | Universidad Catolica de Valparaiso, Chile |
| Cristia Maximiliano | CIFASIS and UNR, Argentina |
| Cunha Gaspar | University of Minho, Portugal |
| Cunha Jácome | University of Minho, Portugal |
| Cutini Valerio | University of Pisa, Italy |
| Danese Maria | IBAM, CNR, Italy |
| Da Silva B. Carlos | University of Lisboa, Portugal |
| De Almeida Regina | University of Trás-os-Montes e Alto Douro, Portugal |
| Debroy Vidroha | Hudson Alley Software Inc., USA |
| De Fino Mariella | Polytechnic of Bari, Italy |
| De Lotto Roberto | University of Pavia, Italy |
| De Paolis Lucio Tommaso | University of Salento, Italy |
| De Rosa Fortuna | University of Napoli Federico II, Italy |
| De Toro Pasquale | University of Napoli Federico II, Italy |
| Decker Hendrik | Instituto Tecnológico de Informática, Spain |
| Delamé Thomas | CNRS, France |
| Demyanov Vasily | Heriot-Watt University, UK |
| Desjardin Eric | University of Reims, France |
| Dwivedi Sanjay Kumar | Babasaheb Bhimrao Ambedkar University, India |
| Di Gangi Massimo | University of Messina, Italy |
| Di Leo Margherita | JRC, European Commission, Belgium |

Di Trani Francesco              University of Basilicata, Italy
Dias Joana                      University of Coimbra, Portugal
Dias d'Almeida Filomena         University of Porto, Portugal
Dilo Arta                       University of Twente, The Netherlands
Dixit Veersain                  Delhi University, India
Doan Anh Vu                     Université Libre de Bruxelles, Belgium
Dorazio Laurent                 ISIMA, France
Dutra Inês                      University of Porto, Portugal
Eichelberger Hanno              University of Tuebingen, Germany
El-Zawawy Mohamed A.            Cairo University, Egypt
Escalona Maria-Jose             University of Seville, Spain
Falcão M. Irene                 University of Minho, Portugal
Farantos Stavros                University of Crete and FORTH, Greece
Faria Susana                    University of Minho, Portugal
Faruq Fatma                     Carnegie Melon University,, USA
Fernandes Edite                 University of Minho, Portugal
Fernandes Rosário               University of Minho, Portugal
Fernandez Joao P                Universidade da Beira Interior, Portugal
Ferreira Fátima                 University of Trás-Os-Montes e Alto Douro,
                                  Portugal
Ferrão Maria                    University of Beira Interior and CEMAPRE,
                                  Portugal
Figueiredo Manuel Carlos        University of Minho, Portugal
Filipe Ana                      University of Minho, Portugal
Flouvat Frederic                University New Caledonia, New Caledonia
Forjaz Maria Antónia            University of Minho, Portugal
Formosa Saviour                 University of Malta, Malta
Fort Marta                      University of Girona, Spain
Franciosa Alfredo               University of Napoli Federico II, Italy
Freitas Adelaide de Fátima
  Baptista Valente              University of Aveiro, Portugal
Frydman Claudia                 Laboratoire des Sciences de l'Information et des
                                  Systèmes, France
Fusco Giovanni                  CNRS - UMR ESPACE, France
Fussel Donald                   University of Texas at Austin, USA
Gao Shang                       Zhongnan University of Economics and Law,
                                  China
Garcia Ernesto                  University of the Basque Country, Spain
Garcia Tobio Javier             Centro de Supercomputación de Galicia
                                  (CESGA), Spain
Gavrilova Marina                University of Calgary, Canada
Gensel Jerome                   IMAG, France
Geraldi Edoardo                 National Research Council, Italy
Gervasi Osvaldo                 University of Perugia, Italy

| | |
|---|---|
| Giaoutzi Maria | National Technical University Athens, Greece |
| Gizzi Fabrizio | National Research Council, Italy |
| Gomes Maria Cecilia | Universidade Nova de Lisboa, Portugal |
| Gomes dos Anjos Eudisley | Federal University of ParaÃba, Brazil |
| Gomez Andres | Centro de Supercomputación de Galicia, CESGA (Spain) |
| Gonçalves Arminda Manuela | University of Minho, Portugal |
| Gravagnuolo Antonia | University of Napoli Federico II, Italy |
| Gregori M. M. H. Rodrigo | Universidade Tecnológica Federal do Paraná, Brazil |
| Guerlebeck Klaus | Bauhaus University Weimar, Germany |
| Guerra Eduardo | National Institute for Space Research, Brazil |
| Hagen-Zanker Alex | University of Surrey, UK |
| Hajou Ali | Utrecht University, The Netherlands |
| Hanzl Malgorzata | University of Lodz, Poland |
| Heijungs Reinout | VU University Amsterdam, The Netherlands |
| Henriques Carla | Escola Superior de Tecnologia e Gestão, Portugal |
| Herawan Tutut | University of Malaya, Malaysia |
| Iglesias Andres | University of Cantabria, Spain |
| Jamal Amna | National University of Singapore, Singapore |
| Jank Gerhard | Aachen University, Germany |
| Jiang Bin | University of Gävle, Sweden |
| Kalogirou Stamatis | Harokopio University of Athens, Greece |
| Kanevski Mikhail | University of Lausanne, Switzerland |
| Kartsaklis Christos | Oak Ridge National Laboratory, USA |
| Kavouras Marinos | National Technical University of Athens, Greece |
| Khan Murtaza | NUST, Pakistan |
| Khurshid Khawar | NUST, Pakistan |
| Kim Deok-Soo | Hanyang University, South Korea |
| Kim Moonseong | KIPO, South Korea |
| Kolingerova Ivana | University of West Bohemia, Czech Republic |
| Kotzinos Dimitrios | Université de Cergy-Pontoise, France |
| Lazzari Maurizio | CNR IBAM, Italy |
| Laganà Antonio | Department of Chemistry, Biology and Biotechnology, Italy |
| Lai Sabrina | University of Cagliari, Italy |
| Lanorte Antonio | CNR-IMAA, Italy |
| Lanza Viviana | Lombardy Regional Institute for Research, Italy |
| Le Duc Tai | Sungkyunkwan University, South Korea |
| Le Duc Thang | Sungkyunkwan University, South Korea |
| Lee Junghoon | Jeju National University, South Korea |

| | |
|---|---|
| Lee KangWoo | Sungkyunkwan University, South Korea |
| Legatiuk Dmitrii | Bauhaus University Weimar, Germany |
| Leonard Kathryn | California State University, USA |
| Lin Calvin | University of Texas at Austin, USA |
| Loconte Pierangela | Technical University of Bari, Italy |
| Lombardi Andrea | University of Perugia, Italy |
| Lopez Cabido Ignacio | Centro de Supercomputación de Galicia, CESGA |
| Lourenço Vanda Marisa | University Nova de Lisboa, Portugal |
| Luaces Miguel | University of A Coruña, Spain |
| Lucertini Giulia | IUAV, Italy |
| Luna Esteban Robles | Universidad Nacional de la Plata, Argentina |
| Machado Gaspar | University of Minho, Portugal |
| Magni Riccardo | Pragma Engineering SrL, Italy, Italy |
| Malonek Helmuth | University of Aveiro, Portugal |
| Manfreda Salvatore | University of Basilicata, Italy |
| Manso Callejo Miguel Angel | Universidad Politécnica de Madrid, Spain |
| Marcheggiani Ernesto | KU Lueven, Belgium |
| Marechal Bernard | Universidade Federal de Rio de Janeiro, Brazil |
| Margalef Tomas | Universitat Autònoma de Barcelona, Spain |
| Martellozzo Federico | University of Rome, Italy |
| Marvuglia Antonino | Public Research Centre Henri Tudor, Luxembourg |
| Matos Jose | Instituto Politecnico do Porto, Portugal |
| Mauro Giovanni | University of Trieste, Italy |
| Mauw Sjouke | University of Luxembourg, Luxembourg |
| Medeiros Pedro | Universidade Nova de Lisboa, Portugal |
| Melle Franco Manuel | University of Minho, Portugal |
| Melo Ana | Universidade de São Paulo, Brazil |
| Millo Giovanni | Generali Assicurazioni, Italy |
| Min-Woo Park | Sungkyunkwan University, South Korea |
| Miranda Fernando | University of Minho, Portugal |
| Misra Sanjay | Covenant University, Nigeria |
| Modica Giuseppe | Università Mediterranea di Reggio Calabria, Italy |
| Morais João | University of Aveiro, Portugal |
| Moreira Adriano | University of Minho, Portugal |
| Mota Alexandre | Universidade Federal de Pernambuco, Brazil |
| Moura Pires João | Universidade Nova de Lisboa - FCT, Portugal |
| Mourelle Luiza de Macedo | UERJ, Brazil |
| Mourão Maria | Polytechnic Institute of Viana do Castelo, Portugal |
| Murgante Beniamino | University of Basilicata, Italy |
| NM Tuan | Ho Chi Minh City University of Technology, Vietnam |

| | |
|---|---|
| Nagy Csaba | University of Szeged, Hungary |
| Nash Andrew | Vienna Transport Strategies, Austria |
| Natário Isabel Cristina Maciel | University Nova de Lisboa, Portugal |
| Nedjah Nadia | State University of Rio de Janeiro, Brazil |
| Nogueira Fernando | University of Coimbra, Portugal |
| Oliveira Irene | University of Trás-Os-Montes e Alto Douro, Portugal |
| Oliveira José A. | University of Minho, Portugal |
| Oliveira e Silva Luis | University of Lisboa, Portugal |
| Osaragi Toshihiro | Tokyo Institute of Technology, Japan |
| Ottomanelli Michele | Polytechnic of Bari, Italy |
| Ozturk Savas | TUBITAK, Turkey |
| Pacifici Leonardo | University of Perugia, Italy |
| Pages Carmen | Universidad de Alcala, Spain |
| Painho Marco | New University of Lisbon, Portugal |
| Pantazis Dimos | Technological Educational Institute of Athens, Greece |
| Paolotti Luisa | University of Perugia, Italy |
| Papa Enrica | University of Amsterdam, The Netherlands |
| Papathanasiou Jason | University of Macedonia, Greece |
| Pardede Eric | La Trobe University, Australia |
| Parissis Ioannis | Grenoble INP - LCIS, France |
| Park Gyung-Leen | Jeju National University, South Korea |
| Park Sooyeon | Korea Polytechnic University, South Korea |
| Pascale Stefania | University of Basilicata, Italy |
| Passaro Pierluigi | University of Bari Aldo Moro, Italy |
| Peixoto Bitencourt Ana Carla | Universidade Estadual de Feira de Santana, Brazil |
| Perchinunno Paola | University of Bari, Italy |
| Pereira Ana | Polytechnic Institute of Bragança, Portugal |
| Pereira Francisco | Instituto Superior de Engenharia, Portugal |
| Pereira Paulo | University of Minho, Portugal |
| Pereira Ricardo | Portugal Telecom Inovacao, Portugal |
| Pietrantuono Roberto | University of Napoli "Federico II", Italy |
| Pimentel Carina | University of Aveiro, Portugal |
| Pina Antonio | University of Minho, Portugal |
| Pinet Francois | IRSTEA, France |
| Piscitelli Claudia | Polytechnic University of Bari, Italy |
| Piñar Miguel | Universidad de Granada, Spain |
| Pollino Maurizio | ENEA, Italy |
| Potena Pasqualina | University of Bergamo, Italy |
| Prata Paula | University of Beira Interior, Portugal |
| Prosperi David | Florida Atlantic University, USA |
| Qaisar Saad | NURST, Pakistan |

| | |
|---|---|
| Quan Tho | Ho Chi Minh City University of Technology, Vietnam |
| Raffaeta Alessandra | University of Venice, Italy |
| Ragni Mirco | Universidade Estadual de Feira de Santana, Brazil |
| Rautenberg Carlos | University of Graz, Austria |
| Ravat Franck | IRIT, France |
| Raza Syed Muhammad | Sungkyunkwan University, South Korea |
| Ribeiro Isabel | University of Porto, Portugal |
| Ribeiro Ligia | University of Porto, Portugal |
| Rinzivillo Salvatore | University of Pisa, Italy |
| Rocha Ana Maria | University of Minho, Portugal |
| Rocha Humberto | University of Coimbra, Portugal |
| Rocha Jorge | University of Minho, Portugal |
| Rocha Maria Clara | ESTES Coimbra, Portugal |
| Rocha Maria | PRODEB, San Salvador, Brazil |
| Rodrigues Armanda | Universidade Nova de Lisboa, Portugal |
| Rodrigues Cristina | DPS, University of Minho, Portugal |
| Rodriguez Daniel | University of Alcala, Spain |
| Roh Yongwan | Korean IP, South Korea |
| Roncaratti Luiz | Instituto de Fisica, University of Brasilia, Brazil |
| Rosi Marzio | University of Perugia, Italy |
| Rossi Gianfranco | University of Parma, Italy |
| Rotondo Francesco | Polytechnic of Bari, Italy |
| Sannicandro Valentina | Polytechnic of Bari, Italy |
| Santos Maribel Yasmina | University of Minho, Portugal |
| Santos Rafael | INPE, Brazil |
| Santos Viviane | Universidade de São Paulo, Brazil |
| Santucci Valentino | University of Perugia, Italy |
| Saracino Gloria | University of Milano-Bicocca, Italy |
| Sarafian Haiduke | Pennsylvania State University, USA |
| Saraiva João | University of Minho, Portugal |
| Sarrazin Renaud | Université Libre de Bruxelles, Belgium |
| Schirone Dario Antonio | University of Bari, Italy |
| Schneider Michel | ISIMA, France |
| Schoier Gabriella | University of Trieste, Italy |
| Schutz Georges | CRP Henri Tudor, Luxembourg |
| Scorza Francesco | University of Basilicata, Italy |
| Selmaoui Nazha | University of New Caledonia, New Caledonia |
| Severino Ricardo Jose | University of Minho, Portugal |
| Shakhov Vladimir | Russian Academy of Sciences, Russia |
| Shen Jie | University of Michigan, USA |
| Shon Minhan | Sungkyunkwan University, South Korea |

| | |
|---|---|
| Shukla Ruchi | University of Johannesburg, South Africa |
| Silva J.C. | IPCA, Portugal |
| Silva de Souza Laudson | Federal University of Rio Grande do Norte, Brazil |
| Silva-Fortes Carina | ESTeSL-IPL, Portugal |
| Simão Adenilso | Universidade de São Paulo, Brazil |
| Singh R K | Delhi University, India |
| Soares Inês | INESC Porto, Portugal |
| Soares Maria Joana | University of Minho, Portugal |
| Soares Michel | Federal University of Sergipe, Brazil |
| Sobral Joao | University of Minho, Portugal |
| Son Changhwan | Sungkyunkwan University, South Korea |
| Sproessig Wolfgang | Technical University Bergakademie Freiberg, Germany |
| Su Le Hoanh | Ho Chi Minh City Technical University, Vietnam |
| Sá Esteves Jorge | University of Aveiro, Portugal |
| Tahar Sofiène | Concordia University, Canada |
| Tanaka Kazuaki | Kyushu Institute of Technology, Japan |
| Taniar David | Monash University, Australia |
| Tarantino Eufemia | Polytechnic of Bari, Italy |
| Tariq Haroon | Connekt Lab, Pakistan |
| Tasso Sergio | University of Perugia, Italy |
| Teixeira Ana Paula | University of Trás-Os-Montes e Alto Douro, Portugal |
| Teixeira Senhorinha | University of Minho, Portugal |
| Tesseire Maguelonne | IRSTEA, France |
| Thorat Pankaj | Sungkyunkwan University, South Korea |
| Tomaz Graça | Polytechnic Institute of Guarda, Portugal |
| Torre Carmelo Maria | Polytechnic of Bari, Italy |
| Trunfio Giuseppe A. | University of Sassari, Italy |
| Urbano Joana | LIACC University of Porto, Portugal |
| Vasconcelos Paulo | University of Porto, Portugal |
| Vella Flavio | University of Rome La Sapienza, Italy |
| Velloso Pedro | Universidade Federal Fluminense, Brazil |
| Viana Ana | INESC Porto, Portugal |
| Vidacs Laszlo | MTA-SZTE, Hungary |
| Vieira Ramadas Gisela | Polytechnic of Porto, Portugal |
| Vijay NLankalapalli | National Institute for Space Research, Brazil |
| Villalba Maite | Universidad Europea de Madrid, Spain |
| Viqueira José R.R. | University of Santiago de Compostela, Spain |
| Vona Marco | University of Basilicata, Italy |

## Sponsoring Organizations

ICCSA 2014 would not have been possible without the tremendous support of many organizations and institutions, for which all organizers and participants of ICCSA 2014 express their sincere gratitude:

**Universidade do Minho**
Escola de Engenharia
Universidade do Minho
(http://www.uminho.pt)

University of Perugia, Italy
(http://www.unipg.it)

University of Basilicata, Italy (http://www.unibas.it)

Monash University, Australia
(http://monash.edu)

Kyushu Sangyo University, Japan
(www.kyusan-u.ac.jp)

Associação Portuguesa de Investigação Operacional
(apdio.pt)

# Table of Contents

## Workshop on Software Quality (SQ 2014)

## Workshop on Big Data Analytics and Management (WBDAM 2014)

## Workshop on Bio-inspired Computing and Applications (BIOCA 2014)

## Workshop on Agricultural and Environmental Information and Decision Support Systems (AEIDSS 2014)

## Erratum

# Proof-Carrying Model for Parsing Techniques

Mohamed A. El-Zawawy[1,2]

[1] College of Computer and Information Sciences,
Al Imam Mohammad Ibn Saud Islamic University (IMSIU)
Riyadh, Kingdom of Saudi Arabia
[2] Department of Mathematics, Faculty of Science, Cairo University
Giza 12613, Egypt
maelzawawy@cu.edu.eg

**Abstract.** This paper presents new approaches to common parsing algorithms. The new approach utilizes the concept of inference rule. Therefore the new approach is simple, yet powerful enough to overcome the performance of traditional techniques. The new approach is basically composed of systems of inference rules.

Mathematical proofs of the equivalence between proposed systems and classical algorithms are outlined in the paper. The proposed technique provides a correctness verification (a derivation of inference rules) for each parsing process. These verifications are required in modern applications such as mobile computing. Results of experimental proving the efficiency of the proposed systems and their produced verifications are shown in the paper.

**Keywords:** Inference rules, LL(1) parsers, Operator-precedence passers, LR parsers, Parsing.

## 1 Introduction

The parsing process [14,5] aims at analyzing and checking the structure of a given input (computer program) with respect to a specific grammar. Each grammar can typically produce an infinite number of *sentences* (programs). However grammars [28,8] have finite sizes. Therefore a grammar abstracts briefly an infinite number of program architectures of a specific type (structural, functional, object-oriented, etc).

Parsing is very critical for several reasons [21]. Parsing process is essential for further processing of parsed objects (programs). For example in natural languages processing, recognizing the verb of a sentence facilitates the sentence translation. Parsing process is also important to understand grammars which summarize our realization of a class of programs. Error-recovering parsers [29] are important tools for correcting some program faults. Theoretical foundations of parsing excuse it from being described as esoteric or as a mathematical discipline. Using string cutting and pasting, parsing can be realized and applied.

Parsing techniques [28,8] are classified into two categories; top-down and bottom-up. A parsing technique that tries to establish a derivation for the input program starting from the start symbol is classified as a top-down algorithm. If a parsing technique starts with the input program and rolls back trying to establish the derivation in the reverse

B. Murgante et al. (Eds.): ICCSA 2014, Part V, LNCS 8583, pp. 1–14, 2014.

order until reaching the start symbol, the technique is then described as bottom-up. Derivations produced by a bottom-up (top-down) algorithm are right-most (left-most). In top-down (bottom-up) parsers, parse trees are constructed from the top (bottom) downwards (upwards). Common examples of top-down and bottom- up parsers are LL(1) and LR(1) [3], respectively.

Some applications like proof-carrying code [23,17] and mobile computing [30] require the parser to associate each parse tree with a correctness proof. This proof ensures the validity of the tree. Unfortunately most of existing parsers are algorithmic in their style and hence build the parse tree but not the correctness proof. Proposing new sittings for common parsing algorithms so that correctness proofs are among outputs of these algorithms (rather than parse trees) is a requirement of many modern computing applications.

This paper introduces new models for common existing parsing techniques such as LL(1), operator-precedence, SLR, LR(1), and LALR. The proposed models are built mainly of inference rules whose outputs in this case are pairs of parse trees and correctness proofs. Rather than providing the required correctness proofs, the new models are faster than their corresponding ones as is confirmed by experimental results. Mathematical proofs of the equivalence between each proposed model and its corresponding original parsing technique are outlined in the paper. The proposed models are supported with illustrative parsing examples that are detailed.

## Motivation

Systems of inference rules were found very useful for optimization phases of compilers [11,12]. This is so as they provide compact correctness proofs for optimizations. These proofs are required by applications like proof-carrying code and mobile computing. The motivation of this paper is the need for parsing techniques that associate each parsing process with a simply-structured correctness proof. The research behind this paper reveals that parsing techniques having the form of inference rules are the perfect choice to build the required proofs. The resulting proofs have the shape of rule derivations as in Figures 2 and 5.

## Contributions

Contributions of the paper are the following:

1. A new approach to LL parsing techniques in the form of system of inference rules. The new approach was found more efficient than the traditional one.
2. A new setting for operator-precedence parsers.
3. A novel technique to achieve LR parsing methods.

## Paper Outline

The rest of the paper is organized as follows. Related work is discussed in Section 2. Section 3 presents the new technique for LL(1) parsers. The systems of inference rules for LR parsers and operator-precedence parsers are shown in Sections 4.1 and 4.2, respectively. Each of Sections 3, 4.1, and 4.2 includes mathematical proofs of the correctness of its proposed technique. Section 5 shows the experimental results. A conclusion to the paper is in Section 6.

## 2 Related Work

Generalized LR (GLR) parsers [25,18] are used to handle ambiguous grammars whose parsing tables compilation may last for minutes for large grammars. These parsers are not convenient for natural-language grammars because parsing tables can become very large. For natural languages more specific parsers [9,20] were designed. Such parsers, typically, support PM context-free grammars. Some attempts [10,2,13] towards parsers for visual languages were done. However problems with parsing visual languages include that parsing tables need to be recomputed with each grammar. Most parsers [24,19] using no parsing tables are inefficient and restricted to a small categories of grammars.

In dependent grammars [22,7], semantic values are associated with variables. Therefore these grammars are convenient for constrain parsing [6,15]. Common attributes not covered by context-free grammars, like intended indentations and field lengths in data types of programming languages are determined by dependent grammars. Researchers have been trying to extend classical parsing algorithms to support dependent parsing [16,32]. For example, in [16] a point-free algorithm language for dependent grammars is proposed. The language algorithms are similar to classical algorithms for context-free parsing. Hence the point-free language acts as a tool to get point-free grammars from classical dependent ones.

Boolean grammars [27,26] are grammars equipped with operations that are explicitly set-theoretic and used to express rules formally defined by equations of the language equations. Boolean grammars are extensions of context-free grammars. Much research was done to extend algorithms of the later grammars to that of the former ones. Other attempts were done to generalize LR parsing algorithm to cover Boolean grammars. This extension included traditional LR operations, Reduce and Shift, and a new third operation called, Invalidate. This third action reverses a prior reduction. The generalization idea was to make the algorithm complectly different from its prototype.

Although classical top-down parsers [21,29] simplicity and ability to treat empty rules, they can not treat many sorts of left-recursions. In the same time, it is not preferred to normalize a left-recursion grammar as the rules semantic-structure gets affected. On the other hand, bottom-up parsers [21,29] are able to treat left-recursion but not empty rules. The research in [31,4] proposed a deterministic cancelation parser whose structure is recursive-descent. This parser can manipulate left-recursions and empty rules via a combination of bottom-up and top-down applications towards building the parse tree. The challenge of these parsers is to manipulate all sorts of left-recursions, such as hidden and indirect left recursions.

Unfortunately, none of the parsers reviewed above associates each parsing process with a justification or a correctness proof. Therefore all these parsers are not applicable in modern applications of mobile computing and proof-carrying code.

## 3  New Approach for Top-Down Parsers: LL(1)

LL(1) [3] is a top-down parsing algorithm. For a given input string and starting from the start symbol of the given grammar, top-down parsers tries to create a parse tree using depth-first methods. Hence top-down parsers can be realized as searching methods for leftmost derivations for input strings. LL(1) is predictive in the sense that it is able to decide the rule to apply based only on the currently processed nonterminal and following input symbol. This is only possibly if the grammar has a specific form called LL(1). The LL(1) parser uses a stack and a parse table called LL(1)-table. The stack stores the productions that the parser is to apply.

The LL(1)-table columns correspond to terminals of the grammar plus an extra column for the $ symbol (the marker for the input symbol end). Rows of LL(1)-table correspond to nonterminals of the LL(1) grammar. Each cell of the LL(1)-table is either empty or including a single grammar production. Therefore the table includes the parser actions to be taken based on the top value of the stack and the current input symbol.

$$\frac{\quad}{\epsilon : \epsilon \to \$} \text{ (Base)} \qquad \frac{a \in \text{first}(T) \qquad (N \to T) \in P}{a\alpha : NM \to \$} \qquad \frac{a\alpha : TM \to \$}{} \text{ (Predict}_1)$$

$$\frac{a \in \text{follow}(N) \qquad N \to \epsilon \qquad a\alpha : M \to \$}{a\alpha : NM \to \$} \text{ (Predict}_2) \qquad \frac{\alpha : \beta \to \$}{a\alpha : a\beta \to \$} \text{ (Match)}$$

**Fig. 1.** Inference rules for LL(1) parser

Figure 1 presents a new model for the LL(1) functionality in the form of a system of inference rules. The following definition is necessary for linking the semantics and the use of inference rules in Figure 1 to functionality of LL(1). The definition is also necessary for proving equivalence between the original model and our new one.

**Definition 1.** *Let* $(S, N, T, P)$ *be a LL(1) grammar. Then* **LL(1)-infer** *is the set of strings* $w \in T^+$ *such that* $w : S \to \$$ *is derivable in the system of inference rules of Figure 1. Moreover* **path(w,ll(1))** *is the string derivation obtained by applying the grammar rules in the derivation of* $w : S \to \$$. *This rule applications is intended to respect the order of the rule appearances in the direction bottom-up of the derivation.*

Our proposed model works as follows. For a given string $w$, we start by trying to derive $w : S \to \$$ using the system of inference rules. This is equivalent to the start of the LL(1) algorithm; reading the first token of input and pushing on the stack the start symbol. In the LL(1) method and using the LL(1)-table, there are three possible cases for the input token, $a$, and the top stack symbol $x$. In the first case $x = a = \$$, meaning that $w$ is correct. This case is modeled through the rule *(Base)*. In the second case, $x = a \neq \$$, the algorithm pops $x$ and moves to next input token. This is achieved via the rule *(Match)*. In the third case $x$ is a nonterminal and $x \neq a$. In this case, the LL(1)-table is referenced at the cell $(x, a)$ towards a rule application. There are three possible sub-cases for this case. The first subcase happens when the cell $(x, a)$ is empty meaning that it is not possible to construct the derivation and hence $w$ is wrong. The second subcase happens when there is a rule at in the cell $(x, a)$ and the left hand side of the rule is not empty. The application of this rule is achieved by the rule $(Predict_1)$. The third subcase happens when the left hand side of the rule is empty. This subcase is treated via the rule $(Predict_2)$.

The following theorem proves the equivalence between our proposed model and the original (one which does not provide the correctness verifications).

**Theorem 1.** *Let* $(S, N, T, P)$ *be a LL(1) grammar and* $w \in T^+$. *Then* $w \in LL(1) - infer$ *if and only if* $w$ *is accepted by the LL(1)-algorithm. Moreover the left-most derivation obtained by the LL(1)-algorithm is the same as* **path(w,ll(1))**.

*Proof.* The LL(1)-algorithm has the following cases:

- The case $(Match)$ meaning a symbol $a$ is on the top of the stack and is also the first symbol of the current input string. In this case, the $LL(1)$ algorithm removes $a$ from the stack and input. Also in this case, the rule $(Match)$ of Figure 1 applies and does the same action as the $LL(1)$-algorithm.
- The case $(Predict_1)$: meaning the top of the stack is a nonterminal symbol $x$ and the first symbol of the current input string is a terminal symbol $a$. In this case, the algorithm references the $LL(1)$-table at location $(x, a)$ towards a rule application and pushes into the stack the r.h.s. of rule. Suppose this rule is $N \rightarrow T$. In this case $a \in$ first$(T)$ and the rule $(Predict_1)$ of Figure 1 applies and does the same action as the $LL(1)$-algorithm.
- The case $(Predict_2)$: this case is similar to the previous one except that $T = \epsilon$. In this case $a \in$ follow$(N)$ and the $LL(1)$-algorithm removes $N$ form the stack. Clearly, in this case, the rule $(Predict_2)$ of Figure 1 applies and does the same action as the $LL(1)$-algorithm.
- The case $(Accept)$, in this case the stack and input have only $\$$. This case is covered by the rule $(Base)$ of Figure 1.

The following two lemmas formalize the relationship between parse trees obtained in our prosed model of inference rules and that obtained using the original LL(1)-algorithm. The lemmas are easily proved by a structure induction on the inference-rule derivations.

**Lemma 1.** *Let* $(S, N, T, P)$ *be a LL(1) grammar and* $w \in LL(1) - infer$. *Then path(w) is left-most.*

**Lemma 2.** *Let* $(S, N, T, P)$ *be an LL(1) grammar. Then for* $w \in T^+$,

$$S \Rightarrow^+ w \iff w : S \rightarrow \$.$$

For the grammar consists of the rules:

$$S \rightarrow aBa, B \rightarrow bB, \text{ and } B \rightarrow \epsilon.$$

The upper part of Figure 2 shows the derivation for proving the correctness of the statement $abba$ using inference rules of Figure 1. The obtained derivation is the following:

$$S \Rightarrow_{lm} aBa \Rightarrow_{lm} abBa \Rightarrow_{lm} abbBa \Rightarrow_{lm} abba.$$

**Fig. 2.** Derivation Examples

# 4 New Approaches for Bottom-Up Parsers

As suggested by their names, bottom-up parsers [3] work oppositely to top-down parsers. Starting with the input string, a bottom-up parser proceeds upwards from leaves to the start symbol. Hence a bottom-up parser works backwards and applies the grammar rules in a reverse direction until arriving at the start symbol. This reverse application amounts to finding a substring of the stack content that coincides with the r.h.s. of some production of the grammar. Then the l.h.s. of this production replaces the found substring. Hence a reduction takes place. The idea is to keep reducing until arriving at the start symbol, therefore the string is correct, otherwise it is not correct. In the following, we show how common bottom-up parsers (LR and operator-precedence parsers) can be modeled using systems of inference rules.

## 4.1 LR Parsers Family

LR parsers [3] work as following. Fixing handles depends on the stack content and hence on the context of the parsing process. The LR parsers do not push tokens only into the stack, but push as well state numbers in alternation with tokens. The state numbers describe the stack content. The state on the top of the stack determines whether the next action is to move (shift) a new state to the stack (for the next symbol of input) or is to reduce.

Two tables are used by LR parsers; an action table and a goto table. For the action table, columns correspond to terminals and rows correspond to state numbers. For the goto table, columns correspond to nonterminals and rows correspond to state numbers. The entry of the action table at the intersection $(s, a)$ determines the action to be taken when the next terminal is $a$ and $s$ is the state on the top of the stack. Figure 3 models possible actions in the form of inference rules. One possible action is to *shift* and is modeled by the rule *(Shift)*. Another action is to *reduce* and is modeled by the rule *(Reduce)*. The *accept* case is modeled by the rule *(Base)*. In case of error, it is not possible to find a derivation for the given input string. The goto-table entry at the intersection $(s, x)$ determines the state to be inserted into the stack after reducing $x$ when the top of stack is $s$.

For a given string $w$, if it is possible to derive $w\$ : 0 \rightarrow S$ using inference rules of Figure 3, then $w$ is correct (accepted), otherwise it is wrong. This is corresponding

$$\frac{(n, \$) = \text{accept}}{\$ : \beta n \rightarrow S} \text{ (Base)} \qquad \frac{\text{goto}(n_1, a) = sn_2 \qquad \alpha : \beta n_1 a n_2 \rightarrow S}{a\alpha : \beta n_1 \rightarrow S} \text{ (Shift)}$$

$$\frac{\begin{array}{c} \text{goto}(n_1, a) = r \text{ by } (T \rightarrow \gamma) \\ \# - \text{terminals}(\gamma) = m \end{array} \qquad \begin{array}{c} \text{deduct}(\beta n_1, 2m) = \delta n_2 \\ \text{goto}(n_2, T) = n_3 \\ a\alpha : \delta T n_3 \rightarrow S \end{array}}{a\alpha : \beta n_1 \rightarrow S} \text{ (Reduce)}$$

**Fig. 3.** Inference rules for SLR, LR(1), and LALR parsers

to pushing the state 0 into the stack and continuing in the shift-reduce process until reaching an accept or a reject.

Definition 2 recalls important concepts of grammars. Definition 3 introduces necessary concepts towards proving soundness of inference rules of Figure 3.

**Definition 2.**   – *A LR parser using SLR parsing table for a grammar G is called a SLR parser for G.*
– *If a grammar G has a SLR parsing table, it is called SLR-grammar.*[1]

**Definition 3.** *Let $(S, N, T, P)$ be a SLR grammar. Then **der-stg** is the set of strings $w \in T^+$ such that $w : 0\$ \to S$ is derivable in the system of inference rules of Figure 3. Moreover **way(w,slr)** is the string derivation obtained by applying the grammar rules in the obtained derivation. This rule applications is intended to respect the order of the rules appearances in the direction bottom-up of the derivation.*

The following theorem proves the soundness of the proposed model and its equivalence to the original SLR model.

**Theorem 2.** *Let $(S, N, T, P)$ be a SLR grammar and $w \in T^+$. Then $w \in$ der-stg if and only if $w$ is accepted by the SLR-algorithm. Moreover the right-most derivation obtained by the SLR-algorithm is the same as **way(w,slr)**.*

*Proof.* The SLR algorithm has the following possible actions cases:

- The *shift* action: in this case the intersection of the number, $n_1$, at the top of the stack and the symbol $a$ at the beginning of the input, in the SLR table, is $sn_2$. In this case the SLR-algorithm pushes $a$ followed by $n_2$ into the stack. Also in this case, the rule *(Shift)* of Figure 3 applies and does the same action as the SLR-algorithm.
- The *reduce* action: in this case the intersection of the number $n_1$ at the top of the stack and the symbol $a$ at the beginning of the input, in the SLR-table, is $r$ by $T \to \gamma$. In this case, a number (twice the number of symbols in $\gamma$) of elements is removed from the stack and $T$ is added to the stack. If $n_2$ is the number prior to $T$ in the stack, then the intersection of $T$ and $n_2$ in the goto table is added to the stack. In this case, the rule *(Reduce)* of Figure 3 applies and does the same action as the SLR-algorithm.
- The case of an acceptance which is equivalent to applying the rule *Base*.

The relationship between parse trees resulting from our prosed system of inference rules and that resulting from original SLR-algorithm are formalized by the following two lemmas. A straightforward structure induction on the inference-rule derivations completes the lemma proofs.

**Lemma 3.** *Let $(S, N, T, P)$ be a SLR-grammar and $w \in$ der-stg. Then **way(w,slr)** is a right-most derivation.*

---

[1] Recall that every SLR grammar is unambiguous, but not every unambiguous grammar is SLR grammar [3].

**Lemma 4.** *Let $(S, N, T, P)$ be a SLR-grammar. Then for $w \in T^{+}$,*

$$S \Rightarrow^{+} w \iff w\$ : 0 \to S.$$

Consider the following grammar.

$$E \to E + E, E \to T, T \to T * F, T \to F, F \to (E), \text{ and } F \to id.$$

Figure 5 presents a derivation for proving the correctness of the statement $id * id + id$ using inference rules of Figure 3. The following is the obtained derivation.

$$E \Longrightarrow_{rm} E + T \Longrightarrow_{rm} E + F \Longrightarrow_{rm} E + id \Longrightarrow_{rm} T * F + id \Longrightarrow_{rm}$$

$$T * id + id \Longrightarrow_{rm} F * id + id \Longrightarrow_{rm} id * id + id.$$

*Remark 1.* All results above about SLR-grammars and parsers are correct and applicable for LR(1)-grammars and parsers and for LALR-grammars and parsers.

## 4.2 Operator-Precedence Parsing

Operator-precedence parsing [29] is a simple parsing technique suitable for particular grammars. This grammars are *operator-grammars* characterized by the absence of $c$ and consecutive nonterminals in l.h.s. of their productions. A precedence relation is defined on the set of terminals: $a \lessdot b$ means that $b$ has higher precedence than $a$, where $a \doteq b$ means that $b$ enjoys the same precedence as $a$. Classical concepts of operators precedence and associativity determine the convenient precedence relations. The handle of a right-sentential form is determined by the precedence relation. Hence the left (right) end of the handle is marked by $\lessdot$ ($\gtrdot$). Figure 4 presents a system of inference rules to carry the operator-precedence parsing. Hence the parsing process amounts to a derivation construction. Theorem 3 illustrates the use of the inference rules and their equivalence to the operator-precedence algorithm.

$$\frac{}{\$ : \$ \to S} \text{ (Base)} \qquad \frac{a \gtrdot\doteq b \qquad a\alpha : b\beta \to S}{\alpha : ab\beta \to S} \text{ (Shift)}$$

$$\frac{(T \to a) \in P \qquad a \lessdot b \qquad a\alpha : b\beta \to S}{a\alpha : \beta \to S} \text{ (Reduce}_1)$$

$$\frac{(T \to \gamma_1 o \gamma_2) \in P \qquad o \lessdot b \qquad o\alpha : b\beta \to S}{o\alpha : \beta \to S} \text{ (Reduce}_2)$$

**Fig. 4.** Inference Rules for the Operator-Precedence Parser

The following definition introduces necessary concepts towards proving soundness of inference rules of Figure 4.

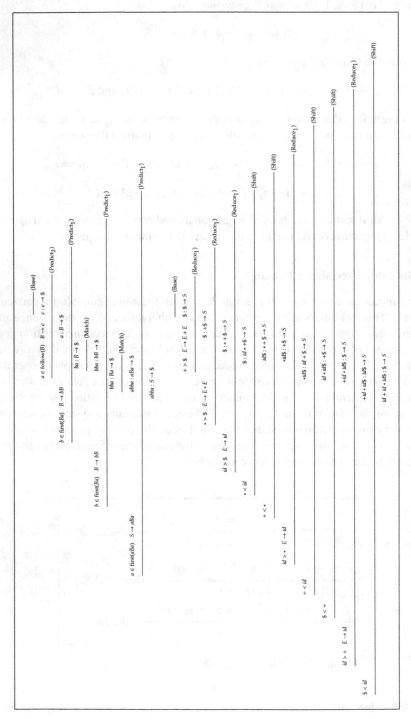

**Fig. 5.** SLR Derivation Example for the statement id*id+id using rules of Figure 3

**Definition 4.** *Let* $(S, N, T, P)$ *be an operator-grammar. Then* **op-infer** *is the set of strings* $w \in T^+$ *such that* $w$ *has a rule derivation for* $w\$ : \$ \rightarrow S$ *in the system of inference rules of Figure 4. The string derivation obtained by applying the grammar rules in the derivation is denoted by* $path(w, op)$. *Respecting the order of the rules appearance bottom-up in the derivation is imperative for rule applications.*

Theorem 3 proves the soundness of the system of inference rules and the original operator-precedence algorithm which has the drawback of not providing correctness proofs.

**Theorem 3.** *Let* $(S, N, T, P)$ *be an operator-grammar and* $w \in T^+$. *Then* $w \in$ *op-infer if and only if* $w$ *is accepted by the operator-precedence algorithm. Moreover the right-most derivation obtained by the operator-precedence algorithm is the same as* **path(w,op)**.

*Proof.* The operator-precedence algorithm has the following action cases:

- The case of a *shift*: this means a symbol $b$ is on the top of the stack and is greater than or equal to the first symbol, $a$, of the current input string. The order relation is obtained via the operator-precedence table. In this case the operator-precedence algorithm pushes $a$ into the stack. Also in this case, the rule *(Shift)* of Figure 4 applies and does the same action as the operator-precedence algorithm.
- The case of a *reduce*: in this case a symbol $b$ is on the top of the stack and is greater than the first symbol of the current input string. If $b$ is a terminal but not an operator, then the operator-precedence algorithm discards $b$ from the stack and records the grammar rule whose right hand side coincides $b$. For this case, the rule *(Reduce₁)* of Figure 4 applies and does the same action as the operator-precedence algorithm. If $b$ is an operator and there is a grammar rule $T \rightarrow \gamma_1 b \gamma_2$, then the operator-precedence discards $b$ from the stack and records the grammar rule. In this case, the rule *(Reduce₂)* of Figure 4 applies and does the same action as the operator-precedence algorithm.
- The case of *accept*, in this case the stack and input have only \$. This case is simulated by the rule *(Base)* of Figure 4.

Parse trees obtained by inference rules of Figure 4 and that obtained by original operator-precedence algorithm are equivalent by the following two lemmas. The lemmas are proved by a straightforward structure induction on the inference-rule derivations.

**Lemma 5.** *Let* $(S, N, T, P)$ *be an operator-precedence grammar and* $w \in$ *op-infer. Then* $path(w, op)$ *is right-most.*

**Lemma 6.** *Let* $(S, N, T, P)$ *be an operator-precedence grammar. Then for* $w \in T^+$,

$$S \Rightarrow^+ w \iff w\$ : \$ \rightarrow S.$$

Consider the grammar:

$$E \rightarrow E + E, E \rightarrow E - E, E \rightarrow E * E, E \rightarrow E/E, E \rightarrow E^E, E \rightarrow (E), E \rightarrow -E, \text{ and } E \rightarrow id$$

| # Symbols | LL(1) | LL(1)-Inf | Proof Size |
|---|---|---|---|
| 15 | 71.0 ms | 52.0 ms | 3.1 KB |
| 30 | 122.0 ms | 98.0 ms | 5.7 KB |
| 50 | 143.0 ms | 112.0 ms | 8.9 KB |
| # Symbols | SLR | SLR-Inf | Proof Size |
| 15 | 83.0 ms | 57.0 ms | 3.5 KB |
| 30 | 143.0 ms | 112.0 ms | 6.5 KB |
| 50 | 195.0 ms | 134.0 ms | 9.8 KB |
| # Symbols | Operator-Precedence | OP-Inf | Proof Size |
| 15 | 62.0 ms | 46.0 ms | 2.0 KB |
| 30 | 93.0 ms | 73.0 ms | 4.3 KB |
| 50 | 147.0 ms | 129.0 ms | 7.6 KB |

**Fig. 6.** Experiential Results

The lower derivation of Figure 2 presents a derivation for proving the correctness of the statement $id + id * id$ using inference rules of Figure 4. The following is the obtained derivation.

$$E \Longrightarrow_{rm} E + E \Longrightarrow_{rm} E + E * E \Longrightarrow_{rm} E + E * id \Longrightarrow_{rm} E + id * id \Longrightarrow_{rm} id * id + id.$$

## 5    Implementation and Evaluation

Timing experimenters were performed to evaluate *LL(1)-Inf, SLR-Inf,* and *OP-Inf* and to compare them to LL(1), SLR, Operator-precedence parsing algorithms, respectively. All the algorithms were implemented in $C + +$. The experiments were run on a Linux system whose processor is Intel(R)-Core2(TM)-i5-CPU(2.53GHz) and whose RAM is 4GB. Computed beforehand parse tables were stored and used internally in all experiments. In order to have realistic comparisons, some modifications were done to proposed methods to simulate overhead of lexical analysis and startup penalty of running compilers. The proposed algorithms as well as the original ones were run on string of different sizes (15, 30, 50 symbols). The same grammars used in Sections 3, 4.1, 4.2 to illustrate *LL(1)-Inf, SLR-Inf,* and *OP-Inf,* respectively, were used to build the test strings.

The experimental results are shown in Figure 6. For the sake of accuracy, all readings are averaged using results of 20 runs. Parameters used to measure the performance are timings and the size of produced correctness proofs. The second parameter, the proof size, is quite an important parameter. One of the main advantages of proposed techniques in this paper over their corresponding ones is the association of each parsing result with a correctness proof in the form of inference-rule derivations. This proof is to be delivered with the parsing result. In applications like proof-carrying code and mobil computing, this proof is to be communicated. Hence it is quite important to ensure that these proofs are of convenient sizes.

As expected and confirmed by experimental results, the proposed techniques are faster than their original corresponding ones. Moreover, the sizes of the proofs are convenient compared to the lengths of the used strings. This guaranties scalability of the proposed techniques.

## 6   Conclusion

This paper revealed that using systems of inference rules for achieving the parsing problem is a convenient choice for many modern applications. The relative simplicity of inference rules required to express parsing algorithms (as it is confirmed in this paper for like LL(1), LR(1), and operator-precedence parsers) attracts compiler designers to use them instead of tractional algorithmic way. The paper showed mathematical proofs for the equivalence between proposed systems of inference rules and their classical corresponding algorithms. Results of carried experiments, endorsing the efficiency of the use of inference rules, were shown in the paper. The paper also presented detailed examples of using proposed systems of inference rules.

**Acknowledgment.** The author acknowledges the support (grant number 330917) of the deanship of scientific research of Al Imam Mohammad Ibn Saud Islamic University (IMSIU).

## References

1. 26th IEEE Canadian Conference on Electrical and Computer Engineering CCECE 2013, Regina, SK, Canada, May 5-8. IEEE (2013)
2. Al-Mulhem, M., Ather, M.: Mrg parser for visual languages. Inf. Sci. 131(1-4), 19–46 (2001)
3. Aho, R.S.A.V., Lam, M.S., Ullman, J.D.: Compilers: Principles, Techniques, and Tools, 2nd edn. Prentice Hall, New Jersey (2007)
4. Bahrololoomi, M.H., Younessi, O., Moghadam, R.A.: Exploration of conflict situations in deterministic cancellation parser. In: CCECE [1], pp. 1–4
5. Bernardy, J.-P., Claessen, K.: Efficient divide-and-conquer parsing of practical context-free languages. In: Morrisett, G., Uustalu, T. (eds.) ICFP, pp. 111–122. ACM (2013)
6. Bodenstab, N., Hollingshead, K., Roark, B.: Unary constraints for efficient context-free parsing. In: ACL (Short Papers), pp. 676–681. The Association for Computer Linguistics (2011)
7. Brink, K., Holdermans, S., Löh, A.: Dependently typed grammars. In: Bolduc, C., Desharnais, J., Ktari, B. (eds.) MPC 2010. LNCS, vol. 6120, pp. 58–79. Springer, Heidelberg (2010)
8. Cooper, K., Torczon, L.: Engineering a Compiler, 2nd edn. Morgan Kaufmann, Waltham (2011)
9. de Piñerez Reyes, R.E.G., Frias, J.F.D.: Building a discourse parser for informal mathematical discourse in the context of a controlled natural language. In: Gelbukh, A. (ed.) CICLing 2013, Part I. LNCS, vol. 7816, pp. 533–544. Springer, Heidelberg (2013)
10. Deufemia, V., Paolino, L., de Lumley, H.: Petroglyph recognition using self-organizing maps and fuzzy visual language parsing. In: ICTAI, pp. 852–859. IEEE (2012)
11. El-Zawawy, M.A.: Detection of probabilistic dangling references in multi-core programs using proof-supported tools. In: Murgante, B., Misra, S., Carlini, M., Torre, C.M., Nguyen, H.-Q., Taniar, D., Apduhan, B.O., Gervasi, O. (eds.) ICCSA 2013, Part V. LNCS, vol. 7975, pp. 516–530. Springer, Heidelberg (2013)
12. El-Zawawy, M.A.: Distributed data and programs slicing. Life Science Journal 10(4), 1361–1369 (2013)

13. Ferrucci, F., Tortora, G., Tucci, M., Vitiello, G.: A predictive parser for visual languages specified by relation grammars. In: VL, pp. 245–252 (1994)
14. Grune, D., Jacobs, C.J.H.: Parsing Techniques: A Practical Guide, 2nd edn. Springer, Heidelberg (2007)
15. Hulden, M.: Constraint grammar parsing with left and right sequential finite transducers. In: Constant, M., Maletti, A., Savary, A. (eds.) FSMNLP, ACL Anthology, pp. 39–47. Association for Computational Linguistics (2011)
16. Jim, T., Mandelbaum, Y.: A new method for dependent parsing. In: Barthe, G. (ed.) ESOP 2011. LNCS, vol. 6602, pp. 378–397. Springer, Heidelberg (2011)
17. Jobredeaux, R., Herencia-Zapana, H., Neogi, N.A., Feron, E.: Developing proof carrying code to formally assure termination in fault tolerant distributed controls systems. In: CDC, pp. 1816–1821. IEEE (2012)
18. Kats, L.C.L., de Jonge, M., Nilsson-Nyman, E., Visser, E.: Providing rapid feedback in generated modular language environments: adding error recovery to scannerless generalized-lr parsing. In: Arora, S., Leavens, G.T. (eds.) OOPSLA, pp. 445–464. ACM (2009)
19. Koo, T., Collins, M.: Efficient third-order dependency parsers. In: Hajic, J., Carberry, S., Clark, S. (eds.) ACL, pp. 1–11. The Association for Computer Linguistics (2010)
20. Lei, T., Long, F., Barzilay, R., Rinard, M.C.: From natural language specifications to program input parsers. In: ACL (1), pp. 1294–1303. The Association for Computer Linguistics (2013)
21. Linz, P.: An Introduction to Formal Languages and Automata, 5th edn. Jones & Bartlett Learning, Burlington, MA 01803, USA (2011)
22. Middelkoop, A., Dijkstra, A., Swierstra, S.D.: Dependently typed attribute grammars. In: Hage, J., Morazán, M.T. (eds.) IFL. LNCS, vol. 6647, pp. 105–120. Springer, Heidelberg (2011)
23. Necula, G.C.: Proof-carrying code. In: Avan Tilborg, H.C., Jajodia, S. (eds.) Encyclopedia of Cryptography and Security, 2nd edn., pp. 984–986. Springer, Heidelberg (2011)
24. Nederhof, M.-J., Bertsch, E.: An innovative finite state concept for recognition and parsing of context-free languages. Natural Language Engineering 2(4), 381–382 (1996)
25. Okhotin, A.: Generalized lr parsing algorithm for boolean grammars. Int. J. Found. Comput. Sci. 17(3), 629–664 (2006)
26. Okhotin, A.: Recursive descent parsing for boolean grammars. Acta Inf. 44(3-4), 167–189 (2007)
27. Okhotin, A.: Parsing by matrix multiplication generalized to boolean grammars. Theor. Comput. Sci. 516, 101–120 (2014)
28. Scott, M.L.: Programming Language Pragmatics, 3rd edn. Morgan Kaufmann, Waltham (2009)
29. Sippu, S., Soisalon-Soininen, E.: Parsing Theory: Volume II LR(k) and LL(k) Parsing, 1st edn. Springer, Heidelberg (2013)
30. Yang, Y., Ma, M.: Sustainable mobile computing. Computing 96(2), 85–86 (2014)
31. Younessi, O., Bahrololoomi, M.H., Yazdani, A.: The extension of deterministic cancellation parser to directly handle indirect and hidden left recursion. In: CCECE [1], pp. 1–4
32. Zhou, G., Zhao, J.: Joint inference for heterogeneous dependency parsing. In: ACL (2), pp. 104–109. The Association for Computer Linguistics (2013)

# An Approach for Unit Testing in OOP
# Based on Random Object Generation

Pablo Andrés Barrientos

Universidad Nacional de La Plata
50 y 120, La Plata, Argentina
pbarrientos@sol.info.unlp.edu.ar

**Abstract.** When unit testing, developers create tests cases to cover all possible states and configurations of the software units in order to discover bugs. However, due to the increasing complexity of modern systems, it is becoming very difficult to test the software using a large combination of inputs and outputs. Furthermore, software testing cannot be used to show the absence of bugs. Nowadays, many software testing techniques exist and most of them rely on the automation of specific execution paths or steps, with hard-coded objects and specific conditions. In this paper, we introduce a unit testing technique along with a tool which allows the tests to make use of randomly generated objects.

The idea of random generation of values for testing is not new. In functional programming there is a tool for testing specifications over functions called *QuickCheck*. This tool with its underlying ideas is used as the foundation for our work. We also made use of features that only exist in the OO programming paradigm. Our main goal is to show that this technique can be used in OO testing as a good complement of existing techniques. In this paper, we present a tool together with the underlying ideas behind it. This tool is a full implementation of the QuickCheck features and is the first step toward the creation of a new random testing technique for OO.

**Keywords:** Testing tool, software testing, OO testing, random testing, random object generation.

## 1 Introduction

Current and future trends for software development include complex requirements and frequent change requests. High complexity implies that a system may have potentially infinite combinations of inputs and outputs. It is very difficult to cover all possible execution paths and object states, within the source code of such a system, using current automated test-case techniques.

When unit testing, it is common practice to hard-code objects in the test cases. The developer implements each test method using three major sections: *arrange*, *act*, and *assert*. In the arrange section, an instance of the class being tested is created, initialized and its state is set. In the act section, the method is invoked. Finally, in the assert section, post-conditions are verified.

B. Murgante et al. (Eds.): ICCSA 2014, Part V, LNCS 8583, pp. 15–31, 2014.

With this approach, the developer implements multiple test methods with as many *configurations* and object states as possible, simulating the object state and behaviour when invoked. This is done to cover as many execution paths as possible within the method, and often the resulting act and assert sections of those tests are the same. It is common that due to the complexity of the unit under test or just due to a lack of time, this work cannot be completed. As a result, many execution paths are left uncovered. In the worst case, the most used paths in runtime are left uncovered, and the written tests are useless in terms of achieving real code coverage and effectiveness.

In this paper we propose the random generation of test objects for unit testing OO software. This idea, taken from QuickCheck [19] is another way of random testing [21]. Within this testing approach, test cases receive random objects and the *arrange* section does not exist or if it exists it becomes very short. Every test method looks like a specification of what the object should do, with no hard-coded values. Moreover, every test method is run many times to try to falsify the specification.

We present a tool called YAQC4J (**Y**et **A**nother **Q**uickCheck for **J**ava) [12], which implements these ideas. This tool aims to easily ensure specifications, so that a substantial number of errors are detected at a lower cost. The tool we present is based on the ideas behind Claessen's QuickCheck, and it also handles any new problems generated that arose from the migration of these ideas to the OO environment, including any new ideas from the OO paradigm. No other similar existing tool addresses the problems that we do, and this is the major technical contribution of our work.

The remainder of this paper is organized as follows. Section 2 gives the background needed to understand our approach. Section 3 introduces the tool. Section 4 also introduces some issues we addressed for OOP that no other similar tool has done before. Section 5 presents two different case studies. These case studies show some advantages of *YAQC4J* over standard JUnit testing. Section 6 presents the related work. Section 7 presents a comparison between our tool and tools following QuickCheck's approach. Finally, section 8 discusses some points about *YAQC4J* tool and draws future work we intend to undertake.

## 2   Background

In the functional paradigm, Claessen and Hughes [19] created a tool (a combinators library) for formulating and testing properties of Haskell programs. That tool was translated into many functional languages like Erlang [15] and Curry [18] due to its success. It was also translated into OO programming languages like Java [2, 11] and C++ [10].

In the original QuickCheck for Haskell, *specifications* are the way in which properties are proven. These tests are specifically generated to attempt to falsify assertions. For example, the property `testreverse` defined as:

```
testreverse xs = reverse (reverse xs) == xs
```

is a simple property over lists of any type. It says that when reversing a list twice, the resulting list must be equal to the original one. QuickCheck tries to falsify this property by proving the specification with randomly generated lists. In order to generate these lists, it restricts xs to be a type of class **Arbitrary**. Instances of this class provide the framework with all it needs to generate random values. QuickCheck provides combinators to define properties. For instance, in case we want to specify that the property will hold under certain conditions, we can set a precondition using the combinator ==> (also known as suchThat), as: ordered xs ==> ordered (insert x xs). QuickCheck provides a way to define test data generators by defining an instance of the **Arbitrary** class and implementing the **arbitrary** function. QuickCheck also provides combinators to classify randomly generated values with the **classify** combinator. Besides, if we want to distinguish which data is generated (for example, we would like to watch the length of generated lists) we can use the **collect** combinator.

Many predefined functions or combinators were written to help us to create custom generators, such as oneof, choose, frequency, vector, and elements; all of them return a Gen a (a generator). For instance, if we have the type data TernaryLogic = Yes | No | Unknown, we can define a generator for it using choose, in the following form:

```
instance Arbitrary TernaryLogic where
   arbitrary = do n <- choose (0, 2) :: Gen Int
               return $ case n of
                     0 -> Yes
                     1 -> No
                     2 -> Unknown
```

Once a property is written, it can be tested by using the quickCheck function.

Most of these features from QuickCheck have been used for our tool, as explained in Section 3. For more details about QuickCheck for Haskell we recommend reading the original paper [19], as this paper is not intended to describe QuickCheck at depth.

## 3   Tool Description

JUnit [31] is the well-known framework used in Java for unit testing. In JUnit 4.x, the annotation @Test identifies test methods inside a class that are used for testing a target class. There are also two additional annotations to manage the general context: @Before (used to set up the test environment), and @After (used to clean up the test environment). The framework runs the class, method by method, executing these methods every time. In Fig. 1 we show the typical execution cycle of JUnit.

Each method is written for a particular *configuration* and state of the object. The object itself must be created, and then the message being tested is sent to the object. The developer has two alternatives for setting up the object: to set it up as part of the general context, or to hardcode it in the test method. A very simple test in JUnit (without general context initialization) is shown below.

Our tool enters at this point to provide the creation of random objects. With *YAQC4J*, the developer does not test a hard-coded object but the method that should be tested. The developer does not need to write as many different tests as the different states an object could have when receiving the method under

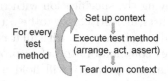

**Fig. 1.** Execution cycle of test methods in JUnit

test. Our tool runs each test method several times (100 by default), and every time different randomly generated objects are provided.

```
public class TestClass {
 @Test
 public void testReverse() {
    StringBuffer buff = new StringBuffer("some hardcoded string");
    assertEquals(str, buff.reverse().reverse().toString());
 }
}
```

We created a special runner which is a subclass of an existing JUnit runner. This runner manages the testing cycle based on metadata given by the @Configuration annotation. The developer only needs to indicate which objects must be generated randomly by putting them as parameters of the test method. In Fig. 2 we show the typical execution cycle of *YAQC4J*. As the runner is based in JUnit, the tool can be used in visual IDEs with support for JUnit.

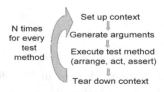

**Fig. 2.** Execution cycle of test methods in YAQC4J

We show below a simple example where the runner is specified with the annotation @RunWith and a random String object is used.

```
@RunWith(QCCheckRunner.class)
public class SimpleTest {
   @Test
   public void testReverse(final String str) {
       StringBuffer buff = new StringBuffer(str);
       assertEquals(str, buff.reverse().reverse().toString());
   }
}
```

The developer can create a custom generator for a specific type. New generators must implement the Gen<T> interface, being T the class of the object to be generated.

```
public interface Gen<T> {
    T arbitrary(Distribution random, long minsize, long maxsize);
}
```

The method arbitrary has three parameters: the random distribution used to get random values (we provide some distribution classes, and the developer could create his own Distribution subclass), and the other two parameters (minsize and maxsize) that can be used as lower and upper bounds for types such as numbers, trees, Strings, collections, etc.

In Fig. 3 it can be seen an example of test configuration (at class and method level), and the use of the @Generator annotation to use a specific generator. Many generators can be defined for the same type using the annotation @UseGenerators. In that case, our tool randomly selects one of the available generators for the type every time the test runs.

```
@Configuration(maxsize = 50)
@Generator(generator = IntegerGen.class, klass = int.class)
public class SimpleTest {
    @Test
    @Configuration(distribution=InvertedNormalDistribution.class)
    public void classConfigurationTest(final int i) {
        assertTrue("@Configuration for class (or int generator) failed", i>=-50 && i<=50);
    }

    @Test
    @Configuration(maxsize = 5, minsize = -5)
    public void methodConfigurationTest(final int i) {
        assertTrue("@Configuration for methods (or in generator) failed", i>=-5 && i<=5);
    }
}
```

Fig. 3. Samples of metadata in *YAQC4J* tests

Sometimes, any precondition or constraint could be required for a random object. The developer can create a custom generator which considers the constraint(s), or use the Implication.imply method. It receives a condition that the generated object must hold, otherwise the test is discarded. The number of failures before the test is declared failed as well as the number of runs could be specified with the properties maxArgumentsFails and tests in the annotation @Configuration).

Our tool provides some generators for several standard types, such as: String, Long, Boolean, Character, Float, BigDecimal, BigInteger, primitives, arrays and collections, Date, enumeratives, etc. We also include many classes equivalent to the combinators from the original QuickCheck, and added more generators for features present only in OO programming: NullGen, IdentityGen, and CloneGen. Finally, a TransformerGen allows mapping from one type to another. That is the way in which we created a Calendar generator by using the generator of Date instances.

We also implemented the QuickCheck features for classifying the objects generated. In Fig. 4 there is a simple example of these annotations. The annotation @Classify receives the name that will be used to report results and the list of classifiers (subclasses of `Classifier` that implement the method `public boolean classify(T object)`). The annotation `@Collect` receives the name that will be used to report results, and the subclass of `Collector` which will classify every object generated in a particular partition defined by the method `public K getCategoryFor(T t)`. The result of executing the given example is:

```
Results for collector "partitionInt":
  0: 25 occurrences (25.0%)
  1: 24 occurrences (24.0%)
  2: 13 occurrences (13.0%)
 -2: 11 occurrences (11.0%)
 -1: 27 occurrences (27.0%)

Results for collector "sampleIntegers":
  even: 48 occurrences (48.0%)
  odd: 52 occurrences (52.0%)

sample1: OK. Passed 100 tests.
```

```java
public class SampleCollector {
    @Test
    @Configuration(minsize=-2, maxsize=2)
    @Generator(klass = Integer.class, generator = IntegerGen.class)
    public void sample1(
      @Classify(name="sampleIntegers",
                classifiers = {EvenClassifier.class, OddClassifier.class}) final Integer a,
      @Collect(name="partitionInt",collector=IntegerPartitionCollector.class) final Integer b){
        ...
    }
}
```

**Fig. 4.** Classifiers

We used JUnit to test our tool, but we also used *YAQC4J* to test itself. We found some interesting bugs, one of which forced us to make a design decision on the tool. When defining tests for the lists generator, we found that *YAQC4J* failed to create instances of ArrayList because one of its constructors receives an integer as argument. When the integer is negative, the constructor fails. We had to remove the integers generator from the list of default generators, and we also defined a PositiveIntegerGen that can be used when generating lists.

## 4   Special issues

Functional and OO programming have big differences. OOP has side effects, class hierarchies, interfaces, abstract classes, and developers can define singletons, etc. In this section, we explain how we deal with these issues.

### 4.1   Singletons

The singleton deign pattern [24] is used to implement the mathematical concept of a singleton, and implies that one class has only one object in the whole system. The problem arises when the instance is stateful. In the context of unit testing, a singleton is shared among all test methods and that, could give unexpected results if one test modifies the state of the singleton. For solving this issue, we created the `SingletonGen`, which is an abstract class that accesses the singleton method in run-time. The method is assumed to be `getInstance`, but it can be specified when subclassifying. This generator resets the class variable that holds the unique instance every time it has to generate an instance of the class, and a new instance is automatically created. An easy way the developer has to use this generator, is to add the annotation IsSingleton to a test parameter. *YAQC4J* will automatically associate a SingletonGen to that type and generate fresh instances every time the test runs.

## 4.2   Abstract Classes and Hierarchies

Our idea is to simplify the amount of code developers have to write for using the tool and test their code. For that purpose, when a developer does not specify or does not have a generator for a class, we try to create an instance by looking all generators defined for subclasses in its hierarchy and selecting randomly one of those generators. That logic is part of the runner. We also defined the generator `ClassHierarchyOrInterfaceGen`, which proceeds in the same way, and can be used when the test parameter is an abstract class or interface.

## 4.3   Interfaces

When an argument for a test is an interface, we have two options. The first one is to look for classes that implement the interface and find generators for those classes and randomly choose one of them (as described previously). The other option we provide is the InterfaceInstancesGen, which generates a proxy object that implements the interface. The proxy is able to receive messages and it returns a randomly generated object if possible. It means that the method will return an object if the return type is not `void` and if the type has a generator. If the return type is the interface itself, the proxy returns `null` to avoid a potential infinite loop. *YAQC4J* uses this generator in case there is no other generator associated to the interface or any class in the classpath which implements it.

## 4.4   Constructor Based Generator

We defined a generic generator which creates instances of any class based on the public constructors the class has. The generator looks up for public constructor of the class, selects one of them randomly, and creates an instance by generating (also randomly) the arguments needed by the constructor. This generator is used automatically in case there is no generator associated to the type, and there is no other class in the hierarchy which has associated generators.

# 5   Discussion

We have used our tool in many open source projects to validate its features. We present in this section two interesting examples with interesting results.

## 5.1 Commons Math

Commons Math v3.2 [1] is a library of mathematics and statistics components. We selected the class Complex and rewrote the existing tests. As a result, the branch coverage[1] increased from 91.2% to 94.1%, and the total lines of code[2] (including generators for double and Complex) decreased from 1038 to 992. There were instance samples that were not covered by the original hand-written test cases that *YAQC4J* generated. We would also highlight how oracles are specified in each case. In the original tests, assertions have hard-coded values, while in the new version they are specification based on calculations that work for any input value. A visual comparison of the tests can be done by observing Fig. 5.

```
//ComplexTest (JUnit)
@Test
public void testConstructor() {
  Complex z = new Complex(3.0, 4.0);
  Assert.assertEquals(3.0,z.getReal(),1.0e-5);
  Assert.assertEquals(4.0, z.getImaginary(), 1.0e-5);
}

@Test
public void testConstructorNaN() {
  Complex z = new Complex(3.0, Double.NaN);
  Assert.assertTrue(z.isNaN());
  z = new Complex(nan, 4.0);
  Assert.assertTrue(z.isNaN());
  z = new Complex(3.0, 4.0);
  Assert.assertFalse(z.isNaN());
}

//YAQC4JComplexTest
@Test
public void testConstructor(Double r,Double i){
  Complex z = new Complex(r, i);
  assertEquals(r, z.getReal(), 1.0e-5);
  assertEquals(i, z.getImaginary(), 1.0e-5);
  assertEquals(Double.isNaN(z.getImaginary()) || Double.isNaN(z.getReal()), z.isNaN());
  assertEquals(!Double.isNaN(z.getReal()) && !Double.isNaN(z.getImaginary()), !z.isNaN());
}
```

**Fig. 5.** Test fragments from *Commons Math* with *J*Unit and *YAQC4J*

## 5.2 JStock

JStock [5] (v1.0.5y) is an open source library for tracking stock investments. We wrote tests for the class org.yccheok.jstock.engine.Duration, without considering the tests written for this class. We found two new bugs in the implementation with less code (test size decreased from 110 to 106), the branch coverage increased from 62.5% to 75%, and the instructions coverage increased from 71.3% to 95.1%.

---

[1] Measured with the Eclipse plugin eCobertura [3].
[2] Measured with the Eclipse plugin Metrics [7], and the default Eclipse code formatter.

We also rewrote a test for the method TSTSearchEngine.searchAll (used to suggest a list of items given a prefix). We found bugs for simple input cases and we wrote 58 lines code less than in the original test. The intention of the developer in the original test was to show that the search worked for a fixed prefix and a fixed number of occurences in upper and lower case of that prefix for certain number of inputs. In the *YAQC4J* version of the test, a generator receives an array of randomly generated strings and inserts the prefix in each string, changing also the cases of the prefix. In Fig. 6 we show one of the many tests written originally, and the new version with *YAQC4J*.

```
public void testSearchAll() {
  TSTSearchEngine<Name> engine = new TSTSearchEngine<Name>();
  engine.put(new Name("Mr Cheok"));
  engine.put(new Name("miss Lim"));
  engine.put(new Name("mRM"));
  engine.put(new Name("mr H"));
  engine.put(new Name("ABCDEFG"));
  assertEquals(3,engine.searchAll("MR").size());
}

public void qcTestSearchAll(String... strings) {
  Distribution dist = Arbitrary.defaultDistribution();
  String substring = stringGen.arbitrary(dist, 1, 5);
  int numOccurrences = positiveIntGen.arbitrary(dist, 1, strings.length);
  Gen<String[]> gen = new TSTSearchEngineStringGen(substring, numOccurrences, strings);
  String[] inputStrings = gen.arbitrary(dist, Integer.MIN_VALUE, Integer.MAX_VALUE);
  TSTSearchEngine<Name> engine = new TSTSearchEngine<Name>();
  for (String input : inputStrings) {
    if (!StringUtils.isEmpty(input)) engine.put(new Name(input));
  }
  assertEquals(numOccurrences, engine.searchAll(substring).size());
}
```

**Fig. 6.** JUnit and *YAQC4J* tests for method TSTSearchEngine.searchAll

# 6   Related Work

The interest in random testing OO software has increased greatly in recent years. Several tools combine random testing with other techniques and strategies in order to generate better results in terms of code coverage, mutation score, faults detection or false positives reported, among other evaluation criteria. The most significant tools are: JCrasher [20] which uses a complete black-box approach generating a parameter's graph to analyze and generate the tests input, Eclat [27] which reuses generated input from an object pool and invokes methods to modify them, Rute-J [14], Randoop [28] which is an improvement over Eclat using feedback-directed random testing [29] and is available for Java and .Net, Jartege [26] which uses JML for its systematic approach of random testing, AutoTest [25] which is written in Eiffel and exploits its *contracts* feature and combine it with adaptive random testing [17], Pex [32] which uses dynamic symbolic execution and a constraint solver for generating .NET tests, EvoSuite [22] which uses metaheuristics (evolutionary algorithms and dynamic symbolic execution) as its search-based approach, TestFul [16] and eToc [33] use an evolutionary algorithm for mutating tests and maximize a given coverage measure, and finally YETI [13] uses a proprietary random strategy, an abstract meta-model which allows implementation for different languages, and a unique feature to change runtime parameters and monitor the evolution of tests.

JCrasher, Eclat, Randoop, Jartege, AutoTest and YETI have automatic or semi-automatic tests generation. The developer has to work at the beginning of the process to configure parameters, and at the end to filter false positives and improve the code legibility which is not an easy task [23]. Finally, JCrasher, eClat, and Evosuite generate oracles based on predefined heuristics or constraints given by the user, and Jartege and Autotest use JML and Eiffel contracts.

As mentioned before, our tool is the first step toward the creation of a new testing technique for OO. We achieved a full implementation of the QuickCheck features for OOP, but it is our aim to add a systematic search-based approach for generating inputs as well. Our future plans will be outlined in section 8.

On the other hand, our tool is not the only one that exists for automatic generation of objects. A few libraries were written before. As *YAQC4J* aims to cover many features that none of these tools did, we present them and do a comparison in section 7.

## 6.1   JCheck

JCheck [2] (v.0.1) provides generators for classes from package `java.lang` and most of the primitive types, a custom JUnit runner, the `imply` operator, test configuration by annotations (such as number of tests or `size` parameter), and the capability of defining new generators. However, this tool uses just the standard class `java.util.Random` and we also found that the execution cycle is not properly managed and the context is reused for different tests in the same class.

## 6.2   QuickCheck for Java

This tool [11] (v0.6) has many features in common with ours. It as generators for primitives, several generators from the original QuickCheck (`frequency`, `oneOf`, etc), transformation generators, different distribution types, size for random generation, and a special JUnit runner. Nevertheless, there is a big difference in the way random data is obtained. In *YAQC4J* the data comes from parameters defined in the test signature, while in QuickCheck the data is provided by helper methods, invoked by the user in the test implementation.

## 6.3   QC4J

QC4J [9] (v0.1) provides a few generators for basic types but provides a base generator for defining custom ones. It also provides a powerful mechanism to map classes with their corresponding generators (our tool has this feature as well). QC4J does not provide any integration with JUnit (it has its own runner, based on a common `main()` method). As JCheck, QC4J uses only `java.lang.Random` rather than parameterizable distributions. Finally, this tool does not provide a mechanism to parameterize the number of tests and size of the generated object.

## 6.4   JUnit-QuickCheck

JUnit-QuickCheck [30] (v0.2) is a testing library written for JUnit whose main achievement is to provide random values to JUnit theories [30]. Theories are a relative new element in JUnit that allows parameterized tests using *data points* (data sources of a given type). Theories provide a bounded universe in which some behaviour is tested.

This tool feeds JUnit theories with random objects, and provides generators for common types and collections. It also provides a way to create new generators and the user can provide the *source of randomness* (subclass of `java.util.Random`). It also has an operator to restrict invalid objects (like the operator `imply` in our tool). It reuses the runner used by JUnit theories. This tool is promising, but it is still in its experimental phase.

## 7   Comparison

In this section, we compare *YAQC4J* against other similar tools described in previous section. We describe the methodology we then show the results and conclusions.

## 7.1   Methodology

We evaluated all tools in different dimensions and under specific conditions which are described in this section.

### Dimensions Covered

- Features: we analyze each tool in terms of QuickCheck and OOP features.
- End-user support: it includes development environment, documentation and examples of use.
- End-user effort: comprehensibility (intelligibility, ease of use, reporting and debugging), coding and execution time[3], size (ELOC, number of classes, methods, and variables [2]).
- Effectiveness: branch and statements coverage[1].

---

[3] Measured using Perf4J [8] in a Dell XPS L501X, with Intel Core i5 M460 processor @ 2.53GHz 64bits, 4Gb RAM, Window 7 Home Premium, JDK v1.6.0_23 (default jvm values). All tools ran around 100 times each test method.

**Rules and Conditions**

- prepare all examples to be compiled and run from the same development environment: a Java 1.6 project for Eclipse Kepler (Build id: 20130614-0229) with all tools, libraries and initial code.
- provide all subjects with the documentation available of each tool, and provide an introductory lesson of the testing approach. A survey was given to each subject to be completed during the session. This survey includes items about the end-user effort and the time required to develop each solution.
- select useful case studies and test cases. These cases must allow the comparison of all tools using the features they all have.
- there were a total of five Java professional developers with different experience in Java and unit testing, and from different countries (China – 12 years of experience, Peru – 2 years of experience, Colombia – 7 years of experience, and Argentina – 8 and 5 years of experience).

To the best of our knowledge, there are no recommendations or guidelines to select case studies for evaluating testing tools. In this context, we used the following criteria:

- open source and documented code, to be accessed and studied.
- easy to understand but having non-trivial logic. All examples belong to simple domains, with branches and conditions to create interesting test cases.
- case study with existing test cases written, given that one of the tasks is to rewrite tests with every tool.

The case studies we selected are:

- IMoney [6]. Simple example used to explain XUnit tools. In this case, the developers had to rewrite the existing test case MoneyTest using every tool.
- javaGeom (v0.11.1) [4]. A library for performing geometric computations. It has a few tests written. The developers were required to write additional tests for the class Box2D. In this case, subjects were asked to develop additional generators for Point2D and Box2D.

### 7.2   Results and Discussion

Due to space restrictions, we will use the following numbers in all tables shown below to refer to the tools: $1 = YAQC4J$, $2 =$ JCheck, $3 =$ QuickCheck for Java, $4 =$ QC4J, $5 =$ JUnit-QuickCheck, $6 =$ QuickCheck, $7 =$ JUnit.

| Features | 1 | 2 | 3 | 4 | 6 | 5 |
|---|---|---|---|---|---|---|
| Execution cycle | Y | - | Y | Y | Y | Y |
| Generator combinators | Y | Y | Y | N | Y | N |
| Primitives generators | Y | Y | Y | Y | Y | Y |
| Input classification | Y | N | N | N | Y | N |
| *imply* operator | Y | Y | Y | N | Y | Y |
| Standard classes generators | Y | Y | Y | N | N/A | N |
| Abstract classes generator | Y | N | N | N | N/A | N |
| Hierarchy generator | Y | N | N | N | N/A | N |
| Interfaces generator | Y | N | N | N | N/A | N |
| Constructor-based generator | Y | N | N | N | N/A | N |
| Singleton generator | Y | N | N | N | N/A | N |
| Addicional generators | Y | Y | Y | N | Y | N |
| Configurable tests | Y | Y | Y | N | Y | Y |
| Configurable random distribution | Y | N | Y | N | N | Y |
| Parametrized tests | Y | Y | N | Y | N | Y |

The previous table shows that *YAQC4J* is the only tool that implements all features. Its features match with the original tool (included in that table for reference) and also with the OO features. The rest of the tools only provide partial implementations, highlighting only the fact that QC4J does not provide tests configuration (key feature in this approach). Another unique contribution of *YAQC4J* is the classification of the input generated during the tests.

| End-user support | 1 | 2 | 3 | 4 | 5 | 7 |
|---|---|---|---|---|---|---|
| Dev. environment | * | * | * | own runner | * | * |
| Documentation | online/javadocs | online | online/javadocs | No | online | online |
| Example of use | online | few online | online | online | online | online |

\* = any environment with UI and support for JUnit.

From the end-user support table it can be observer that QC4J has a proper runner (i.e. a different way to execute tests). It was pointed out by the developers that it has been the cause of its negative qualifications in the end-user effort dimension.

All tools have documentation available either online or as javadocs. Only QC4J has no documentation, except for a few examples of use. This weakness and the runner played against it from an ease of use point of view.

| End-user effort | 1 | 2 | 3 | 4 | 5 | 7 |
|---|---|---|---|---|---|---|
| Intelligibility(1 to 5) | 4.2 | 3.2 | 2.8 | 2.8 | 3.6 | 4 |
| Ease of use (1 to 5) | 4.2 | 3.4 | 2.2 | 2.8 | 2.8 | 4.2 |
| Reporting (1 to 5) | 4 | 2.8 | 2.8 | 3.8 | 3 | 3.6 |
| Debugging (1 to 5) | 2.8 | 3 | 2.2 | 3 | 3.6 | 4.2 |
| Overall value (1 to 10) | 8.2 | 6.4 | 5 | 6.6 | 6.6 | 7.4 |

The table above and the ones below show the average values for each item. From an end-user effort point of view, JUnit-Quickcheck and *YAQC4J* are the easiest to understand, but our tool is the easiest to use. Only JUnit (added as a reference) obtains similar results. QuickCheck for Java was hard to use and the cause was the tool problems occurred during test executions. This issue also impacted negatively in other dimensions of the comparison. QuickCheck for Java was also hard to use and understand due to the way the random data must be generated. Regarding the reporting, *YAQC4J* and QC4J obtained good qualifications. They report the results in a similar manner as does the original tool. A weakness in *YAQC4J* seems to be the debugging. The problem indicated by some subjects was that once an error occurs, the stacktrace refers firstly to the tool code and not to the test or unit itself. This weakness will be worked out in a future release of *YAQC4J*. Finally, in the overall value, *YAQC4J*, QuickCheck for Java and JUnit-QuickCheck received the highest qualifications, being *YAQC4J* clearly the best qualified.

## Time

| IMoney | 1 | 2 | 3 | 4 | 5 | 7 | javaGeom | 1 | 2 | 3 | 4 | 5 | 7 |
|---|---|---|---|---|---|---|---|---|---|---|---|---|---|
| Total coding (m) | 107 | 247 | 260 | 99 | 194 | N/A | Total coding (m) | 32 | 46 | 46 | 36 | 47 | N/A |
| Generators (m) | 7 | 16 | 5 | 15 | 0 | N/A | Generators (m) | 11 | 12 | 18 | 14 | 13 | 25 |
| Execution (ms) | 571 | 897 | 963 | 321 | 6027 | 293 | Execution (ms) | 377 | 457 | 496 | 225 | 2262 | 243 |

From the tables above, it can be observed that the time used to code generators differs from tool to tool, but our tool consumes the shortest amount of time in most of the cases. In some cases there was needed some extra time for coding new generators, and it coincides with the tools that do not have generators for standard types (JCheck) or the way to define them is not intuitive (QC4J). The time to code tests is significantly greater in JCheck, and always above the average for JUnit-QuickCheck. *YAQC4J* obtained the lowest times in both cases. Regarding the execution time, all tools registered good results except Junit-QuickCheck. Unfortunately, for this tool and for QuickCheck for Java the number of times a method is run depends on the random data to be generated. We used default values for the other tools (100) and tried to approximate the numbers of executions for them. We observed that JUnit-QuickCheck is the least efficient tool, and *YAQC4J* and QC4J are the most efficient ones.

## Size

| IMoney | 1 | 2 | 3 | 4 | 5 | 7 | javaGeom | 1 | 2 | 3 | 4 | 5 | 7 |
|---|---|---|---|---|---|---|---|---|---|---|---|---|---|
| ELOC | 269 | 279 | 371 | 231 | 277 | 157 | ELOC | 102 | 104 | 116 | 92 | 105 | 81 |
| # classes | 2 | 2 | 2 | 3 | 1 | 1 | # classes | 2 | 3 | 3 | 3 | 3 | 1 |
| # methods | 23 | 23 | 22 | 25 | 22 | 23 | # methods | 6 | 7 | 7 | 8 | 8 | 6 |
| # variables | 0 | 5 | 1 | 2 | 0 | 6 | # variables | 0 | 0 | 2 | 1 | 0 | 1 |

An analysis of size metrics does not show that any tool is superior to other. None of the tools could test the same functionality than JUnit using less code. QuickCheck for Java always required more code to cover the same functionality, due to the way generators must be used inside each method (using nested loops). The variation on the number of classes, methods and variables does not allow any conclusions to be made.

**Effectiveness**

| IMoney | Class | 1 | 2 | 3 | 4 | 5 | 7 |
|---|---|---|---|---|---|---|---|
| Inst. coverage(%) | Money | 95.44 | 98.54 | 94 | 98.74 | 91.98 | 94 |
| Branch coverage(%) | Money | 82.94 | 88.78 | 70.3 | 91.26 | 70.04 | 68.8 |
| Inst. coverage(%) | MoneyBag | 87.8 | 86.76 | 85.93 | 88.72 | 88,24 | 86.8 |
| Branch coverage(%) | MoneyBag | 86.18 | 83.34 | 80.35 | 89.44 | 81.6 | 85.7 |

| javaGeom | 1 | 2 | 3 | 4 | 5 | 7 |
|---|---|---|---|---|---|---|
| Inst. coverage(%) | 13,88 | 13,88 | 13,98 | 15,22 | 15 | 12,52 |
| Branch coverage(%) | 4,22 | 4,22 | 4,62 | 8,4 | 4,62 | 2,52 |

The coverage metrics indicate that *YAQC4J* reaches better instruction and branch coverage than JUnit and QuickCheck for Java. However, it is sometimes below other tools and QC4J always scored the highest coverage rates. As mentioned before, one of our future goals will be to improve the code coverage by following a strategic search based input generation.

## 7.3 Threats of Validity

The number of subjects for conducting the comparison may not be enough to generalize the results. However, all subjects are experiences professionals with different cultural and educational backgrounds, which could not be found in other comparison experiments for random testing tools.

Due to the lack of standard guidelines to select case studies and examples, the ones we used may not cover all aspects of the testing tools. Therefore, the case studies may not be representative for doing the comparison, and additional criteria and case studies could have been used.

## 8 Conclusions and Future Work

QuickCheck has been a successful tool and random testing approach in functional programming. Many tools similar to QuickCheck have also been written for different languages. OO programming has some examples of QuickCheck-like tools. In this paper we have presented the testing approach and a testing tool called *YAQC4J*, that follows the ideas behind QuickCheck and sets about OOP problem like no other existent tools. This tool is the first step toward a new way to define unit test cases. We showed that the approach increases the code coverage and effectiveness when compared to JUnit, and the tool is superior in various dimensions when compared to existing similar tools.

We believe there is still a lot of work to do. Two lines of research we will work in the near future are:

- Search-based object generation – As mentioned before, we will analyze different search-based approaches used in other random testing tools and extend our tool with a more systematic way of choosing the input values. For instance, using an objects pool for reusing existent objects or combining pure random generation with dynamic symbolic execution could improve actual code coverage metrics.

– Environment with IoC – We implemented a special runner for our tool which manages the execution cycle, but it would be interesting to write another runner which has IoC capabilities. This runner could inject random and fixed dependencies to the object being tested.

**Acknowledgments.** We want to thank to all developers who helped us to compare *YAQC4J* with other testing tools: Mandy Siu, David Vara, Andrés Sayago, Mariano Pilotto and Victor Toledo.

# References

[1] Commons math: The apache commons mathematics library, http://commons.apache.org/proper/commons-math/ (last accessed January 3, 2014)

[2] EasyCheck - test data for free, http://www.jcheck.org/ (last accessed December 27, 2013)

[3] Eclipse plugin for cobertura, http://ecobertura.johoop.de (last accessed August 27, 2013)

[4] Javageom, a geometry library for java, http://sourceforge.net/projects/geom-java/ (last accessed August 27, 2013)

[5] JStock, http://jstock.sourceforge.net/ (last accessed August 27, 2013)

[6] Junit test infected: Programmers love writing tests, http://junit.sourceforge.net/doc/testinfected/testing.htm (last accessed August 27, 2013)

[7] Metrics, http://metrics.sourceforge.net/ (last accessed August 27, 2013)

[8] Perf4j v0.9.16, http://perf4j.codehaus.org/ (last accessed December 16, 2013)

[9] QC4J, http://sourceforge.net/projects/qc4j/ (last accessed December 27, 2013)

[10] QuickCheck++, http://software.legiasoft.com/quickcheck/ (last accessed December 27, 2013)

[11] QuickCheck for Java, http://java.net/projects/quickcheck/pages/Home/ (last accessed December 27, 2013)

[12] YAQC4J, Yet Another QuickCheck for Java, http://sourceforge.net/projects/yaqc4j/ (last accessed on December 28, 2013)

[13] Yeti site, http://code.google.com/p/yeti-test/ (last accessed December 27, 2013)

[14] Andrews, J.H., Haldar, S., Lei, Y., Li, F.C.H.: Tool support for randomized unit testing. In: Proceedings of the 1st international workshop on Random testing, RT 2006, pp. 36–45. ACM, New York (2006)

[15] Arts, T., Hughes, J., Johansson, J., Wiger, U.: Testing telecoms software with quviq quickcheck. In: Proceedings of the 2006 ACM SIGPLAN workshop on Erlang, ERLANG 2006, pp. 2–10. ACM, New York (2006)

[16] Baresi, L., Lanzi, P.-L., Miraz, M.: Testful: An evolutionary test approach for java. In: 2010 Third International Conference on Software Testing, Verification and Validation (ICST), pp. 185–194 (2010)

[17] Chen, T.Y., Leung, H., Mak, I.K.: Adaptive random testing. In: Maher, M.J. (ed.) ASIAN 2004. LNCS, vol. 3321, pp. 320–329. Springer, Heidelberg (2004)

[18] Christiansen, J., Fischer, S.: EasyCheck — test data for free. In: Garrigue, J., Hermenegildo, M.V. (eds.) FLOPS 2008. LNCS, vol. 4989, pp. 322–336. Springer, Heidelberg (2008)

[19] Claessen, K., Hughes, J.: Quickcheck: a lightweight tool for random testing of haskell programs. In: Proceedings of the Fifth ACM SIGPLAN International Conference on Functional Programming, ICFP 2000, pp. 268–279. ACM, New York (2000)

[20] Csallner, C., Smaragdakis, Y.: Jcrasher: an automatic robustness tester for java. Softw. Pract. Exper. 34(11), 1025–1050 (2004)

[21] Duran, J.W., Ntafos, S.C.: An evaluation of random testing. IEEE Trans. Softw. Eng. 10(4), 438–444 (1984)

[22] Fraser, G., Arcuri, A.: Evosuite: Automatic test suite generation for object-oriented software. In: Proceedings of the 19th ACM SIGSOFT Symposium and the 13th European Conference on Foundations of Software Engineering, ESEC/FSE 2011, pp. 416–419. ACM, New York (2011)

[23] Fraser, G., Staats, M., McMinn, P., Arcuri, A., Padberg, F.: Does automated white-box test generation really help software testers? In: Proceedings of the 2013 International Symposium on Software Testing and Analysis, ISSTA 2013. ACM, New York (2013)

[24] Gamma, E., Helm, R., Johnson, R., Vlissides, J.: Design patterns: elements of reusable object-oriented software. Addison-Wesley Longman Publishing Co., Inc., Boston (1995)

[25] Meyer, B., Ciupa, I., Leitner, A., Liu, L.L.: Automatic testing of object-oriented software. In: van Leeuwen, J., Italiano, G.F., van der Hoek, W., Meinel, C., Sack, H., Plášil, F. (eds.) SOFSEM 2007. LNCS, vol. 4362, pp. 114–129. Springer, Heidelberg (2007)

[26] Oriat, C.: Jartege: a tool for random generation of unit tests for java classes. In: Reussner, R., Mayer, J., Stafford, J.A., Overhage, S., Becker, S., Schroeder, P.J. (eds.) QoSA 2005 and SOQUA 2005. LNCS, vol. 3712, pp. 242–256. Springer, Heidelberg (2005)

[27] Pacheco, C., Awasthi, P.: Eclat: Automatic generation and classification of test inputs. In: Gao, X.-X. (ed.) ECOOP 2005. LNCS, vol. 3586, pp. 504–527. Springer, Heidelberg (2005)

[28] Pacheco, C., Ernst, M.D.: Randoop: feedback-directed random testing for java. In: Companion to the 22nd ACM SIGPLAN Conference on Object-Oriented Programming Systems and Applications Companion, OOPSLA 2007, pp. 815–816. ACM, New York (2007)

[29] Pacheco, C., Lahiri, S.K., Ball, T.: Finding errors in.net with feedback-directed random testing. In: Proceedings of the 2008 International Symposium on Software Testing and Analysis, ISSTA 2008, pp. 87–96. ACM, New York (2008)

[30] Saff, D.: Theory-infected: or how i learned to stop worrying and love universal quantification. In: Companion to the 22nd ACM SIGPLAN Conference on Object-Oriented Programming Systems and Applications Companion, OOPSLA 2007, pp. 846–847. ACM, New York (2007)

[31] Tahchiev, P., Leme, F., Massol, V., Gregory, G.: JUnit in Action, 2nd edn. Manning Publications Co., Greenwich (2010)

[32] Tillmann, N., de Halleux, J.: Pex–white box test generation for.NET. In: Beckert, B., Hähnle, R. (eds.) TAP 2008. LNCS, vol. 4966, pp. 134–153. Springer, Heidelberg (2008)

[33] Tonella, P.: Evolutionary testing of classes. In: Proceedings of the 2004 ACM SIGSOFT International Symposium on Software Testing and Analysis, ISSTA 2004, pp. 119–128. ACM, New York (2004)

# Weighted-Frequent Itemset Refinement Methodology (W-FIRM) of Usage Clusters

Veer Sain Dixit[1,*], Shveta Kundra Bhatia[2], and Sarabjeet Kaur[3]

[1] Department of Computer Science,
Atma Ram Sanatan Dharma College,
University of Delhi,
New Delhi, India
veersaindixit@rediffmail.com
[2] Department of Computer Science, Research Scholar,
University of Delhi,
New Delhi, India
shvetakundra@gmail.com
[3] Department of Computer Science,
Indraprastha College for Women,
University of Delhi,
New Delhi, India
sarabjeet.kochhar@gmail.com

**Abstract.** Due to information overload on the Internet a large number of systems have been developed for extracting user behavior. This paper presents mining of Frequent Itemsets and refinement of usage clusters for web based applications. Here a particular case is under consideration where sessions in a cluster are in abundance, consequently leading to a very large number of not-so interesting recommendations for the user. To solve such problems we intend to refine clusters on the basis of Weighted Frequent Itemsets that in turn help to generate improved quality refined clusters. In the proposed work, Frequent Itemsets are sets of web pages that occur in sessions more than a given threshold known as the minimum support. Motivation for adapting Frequent Itemsets for refinement is the demand of dimensionality reduction. Experimental results show that the cluster quality using the proposed approach is better than the existing approaches (DBS, 2011 and HITS, 2010). After getting refined clusters the same can be used for number of applications such as Web Personalization, improvement in Web Site Structure, Analysis of Users' Online Behavior and the services of a Recommender System.

**Keywords:** Web Usage Mining, Local Frequent Itemsets (LFI), Global Frequent Itemsets (GFI), Daveis Bouldin (DB) Index, and Silhouette Coefficient.

---

[*] Corresponding author.

B. Murgante et al. (Eds.): ICCSA 2014, Part V, LNCS 8583, pp. 32–50, 2014.

# 1   Introduction

In the World Wide Web context, Web Usage Mining process is divided into three main sub processes: pre-processing, discovery and analysis. Pre-processing of web log entries is followed by discovery of patterns using Data Mining tasks such as Association Rules, Sequential Pattern Analysis, Clustering and Classification [1, 23] which in turn lead to the discovery of useful patterns and groups. Analysis of the patterns and groups leads to applications such as Web Personalization, Web Page Recommendation, improvement in Customer Behavior and Web Site Structure [23, 24].

In the constantly developing web based environment abundant amount of information is generated and delivered to the web user. Many times user gets trapped in a difficult situation while searching information as per his requirements. Recommender Systems facilitate and help the users to locate items of their interest by changing the way people find products or information on the web. A large number of Recommender Systems [2, 3] are available (content, structure and usage) in the literature. These systems apply Knowledge Discovery Techniques from the fields of Artificial Intelligence and Soft Computing (Information Retrieval, Clustering, Association Rules, Classification, etc) [25, 26].

We shall focus on Clustering that can be performed by using various different algorithms based on Distance Measures [4], Neural Networks [5], Fuzzy [6], Hierarchical and Agglomerative Clustering Techniques [7]. The traditional clustering techniques group the sessions or pages by analyzing clusters, and then recommend them to one or more users in accordance to the users' cluster. Clustering of web sessions generates group of web pages; these group of web pages can be used for web recommendations. Web Recommendation is an application which serves the need of the web users in a better manner. The issue of clustering web sessions is a part of Web Usage Mining for discovering access behavior from large web log files. It is a technique of grouping sessions such that the sessions within a single cluster have similar page views. The core question of the clustering technique is to find clusters with good quality along with the reduction in dimensionality of the attributes for good recommendations. In the past, session clustering has been performed on numerical data using various clustering algorithms [27, 28] and on categorical data by using algorithms such as ROCK [29], CHAMELEON [30] and TURN [17]; where, sessions have been treated as unordered sets of clicks for various web applications. With the massive growth of the World Wide Web in the recent years the size of log files has increased exponentially due to which size of clusters incorporating similar sessions is very large. This leads to poor quality recommendations. A possible solution to this problem is to cluster similar sessions and refine them. Refinement is the technique of eliminating sessions from the cluster having a factor of dissimilarity with other sessions to improve the Cohesion and in turn the quality of the cluster. The advantages of refinement are to reduce and improve the set of recommendations being provided to the target user.

The goal of the Frequent Itemsets Mining [8, 9] is to discover correlations between existing elements in a dataset that the author shall use for refinement of clusters. Apriori [32], Eclat [33], Clique [34], FP-Growth [35] are some classical algorithms used to discover Frequent Itemsets.

In this paper, the main focus is on the clustering of web sessions, refinement of clusters and use of refined clusters for recommendation with reduced dimensionality. Main intension here is to generate refined clusters on the basis of Frequent Itemsets. A Frequent Itemset is a set of web pages that occur in a cluster of sessions. The refined cluster's quality is evaluated by using Davies Bouldin index [31, 36] and Silhouette Coefficient [37] quality measures. Test results confirm the effectiveness of our proposed method.

The rest of the paper is organized as follows. Section 2 describes about related work. Section 3 includes a description of the proposed framework. Section 4 describes the performance evaluation metrics. Section 5 presents the experiments conducted and the results obtained. Section 6 concludes the work and briefly discusses future work.

## 2      Related Work

A large number of real world applications such as Medicine, Image Processing, Biology, etc [10] use clustering as one of the techniques to extract useful information and knowledge. Clustering of web data is the process of grouping web data into clusters so that similar objects are in a single group and dissimilar objects are in different groups [11, 12]. The goal is to meet preferences of the target user and organize the data into groups so as to increase web Information Accessibility, improve Information Retrieval and Content Delivery on the Web. Web data clustering can be link based [13, 14, 15] and session-based [19, 5]. Link based clustering [16, 17, 18] groups together sessions on the basis of access patterns and session based clustering [21] on the viewing of web pages. The records of user's activities on a Web site are stored in a log file supported by various attributes described in [37]. Literature on various clustering algorithms can be found in [4, 25, 26].

The process of clustering of web sessions is an important research area in Web Mining. Banerjee and Ghosh [13] proposed similarity graph in conjunction with the time spent on web pages to estimate group similarity where the similarity is defined by the relative time spent on the longest common sub-sequences. Hay, Vanhoof and Wester [16] measured similarity on the basis of sequence alignment. The BIRCH algorithm was developed by Fu [15] that generalized the session in attribute oriented induction with respect to a data structure called page hierarchy-partial ordering. Xie and Phoha [39] suggested web user clustering from access log using the belief function. Shahabi and Kashani [40] suggested path similarity measures used to group the sessions. Ypma, T. Heskes [41] presented the clustering based on a mixture of the Markov models, information of the web user and the navigation patterns. Nasraoui

[42] correlates Fuzzy Artificial Immune System and clustering techniques to improve the profiles of users obtained through clustering. Xu and Liu [6] performed web user clustering based on K-Means Algorithm by introducing a cosine similarity measure based on the number of hits by a user on a web page in a session. Castellano [6] proposed a relational fuzzy clustering algorithm to discover profiles and extracting real user preferences. Oyanagi [44] presents an application of matrix clustering to Web usage data. Liu and Li [7] suggested a no-Euclidean distance measure on the access sequences of a pair of sessions. Kivi and Azmi [45] proposed a web page similarity measure for web session clustering using sequence alignment. Along with the above clustering techniques based on different features K-means is one of the most widely used clustering algorithms. Due to certain limitations of the K-means algorithm several modification are done with it time to time for improvement in cluster cohesion. Bentley [46] suggested kd-trees to improve triangle inequality to enhance K-Means. Bradley and Fayyad [47] presented a procedure for computing a refined starting condition that is based on a technique for estimating the modes of a distribution. A Kernel-Means algorithm was established in [48], mapping data points from the input space to a feature space through a non linear transformation by minimizing the clustering error in feature space. Dhillon [49] suggested refining clusters in high dimensional text data by combining the first variation principal and spherical K-Means. Elkan [50] suggested the use of triangle inequality to accelerate K-Means.

Frequent Itemsets Mining [20, 22, 38] plays an essential role in data mining tasks trying to find interesting patterns from databases. The original motivation for searching Frequent Itemsets came from the need to analyze supermarket transaction data [9, 43]. Frequent itemsets can been divided into local and global frequent items for which a number of algorithms [51, 52, 53] have been designed. Weighed Frequent Itemsets mining has also been suggested to find important Frequent Itemsets by considering the weights of itemsets. Some weighted frequent pattern mining algorithms MINWAL [54], WARM [55], WAR [56] WFIM [57], WIP [58] have been developed.

In this paper, refining of clusters is performed based on the Weighted Local Frequent Itemsets which are not a part of the Weighted Global Frequent Itemsets for improvement in quality of clusters.

# 3    Proposed Approach

The proposed approach is divided into four phases. The first phase includes the pre-processing of the web log data. The second phase extracts clusters from the pre-processed web log data. The third phase extracts global and local frequent itemsets. The fourth phase refines the clusters using global and local frequent itemsets as shown in Figure 1.

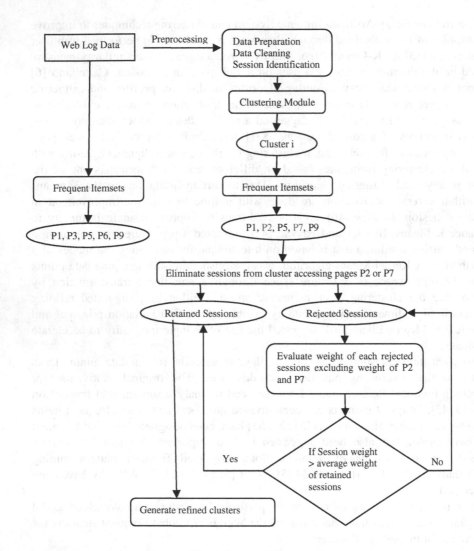

**Fig. 1.** Proposed Framework

## 3.1 Data Preparation

Data preparation section prepares the raw web log data for mining patterns, clusters and rules for recommendations. Several steps are needed to change the web data to a reliable format by eliminating background image files and unsuccessful requests. The pre-processing steps are the same for any web mining problem studied in [37]. The logs are pre-processed to group requests from the same user into sessions. A session contains the requests from a single visit of a user to the Web site. The users are identified by the IP addresses in the log and all requests from the same IP address within a certain time-window are put into a session.

Suppose for the given web site, there are n sessions S= {$S_1$, $S_2$, $S_3$... $S_n$} with the set of pages P= {$p_1$, $p_2$, $p_3$... $p_m$}. The pages are associated with the sessions as follows:

$$\text{Access } (p_i, S_j) = \begin{cases} 1 \text{ if } p_i \text{ is viewed by } S_j \\ 0 \text{ otherwise} \end{cases}$$

## 3.2    Session Cluster Formation

Clustering of similar sessions is the way to group the sessions with similar interests. Clustering has been performed by using the K-Means algorithm. K-Means, an unsupervised learning technique solves the well known clustering problems with the advantage of time and space complexity. The procedure is to classify a given data set through a certain number of clusters (assume K clusters) fixed a priori. The main idea is to define K centroids, one for each cluster. These centroids should be placed far away from each others because a different location causes different results. The next step is to take each point belonging to a given data set and associate it to the nearest centroid. When no point remains, the first step is completed and an early grouping is done. At this point we need to re-calculate K new centroids as of the clusters resulting from the previous step. After getting K new centroids, a new binding has to be done between the same data set points and the nearest new centroid. A loop is continued until no more changes are done in the centroid values.

The original clusters are generated using the above steps. The clusters are defined as follows:

$$C_i = \{ S_i \}, \quad i=1, 2..., x \tag{1}$$

Where $S_1$, $S_2$ ..., $S_x$ are similar sessions.

## 3.3    Frequent Itemsets Extraction

Frequent itemsets represent correlations among a set of items. Let I = {$i_1$, $i_2$, $i_3$... $i_n$} be a set of binary attributes representing web pages. Let D be a set of transactions representing the sessions and each session transaction T is a set of items such that T⊆ I. X is a set of items. A transaction T is said to support X, if and only if, X⊆ T. The support of an itemset X in D is the number of transactions in D that contain X. An itemset X is said to be frequent if its frequency is above a user defined threshold where frequency is defined as its relative support to the database size.

The FP growth algorithm was used for frequent itemsets generation as it overcomes the problem of number of passes over the database along with a situation where, as the support goes down the number and length of the frequent itemsets increases dramatically scaling much better than the previous algorithms. The authors shall focus and use the frequent itemsets for the process of refining clusters.

The above techniques of clustering and frequent itemsets mining have been used individually in an efficient and effective manner to generate recommendations for the end user. When a request is accepted from the current user a recommendation set of web pages is generated containing n number of pages which is in abundance leading to difficulty in selection for the end user. We intend to combine both the techniques by refining clusters on the basis of weighted frequent item sets from the complete log file (Global Frequent Itemsets (GFI)) and the individual clusters (Local Frequent Itemsets for Cluster $C_i$ (LFI($C_i$)).

## 3.4   Refinement of Clusters

The developed approach adopts the clusters generated using K-Means and frequent itemsets using the FP-growth algorithm from the Complete Log File (CLF) termed as Global Frequent Itemsets (GFI). Groups of Local Frequent Itemsets (LFI ($C_i$)) have been extracted from the individual clusters. The idea of generating clusters is to group similar sessions together having similar page views. To refine the obtained clusters we propose to use the GFI and LFI ($C_i$).

From the frequent itemsets obtained the items from the complete log file represent a fact that shall be true always, whereas the items obtained from the individual clusters have been generated in accordance to the sessions in a cluster. We first extract the itemsets from the cluster that are not in the CLF and populate the CHeck IMPortance (CHIMP []) array using the algorithm given in Figure 2:

```
INPUT: Cᵢ - cluster of sessions
       LFI (Cᵢ) - Itemsets of Cluster
       GFI- Itemsets of Complete log file
    // Finding itemsets from the cluster not existing in GFI
       NC = Number of itemsets in GFI
       For each frequent itemset in LFI (Cᵢ)
           If does not exist in GFI
               Append in CHeck IMPortance (CHIMP []) array
           End If
       End For
```

**Fig. 2.** Algorithm for searching itemsets existing in cluster $C_i$ but not existing in the complete log file

The cluster is divided into rejected and retained sessions by adding sessions to the rejected file (REJ) if web pages from CHIMP [] have been viewed otherwise the sessions are added to retained file (RET). For example, if the web page $I_9$ is a frequent itemset of the cluster $C_2$ but not a part of GFI, the cluster $C_2$ is divided into the rejected and retained files as follows:

$$REJ= \{S_1, S_2, S_5, S_7 \dots S_{24}\} \text{ Where, for each } S_i, I_9=1;$$

$$RET= \{S_3, S_4, S_6, S_9 \dots S_{15}\} \text{ Where, for each } S_i, I_9=0;$$

The algorithm for the division of a cluster is given in Figure 3:

```
INPUT: Cᵢ- Cluster of sessions
          CHIMP [] - Array of itemsets existing in cluster but not existing
          in the complete log file
          NCH- Number of elements in CHIMP []
                For each session in cluster Cᵢ
                      For k=1 to NCH
                            If Pₖ ==1
                                  Transfer session to Rejected file (REJ)
                            Else
                                  Transfer session to Retained file (RET)
                            End If
                      End For
                End For
```

**Fig. 3.** Algorithm for eliminating sessions from the cluster where elements of CHIMP [] have been accessed

The sessions eliminated from the cluster are based on the uncommon itemsets between GFI and LFI($C_i$) with an anomaly that the same Itemset might contribute to the formation of the cluster. Hence, here we introduce a factor of importance of a web page corresponding to the weight of the web page with respect to the cluster by Eq. (2). Algorithm is given in Figure 4.

$$Wt_{ij} = f_j/N_c * log(Q/n) \tag{2}$$

Where,

$f_j$ is the frequency of access of item/page $I_j$ in cluster C.

$N_c$ is the number of sessions in cluster C.

Q is the total number of sessions in the preprocessed log that we term as CLF

n is the number of sessions that accessed item/page $I_j$ in the CLF

```
INPUT: Fᵢ -Frequency of access of web page pᵢ
          FCᵢ -Frequency of access of web pᵢ in cluster Cᵢ
          Q- Number of sessions in Complete Log File (CLF)
          Nᵢ - Number of sessions in cluster Cᵢ
          For each page pᵢ
                If (FCᵢ~=0)
                      Wtₚᵢ = (FCᵢ / Nᵢ) * log (Q/Fᵢ)
                End If
          End For
          For each session in cluster Cᵢ
                For each page accessed in session Sᵢ
                      Wtₛᵢ = Σᵢ₌₁ˣ Wtₚᵢ
                      Where x is the number of pages viewed in the
session
                      Wtₛᵢ' = Σᵢ₌₁ʸ Wtₚᵢ
                      Where y is the number of pages in CHIMP[]
                      Wtₛᵢ'' = Wtₛᵢ − Wtₛᵢ'
                End For
          End For
```

**Fig. 4.** Algorithm for calculating importance of web page on the basis of weight of a web page corresponding to the cluster

From the above algorithm the weights for the web pages corresponding to a cluster are defined as follows:

$$Wt_C = \{ w_1, w_2, w_3, \dots w_9, \dots w_{366} \}$$

Using the evaluated weights we calculate the total weight of each session in the REJ file by Eq. (3):

$$Wt_{s_i} = \sum_{j=1}^{x} w_j \text{ where } x \text{ is the number of pages in session } s_i \qquad (3)$$

In reference to the above example we calculate the new weight of the session by excluding the weight of page $I_9$ and defining a modified weight of the session by Eq. (4).

$$Wt'_{s_i} = \sum_{j=1}^{x} w_j - w_9 \text{ where } x \text{ is the number of pages in session } s_i \qquad (4)$$

Using the modified weight we check the importance of the session and shift it to the retained file if the new weight is greater than the average weight of all the sessions in the retained file. The average weight is defined by Eq. (5).

$$Avg\_wt(RET) = \sum_{i=1}^{N} Wt_{si} / N \qquad (5)$$

The algorithm for retaining sessions from rejected file on the basis of weight is given in Figure 5.

```
INPUT: REJ- Set of rejected sessions
            RET - Set of retained sessions
            Wtsi''- Modified session weight
            For each session Si in RET
                Wtsi = ∑ˣᵢ₌₁ Wtpi
                      Where x is the number of pages viewed in the
                      session
                Avg_wt(RET) = ∑ᴺᵢ₌₁ Wtsi /N
            End For
            For each session Si in REJ
                If Wtsi''> Avg_wt(RET)
                    Then append Si to REJ
                End If
            End For
```

**Fig. 5.** Algorithm for retaining sessions from rejected file on the basis of weight of a session

The advantage of the proposed refinement technique is to filter clusters and improve their quality in terms of cohesion of the cluster. The overall refinement technique is displayed as a flowchart in Figure 1. The results of the above refinement are compared with the Distance between Sequences (DBS) algorithm [60] finding the difference between the sequences of access on the basis of the number of replacements, insertions and deletions required to make the session sequences equal. Comparison of W-FIRM algorithm with the number of hits on a web page in a session [59] is also performed.

# 4    Performance Evaluation

## 4.1    Daveis Bouldin Index

This index aims to identify sets of clusters that are cohesive and well separated. The Davies-Bouldin index is defined by Eq. (6):

$$DB = 1/K \sum_{i,j=1}^{K} \max_{i \neq j} \left[ \frac{diam(C_i) + diam(C_j)}{d(C_i, C_j)} \right] \tag{6}$$

Where K denotes the number of clusters, $i$, $j$ are cluster labels, $diam(C_i)$ and $diam(C_j)$ are the diameters of the clusters $C_i$ and $C_j$, $d(C_i, C_j)$ is the average distance between the clusters. Smaller values of average similarity between each cluster and its most similar one indicate a "better" clustering solution.

## 4.2    Silhouette Coefficient

This coefficient combines the ideas of cohesion and separation for clustering techniques where cohesion measures how closely are the objects related in the cluster and separation measures how distinct or well separated a cluster is from other clusters. The silhouette coefficient is defined by Eq. (7):

$$SC = \frac{1}{N} \sum_{i=1}^{N} s(x) \tag{7}$$

Where,

$$s(x) = \frac{(b(x) - a(x))}{\max\{a(x), b(x)\}} \tag{8}$$

Where,

a(x) is the Cohesion defined as the average distance of x to all other vectors in the same cluster and b(x) is the Separation defined as the average distance of x to the vectors in other clusters. The silhouette plot displays a measure of how close each point in one cluster is to points in the neighboring clusters. This measure ranges from [+1,-1]. +1 indicating points that are very distant from neighboring clusters; 0 indicating points that are not distinctly in one cluster or another; -1 indicating points that are probably assigned to the wrong cluster.

## 5    Experimental Setup and Results

Data used in the experiment is taken from different log files from the internet traffic archive containing information about all web requests to the websites. The datasets used are as follows:

The NASA dataset has been taken from the Internet Traffic Archive. It is a moderated repository to support widespread access to traces of Internet network traffic; sponsored by ACM SIGCOMM. The NASA dataset is a benchmark dataset which has been used for research in a number of web based applications. The HTTP requests of one month to the NASA Kennedy Space Center server in Florida were recorded and used. The logs are an ASCII file with one line per request along with the following attributes: host making the request, Timestamp of the request, request given in quotes, HTTP reply code and bytes in the reply. The log was collected from 00:00:00 July 1, 1995 through 23:59:59 July 31, 1995, a total of 31 days. The raw web log file used for the experiment contained 131031 web requests. Two synthetic datasets DS1and DS2 have been designed from the NASA dataset having 6000 and 5000 sessions respectively for evaluating results of the proposed work.

The data sets used for the experiments are listed as follows:

**Table 1.** Data Sets used

| Real Data Sets | Number of Sessions |
|:---:|:---:|
| NASA | 131031 |
| **Synthetic Data Sets** | |
| DS1 | 6000 |
| DS2 | 5000 |

Pre-processing of a web log file is to filter out log entries that are not required for our task. These include entries including the error code, image files, request methods other than "GET" that are transmitted to the user's machine. For creation of sessions, an algorithm is designed which is implemented in C++ language. The program computes session information by tracking the page, date/time, and visitor id for each page view. When a session view is requested, it processes all of these page views at the time of the request. The program groups the hits into initial sessions based on the visitors IP address. Subsequent requests from the same IP address are added to the session as long as the time difference between two consecutive requests is 30 minutes (standard). Otherwise, the running current session is closed and a new session is created; where, the session is defined as a sequence of accesses by a user. In general, a user session is defined as accesses from the same IP address such that the duration of time elapsed between any two consecutive accesses in the session within a pre-specified threshold.

The NASA dataset after pre-processing and optimization was reduced to 28213 sessions and 366 web pages. The URLs in the site are assigned a unique name $X_i$, $i \in \{1... n\}$, where n is the total number of valid URLs and the sessions are identified as $S_j$, $j \in \{1 \dots m\}$. The synthetic datasets DS1 and DS2 extracted from the NASA dataset are used for experimentation having 6000 and 5000 entries respectively.

K-Means is applied on the Page View Matrix to form original clusters; where rows represent sessions and columns represent the web pages. The results of W-FIRM are compared with Distance Between Sequences (DBS) [60] and distance on the basis of Hits [59] algorithms. The results are discussed as follows:

**Table 2.** Comparison of Davies Bouldin Index for different refinement techniques

| Data Set | Number of Clusters | Original Clusters | Refined Clusters (W-FIRM) | Refined Clusters (DBS) | Refined Clusters (Hits) |
|---|---|---|---|---|---|
| | | | **Daveis Bouldin Index** | | |
| NASA | 10 | 2.9919 | 2.3784 | 2.9512 | 2.8428 |
| | 15 | 2.6317 | 2.1528 | 2.5862 | 2.3717 |
| | 20 | 2.4674 | 2.1705 | 2.3459 | 2.2321 |
| DS1 | 10 | 2.7734 | 2.2434 | 2.5404 | 2.677 |
| | 15 | 2.7154 | 1.8436 | 2.4985 | 2.4381 |
| | 20 | 3.0535 | 1.9758 | 2.8827 | 2.8918 |
| DS2 | 10 | 2.9493 | 2.5671 | 2.7818 | 2.6094 |
| | 15 | 2.9532 | 1.4022 | 2.8357 | 2.7337 |
| | 20 | 3.2713 | 1.5647 | 3.2296 | 3.1768 |

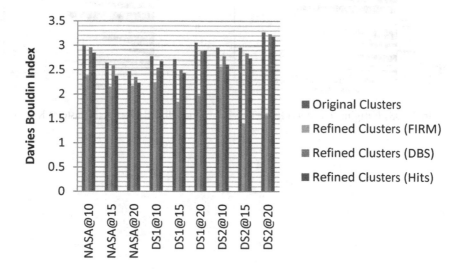

**Fig. 6.** Comparison of Davies Bouldin Index

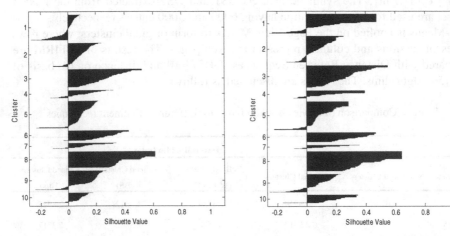

**Fig. 7.** DS1 (Original Clusters) K@10        **Fig. 8.** DS1 (Refined Clusters) K@10

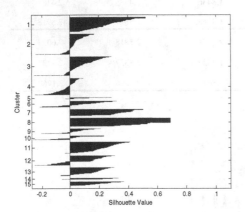
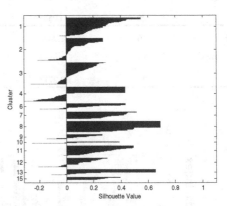

**Fig. 9.** DS1 (Original Clusters) K@15        **Fig. 10.** DS1(Refined Clusters) K@15

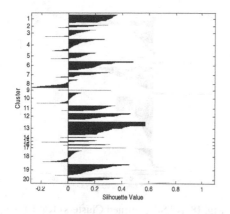

**Fig. 11.** DS1 (Original Clusters) K@20

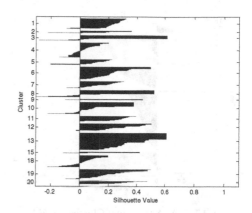

**Fig. 12.** DS1 (Refined Clusters) K@20

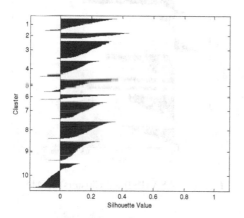

**Fig. 13.** DS2 (Original Clusters) K@10

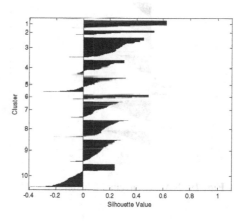

**Fig. 14.** DS2 (Refined Clusters) K@10

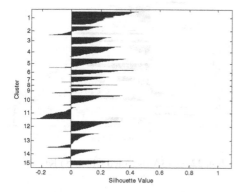

**Fig. 15.** DS2 (Original Clusters) K@15

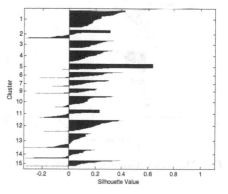

**Fig. 16.** DS2 (Refined Clusters) K@15

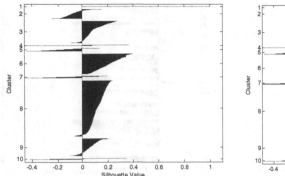

**Fig. 17.** NASA (Original Clusters) K@10

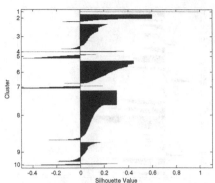

**Fig. 18.** NASA (Refined Clusters) K@10

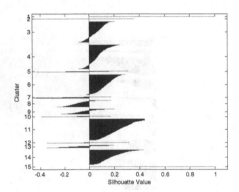

**Fig. 19.** NASA (Original Clusters) K@15

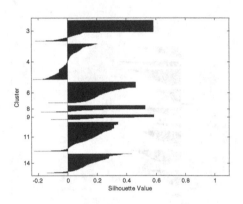

**Fig. 20.** NASA (Refined Clusters) K@15

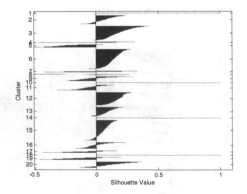

**Fig. 21.** NASA (Original Clusters) K@20

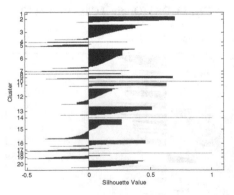

**Fig. 22.** NASA (Refined Clusters) K@20

Figure 6 exhibits the change in the Davies Bouldin Index for clusters using the various refinement techniques. In the graph lower value of DB Index represents better clusters while values on the higher side represent poor quality clusters. It is evident that refinement using W-FIRM produces results with better quality clusters as compared to other refinement techniques. Figure 7 and 8 represent the silhouette coefficients for the original and refined clusters of the synthetic dataset (DS1); it is visibly evident that the quality of cluster number 1, 4 and 10 has improved significantly. Similarly, Figure 20 and 21 represent the silhouette coefficients for the original and refined clusters of the real dataset (NASA); it is visibly evident that the quality of cluster number 3, 6, 8 and 9 has improved significantly. On the same lines Figure 9 to 18, 21 and 22 depict the Silhouette values for the original and refined clusters for varying values of K as 10, 15 and 20.From the graph results it is clear that the quality of refined clusters using W-FIRM have improved significantly.

## 6    Conclusion and Future Work

In Web Usage Mining, the clustering technique generates clusters helping to understand the similarity in sessions and hence using the cluster information for various web based applications such as Web Personalization, Web Page Recommendation, improvement in Customer Behavior and Web Site Structure. This paper discusses the problem of web usage clusters having large number of web sessions leading to poor quality of clusters. A refinement technique for improving the quality of clusters has been proposed on the basis of Weighted Frequent Itemsets extracted from the log file and clusters obtained from the log file. The work was tested on web log files and it was shown how the quality of clusters improved by evaluating the Davies Bouldin index and Silhouette Coefficient values on original and refined clusters.

In future research, a Recommender System shall be proposed to evaluate the effectiveness of recommendations to the end user based on W-FIRM.

## References

[1] Mobasher: Discovery of aggregate usage profiles for web personalization. WebKDD, Boston (2009)

[2] Bobadilla, J., Ortega, F., Hernando, A., Gutiérrez, A.: Recommender systems survey. Knowledge-Based Systems 46, 109–132 (2013)

[3] Ziegler, C.N.: On Recommender Systems. In: Ziegler, C.N. (ed.) Social Web Artifacts for Boosting Recommenders. SCI, vol. 487, pp. 11–22. Springer, Heidelberg (2013)

[4] Berkhin, P.: Survey of clustering data mining techniques. Springer, Heidelberg

[5] Flake, G., Lawrence, S., Giles, C.L., Coetzee, F.: Self-organization and identification of Web Communities. IEEE Computer 35(3) (2002)

[6] Castellano, G., Fanelli, A.M., Mencar, C., Torsello, M.A.: Similarity based Fuzzy clustering for user profiling. In: Proceedings of International Conference on Web Intelligence and Intelligent Agent Technology, IEEE/WIC/ACM (2007)

[7] Xu, R., Wunsch, D.: Survey of clustering algorithms. IEEE Trans. Neural Networks 16(3), 645–678 (2005)

[8] Mobasher, B., Cooley, R., Srivastava, J.: Automatic Personalization based on Web Usage Mining. Communications of the ACM 43(8), 142–151 (2000)

[9] Agrawal, R., Imieliński, T., Swami, A.: Mining association rules between sets of items in large databases. In: Proceedings of the 1993 ACM SIGMOD International Conference on Management of Data, SIGMOD 1993, p. 207 (1993)

[10] Nock, R., Nielsen, F.: On Weighting Clustering. IEEE Transactions and Pattern Analysis and Machine Intelligence 28(8), 1223–1235 (2006)

[11] Baldi, P., Frasconi, P., Smyth, P.: Modeling the Internet and the Web. Wiley (2003)

[12] Chakrabarti, S.: Mining the Web. Morgan Kaufmann Publishers (2003)

[13] Banerjee, A., Ghosh, J.: Click stream clustering using weighted longest common subsequences. In: Proceedings of the Web Mining Workshop at the 1st SIAM Conference on Data Mining (2001)

[14] Cadez, H.D., Meek, C., Smyth, P., White, S.: Model-based clustering and visualization of navigation patterns on a Web site. Data Mining and Knowledge Discovery 7(4), 399–424 (2003)

[15] Fu, Y., Sandhu, K., Shih, M.Y.: Clustering of Web users based on access patterns. In: Proceedings of WEBKDD (1999)

[16] Hay, B., Vanhoof, K., Wetsr, G.: Clustering navigation patterns on a Website using a sequence alignment method. In: Proceedings of 17th International Joint Conference on Artificial Intelligence, Seattle, Washington, USA (2001)

[17] Wang, W., Zaane, O.R.: Clustering Web sessions by sequence alignment. In: Proceedings of the 13th International Workshop on Database and Expert Systems Applications, pp. 394–398. IEEE Computer Society, Washington, DC (2002)

[18] Shahabe, C., Zarkesh, A.M., Abidi, J., Shah, V.: Knowledge discovery from user's web-page navigation. In: Proceedings Seventh IEEE International Workshop on Research Issues in Data Engineering (RIDE), pp. 20–29 (1997)

[19] Eiron, N., McCurley, K.S.: Untangling compound documents on the Web. In: Proceedings of the Fourteenth ACM Conference on Hypertext and Hypermedia (2003)

[20] Agrawal, R., Srikant, R.: Fast algorithms for mining association rules. In: VLDB 1994, pp. 487–499 (1994)

[21] Greco, G., Greco, S., Zumpano, E.: Web communities: models and algorithms. Journal of World Wide Web 7(1), 58–82 (2004)

[22] Cheng, J., Ke, Y., Ng, Q.: A survey on algorithms for mining frequent itemsets over data streams. Knowledge and Information Systems 16, 1–27 (2008)

[23] Srivastava, J., Cooley, R., Deshpande, M., Tan, P.: Web usage mining: discovery and applications of usage patterns from Web data. ACM SIGKDD Explorations Newsletter 1(2), 12–23 (2000)

[24] Munk, M., Kapusta, J., Svec, P.: Data preprocessing evaluation for web log mining: reconstruction of activities of a web visitor. Procedia Computer Science 1(1), 2273–2280 (2010)

[25] Kosala, R., Blockeel, H.: Web Mining Research: A Survey. ACM SIGKDD Explorations 2(1), 1–15 (2000)

[26] Liao, S.H., Chu, P.H., Hsiao, P.Y.: Data mining techniques and applications – A decade review from 2000 to 2011. Expert Systems with Applications 39(12), 11303–11311 (2011)

[27] Jain, A.K.: Data clustering: 50 years beyond K-means. Pattern Recognition Letters 31(8), 651–666 (2010), Award Winning Papers from the 19th International Conference on Pattern Recognition (ICPR) (2010)

[28] Boyinbode, O., Le, H., Takizawa, M.: A survey on clustering algorithms for wireless sensor networks. International Journal of Space-Based and Situated Computing 1(2-3), 130–136 (2011)

[29] Prusiewicz, A., Zięba, M.: Services Recommendation in Systems Based on Service Oriented Architecture by Applying Modified ROCK Algorithm. In: Networked Digital Technologies Communications in Computer and Information Science, vol. 88, pp. 226–238. Springer (2010)

[30] Karypis, G., Han, E.H., Kumar, V.: Chameleon: hierarchical clustering using dynamic modeling. Computer 32(8), 68–75 (1999)

[31] Davies, D.L., Bouldin, D.W.A.: Cluster Separation Measure. Pattern Analysis and Machine Intelligence. IEEE Transactions PAMI-1(2), 224–227 (1979)

[32] Chen, M.S., Han, J., Yu, P.S.: Data mining: an overview from a database perspective. Knowledge and Data Engineering 8(6), 866–883 (1996)

[33] Goethals, B.: Frequent Set Mining. In: Data Mining and Knowledge Discovery Handbook, pp. 377–397. Springer (2005)

[34] Berkhin, P.: A Survey of Clustering Data Mining Techniques. In: Grouping Multidimensional Data, pp. 25–71. Springer (2006)

[35] Borgelt, C.: An implementation of the FP-growth algorithm. In: Proceedings of the 1st International Workshop on Open Source Data Mining: Frequent Pattern Mining Implementations, pp. 1–5. ACM (2005)

[36] Maulik, U., Bandyopadhyay, S.: Performance evaluation of some clustering algorithms and validity indices. Pattern Analysis and Machine Intelligence 24(12), 1650–1654 (2002)

[37] Cooley, R., Mobasher, B., Srivastava, J.: Data Preparation for mining World Wide Web Browsing Patterns. In: Knowledge and Information Systems, pp. 1–25. Springer (1999)

[38] Cormode, G., Hadjieleftheriou, M.: Methods for finding frequent items in data streams. The VLDB Journal 19, 3–20 (2010)

[39] Xie, Y., Phoha, V.V.: Web user clustering from access log using belief function. In: Proceedings of the First International Conference on Knowledge Capture (K-CAP 2001), pp. 202–208. ACM Press (2001)

[40] Shahabi, C., Banaei-Kashani, F.: A framework for efficient and anonymous web usage mining based on client-side tracking. In: Kohavi, R., et al. (eds.) WebKDD 2001. LNCS (LNAI), vol. 2356, pp. 113–144. Springer, Heidelberg (2002)

[41] Ypma, A., Heskes, T.: Clustering web surfers with mixtures of hidden markov models. In: Proceedings of the 14th Belgian–Dutch Conference on AI (BNAIC 2002) (2002)

[42] Nasraoui, O., Frigui, H., Joshi, A., Krishnapuram, R.: Mining Web Access Logs Using Relational Competitive Fuzzy Clustering. Presented at the Eight International Fuzzy Systems Association World Congress, IFSA 1999, Taipei (1999)

[43] Tseng, F.C.: Mining frequent itemsets in large databases: The hierarchical partitioning approach. Expert Systems with Applications 40(5), 1654–1661 (2013)

[44] Oyanagi, S., Kubota, K., Nakase, A.: Application of matrix clustering to web log analysis and access prediction. In: Third International Workshop EBKDD 2001—Mining Web Log Data Across All Customers Touch Points (2001)

[45] Kivi, M., Azmi, R.: A webpage similarity measure for web sessions clustering using sequence alignment. In: Proceedings of 2011 International Symposium Artificial Intelligence and Signal Processing (AISP). IEEE Press (2011)

[46] Bentley, J.: Multidimensional Binary Search Trees Used for Associative Searching. ACM 18(9), 509–517 (1975)

[47] Bradley, P.S., Fayyad, U., Reina, C.: Scaling Clustering Algorithms to Large Databases. In: 4th International Conference on Knowledge Discovery and Data Mining (KDD 1998). AAAI Press (1998)

[48] Scholkopf, B., Smola, J., Muller, R.: Technical Report: Nonlinear component analysis as a kernel eigen value problem. Neural Comput. 10(5), 1299–1319 (1998)

[49] Dhillon, I.S., Fan, J., Guan, Y.: Efficient clustering of very large document collections. In: Data Mining for Scientific and Engineering Applications, pp. 357–381. Kluwer Academic Publishers (2001)

[50] Elkan, C.: Using the Triangle Inequality to Accelerate k-Means. In: Proceedings of the Twentieth International Conference on Machine Learning (ICML 2003), pp. 609–616 (2003)

[51] Yin, K.C., Hsieh, Y.L., Yang, D.L.: GLFMiner: Global and local frequent pattern mining with temporal intervals. In: 2010 the 5th IEEE Conference Industrial Electronics and Applications (ICIEA), pp. 2248–2253 (2010)

[52] Baralis, E., Cerquitelli, T., Chiusano, S., Grand, A., Grimaudo, L.: An Efficient Itemset Mining Approach for Data Streams. In: König, A., Dengel, A., Hinkelmann, K., Kise, K., Howlett, R.J., Jain, L.C. (eds.) KES 2011, Part II. LNCS, vol. 6882, pp. 515–523. Springer, Heidelberg (2011)

[53] Zhao, C., Jia, B., Liu, Y., Chen, L.: Mining global frequent sub trees. In: 2010 Seventh International Conference on Fuzzy Systems and Knowledge Discovery (FSKD), vol. 5, pp. 2275–2279 (2010)

[54] Cai, C.H., Fu, A.W., Cheng, C.H., Kwong, W.W.: Mining association rules with weighted items. In: Proceedings of the International Database Engineering and Applications Symposium, IDEAS 1998, Cardiff, Wales, UK, pp. 68–77 (1998)

[55] Tao, F.: Weighted association rule mining using weighted support and significant framework. In: Proceedings of the 9th ACM SIGKDD, Knowledge Discovery and Data Mining, pp. 661–666 (2003)

[56] Wang, W., Yang, J., Yu, P.S.: WAR: weighted association rules for item intensities. Knowledge Information and Systems 6, 203–229 (2004)

[57] Yun, U., Leggett, J.J.: WFIM: weighted frequent itemset mining with a weight range and a minimum weight. In: Proceedings of the 15th SIAM International Conference on Data Mining (SDM 2005), pp. 636–640 (2005)

[58] Yun, U.: Efficient Mining of weighted interesting patterns with a strong weight and/or support affinity. Information Sciences 177, 3477–3499 (2007)

[59] Xu, J., Liu, H.: Web User Clustering Analysis based on K-Means Algorithm. In: International Conference on Information Networking and Automation. IEEE (2010)

[60] Liu, P., Li, W.: Navigation Pattern Discovery on Web Site Based on the Distance Between Sequences. Artificial Intelligence. In: Artificial Intelligence, Management Science and Electronic Commerce (AIMSEC). IEEE Press (2011)

# An Approach to Creating a Personified Job Description of a Designer in Designing a Family of Software Intensive Systems

Petr Sosnin and Andrey Pertsev

Ulyanovsk State Technical University, Severny Venetc str. 32,
432027 Ulyanovsk, Russia
sosnin@ulstu.ru

**Abstract.** This study is bound with rational management of workforces that are used in the design company for developing the software intensive systems. The approach applied in the study is based on modeling the experience of any designer of the company team in the form of a personified job description which is bound with precedents' models used in solutions of all appointed tasks. Question-answer reflections of tasks being solved by the designer on models of precedents define the specificity of the suggested approach. The personified job description is built on the base of measured competencies registered in the Experience Base. The used realization of the approach facilitates increasing the efficiency of designing the family of software intensive systems.

**Keywords:** Automated designing, competency, precedent, question-answering, software intensive system.

## 1 Introduction

In the theory and practice of developing the Software Intensive Systems (SIS), the concept «occupational maturity» is widely used in different aims. This notion has found the normative representation in a number of standards, most popular of which are CMMI-1.3 (Capability Maturity Model Integrated for Development) [1] and P-CMM 2.0 (People Capability Maturity Model) [2]. The first of these standards is focused on specifications of occupationally mature processes used in designing the SIS while the second standardizes occupational maturity of their developers. Both of named standards are additional everyone to another, and this fact should be taken into account in their use especially for continuous improvement of the project activity.

In this case value of P-CMM 2.0 standard is caused that it helps to build the answer this question: How the work with a team of designers should be organized when it is coordinated with CMMI and is aimed at the improvement of managing the workforces. In turn, effectiveness of the answer depends on models of design processes and designers.

Till now, the search of new indicated models continues. The main reason of such condition is extremely low degree of success (approximately 35 % all last 20 years) in

B. Murgante et al. (Eds.): ICCSA 2014, Part V, LNCS 8583, pp. 51–62, 2014.

the development of the SIS [3]. Perspective ways of search are opened by the initiative called SEMAT (Software Engineering Methods and Theory) where necessity of reshaping the bases of software engineering is declared [4].

The approach proposed in this paper is oriented on the use of empirical precedent forms for personified modeling of designers. "Precedents are actions or decisions that have already happened in the past and which can be referred to and justified as an example that can be followed when the similar situation arises"    (http: //dictionary. cambridge.org/dictionary/british/ precedent).

It is realized in an environment of the specialized toolkit WIQA (Working In Questions and Answers) which was also used for creating the precedent-oriented Experience Base [5]. Such a solution allows binding the models of processes and models of designers. Relations between models of both types is provided by models of practices and used in processes and embedded to the Experience Base. It opens the possibility for the management of these relations or, by the other words, for a rational distribution of work forces in the development of the SIS.

The remainder of the paper is structured as follows. Section 2 discusses the brief works related to the designer model. In Section 3, an operational space of designer's activity is shown. In Section 4, we introduce the precedent-oriented approach to the simulation of design practices. Understanding and specifying the competency is presented in Section 5. In Section 6, the creation of the personified job description is described. Finally, Section 7 makes conclusions.

## 2    Related Works

The Personified Job Description (PJD) is specified for its use in conditions of the continuous improvement of the project activity in accordance with ISO/MEK standard 9004-2009 [6]. It is provided by inheritance from this standard set of recommendation for achieving the sustained success of a design company. Requirements inherited from this standard were detailed in accordance with standards [1] and [2].

Very useful information was extracted from a number of publications devoted to specialized areas of the occupational activity. In this set, we mark Project Management Maturity Model specifying ways of the perfection of the project management [7]; Business Process Maturity Model opening forms and means of the business-processes perfection (without the accent on the development of the SIS) [8]; Business Intelligence Maturity Model, which focuses on the perfection of practices for intellectual support of the occupational action [9]; Enterprise Architecture Maturity Model including practices for creating an enterprise architecture and ways for their perfection [10].

The next group of related papers is bound with competences that mastering by designers is necessary for designing the SIS. First of all this group includes definitions of competencies [11], systems engineering competencies framework [12], competency wheel [13] and improving software architecture competence [14]. This group also includes specifying the roles in Rational Unified Process [15].

One more group of related works focuses on modeling the experience used in designing the SIS. In this group, we mark papers [16] and [17] where question-answering is also applied but for the other aims.

# 3     Operational Space of Designer's Activity

As told above, the proposed approach is aimed at the creation and use of the designer model in collective designing of the SIS family. In this case, the designer should operate in operational space the general scheme of which is presented in Fig. 1.

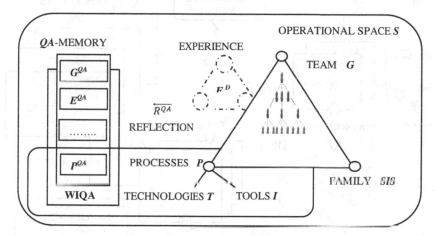

**Fig. 1.** Operational space

Specificity of named actions is expressed by a reflection $R^{QA}$ of the operational space to a question-answer memory (QA-memory) of the WIQA toolkits. On the scheme, this reflection shows that all what involving in designing of the family of SIS is found their expression as models in the QA-memory. It can be written by the following expression:

$$WW(P, G, E^D, E^{QA}, E^{Pr}, \{SIS_i\}, t) \xrightarrow{R^{QA}} G^{QA}(t+\Delta t) \cup$$
$$\cup\ E^{QA}(t+\Delta t) \cup P^{QA}(t+\Delta t) \cup SIS^{QA}(t+\Delta t), \tag{1}$$

where $WW$ is a Way-of-Working used by designers and all other symbolic designations corresponds to the names of essences in Fig. 1. Let us additionally note that results of the reflection $R^{QA}$ are dynamic objects, $S^{QA}(t_j)$ models all relations among essences, $E^{QA}(t_j)$ presents models corresponding to the used occupational experience $E^D$ which is mastered by members $\{D_k\}$ of the team $G(\{D_k\}$.

At the conceptual stage of designing, means of WIQA are used by designers for the following aims:

- registering of the set of created projects each of which is presented by the tree of its tasks in the real time;
- parallel implementing of the set of workflows $\{W_m\}$ each of which includes subordinated workflows and/or project tasks $\{Z_n\}$;
- pseudo-parallel solving of the project tasks on each workplace in the corporate network;
- simulating the typical units of designers' behavior with using the precedent framework.

These details of operational space are opened on the scheme presented in Fig. 2 where relations with an operational environment are detailed only for one of designers.

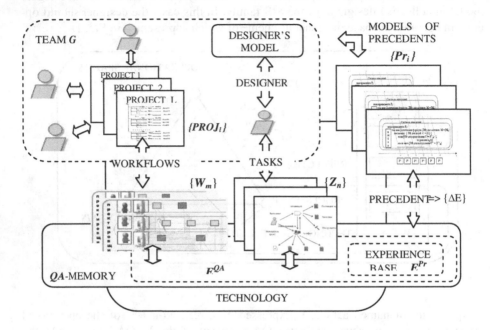

**Fig. 2.** Details of operational space

For the process $P_l$ of designing the project $PROJ_l$, all indicated aims are achieved by the following reflections:

$$WW(P_l) = WW(PROJ_l, \{W_m\}, \{Z_n\}, t) \xrightarrow{R^{QA}(P)} ZP_l{}^{QA}(t+\Delta t) \bigcup$$

$$\bigcup \{ZW_m{}^{QA}(t+\Delta t)\} \bigcup \{Z_n{}^{QA}(t+\Delta t)\} , \tag{2}$$

$$\{Z_n{}^{QA}(t) \xrightarrow{R^{QA}(Z^{QA})} Pr_n{}^{QA}(t+\Delta t)\},$$

where symbol $Z$ underlines that models have a task type, $Pr^{QA}(t)$ designates a model of precedent for the corresponding task, $R^{QA}(X)$ indicates that the reflection $R^{QA}$ is applied to the essence or artifact $X$. For example, $R^{QA}(Z^{QA})$ designates applying this reflection to the model $Z^{QA}$ of the task $Z$. In WIQA-environment, the model of such kind is named as "QA-model of the task".

The second scheme includes a designer's model with feedback which shows that, in the proposed approach, a set of precedents mastered by the designer can be used for the creation of the designer model.

The above not only indicates reflections of projects on the QA-memory, but it also demonstrates structures that should find their presentations in such memory. Below, for specifications of question-answer objects (QA-objects) in the QA-memory, a formal grammar $GR^{QA}$ with extended BNF-notations is used. For example, structures of QA-objects from (2) should correspond to the following rules of $GR^{QA}$:

$$
\left.
\begin{aligned}
&PROJ = ZP; \\
&ZP = (Z, \text{``}\downarrow\text{''}, \{Workflows\ \}; \\
&Workflows = ZW \mid (Workflows, \text{``}\downarrow\text{''}, \{ZW\}); \\
&ZW = \{Task\}; \\
&Task = Z \mid (Task, \text{``}\downarrow\text{''}, \{Z\}); \\
&Z = QA\text{-}model \mid (Z, \text{``}\downarrow\text{''}, \{QA\text{-}model\}); \\
&QA\text{-}model = \{QA\} \mid (QA\text{-}model, \text{``}\downarrow\text{''}, \{QA\}); \\
&QA = (Question, Answer), \\
&Question = Q \mid (Question, \text{``}\downarrow\text{''}, \{Q\}); \\
&Answer = A \mid (Answer, \text{``}\downarrow\text{''}, \{A\}),
\end{aligned}
\right\}
\tag{3}
$$

where "$Z$", "$Q$" and "$A$" are typical visualized objects stored in cells of QA-memory, symbol "$\downarrow$" designates an operation of "subordinating". These objects have the richest attribute descriptions [5]. For example, a set of attributes includes the textual description, index label, type of objects in the QA-memory, name of a responsible person and the time of last modifying. Any designer can add necessary attributes to the chosen object by the use of the special plug-ins "Additional attributes" (object-relational mapping to C# classes).

## 4    Simulation of Project Practices

As told above, interests of the proposed approach are connected with mastering the occupational experience the important part of which is specified in standards CMMI-1.3 and P-CMM 2.0. Therefore, practices of these standards should occupy a central place in the Experience Base. These practices are invariant to their application for designing in definite subject areas consequently they should be adjusted on specificity of definite conditions of designing the SIS. Successful designing demands to widen the invariant practices by practices from the subject area of the designed SIS.

Thus, the team competence should be enough for the real time work with following sets of tasks: subject tasks $Z^S = \{Z^S_i\}$ of SIS subject area; normative tasks $Z^N = \{Z^N_j\}$ of technology used by designers; adaptation tasks $Z^A = \{Z^A_j\}$ providing an adjustment of tasks $\{Z^N_j\}$ for solving the tasks $\{Z^S_i\}$; workflow tasks $\{Z^W_m\}$ providing the works with tasks of $Z^S$-type in workflows $\{W_m\}$ in SIS; workflow tasks $\{Z^W_n\}$ providing the works with tasks of $Z^N$-type in corresponding workflows $\{W_n\}$ in the used technology; workflow tasks $\{Z^G_p\}$ and $\{Z^G_r\}$ any of which corresponds to the definite group of workflows in SIS or technology.

The indicated diversity of tasks emphasizes that designers should be very qualified specialists in the technology domain, but that is not sufficient for successful designing. Normative tasks are invariant to the SIS domain and, therefore, designers should gain particular experience needed for solving the definite tasks of the SIS subject area. The most part of the additional experience is being acquired by designers in experiential learning when tasks of $Z^S$-type are being solved. Solving of any task $Z^S_i$ is similar to its expanding into a series on the base of normative tasks.

Let us note that interests of the proposed approach are related only with the conceptual stage of designing and, therefore, with conceptual reflections of the denoted tasks on the QA-memory. Below, in rules of grammar $GR^{QA}$, the symbol "$Z$" is used in designations for tasks of any kind indicated above.

In designing the definite SIS, all tasks ZP of its project is divided on two classes. The first class includes the tasks $\{ZPr_s\}$ which are reusable as precedents in the other projects of SISs. The second class $ZO$ comprises the other tasks $\{ZO_v\}$.

For differentiating of these classes in grammar $GR^{QA}$, the rule for the task

$$Z = QA\text{-}model \mid (Z, ``\downarrow", \{QA\text{-}model\}),  \qquad (4)$$

should be changed at the following set of rules:

$$\left. \begin{array}{l} Z = ZPr \mid ZO; \\ ZPr = QA\text{-}model \mid (ZPr, ``\downarrow", \{QA\text{-}model\}); \\ ZO = QA\text{-}model \mid (ZO, ``\downarrow", \{QA\text{-}model\}). \end{array} \right\} \qquad (5)$$

Implementations of RQA(ZQA)-reflections are oriented on the use of the framework of the precedent model ($FPr$), the scheme of which is presented in Fig. 3.

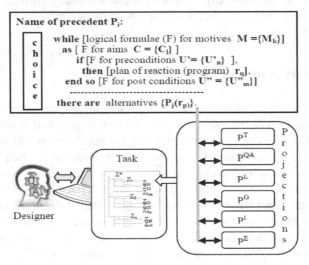

**Fig. 3.** Precedent framework

The central place in this framework is occupied the logical scheme of the precedent. The scheme explicitly formulates "cause-effect regularity" of the simulated behavior of the designer. The structure of the integrated precedent model $FPr$ is coordinated with the process of task-solving and preparing the solution of the task for the reuse. This structure includes a textual model $TPr$ of the solved task $ZPr$, its model $QAPr$ in the form of the registered QA-reasoning, the logical formulae $LPr$ of the precedent regularity, a graphical (diagram) representation $GPr$ of the precedent, its pseudo-code model $IPr$ in a form of a pseudo-code program and model which presents its executable code $EP$.

In the grammar $GR^{QA}$, the precedent framework is described by the following set of rules:

$$
\left.\begin{array}{ll}
Fpr = \text{``}\rho\text{''}, ZPr; \\
FPr \quad = (Keys, TPr, LPr, QAPr, GPr, IPr, EPr); \\
VFPr \ = FPr - [QAPr] - [GPr] - [IPr] - [EPr)]; \\
VFPr \ = \text{``}\pi\text{''}, FPr; \\
Keys \ = \{Key\},
\end{array}\right\} \quad (6)
$$

where "$\rho$" presents the reflection $R^{QA}(Z^{QA})$ in the operation form, *VFP* designates a variant or projection of the precedent use, and $\pi$ is an operation of the projection. Possibility of projecting is included in the potential of reflections for adjusting the precedent models on conditions of their reuse.

At the level of tasks, the rules of the set (5) can be detailed with the help of the following grammar rules:

$$
\left.\begin{array}{ll}
FPr \quad = \ ((ZPr, \text{``}\downarrow\text{''},(TZ, LZ, QAZ, GZ, IZ, EZ)); \\
TPr \quad = TZ; \ (*\ specialized\ Z*) \\
QAPr = QAZ; \ (*\ specialized\ Z*) \\
LPr \quad = LZ; \ (*\ specialized\ Z*) \\
GPr \quad = GZ \ (*\ specialized\ Z*) \\
IPr \quad = IZ \ (*\ specialized\ Z*) \\
EPr \quad - EZ \ (*\ specialized\ Z*).
\end{array}\right\} \quad (7)
$$

Thus, the task structure (*ZPr, TZ, QAZ, LZ, GZ, IZ, EZ*) is represented the precedent model by means of QA-objects.

## 5   Specification of Competencies

In the proposed approach models of precedents are used for modeling of assets applied by the team in designing the family of SISs. Any asset is understood from the viewpoint of its application, and; therefore, it binds with the corresponding task of *ZPr*-type.

Assets are divided on classes of different types the specifications of which are described by a set of attributes including "the name of type", "description", "kind of precedent version *VFPr* used for its storing in QA-memory" and the others. Some of the types specify the models of precedents for human resources, for example, they should include the type specifying the personified models of designers.

In any current state, the Experience Base contains:

- a model of an organizational structure (a team description) of the project company;
- a set of tasks' trees for completed and implemented projects of SISs;
- a systematized set of precedents' models coding the assets of the company.

As told above, set of attributes of any Z-object of any tasks' tree includes the personified name of the designer who has been appointed as the solver of the corresponding task. It gives the opportunity for extracting from indicated structures for any designer:

- lists of the solved task as for tasks of *ZPr*-type so for tasks of *ZO*-type;
- a list of the used precedents models;
- a list of precedents models created by the designer;

- the other occupational information that reflects the occupational qualification of the designer.

The denoted opportunity is used for creating the designer model on the base of precedents' models, which have been successfully applied by the designer in the solved tasks.

The first step of such work is bound with grouping the precedents mastered by the designer. The grouping is oriented on the use of mastered competencies and roles being executed. In the suggested approach, a competency is understood as "a measurable pattern of knowledge, skills, abilities, behaviors, and other characteristics that individual needs to perform work roles or occupational functions successfully" (http://apps.opm.gov/ADT/Content.aspx?page=1-03&AspxAutoDetectCookieSupport =1 &JScript=1). The base of measurability is a set of successfully solved tasks that have been appointed to the designer who used precedents' models.

This understanding has led to the solution of using the precedents' models as units for "measurement" of competencies. Therefore, competencies and roles have been specified in grammar $GR^{QA}$. Competency $K$ of the designer is expressed structurally by the following grammar rules:

$$\left.\begin{aligned}
&K = (Name, Definition, FPr); \\
&K = (Name, Definition, ``\gamma1", \{FPr\}); \\
&K = (Name, Definition, ``\gamma2", \{K\}); \\
&Competencies = \{K\}; \\
&Role = (Name, Definition, ``\gamma2", \{K\}); \\
&Roles = \{Role\},
\end{aligned}\right\} \qquad (8)$$

where **Definition** is a verbal description for the corresponding **Name**, $\gamma1$ and $\gamma2$ are operations of grouping.

These rules correspond to relations between a net of competencies and a set of precedents' models that are generally presented in Fig. 4.

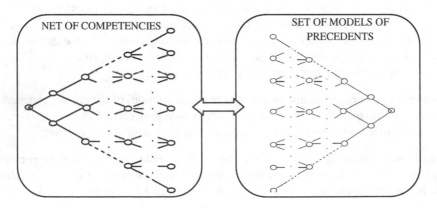

**Fig. 4.** Scheme of relations between competencies and precedents

Rules underline that some competencies can be expressed through a number of subordinated competencies. It helps the use of generalization in brief describing the occupational experience mastered by the designer. The rule for the role fulfills the similar function, but roles are usually used for qualifying the occupational area of designer responsibilities, for example, architect, programmer, tester.

# 6     Personified Job Description

Opportunity of the precedent-oriented description of designer competencies opens the question about their systematization. In the management practice of workforces, the job descriptions are widely used. "Job descriptions are written statements that describe the duties, responsibilities, most important contributions and outcomes needed from a position, required qualifications of candidates, and reporting relationship and coworkers of a particular job" (http://humanresources.about.com/od/ jobdescrip-tions/a/develop_job_des.htm).

Documents of this kind can be written as a result of job analysis in different forms. In our case, these documents should be oriented on the personified modeling of any member of the team designing the family of SISs. Moreover, the personified job description should systematize the measurable competencies of the designer.

Traditionally, a normative text of the job description *JD* includes the following sections: Job title; General summary of job; Key relationships; Education and *Experience (Minimum qualifications); Knowledge, skills and abilities; Principal duties and essential functions;* Major challenges; Physical, mental, sensory requirements; and Working conditions.

In the section structure, the italic indicates elements the main content of which can be presented by names of competencies and references to models of precedents. For this reason, JD-document has been chosen as a kernel of the designer model. This choice requires the use of reflecting the documents on the QA-memory

$$JD = (DName, \{Section\});$$
$$Section = Text -\{[Role]\} -\{[K]\}- \{[AF]\}, \tag{9}$$

where *DName* is the identifier of the designer, *{AF}* indicates the opportunity including the additional feature of the designer in the *JD*.

In the WIQA-environment, the work with documents is supported with using their QA-patterns, which should be previously developed and stored in the specialized library. The typical model of each document is presented by two its patterns. The first pattern reflects the document structure and the second part defines its printed version. The specialized plug-ins "Documenting" provides adjusting of both patterns on conditions of their use.

The denoted plug-ins is modified on conditions when information about competencies and models of precedents can be included to the *JD*-documents from the Experience Base. Such creation of the personified *JD* is generally presented in Fig. 5.

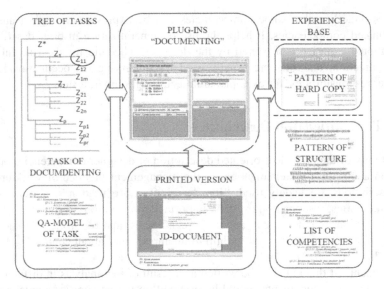

**Fig. 5.** Creating Job Description

As told above, the **JD**-documents are created in the process of the job analysis. In the described case, one stage of this analysis should be aimed at forming the list of competencies. Such work should be implemented with using the generalization for the net of competencies.

# 7    The Use of Personified Job Descriptions

The **JD** is not separated from its using in solving a set of tasks related with the management of human resources including the management of workforces. The representative systems of such tasks are described and specified in standards CMMI-1.3 and P-CMM 2.0 that are supplementary each other. The first of them indicates on the collection of best practices of mature processes while the second is based on state-of-the-art workforce practices that are used in the design company, for example, for:

- Developing the workforce required to effective create the family of SIS.
- Characterizing the maturity of workforce practices.
- Setting the priorities for improving workforce capability.
- Integrating the improvements in process and workforce.

In P-CMM 2.0, the allocated workforce practices are distributed on 22 process areas which include a number of areas that are directly connected with competencies of designers, for example, with such areas as Competency Analysis, Development Competency, Competency-Based Practices, Competency Integration and Competency-Based Assets. The use of practices from the other process areas is also based on competencies of designers, for example, practices that help in Staffing, Development Workforce Planning, Workgroup Development, Quantitative Performance

Management and Empowered Workgroups. Furthermore, a number of process areas define the steps of continuous improving the work with human resources including the continuous workforce innovations.

The described *JD* opens the opportunity for more rational and accurate solving the tasks that correspond to workforces practices. Such opportunity is caused by not only reflections of mature practices on competencies (and roles) but also following features:

- A set of reflected practices consists of process practices and workforce practices, and all of them are specified by the uniform way as models of precedents.
- Workforce practices can be constructively specified as meta-practices for process practices as competency-based practices.
- The pseudo-code of any practice is implemented by the designer who fulfils the role of the intellectual processor [18].
- Any member of the company staff is personally ready and responsible for the measurable programmable implementation of practices that are included in his/her model in the *JD*-form.
- Measurability of competencies that are included into any JD-model is constructively expressed by a set of corresponding precedents extracted from the Experience Base.

The *JD* has been used in solutions of a number of tasks that correspond to the standard P-CMM 2.0. The set of solved tasks includes practices that support staffing, development workforce planning and workgroup development (for temporal groups that solve communicative tasks). The introduction of *JD*-models of designers into the work with these practices has led to increasing the level of their automation. Furthermore, programmable components of these models support processes of experimenting that facilitate experiential learning in the real time process of designing.

It is necessary to note that the offered *JD* is developed as a specialized type of assets intended for solving the indicated tasks. Assets of this type are stored in the Experience Base of the design company that use *JD*-models for the management of own workforces.

# 8    Conclusion

The approach described in this paper has led to the system of means providing the creation and use of the personified *JD*. Specificity of the approach is defined by the reflection of occupational actions of the designer and other members of their team on the QA-memory. It opens the opportunity for simulating the accessible experience in the form of intellectually processed precedents.

The personified *JD* is the interactive object the textual description of which includes the names of occupational competencies and roles. Explicitly or implicitly, all these names are lead to corresponding models of precedents in the Experience Base. The *JD* is intended for solving the tasks oriented on management of human resources and workforces in designing the family of SIS. For example, it helps to group the designers and to distribute the tasks between group members. Our future work is aimed at improving the designer model that is based on the **JD**. The next step of the improvment is bound with creating the library of **JD**-applications that will cover practices of P-CMM 2.0.

# References

1. Capability Maturity Model Integrated for Development, Version 1.3 (2010), http://www.sei.cmu.edu/reports/10tr033.pdf
2. Curtis, B.: People Capability Maturity Model (P-CMM) Version 2.0. Curtis, B., Hefley, B., Miller, S. (eds.) Technical Report CMU/SEI-2009-TR-003 P (2009)
3. El Emam, K., Koru, A.G.: A Replicated Survey of IT Software Project Failures. IEEE Software 25(5), 84–90 (2008)
4. Jacobson, I., Ng, P.-W., McMahon, P., Spence, I., Lidman, S.: The Essence of Software Engineering: The SEMAT Kernel. Queue (10) (2012)
5. Sosnin, P.: Experiential Human-Computer Interaction in Collaborative Designing of Software Intensive Systems. In: Proc.of the 11th International conference on Software Methodology and Technics (SoMeT), pp. 180–197 (2012)
6. Managing for the sustained success of an organization – A quality management, ISO 9004:2009 (2009), http://www.iso.org/iso/catalogue_detail?csnumber=41014
7. Robertson, K.: Project Management Maturity Model, http://www.klr.com/white_papers/pdfs/pm_maturity_model.pdf
8. Roglinger, M., Poppelbuth, J., Becker, J.: Maturity models in business process management. Business Process Management 18(2), 328–346 (2012)
9. Lahrmann, G., Marx, F., Winter, R., Wortmann, F.: Business Intelligence Maturity Models: An Overview, http://www.alexandria.unisg.ch/Publikationen/72444/L-en
10. Enterprise Architecture Maturity Model, http://www.nascio.org/publications/documents/nascio-eamm.pdf
11. Delamare Le Deist, F., Winterton, J.: What is competence? Human Resource Development International 8(1), 17–36 (2005)
12. Systems Engineering Competency Framework (2010), http://www.incoseonline.org.uk/Documents/zGuides/Z6_Competency_WEB.pdf
13. Von Rosing, M., Moshiri, S., Gräslund, K., Rosenberg, A.: Competency Maturity Model Wheel, http://www.valueteam.biz/downloads/model_cmm_wheel.pdf
14. Bass, L., Clements, P., Kazman, R., Klein, M.: Models for Evaluating and Improving Architecture Competence, Technical Report, CMU/SEI-2008-TR-006, ESC-TR-2008-006 (2008)
15. Borges, P., Monteiro, P., Machado, R.J.: Mapping RUP roles to small software development teams. In: Biffl, S., Winkler, D., Bergsmann, J. (eds.) SWQD 2012. Lecture Notes in Business Information Processing, vol. 94, pp. 59–70. Springer, Heidelberg (2012)
16. Basili, V.R., Rombach, H.D., Schneider, K., Kitchenham, B., Pfahl, D., Selby, R.W. (eds.): Empirical Software Engineering Issues. LNCS, vol. 4336. Springer, Heidelberg (2007)
17. Henninger, S.: Tool Support for Experience-based Software Development Methodologies. Advances in Computers 59, 29–82 (2003)
18. Sosnin, P.: Role "Intellectual Processor" in Conceptual Designing of Software Intensive Systems. In: Murgante, B., Misra, S., Carlini, M., Torre, C.M., Nguyen, H.-Q., Taniar, D., Apduhan, B.O., Gervasi, O. (eds.) ICCSA 2013, Part III. LNCS, vol. 7973, pp. 1–16. Springer, Heidelberg (2013)

# Mobile Application Development:
# How to Estimate the Effort?

Laudson Silva de Souza and Gibeon Soares de Aquino

Department of Informatics and Applied Mathematics,
Federal University of Rio Grande do Norte,
59078-970 Natal, Brazil
{laudyson,gibeo}@gmail.com

**Abstract.** The context of mobile applications, is a technological scenario that is emerging with new requirements and restrictions requires a
reevaluation of current knowledge about the processes of planning and
building software systems. These new systems have different characteristics and, therefore, an area in particular that demands such adaptation
is software estimation. The estimation processes, in general, are based on
characteristics of the systems, trying to quantify the complexity of implementing them. For this reason, it is important to analyze the methods
currently proposed for software projects estimation and evaluate their
applicability to this new context of mobile computing. Hence, the main
objective of this paper is to present a partial validation of the proposed
model estimate.

**Keywords:** Software Engineering, Software Quality, Estimating Software, Systematic Review, Mobile Applications, Mobile Computing.

## 1  Introduction

It is notable that these days, that traditional information systems are undergoing
a process of adaptation to this new computing context. Current developments,
including the increase of the computational power of these new devices, in addition to the integration of multiple devices on a single one and lined up with
the change of the users' behavior, actually create a new environment for the
development of computing solutions. However, it is important to note that the
characteristics of this new context are different. They present new resources
and, thereafter, new possibilities [22], [13], [23] and [16], as well as introduce
non-existing restrictions in conventional systems [12] and [19].

These new systems have different characteristics and, therefore, an area in
particular that demands such adaptation is software estimation. The estimation processes, in general, are based on characteristics of the systems, trying to
quantify the complexity of implementing them. For this reason, it is important
to analyze the methods currently proposed for software projects estimation and
evaluate their applicability to this new context of mobile computing. The fact is
that this new technological scenario that is emerging with new requirements and

B. Murgante et al. (Eds.): ICCSA 2014, Part V, LNCS 8583, pp. 63–72, 2014.

restrictions requires a reevaluation of current knowledge about the processes of planning and building software systems. Hence, the main objective of this paper is to present a partial validation of the proposed model estimate.

## 2    Estimation Methods in Accordance with ISO

In order to identify how the estimation methods in accordance with ISO could address the characteristics of the systems, a literature review on the estimation methods was performed. The methods identified in the survey can be seen in Table 1. All methods identified with their features can be accessed at the Following address: http://www.laudson.com/methods.pdf.

**Table 1.** Estimation Methods in accordance with ISO

| Year | Method | Author |
|------|--------|--------|
| 1979 | Function Point Analysis (FPA) | Albrecht [18] |
| 1988 | Mark II FPA | Charles Symons [21] |
| 1990 | Netherlands Software Metrics Users Association (NESMA) FPA | The Netherlands Software Metrics Users Association [4] |
| 1999 | Common Software Measurement International Consortium (COSMIC) FFP | Common Software Measurement International Consortium (COSMIC) [3] |
| 2004 | Finnish Software Metrics Association FSM | The Finnish Software Metrics Association (FiSMA) [7] |

Table 1 displays in chronological order the estimation methods in accordance with ISO, showing the year of creation, the name of the method and the author of it. At first glance, one realizes that the main existing methods were not designed to consider the requirements of mobile applications. Indeed, the very creation of most of them precedes the emergence of mobile devices as we know today. This suggests that the use of these methods to estimate the effort of the development of projects involving systems or applications for mobile devices would cause a possible failure to quantify the complexity of some features and, therefore, would not produce adequate estimates.

## 3    Characteristics of Mobile Applications

In order to identify characteristics that are inherent to systems and mobile applications, a surveying of the characteristics of these types of software was accomplished through a systematic review. Conducting a systematic review is relevant because most searches begin with some kind of review of the literature, and a systematic review summarizes the existing work fairly, without inclinations. So the surveys were conducted according to a predefined search strategy, in which the search strategy should allow the integrity of the research to be evaluated.

The planning and accomplishment of the methodology discussed were directed by *Procedures for Performing Systematic Reviews* [14].

In the context of the research questions, the following research questions were formulated: "What are the characteristics of mobile applications" and "What are the main differences between the Mobile Applications and other Applications"?. Procedures for The Evaluation of the Articles: the articles will be analyzed considering its relation with the issues addressed in the research questions, inclusion criteria and exclusion criteria, and their respective situation will be assigned with either "Accepted" or "Rejected". The evaluation will follow the following procedure: read the title and abstract and, should it be related with the research question, also read the whole article. The implementation of the systematic review was performed almost in line with its planning, except for the need to adjust the syntax of the proposed search string due to the particularities of the research bases. 234 articles were analyzed, of which 40 were selected and considered "Accepted" according to the inclusion criteria; 194 were considered "Rejected" according to the exclusion criteria. The list with all the articles Inclusion and Exclusion Criteria and Criteria can be accessed at the following address: http://www.laudson.com/sr-articles.pdf. The 40 articles that were accepted were fully read, thus performing the data extraction. All of the features found during this phase extraction are described below.

Given the results extracted from the systematic review, it's is possible to identify 29 kinds of characteristics in 100% of the articles evaluated and considered accepted in accordance with the inclusion criteria. However some of these are a mixture of characteristics of mobile devices and characteristics of mobile applications, such as the characteristic called "Limited Energy", which is a characteristic of the device and not the application, however the articles that mention this type of characteristic emphasize that in the development of a mobile application, this "limitation" must be taken into account since all the mobile devices are powered by batteries, which have a limited life, depending completely on what the user operates daily. Applications requiring more hardware or software resources will consume more energy. The 23 types of characteristics mentioned the most in the selected articles can be observed following. There is a description of each characteristic identified in the review:

o Limited energy [20]; Small screen [20]; Limited performance [17]; Bandwidth [17]; Change of context [17]; Reduced memory [20]; Connectivity [5]; Interactivity [17]; Storage [17]; Software portability [17]; Hardware portability [17]; Usability [5]; 24/7 availability [5]; Security [5]; Reliability [15]; Efficiency [1]; Native vs. Web Mobile [5]; Interoperability [17]; Response time [10]; Privacy [5]; Short term activities [2]; Data integrity [10]; Key characteristics [2]; Complex integration of real-time tasks [11]; Constant interruption of activities [2]; Functional area [9]; Price [9]; Target audience [9]; Provider type [9].

After this survey, a refinement was made and a mix of characteristics was elicited with the purpose of defining which characteristics would be emphasized. Of a total of 23 types of characteristics that were most mentioned in the selected articles, a common denominator of 13 characteristics was reached, some of which had their names redefined, like "Interactivity", which became "Input Interface".

With the conclusion of the systematic review, a survey was carried out among experts in mobile development with the purpose of ratifying the characteristics previously raised and to prove their respective influence on mobile development. The disclosure of the survey was conducted in more than 70 locations, among them universities and businesses, through e-mails, study groups and social groups. In general, of all 117 feedbacks received through the survey, 100% of the experts confirmed the characteristics; among them, an average of 72% indicated a greater effort and complexity regarding the characteristics during development, an average of 12% indicated less effort and complexity and, finally, an average of 16% indicated they did not perceive any difference in mobile development, even though they confirmed the presence of the characteristics.

## 4   Problem Addressed

As noted in Section 2, there is no estimation method developed for mobile applications projects. Moreover, some of the characteristics elicited in Section 3 aggravate the complexity and, thereafter, the effort in the development of mobile applications. From the analysis that follows, with the characteristics of applications on mobile devices elicited in Section 3, it is clear that they are different from the characteristics of traditional systems and directly influence its development. A clear example, which is different from the information or desktop systems, is the characteristic that the mobile devices have "Limited Energy". As mobile devices are powered by battery, which have a limited lifetime period, the applications must be programmed to require the minimal amount of hardware resources possible, since the more resources consumed, the greater amount of energy expended. This characteristic makes it necessary for the solution project to address this concern, generating a higher complexity of development and, thereafter, a greater effort and cost. All other characteristics that tend to influence the development of a mobile application and its attached thereto analyzes can be accessed at: http://www.laudson.com/characteristics.pdf.

From the survey of the most popular estimation methods cited in Section 2, it was found that these characteristics are not covered by the current estimation methods for two explicit reasons: first, none of the existing methods was designed to perform project estimation in mobile applications development; and second, all the characteristics discussed in this section are exclusive to mobile applications, with direct interference in their development, thereby generating a greater complexity and, thereafter, a greater effort. However, to consider any of the existing estimation methods to apply to the process of development of mobile applications is to assume that this kind of development is no different than the project of developing desktop applications, in other words, an eminent risk is assumed.

# 5 Proposal: Estimation in Mobile Application Development Project

The approached proposed is an adaptation of an existing method, which was named the "MEstiAM (Estimation Model for Mobile Applications)", based exclusively on methods recognized as international standards by ISO. Among the most popular estimation methods mentioned in Section 2, the method used to base the proposal below on is known as *"Finnish Software Metrics Association* (FISMA)". The model is one of the five methods for measuring software that complies with the ISO/IEC 14143-1 standard, is accepted as an international standard for software measuring [7] and nowadays over 750 software projects are completed being estimated by FISMA. However, the difference between this and other methods that are in accordance with the above standard, which are the *Common Software Measurement International Consortium Function Points* (COSMIC FP) [3], the *International Function Point Users Group* (IFPUG) FPA [18], *MarkII FPA* [21] and the *Netherlands Software Metrics Association* (NESMA) WSF [8], is that the method used is based in functionality but is service-oriented. It also proposes in its definition that it can be applied to all types of software, but this statement is lightly wrong since in its application, the method does not take into account the characteristics elicited in Section 3. Finally, other methods were analyzed tested, become unfeasible a possible adapto of them because mostly has its year of creation before the FISMA itself.

The COMISC FP [3], the MarkII FPA [21] and the NESMA [8] were created based on the FPA [18], in other words, they assume the counting of Function Point (FP), but considering the implemented functionality from the user's point of view. With this, it is clear that the methods mentioned above do not take into account the characteristics of mobile applications because they are not noticed by the user. The methods are independent of the programming language or technology used. And, unlike FISMA, they do not bring in their literature the information that they can be applied to all types of software.

Overall, the FISMA method proposes that all services provided by the application are identified. It previously defines some services, among which stands out the user's interactive navigation, consulting services, user input interactive services, interface services for other applications, data storage services, algorithmic services and handling services. Finally, after identifying all the services, the size of each service is calculated using the same method and thus obtaining a total functional size of the application by adding the size of each service found [6].

## 5.1 Approaching the Chosen Model

The FiSMA method in its original usage proposes a structure of seven classes of the Base Functional Component or BFC (Base Functional Component) type, which is defined as a basic component of functional requirement. The seven classes used to account for the services during the application of the method are [6]: interactive navigation of the end user and query services (q); interactive input services from end users (i); non-interactive outbound services for the end user (o); interface services for another application (t); interface services for other applications (f); data storage services (d) and algorithmic manipulation services (a).

The identification for each class name BFC previously mentioned, with a letter in parenthesis, is used to facilitate the application of the method during the counting process, because each of the seven classes BFCs are composed of other BFC classes which, at the time of calculating, these BFCs "daughter" classes are identified by the letter of their BFC "mother" class followed by a numeral. The unit of measurement is the point of function with the letter "F" added to its nomenclature to identify the "FiSMA", resulting in FfP (*FiSMA Function Point*) or Ffsu (*FiSMA functional size unit*). The measurement process generally consists of measuring the services and end-user interface and the services considered indirect [6]. Briefly, the process of counting should be done as follows. Identify: ○ How many types of BFCs does the software have? ○ Which are they? (identify all) ○ What are they? (provide details of each BFC identified). After doing this, it is necessary to add each BFC root using the formulas pre-defined by the method and their assignments. Finally, the formula of the final result of the sum is the general sum of all the BFCs classes.

## 5.2    Applying the Chosen Model

The FiSMA method can be applied manually or with the aid of the Experience Service[1] tool, which was the case, provided by FiSMA itself through contact made with senior consultant Pekka Forselius and with the chairman of the board Hannu Lappalainen.

When using the tool, it is necessary to perform all the steps of the previous subsection to obtain the functional size. Figure 1 shows the final report after the implementation of the FiSMA on a real system, the Management of Academic Activities Integrated System (Sigaa) in its Mobile version, developed by the Superintendence of Computing (SINFO) of the Federal University of Rio Grande do Norte (UFRN).

After the application of FiSMA, the functional size of the software is obtained and from this it is possible to find the effort using the formula: Estimated effort (h) = size (fp) x reuse x rate of delivery (h/fp) x project status; the latter is related to productivity factors that are taken into account for the calculation of the effort. However, of the factors predefined by the FiSMA regarding the product, only 6 (six) are proposed, in which the basic idea of the evaluation is that "the better the circumstances of the project, the more positive the assessment". The weighting goes from - - to + +, as follows: **Caption:** ○ (+ +) = [1.10] Excellent situation, much better circumstances than in the average case; ○ (+) = [1.05] Good situation, better circumstances than in the average case; ○ (+ / -) = [1.0] Normal situation; ○ (-) = [0.95] Bad situation, worse circumstances than in

---

[1] http://www.experiencesaas.com/

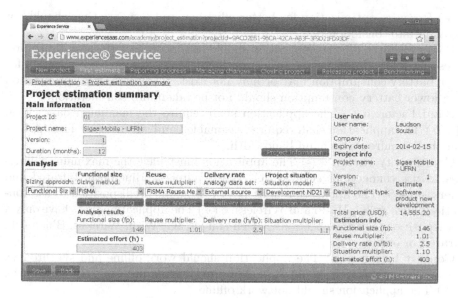

**Fig. 1.** Final Report of FiSMA applied to Sigaa Mobile

the average case; ○ (- -) = [0.90] Very bad situation, much worse circumstances than in the average case.

**Productivity Factors:** ○ Functionality requirements → compatibility with the needs of the end user, the complexity of the requirements; ○ Reliability requirements → maturity, tolerance to faults and recovery for different types of use cases; ○ Usability requirements → understandability and easiness to learn the user interface and workflow logic; ○ Efficiency requirements → effective use of resources and adequate performance in each use case and under a reasonable workload; ○ Maintainability requirements → lifetime of the application, criticality of fault diagnosis and test performance; ○ Portability requirements → adaptability and instability to different environments, to the architecture and to structural components.

Among the productivity factors mentioned above, only the "Portability Requirement" factor fits in harmony with the "Portability" characteristic regarding both hardware and software. However, none of the other factors discusses the characteristics of mobile application, in other words, after obtaining the functional size of the software and applying the productivity factors related to the product to estimate the effort, this estimate ignores all of the characteristics of mobile applications, judging that the estimate of traditional information systems is equal to the mobile application. However, with the proposal of the creation of new productivity factors, which would be the specific characteristics of mobile applications, this problem will be solved, as presented below.

Performance Factor: ○ (-) The application should be concerned with the optimization of resources for a better efficiency and response time. ○ (+/-) Resource optimization for better efficiency and response time may or may not exist.

o (+) Resource optimization for better efficiency and response time should not be taken into consideration.

Power Factor: o (-) The application should be concerned with the optimization of resources for a lower battery consumption. o (+/-) Resource optimization for lower battery consumption may or may not exist. o (+) Resource optimization for a lower battery consumption should not be taken into consideration.

Band Factor: o (-) The application shall require the maximum bandwidth. o (+/-) The application shall require reasonable bandwidth. o (+) The application shall require a minimum bandwidth.

Connectivity Factor: o (-) The application must have the maximum willingness to use connections such as 3G, Wi-fi, Wireless, Bluetooth, Infrared and others. o (+/-) The application must have reasonable predisposition to use connections such as 3G, Wi-Fi and Wireless. o (+) The application must have only a predisposition to use connections, which can be: 3G, Wi-fi, Wireless, Bluetooth, Infrared or others.

Context Factor: o (-) The application should work offline and synchronize. o (+/-) The application should work offline and it is not necessary to synchronize. o (+) The application should not work offline.

Graphic Interface Factor: o (-) The application has limitations due to the screen size because it will be mainly used by cell phone users. o (+/-) The application has reasonable limitation due to the screen size because it will be used both by cell phone and tablet users. o (+) The application has little limitation due to the screen size because it will be mainly used by tablet users.

Input Interface Factor: o (-) The application must have input interfaces for touch screen, voice, video, keyboard and others. o (+/-) The application must have standard input interfaces for keyboard. o (+) The application must have any one of the types of interfaces, such as: touch screen, voice, video, keyboard or others.

The proposed factors take into account the same weighting proposed by FiSMA, but only ranging from - to +, in other words: o (+) = [1.05] Good situation, better circumstances than in the average case; o (+ / -) = [1.0] Normal Situation; o (-) = [0.95] Bad situation, worse circumstances than in the average case. The functional size remains the same, thus affecting only the formula used to obtain the effort, which will now consider in its "project situation" variable the new productivity factors specific for mobile applications.

The validation process was as follows, was raised the total effort expended in developing the Sigaa Mobile project, ie, we obtained the actual effort. After we applied the method of estimation FISMA, in his original proposal thus obtaining an estimate of effort. Then we applied the method MEstiAM also generating an effort estimate finally the comparative analysis between the three estimates generated was performed to verify which method is closer to the actual effort spent. As can be seen in Table 2.

As can be seen in Table 2, the proposed method, MEstiAM, which is closest to the actual effort spent. You FISMA model, it was very much desired for the new model and the actual effort expended.

**Table 2.** Analysis of Estimates of Sigaa Mobile

| Real Effort Spent | MEstiAM Model | FiSMA Model |
|:---:|:---:|:---:|
| 860 h | 792 h | 403 h |

# 6   Conclusion

Given the results presented, based on the literature review of estimation methods and on the systematic review of the characteristics of mobile applications, it was observed that this sub-area of software engineering still falls short. Basically, it's risky to use any existing estimation method in development projects for mobile applications, as much as there are some models already widespread in industry, such as the Function Point Analysis, the Mark II and the COSMIC-FFP, which are even approved by ISO as international standards. They all fall short by not taking into account the particularities of mobile applications, which makes the method partially ineffective in this situation. Based on this study, it is concluded that the proposal presented in this work is entirely appropriate and viable and that this proposal should take into account all the peculiarities of such applications, finally creating a belief that there actually are considerable differences in the development project for mobile applications.

# References

1. Al-Jaroodi, J., Al-Dhaheri, A., Al-Abdouli, F., Mohamed, N.: A survey of security middleware for pervasive and ubiquitous systems. In: International Conference on Network-Based Information Systems, NBIS 2009, pp. 188–193. IEEE (2009)
2. Rogov, I., Erlick, D., Gerbert, A., Mandadi, A., Mudegowder, D.: Mobile applications: Characteristics & group project summary. In: Mobile Application Development. Google (2009)
3. COSMIC-Common Software Measurement International Consortium, et al.: The cosmic functional size measurement method-version 3.0 measurement manual (the cosmic implementation guide for iso/iec 19761: 2003) (2007)
4. Engelhart, J., Langbroek, P., et al.: Function Point Analysis (FPA) for Software Enhancement. In: NESMA (2001)
5. Feng, H.: A literature analysis on the adoption of mobile commerce. In: IEEE International Conference on Grey Systems and Intelligent Services, GSIS 2009, pp. 1353–1358. IEEE (2009)
6. Finnish Software Measurement Association FiSMA. Fisma functional size measurement method version 1-1 (2004)
7. Forselius, P.: Finnish software measurement association (fisma), fsm working group: Fisma functional size measurement method v. 1.1 (2004)
8. Gencel, C., Heldal, R., Lind, K.: On the conversion between the sizes of software products in the life cycle
9. Giessmann, A., Stanoevska-Slabeva, K., de Visser, B.: Mobile enterprise applications–current state and future directions. In: 2012 45th Hawaii International Conference on System Science (HICSS), pp. 1363–1372. Google (2012)

10. Hameed, K., et al.: Mobile applications and systems (2010)
11. Hayenga, M., Sudanthi, C., Ghosh, M., Ramrakhyani, P., Paver, N.: Accurate system-level performance modeling and workload characterization for mobile internet devices. In: Proceedings of the 9th Workshop on MEmory Performance: DEaling with Applications, Systems and Architecture, MEDEA 2008, pp. 54–60. ACM, New York (2008)
12. Husted, N., Saïdi, H., Gehani, A.: Smartphone security limitations: conflicting traditions. In: Proceedings of the 2011 Workshop on Governance of Technology, Information, and Policies, GTIP 2011, pp. 5–12. ACM, New York (2011)
13. Ketykó, I., Moor, K.D., Pessemier, T.D., Verdejo, A.J., Vanhecke, K., Joseph, W., Martens, L., Marez, L.D.: Qoe measurement of mobile youtube video streaming. In: Proceedings of the 3rd Workshop on Mobile Video Delivery, MoViD 2010, pp. 27–32. ACM, New York (2010)
14. Kitchenham, B.: Procedures for performing systematic reviews. Keele, UK, Keele University 33 (2004)
15. Maji, A.K., Hao, K., Sultana, S., Bagchi, S.: Characterizing failures in mobile oses: A case study with android and symbian. In: 2010 IEEE 21st International Symposium on Software Reliability Engineering (ISSRE), pp. 249–258. IEEE (2010)
16. Lowe, R., Mandl, P., Weber, M.: Context directory: A context-aware service for mobile context-aware computing applications by the example of google android. In: 2012 IEEE International Conference on Pervasive Computing and Communications Workshops (PERCOM Workshops), pp. 76–81 (2012)
17. Mukhtar, H., Belaïd, D., Bernard, G.: A model for resource specification in mobile services. In: Proceedings of the 3rd International Workshop on Services Integration in Pervasive Environments, SIPE 2008, pp. 37–42. ACM, New York (2008)
18. Oligny, S., Desharnais, J.-M., Abran, A.: A method for measuring the functional size of embedded software. In: 3rd International Conference on Industrial Automation, pp. 7–9 (1999)
19. Shabtai, A., Fledel, Y., Kanonov, U., Elovici, Y., Dolev, S., Glezer, C.: Google android: A comprehensive security assessment. IEEE Security Privacy 8(2), 35–44 (2010)
20. Sohn, J.-H., Woo, J.-H., Lee, M.-W., Kim, H.-J., Woo, R., Yoo, H.-J.: A 50 mvertices/s graphics processor with fixed-point programmable vertex shader for mobile applications. In: 2005 IEEE International Solid-State Circuits Conference, ISSCC. Digest of Technical Papers, vol. 1, pp. 192–592. Google (2005)
21. Symons, C.: Come back function point analysis (modernized)–all is forgiven!). In: Proc. of the 4th European Conference on Software Measurement and ICT Control, FESMA-DASMA, pp. 413–426 (2001)
22. Yang, C.-C., Yang, H.-W., Huang, H.-C.: A robust and secure data transmission scheme based on identity-based cryptosystem for ad hoc networks. In: Proceedings of the 6th International Wireless Communications and Mobile Computing Conference, IWCMC 2010, pp. 1198–1202. ACM, New York (2010)
23. Yang, S.-Y., Lee, D.L., Chen, K.-Y.: A new ubiquitous information agent system for cloud computing - example on gps and bluetooth techniques in google android platform. In: 2011 International Conference on Electric Information and Control Engineering (ICEICE), pp. 1929–1932 (2011)

# Support for Refactoring an Application towards an Adaptive Object Model

Eduardo Guerra[1] and Ademar Aguiar[2]

[1] National Institute for Space Research (INPE) - Laboratory of Computing and Applied
Mathematics (LAC) - P.O. Box 515 – 12227-010 - São José dos Campos, SP, Brazil
[2] Departamento de Engenharia Informática, Faculdade de Engenharia
Universidade do Porto (FEUP), Rua Dr. Roberto Frias, 4200-465 Porto, Portugal

**Abstract.** Flexibility requirements can appear in the middle of a software
development, perceived by several client requests to change the application. A
flexible domain model, usually implemented with using the adaptive object
model (AOM) architectural style, required custom-made components to handle
the current implementation of the domain entities. The problem is that by
evolving an AOM model, the components need to be evolved as well, which
generates constant rework. This work studied the possible AOM evolution
paths, in order to provide support in the components for model changing. An
evolution of the Esfinge AOM RoleMapper framework were developed to
provide this functionality, allowing AOM models in different stages to be
mapped to a single structure. The study was evaluated using a set of tests that
were applied in each possible structure for the model.

**Keywords:** framework, refactoring, adaptive object model, metadata, software
design, software architecture, and software evolution.

## 1 Introduction

An adaptive object model (AOM) is an architectural style where the types are defined
as instances, allowing their change at runtime [1, 2]. Metadata about the types are
stored in external sources, such as XML files or databases, and used to create the
application domain model. By using this architectural style, new entity types can be
created and existing entity types can be changed by application users. This kind of
architecture is suitable for systems in which the evolution of the domain model is part
of the business processes. There are documented case studies that use AOM in fields
like insurance [3], health [2] and archeology [4].

The requirement for domain model flexibility is usually perceived in the middle of
a software project. The initial domain model usually reflects the current user needs
when the system was requested. However, during the development, when several
customer requests are made to change the domain model, some flexibility should be
introduced in the entities to allow changes to be made at runtime, without having to
change the software code. Following this approach, some information that was
defined as domain classes' code is now defined as data that describes the domain
entities. This way, it is easier to be changed at runtime.

B. Murgante et al. (Eds.): ICCSA 2014, Part V, LNCS 8583, pp. 73–89, 2014.

When flexibility requirements appear in the middle of the project, the AOM patterns [5] are usually applied gradually through the iterations. A problem that happens with this evolution is that the other application components need to be change accordingly to the domain model implementation. For instance, architectural components, such as for persistence and to manage graphical interfaces, which are created for a fixed domain model, are usually not able to handle a flexible domain model. Even for different degrees of flexibility, the implementation of such components needs to change.

The goal of the research work presented in this paper is to map the paths for refactoring a domain model to an AOM, providing a framework support to allow this evolution without needing to change the existing components. This work is based on the Esfinge AOM RoleMapper framework [6], which uses annotations to map a domain specific AOM implementation to a general AOM implementation. The research work performed included an evolution of this framework to support the mapping of AOM patterns implementations in different stages. Several models on different stages of evolution were mapped using Esfinge AOM RoleMapper to validate the viability of this model implementation.

This paper is organized as follows: section 2 presents the AOM model and the patterns in which it is based on; section 3 talks about metadata-based frameworks, which are used in the framework that implemented the proposed model; section 4 presents the framework Esfinge AOM Role Mapper; section 5 introduces the possible evolution paths that a domain model can have towards an AOM; section 6 presents the support for the evolution path implemented in Esfinge AOM Role Mapper and how each stage can be mapped using metadata; section 7 presents an evaluation of the proposed solution; and, finally, section 8 concludes this paper and presents some future directions.

## 2    Adaptive Object Models

In a scenario in which business rules are constantly changing, implementing up-to-date software requirements has been a challenge. Currently, this kind of scenario has been very common and requirements usually end up changing faster than their implementations, resulting in systems that do not fulfill the customer needs and projects that have high rates of failure [4].

Adaptive Object Model (AOM) is an architectural style where the types, attributes, relationships and behaviors are described as instances, allowing them to be changed at runtime [1, 2]. The information about the types, which is the entities metadata, is stored in an external source, such as a database or an XML file, for the application to be able to easily read and change it. This model is described through a set of patterns, which can be combined in the same application to implement a flexible domain model. An important point is that it is not mandatory to implement always all the patterns, and only the ones that are suitable to the application requirements should be used.

The following describes briefly the core patterns of an AOM architecture. The names highlighted in bold and italic represent important elements of an AOM.

- **Type Object** [7]: This pattern should be used in scenarios where the number of subtypes of an *Entity* cannot be determined at development time. This pattern solves this issue by creating a class whose instances represent the subtypes at runtime, which is called the *Entity Type*. Then, to determine the type, *Entity* instances are composed by instances of *Entity Type*.
- **Properties** [8]: This pattern should be used when different instances of an *Entity* can have different kinds of *Properties*. In this scenario it is not viable to define attributes for all possibilities or subclasses for all possible variations. Implementing this pattern, *Properties* are represented as a list of named values in an instance. As a consequence, each instance can have only the necessary *Properties* needed, and new ones can be easily added at runtime to an *Entity*.
- **Type Square** [1, 2]: This pattern join the implementation of the two previous patterns, implementing **Type Object** twice, for the *Entity* and for the *Property*, which now has a *Property Type*. This pattern should be used when each *Entity* can have different *Properties*, but depending on its *Entity Type* there are certain *Property Types* allowed. As a consequence, new *Entity Types* can be added at runtime, as *Property Types* can be added and changed in existing *Entity Types*. Figure 1 depicts the structure of the **Type Square** pattern.

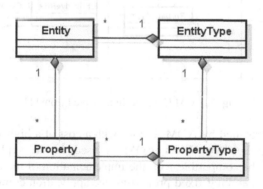

**Fig. 1.** The structure of the **Type Square** pattern

- **Accountability** [1, 8]: This pattern is used to represent a relationship or an association between Entity Types. Following this approach, an instance of a class is used to represent this *Accountability*, and an *Accountability Type* represents the allowed relationships. Despite this approach is often applied in an AOM architectural style; there are other approaches to represent the relations between *Entities*. Examples of these other approaches are: to create two subclasses for representing a *Property*, in which one of them are to represent associations; or to check the type of the value of a *Property* instance, to verify if it is an *Entity* or an instance of a language-native class.

The research work presented in this paper focus on these core structural patterns, however there are other patterns to represent the behavioral level of an AOM. Usually, `Strategy` [9] and `Rule Object` [10] are used to represent business rules. The core design of an AOM system is depicted in Fig. 2. The **Operational Level** is used to represent the application instances, which contain the information that is from the direct interest of the application, and the **Knowledge Level** represent the application metadata, which describe the application entities.

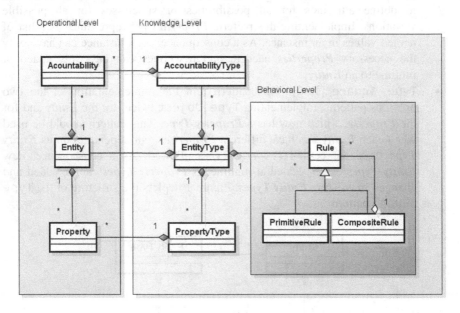

**Fig. 2.** AOM Core Design, adapted from [1]

The flexibility provided by AOMs comes with a cost of a higher complexity when developing the application. When AOM is used in an application, the other components should be compatible with the implementation of the domain model. For instance, instead of handling fixed properties, the application components need to be able to handle dynamic properties. The implementation of such components is usually coupled with the AOM implementation.

Two concerns that often should be implemented in an AOM application are persistence and graphical interfaces. The persistence should consider, not only the persistence of entities with a variable number of properties and relationships, but also the persistence of the model. The pattern AOM Builder [11] should be considered in this implementation. Graphical interfaces due to the dynamic nature of AOMs, should be able to adapt to changes in the model, and in order to implement it, rendering patterns for AOMs [12] should be implemented. Besides the issues presented, there are many other points to be considered, such as security and instance validation.

# 3  Metadata-Based Frameworks

A framework can be considered incomplete software with some points that can be specialized to add application-specific behavior, consisting in a set of classes that represents an abstract design for a family of related problems. It provides a set of abstract classes that must be extended and composed with others to create a concrete and executable application. The specialized classes can be application-specific or taken from a class library, usually provided along with the framework [13].

More recent frameworks make use of introspection [14, 15] to access application classes metadata at runtime, like their superclasses, methods and attributes. As a result, it eliminates the coupling between application classes and framework abstract classes and interfaces. For instance, following this approach the framework can search in the class structure for the right method to invoke. The use of this technique provides more flexibility to the application, since the framework reads dynamically the classes structure, allowing them to evolve more easily [16].

When a framework uses reflection [16, 17] to access and find the class elements, sometimes the class intrinsic information is not enough. If framework behavior should differ for different classes, methods or attributes, it is necessary to add a more specific meta-information to enable differentiation. For some domains, it is possible to use marking interfaces, like Serializable in Java Platform, or naming conventions [18], like in Ruby on Rails [19]. But those strategies can be used only for a limited information amount and are not suitable for scenarios that need more data.

Metadata-based frameworks can be defined as frameworks that process their logic based on instances and classes metadata [20]. In those, the developer must define additional domain-specific or application-specific metadata into application classes to be consumed and processed by the framework. The use of metadata changes the way frameworks are build and their usage by software developers [21].

From the developer's perspective in the use of those frameworks, there is a stronger interaction with metadata configuration, than with method invocation or class specialization. In traditional frameworks, the developer must extend its classes, implement its interfaces and create hook classes for behavior adaptation. He also has to create instances of those classes, setting information and hook class instances. Using metadata-based frameworks, programming activity is focused on declarative metadata configuration; and the method invocation in framework classes is smaller and localized.

Despite the use of code conventions, metadata can be defined on external sources like databases and XML files, but another alternative that is becoming popular in the software community is the use of code annotations. Using this technique the developer can add custom metadata elements directly into the class source code, keeping this definition less verbose and closer to the source code. The use of code annotations is called attribute-oriented programing [22]. Some programming languages, such as Java [23] and C# [24], have native support for annotations.

The pattern Entity Mapping [25] documented a common usage of metadata-based frameworks, which is to map between different representations of the same entity in a system. For instance, an application class can be mapped to a database to allow the framework to handle its persistence. Similarly, entities can be mapped to another class schema or to an XML format, and the framework can translate between the two representations based on the class metadata. The idea of metadata mapping is used by the Esfinge AOM RoleMapper Framework, which is presented on the next section.

## 4  Esfinge AOM RoleMapper Framework

Esfinge AOM RoleMapper framework was created in the context of the Esfinge project (http://esfinge.sf.net), which is an open source project that comprises several metadata based frameworks for different domains. Examples of other frameworks developed in this project were Esfinge QueryBuilder [6] for query generation based on method metadata, Esfinge Guardian [26] for access control, and Esfinge SystemGlue [27] for application integration. Despite each framework provide innovations in its domain, they are also used in a broader research to identify models, patterns and best practices for metadata-based frameworks.

The main goal of this framework is to map domain-specific AOM models to a general AOM model. With this mapping it is possible to have software components developed for the general AOM model, and reused for several specific models. The framework uses adapters that have the general model interfaces, but invoke methods on the application-specific model. The map between the models, the developer should add annotations to the domain-specific AOM model, and use the factory method on the adapter class to encapsulate the original class.

The common AOM core structure provided by the framework consists of the following interfaces: *IEntityType*, *IEntity*, *IPropertyType* and *IProperty*, which represent an API for the `Type Square` pattern. One implementation of these interfaces has a general implementation of an AOM, and other one has adapters to encapsulate the access to a domain-specific AOM model. The interface *IProperty* has two adapter implementations, *AdapterFixedProperty* for fixed entity attributes and *AdapterProperty* for dynamic attributes.

The framework uses code annotations, such as *@Entity*, *@EntityType*, *@PropertyType* and *@EntityProperties* to identify the roles of the classes in the AOM model. These annotations can be added in the class definition and in the attribute the stores such information. For instance, the *@EntityType* annotation should be added in the class that represents it, and in the entity attribute that stores the entity type. There are also other annotations, such as *@Name* and *@PropertyValue* that identifies the attributes that store such properties of the AOM elements.

A class named *ModelManager* manages the AOM instances created by the framework. This class is responsible to all the operations involving the manipulation of the model, including model persistence, loading and querying. For accessing the database, the *ModelManager* makes use of the *IModelRetriever* interface, which should be implemented to allow the persistence of the model metadata. The Service Locator pattern is used to locate the persistence component.

One of the main responsibilities of the *ModelManager* class is to guarantee that a logical element is not instantiated twice in the framework. In order to control that, the *ModelManager* contains two *Map* objects – one for storing the loaded Entities by their IDs and one for storing the loaded Entity Types by their IDs. Whenever a method that loads an Entity or an Entity Type is called, the *ModelManager* checks whether the ID of the instance to be loaded is already found in the corresponding map. If so, it returns the previously loaded object. Otherwise, it calls the *IModelRetriever* object for loading the object into the memory and saves it into the map.

A problem of the previous implementation of Esfinge AOM RoleMapper is that the mapping is only possible if the domain-specific AOM model implements the Type Square pattern. That can prevent the use of this framework in applications that requires a lower level of flexibility, but still need to use some of the AOM patterns. This scenario usually happens in applications that are evolving from classes with fixed properties towards an AOM model. It would be important for the framework to give support to the AOM model evolution, because the flexibility requirements that drive the design in the direction of an AOM are often discovered in the middle of the project.

## 5    AOM Evolution Paths

In order to find the evolution paths that the implementation of an AOM can have from a set of fixed classes to the Type Object pattern implementation, an empirical study was performed with developers of AOM systems. In this study were included some papers [1, 3, 4] and case studies that narrates the implementation of an AOM step by step. The goal is to find the intermediate solutions that can be implemented. Figure 3 presents the evolution path that resulted from this study.

The initial point is presented as **Fixed Entity**. In this stage classes with fixed properties compose the application model. To change or add a new property, the class source code should be changed, and to add a new type a new class should be created. In Java language, this kind of entity is usually implemented as JavaBeans [], in which each property is implemented as a private attribute with get/set access methods.

A possible evolution to add flexibility in properties is the model presented as **Property Implementation**. This change is usually motivated by several changes in the entity properties, demanding changes in the source code and in other components that depends on it. Following this model, the entity can still have some fixed properties, but it also has a list where new properties can be included. The class that represents a property usually has a name, a type and a value. The entities with a property list can receive new values as properties with no restrictions.

Another path that can be followed from a class with fixed properties is the model named **Type Object Implementation**. This path is followed when there is an explosion of subclasses to characterize different types, and all existing types can not be predicted at compile time. In this model, the type is represented by an instance that composes the entity. As a consequence, simply creating new instances of the Entity

Type class it is possible to create new types at runtime. Despite flexibility was introduced regarding the type definition, the entity properties are still fixed.

The next step, which can be implemented after the both previous models, is the **Flexible Entity Without Property Enforcement**. This model joins the implementation of Type Object and Properties patterns. New properties can be added in the entity and entity types can also be determined and set at runtime. However, the entity type does not enforce the allowed properties in a given entity. Consequently, properties can be added independent of the entity type.

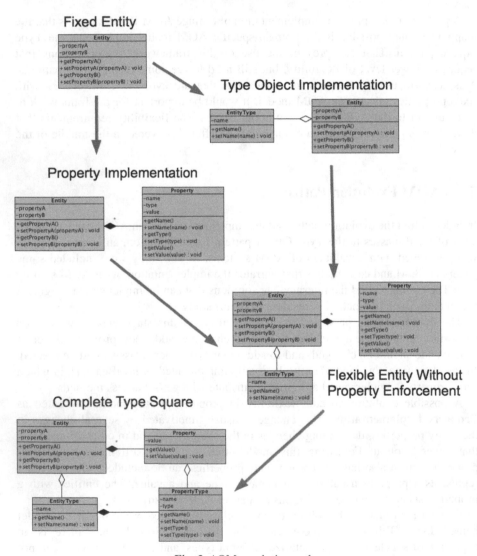

**Fig. 3.** AOMs evolution path

Finally, the last step covered by this evolution path is the **Complete Type Square**. Now the Type Object pattern is applied twice, one for the Entity, which has the Entity Type, and other for the Property, which have the Property Type. Now, the Entity Type defines the allowed Properties with the respective Property Types. Based on this information, the Entity only accepts Properties whose Property Type exists on its Entity Type. This last evolution does not add flexibility in comparison to the previous model, however it adds mode safety by allowing a more precise definition of the types and validating the Entity according to its Entity Type.

This evolution model considered the structural refactoring path to evolve from a model with fixed entities to a complete Type Square implementation. The other aspects of an AOM related to behavior were considered out of the scope of this research study. Other evolutions after the Type Square implementation are also possible, such as the introduction of inheritance and the introduction of custom metadata types, but they were also out of the scope of this work.

# 6    Implemented Support to AOM Evolution

After mapping the possible evolution paths towards an AOM model, the next step is to evolve the Esfinge AOM Role Mapper framework [6] to allow mapping for each possible implementation. No additional annotation was introduced to support the intermediary mappings. Most of the effort was employed to handle incomplete models, providing their mapping to a complete generic AOM used as API by the framework.

This section presents how each model stage can be mapped using the framework. Each subsection presents how the metadata should be configured and how the general AOM is expected to work when mapped to that model stage.

## 6.1    Fixed Entity

The first model to be supported is the fixed entity, which is a class whose properties are represented directly by attributes. Listing 1 presents how this class should be mapped using annotations. The class should receive the *@Entity* annotation and each property should receive the *@EntityProperty* annotation. Despite the annotations are added to the attributes, the access is performed using accessor methods following the JavaBeans standard. These methods can be used to add some additional logic, such as validations or transformation.

```
@Entity
public class Person {

        @EntityProperty private String name;
        @EntityProperty private String lastName;
        @EntityProperty private int age;
        @EntityProperty private Date bithday;
        //getters and setters omitted
}
```

**Listing 1.** Mapping a Fixed Model

The main restriction of this model is that new properties cannot be added to the entity, since it does not provide this kind of flexibility. When mapped to an AOM Entity by the framework, when the Entity Type is retrieved, it returns an instance of the class *GenericEntityType* created during mapping. It will have the same name as the mapped class and will have a Property Type list that represents the existing attributes in the Entity class.

## 6.2    Property Implementation

This model adds to the previous model the possibility to add dynamic properties. The entity needs to have a list of a class that represents a property. This list should be annotated with the *@EntityProperties* annotation. Additionally the entity can still have fixed properties, like in the previous model. An example of the resulting mapping can be found on Listing 2.

```
@Entity
public class Person {

        @EntityProperties
        private List<PersonInfo> infos = new ArrayList<>();
        @EntityProperty private String name;
        //getters and setters omitted
}
```

**Listing 2.** Mapping a model with Properties implementation

The class that represents a property should also be annotated with some metadata to indicate the meaning of the attributes in the AOM context. The class itself should have the *@EntityProperties* annotation, and should have, at least, one attribute with the annotation *@Name* and other with the annotation *@PropertyValue*. These attributes represent respectively the property name and value. A mapping example is presented on Listing 3.

```
@EntityProperties
public class PersonInfo {

        @Name private String name;
        @PropertyValue private Object info
        //getters and setters omitted
}
```

**Listing 3.** Property representation

Similarly to the previous model, the invocation of the method getEntityType() on a mapped instance of the class Entity, will return an instance of GenericEntityType. There is no restriction to each properties can be added in an entity. Because of that, if an invocation was performed in the entity type to retrieve the property types, an empty list will be returned, unless the entity class has also some fixed property, which should be included on the list.

## 6.3 Type Object Implementation

This model is an evolution from the model with fixed properties, presented in section 6.1, and it is a parallel evolution to the model presented on section 6.2. This approach introduces the Entity Type as an instance that composes the Entity. As presented on Listing 4, the attribute that represents the Entity Type should receive the annotation *@EntityType*. In this model, the properties are fixed in the Entity, so each one should be annotated with *@EntityProperty*.

Listing 5 presents the class that represents the Entity Type. The annotation *@EntityType* is also used in the class to configure the role it represents. The only required field annotation in this class is *@Name* that is used to differentiate between different types.

The greatest difference from the fixed domain model is that the retrieved Entity Type is an instance of *AdapterEntityType*, which encapsulates an instance of the mapped class. Consequently, while previously the name of the type was always the same, here it depends on the instance configured as the Entity Type.

```
@Entity
public class Person {

        @EntityType private PersonType type;
        @EntityProperty private String name;
        //getters and setters omitted
}
```

**Listing 4.** Entity type mapping

```
@EntityType
public class PersonType {

        @Name private String typeName;
        //getters and setters omitted
}
```

**Listing 5.** Class that represents the Entity Type

## 6.4    Flexible Entity without Property Enforcement

This model joins the independent evolutions of both previous models: the Entity has a list of Properties and a configured Entity Type as an attribute. Listing 6 presents an example of an entity implementation, that contains an attribute annotated with *@EntityType* and another one with *@EntityProperties*. Independent of that, it is still possible to have fixed properties that receive the *@EntityProperty* annotation.

```
@Entity
public class Person {

        @EntityType private PersonType type;
        @EntityProperties
        private List<PersonInfo> infos = new ArrayList<>();
        @EntityProperty private String name;
        //getters and setters omitted
}
```

**Listing 6.** Entity type and properties mapping

In this implementation, the Property class is similar to the one is Listing 3 and the Entity Type is similar to the one in Listing 5. An important characteristic in this model is that there is no restriction on the properties inserted in the entity, despite is has a defined Entity Type.

## 6.5    Complete Type Square

In this last model of the studied path there is a complete implementation of the Type Square pattern. The main difference is that the Property also has a defined type, implementing again the Type Object pattern. Listing 7 presents an example of how a class that represents the Property Type should be mapped. The class itself should receive a *@PropertyType* annotation, and additionally it should have attributes to define the property name and the property type respectively annotated with *@Name* and *@PropertyTypeType*. It is important to notice that the type is represented by an *Object*, because it can be an instance of *Class* in case of a simple property or it can be an *EntityType* in case of a relationship.

```
@PropertyType
public class InfoType {

        @Name private String name
        @PropertyTypeType private Object type;
        //getters and setters omitted
}
```

**Listing 7.** The Property Type mapping

To enable the mapping of this kind of model other classes should also be changed. As presented in Listing 8, the Entity Type should define a list of Property Types that are allowed for its respective Entities. This list should be mapped using the *@PropertyType* annotation. Additionally, as presented in Listing 9, the Property representation instead of being identified by a String with the *@Name* annotation, it is identified by a reference to its property type instance, which should also receive the *@PropertyType* annotation.

```
@EntityType
public class PersonType {

        @Name private String typeName;
        @PropertyType private List<InfoType> list = new ArrayList<>();
        //getters and setters omitted
}
```

**Listing 8.** Class that represents the Entity Type

```
@EntityProperties
public class PersonInfo {

        @PropertyType private InformationType type;
        @PropertyValue private Object info
        //getters and setters omitted
}
```

**Listing 9.** Property representation

An important characteristic of this final model is that independent of the mapped implementation, it will validate if the Properties inserted in an Entity actually exist in its Property Type. Consequently, by configuring an Entity Type it is defined which types of properties an Entity can have, in addition to its fixed properties.

# 7     Evaluation

In order to evaluate if the proposed model fulfill its goals, a small experiment was performed. The aim of this section is to describe this experiment and present the obtained results.

## 7.1     Experiment Goal

The goal of this experiment is to verify if by using the proposed mapping model it is possible to evolve an AOM model increasing its flexibility without changing the code that handles it. In this context, two hypotheses were formulated:

*H1 – The code that handles an AOM model can be decoupled from its domain-specific implementation.*
*H2 – An AOM model can be evolved without changing the code that handles it.*

## 7.2    Experiment Description

The experiment consists in a set of tests that manipulate the model through the Esfinge AOM Role Mapper API. These tests consist in the execution of operations that access the property list, insert a new property value, access a property and change a property value. The same tests are executed in all the model stages proposed in evolution model.

Figure 4 presents a graphic representation of how the software components were organized to implement the experiment. The tests use an entity factory that return an instance of the interface *IEntity* that represents the entity that should be used for the test. The factory creates the domain specific entities and uses the Esfinge Role Mapper framework to adapt it to the general AOM API, returning it. For each different model, only the Entity Factory and the Domain-specific AOM were changed.

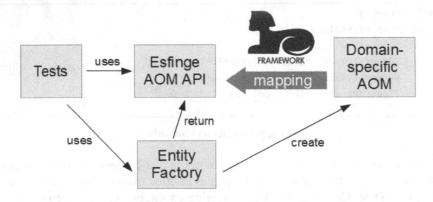

**Fig. 4.** Experiment representation

## 7.3    Results and Analysis

The elaborated tests executed successfully in all different model implementations. Based on that it is possible to confirm the first hypothesis H1, because the test code does not depend directly on the domain-specific model and could be reused and executed for different implementations of the same model.

The second hypothesis H2 can also be confirmed to be true, because the test could be executed successfully by evolving the domain-specific model, which vary from the less flexible format with fixed properties to the more flexible format with a complete type square implementation.

## 8    Related Works

In [2], many examples of systems that use the AOM architectural style are presented. While these systems aim at solving specific issues in specific domains, other frameworks, such as Oghma [4], ModelTalk [3] and its descendant, Ink [28] aim at providing generic AOM frameworks for easing the creation of adaptive systems, mainly through the use of a Domain-Specific Language (DSL).

Oghma is an AOM-based framework written in C#, which aims to address several issues found when building AOM systems, namely: integrity, runtime co-evolution, persistence, user-interface generation, communication and concurrency [Ferreira et al. 2009]. The modules that handle each of these concerns reference the AOM core structure of the framework, which was developed to be self-compliant by using the Everything is a Thing pattern [29].

ModelTalk and Ink are AOM frameworks that rely on a DSL interpreter to add adaptability to the model. At runtime, instances of DSL classes are instantiated and used as meta-objects for their corresponding Java instances through a technique called model-driven dependency injection [Hen-Tov et al. 2009]. Developers are able to change the model by editing the ModelTalk/Ink configuration in an Eclipse IDE plug-in specially developed to handle the framework DSL. When changes in the model are saved, the plug-in automatically invokes the framework's DSL analyzer, performing incremental cross-system validation similar to background compilation in Java.

Since these frameworks do not handle domain specific AOM models, it does not make sense to support the evolution of AOMs in different stages. To refactor a system towards the AOM architectural style by using these alternatives, the general AOM model provided by the framework should replace the original application model. Considering this fact, the solution proposed by this work provides a more gentle curve for application refactoring.

## 9    Conclusions

This work proposes the usage of metadata mapping to support the evolution of an AOM model without changing the code that handles it. By studding different cases in AOM implementation, it was identified the possible evolution paths that the AOM flexibility can be introduced. An implementation using the Esfinge AOM Role Mapper was incremented to enable the mapping between a general AOM Model and each one of the evolution stages. An evaluation was performed in order to verify if the goal to map different stages of the model was achieved.

This work can be considered the first step to achieve the support for mapping hybrid models, which can have different implementations in different stages co-existing on the same model and mapped to the same structure. To achieve this final stage, a further study should focus on the relationship between different entities in different evolution stages. Additionally, another future work could provide the inverse mapping, allowing an AOM model to be mapped to fixed classes with properties in Java Beans style.

We thank for the essential support of Institutional Capacitation Program (PCI-MCTI – modality BSP), which enabled a technical visit when this work was developed.

# References

1. Yoder, J.W., Balaguer, F., Johnson, R.: Architecture and design of Adaptive Object-Models. In: Proceedings of the 16th Object-Oriented Programming, Systems, Languages & Applications (2001)
2. Yoder, J.W., Johnson, R.: The Adaptive Object-Model architectural style. In: Proc. of 3rd IEE/IFIP Conference on Software Architecture: System Design, Development and Maintenance (2002)
3. Hen-Tov, A., Lorenz, D.H., Pinhasi, A., Schachter, L.: ModelTalk: when everything is a domain-specific language. IEEE Software 26(4), 39–46 (2009)
4. Ferreira, H.S.: Adaptive-Object Modeling: Patterns, Tools and Applications. PhD Thesis, Faculdade de Engenharia da Universidade do Porto (2010)
5. Welicki, L., Yoder, J.W., Wirfs-Brock, R., Johnson, R.E.: Towards a pattern language for Adaptive Object-Models. In: Proc. of 22th Object-Oriented Programming, Systems, Languages & Applications (2007)
6. Matsumoto, P., Guerra, E.M.: An Architectural Model for Adapting Domain-Specific AOM Applications. In: SBCARS- Simpósio Brasileiro de Componentes, Arquitetura e Reutilizacação de Software, Natal (2012)
7. Johnson, R., Wolf, B.: Type Object. In: Pattern Languages of Program Design, vol. 3, pp. 47–65. Addison-Wesley (1997)
8. Fowler, M.: Analysis patterns: reusable object models. Addison-Wesley Professional (1996)
9. Gamma, E., Helm, R., Johnson, R., Vlissides, J.: Design Patterns: elements of reusable object oriented software. Addison-Wesley (1994)
10. Arsanjani, A.: Rule Object: a pattern language for adaptive and scalable business rule construction. In: Proc. of 8th Conference on Pattern Languages of Programs (PLoP) (2001)
11. Welicki, L., Yoder, J.W., Wirfs-Brock, R.: Adaptive Object-Model Builder. In: Proceedings of the 16th PLoP (2009)
12. Welicki, L., Yoder, J.W., Wirfs-Brock, R.: A pattern language for Adaptive Object Models - rendering patterns. In: Proc. of 14th Conference on Pattern Languages of Programs, PLoP (2007)
13. Johnson, R., Foote, B.: Designing reusable classes. Journal Of Object-Oriented Programming 1(2), 22–35 (1988)
14. Doucet, F., Shukla, S., Gupta, R.: Introspection in system-level language frameworks: meta-level vs. Integrated. In: Source Design, Automation, and Test in Europe. Proceedings... [S.l.: s.n], pp. 382–387 (2003)
15. Forman, I., Forman, N.: Java reflection in action. Manning Publ., Greenwich (2005)
16. Foote, B., Yoder, J.: Evolution, architecture, and metamorphosis. In: Pattern Languages of Program Design 2, ch. 13, pp. 295–314. Addison-Wesley Longman, Boston (1996)
17. Maes, P.: Concepts and Experiments in Computational Reflection. In: Proceedings of the International Conference on Object-Oriented Programming, Systems, Languages and Applications, OOPSLA 1987, pp. 147–169 (1987)

18. Chen, N.: Convention over configuration (2006),
    http://softwareengineering.vazexqi.com/files/pattern.htm
    (accessed on December 17, 2009)
19. Ruby, S., Thomas, D., Hansson, D.: Agile Web Development with Rails, 3rd edn.
    Pragmatic Bookshelf (2009)
20. Guerra, E., Souza, J., Fernandes, C.: A pattern language for metadata-based frameworks.
    In: Proceedings of the Conference on Pattern Languages of Programs, Chicago, vol. 16
    (2009)
21. O'Brien, L.: Design patterns 15 years later: an interview with Erich Gamma, Richard Helm
    and Ralph Johnson. InformIT (October 22, 2009),
    http://www.informit.com/articles/article.aspx?p=1404056
    (accessed on December 26, 2009)
22. Miller, J., Ragsdale, S.: Common language infrastructure annotated standard. Addison-
    Wesley, Boston (2003)
23. Schwarz, D.: Peeking inside the box: attribute-oriented programming with Java 1.5 [S.n.t.]
    (2004),
    http://missingmanuals.com/pub/a/onjava/2004/
    06/30/insidebox1.html (accessed on December 17, 2009)
24. JSR 175: a metadata facility for the java programming language (2003),
    http://www.jcp.org/en/jsr/detail?id=175
    (accessed on December 17, 2009)
25. Guerra, E., Fernandes, C., Silveira, F.: Architectural Patterns for Metadata-based
    Frameworks Usage. In: Proceedings of Conference on Pattern Languages of Programs,
    Reno, vol. 17 (2010)
26. Silva, J., Guerra, E., Fernandes, C.: An Extensible and Decoupled Architectural Model for
    Authorization Frameworks. In: Murgante, B., Misra, S., Carlini, M., Torre, C.M., Nguyen,
    H.-Q., Taniar, D., Apduhan, B.O, Gervasi, O. (eds.) ICCSA 2013, Part IV. LNCS,
    vol. 7974, pp. 614–628. Springer, Heidelberg (2013)
27. Guerra, E., Buarque, E., Fernandes, C., Silveira, F.: A Flexible Model for Crosscutting
    Metadata-Based Frameworks. In: Murgante, B., Misra, S., Carlini, M., Torre, C.M.,
    Nguyen, H.-Q., Taniar, D., Apduhan, B.O., Gervasi, O. (eds.) ICCSA 2013, Part II. LNCS,
    vol. 7972, pp. 391–407. Springer, Heidelberg (2013)
28. Acherkan, E., Hen-Tov, A., Lorenz, D.H., Schachter, L.: The ink language meta-
    metamodel for adaptive object-model frameworks. In: Proc. of 26th ACM International
    Conference Companion on OOPSLA Companion (2011)
29. Ferreira, H.S., Correia, F.F., Yoder, J., Aguiar, A.: Core patterns of object-oriented meta-
    architectures. In: Proc. of 17th Conference on Pattern Languages of Programs (PLoP)
    (2010)

# A Strongly Consistent Transformation from UML Interactions to PEPA Nets

Juliana Bowles and Leïla Kloul

[1] School of Computer Science, University of St. Andrews, St. Andrews, UK
jkfb@st-andrews.ac.uk
[2] PRiSM, Université de Versailles, 45, Av. des Etats-Unis, 78000 Versailles
kle@prism.uvsq.fr

**Abstract.** A seamless approach suitable for both design and analysis of mobile and distributed software systems is a challenge. In this paper we provide a new constructive approach that links interaction diagrams in UML to PEPA nets, a performance modelling technique which offers capabilities for capturing notions such as location, synchronisation and message passing. Our formally defined transformation is defined in such a way that a PEPA net model realises the same language as a given Interaction Overview Diagram in UML. Furthermore, the languages are *strongly consistent*, in other words, there is a one-to-one correspondence between the traces of both models.

## 1 Introduction

The increasing complexity of modern distributed mobile systems requires a careful and sophisticated design approach for successful implementation. UML can describe the structural and behavioural aspects of these systems and is a popular medium for effective design. However, to demonstrate whether the system being developed meets the performance needs and the resources constraints specified by the client, a performance analysis of the UML-based design should be carried out.

In order to carry out performance studies of modern systems a new generation of performance modelling techniques has been introduced over the last decades. These techniques have aimed to provide mechanisms to represent the increasing complexity of the synchronisation constraints of modern systems whilst retaining the compositional structure explicitly within the model. Examples of performance modelling techniques include Stochastic Automata Networks (SAN) [16], modified Stochastic Petri net formalisms such as GSPN [6] and stochastic process algebras such as EMPA [2] and IMC [10]. However, these techniques remain unsuitable for a correct representation of many modern software systems such as mobile distributed systems which require the distinction between several contexts of computation which may depend on physical location, operating conditions, or both. For example, these techniques do not provide means for differentiating between simple local communication and the movement (migration) of processes which can change the allowable pattern of communication.

By contrast, PEPA nets [7] allows us to make a clear distinction between such mechanisms. PEPA nets are the result of the combination of coloured stochastic Petri nets and the stochastic process algebra PEPA (Performance Evaluation Process Algebra) [11].

B. Murgante et al. (Eds.): ICCSA 2014, Part V, LNCS 8583, pp. 90–105, 2014.

The Petri net infrastructure allows us to represent system locations as places in the net. The tokens that can move between locations in the net correspond to PEPA components and denote the objects in the system which carry out activities within a place either on their own or through synchronisation with other available tokens. Thus PEPA nets have capabilities for capturing notions such as location, synchronisation and message passing, and are consequently ideally suited for performance analysis of applications such as mobile code systems where the program code moves between network hosts.

We combine UML-based design with formal modelling technique PEPA nets for performance analysis of mobile distributed software systems. In particular, we are interested in a combination which can be completely seamless and transparent to software developers. In [13], we have shown how to model mobility and performance information at the design level using UML 2 interactions diagrams - sequence diagrams (SDs) and interaction overview diagrams (IODs) - and PEPA nets. While SDs capture the behaviour of objects at system locations, the IOD is used to model the overall distributed system and how mobile objects move between locations. In [3], we have described some aspects of IODs formally and defined languages associated to IODs and PEPA nets. The paper also discussed the equivalence of these languages and provided a preliminary translation algorithm of an IOD into a PEPA net model. Here, we extend the previous results to capture complex behaviour both at the IOD level and within a node. The performance annotations are extended to capture priorities and branching probabilities, and a complete formalisation of IODs is given. We obtain a framework to generate the exact corresponding PEPA expressions for any location in the net. We describe the idea of an algorithm to synthesise a PEPA net model from an IOD model, and how it guarantees that our languages are *strongly consistent*. In other words, the legal traces of both models have a one-to-one correspondence.

*Structure of this paper:* In Section 2, we present the UML2 interaction diagrams and extended notation used to model mobile distributed systems. Section 3 is dedicated to PEPA nets formalism. Section 4 provides a description of the formal model for IODs and IOD nodes. Section 5 describes the languages associated with both models. In Section 6 the synthesis algorithm and a proof for the equivalence of the languages are presented. Section 7 describes related work, and Section 8 concludes the paper.

## 2   Interaction Overview Diagrams

IODs constitute a high-level structuring mechanism that is used to compose scenarios through sequence, iteration, concurrency or choice. They are a special and restricted kind of activity diagrams (ADs) in UML where nodes are interactions or interaction uses, and edges indicate the order in which these interactions occur (cf. [15]).

Even though IODs only describe control flow and cannot show object flow, the notion of object flow is implicitly present. A node in an IOD is a SD containing objects that can progress to a further interaction according to the edges at the IOD level. Moreover, from an IOD we can derive the expected traces of behaviour for each of the instances involved. Take the example in Fig. 1(a) showing an IOD with two inline interactions. The first interaction (sd1) shows object o1 sending a message m1 to object o2 and independently object o3 sending a message m2 to object o4. The second interaction (sd2) shows just object o1 sending a message m3 to object o2. At the higher level, there are two ways of understanding the edge from sd1 to sd2 which correspond to the two possible interpretations of sequential composition of interactions in an IOD. The first interpretation could be that interaction sd1 has to complete before the behaviour described in interaction sd2 can start. This is the typical interpretation of transitions in an AD and corresponds to the notion of *strong sequential composition*. However, it is not entirely justified in the case of IODs as explained below.

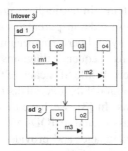   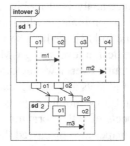

**Fig. 1.** (a) Simple IOD with two inline interactions      (b) Independent object progression

A second and weaker interpretation could be that since only objects o1 and o2 are involved in the second interaction. Both objects can move from the first interaction to the second after completing their behaviour in the first one. Thus it should be possible for o1 and o2 to proceed to the second interaction after having synchronised on message m1 independently of whether o3 and o4 have synchronised on message m2 or not. We are primarily interested in explicitly modelling the mobility of objects, and thus we borrow the notation of object flow and pins from ADs as in Fig. 1(b). This means that we assume the second interpretation (weak sequential composition) by default.

Objects that want to progress from one interaction to another must have an output pin with the name and type of the object, and an input pin with the same name and type in the following interaction. As soon as an object completes its behaviour in an interaction, a token is put in the corresponding output pin and the edge can fire provided target pin has enough room. Whether or not the following interaction can execute depends on how many input tokens are required. In Fig.1(b), n sd2 can only start once both tokens are available in respective input pins, regardless of whether m2 has been sent or not.

To allow both interpretations of sequential composition, we can represent strong sequential composition using a fork (Fig. 2). Here the edges for objects o1 and o2 cannot fire independently and are synchronised with those for objects o3 and o4. Only when all tokens are available on the fork execution proceeds, with o1 and o2 moving to node sd2 and o3 and o4 returning to node sd1. As in this case, a fork can be used to synchronise the objects associated with the edges it cuts across. Both interpretations of sequential composition give us a powerful language to model and structure interactions.

As in Fig. 2, we use a tagged value $\{initBound = n\}$ next to a pin to indicate the initial number of tokens $n$ associated with that pin (initial marking of the IOD). By default $\{initBound = 0\}$. Since we model *object flow* and not control flow, we do not

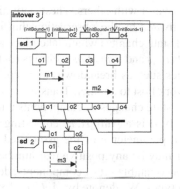

**Fig. 2.** Strong sequential composition in an IOD

need joins, and forks are used as a way to enforce object synchronisation along edges. Decision points are not needed and we represent guards directly on the edges.

For modelling mobility through edges in an IOD, it is useful to be able to indicate, if intended, the explicit activity (an action type with its occurrence rate) that corresponds to the movement of an object from one node (location) to another. We can indicate this activity at the source pin of an IOD edge. According to UML specification, a pin has a name and type (one or the other may be omitted). We assume that all information on an edge is given at it's source pin, that is, the source pin of an edge carries the information on the associated activity by giving the corresponding *action type* and *rate*. Additional optional information includes a condition and information on the priority and/or the probability of the edge. The condition is a boolean expression that determines whether an edge is enabled or not. An edge may also be temporarily disabled if the target pin is full (given by the upperBound tag mentioned earlier). A priority on an edge, given by a natural number, is used if we have several possible edges leaving a node, and we want to model explicitly that edges have different priorities. The probability on an edge denotes the probability of the associated target IOD node. The textual label of a source pin is given by `<[c]>pin_type;action_type,<q>/rate,<p>` where c denotes a condition or guard, q the priority and p the probability of the edge (see Fig. 3). Priorities and probabilities are only given at the IOD level, not on messages.

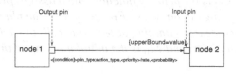

**Fig. 3.** Input and output pins

# 3   PEPA Nets

PEPA nets [7] combine the process algebra PEPA with stochastic coloured Petri nets. In this hybrid formalism, a system is described as an interaction of *components* which engage in *activities* on their own or with others. These activities represent system state changes. PEPA nets are motivated by the observation that in many systems we can identify two distinct types of state changes which take place at different levels. Consequently PEPA net considers *firings* and *transitions*. Firings (of the net) are typically used to model *global* state changes such as breakdowns and repairs, or a mobile software agent moving from one network host to another. Transitions (of PEPA components) are typically used to model *local* state changes as components undertake activities.

In PEPA nets, each activity $a$ is a pair $(\alpha, r)$ consisting of an *action type* $\alpha$ and its duration represented by a parameter $r$ of the associated exponential distribution: *activity rate*. This parameter may be any positive real number, or the symbol $\top$ (read as *unspecified*). We assume a countable set of components, denoted $\mathcal{C}$, and a countable set, $\mathcal{Y}$, of all possible action types. We denote by $\mathcal{A}ct \subseteq \mathcal{Y} \times \mathbb{R}^*$, the set of activities, where $\mathbb{R}^*$ is the set of positive real numbers and zero, together with the symbol $\top$. As the firings and the transitions are special cases of PEPA activities, we denote by $\mathcal{Y}_f$ the set of action types at the net level and by $\mathcal{Y}_t$ the set of action types inside the places such that $\mathcal{Y} = \mathcal{Y}_f \cup \mathcal{Y}_t$. Similarly, we denote by $\mathcal{A}ct_t \subseteq \mathcal{Y}_t \times \mathbb{R}^*$ the set of activities undertaken by the components inside the places and by $\mathcal{A}ct_f \subseteq \mathcal{Y}_f \times \mathbb{R}^+$ the set of activities at the net level such that $\mathcal{A}ct = \mathcal{A}ct_f \cup \mathcal{A}ct_t$. Thus, unlike activities labelling the firing transitions, local activities to a place can have a rate equal to zero.

A PEPA net is made up of PEPA *contexts*, one at each place in the net. A context consists of zero or more *static* components and one or more *cells*. A cell is a storage area dedicated to a *mobile* PEPA component which circulate as the *tokens* of the net. By contrast, the static components cannot move. Mobile components are terms of PEPA which define the components behaviour. Thus each token has a type given by its definition. This type determines the transitions and firings which a token can engage in; it also restricts the places in which it may be, since it may only enter a cell of the corresponding type. We assume a countable set of static components $\mathcal{C}_S$ and a countable set of mobile components or tokens $\mathcal{C}_M$ such that $\mathcal{C}_S \cup \mathcal{C}_M = \mathcal{C}$.

**Definition 31.** *A PEPA net $\mathcal{V}$ is a tuple $\mathcal{V} = (\mathcal{P}, \mathcal{T}, I, O, \ell, \pi, \mathcal{F}_P, K, M_0)$ such that $\mathcal{P}$ is a finite set of places; $\mathcal{T}$ is a finite set of net transitions; $I : \mathcal{T} \to \mathcal{P}$ is the input function; $O : \mathcal{T} \to \mathcal{P}$ is the output function; $\ell : \mathcal{T} \to (\mathcal{Y}_f, \mathbb{R}^+ \cup \{\top\})$ is the labelling function, which assigns a PEPA activity ((type, rate) pair) to each transition. The rate determines the negative exponential distribution governing the delay associated with the transition; $\pi : \mathcal{Y}_f \to \mathbb{N}$ is the priority function which assigns priorities (represented by natural numbers) to firing action types; $\mathcal{F}_P : \mathcal{P} \to \mathbf{P}$ is the place definition function which assigns a PEPA context, containing at least one cell, to each place; $K$ is the set of token component definitions; and $M_0$ is the initial marking of the net.*

In PEPA nets, conflicts may be solved by the race policy but it is also possible to assign different priorities to net transitions using priority function $\pi$. As with classical Petri nets with priorities, a transition $t$ will be enabled for a firing if there is no other firing transitions of higher priority that can be legally fired according to the net structure and the current marking. The syntax of PEPA nets is given in the following where $S$ denotes a *sequential component* and $P$ a *concurrent component* which executes in parallel. $I$ stands for a constant denoting either a sequential or a concurrent component.

$$N ::= K^+ M \qquad \text{(net)}$$

**(definitions and marking)**

$$M ::= (M_{\mathbf{P}}, \ldots) \text{ (marking)} \qquad K ::= I \overset{def}{=} S \qquad \text{(component defn)}$$

$$M_{\mathbf{P}} ::= \mathbf{P}[X, \ldots] \text{ (place marking)} \qquad | \quad \mathbf{P}[X] \overset{def}{=} P[X] \qquad \text{(place defn)}$$

$$| \quad \mathbf{P}[X, \ldots] \overset{def}{=} P[X] \underset{L}{\S} C \text{ (place defn)}$$

**(marking vectors)**              **(identifier declarations)**

$$S ::= (\alpha, r).S \text{ (prefix)} \qquad P ::= P \underset{L}{\S} P \text{ (cooperation)} \qquad X ::= \text{`\_' (empty)}$$

$$| \quad S + S \quad \text{(choice)} \qquad | \quad P/L \quad \text{(hiding)} \qquad | \quad S \quad \text{(full)}$$

$$| \quad I \qquad \text{(identifier)} \qquad | \quad P[X] \quad \text{(cell)}$$

$$| \quad I \qquad \text{(identifier)}$$

**(sequential components)**    **(concurrent components)**    **(cell term expressions)**

PEPA net behaviour is governed by structured operational semantic rules [7]. The states of the model are the marking vectors, which have one entry for each place of the PEPA net. The semantic rules govern the possible evolution of a state, giving rise to a multi-labelled transition system or derivation graph. The nodes of the graph are the marking vectors and the activities give the arcs of the graph. This one gives rise to a Continuous-Time Markov Chain (CTMC) which can be solved to obtain a steady-state probability distribution from which performance measures can be derived.

## 4 A Formal IOD Model

In this section we describe IODs and IOD nodes formally.

**Definition 41.** *An IOD $\mathcal{D}$ is a tuple defined by $\mathcal{D} = (\mathcal{N}, \mathcal{S}, \mathcal{T}, \mathcal{P}, \mathcal{A}ct, \mathcal{L}_O, \mathcal{L}_I, \mathcal{K}, \mathcal{F}, \mathcal{C}, \mathcal{B})$ where*

- *$\mathcal{N}$ is a finite set of nodes;*
- *$\mathcal{S}$ is a finite set of fork nodes;*
- *$\mathcal{T}$ is a finite set of transitions;*
- *$\mathcal{P}$ is a set of pin types such that $\mathcal{P} = \mathcal{P}_I \cup \mathcal{P}_O$ and $\mathcal{P}_I \cap \mathcal{P}_O = \emptyset$, where $\mathcal{P}_I$ is the set of input pin types and $\mathcal{P}_O$ is the set of output pin types in $\mathcal{D}$;*
- *$\mathcal{A}ct$ is a set of activities such that $\mathcal{A}ct = \mathcal{A}ct_n \cup \mathcal{A}ct_p$ where $\mathcal{A}ct_n$ is the set of activities in the nodes and $\mathcal{A}ct_p$ is the set of activities at the IOD level. Each activity in $\mathcal{A}ct$ is a pair $(a, r)$ consisting of an action type $a$ and a rate $r \in \mathbb{R}^+ \cup \{\top\}$ if $a \in \mathcal{A}ct_p$ and $r \in \mathbb{R}^* \cup \{\top\}$ if $a \in \mathcal{A}ct_n$;*
- *$\mathcal{L}_O \colon \mathcal{T} \to \{\mathcal{P}_O, \mathcal{A}ct_p\}$ is a total labelling function which assigns an output pin type and an activity to the source pin of a transition;*
- *$\mathcal{L}_I \colon \mathcal{T} \to \{\mathcal{P}_I, \mathbb{N}^+\}$ is a total labelling function which assigns an input pin type and an upper bound to the target pin of a transition;*
- *$\mathcal{K} \colon \mathcal{T} \to \{Exp, \mathbb{N}, [0, 1]\}$ is a partial labelling function which associates a triple $(g, q, p)$ to the source pin of a transition such that $g$ is an expression denoting a guard, $q$ is a natural number denoting the priority of the transition, and $p$ is a number between $[0, 1]$ denoting the probability of the transition. $g$, $q$ and $p$ are optional;*

- $\mathcal{F} \colon \mathcal{T} \to \mathcal{N} \times \mathcal{N}$ is a total function which assigns a pair of nodes (a source node and a target node) to a transition;
- $\mathcal{C} \colon \mathcal{S} \to 2^{\mathcal{T}}$ is a total function which assigns a set of transitions to a fork node;
- $\mathcal{B} \colon \mathcal{P} \to \mathbb{N}$ is the initial marking of the IOD.

An IOD $\mathcal{D}$ is described by a set of nodes $\mathcal{N}$ and edges $\mathcal{T}$ called transitions between the nodes. An IOD contains an additional (possibly empty) set of fork nodes given by $\mathcal{S}$. To capture object mobility, a transition in an IOD is associated with a unique object and indicates how it moves from one node to another. To indicate which object is associated with a transition we use a set of pin types $\mathcal{P}$ distinguishing between input pin types $\mathcal{P}_I$ and output pin types $\mathcal{P}_O$. We use a set of activities $\mathcal{A}ct_p$ to indicate the action and rate associated with the object move (transition). All transitions are associated with two pin types: one output pin type (the source pin of the transition) and one input pin type (the target pin of the transition). We use functions $\mathcal{L}_O$ and $\mathcal{L}_I$ to associate the specific pins with a transition. The source pin carries the activity associated with the object move. In addition, it can also have optional information such as a guard (a boolean expression in $Exp$), a natural number denoting the priority of the transition and/or a number denoting the probability of the transition. Function $\mathcal{K}$ is used to add this optional information to the source pin. The target pin has a natural number indicating the number of tokens allowed (*upperBound* - see Fig. 3). If the target pin has reached its maximum number of tokens the transition is not enabled. A fork node in $\mathcal{S}$ cuts across several transitions to synchronise the objects associated with the transitions. The set of transitions affected by a fork is given by function $\mathcal{C}$. Finally, the initial marking $\mathcal{B}$ defines how many tokens are available at pin types. When a transition fires one token from the source pin of the transition is removed and placed at the associated target pin. Take example of Fig. 2. The IOD is given by the set of nodes $\mathcal{N} = \{sd1, sd2\}$, one fork node $\mathcal{S} = \{s1\}$, transitions $\mathcal{T} = \{t_1, t_2, t_3, t_4\}$, input pins $\mathcal{P}_I = \{o_{1\,isd1}, o_{2\,isd1}, o_{3\,isd1}, o_{4\,isd1}, o_{1\,isd2}, o_{2\,isd2}\}$. (compact notation $o_{1\,isd1}$ represents input pin for object o1 in node sd1), output pins $\mathcal{P}_O = \{o_{1\,osd1}, o_{2\,osd1}, o_{3\,osd1}, o_{4\,osd1}\}$, empty set of activities $Act$ (the example does not show activities), and for instance $\mathcal{L}_O(t_1) = (o_{1\,osd1}, act_{o_1})$, $\mathcal{K}$ is undefined as this example does not contain guards, priorities or probabilities on transitions, $\mathcal{L}_I(t_1) = (o_{1\,isd2}, 1)$, $\mathcal{F}(t_1) = (sd1, sd2)$, $\mathcal{C}(s1) = \mathcal{T}$, $\mathcal{B}(o_{1\,isd1}) = 1$, $\mathcal{B}(o_{1\,osd1}) = 0$, etc.

An IOD defines the system behaviour of whereas each individual node (SD) in the IOD describes the behaviour of a location in the system. We define a node as follows.

**Definition 42.** *A node $\mathcal{A}$ in an IOD $\mathcal{D}$ where $\mathcal{A} \in \mathcal{N}$ is a tuple $\mathcal{A} = (\mathcal{O}, \mathcal{E}, <, \mathcal{M}_A, \mathcal{T}_A, \mathcal{P}_A, \mu_A, \mathcal{I}_A, \mathcal{U}_A)$ such that*

- $\mathcal{O}$ *is a finite set of object types such that $\mathcal{O} = \mathcal{O}_M \bigcup \mathcal{O}_S$ where $\mathcal{O}_M$ is the set of mobile object types and $\mathcal{O}_S$ is the set of static object types;*
- $\mathcal{E}$ *is a set of events such that $\mathcal{E} = \mathcal{E}_S \bigcup \mathcal{E}_R$ where $\mathcal{E}_S$ is the set of send events and $\mathcal{E}_R$ is the set of receive events, and $\mathcal{E} = \bigcup_{o \in \mathcal{O}} \mathcal{E}_o$ such that for any $o_1, o_2 \in \mathcal{O}$, if $o_1 \neq o_2$ then $\mathcal{E}_{o_1} \bigcap \mathcal{E}_{o_2} = \emptyset$;*
- $<$ *is a set of partial orders $<_o \subseteq \mathcal{E}_o \times \mathcal{E}_o$ with $o \in \mathcal{O}$;*
- $\mathcal{M}_A$ *is a finite set of local labels (messages). Each label $m \in \mathcal{M}_A$ is defined as $m = a/r_1; r_2$ where $(a, r_1) \in Act_n$ and $(a, r_2) \in Act_n$;*
- $\mathcal{T}_A$ *is the set of local transitions such as $\mathcal{T}_A \subseteq \mathcal{E}_S \times \mathcal{M}_A \times \mathcal{E}_R$;*

- $\mathcal{P}_A$ is the set of pin types of $A$ such that $\mathcal{P}_A \subseteq \mathcal{P}$;
- $\mu_A : \mathcal{P}_A \to \mathcal{O}_M$ is total function which associates a mobile object type to a pin type;
- $\mathcal{I}_A$ is set of inputs to $A$ such that each input $I \in \mathcal{I}_A$ is a set of pairs $\{(p, n)/p \in \mathcal{P}_{I_A}, n \in \mathbb{N}^+\}$ where $\mathcal{P}_{I_A}$ is set of input pin types to $A$ and $n$ is a number of tokens;
- $\mathcal{U}_A$ is set of outputs from $A$ such that each output $U \in \mathcal{U}_A$ is a set of pairs $\{(p, n)/p \in \mathcal{P}_{O_A}, n \in \mathbb{N}^+\}$ where $\mathcal{P}_{O_A}$ is set of output pin types of $A$ and $n$ is number of tokens.

A node $A$ in an IOD is a SD describing an interaction between objects in $\mathcal{O}$. Some of the objects enter/leave the node through input/output pins and are the *mobile objects* given by the set $\mathcal{O}_M$ (the exact mapping of pin types to object types is given by the total function $\mu_A$). Additional objects involved in the interaction described by the diagram are *static* and given by the set $\mathcal{O}_S$. Static objects reside in an IOD node and do not participate in any other interaction (node) elsewhere in the IOD. The behaviour of the node is described by a set of events $\mathcal{E}$ corresponding to the sending/receiving of messages ($\mathcal{E}_S$ and $\mathcal{E}_R$ respectively). Each event is associated with one unique object. A partial order $<$ is defined over the set of events and based on local partial orders. Given a set of events and message labels, transitions in the node correspond to triples $(e_1, m, e_2)$ whereby $e_1$ (respectively $e_2$) is an event associated with the sending (receipt) of message $m$. Each message $m = a/r_1; r_2$ consists of an action type $a$, and two rates $r_1$ and $r_2$. If one of the rates is unspecified ($r_1 = \top$ or $r_2 = \top$), then it is omitted leading to a message $m = a/r$ where $r = r_1$ or $r = r_2$. In this case the rate is associated with the object sending the message. The rate gives the frequency at which the activity is performed.

A node $A$ has a set of pin types $\mathcal{P}_A$ which is a subset of the pin types. It consists of a disjoint set of input and output pin types. For a node to execute, it needs to have a set $\mathcal{I}_A$ of tokens available at its input pins. In particular, a node can have *alternative* inputs. For example, $\mathcal{I}_A = \{\{(p_1, 1), (p_2, 2)\}, \{(p_3, 1)\}\}$ indicates that node $A$ has three input pin types $p_1, p_2$ and $p_3$, but $p_1, p_2$ are an alternative input to $p_3$. Further, for the node to execute, we need one token of type $p_1$ and two tokens of type $p_2$ or alternatively one token of type $p_3$. Similarly, once a node has executed, it generates a set of tokens at its output pins. Outputs correspond to a family of sets of output pins $\mathcal{U}_A$. Consider node $sd1$ in Fig. 4. It is defined by objects $\mathcal{O}_M = \{o_1, o_3\}$ and $\mathcal{O}_S = \{o_2\}$, events $\mathcal{E} = \{e_1, \ldots e_6\}$ where local object events are $\mathcal{E}_{o_1} = \{e_1, e_3\}$, $\mathcal{E}_{o_2} = \{e_2, e_6\}$, $\mathcal{E}_{o_3} = \{e_4, e_5\}$ and the partial order is such that $e_4 <_{o_3} e_5$. The set of messages is $\mathcal{M}_{sd3} = \{m_1, m_2, m_3\}$, and the local transitions $\mathcal{T}_{sd3} = \{t_1, t_2, t_3\}$ are such that $t_1 = (e_1, m_1, e_2)$, $t_2 = (e_3, m_2, e_4)$ and $t_3 = (e_5, m_3, e_6)$. The pins to the node are $\mathcal{P}_{sd3} = \{p_1, p_2, p_3, p_4\}$ and $\mu_{sd3}(p_1) = \mu_{sd3}(p_3) = o_1$, $\mu_{sd3}(p_2) = \mu_{sd3}(p_4) = o_3$. Activities are such that, i.e., $(a_4, r_4) \in Act_p$ and $(a_1, r_{11}) \in Act_n$. There is one possible input and one possible output: $\mathcal{I}_{sd3} = \{\{(p_1, 1), (p_2, 1)\}\}$ and $\mathcal{U}_{sd3} = \{\{(p_3, 1), (p_4, 1)\}\}$.

Since we can describe a variety of behaviours in a node using interaction fragments such as parallel, alternative and iterative behaviour, we need to capture the fragments associated with an IOD node. In the following, we consider a *region* as a subset of events and define a so-called *basic region* next.

**Definition 43.** $\mathcal{R}$ is a basic region over $(\mathcal{E}, <)$ if $\mathcal{R} \subseteq \mathcal{E}$ and $\mathcal{R} = \bigcup_{o \in \mathcal{O}} \mathcal{R}_o$, where $\mathcal{R}_o$ is a totally ordered set of events for object $o$. The minimal and the maximal events of $\mathcal{R}_o$ are denoted $first_{\mathcal{R}_o}$ and $last_{\mathcal{R}_o}$ respectively.

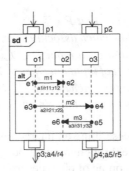

**Fig. 4.** An IOD node

A basic region is a subset of events where all events belonging to the same object are totally ordered and hence we can refer to the first and last events for that object. One example of a basic region is a node without any interaction fragments in it. Nodes $sd1$ and $sd2$ in Fig. 2 are basic regions. Notice that a basic region can be empty.

**Definition 44.** *Let $\mathcal{R}$ be a basic region over $(\mathcal{E}, <)$ defined over IOD node $\mathcal{A}$. $\mathcal{R}$ is closed iff for any $e \in \mathcal{R}$ if there is a local transition $t \in \mathcal{T}_\mathcal{A}$ with $t = (e, m, e')$ or $t = (e', m, e)$ then $e' \in \mathcal{R}$.*

A closed basic region does not cut across local transitions. Thus, if one event involved in a transition belongs to a basic region so does its corresponding send/receive event. In the sequel, we assume that all our basic regions are closed. Once we include alternative or parallel behaviour, we are no longer able to characterise the events of an object as being totally ordered. In Fig. 4, the whole set of events $\mathcal{E} = \{e_1, \ldots e_6\}$ does not define a region as $e_1 \not<_{o_1} e_3$. We have two basic regions which correspond to the operands of the alt fragment: $\mathcal{R}_1 = \{e_1, e_2\}$ and $\mathcal{R}_2 = \{e_3, e_4, e_5, e_6\}$.

**Definition 45.** *Let $\mathcal{R}$ be a basic region over $(\mathcal{E}, <)$ defined over IOD node $\mathcal{A}$. The associated set of local transitions for $\mathcal{R}$ is given by $\mathcal{T}_{\mathcal{A}_\mathcal{R}}$ and is such that for each $l \in \mathcal{T}_{\mathcal{A}_\mathcal{R}}, l = (e_1, m, e_2)$ with $e_1, e_2 \in \mathcal{R}$.*

In Fig. 4, the local transitions are given by $\mathcal{T}_{sd3_{\mathcal{R}_1}} = \{t_1\}$ and $\mathcal{T}_{sd3_{\mathcal{R}_2}} = \{t_2, t_3\}$.

**Definition 46.** *Let $\mathcal{R}$ be a basic region over $(\mathcal{E}, <)$ defined over IOD node $\mathcal{A}$, and $t_1, t_2 \in \mathcal{T}_{\mathcal{A}_\mathcal{R}}$. $t_1$ precedes $t_2$ in the set of local transitions (written $t_1 \ll t_2$) for $t_1 = (e_{11}, m_1, e_{12}), t_2 = (e_{21}, m_2, e_{22})$, and some $o \in \mathcal{O}$ iff*

*1. at least one of the following holds*
   *(a) $e_{11} <_o e_{21}$, where $e_{11}, e_{21} \in \mathcal{E}_o$;*     *(b) $e_{12} <_o e_{22}$, where $e_{12}, e_{22} \in \mathcal{E}_o$;*
   *(c) $e_{12} <_o e_{21}$, where $e_{12}, e_{21} \in \mathcal{E}_o$;*     *(d) $e_{11} <_o e_{22}$, where $e_{11}, e_{22} \in \mathcal{E}_o$;*
*2. or the transitions are independent, i.e., $o_1 \neq o_3$, $o_1 \neq o_4$, $o_2 \neq o_3$ and $o_2 \neq o_4$, for $e_{11} \in \mathcal{E}_{o_1}, e_{12} \in \mathcal{E}_{o_2}, e_{21} \in \mathcal{E}_{o_3}, e_{22} \in \mathcal{E}_{o_4}$.*

According to definition 46, $t_1 \ll t_2$ if the transitions share at least one object and the associated events for that object are ordered, or the transitions are independent. Back in our example, for basic region $\mathcal{R}_2$ with $t_2, t_3 \in \mathcal{T}_{sd3_{\mathcal{R}_2}}$, $t_2 \ll t_3$ as $e_4 <_{o_3} e_5$. In the case of independent transitions we always have both $t_1 \ll t_2$ and $t_2 \ll t_1$.

**Definition 47.** *Let $\mathcal{R}$ be a basic region and $\mathcal{T}_{A_\mathcal{R}}$ the associated set of local transitions over $\mathcal{R}$. The concurrency level of $\mathcal{T}_{A_\mathcal{R}}$ is $l$, if $\mathcal{T}_{A_\mathcal{R}}$ contains $l$ totally ordered subsets of transitions $\mathcal{T}_{A_\mathcal{R}} = \mathcal{T}_{A_{\mathcal{R}1}} \cup \ldots \cup \mathcal{T}_{A_{\mathcal{R}l}}$ such that, for two arbitrary distinct transitions $t_1$ and $t_2$, $t_1 \in \mathcal{T}_{A_{\mathcal{R}i}}$, $t_2 \in \mathcal{T}_{A_{\mathcal{R}j}}$ and $i \neq j \in [1, \ldots, l]$, $t_1$ and $t_2$ are independent.*

A basic region $\mathcal{R}$ where $\mathcal{T}_{A_\mathcal{R}}$ has concurrency level 1 is such that $\mathcal{T}_{A_\mathcal{R}}$ is a totally ordered set of local transitions. In our example, $\mathcal{T}_{sd3_{\mathcal{R}2}}$ is a totally ordered set of local transitions and has concurrency level 1. In Fig. 2, basic region in node $sd1$ containing both transitions labelled $m_1$ and $m_2$ has concurrency level 2 as $m_1$ and $m_2$ are independent. We can thus consider two basic regions one associated with the set of transitions containing $m_1$ and the other associated with the set of transitions containing $m_2$.

**Definition 48.** *For basic regions $\mathcal{R}_1$ and $\mathcal{R}_2$, $\mathcal{R}_1.\mathcal{R}_2$ is a basic region denoting the sequential composition of the regions satisfying $last_{\mathcal{R}_{1o}} <_o first_{\mathcal{R}_{2o}}$ for any $o \in \mathcal{O}$.*

The sequential composition of regions is as expected a way of ordering the events of the respective regions sequentially.

**Definition 49.** *$\mathcal{G}$ is a basic arbitrary region if :*

- *$\mathcal{G}$ is a basic alt region, that is $\mathcal{G} = \mathcal{R}_1 \cup \mathcal{R}_2 \cup \ldots \cup \mathcal{R}_N$, $N \subset \mathbb{N}$, where each $\mathcal{R}_n$, $n = 1, \ldots, N$, is a basic region;*
- *$\mathcal{G}$ is a basic loop region, that is $\mathcal{G} = \mathcal{R}^N$ where $\mathcal{R}$ is a basic region and $N$, $N \in \mathbb{N}$, is the loop index; or*
- *$\mathcal{G}$ is a basic par region, that is $\mathcal{G} = [\mathcal{R}_1, \mathcal{R}_2, \ldots, \mathcal{R}_N]$, $N \in \mathbb{N}$, where each $\mathcal{R}_n$, $n = 1, \ldots, N$, is a basic region and the associated set of local transitions $\mathcal{T}_{A_{\mathcal{R}n}}$ is of concurrency level one.*

Interaction fragments without nesting constitute basic arbitrary regions. A basic alternative fragment, called basic alt region, can be seen as a finite union $\mathcal{G}$ of regions where each region corresponds to one of the operands in the alternative fragment and these regions are basic (see Fig. 4). Similarly, a basic loop region and a basic par region are made from basic regions: one in the case of the loop where there is an iteration over that basic region, and as many as there are operands in the case of the par. We refine the definition of transition order for sequential composition of regions.

**Definition 410.** *Let $\mathcal{R}_0.\mathcal{G}$ be the sequential composition of basic region $\mathcal{R}_0$ and basic arbitrary region $\mathcal{G}$.*

1. *If $\mathcal{G}$ is a basic alt region, that is $\mathcal{G} = \mathcal{R}_1 \cup \mathcal{R}_2 \cup \ldots \cup \mathcal{R}_N$:*
   (a) *if $t_1, t_2 \in \mathcal{T}_{A_{\mathcal{R}i}}$ for some $i = 1, ..., N$, $t_1 \ll t_2$ if Definition 46 applies,*
   (b) *if $t_1 \in \mathcal{R}_0$ and $t_2 \in \mathcal{T}_{A_{\mathcal{R}i}}$ for some $i = 1, ..., N$, $t_1 \ll t_2$ if Definition 46 applies,*
   (c) *if $t_1 \in \mathcal{T}_{A_{\mathcal{R}i}}$ and $t_2 \in \mathcal{T}_{A_{\mathcal{R}j}}$ with $i \neq j \geq 1$ then $t_1 \not\ll t_2$ and $t_2 \not\ll t_1$.*
2. *If $\mathcal{G}$ is a basic par region, that is $\mathcal{G} = [\mathcal{R}_1, \mathcal{R}_2, \ldots, \mathcal{R}_N]$:*
   (a) *if $t_1, t_2 \in \mathcal{T}_{A_{\mathcal{R}i}}$ for some $i = 1, ..., N$, $t_1 \ll t_2$ if Definition 46 applies,*
   (b) *if $t_1 \in \mathcal{R}_0$ and $t_2 \in \mathcal{T}_{A_{\mathcal{R}i}}$ for some $i = 1, ..., N$, $t_1 \ll t_2$ if Definition 46 applies,*
   (c) *if $t_1 \in \mathcal{T}_{A_{\mathcal{R}i}}$ and $t_2 \in \mathcal{T}_{A_{\mathcal{R}j}}$ with $i \neq j \geq 1$ then $t_1 \ll t_2$ and $t_2 \ll t_1$ (independent transitions).*
3. *If $\mathcal{G}$ is a basic loop region, that is $\mathcal{G} = \mathcal{R}_1.\mathcal{R}_2.\ldots.\mathcal{R}_N$:*
   (a) *if $t_1, t_2 \in \mathcal{T}_{A_{\mathcal{R}i}}$ for some $i = 1, ..., N$, $t_1 \ll t_2$ if Definition 46 applies,*
   (b) *if $t_1 \in \mathcal{R}_0$ and $t_2 \in \mathcal{T}_{A_{\mathcal{R}1}}$, $t_1 \ll t_2$ if Definition 46 applies,*

(c) if $t_1 \in \mathcal{T}_{A_{\mathcal{R}_i}}$ and $t_2 \in \mathcal{T}_{A_{\mathcal{R}_{i+1}}}$ for some $i \geq 1$ then $t_1 \ll t_2$ if $t_1$ is the last transition in $\mathcal{R}_i$ and $t_2$ is the first transition in $\mathcal{R}_{i+1}$.

The above definition only considers sequential composition of regions where $\mathcal{R}_0$ is a basic region and $\mathcal{G}$ is a basic arbitrary region. For other possible cases ($\mathcal{R}_0$ and $\mathcal{G}$ are basic arbitrary regions, or $\mathcal{R}_0$ is a basic arbitrary region and $\mathcal{G}$ is a basic region) we can derive similar orderings. For a basic alt region, transitions from different operands must not be ordered in any way. For a basic par region, transitions from different operands are considered independent. For a basic loop region, the basic region of the loop is composed sequentially with $N$ copies of itself and transitions are ordered between them as expected. Transitions from $\mathcal{R}_0$ precede the transitions from each one of the operands of the interaction fragments. If we have a basic region with concurrency level greater than 1 (say $l$) we can see it as a basic par fragment with $l$ operands where each operand is a basic region of concurrency level 1. This is stated in the following lemma.

**Lemma 1.** *Let $\mathcal{R}$ be a basic region such that the associated set of local transitions $\mathcal{T}_{A_{\mathcal{R}}}$ has concurrency level $l$. Then there is an equivalent basic par region $\mathcal{G}$ with $l$ operands such that $\mathcal{G} = [\mathcal{R}_1, \mathcal{R}_2, \ldots, \mathcal{R}_l]$ and where each $\mathcal{R}_p$, for $p \in [1, \ldots, l]$, is a basic region with associated set of local transitions $\mathcal{T}_{A_{\mathcal{R}_p}}$ of concurrency level 1.*

*Proof.* According to Definition 47, if $\mathcal{T}_{A_{\mathcal{R}}}$ has concurrency level $l$ then $\mathcal{T}_{A_{\mathcal{R}}} = \mathcal{T}_{A_{\mathcal{R}_1}} \cup \ldots \cup \mathcal{T}_{A_{\mathcal{R}_l}}$ each totally ordered and such that two arbitrary transitions $t_1 \in \mathcal{T}_{A_{\mathcal{R}_i}}$ and $t_2 \in \mathcal{T}_{A_{\mathcal{R}_j}}$ with $i \neq j$ are independent. Since each $\mathcal{T}_{A_{\mathcal{R}_i}}$ is totally ordered it has concurrency level 1. Further, $\mathcal{G} = [\mathcal{R}_1, \mathcal{R}_2, \ldots, \mathcal{R}_l]$ satisfies Definition 410 2.c) and constitutes a basic alt region.                                                      □

The idea is that a basic region with concurrency level greater than one can always be understood as a basic par region where the level of concurrency gives us the number of operands of the par region. Given the lemma and without loss of generality, from now on we only consider basic regions with associated set of local transitions of concurrency level 1. The next definition deals with more general parallel fragments where nesting is allowed but restricted to a finite number of times given by $k$.

**Definition 411.** *$\mathcal{G}_k$ is a par region of level $k > 1$ with $N$ operands, $N \in \mathbb{N}^+$, if $\mathcal{G}_k = [\mathcal{R}_1, \mathcal{R}_2, \mathcal{R}_3, \ldots, \mathcal{R}_N]$ where $\mathcal{R}_n$, $1 \leq n \leq N$, is either a basic region or, for at least one value $n_1$, $\mathcal{R}_{n_1} = [\mathcal{P}re.\mathcal{G}_{k-1}.\mathcal{P}ost]$. Both $\mathcal{P}re$ and $\mathcal{P}ost$ are basic regions and $\mathcal{G}_{k-1}$ is an arbitrary region of level $k - 1$. If $k = 1$ then $\mathcal{G}_1$ is a basic arbitrary region.*

Consider the example of Fig. 5(a). It describes a par region of level $k = 2$ with two operands ($N = 2$). Thus we have $\mathcal{G}_2 = [\mathcal{R}_1, \mathcal{R}_2]$ where $\mathcal{R}_1 = [\mathcal{P}re.\mathcal{G}_1.\mathcal{P}ost]$ and $\mathcal{R}_2 = [\mathcal{P}re'.\mathcal{G}_1'.\mathcal{P}ost']$. $\mathcal{P}re$ is a basic region with one message $m_0$ whereas $\mathcal{P}ost$, $\mathcal{P}re'$ and $\mathcal{P}ost'$ are empty basic regions. Both $\mathcal{G}_1$ and $\mathcal{G}_1'$ are basic par regions.

**Definition 412.** *$\mathcal{G}_k$ is an alt region of level $k > 1$ with $N$ operands, $N \in \mathbb{N}^+$, if $\mathcal{G}_k = \bigcup_{n=1}^{N} \mathcal{R}_n$, where each $\mathcal{R}_n$ is either a basic region or, for at least one value $n_1$ of $n$, $\mathcal{R}_{n_1} = \mathcal{P}re.\mathcal{G}_{k-1}.\mathcal{P}ost$. Both $\mathcal{P}re$ and $\mathcal{P}ost$ are basic regions and $\mathcal{G}_{k-1}$ is an arbitrary region of level $k - 1$. If $k = 1$ then $\mathcal{G}_1$ is a basic arbitrary region.*

If there is nesting in one of the operands of an alternative fragment, the operand is not basic and can be seen as the sequential composition of three regions given by

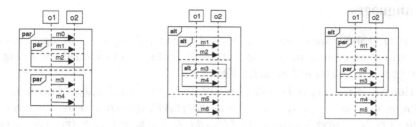

**Fig. 5.** (a) Par region ($k = 2$)      (b) Alt region ($k = 3$)      (c) Nested alt and par region ($k = 3$)

($Pre.P.Post$) where the first and the last are basic and $P$ is again an interaction fragment of some kind (alternative, parallel, or loop). The example of Fig. 5(b) describes an alt region of level $k = 3$ with two operands ($N = 2$). Thus we have $\mathcal{G}_3 = \mathcal{R}_1 \cup \mathcal{R}_2$ where $\mathcal{R}_1 = Pre.\mathcal{G}_2.Post$ and $\mathcal{R}_2$ is a basic region with associated set of local transitions $\mathcal{T}_{A_{\mathcal{R}_2}} = \{m_5, m_6\}$. In $\mathcal{R}_1$, $\mathcal{G}_2 = \mathcal{R}_1' \cup \mathcal{R}_2'$, and both $Pre$ and $Post$ are empty. At the second level of nesting $\mathcal{R}_1'$ is a basic region with associated set of local transitions $\mathcal{T}_{A_{\mathcal{R}_1'}} = \{m_1, m_2\}$ whereas $\mathcal{R}_2' = Pre'.\mathcal{G}_1.Post'$. Both $Pre'$ and $Post'$ are empty and $\mathcal{G}_1$ is a basic alt region. Finally, $\mathcal{G}_1 = \mathcal{R}_1'' \cup \mathcal{R}_2''$ where $\mathcal{R}_1''$ and $\mathcal{R}_2''$ are basic regions, and $\mathcal{T}_{A_{\mathcal{R}_1''}} = \{m_3\}$ and $\mathcal{T}_{A_{\mathcal{R}_2''}} = \{m_4\}$. Further, $\mathcal{G}_3 = \mathcal{R}_1 \cup \mathcal{R}_1'' \cup \mathcal{R}_2'' \cup \mathcal{R}_2$ and we can obtain an equivalent basic alt region of level 1 with these four operands. The only ordered local transitions we can have are those that satisfy Definition 410 1.a).

**Definition 413.** *$\mathcal{G}$ is a loop region if $\mathcal{G} = Pre.(\mathcal{R})^n.Post$ where $n, n \in \mathbb{N}$, is the loop index, $Pre$ and $Post$ are basic regions and $\mathcal{R}$ is an arbitrary region.*

As defined next, an arbitrary region can be any possible region (basic, alt, par). Fig. 5(c) describes an example of nested alt and par regions of level $k = 3$ with two operands ($N = 2$). As the region at the third level is an alt region we have $\mathcal{G}_3 = \mathcal{R}_1 \cup \mathcal{R}_2$ where $\mathcal{R}_1$ is a par region of level 2 and thus $\mathcal{R}_1 = [\mathcal{R}_1', \mathcal{R}_2']$ and $\mathcal{R}_2$ is a basic region with associated set of local transitions $\mathcal{T}_{A_{\mathcal{R}_2}} = \{m_4, m_5\}$. $\mathcal{R}_1'$ is a basic region with $\mathcal{T}_{A_{\mathcal{R}_1'}} = \{m_1\}$ and $\mathcal{R'}_2 = [Pre.\mathcal{G}_1.Post]$ where $Pre$ and $Post$ are empty and $\mathcal{G}_1$ is a basic par region such that $\mathcal{G}_1 = [\mathcal{R}_1'', \mathcal{R}_2'']$ with $\mathcal{T}_{A_{\mathcal{R}_1''}} = \{m_2\}$ and $\mathcal{T}_{A_{\mathcal{R}_2''}} = \{m_3\}$.

Given the framework described above to deal with interaction fragments, we define an IOD node fragment specification. Assume given $\Omega$ with $par, alt, loop \in \Omega$.

**Definition 414.** *An IOD node $\mathcal{A}$ fragment specification $Spec_{\mathcal{A}} = (Int_{\mathcal{A}}, f_{\mathcal{A}}, g_{\mathcal{A}})$ is such that $Int_{\mathcal{A}}$ is a set of interaction fragment identifiers in $\mathcal{A}$; $f_{\mathcal{A}} : Int_{\mathcal{A}} \to \Omega \times \mathbb{N} \times \mathbb{N}$ is a total function that assigns a triple $(d, l, N)$ to an interaction fragment identifier where $d$ is an operator, $l$ a natural number indicating the level of the fragment and $N$ the number of operands; and $g_{\mathcal{A}} : Int_{\mathcal{A}} \to 2^{\mathcal{E}}$ is an injective function that returns a set of events (region) for $i$, and such that if $f_{\mathcal{A}}(i) = (d, l, N)$ it returns a $d$ region of level $l$ with $N$ operands.*

From $Spec_{\mathcal{A}}$ we can infer the possible orderings of transitions in the node $\mathcal{A}$ and as such its associated possible traces.

# 5   Languages

For an IOD $\mathcal{D}$ above we define its associated language $\mathcal{L}(D)$ as set of legal traces of $\mathcal{D}$. The traces are defined by the order of events in the IOD nodes and respecting the ordering given by the transitions at the IOD level.

**Definition 51.** *A trace of IOD node* $\mathcal{A} = (\mathcal{O}, \mathcal{E}, <, \mathcal{M}_A, \mathcal{T}_A, \mathcal{P}_A, \mathcal{I}_A, \mathcal{U}_A)$ *is a (possibly infinite) word* $w = c_1.c_2 \ldots$ *over the alphabet* $\mathcal{M}_A$ *iff there is a sequence of local transitions* $t_1.t_2 \ldots$ *over* $\mathcal{T}_A$, *such that* $t_1 \ll t_2 \ll \ldots, t_i = (e_{si}, a_i/r_{i1}; r_{i2}, e_{ri})$ *and* $c_i = (a_i, min(r_{i1}, r_{i2}))$ *for* $0 < i \le |w|$, $e_{si} \in \mathcal{E}_S$ *and* $e_{ri} \in \mathcal{E}_R$.

We define $L_1$ as the IOD alphabet such that $L_1 = Act_p \cup Act_t$.

**Definition 52.** *A trace of IOD* $\mathcal{D} = (\mathcal{N}, \mathcal{S}, \mathcal{T}, \mathcal{P}, Act, \mathcal{L}_O, \mathcal{L}_I, \mathcal{K}, \mathcal{F}, \mathcal{C}, \mathcal{B})$ *is a (possibly infinite) word* $W = w_1.c_1.w_2.c_2 \ldots$ *over the alphabet* $L_1$ *iff there is a sequence of transitions* $t_1.t_2 \ldots$ *over* $\mathcal{T}$ *such that, for* $0 < i \le |W|$, $w_i$ *is a trace of IOD node* $\mathcal{A}_i$, $\mathcal{L}_O(t_i) = (p_i, c_i)$ *where* $p_i \in \mathcal{P}_O$ *and* $c_i = (a_i, r_i) \in Act_p$, $\mathcal{F}(t_i) = (\mathcal{A}_i, \mathcal{A}_{i+1})$ *where* $\mathcal{A}_i, \mathcal{A}_{i+1} \in \mathcal{N}$, *and* $t_1 \in \mathcal{T}_\mathcal{B}$, *the set of possible initial transitions obtained from the initial marking* $\mathcal{B}$.

**Definition 53.** *Let a maximal trace be a trace which is not a proper prefix of any other trace. The language of IOD* $\mathcal{D}$ *is the set* $L_1(\mathcal{D})$ *of words over the alphabet* $L_1$ *where* $L_1(\mathcal{D}) = \{W \mid W \text{ is a maximal trace of } \mathcal{D}\}$.

We define the set of legal traces of a PEPA net model next. Let $V$ be the labelled transition system of a place $P \in \mathcal{P}$ and let $\mathcal{T}_V$ be the set of all transitions in $V$. Let $h$ be the labelling function which assigns a PEPA activity to each transition in $\mathcal{T}_V$.

**Definition 54.** *Let* $t_1, t_2 \in \mathcal{T}_V$. $t_1$ *precedes* $t_2$ *in the set of transitions (written* $t_1 \ll t_2$) *iff there is a sequence of activities* $h(t_1).h(t_2)$ *where* $h(t_1) = (a_1, r_1)$ *and* $h(t_2) = (a_2, r_2)$, $r_1, r_2 \in \mathbb{R}^* \cup \{\top\}$.

To define the PEPA net language $\mathcal{V}$, we first define the trace of the net place $P \in \mathcal{P}$.

**Definition 55.** *A trace of a PEPA net place* $P$ *is a (possibly infinite) word* $w = c_1.c_2.\ldots$ *over the alphabet* $Act_t$ *iff there is a sequence of transitions* $t_1, t_2, \ldots$ *over* $\mathcal{T}_V$ *such that, for* $0 < i \le |w|$, $t_1 \ll t_2 \ll \ldots$ *and* $c_i = h(t_i) = (a_i, r_i)$ *where* $c_i$ *is an individual or shared activity between two components* $C_1$ *and* $C_2$ *with rate* $r_i = min(r_{i1}, r_{i2})$ *where* $r_{i1}$ *and* $r_{i2}$ *are the rates of the activity in components* $C_1$ *and* $C_2$, *respectively.*

We define $L_2 = Act_t \cup Act_f$ as the PEPA net alphabet. Using the definition of the trace $w_i$ of each place $P_i \in \mathcal{P}$, the trace of a PEPA net $\mathcal{V}$ is defined as follows.

**Definition 56.** *A trace of a PEPA net* $\mathcal{V} = (\mathcal{P}, \mathcal{T}, I, O, \ell, \pi, \mathcal{C}, K, M_0)$ *is a (possibly infinite) word* $W = w_1.c_1.w_2.c_2 \ldots$ *over the alphabet* $L_2$ *iff there is a sequence of transitions* $t_1.t_2 \ldots$ *over* $\mathcal{T}_f$ *such that, for* $0 < i \le |W|$, $w_i$ *is a trace of the PEPA net place* $P_i \in \mathcal{P}$, $\mathcal{O}(t_i) = P_i$, $\mathcal{I}(t_i) = P_i'$ *where* $P_i' \in \mathcal{P}$, $c_i = l(t_i) = (a_i, r_i)$ *where* $c_i \in Act_f$, *and* $t_1 \in \mathcal{T}_{M_0}$ *where* $\mathcal{T}_{M_0}$ *is the set of possible initial transitions obtained from the initial marking* $M_0$.

Finally, we define the language of a PEPA net $\mathcal{V}$, noted $L_2(\mathcal{V})$, as follows:

**Definition 57.** *Let a maximal trace be a trace which is not a proper prefix of any other trace. The language of the PEPA net* $\mathcal{V}$ *is the set* $L_2(\mathcal{V})$ *of words over the alphabet* $L_2$ *such that* $L_2(\mathcal{V}) = \{W \mid W \text{ is a maximal trace of } \mathcal{V}\}$.

Note that the set of legal traces of a PEPA net model that can be derived from the associated CTMC is different from the one defined above. The difference is due to model semantics that is taken into account in the former to build the CTMC. Indeed, in our definition of a word $W$, the order of two parallel activities occurring in two different places is not taken into account. However, from the legal traces we have defined, we can derive the legal set of traces from the underlying CTMC.

## 6  The Strongly Consistent Transformation

In this section, we describe the IOD-to-PEPA net model transformation and its correctness by sketching the proof of the equivalence of the underlying languages.

Each IOD node and object in the UML model is associated with a place and a component in the PEPA net model. Both models use activities and there is a one-to-one correspondence between activities in IOD edges or IOD node messages, and PEPA net firing transitions or PEPA transitions respectively. In other words, an IOD can be viewed as a PEPA net model where each IOD node corresponds to a place in the PEPA net. An edge or transition between two IOD nodes is transformed into a firing transition between two places in the net with the same label. The table below describes the correspondence between the elements of an IOD and those of a PEPA net, in accordance with our definitions. For space reasons, the complete details of the algorithm are omitted. A definition of a formal semantics for UML sequence diagrams and operators is given in [12]. This semantics is based on PEPA's structural operational semantics.

| IODs | PEPA nets |
| --- | --- |
| IOD $\mathcal{D}$ (Definition 41) | PEPA net $\mathcal{V}$ (Definition 31) |
| IOD node $\mathcal{A} \in \mathcal{N}$ | Place $P \in \mathcal{P}$ |
| IOD transition $t \in \mathcal{T}$ | Firing transition $t \in \mathcal{T}_f$ |
| IOD activity $c \in Act_p$ | Firing activity $c \in Act_f$ |
| IOD node local transition $t \in \mathcal{T}_A$ | Transition $t \in \mathcal{T}_t$ |
| Static object $O \in \mathcal{O}_S$ | Static component $C \in \mathcal{C}_S$ |
| Mobile object, token $O' \in \mathcal{O}_M$ | PEPA net token $C' \in \mathcal{C}_M$ |
| IOD node activity $c \in Act_n$ | PEPA activity $c \in Act_t$ |
| Set of inputs to IOD node $\mathcal{A}$ $(p, n) \in \mathcal{I}_A$ | Number of cells $n$ in place $P$ for corresponding token |
| IOD fork node $s \in \mathcal{S}$ | PEPA component synchronisation in the source place |

The set of legal traces (words) determines the language of an IOD $\mathcal{D}$ given by $L_1(\mathcal{D})$ or a PEPA net $\mathcal{V}$ given by $L_2(\mathcal{V})$. Given the translation algorithm, the languages are equivalent, also known as *strongly consistent*.

**Theorem 61.** *Let $\mathcal{D}$ be an IOD and $\mathcal{V}$ the PEPA net derived from $\mathcal{D}$. If $L_1(\mathcal{D})$ is the set of words over the alphabet $L_1$ of $\mathcal{D}$ and $L_2(\mathcal{V})$ is the set of words over the alphabet $L_2$ of $\mathcal{V}$ then 1) $L_1 = L_2$ and 2) $L_1(\mathcal{D}) = L_2(\mathcal{V})$.*

*Proof.* 1) is true by definition as $L_1 = Act_n \cup Act_p = Act_t \cup Act_f = L_2$. The language equality can be proven in two steps: (1) $L_1(\mathcal{D}) \subseteq L_2(\mathcal{V})$ and (2) $L_2(\mathcal{V}) \subseteq L_1(\mathcal{D})$. For space reasons we only give an idea of the proof. We prove (1) by contradiction and assume there is a word $W$ such that $W \in L_1(\mathcal{D})$ and $W \notin L_2(\mathcal{V})$. Since strong consistency is assumed by hypothesis, the trace violation occurs at length $i + 1$, i.e., there is a trace $W = w_1.c_1.w_2.c_2 \ldots w_i.c_i.w_{i+1}.c_{i+1}$ such that $w_1.c_1.w_2.c_2 \ldots w_i.c_i \in L_2(\mathcal{V})$ but there is no trace in $L_2(\mathcal{V})$ which would contain the continuation $w_{i+1}.c_{i+1}$ and thus there is either no word $w_{i+1}$ in the PEPA place $P_{i+1}$ associated with node $\mathcal{A}_{i+1}$ or there is no net transition $c_{i+1} \in \mathcal{A}_f$. The first assumption contradicts the one-to-one correspondence between the event ordering in an IOD node (and thus the local transition ordering) and the sequences of activities possible for the components in place $P_{i+1}$. The second assumption contradicts the one-to-one correspondence between the IOD transitions and the PEPA net transitions (i.e., $L_1 = L_2$). Similarly for (2).    □

# 7    Related Work

An increased number of approaches have emerged for the design and/or formal verification of mobile systems. At the software design level this includes extensions of UML for mobility (e.g. [1, 9]). In [1], the authors extend UML activity diagrams to capture mobile systems. However, activity diagrams are not adequate to capture at the same time the structure of the system (locations), how objects move between locations, *and* how objects behave/interact within locations. Several performance modelling approaches using UML and an underlying formal model for performance analysis have been developed including [4, 5, 8, 14]. Some of the work using UML for performance analysis has different motivation than ours. In this context [14] uses activity diagrams to refine do activities in state machines and then obtain predictive performance measures from the performance model obtained from these diagrams. In [8] the authors introduce a mobility profile for the performance analysis domain, but do not focus on new notations available in UML2.3. In [5] the authors report on a toolset for modelling systems with performance information using UML. This approach does not consider mobility, and assumes an underlying translation of mainly UML1.x notation into the process algebra PEPA. In [17] the authors are also concerned with mobility. However they propose a translation of UML1.x specifications made up of sequence and state diagrams into $\pi$-calculus processes. Other synthesis approaches, e.g. [19], often have the problem of implied (additional) scenarios as the models used are very different in nature and essentially capture different views of the system. Such approaches have to focus on mechanisms to detect such unwanted and unacceptable additional behaviours.

# 8    Conclusion

We have shown how to formalise performance annotated IODs taking into account complex behaviour. The legal traces of an IOD have a one-to-one correspondence to the legal traces of the underlying PEPA net model. An advantage of strongly consistent models is the absence of implied scenarios in the synthesised model, which ensures an accurate performance analysis on the given UML design models. A challenging aspect of our future work concerns the performance analysis itself. PEPA nets mainly rely on the performance techniques available for PEPA and these ignore the location or mobility information of the PEPA net. By contrast we want to exploit the design structure of our IOD and PEPA nets to enhance verification and scalability and thus have a more suitable approach for performance evaluation of complex mobile distributed applications.

# References

1. Baumeister, H., Koch, N., Kosiuczenko, P., Wirsing, M.: Extending activity diagrams to model mobile systems. In: Akşit, M., Mezini, M., Unland, R. (eds.) NODe 2002. LNCS, vol. 2591, pp. 278–293. Springer, Heidelberg (2003)
2. Bernardo, M., Gorrieri, R.: A tutorial on EMPA: a theory of concurrent processes with non-determinism, priorities, probabilities and time. TCS 202, 1–54 (1998)
3. Bowles, J., Kloul, L.: Synthesising PEPA nets from IODs for performance analysis. In: WOSP/SIPEW 2010, pp. 195–200. ACM (2010)
4. Canevet, C., Gilmore, S., Hillston, J., Kloul, L., Stevens, P.: Analysing UML 2.0 activity diagrams in the software performance engineering process. In: WOSP 2004, Short Papers, pp. 74–78. ACM (2004)
5. Canevet, C., Gilmore, S., Hillston, J., Prowse, M., Stevens, P.: Performance modelling with UML and stochastic process algebras. IEE Proceedings: Computers and Digital Techniques 150(2), 107–120 (2003)
6. Donatelli, S.: Superposed Generalised Stochastic Petri Nets: Definition and Efficient Solution. In: Valette, R. (ed.) ICATPN 1994. LNCS, vol. 815, pp. 258–277. Springer, Heidelberg (1994)
7. Gilmore, S., Hillston, J., Kloul, L., Ribaudo, M.: PEPA nets: A structured performance modelling formalism. Performance Evaluation 54, 79–104 (2003)
8. Grassi, V., Mirandola, R., Sabetta, A.: UML based modeling and performance analysis of mobile systems. In: MSWIM 2004, pp. 95–104. ACM (2004)
9. Grassi, V., Mirandola, R., Sabetta, A.: A UML profile to model mobile systems. In: Baar, T., Strohmeier, A., Moreira, A., Mellor, S.J. (eds.) UML 2004. LNCS, vol. 3273, pp. 128–142. Springer, Heidelberg (2004)
10. Hermanns, H.: Chapter 4: Interactive Markov Chains. In: Hermanns, H. (ed.) Interactive Markov Chains. LNCS, vol. 2428, pp. 57–88. Springer, Heidelberg (2002)
11. Hillston, J.: A compositional approach to performance modelling. Cambridge University Press (1996)
12. Kloul, L.: Blending UML2.0 and PEPA nets. Technical Report n.2006/102, PRiSM, Université de Versailles (2006),
http://wwwex.prism.uvsq.fr/recherche/rapports
13. Kloul, L., Küster-Filipe, J.: Modelling Mobility with UML 2.0 and PEPA Nets. In: ACSD 2006, pp. 153–162. IEEE Computer Society (2006)
14. López-Grao, J., Merseguer, J., Campos, J.: From UML Activity Diagrams to Stochastic Petri Nets: Application to Software Performance Engineering. In: WOSP 2004, pp. 25–36. ACM (2004)
15. OMG. UML Superstructure Version 2.4.1. Document id:formal/2011-08-06 (2011)
16. Plateau, B.: De l'Evolution du Parallélisme et de la Synchronisation. PhD Thesis, Université de Paris-Sud, Orsay (1984)
17. Pokozy-Korenblat, K., Priami, C.: Towards extracting $\pi$-calculus from UML sequence and state diagrams. Electronical Notes in Theoretical Computer Science 101, 51–72 (2004)
18. Rumbaugh, J., Jacobson, I., Booch, G.: The Unified Modelling Language Reference Manual, 2nd edn. Addison-Wesley (2005)
19. Uchitel, S., Kramer, J., Magee, J.: Detecting implied scenarios in message sequence chart specifications. In: ESEC/FSE 2001, pp. 74–82. ACM (2001)

# Early Effort Estimation for Quality Requirements by AHP

Mohamad Kassab

The Pennsylvania State University, Engineering Division
Malvern, PA, U.S.A
muk36@psu.edu

**Abstract.** The increased awareness of the quality requirements as a key to software project and product success makes explicit the need to include them in any software project effort estimation activity. However, the existing approaches to defining size-based effort relationships still pay insufficient attention to this need. Furthermore, existing functional size measurement methods still remain unpopular in industry. In this paper, we propose the usage of the Analytic Hierarchy Process (AHP) technique in the effort estimation for quality requirements. The paper demonstrates the applicability of the approach through a case study.

**Keywords:** Quality Requirements, AHP, Requirements Engineering, Effort Estimation, Functional Size.

## 1    Introduction

Early in a project, specific details of the nature of the software to be built, details of specific requirements, of the solution, of the staffing needs, and other project variables, are unclear. The variability in these factors contributes to the uncertainty of project effort estimates. As the sources of variability are further investigated and pinned down, the variability in the project diminishes, and so the variability in the project effort estimates can also diminish. This phenomenon is known as the Cone of Uncertainty [1].

In practice, the software development industry, as a whole, has a disappointing track record when it comes to completing a project on time and within budget. From April 2013 through July 2013, we conducted a survey study on the requirements engineering (RE) current state of practice [2]. The survey drew 247 professional participants from 23 countries and from wide range of industries. Respondents were asked to base their responses on one project that they were either currently involved with or had taken part in during the past five years. Questions relating to delivery timeline, schedule and costs indicated that the majority of projects represented in this study took longer than the respondents had expected to deliver. Only 48% of the

B. Murgante et al. (Eds.): ICCSA 2014, Part V, LNCS 8583, pp. 106–118, 2014.
© Springer International Publishing Switzerland 2014

respondents agreed that the duration of the project was within schedule; and only 21% agreed that the project goals were achieved earlier than predicted. Also only 45% agreed that project costs were within budget estimates.

While experiences show that quality requirements may represent more than 50% of the total effort to produce a software product [3]; software developers are constantly under pressure to deliver on time and on budget. As a result, many projects focus on delivering functionalities at the expense of meeting quality requirements such as reliability, security, maintainability, portability, accuracy, among others. As software complexity grows and clients' demands on software quality requirements increase, these qualities can no longer be considered of secondary importance. Many systems fail or fall into disuse precisely because of inadequacies in quality requirements [4].

While effort is a function of size [5]; one way to respond to the need to deal comprehensively and objectively with the effect of quality requirements on the scope of a software project is in terms of their corresponding functional size when applicable. Nevertheless, using the functional size measurement (FSMs) methods still remain unpopular in industry. In the RE state of practice survey that we conducted, Out of the 60% of those who reported on performing estimation for the size of requirements or the effort of building them, less than 7% reported on the usage of any FSM method [2].

In addition, existing FSM methods have been primarily focused on sizing the functionality of a software system. Size measures are expressed as single numbers (function points (FP) [6, 7, 8, 9]), or multidimensional 'arrays' designed to reflect how many of certain types of items there are in a system [10]. The existing function-point-based FSM techniques have so far addressed the topic of quality requirements only with respect to the task of adjusting the (unadjusted) function point counts to the project context or the environment in which the system is supposed to work.

The lack of effort estimation approaches which take into account the effect of the quality requirements on early effort estimation contributes to the Cone of Uncertainty phenomenon.

The goal of this research is to investigate requirements-based tuned early estimation of the software effort. In particular, we propose the usage of the Analytic Hierarchy Process (AHP) technique in the effort estimation for quality requirements. These requirements are subjective and usually captured in qualitative format at the early stages of RE. Since AHP integrates qualitative approach with quantitative one, and subjective approach with objective one, it is appropriate for estimating the quality requirements at the beginning of development.

In the rest of this paper, Section 2 introduces the background of quality-tactics relation and the AHP technique. In Section 3, we present our approach of incorporating the AHP for the quality effort estimation; and we demonstrate it through a case study. Section 4 reviews related work and finally we summarize and conclude the paper in the Section 5.

# 2    Background

## 2.1    Quality Requirements and Tactics

Quality is "the totality of characteristics of an entity that bear on its ability to satisfy stated and implied needs" [11]. Software Quality is an essential and distinguishing attribute of the final product. Tactics on the other hand are measures taken to implement the quality attributes [12]. For example, introducing concurrency for a better resource management is a tactic to improve system's Performance. Similarly, Authentication and Authorization are popular tactics to resist unwanted attacks on the system and improve the overall Security. In [12], the authors list the common tactics for the qualities: Availability, Modifiability, Performance, Security, Testability and Usability.

Tactics are considered as the building blocks from which software architectures are composed [12]; and the meeting point between requirements and architecture. Because qualities are being satisfied by implementing their corresponding set of tactics; the effort of building the quality requirements is in fact the effort of implementing their derived tactics. In this paper, our aim is to estimate the effort of building tactics that aim at satisfying the qualities.

## 2.2    The AHP Technique

The AHP [13, 14] is a technique for modeling complex and multi-criteria problems and solving them using a pairwise comparison process. Based on mathematics and psychology, it was developed by Thomas L. Saaty in the 1970s and has been extensively studied and refined since then. AHP was refined through its application to a wide variety of decision areas, including transport planning, product portfolio selection, benchmarking and resource allocation and energy rationing.

Simply described, AHP breaks down a complex and unstructured problems into a hierarchy of factors. A super-factor may include sub-factors. By pairwise comparison of the factors in the lowest level, we can obtain a prior order of factors under a certain decision criterion. The prior order of super-factors can be deduced from the prior order of sub-factors according to the hierarchy relations.

The AHP process starts by a detailed definition of the problem; goals, all relevant factors and alternative actions are identified. The identified elements are then structured into a hierarchy of levels where goals are put at the highest level and alternative actions are put at the lowest level. Usually, an AHP hierarchy has at least three levels: the goal level, the criteria level, and the alternatives level. This hierarchy highlights relevant factors of the problem and their relationships to each other and to the system as a whole.

Once the hierarchy is built, involved stakeholders (i.e., decision makers) judge and specify importance of the elements of the hierarchy. To establish the importance of elements of the problem, a pairwise comparison process is used. This process starts at the top of the hierarchy by selecting an element (e.g., a goal) and then the elements of the level immediately below are compared in pairs against the selected element. A pairwise matrix is built for each element of the problem; this matrix reflects the relative importance of elements of a given level with respect to a property of the next higher level. Saaty proposed the scale [1...9] to rate the relative importance of one criterion over another (See Table 1). Based on experience, a scale of 9 units is reasonable for humans to discriminate between preferences for two items [13, 14].

One important advantage of using AHP technique is that it can measure the degree to which manager's judgments are consistent. In the real world, some inconsistency is acceptable, and even natural. For example, in a sporting contest, if team A usually beats team B, and if team B usually beats team C, this does not imply that team A usually beats team C. The slight inconsistency may result because of the way the teams match up overall. The point is to make sure that inconsistency remains within some reasonable limits. If it exceeds a specific limit, some revision of judgments may be required. AHP technique provides a method to compute the consistency of the pairwise comparisons [13, 14].

**Table 1.** Pairwise comparison scale for AHP [13]

| Intensity of judgment | Numerical Rating |
| --- | --- |
| Extreme Importance | 9 |
| Very Strong Importance | 7 |
| Strong Importance | 5 |
| Moderate Importance | 3 |
| Equal Importance | 1 |
| For compromise between the above values | 2, 4, 6, and 8 |

# 3    Incorporating the AHP into Tactics Effort Estimation

## 3.1    AHP Hierarchy for Effort Estimation

The first step in the AHP process is to construct the hierarchy model. One challenge was to identify the elements of the criteria levels of the effort estimation AHP hierarchy.

We conducted a workshop that drew 24 professionals. During the workshop, professionals participated in a questionnaire and brainstorming session aiming at identifying the set of criteria that contributes to the effort of implementing the tactics. The participants reflected a diverse range of positions; describing themselves as programmers / developers, software / system engineers, or testers 46% of the time. Architects, project / product managers, analysts, and consultants comprised the remaining 54% of respondents; positions typically involved in the higher-level aspects of computerized system's technical design. Given this population, responses to the questionnaire are more likely to reflect the opinions and biases of any given project's development team rather than those of other groups represented in a software development effort.

The outcome of the workshop was the generation of Figure 1 which presents the proposed AHP hierarchy model for the tactics effort estimation.

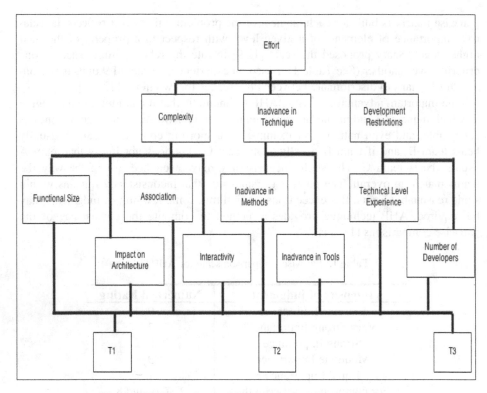

**Fig. 1.** AHP hierarchy model for the tactics effort estimation

In the hierarchy, the effort as the object of decision-making appeared at the top level. The alternatives level, namely the bottom level, was composed of the tactics. In the middle, there were two criteria levels. One level had three criteria (level 2), i.e., Complexity, Inadvance in Technique, and Restriction from Developers. There was a convention that the weight of a super-criterion would increase with the increase of the sub-criterion's weights. Thus, we used "inadvance" as a criterion, but not "advance".

In the second criteria level (level 3), there were eight criteria, where functional size, Impact on architecture, association and interactivity are the sub-criteria of complexity, and so on. The contribution of sub-criteria of "Inadvance in Technique" and "Development Restrictions" towards the effort is trivial. The sub-criteria of "Complexity" are briefly explained below:

— **Functional size:** Functional Size Methods have shifted the focus from measuring the technical characteristics of the software towards measuring the functionality of the software that is required by the intended users of the software. If the tactic corresponds to a functionality; then it will have a functional size that can be estimated and compared; otherwise if the tactic corresponds to other types of architectural decisions; then it will have no functional size. The effort is a function of size [5] where the increase in functional size increases the effort.

— **Impact on Architecture:** In [15], the authors identified six types of changes that an architectural structure or behavior might undergo when a tactic is implemented within the structure and they define a scale to rate these changes. For example, a tactic can have a minimum impact on the architecture when there are no major changes to be implemented to the current structure (e.g. minimal changes to be implemented within existing components); while it may have high impact on the architecture if the current structure require major changes (e.g. requiring the addition of 3 or more new components that will break the current structure). The higher the impact on the architecture is; the higher contribution to the effort. This criterion suggests also to take into account the order in which the tactics to implemented. That is, if tactic B implemented after tactic A, then its implementation may require modification to the existing structure of A - a different effort from incorporating tactic B from scratch.

— **Association:** This criterion suggests considering the range of items a tactic is associated to (e.g. functions, resources, processes, or the whole product). The wider this range is, the higher the effort will be.

— **Interactivity:** Typically, systems have multiple important quality attributes, and decisions made to satisfy a particular quality may help or hinder the achievement of another quality attribute. The best-known cases of conflicts occur when the choice of a tactic to implement certain quality attribute contributes negatively towards the achievement of another quality. For example, decisions to maximize the system reusability and maintainability through the usage of "abstracting common services" tactic may come at the cost of the "response time". If a tactic contributes negatively to satisfy other qualities; then it contributes towards the increase in the effort and if it contributes positively towards the satisfaction of other qualities; then it contributes towards the decrease in the total effort.

## 3.2 Case Study Description

We demonstrate the incorporation of the AHP into quality/tactics effort estimation through an automated building system case study.

A company manufactures devices for the building automation domain and software applications that manage a network of these devices. With the hardware being commoditized, its profit margins have been shrinking. The internal development costs for the software applications that manage different devices have also been rising. To sustain their business long term, the company decides to create a new integrated building automation system. The intended system would broadly perform the following functions: Manage field devices currently used for controlling building functions; Define rules based on values of field device properties that trigger reactions; Issue commands to set values of field device properties; and for life critical situations, trigger alarms notifying appropriate users.

Taking this approach would allow the company to reduce internal development costs – several existing applications will be replaced with the new system. The company could also achieve market expansion by entering new and emerging geographic markets and opening new sales channel in the form of Value Added Resellers (VARs).

In order to support a multitude of hardware devices and consider different languages and cultures, the system must be modifiable (a modifiability requirement). In order to support different regulations in different geographic markets, the system must respond to life threatening events in a timely manner (a performance requirement).

To apply the modifiability tactics, we aim to limit the impact of change and minimize the number of dependencies on the part of the system responsible for integrating new hardware devices. There are three design concerns related with modifiability: 1) Localize changes: this relates to adding a new field device; 2) Prevention of ripple effects: this relates to minimizing the number of modules affected as a result of adding a new field device; and 3) Defer binding time: this relates to the time when a new field device is deployed and the ability of non-programmers to manage such deployment.

We address these concerns by creating adaptors for field devices, an "anticipation of expected changes" tactic. We use two additional architectural tactics to minimize propagation of change. First we specify a standard interface to be exposed by all adaptors ("maintain existing interfaces"). Second, we use the adaptor as an "intermediary" responsible for semantic translation into a standard format, of all the data received from different field devices.

As of the performance quality attribute of the building automation system, there are two design concerns: 1) Resource Demand: the arrival of change of property value events from the various field devices and the evaluation of automation rules in response to these events are source of resource demand; and 2) Resource Management: the demand on resources may have to managed in order to reduce the latency of event and alarm propagation.

To address these concerns, we move the responsibility of rule evaluation and execution, and alarm generation, respectively to a newly added separate Logic & Reaction (L&R) component and an Alarm component. These components running outside the automation server can now be easily moved to dedicated execution nodes if necessary. In doing so, we are making use of the "increase available resources" tactic to address the resource management concern and the "reduce computational overhead" tactic to address the resource demand concern. We use an additional tactic to address the resource management concern. This tactic relies on introducing "concurrency" to reduce delays in processing time. Concurrency is used inside the L&R and Alarm components to perform simultaneous rule evaluations.

So to satisfy modifiability and performance qualities in the building automated systems; we introduced the five tactics: 1) an anticipation of expected changes; 2) maintain existing interfaces; 3) usage of an intermediary; 4) increase available resources; and 5) introducing concurrency.

## 3.3    AHP in Action

Construction of the hierarchy model (as shown in section 3.1) is the first step in the problem solving process of the AHP technique. In the hierarchy, each of the eight criteria from level 3 is related to all tactics. In the automated building system case study, this means that each of the criteria: functional size, impact on architecture, association, interactivity, inadvance in methods, inadvance in tools, technical level experience, and number of developers is related to all tactics: an anticipation of expected changes; maintain existing interfaces; usage of an intermediary; increase available resources; and introducing concurrency.

Each of these relations between a level 3 criteria and tactics will be assessed via the pairwise comparison. The comparisons for this effort estimation problem applied on the building automation system case study are shown below (this is an actual execution of our approach by one of the architects participated in the workshop described in 3.1):

1. We start by the pairwise comparisons of evaluation criteria (level-2 elements in the decision hierarchy) – Table 2. This comparison represents the prioritization of the criteria in that level in respect to their impact on the effort. The weights values are calculated by; first calculating the geometric mean for each row; then dividing the geometric mean of each row by the total summation of geometric mean values from all rows. The geometric mean of n numbers, say, $X_1$, $X_2$, ... $X_n$ is given by: $(X_1 * X_2 * ... * X_n)^{1/n}$

**Table 2.** Pairwise comparisons matric for level-2 criteria in the automated home system

|  | Complexity | Inadvace in Technique | Development Restrictions | Geometric Mean | Weight |
|---|---|---|---|---|---|
| Complexity | 1 | 3 | 5 | 2.47 | 0.6 |
| Inadvace in Technique | 0.33 | 1 | 0.2 | 0.4 | 0.1 |
| Development Restrictions | 0.2 | 5 | 1 | 1 | 0.3 |

2. Similarly, we complete the pairwise comparisons of sub-criteria from the third levels with respect to level-2 criteria of the decision hierarchy (See tables 3, 4 and 5).

**Table 3.** Pairwise comparisons matric for "Complexity" criterion

|  | Functional Size | Impact on Architecture | Association | Interactivity | Weight |
|---|---|---|---|---|---|
| Functional Size | 1 | 5 | 5 | 5 | 0.6 |
| Impact on Architecture | 0.2 | 1 | 3 | 3 | 0.2 |
| Association | 0.2 | 0.33 | 1 | 1 | 0.1 |
| Interactivity | 0.2 | 0.33 | 1 | 1 | 0.1 |

**Table 4.** Pairwise comparisons matric for "Inadvance in Technique" criterion

|  | Inadvance in Methods | Inadvance in Tools | Weight |
|---|---|---|---|
| Inadvance in Methods | 1 | 1 | 0.5 |
| Inadvance in Tools | 1 | 1 | 0.5 |

**Table 5.** Pairwise comparisons matric for "Development Restrictions" criterion

|  | Technical Experience Level | Number of Developers | Weight |
|---|---|---|---|
| Technical Experience Level | 1 | 3 | 0.75 |
| Number of Developers | 0.33 | 1 | 0.25 |

3. We then complete the pairwise comparisons of the tactics (elements of the lowest level in the hierarchy with respect to every crtiterion from level 3). Table 6 shows the pairwise comparison of the tactics with respect to "functional size" criterion. We will not show the pairwise computations with respect to other criteria in this paper due to the space constraints.

**Table 6.** Pairwise comparisons of tactics with respect to "functional size" criterion: (**T1**: Anticipation Expected Changes; **T2**: Maintain Existing Interfaces; **T3**: Usage of an Intermediary; **T4**: Increase Available Resources; **T5**: Introducing Concurrency)

|  | T1 | T2 | T3 | T4 | T5 | Weight |
|---|---|---|---|---|---|---|
| T1 | 1 | 5 | 3 | 1 | 3 | 0.34 |
| T2 | 0.2 | 1 | 0.33 | 0.2 | 0.33 | 0.05 |
| T3 | 0.33 | 3 | 1 | 0.33 | 1 | 0.13 |
| T4 | 1 | 5 | 3 | 1 | 3 | 0.34 |
| T5 | 0.33 | 3 | 1 | 0.33 | 1 | 0.13 |

4. Once the normalized are computed for all levels of the hierarchy, they are combined by moving through the hierarchy starting at the lowest level. Figure 2 illustrates this procedure. For example, after one level of composition the average weights of the tactics (anticipation of expected changes, maintain existing interfaces, usage of an intermediary, increase available resources, introducing concurrency) with respect to "Development restrictions" are:

(0.323, 0.165, 0.12, 0.325, 0.068)  =  0.75 * (0.34, 0.13, 0.13, 0.34, 0.06)  + 0.25 * (0.27, 0.27, 0.09, 0.28, 0.09).

Following this procedure, the overall weights for the tactics: (anticipation of expected changes, maintain existing interfaces, usage of an intermediary, increase available resources, introducing concurrency) are calculated to be (0.33, 0.11, 0.12, 0.34, 0.1).

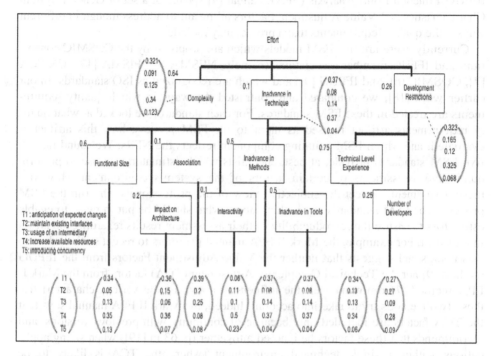

**Fig. 2.** Hierarchy composition of weights for the Home Automated system case study

5. Finally, if we know the effort of anyone among the five tactics from a historical project, then the effort of the others can be calculated. For example, if had known the effort of "introducing concurrency" from a previous project to be 3 person days; then the effort of "anticipation of expected changes" would be: (0.33/0.1) * 3 = 11 person days.

# 4    Related Work

Over the years, different estimation techniques have been developed in industry and academia, primarily with the objective of improving the accuracy of schedule, effort and cost estimation. These estimation techniques can primarily be subdivided into two major categories: formal methods and expert-judgment based methods. Overall, quality requirements received little attention compared to functionalities from these effort estimation techniques.

The existing function-point-based FSM techniques have so far addressed the topic of quality requirements only with respect to the task of adjusting the (unadjusted) FP counts to the project context or the environment in which the system is supposed to work. For example, the International Function Point Users Group (IFPUG) [16] has been approaching the inclusion of quality requirements in the final FP count by using qualitative judgments about the system's environment. The current version of the IFPUG Function Point Analysis (FPA) manual [9] speaks of a set of General System Characteristics and Value Adjustment Factors all meant to address though in different ways – the quality requirements that a project may include.

Currently, there are five FSM models which are proposed by the COSMIC consortium and IFPUG member associations (namely, NESMA [8], FISMA [17], UKSMA [7], COSMIC [6], and IFPUG [9]) and which are recognized as ISO standards. In our earlier work [18], we compared and contrasted the ways in which quality requirements are treated in these FSM standards. For each standard, we looked at what quality requirements artifact is used as input to the FSM process, how this artifact is evaluated, and which FSM counting component reflects the NFRs. We found that all five FSM standards provide, at best, checklists which estimators can use to perform qualitative assessments of certain factors of the system's environment. However, these assessments reflect the subjective view of the professionals who run the FSM process. The FSM standards say nothing about what should be put in place to enable estimators to ensure the reproducibility of their assessment results regarding the NFRs in a project. For example, the Mark II FPA manual [7] refers to recent statistical analysis results and suggests that neither the Value Adjustment Factors from the IFPUG method [9] nor the Technical Complexity Adjustment (TCA) factors from the Mark II FPA method [7] represent well the influence on size of the various characteristics these two methods try to take into account. Indeed, the Mark II FPA manual says that the TCA factors are included only because of continuity with previous versions, and recommends that these factors be ignored altogether (p. 63 in [19]) when sizing applications within a single technical environment (where the TCA is likely to be constant).

Recently, "Software Non-functional Assessment Process" (SNAP) [20] was introduced as a measurement of non-functional software size. SNAP point sizing is a complement to a function point sizing, which measures functional software size. Nevertheless, as we pointed earlier, the usage of any FSM for estimation remains unpopular in the industry. In the RE state of practice survey we conducted; Out of the 60% of those who reported on performing any estimation for the size / effort of requirements or the effort of building them, 62% reported on taking into account the Non-Functional Requirements (NFRs). That group of respondents reported that "Expert Judgments" was the most popular technique (24%) for estimating NFRs; followed by "Group estimation" techniques (e.g. planning poker, wideband Delphi) and "story points". The usage of FSM was reported to be less than 7% for estimating NFRs.

## 5    Conclusion

The quality effort estimation methodology presented in this paper aims at improving the predictive quality of the software industry's effort estimation models. The paper demonstrates the feasibility of the proposed approach on a case study.

The research we presented is multidisciplinary in nature, which opened multiple avenues of future work that we could effectively pursue. In our immediate next steps; we plan to validate our work further in real industrial settings. In addition, we plan on extending the effort estimation model to consider the cost estimation for the project.

We acknowledge that our solution proposal may sound complex for implementation by practitioners when large number of tactics are selected. Thus, we are also looking to automate the process of collecting the pairwise judgments and the final weights' calculations through implementing an automating tool for the process.

# References

1. McConnell, S.: Software Estimation: Demystifying the Black Art. Microsoft Press (2006)
2. Kassab, M., Neill, C., Laplante, P.: State of Practice in Requirements Engineering: Contemporary Data. Innovations in Systems and Software Engineering; A NASA Journal (2014), doi:10.1007/s11334-014-0232-4
3. IBM website: SAS Hub Non Functional Requirements (NFRs),
   http://www.haifa.ibm.com/projects/software/nfr/index.html
   (last visited January 2014)
4. Kassab, M.: Non-Functional Requirements: Modeling and Assessment. VDM Verlag Dr. Mueller (2009) ISBN 978-3-639-20617-3
5. Pfleeger, S.L., Wu, F., Lewis, R.: Software Cost Estimation and Sizing Methods: Issues and Guidelines. RAND Corporation (2005)
6. Abran, A., Desharnais, J.-M., Oligny, S., St-Pierre, D., Symons, C.: COSMIC FFP – Measurement Manual (COSMIC Implementation Guide to ISO/IEC 19761:2003). École detechnologie supérieure – Université du Québec, Montréal (2003),
   http://www.gelog.etsmtl.ca/cosmic-ffp/manual.jsp
7. UKSMA: Estimating with Mark II, v.1.3.1., ISO/IEC 20968:2002(E),
   http://www.uksma.co.uk (2002)
8. NESMA: NESMA Functional Size Measurement method compliant to ISO/IEC 24570 (2006), http://www.nesma.nl
9. IFPUG 4.1 Unadjusted Functional Size Measurement Method - Counting Practices Manual, ISO/IEC 20926:2003, 1st edn. (September 1, 2003), http://www.ifpug.org
10. Stensrud, E.: Alternative Approaches to Effort Prediction of ERP projects. Journal of Information and Software Technology 43(7), 413–423 (2001)
11. Glinz, M.: On Non-Functional Requirements. In: 15th IEEE International Requirements Engineering Conference (RE 2007), Delhi, India, pp. 21–26 (2007)
12. Bass, L., Clements, P., Kazman, R.: Software architecture in practice, 3rd edn. Addison-Wesley (2013)
13. Saaty, T.L.: The analytic hierarchy process. McGraw-Hill, New York (1980)
14. Saaty, T.L.: Decision making for leaders. LifeTime Leaning Publications, Belmont (1985)
15. Harrison, N.B., Avgeriou, P.: How do architecture patterns and tactics interact? A model and annotation. Journal of Systems and Software 83(10), 1735–1758 (2010)
16. FP Users Group, http://www.ifpug.org
17. FISMA: FiSMA 1.1 Functional Size Measurement Method, ISO/IEC 29881 (2008),
    http://www.fisma.fi/
    wp-content/uploads/2008/07/fisma_fsmm_11_for_web.pdf

18. Kassab, M., Daneva, M., Ormandjieva, O.: A Meta-model for the Assessment of Non-Functional Requirement Size. In: Proceedings of the 34th Euromicro Conference Software Engineering and Advanced Applications, SEAA 2008, pp. 411–418 (2008)
19. ISO 14143-1: Functional size measurement – Definitions of concepts, International Organization for Standardization – ISO, Geneva (1988)
20. SNAP,
    http://www.ifpug.org/ISMA6/
    ITPC%20SNAP-SW%20Non-Functional%20Assessment%
    20Process-Sept13.pdf (last visited May 2014)

# A Case Study to Enable and Monitor Real IT Companies Migrating from Waterfall to Agile

Antonio Capodieci, Luca Mainetti, and Luigi Manco

Department of Innovation Engineering,
University of Salento, Monteroni Street, 73100 Lecce, Italy
{Antonio.capodieci,luca.mainetti,luigi.manco}@unisalento.it

**Abstract.** Agile development methods are becoming increasingly important to face continuously changing requirements. Nevertheless, the adoption of such methods in industrial environments still needs to be fostered. Companies call for tools to keep under control both agility and coordination of IT teams.

In this paper, we report on an empirical case study aiming at enabling real companies migrating from Waterfall to Agile. Our research effort has been spent in introducing 11 different IT small and medium-sized enterprises to Agile, and to observe them executing projects. To have a common evaluation framework, we selected a set of 61 metrics, with the purpose of measuring the evolution towards Agile. We provide readers with empirical data on two categories of companies' feedbacks: (i) the metrics they considered to be useful beyond the theoretical definitions; (ii) the tools they integrated with existing development environments to collect data from metrics, and evaluate quantitative improvements of Agile.

**Keywords:** Agile, Software Metrics, Software Engineering, Waterfall, Migration towards Agile.

## 1 Introduction

Producing software artefacts is a factory in which technologies and productive methodologies are involved in the production chain to create an output. In recent years, there has been a steady transformation of the software development methodologies, from the classical type, such as Waterfall, to the newer Agile ones. Such a trend is due to the several benefits introduced by these methods, such as increasing customer and developer satisfaction, decreasing effort spent (in terms of cost and time) in marketing under uncertain conditions, flexibility, highly iterative development with a strong emphasis on stakeholder involvement, a general reduction in the defect density of the code, adaptability to changeable requirements and complex specifications, iterative development on smaller size projects, and so forth.

The present paper describes a study related to the adoption of Agile methods by some SMEs[1] belonging to the Apulian IT District, an Italian Industrial District,

---

[1] Small and medium enterprises (SMEs): companies whose personnel numbers fall below certain limits.

B. Murgante et al. (Eds.): ICCSA 2014, Part V, LNCS 8583, pp. 119–134, 2014.
© Springer International Publishing Switzerland 2014

founded in 2009. The District is composed of 94 Apulian software companies, four research centres and universities, and 11 associations and trade unions, with a total of more than 4,000 employees.

The presented study is part of the SMART project (Strategies, Methodologies and technologies for Agile Review and Transformation), funded by the Apulia Region Government. The SMART project involves 11 enterprises, two universities and two research centres, all belonging to the District. The project mission is to create a methodological framework to guide and monitor the adoption of Agile methods within projects already started with classical software engineering techniques or without any methodological approach in their development. Agile methods precisely describe the steps and actions to be undertaken with respect only to projects started from scratch [1]. SMART, therefore, tries to extend Agile methods and to overcome their gaps, working on a set of experimental software projects provided by the involved companies.

The study presented in this paper concerns the first stage of the SMART Project. In this phase, we have selected a wide set of software metrics in order to monitor and to successively analyse the transformation from classical development to Agile methodologies in the involved experimental projects. These metrics provide a complete control dashboard for a software project developed with Agile methods and promote the adoption of Agile practices.

Moreover, further refinements of the metrics framework were based on the needs of the enterprises involved in SMART. We will show how the business needs have conditioned the metrics selection and how they have contributed to the synthesis of an easy-to-use, minimally invasive and efficient monitoring system for software development projects. The enterprises' intervention in the metrics selection has resulted in a monitoring system transparent to the developers and suitable for coexistence with the business processes.

So, with this study we achieve two goals: providing a complete list of software metrics that best fit Agile methodologies and facilitating the adoption of the methods by real companies. In this way, we can support the transition process among development methods by monitoring metrics and provide a broad framework, which needs to be tailored to specific situations.

Selecting a broad-spectrum metrics set, we have produced a general-purpose monitoring framework for Agile software projects, flexible and adaptable to different kinds of managerial and technological conditions.

In the next section we describe the context of the study and the adopted research method. We analyse the systematic approach used to select the metrics according to the enterprises' needs and the rescarch environment. In Section 3 we show the selected metrics. Section 4 describes the role of the enterprises in the selection of the metrics and the tools useful in retrieving the measurements. Here, we analyse their feedback in response to the metrics' submission and the resulting metrics framework changes. In Section 5 we present the related work.

# 2     Research Context

The 11 SMEs linked to the SMART project produce a global sales volume of about 32 million euros (updated 31st December 2010) and provide direct employment for more than 600 workers. Their tasks deal with ICT in both public and private fields, in many countries around the World. So, their business turnover and employment commitments take on a remarkable relevance among the Apulian IT District.

Although well connected and integrated within the Italian ICT business environment, the enterprises involved in the study lack development approaches and methodologies, as well as the specific skills and experience necessary to perform a migration towards modern development practices. For these reasons, these enterprises still use classical but outmoded development methodologies or do not use any well-defined software production strategy and this deficit represents a limit on the current time-to-market and product quality needs. Despite that, they represent an ideal platform to introduce Agile methodologies, because they work on projects conducted by small teams, tightly connected with customers and subject to variable conditions and requirements.

Here-hence the origin of the SMART project. It aims to provide the involved enterprises with the necessary theoretical and practical competencies in the methodological migration by universities and research centres intervention. The experimental projects used as the base for the metrics synthesis were started with a certain amount of time and they differ from each other in the involved software technologies, requirements, and team compositions. So, they spawned a very heterogeneous research background.

## 2.1     Research Method

The search for the metrics was based on a systematic research approach, necessarily adapted to the background environment described in the previous section: (i) the training of the companies on Agile methods to achieve vocabulary and know-how alignment; (ii) analysis of the metrics in the scientific literature, with emphasis on those specific to the transition from Waterfall to Agile; (iii) focus groups for submission of the metrics to the partners; (iv) collection of the feedbacks from the partners by questionnaire; (v) metrics refining. The analysis of the state of the art aimed to discover the work relating to the monitored introduction of the Agile methodologies in different software factories. The result of the search was a set of metrics comprising both by a certain number of classical metrics and by several metrics closely suited to Agile methods. The business partners reviewed the selected metrics to prevent the selection of hard-to-retrieve metrics and to finally select really adoptable metrics from the enterprises' point of view. This allowed us to refine the initial set of metrics in order to strike a balance between a solid scientific metrics framework and a really applicable monitoring system. First of all, several meetings and workshops were organized to promote information sharing and to encourage the acceptance of the metrics. Secondly, the partners gave feedback, in the form of answers to questions, such as thaose shown in Table 1.

**Table 1.** Metrics acceptance questionnaire

| Question | Answer values |
|---|---|
| Metric understanding | From 1=not at all, to 5=completely |
| Willingness to collect data | From 1=not at all, to 5=completely |
| Ease of access to data collection tools | From 1=not at all, to 5=completely |
| Data collection scheduling | Each task, sprint, release, etc. |
| Comments | Open comments about the metric |

## 3    The Selected Metrics

In this section, we analyse the assembled metrics framework. The selection of the metrics has been based on two main goals:

- Monitoring the migration from sequential software development methods, such as Waterfall, to Agile methods;
- Promoting the practices strictly related to Agile methods, especially Test-Driven Development (TDD) and Refactoring.

Therefore, the measurements collection activity enables the adoption of Agile methodologies, promoting the use of Test-Driven Development  and Refactoring practices. In this paper, we do not give the definition for each metric. The definition are referenced in the bibliography. We point out especially to metric classification and contextualization.

### 3.1    Metrics Classification

As stated in [2], the broad metrics classification includes items such as product, process, objective and subjective, direct and indirect, resource and project metrics. Taking inspiration from the aforementioned article, we use a more restricted categorization to avoid class intersections in the metrics classification, as shown in Table 2.

**Table 2.** Metrics classification

| Classification Set | Metrics categories |
|---|---|
| Product Metrics | Product size; product complexity; derived product complexity; object oriented; usability; code coverage; quality; refactoring; continuous integration. |
| Process Metrics | Management metrics and life cycle metrics |
| Resource Metrics | Personnel metrics (effort metrics, etc.), software metrics and hardware metrics, performance metrics |
| Project Metrics | Cost, time, quality, risk metrics |

## 3.2     Metrics Definition

In this section we introduce the selected metrics. Each of them is well defined in the literature and here they are only gathered according to the classification specified in the previous paragraph. The only exception is represented by the quality metrics, reworked by us to adapt them to our purposes, starting from their definition.

**Product Metrics.** The metrics are categorized by the subsets specified in the Table 2.

*Product Size.* The selected product size metrics are:

- **Logical Lines Of Code (LLOC);**
- **Test Case Point analysis (TCP)** [3–5];
- **Number of classes;**
- **Number of abstract classes/interfaces;**
- **Abstractness (A)** [6];

In the literature there are several different product size metrics. Therefore, we have chosen the most useful from the point of view of the Agile methods. For example, we have selected the Test Case Point analysis instead of the more common Function Point Analysis (FPA), since the first fits better the problems concerning the TDD paradigm and promotes the use of this important practice.

*Product Complexity*

- **Cyclomatic Complexity (v(G));**
- **Halstead Complexity (V).**

*Derived Product Complexity.* The metrics below extend the Cyclomatic definition:

- **Essential Complexity;**
- **Integration Complexity;**
- **Design Complexity.**

*Object Oriented.* The metrics are classified according to four of the basic principles of the object-oriented technique, namely coupling, cohesion, inheritance and polymorphism [6]:

*Coupling*

- **Afferent Couplings (Ca);**
- **Efferent Couplings (Ce);**
- **Coupling Between Objects (CBO);**
- **Instability (I).**

*Cohesion*

- **Lack Of Cohesion of Methods (LOCM).**

*Inheritance*

- **Number of Overridden Methods (NORM);**
- **Depth of Inheritance Tree (DIT);**
- **Number Of Children (NOC);**
- **Specialization Index (SIX).**

*Polymorphism*

- **Weighted Methods per Class (WMC);**
- **Response For Class (RFC).**

*Usability.* The selected usability metric is based on MiLE+, a framework that enables a rigorous evaluation of the software application's usability [7]. It aims to analyse the whole elements involved in the user interaction with the application.

The framework proposes two types of application inspection:

- **Technical Inspection (TI):** it is based on some heuristics, which evaluate the quality of design and highlight the implementation defects from the point of view of the designers and work personnel.
- **User Experience Inspection (UEI):** checking the usability aspects strongly related to the user needs. In the UEI several usability scenarios are generated and each of them is divided into different tasks. For each task various heuristics evaluate the quality of the interaction between the user and the content of the application. The UEI allows to recognize in advance the potential problems that may arise when a user uses the application.

*Code Coverage.* Literature provides well-known unit test coverage criteria [8]:

- **Statement Coverage;**
- **Branch/Decision Coverage;**
- **Boolean/Condition Coverage;**
- **Loop Coverage;**
- **Path Coverage.**

*Quality.* The quality metrics here proposed are based on the amount and the criticality of the defects found during the development phase. In drawing up them, we have taken into account the difference between the *reworking* and *refactoring* phases.

The reworking phase concerns *an improvement of the external quality of the software*, not ensuring the functional equivalence between the code preceding the change and the next. In reworking, code change is due to its critical defects or to requests for improvements and extensions for already implemented functionalities (a sign of the lack of understanding between the team and the stakeholders).

In contrast, the refactoring phase aims at *improving the internal structure* of the software, without changing the external behaviour [9]. Refactoring procedures are intended to make the software more slender, lighter, and elegant, so as to ensure the maintainability, reusability and performance. Starting from these differences, we have selected the quality metrics from the scientific literature (i.e. [10]), adapting them to the demands imposed by Agile methods. Accordingly they are classified as reworking, refactoring and quality parameters metrics. The latter are quality expressions derived from the union between the first ones..

*Reworking*

- **Critical Defects ($SCO_1{}^2$):** number of critical defects;
- **Improvement Requests ($SCO_2$):** number of requested improvements and extensions for already implemented functionalities;
- **Open Reworks ($RW_O$):** cumulative defective LLOC due to $SCO_1$ and $SCO_2$, not yet fixed;
- **Closed Reworks ($RW_C$):** cumulative LLOC related to fixed $SCO_1$ and $SCO_2$;
- **Rework Effort ($RW_E$):** cumulative effort to fix spent to fix $SCO_1$ and $SCO_2$;
- **Total Reworks ($RW_T$):**

$$RW_T = RW_O + RW_C \tag{1}$$

- **Rework Ratio ($RW_R$)**

$$RW_R = \frac{RW_T}{LLOC} \tag{2}$$

- **Rework Backlog ($RW_B$)**

$$RW_B = \frac{RW_C}{LLOC} \tag{3}$$

- **Rework Stability ($RW_S$)**

$$RW_S = RW_T - RW_C \tag{4}$$

- **Rework Effort Ratio ($RW_{ER}$)**

$$RW_{ER} = \frac{RW_E}{TotalEffort} \tag{5}$$

*Refactoring*

- **Normal Defects ($SCO_3$):** number of non-critical defects;
- **Open Refactors ($RF_O$):** cumulative defective LLOC due to $SCO_3$, not yet fixed;
- **Closed Refactors ($RF_C$):** cumulative LLOC related to fixed $SCO_3$;
- **Refactor Effort ($RF_E$):** cumulative effort spent to fix $SCO_3$;
- **Total Refactor ($RF_T$):**

$$RF_T = RF_O + RF_C \tag{6}$$

---

[2] Software Change Order.

- **Refactor Ratio (RF$_R$)**

$$RF_R = \frac{RF_T}{LLOC} \tag{7}$$

- **Refactor Backlog (RF$_B$)**

$$RF_B = \frac{RF_C}{LLOC} \tag{8}$$

- **Refactor Stability (RF$_S$)**

$$RF_S = RF_T - RF_C \tag{9}$$

- **Refactor Effort Ratio(RF$_{ER}$)**

$$RF_{ER} = \frac{RF_E}{TotalEffort} \tag{10}$$

Combining the shown quality metrics, we can obtain a new set of quality parameters depending on both the rework and refactor procedures. In some of them, an appropriate weight $0 \le \alpha \le 1$ gives more or less importance to the two phases, providing different perspectives of the quality parameters:

*Quality parameters*

- **Number of SCOs (N):**

$$N = SCO_1 + SCO_2 + SCO_3 \tag{11}$$

- **Modularity (Q$_{MOD}$):** this value identifies the average broken LLOC per SCO, which reflects the inherent ability of the integrated product to localize the impact of change. The best case should ensure that SCOs are written for singular source changes:

$$Q_{MOD} = \alpha \frac{RW_T}{SCO_1 + SCO_2} + (1 - \alpha) \frac{RF_T}{SCO_3} \tag{12}$$

- **Changeability (Q$_C$):** this value reflects the ease with which the products can be changed. While a low number of changes is generally a good indicator of a quality process, the magnitude of effort per change is sometimes even more important:

$$Q_C = \alpha \frac{RW_{ER}}{SCO_1 + SCO_2} + (1 - \alpha) \frac{RF_{ER}}{SCO_3} \tag{13}$$

- **Maintainability (Q$_M$):** this value identifies the relative cost of maintaining the product with respect to its development cost by relating the effort ratio with the stability parameter. A value of Q$_M$ much less than 1 would tends to indicate a very maintainable product, at least with respect to development cost. Since we would intuitively expect the maintenance cost of a product to be proportional to its development cost, this ratio provides a fair normalization parameter for comparison between different projects:

$$Q_M = \alpha \frac{RW_{ER}}{RW_S} + (1-\alpha)\frac{RF_{ER}}{RF_S} \qquad (14)$$

Since the numerator of $Q_M$ is in terms of effort and its denominator is in terms of SLOC, it is a ratio of productivities (i.e., effort per SLOC). Some simple mathematical rearrangement will show that $Q_M$ is equivalent to:

$$Q_M = \frac{Productivity_{Maintenance}}{Productivity_{Developement}} \qquad (15)$$

- **Maintainability Index (MI)** [11][12]: It is based on the product complexity and it is obtained from the following parameters:
  - **aveV**: average Halstead Volume V per module;
  - **aveV (G)**: average Cyclomatic Complexity per module;
  - **aveLOC:** LLOC average per module;
  - **perCM:** average percentage of comments per module;

$$MI - [171 - 5.2 * \log_2(aveV) \quad 0.23 * aveG - 16.2 * \log_2(aveLOC)$$
$$+ 50 * \sin(\sqrt{2.4 * perCM})] \qquad (16)$$

- **Defect density (D$_D$)**: represents the defects density within the source code:

$$D_D = \alpha \cdot RW_R + (1-\alpha) \cdot RF_R \qquad (17)$$

*Refactoring.* Since the refactoring does not alter the external behaviour of the software, monitoring the quality of the refactoring phase is not a trivial issue. This metric tracks the time course of the refactoring based on the acceptance tests [9]:

- **Running Tested Features (RTF)** [13]: the metric is defined as the number of features that passed their acceptance test.

Agile methodologies do not allow the application to be designed too much in advance, planning the use of a work-in-progress refactoring. Bad refactoring might damage some application functionalities, invalidating the related acceptance tests. In order to plot a growing straight RTF line over time, an Agile team has to follow good refactoring practices along with the use of an automatic acceptance testing framework. So, at the same time, the RTF metric monitors the quality of the refactoring practice and promotes the use of both these Agile techniques.

*Continuous Integration.* In this section, we describe a simple metric that allows the detection of the Continuous Integration degree implemented in a project:

- **Pulse** [3]: the metric counts the number of commits towards the versioning repository.

As with the RTF metric, the Pulse metric combines monitoring with the promotion of a particular Agile technique.

**Process Metrics.** Process metrics are useful in monitoring the sequence of activities invoked to produce the software product [14].

- **Progress:** it shows the progress of the development as the number of made tasks.

**Resource Metrics.** Resource metrics aim to monitor people, methods and tools time, effort and budget [14]:

- **Effort:** it measures the effort required and planned for each staff member (in terms of time, story points, resources, etc.) compared with that actually provided;
- **Team satisfaction:** it is a survey that provides information about the perception of the Agile transformation from the point of view of the staff involved. Examples of questions can be retrieved from [15].
- **Customer satisfaction:** in the Agile methodologies, customer role is essential to provide a valid software product. So that customer can be considered as a resource of the software life cycle. A questionnaire can be useful for collecting customer considerations about team and software product quality.

**Project Metrics.** Project metrics offer a set of measurements useful in contributing to project control, risk mitigation and to managing team performance [14]:

- **Cost:** cumulative interpretation of the spent effort;
- **Burn-down** [3]: is the opposite of the Cost metric, since it indicates the amount of work still to be done to complete the project (or sprint).

## 4    Feedback from the Companies

As specified in the previous sections of this paper, the selected metrics were subject to the reviews of the business partners in order to refine the set of metrics and to adapt it to the enterprises' needs. The feedback from the companies was based on both meetings and workshops and on a survey about the companies' viewpoints regarding the proposed metrics. The survey was composed with the questions shown in Table 1. Fig. 1 shows three charts representing the results of the survey. Each chart shows the averaged values of the points given by companies to the questions concerning their opinions about each metric. As illustrated in the first chart, the comprehension of the metrics was generally high. However, the second chart shows that there were some difficulties in the willingness to retrieve them. As it can be gathered from the third chart, this problem had to be related above all to the lack of adequate data collection tools for the enterprises. The issue was also confirmed in the workshops with the partners and in the comments attached to the survey feedbacks.

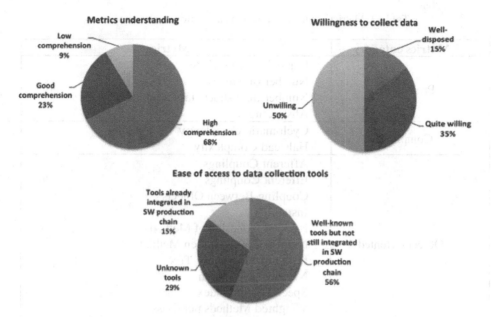

**Fig. 1.** Survey Result

Therefore, at the suggestion of the companies, the problem was overcome by identifying a set of tools and strategies in support of the collection of the measurements. Through a synergistic exchange of ideas between the companies and the research groups, the metrics set was split into three macro-groups, depending on the different methods and tools used in collecting data. The groups do not cover all the metrics set, because the expressions of some of them are derived and then estimable not directly in the monitoring phase.

The first group concerns the metrics whose measurements have to be obtained through automatic calculation tools, due to their computational complexity (Table 3). The only constraint on the tools choice is of allowing the complete export of the collected data, for later analysis. An example of an automatic calculation tool selected by the companies is Eclipse Metrics, a Java plugin for Eclipse. In the .NET environment, a recommend tool is Ndepend. However, the companies adopted various tools concurrently, each one providing different categories of metrics.

The second group comprises the metrics related to the measurements currently not supported by any tool, but achievable by manual procedures. Table 4 shows the metrics belonging to this group. To support the collection of data, spreadsheet templates have been set up for each one of these metrics, as a sort of wizard.

The third and last group affects those metrics achievable by handling data provided by project management and ATDD tools. These metrics are shown in Table 5.

The ATDD tools provide the information needed for the RTF metric. Storing this simple information in a spreadsheet allows an historical vision of the feature test results. Project management tools, adequately used, represent the data source for quality, process, resource and project metrics. They permit to specify for each task the involved teammate, the estimated and effective effort spent and, in the case of SCOs,

**Table 3.** Metrics achievable through automatic calculation tools

| Metrics category | Metrics |
|---|---|
| Product Size | Logical Line Of Code<br>Number of classes<br>Number of abstract classes/interface<br>Abstractness |
| Complexity | Cyclomatic Complexity<br>Halstead Complexity |
| Object Oriented | Afferent Couplings<br>Efferent Couplings<br>Coupling Between Objects<br>Instability<br>Lack Of Cohesion of Methods<br>Number of Overridden Methods<br>Depth of Inheritance Tree<br>Number of Children<br>Specialization Index<br>Weighted Methods per Class<br>Response for a Class |
| Code Coverage | Statement<br>Branch/decision<br>Boolean/Condition<br>Loop<br>Path |
| Quality | Maintainability Index |

**Table 4.** Metrics currently achievable through non-automated procedures

| Metrics category | Metrics |
|---|---|
| Product Size | Test Case Points analysis |
| Usability | MiLE+ |
| Continuous Integration | Pulse |
| Resource | Team satisfaction<br>Customer satisfaction |

the defect type and the involved lines of code. Some of the companies have used Redmine, a project management web application, extended with Backlogs plugin.

A clever answer to SMART requests has been Police, a flexible SVN plugin developed by one of the companies involved in the project, I.T.S. S.r.l. (Informatica, Tecnologie e Servizi S.r.l.). The plugin allows the definition of custom rules to be applied to the code before each commit towards the SVN repository. The company chose this solution to synchronize agile tasks with its internal project management tool, not currently suitable for agile paradigm.

**Table 5.** Metrics achievable from project management and ATDD tools

| Metrics category | Metrics |
|---|---|
| Refactoring | Running Tested Features |
| Quality | Critical Defects ($SCO_1$) |
| | Improvement Requests ($SCO_2$) |
| | Normal Defects ($SCO_3$) |
| | Open Reworks |
| | Closed Reworks |
| | Rework Effort |
| | Open Refactors |
| | Closed Refactors |
| | Refactor Effort |
| Process | Progress |
| Resource | Effort |
| Project | Cost |
| | Burn-down |

# 5     Related Work

The selection of the metrics is based on a review of the already completed work on the monitored migration from classical development methodologies to Agile practices. Below we describe some of them, highlighting the key points that have been useful for our work. Furthermore, in the data analysis the measurements related to the metrics will assume the form of a complex data source, composed of a heterogeneous set of information repositories. So, data mining processes should be applied to the latter in order to obtain consistent information about the trend of experimental projects. For this reason, in this section, we provide a brief introduction to the state of the art of the practices of mining software repositories.

A survey of the Agile metrics and their comparison with traditional development metrics is provided in [2]. Several metrics are listed according to the different categories they belong to, claiming that both methodologies utilize basically the same sets of metrics and classifications. Simply, some differences lie in the use of metrics more closely describing teamwork and Agile aspects, such as team effort, velocity, story points, and so on.

An innovative project, described in [5], has implemented Agile processes and XP practices in a software development team belonging to the Israeli Air Force. As the army is a large and hard-to-change organization with a rigid organizational and managerial structure, the transition to Agile methods was challenging. Thus, such a change was performed in a gradual carefully planned process and it was adapted to the team involved in the study, comprising 60 skilled developers and testers with different individual interests. The project affected by the transition was large-scale, enterprise-critical software, in which quality and fit to customers' needs were not to be compromised by the study. To ensure this and to communicate information on the project's trend to the team, a set of four metrics was selected. They indicated the amount and the quality of work performed and the work status and progress. The four metrics discussed are: (i) *Product Size*, presenting the amount of completed work based on tests written for the application; (ii) *Pulse*, counting how many check-in operations occur per day; (iii) *Burn-down*, providing information about the remaining project work versus the remaining human resources; and (iv) *Faults*, giving the faults per iteration. The use of the aforementioned metrics mechanism increased the confidence of the team members as well of the unit's management with respect to using agile methods.

In [16], Waterfall and Extreme Programming techniques are empirically compared with the aim of looking at the advantages and disadvantages of the methods. The experimental period was five years and the results of the study are based upon the outcomes, generated artefacts and metrics produced in reality by different teams involved in the same project. Some of them used Waterfall development methodologies and the others Agile methodologies. To learn about the effective transition from traditional to agile development methods, the study was conducted at Carnegie Mellon University in Silicon Valley. The metrics used to compare the two methodologies have been divided into three categories: (i) *requirements metrics*; (ii) *design metrics*; and (iii) *implementation metrics*.

First of all, the study points out the fact that setting up this kind of experiment so much in advance is challenging. The paper shows that Waterfall teams spent more time creating high formal documents whereas Extreme Programming teams spent more time writing code and documenting the design and the code. Also, the amount of code and features completed were roughly the same for both methods.

The study described in [15] tries to provide evidence of the impacts of Agile adoption in a very large software development environment, Nokia, from the point of view of the teams involved in the development. A population of more than 1,000 respondents in seven different countries in Europe, North America, and Asia was subjected to a questionnaire. The results reveal that most respondents agree on all accounts with the generally claimed benefits of agile methods. These benefits include higher satisfaction, a feeling of effectiveness, increased quality and transparency, increased autonomy and happiness, and earlier detection of defects. Also, 60% of respondents would not like to return to the old way of working. Beyond this further evidence of the benefits introduced by Agile methods, the study provides an innovative kind of test for evaluating a software development project.

Hereafter we show some examples about the mining software repository practice. It will be essential in order to analyse the measurements retrieved by the proposed metrics framework. In the literature, various methodological frameworks there exist for the application of this practice to software engineering, for example [17].

Mining software repositories has several goals and can lead to different results. For example, in [18], the authors implemented a static source code checker to drive and to help refine the search for bugs. The goals of the study rested upon the data retrieved from the source code repository. In order to refine its results, the system searched for a commonly fixed bug and used information automatically mined from the repository.

The application of this technique on real projects, like Apache web server and Wine, shows that it is more effective than the common static analysis.

The authors of [19] present a framework with the purpose of combining different software repositories in order to ease mining process applications. The framework is called FRASR (FRamework for Analysing Software Repositories). It applies systematic data mining pre-processing procedures, combining data from different sources and extracting logs useful for process mining applications. The study involved various data sources, like versioning repositories, task managers, mailbox archives, and so forth.

Our study aims to differentiate itself from the related work by two main key points: the richness of the software metrics set and the tightly collaborative approach instantiated with the IT business world.

## 6 Conclusion and Future Work

The presented study is part of the SMART project, which aims to enable a systematic approach with real software factories in order to help them in their transition from Waterfall to Agile methodologies. With this paper, we have described a complete set of metrics useful in monitoring different aspects of the software life cycle. We did not propose new metrics, but we have selected the more suitable ones to monitor software projects developed with Agile methods. Empirical data provided by enterprises involved in the project allowed us to select only the more interesting and applicable metrics from the businesses' points of view, discarding those too difficult to integrate into the software development processes. This software metrics framework will be applied to several experimental projects in transition towards Agile methods in the forthcoming period, so monitoring their trends.

The study started from the analysis of the research context. First of all, we examined the experimental software project and the involved teams to give a complete perspective of the technological environment. This analysis revealed the strong heterogeneity of the technologies used and the lack of any kind of Agile practices in the projects.

Subsequently, we presented a large set of metrics, suitable to monitor each aspect of the software development phases, fitting it to the different technological conditions imposed by the heterogeneous projects. The metrics have been selected to monitor the migration towards Agile development methodologies, promoting the practices strictly related to them. The set of metrics was subjected to the project partners review in different meetings and by a survey. We analysed the feedback in the paper, focusing on the difficulties in retrieving the measurements by the enterprises due to a lack of known retrieval tools and strategies. The partners review revealed their own importance also in selecting some of the presented metrics.

Subsequent work will be related to the analysis of the measurements retrieved, applying the presented plan of measures to the experimental projects. The time traces of the collected data will show the effective impact of Agile methods in relation to software projects already started with classical development methods. Therefore, future studies will aim to show the benefits and disadvantages of Agile, introducing scientific innovation because of the numerous monitored aspects and the specific kind of environment under study.

# References

1. Beck, K., Beedle, M., Bennekum, A., Van, C.A., Cunningham, W., Fowler, M., Grenning, J., Highsmith, J., Hunt, A., Jeffries, R., Kern, J., Marick, B., Martin, R.C., Mellor, S., Schwaber, K., Sutherland, J., Thomas, D.: Manifesto for Agile Software Development, http://agilemanifesto.org/
2. Misra, S., Omorodion, M.: Survey on agile metrics and their inter-relationship with other traditional development metrics. ACM SIGSOFT Softw. Eng. Notes 36, 1 (2011)
3. Nguyen, V., Pham, V., Lam, V.: qEstimation: a process for estimating size and effort of software testing. In: Proceedings of the 2013 International Conference on Software and System Process, ICSSP 2013, p. 20. ACM Press, New York (2013)
4. Nguyen, V., Pham, V., Lam, V.: Test Case Point Analysis: An Approach to Estimating Software Testing Size, http://www-scf.usc.edu
5. Martin, R.C.: Agile Software Development: Principles, Patterns, and Practices. Prentice Hall PTR (2003)
6. Dubinsky, Y., Talby, D., Hazzan, O., Keren, A.: Agile metrics at the Israeli Air Force. In: Agile Development Conference (ADC 2005), pp. 12–19. IEEE Comput. Soc. (2005)
7. Triacca, L., Bolchini, D., Botturi, L., Inversini, A.: MiLE: Systematic usability evaluation for e-learning web applications. In: Proceedings of World Conference on Educational Multimedia, Hypermedia and Telecommunications, Chesapeake, VA, pp. 4398–4405 (2004)
8. Zhu, H., Hall, P.A.V., May, J.H.R.: Software unit test coverage and adequacy. ACM Comput. Surv. 29, 366–427 (1997)
9. Kunz, M., Dumke, R.R., Zenker, N.: Software Metrics for Agile Software Development, pp. 673–678 (2008)
10. Royce, W.: Pragmatic Quality Metrics for Evolutionary Software, TRW Space and Defence Sector, Redondo Beach, California (1990)
11. Khan, R.A., Mustafa, K., Ahson, S.I.: Software Quality: Concepts and Practices (2006)
12. Ganpati, A., Kalia, A., Singh, H.: A Comparative Study of Maintainability Index of Open Source Software. Int. J. Emerg. Technol. Adv. Eng. 2, 228–230 (2012)
13. A Metric Leading to Agility, http://xprogramming.com/articles/jatrtsmetric/
14. Royce, W.: Software Project Management: A Unified Framework (1998)
15. Laanti, M., Salo, O., Abrahamsson, P.: Agile methods rapidly replacing traditional methods at Nokia: A survey of opinions on agile transformation. Inf. Softw. Technol. 53, 276–290 (2011)
16. Ji, F., Sedano, T.: Comparing extreme programming and Waterfall project results. In: 24th IEEE-CS Conference on Software Engineering Education and Training (CSEE & T), pp. 482–486. IEEE (2011)
17. Hassan, A.E., Xie, T.: Mining software engineering data. In: Proceedings of the 32nd ACM/IEEE International Conference on Software Engineering, ICSE 2010, p. 503. ACM Press, New York (2010)
18. Williams, C., Hollingsworth, J.: Automatic mining of source code repositories to improve bug finding techniques. IEEE Trans. Softw. Eng. 31, 466–480 (2005)
19. Poncin, W., Serebrenik, A., Van Den Brand, M.: Process Mining Software Repositories. In: 15th European Conference on Software Maintenance and Reengineering, pp. 5–14. IEEE (2011)

# A Solution Proposal for Complex Web Application Modeling with the I-Star Framework

José Alfonso Aguilar[1], Anibal Zaldívar[1], Carolina Tripp[1], Sanjay Misra[2], Salvador Sánchez[1], Miguel Martínez[1], and Omar Vicente García[1]

[1] Señales y Sistemas (SESIS) Facultad de Informática Mazatlán
Universidad Autónoma de Sinaloa, México
82120 Mazatlán, Mexico
[2] Department of Computer and Information Sciences
Covenant University, Nigeria
{ja.aguilar,azaldivar,ctripp,m.martinez}@uas.edu.mx,
ssopam@gmail.com

**Abstract.** In Web Engineering (WE), several Goal-oriented Requirements Engineering (GORE) approaches have emerged using its advantages, such as the representation of actors, their intentions, goals and the tasks needed to achieve the goal, for requirements specification with promising results. Regrettably, the use of GORE approaches has one, among others, gap detected, the scalability. In these modeling frameworks, when the designer performs the requirements specification, the requirements diagram (model) trends to rapidly grow, becoming very difficult to use in projects with a considerable amount of requirements changing and growing constantly. In this paper, we propose an association form for the $i^*$ goal-oriented modeling framework in order to define the creation of two type of modules: Navigational and Service modules, since these are the two types of functional requirements more used for requirements specification in our proposal. Furthermore, we provide an example of application. Finally, with this approach, the benefits are: firstly, the scalability of the Web requirements model will be increased, therefore the model will be less complex and easier to understand and maintain, and secondly, the construction of modeling tools improving the user experience, the maintainability of the models and its reuse.

**Keywords:** Web Engineering, Goal-oriented Requirements Engineering, i-star, Scalability, Requirements Modeling.

## 1 Introduction

Recently, it has emphasized in the success that the Goal-oriented Requirements Engineering (GORE), specially the goal-oriented languages have had for Requirements Engineering (RE) in Software (SE) and Web development, this is mainly because with the use of this kind of languages it is possible to represent (modeling) the user needs (goals) without neglecting the organizational objectives (goals), the software architecture and the business process. One of the most

B. Murgante et al. (Eds.): ICCSA 2014, Part V, LNCS 8583, pp. 135–145, 2014.
© Springer International Publishing Switzerland 2014

widespread GORE approach used these days is the $i^*$ framework [1, 2], used for modeling organization, agent-oriented approaches, business goals among others.

In the context of WE, several approaches currently exist that use different goal-oriented requirements analysis languages for requirements stage, being the most commonly used the $i^*$ modeling framework [3–5] by the reason of it focuses on the description and evaluation of alternatives and their relationships to the organizational objectives. Unfortunately, the robustness of the Web application modeling, i.e., the requirements model, becomes very complex (big size). Therefore, the application of the $i^*$ framework has left much to be desired in order to be applied in real-world projects in view of the fact that the $i^*$ framework lacks scalability due to the absence of modularity [6]. Furthermote, the $i^*$ framework cannot effectively model crosscutting concerns which compromise the modularity and the reusability the results models. This is mainly because of their syntax, for the reason that it does not allow an optimal way to organize the requirements modeling causing usability problems in existing modeling tools which may lead the requirements model difficult to correct when a requirements change it's presented.

In this work, a proposed solution for the modeling of complex Web applications with goal-oriented languages is presented. The proposed solution is applied in the Web Requirements Metamodel, a metamodel based on the $i^*$ modeling language [7]. The main goal of this work is to provide a solid metamodel for requirements specification, in order to help to the constructions of tools (graphical editors) based on this metamodel with which helping the user to more effectively and efficiently modeling the requirements of the Web application to develop. Importantly, we do not propose an extension for the $i^*$ framework, only an association form in order to improve the organization of the requirements model.

The rest of the paper is organized as follows: Section 2 presents some related work relevant to the context of this work. Section 3 describes our GORE proposal where is found the contribution of this work. In Section 4, proposal is explained. The application example is described in Section 5. Finally, the conclusion and future work is presented in Section 6.

## 2   Background

Several proposals have been presented with regard to the use of the $i^*$ framework in the industry, focusing only in software engineering applications. On these works, the authors highlights some gaps with respect the requirements modeling of big projects [8], [9], the most remarkable is the lack of modularity, without it, the $i^*$ goal-oriented diagrams trend to grow up to much making it difficult to maintain. Two of the most widespread contributions in this respect are [10] and [11], in the first proposal the authors uses a specific notation to represent $i^*$ models using aspect-orientation in order to address modularity and composition of crosscutting concerns; in the second one, the authors presents an empirical evaluation to identify and to understand what the practical problems of $i^*$ are in terms of the strengths and weakness that need to be overcome based on a service-oriented architecture. Recently, in [12], the authors present a modularization proposal in the context of Business Intelligence, specifically for Data Warehouse (DW). In this proposal, the requirements elicitation stage is carried out by means of the $i*$ modeling framework trough the modularization of goals, to do this, the authors propose a set of guidelines. Regrettably, this proposal is applied out of the context of Web Engineering and their tool has a very narrow graphical editor.

Since the $i*$ framework is one of the most widespread goal-oriented modeling languages in software engineering it's not surprising that its improvements have been applied only in this field. However, goal-oriented modeling languages have been used recently in the Web Engineering area [3, 13, 14]. In the proposal of [4], the authors illustrates how to use goal-oriented requirements analysis to define hypermedia specific requirements that may be effectively used in Web conceptual design and usability evaluation. Another proposal for using a goal-oriented requirement analysis language is the one presented in [5], this proposal uses the WebGRL and, similar to our previous work [3], the A-OOH Web engineering method and proposes a model transformation approach extending the GOREWEB framework from the requirements phase to the design phase. Unfortunately, any of these works have the mature level for being used in industrial projects since the modularity issue is not covered, thus, the implementation of its tools does not provide well support (i.e., at usability level) for the Web application designer, making the requirements analysis stage difficult to maintain due to the goal-oriented diagrams in real projects trends to grow up accordingly the requirements evolution.

To sum up, there have been many attempts to apply goal-oriented techniques and methods to deal with some aspects of the Requirements Engineering process for Web applications. Nevertheless, there is still a need for solutions in order to bring the gap between scientific/research proposals and their industrial application (real world application projects).

## 3  A Goal-Oriented Modeling Framework in Web Engineering: Web Requirements Metamodel

The $i*$ (pronunced *eye-star*) [1, 2] is one of the most widespread goal-oriented requirements engineering (GORE) frameworks, its has been applied for modeling organizations, business processes, requirements specifications and requirements analysis, among others. As a goal-oriented analysis technique, the $i*$ framework focuses on the description and evaluation of alternatives and their relationships to the organizational objectives [3]. The $i*$ framework consists of two models: the strategic dependency (SD) model, to describe the dependency relationships (represented as $-\!\!\!D\!\!\!-$) among various actors in an organizational context, and the strategic rationale (SR) model, used to describe actor goals and interests and how they might be achieved. Therefore, the SR model (represented as a dashed circle $\bigcirc$) provides adetailed way of modeling the intentions of each actor (represented

**Table 1.** The i* basic elements

| i* Element | Representation | Description |
|---|---|---|
| Goal | elipse ⬭ | Represents adesire of an actor. Goals provide a rationale for requirements, their satisfaction can be described through means-end links (⇾) representing alternative ways for fulfilling goals. |
| Task | hexagon ⬡ | Describes some work to be performed in a particular way. Decomposition links (─┼─) are useful for representing the necessary intentional elements for a task to be performed. |
| Resource | rectangle ▭ | Represents some physical or informational entity required for the actor. |
| Softgoal | eight-shape ◠◡ | Is a goal whose satisfaction criteria is not clearcut. How an intentional element contributes to the satisfaction or fulfillment of a softgoal is determined via contribution links (⟶). Possible labels for a contribution link are make, some+, help, hurt, some-, break, unknown, indicating the (positive, negative or unknown) strength of the contribution. |

as a circle ⭘), i.e., internal intentional elements and their relationships (see table 1).

One important factor to poin out is that the $i*$, as goal-oriented traditional requirement analysis approach, does not cover the special characteristics of Web aplications such as the navigation, and their different type of Web Requirements, Functional and Non-Functional. In order to ameliorate this gap, we adapt the $i*$ modeling framework into the Web Engineering field by using the Web requirements taxonomy proposed in [15]. The authors define six types of Web requirements, these are: (i) Content, defines the information that is useful for the Web application which should be presented to users, in an online book store some examples might be the information about a "book" or a "book category", (ii) Service, it concerns to the internal functionality that the Web application should provide to its users. Following the online book store example, service requirements might be: "register a new client" or "add book to shopping cart", (iii) Navigational, defines the navigational paths for the user (e.g., the user navigation from "index page" to "consult shopping cart"), (iv) Layout, they define the visual interface for users (e.g., "present a colour style"), (v) Personalization, personalization actions to be performed in the Web application (e.g., "show recommendations based on interest in previously acquired books", "adapt font for visually impaired users", etc.), and (vi) Non-Functional Requirements, they are related to quality criteria that the intended Web application should achieve e.g., "good browsing experience", "attract more users", "improve efficiency". To do so, in first place, we defined a profile to formally represent the adaptation of each one of the $i*$ elements with each requirement type from the Web requirements clasification adopted [16]. Then, we implemented this profile in an EMF (Eclipse Modeling Framework) [17] metamodel adding EMF clases according to the different type of Web requirements: the Navigational, Service, Personalization and

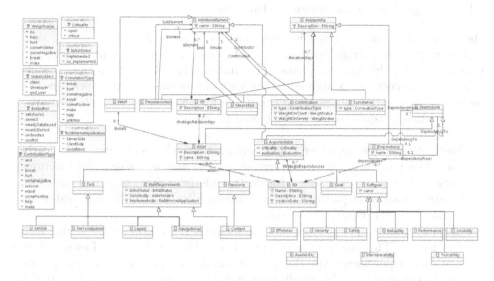

**Fig. 1.** An overview of the $i^*$ Web Requirements Metamodel

Layout requirements extends the $i^*$ Task element, the Content requirement extends the $i^*$ Resource element. It is worth noting that NFRs, until now, can be modeled by directly using the $i^*$ softgoal element. In Figure 1 can be seen an extract of the Web Requirements Metamodel (WRM). The metamodel implements all the $i^*$ concepts such as Actor, Beliefs, Correlations, Dependency and so on. It has been implemented in the Eclipse [18] IDE (Integrated Development Enviroment). For a more detailed information about the $i^*$ modeling framework see [1, 2].

## 4    A Solution-Proposal for Complex Web Application Modeling with the I-Star in Web Engineering

In this section, we shortly describe how our $i^*$ requirements specification approach is complemented improving the scalability of our Web Requirements Metamodel by adding modularity support. To do so, two type of modules are included in the Web Requirements Metamodel: the Navigational Module and the Service Module. Furthermore, we provide a set of rules (considering the fundamental concepts from the work presented in [12]) in order to the correct implementation and use of the modules. The main purpose is to increase the scalability adding modularity in order to improve the user experience making the requirements specification (model or diagram) easy to understand for the designer as well as avoid errors.

## 4.1   Navigational and Service Modules

The *i** requirements specification approach is conformed by five type of Web requirements: the Navigational, Service, Personalization, Content, Layout requirements and the softgoal element for non-functional requirements. At this point, it is important to highlight that we only include two modules, one for navigational requirements and the other one for service requirements since the other fucntional requirements are not used in much.

- Navigational Module. The navigational module includes requirements mostly used to represent the path through which the user will access the functionality of the web application. Inside this module, Service, Personalization, Content, Layout requirements and Softgoal elements can be included, even so navigational requirements too.
- Service Module. The service module includes the functionality of the Web application. This module can include Personalization, Content, Layout, Navigational requirements and Softgoal. It was defined in order to group specific functionality of the Web application, i.e., the module for a record form of a product.

The modules extends two of the core classes from the *i** Web Requirements Metamodel, specifically the *Relationships* class and the *IntentionalElement* class (see Fig. 1). The *Relationships* class is used in the Web Requirements Metamodel to represent all kind of relationships between *i** elements, i.e., decomposition, means-ends, and the *IntentionalElement* class is the core class for all the *i** intentional elements such as Task, Softgoal, Goal, Resource and so on (see Fig. 2). Once the modules definition its done, we present next a set of rules for the correct application of the Navigational and Service modules.

1. Rule No.1. The relationships between two Navigational Modules are valid if at least one *i** intentional element is shared by them.

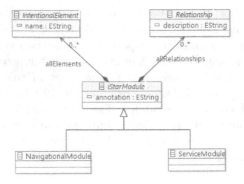

**Fig. 2.** The *i** Modules (extract from the *i** Web Requirements Metamodel)

2. Rule No.2. One module can contain one or more modules navigational or service, it depends of the intentional elements to be decomposed.
3. Rule No.3. If the decomposition of task (functional requirements) is not superior to two levels, it might be not necessary to use their respectively module (Navigational or Service).
4. Rule No.4. For any module created there will be only, as a root element, a Navigational Requirement. This is because the Navigational Requirement is the path with which the user can move from one place to another in the Web application, i.e., browsing through it.
5. Rule No.5. The name of the module, as a mandatory rule, must be, for better identification, the same as the Navigational Requirement in order to help with the organization of the diagram (model).

Although this work was perceived in the context of our Model-Driven Web Engineering approach [3, 16, 19, 7], it is in fact a stand-alone, independent approach that can thus be used in any Web Engineering method. Furthermore, the concepts presented in this proposal ca be easly applied in others GORE approaches. The application example is described in the next section.

## 5   Application Example

In this section, we provide an example of our approach based on a company that sells books on-line. In this case study, a company would like to manage book sales via an online bookstore, thus attracting as many clients as possible. Also there is an administrator of the Web to manage clients.

The requirements specifications was made by using of our $i^*$ framework for Web requirements (see [3]). For the *Online Bookstore*, three actors are detected that depend on each other, namely *"Client"*, *"Administrator"*, and *"Online Bookstore"*. A client depends on the online bookstore in order to *"choose a book to buy"*. The administrator needs to use the online bookstore to *"manage clients"*, while the *"client data"* is provided by the client. These dependencies are modeled by an SD model. Once the actors have been modeled in an SD model, their intentions are specified in SR models.

The SR model of the online bookstore is shown in Fig. 3. The main goal of this actor is to *"manage book sales"*. To fulfill this goal the SR model specifies that two tasks should be performed: *"books should be sold online"* and *"clients should be managed"*. We can see in the SR model that the first of the tasks affects positively the softgoal *"attract more users"*. Moreover, to complete this task four subtasks should be obtained: *"provide book info"* (which is a navigational requirement), *"provide recommended books"* (which is a personalization requirement), *"search engine for books"*, and *"provide a shopping cart"*. We can observe that some of these tasks affect positively or negatively to the non-functional requirement *"easy to maintain"*: *"Provide book information"* is easy to maintain, unlike *"provide recommended books"* and *"use a search engine for books"*. The navigational requirement *"provide book information"* can be decomposed into

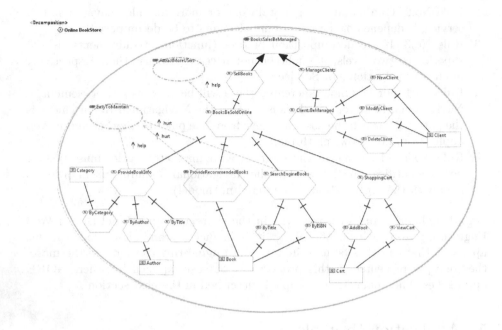

**Fig. 3.** Modeling the online bookstore in an SR model

several navigational requirements according to the criteria used to sort the data. These data is specified by means of content requirements: *"book"*, *"author"* and *"category"*. The personalization requirement *"provide recommended books"* is related to the content requirement *"book"* because it needs the book information to be fulfilled. The task *"search engine for books"* is decomposed into a couple of service requirements: *"search book by title"* and *"search book by ISBN"*, which are also related to the content requirement *"book"*. In the same way, the task *"provide a shopping cart"* is decomposed into two service requirements: *"add book to cart"* and *"view cart content"*. These service requirements are related to the content requirement *"cart"*. Finally, the task *"clients be managed"* is decomposed into three service requirements: *"new client"*, *"modify client"* and *"delete client"*, which are related to the content requirement *"client"*.

This Web requirements specification can be modularized applying the Rules from section 4, thus decreasing the model complexity and providing support to the Web application designer. Therefore, after applying the rules, the requirements model (see Fig. 3) would be organized as follows:

The Fig. 4 shows five modules created at different abstraction levels, i.e., the module *BooksBeSoldOnline* was created with the rules numbre 2, 4 and 5. On the other hand, the Personalization Requirement *ProvideRecommendedBooks* is not a module due to it uses only the rules 3 and 5 which mean that the decomposition of task (functional requirements) is not superior to two levels. Finally, it is important to point out that with this new version of the model

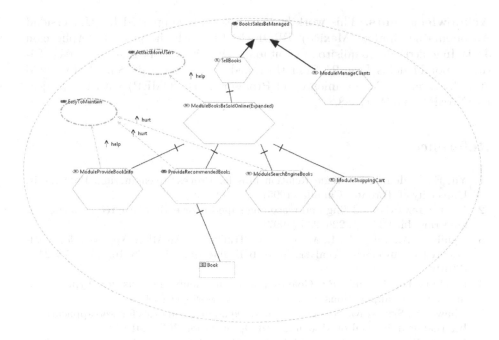

**Fig. 4.** The Requirements Model after the requirements modularization

(after appliying the modularization rules) (see Fig. 4) the user experience is improved because the modularization makes the requirements specification easy to understand for the designer as well as it is helpful in order to avoid errors.

## 6    Conclusions and Future Work

In this work, we have presented an extension to improve scalability in our goal-oriented requirements engineering approach for the development of Web applications [20]. Our proposal offers several advantages such as the complexity management, this is, according to [11], the capability of the modeling method to provide hierarchical structure for its models, thanks to the modularity support integrated to our Web Requirements Metamodel which is not supported natively in the $i^*$ framework. Model reusablity, with the implementation of modules, according to [6], its easy to organize repositories of modules. Moreover, with regard to tool support, with the implementation of modularity in the Web Requirements Metamodel, the usability is improved, because modularization makes the requirements specification (model) easy to construct and to understand for the designer.

Future work consist in the elaboration of a set of experiments in order to validate in real-world cases the improvement level of our tool named WebREd-Tool [21], used to specify Web requirements appliying the $i^*$ framework with the modularization supporting feature.

**Acknowledgments.** This work has been partially supported by: Universidad Autónoma de Sinaloa (México), PROFAPI2013 with the project: "Aplicación de la Ingeniería de Requisitos Orientada a Objetivos en la Ingeniería Web Dirigida por Modelos Fase 2" from Universidad Autónoma de Sinaloa (México) and Programa de Mejoramiento del Profesorado (PROMEP) with the project PROMEP/103.5/13/7588.

# References

1. Yu, E.: Modelling Strategic Relationships for Process Reenginering. PhD thesis, University of Toronto, Canada (1995)
2. Yu, E.: Towards modeling and reasoning support for early-phase requirements engineering. In: RE, pp. 226–235 (1997)
3. Aguilar, J.A., Garrigós, I., Mazón, J.N., Trujillo, J.: An MDA Approach for Goal-oriented Requirement Analysis in Web Engineering. J. UCS 16(17), 2475–2494 (2010)
4. Bolchini, D., Paolini, P.: Goal-driven requirements analysis for hypermedia-intensive web applications, vol. 9, pp. 85–103. Springer (2004)
5. Chawla, S., Srivastava, S.: Goal oriented requirement analysis for web applications. International Journal of Modeling and Optimization 2(3) (2012)
6. Franch, X.: Incorporating modules into the i* framework. In: Pernici, B. (ed.) CAiSE 2010. LNCS, vol. 6051, pp. 439–454. Springer, Heidelberg (2010)
7. Aguilar, J.A.: A Goal-oriented Approach for Managing Requirements in the Development of Web Applications. PhD thesis, University of Alicante, Spain (2011)
8. Maiden, N., Jones, S., Ncube, C., Lockerbie, J.: Using i* in requirements projects: some experiences and lessons. In: Social Modeling for Requirements Engineering, pp. 155–185 (2007)
9. Carvallo, J.P., Franch, X.: On the use of $i^*$ for architecting hybrid systems: A method and an evaluation report. In: Persson, A., Stirna, J. (eds.) PoEM 2009. LNBIP, vol. 39, pp. 38–53. Springer, Heidelberg (2009)
10. Alencar, F., Castro, J., Lucena, M., Santos, E., Silva, C., Araújo, J., Moreira, A.: Towards modular i* models. In: Proceedings of the 2010 ACM Symposium on Applied Computing, pp. 292–297. ACM (2010)
11. Estrada, H., Martínez, A., Pastor, O., Mylopoulos, J., Giorgini, P.: A service oriented approach for the i* framework. In: iStar 3rd International Workshop, p. 21 (2008)
12. Maté, A., Trujillo, J., Franch, X.: A modularization proposal for goal-oriented analysis of data warehouses using I-star. In: Jeusfeld, M., Delcambre, L., Ling, T.-W. (eds.) ER 2011. LNCS, vol. 6998, pp. 421–428. Springer, Heidelberg (2011)
13. Bolchini, D., Mylopoulos, J.: From task-oriented to goal-oriented web requirements analysis. In: Proceedings of the Fourth International Conference on Web Information Systems Engineering (WISE 2003), p. 166. IEEE Computer Society, Washington, DC (2003)
14. Srivastava, S., Chawla, S.: Goal oriented requirements engineering for web applications: A comparative study. International Journal on Recent Trends in Engineering & Technology 4(2) (2010)
15. Escalona, M.J., Koch, N.: Requirements engineering for web applications - a comparative study. J. Web Eng. 2(3), 193–212 (2004)

16. Garrigós, I., Mazón, J.-N., Trujillo, J.: A requirement analysis approach for using i* in web engineering. In: Gaedke, M., Grossniklaus, M., Díaz, O. (eds.) ICWE 2009. LNCS, vol. 5648, pp. 151–165. Springer, Heidelberg (2009)
17. EMF, http://www.eclipse.org/emf/
18. Eclipse (2012), http://www.eclipse.org/
19. Aguilar, J.A., Garrigós, I., Mazón, J.-N.: A goal-oriented approach for optimizing non-functional requirements in web applications. In: De Troyer, O., Bauzer Medeiros, C., Billen, R., Hallot, P., Simitsis, A., Van Mingroot, H. (eds.) ER Workshops 2011. LNCS, vol. 6999, pp. 14–23. Springer, Heidelberg (2011)
20. Aguilar, J.A.: A Goal-oriented Approach for the Development of Web Applications: Goal-oriented Requirements Engineering (GORE) and Model-Driven Architecture (MDA) in the Development of Web Applications. LAP LAMBERT Academic Publishing (2012)
21. Aguilar Calderon, J.A., Garrigós, I., Casteleyn, S., Mazón, J.-N.: Webred: A model-driven tool for web requirements specification and optimization. In: Brambilla, M., Tokuda, T., Tolksdorf, R. (eds.) ICWE 2012. LNCS, vol. 7387, pp. 452–455. Springer, Heidelberg (2012)

# Apply Wiki for Improving Intellectual Capital and Effectiveness of Project Management at Cideco Company

Sanjay Misra[1], Quoc Trung Pham[2], and Tra Nuong Tran[2]

[1] Department of Computer and Information Sciences, Covenant University, Nigeria
[2] School of Industrial Management, HCMC University of Technology, Vietnam
sanjay.misra@covenantuniversity.edu.ng, pqtrung@hcmut.edu.vn,
trantranuong@gmail.com

**Abstract.** Today, knowledge is considered the only source for creating the competitive advantages of modern organizations. However, managing intellectual capital is challenged, especially for SMEs in developing countries like Vietnam. In order to help SMEs to build KMS and to stimulate their intellectual capital, a suitable technical platform for collaboration is needed. Wiki is a cheap technology for improving both intellectual capital and effectiveness of project management. However, there is a lack of proof about real benefit of applying wiki in Vietnamese SMEs. Cideco Company, a Vietnamese SME in construction design & consulting industry, is finding a solution to manage its intellectual capital for improving the effectiveness of project management. In this research, wiki is applied and tested to check whether it can be a suitable technology for Cideco to stimulate its intellectual capital and to improve the effectiveness of project management activities. Besides, a demo wiki is also implemented for 2 pilot projects to evaluate its real benefit. Analysis results showed that wiki can help to increase both intellectual capital and effectiveness of project management at Cideco.

**Keywords:** Knowledge management, Wiki, Intellectual capital, Project management, Cideco Company.

## 1    Introduction

According to [14], in economic crisis, SMEs can only survive if they focus on shared resources, knowledge and innovation. Currently, there are 2 main problems for SMEs including cutting costs and managing intellectual capital effectively.

Cideo [3] is a Vietnamese SME in construction industry. Main business area of Cideco is consulting and designing service to other partners. In the context of economic crisis in Vietnam, Cideco is cutting its cost and facing difficulties in maintaining its human resource. Recently, Cideco applied new policy of cutting salary and working hours by 50%. This causes big problems to Cideco for holding intellectual capitals as well as maintaining the effectiveness of project management. There is a need for Cideco to find a solution for these problems.

Moreover, web 2.0 technologies opened opportunities for SMEs to apply new technology for overcoming their problems [19]. Wiki is one of these technologies,

B. Murgante et al. (Eds.): ICCSA 2014, Part V, LNCS 8583, pp. 146–158, 2014.

which could help to capture knowledge resources, to support collaboration and to improve the effectiveness of project management. However, benefits of applying wiki in a real SME in Vietnam is not realized and checked yet.

Therefore, this research tries to apply Wiki for improving intellectual capital and effectiveness of project management at Cideco company, and then, to evaluate the solution in practice. This research's objectives include: (1) Identify problems in managing intellectual capital at Cideco company, (2) Explore the ability to apply wiki as a solution for Cideco problems, (3) Select a suitable wiki and implement a demo wiki for Cideco, and (4) Conduct experiment for evaluating the demo wiki.

The structure of this paper is organized as follows: (2) Research method; (3) Literature review & research model; (4) Problems of project management at Cideco; (5) Select and Implement a demo wiki; (6) Experimentation and results; and finally, (7) Conclusion.

## 2    Research Method

### 2.1    Data Collection

- Secondary data: literature reviews, scientific journals, papers, related materials from the internet, internal documents of Cidecocompany.
- Primary data: expert interviews, questionnaires, discussing about possibility and feasibility of suggested solution in practice.

### 2.2    Research Process

- Qualitative analysis: projects' document analysis, group discussion with project members, depth interviews with project managers.
- Quantitative analysis: conducting survey for data collection,data analysis (descriptive statistics) for understanding problems at Cideco.
- Experimentation: AHP method (Analytical Hierarchy Process) with the support of Expert Choice software for selecting a suitable wiki, designing & implementing a demo wiki, testing demo wikifor 2 projects, evaluating the solutionbased on users' feedback.

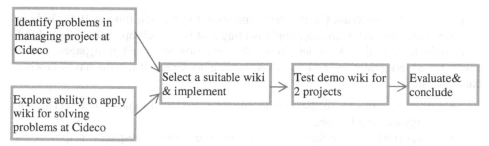

**Fig. 1.** Research process

# 3     Literature Review and Research Model

## 3.1     Knowledge and Knowledge Management Process

Knowledge is defined as "justified belief" [17]. Nonaka [17] also clarified that knowledge is a dynamic process of human beings for proving personal belief and truth.From the viewpoint of science of cognition, knowledge is related to data and information by 2 dimensions: understanding and context independence [25]. Besides, Polanyi [21] classified knowledge into 2 groups: (1) tacit knowledge, which is located in human brain and difficult to capture, and (2) explicit knowledge, which is easier to capture and to present in some form.

Knowledge management is a process of realizing, sharing, using and practicing knowledge inside of an organization [4]. For managing knowledge effectively, a knowledge management process should be established. Kimiz [8] combined previous KM cycles and introduced an integrated KM cycle, including 3 steps: (1) knowledge capture and creation, (2) knowledge sharing and dissemination, and (3) knowledge acquisition and application.

## 3.2     Intellectual Capital

According to Brooking [2], intellectual capital is a collection of intangible assets of an organization. Based on another research of Luthy [13], intellectual capital is anything useful to organization, including business processes, technologies, patents, employees' skills, information of customers, suppliers and stakeholders.

Today, intellectual capital is more and more realized an important resource for improving competitiveness and performance of an organization. For measuring intellectual capital, Skandia, a Sweden consulting company, using a measurement similar to Balanced score card, which includes 4 main dimensions: finance, customer, process and innovation.

According to Roos [23], intellectual capital includes: human capital, relational capital, and structural capital (business process, learning and innovation capability). In this research, measurement scale of Roos is used for measuring current intellectual capital at Cideco company.

## 3.3     Wiki

Wiki was created by Ward Cunningham and appeared the first time in 1995. Wiki is an open source project, therefore, implementing cost is very cheap. According to Leuf & Cunningham [10], wiki is an extensible collection of linked web pages, which could be edited easily by end-users. According to Grace [6], wiki has following characteristics:

- Easy to edit by end-users, who are not necessary to know about HTML or programming language.
- Hyperlinks and references to internal/ external web-pages help readers understanding the topic clearly.

- Version management supports multi-users collaboration by recording different versions and tracking "who change what".
- Supporting search function for finding contents inside of wiki.

According to Amalia [1], wiki helps to increase the readiness of project participation, to stimulate self-learning, and to improve business performance. Besides, according to Pham [19], web 2.0 technologies could be a suitable platform for managing knowledge successfully and improving labor productivity of Vietnamese SME.

Currently, most popular wiki in Vietnam is MediaWiki because it has a user graphic interface similar to Wikipedia, a famous wiki in the world. Other wikis that could be used inside of an organization are: TWiki, MediaWiki, MoinMoin, and PmWiki. In order to select a suitable wiki, some features should be considered, including: general information, system requirement, data storage, security, supported functions, web links, statistics, availability... WikiMatrix (http://www.wikimatrix.org/) may help to choose a suitable wiki platform for an organization.

### 3.4    Effectiveness of Project Management

According to Lock [12], traditional way of thinking about success of a project is based on 3 constraints: on time, within budget and high quality. In this research, effectiveness of a project is measured using measurement of Lei & Skitmore [9], which is based on feedback of project manager and recorded data of the project about: reduce time, reduce cost (training, redesigning, managing, outsourcing), and reduce problem occurred (internal, external).

### 3.5    Research Model

Based on above researches [1], [19] & [23], applying wiki as a collaboration platform could improve intellectual capital of a company; As a result, the effectiveness of project management could also be improved. Therefore, a research model about the impact of applying wiki on intellectual capital, and then, on project management is proposed, which could be summarized in the following figure.

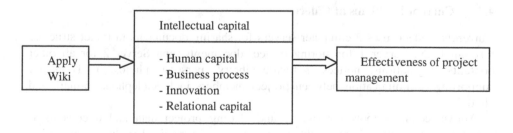

**Fig. 2.** Research model

According to this model, applying wiki could be a solution for solving problems of project management at Cideco. Implementing a demo wiki is necessary for evaluating its real effect on project management activities at Cideco. Experiments with demo wiki are very useful for generalizing this model at an enterprise-wide level.

## 4    Problems of Project Management at Cideco

### 4.1    Cideco Company and Designing Department

Cideco [3] is a Vietnamese Construction Consulting & Designing Company, which is located in BinhThanh District, Ho Chi Minh City. Main business services of Cideco are: Consulting and managing construction projects; Project planning and evaluating; Designing and constructing. Business scope of Cideco include: resident area, metropolitan region, industrial zone, construction projects in industry, agriculture, fishery, telecommunication, transportation, energy, water, and national infrastructure...

Designing department of Cideco has 5 groups (project creating, project planning, architecture, infrastructure, component), with about 50 employees, who are well-educated (greater than 90% got a bachelor degree or higher) and experienced in different areas.

Until now, Cideco has managed more than 400 construction projects of various scales in many areas in Vietnam. The strategy of Cideco is completing projects in a shortest time with high quality.

Construction industry is a knowledge intensive area, which requires efforts in managing knowledge resources and intellectual capital. Similarly, designing construction projects related to various stakeholders with different purposes. For example, customers often change their requirements, while investors would like to reduce the total cost. Inside of designing group, conflicts may occur because of different ideas or viewpoints of project members. Therefore, managing construction projects requires a shared space for sharing group works, communication and collaboration tools for connecting stakeholders and improving the effectiveness of project management.

### 4.2    Current Problems at Cideco

Currently, Cideco uses 2 computer servers for sharing documents in folder structure through LAN. Server 1 for storing project documents, and Server 2 for all other contents (regulations, notices, software library, personal photos...). Tools for supporting communication between project members includetelephone, e-mail, and chat.

Above technical tools for storing and sharing project data and documents is ineffective in practice. Some problems could be easily realized in discussing with a few project members, such as: (1) difficult for searching related documents, (2) high probability of duplicate and overwritten, (3) difficult in protecting project documents from unexpected readers.So, there is a need for a better technical platform for communication and collaboration between project members.

Moreover, in this research, interviews with employees and managers of Designing department are also conducted for understanding problems in managing knowledge resources and intellectual capital. Some main problems are realized and summarized as follows: (1) Storage architecture are complicated; (2) High cost for re-printing documents; (3) Waste of time for finding knowledge; (4) Ineffective in knowledge sharing; (5) Data storage consumes a lot of memory.

Besides, a survey using 5 points Likert measurement scale had been conducted for measuring intellectual capital and effectiveness of project management at Cideco. Questionnaire for the survey was built based on measurement scale of Roos [23] and Lei & Skitmore [9]. Respondents are 20 employees of Designing department. The results are summarized in following tables.

**Table 1.** Current status of intellectual capital at Cideco

| Intellectual capital | Mean | Stdev. |
|---|---|---|
| Knowledge captured | 1.3 | 0.47 |
| Knowledge sharing group | 1.5 | 0.76 |
| Network partner relationship | 1.7 | 0.92 |
| Number of experts | 1.8 | 0.77 |
| Co-operation | 2.2 | 0.95 |
| Information flow | 2.3 | 1.17 |
| Specialization | 2.4 | 0.94 |
| Customer relationship | 4.1 | 0.97 |
| **Average** | 2.1 | 0.87 |

**Table 2.** Effectiveness of project management at Cideco

| Effectiveness of PM | Mean | Stdev. |
|---|---|---|
| Reduce management cost | 2 | 0.79 |
| Reduce training cost | 2.1 | 0.72 |
| Reduce internal conflicting | 2.2 | 0.83 |
| Reduce cost for redesigning | 2.6 | 0.68 |
| Reduce cost for outsourcing | 2.9 | 0.85 |
| Reduce external problems | 3.5 | 1.00 |
| Finish project on time/ early | 3.9 | 0.72 |
| **Average** | 2.7 | 0.80 |

From above analysis, current status of intellectual capital management and effectiveness of project management at Cideco are at a low level (less than 3). Company should find solutions for effectively managing its intellectual capital (especially focus on knowledge captured, knowledge sharing group, and partner relationship), and improving the effectiveness of project management (especially focus on reducing management cost, training cost, and internal conflicting).

In summary, the most problem of Cideco at this time is a lack of suitable technical platform for managing intellectual capital and supporting project management activities. In this research, approach for solving above problems of Cideco is applying a wiki for capturing knowledge resources, facilitating knowledge sharing and supporting project management activities.

## 5     Select and Implement a Demo Wiki

### 5.1     Select a Suitable Wiki

In order to select a suitable wiki for testing, a list of most popular wikis will be compared based on some criteria. This list includes MediaWiki, MoinMoin, PmWiki and Twiki. AHP method (with the support of Expert Choice software) is used for deciding which wiki is the most suitable one. Criteria and weighting of each factor are set by managers of IT department and Designing department. Three main criteria include Specification (System requirement, Data storage, Security), Collaboration support (Group works, Development, Linkage, Media), and User Interface (Syntax, Utilities). Finally, MediaWiki is chosen as the most suitable platform for supporting Cideco in managing its intellectual capital and improving effectiveness of project management (See more in the appendix).

### 5.2     Implement a Demo Wiki

A demo wiki will be built based on MediaWiki, using Xampp [24]. Xampp integrated necessary tools for managing an intranet, such as: Apache web server, PHP, Perl programming language, MySQL DBMS, and phpMyAdmin tool for managing database and website. Main functions of demo wiki are: (1) Access information about project (roles, plan, progress, documents...), (2) Support learning & storing lessons in different expertise (expert profile, case base, forum, experience), and (3) Access general information (regulations, notices, events...).

To support 3 main functions, Cideco wiki contents must be classified into 3 groups, including (1) Homepage, (2) Knowledge portal, and (3) Events. The structure of demo wiki is summarized in following figure.

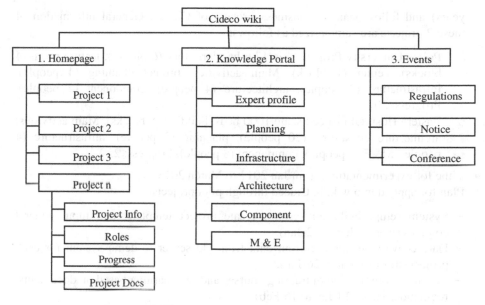

**Fig. 3.** Structure of Cideco wiki

A snapshot screen of Cideco wiki is shown in the following figure.

**Fig. 4.** A screen of Cideco wiki for Eastern Hospital project

## 6    Experimentation and Results

### 6.1    Experimentation Design

- The demo wiki will be applied for 2 projects (Police University, Eastern Hospital) for testing its effectiveness. These projects are long-term projects (more than 3 years) and follow standard business process of Cideco. General information of these 2 projects are summarized as follows:

o  Police University Project: area 17.6 ha, dormitory (6 blocks), lecture room (4 blocks), center (1 block). Main activities: project planning (4 people), infrastructure (4 people), architecture (4 people), component (2 people). Progress: 50%.

o  Eastern Hospital Project: scale 1000 beds, building (8 blocks). Main activities: architecture consulting (20 people), planning (3 people), infrastructure (4 people), M&E (4 people), component (8 people). Progress: 25%.

- Time for experimentation: from Jan 2013 to March 2013
- Plan for apply demo wiki in Cideco through pilot projects

  — System setup: build up system, create project template, and input general contents (from 7-Jan to 12-Jan).
  — Data conversion: input current data from old server to demo system for each project (from 14-Jan to 26-Jan).
  — Training: combination of training courses and self-learning based on documents/ regulations (from 14-Jan to 16-Feb).
  — Testing: using demo wiki for project works (18-Feb to 18-March)

- Evaluation: getting feedbacks from employees and managers of 2 pilot projects (interviews, questionnaires). Measuring 2 dimensions: intellectual capital and effectiveness of project management. Comparing these results with the old ones (before applying wiki).

## 6.2    Experimentation Results

After applying demo wiki for 2 pilot projects, a survey is conducted for collecting feedback data from project members for comparison. Some positive results in improving intellectual capital and effectiveness of project management at Cideco (before and after applying wiki) could be summarized in following table.

**Table 3.** Comparison results before and after applying wiki

| Intellectual capital | Before | | After | |
|---|---|---|---|---|
| | Mean | Stdev. | Mean | Stdev. |
| Knowledge captured | 1.30 | 0.47 | 2.85 | 1.04 |
| Knowledge sharing group | 1.50 | 0.76 | 2.55 | 0.89 |
| Number of experts | 1.80 | 0.77 | 2.65 | 0.75 |
| Information flow | 2.30 | 1.17 | 3.15 | 1.14 |
| Co-operation | 2.20 | 0.95 | 3.20 | 0.95 |
| Specialization | 2.40 | 0.94 | 2.50 | 0.83 |
| Customer relationship | 4.10 | 0.97 | 4.00 | 0.86 |
| Network partner relationship | 1.70 | 0.92 | 3.15 | 0.88 |
| Average: | 2.16 | 0.87 | 3.01 | 0.92 |

**Table 3.** (*continued*)

| Effectiveness of PM | Mean | Stdev. | Mean | Stdev. |
|---|---|---|---|---|
| Reduce management cost | 2.00 | 0.79 | 2.95 | 1.10 |
| Reduce training cost | 2.10 | 0.72 | 2.75 | 1.21 |
| Finish project on time/ early | 3.90 | 0.72 | 3.95 | 0.94 |
| Reduce cost for outsourcing | 2.90 | 0.85 | 2.95 | 0.94 |
| Reduce cost for redesigning | 2.60 | 0.68 | 3.25 | 0.91 |
| Reduce internal conflicting | 2.20 | 0.83 | 3.05 | 1.10 |
| Reduce external problems | 3.50 | 1.00 | 3.35 | 0.88 |
| Average: | 2.74 | 0.80 | 3.18 | 1.01 |

In summary, above numbers showed that wiki helps to increase both intellectual capital and effectiveness of project management (>3).

In comparison between 2 projects, effectiveness of project management of Police University project is better than of Eastern Hospital project. The number of data accessing and discussing through demo wiki of the 1st project is 3 times higher than the 2nd one. The main reason could be in the motivation of project members. The manager of the 1st project is realized more actively in encouraging his members. Therefore, beside technical issues, management policy for encouraging participation should be considered in applying this solution in practice.

Although there are some limitations of this result, such as: slightly decrease in customer relationship (due to technical change), small sample size (20 employees), subjective feedbacks (rating higher than reality), short testing time (3 months)... These limitations could be overcome easily when applying wiki in practice for a long time and doing more experiments. In general, testing results proved the possibility and feasibility of applying wiki for improving both intellectual capital and effectiveness of project management of Cideco in a long-term. It can also be considered to apply for the whole company for a better result.

# 7    Conclusion

Firstly, this research found some difficulties of project management of Design department at Cideco company through internal data analysis and interviews with projects' members. The most problem is found to be a lack of technical platform for managing intellectual capital. Quantitative results showed that intellectual capital and effectiveness of project management at Cideco currently are at a low level (<3).

Secondly, from previous researches and analysis results, wiki is found to be a suitable solution for solving current problems at Cideco Company because of its ability in capturing knowledge resources, sharing project works, and facilitating collaboration.

Based on user requirements and technical criteria, MediaWiki is selected as a suitable platform for building a demo wiki. This demo wiki is implemented and tested with 2 pilot projects for evaluating the real effects of suggested solution.

Experimentation results showed that demo wiki can help to improve intellectual capital and effectiveness of project management at Cideco company to a higher level (>3).

Although applying wiki in practice requires some other attentions, such as: training, creating motivation, and changing organizational culture... The pilot projects proved that wiki could be applied at enterprise-wide level for solving current problems in managing intellectual capital and project management at Cideco company.

However, this research also has some limitations, such as: small sample size, subjective feedback from respondents, short testing time...These limitations should be improved for a better result. Some directions for future researches include: (1) apply wiki at an enterprise-wide level of Cideco company, or test the demo wiki in various companies; (2) change the measurement scale for a more accurate feedback; (3) collect more data for testing the cause and effect relationship between wiki use, intellectual capital, and effectiveness of project management.

**Acknowledgements.** Many thanks to employees and managers of Cideco, who provided data for analysis or participated in several interviews of this research.

# References

1. Amalia, M.: Knowledge management: a case of a multinational company subsidiary in Indonesia. School of Environment and Development (2009)
2. Brooking, A.: Intellectual capital. International Thompson Business Press, UK (1996)
3. Cideco: Company profiles & business processes. Internal Company Publishing (2013)
4. Choi, B., Lee, H.: Knowledge management strategy and its link to knowledge creation process. Expert Systems with Applications 23, 173–187 (2002)
5. Dougherty, D.: The open source web platform, http://onlamp.com/pub/a/onlamp/2001/01/25/lamp.html (retrieve on October 26, 2012)
6. Grace, T.P.L.: Wiki as a knowledge management tool. Journal of Knowledge Management 13(4), 64–74 (2009)
7. Kasvi, J.J.J., Vartiaine, M., Hailikari, M.: Managing knowledge and knowledge competences in projects and project organizations. International Journal of Project Management (2002)
8. Kimiz, D.: Knowledge management in theory and practice. Elsevier Inc. (2005)
9. Lei, W.W.S., Skitmore, M.: Project management competencies: a survey of perspectives from project managers in South East Queensland. Journal of Building and Construction Management (2004)
10. Leuf, B., Cunningham, W.: The wiki way: Collaboration and sharing on the Internet. Addison-Wesley Professional (2001)
11. Levy, M.: Web 2.0 - implications on knowledge management. Journal of Knowledge Management (2007)
12. Lock, D.: Project Management, 9th edn. Gower Publishing (2007)
13. Luthy, D.H.: Intellectual capital and its measurement. Citeseer (1998)

14. Mladen, O., Mladen, V.: Using wiki as knowledge management tools for knowledge sharing and innovation in the times of economic crisis (2012), http://www.symorg.fon.bg.ac.rs/ (retrieve on October 13, 2012)
15. Nguyen, T.H.: Management Information System. VNU-HCM (2010)
16. Nonaka, I., Takeuchi, H.: The knowledge-creating company. Oxford University Press, New York (1995)
17. Pardi, P.: What is knowledge (September 22, 2011), http://www.philosophynews.com (retrieve on October 22, 2011)
18. Pham, Q.T.: Apply KM and SNS for Improving Labor Productivity of Vietnamese SME. Covenant Journal of Informatics and Communication Technology 1(1) (2013)
19. Pham, Q.T., Hara, Y.: KM approach for improving the labor productivity of Vietnamese enterprises. International Journal of Knowledge Management 7(3) (2011)
20. Polanyi, M.: The tacit dimension. Peter Smith, Gloucester (1966)
21. Raman, M., Ryan, T., Olfman, L.: Designing knowledge management systems for teaching and learning with wiki technology. Journal of Information (2005)
22. Roos, G., Roos, J.: Measuring your company's intellectual performance. Long Range Planning, Special Issue on Intellectual Capital. Elsevier Inc. (1997)
23. Seidler, K.O.: Xampp Tool (March 1, 2013), http://www.apachefriends.org/en/xampp.html (retrieve on March 18, 2013)
24. Serban, A.M., Luan, J.: Overview of knowledge management. New Direction for Institutional Research 113, 5–16 (2002)
25. Wagner, C.: Wiki-a technology for conversational knowledge management and group collaboration. Communications of the Association for information system (2004)

## Appendix

Analytic hierarchy for selecting suitable wiki for Cideco is shown in following figure (AHP method – Expert Choice Software)

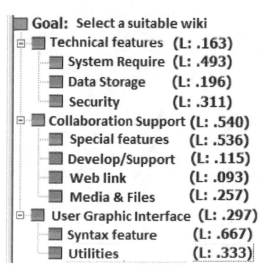

Evaluating result for selecting wiki is summarized in following table.

| Solutions | Overall score | System requirement (L: .493) | Data Storage (L: .196) | Security (L: .311) | Special features (L: .536) | Developing/Supporting (L: .115) | Web link (L: .093) | Media & Files (L: .257) | Syntax feature (L: .667) | Utilities (L: .333) |
|---|---|---|---|---|---|---|---|---|---|---|
| MediaWiki | 0.35 | 1.00 | 0.78 | 1.00 | 0.78 | 1.00 | 0.32 | 0.46 | 1.00 | 1.00 |
| TWiki | 0.29 | 1.00 | 0.46 | 0.81 | 1.00 | 0.47 | 1.00 | 0.37 | 0.52 | 0.31 |
| MoinMoin | 0.23 | 1.00 | 1.00 | 0.51 | 0.31 | 0.21 | 0.77 | 1.00 | 0.40 | 0.58 |
| PmWiki | 0.13 | 1.00 | 0.56 | 0.58 | 0.23 | 0.18 | 0.48 | 0.14 | 0.25 | 0.31 |

# Acceptance and Use of E-Learning Based on Cloud Computing: The Role of Consumer Innovativeness

Thanh D. Nguyen[2], Tuan M. Nguyen[2], Quoc-Trung Pham[2], and Sanjay Misra[1]

[1] Covenant University, Nigeria
[2] HCMC University of Technology, VNU-HCM, Vietnam
`sanjay.misra@covenantuniversity.edu.ng`
`thanh.nguyenduy@gmail.com`, `{n.m.tuan,pqtrung}@hcmut.edu.vn`

**Abstract.** Cloud computing and E-learning are the inevitable trend of computational science in general, and information systems and technologies in specific. However, there are not many studies on the adoption of cloud-based E-learning systems. Moreover, while there are many papers on information system adoption as well as customer innovativeness, the innovativeness and adoption in the same model seems to be rare in the literature. The study combines the extended Unified Theory of Acceptance and Use of Technology (UTAUT2) and consumer innovativeness on the adoption of E-learning systems based on cloud computing. A survey was conducted among 282 cloud-based E-learning participants and analyzed by structural equation modeling (SEM). The findings indicate that the adoption of cloud-based E-learning is influenced by performance expectancy, social influence, hedonic motivation, and habit. Interestingly, although innovativeness is not significant to use intention, it has a positive effect on E-learning usage which is relatively new in Vietnam.

**Keywords:** Adoption, cloud computing, consumer innovativeness, E-learning, information system, Unified Theory of Acceptance and Use of Technology.

## 1 Introduction

In contemporary society, the learning process is becoming a vital factor in business and socioeconomic growth [32]. The first E-learning courses were launched in 1998. Since then the E-learning business has gone global and the competition is fierce. Now, 70% of E-learning courses take place in the United State and Europe, but Asia Pacific is catching up fast, with Vietnam and Malaysia growing the fastest [12]. Vietnam is ranked first within the top ten countries in the world in terms of high-growth in E-learning revenues over the next few years (2011-2016), the Vietnam projected growth rate in E-learning of 44.3% [8]. A large expansion of online higher education possibilities and a growing demand for E-learning in the corporate sector will drive the educational growth. Recently, cloud computing has changed the nature of the Internet from the static environment to a highly dynamic environment, which allows users to run software applications collaborate, share information, create application

B. Murgante et al. (Eds.): ICCSA 2014, Part V, LNCS 8583, pp. 159–174, 2014.

virtual, learn online. Ercan [23]; Masud and Huang [40] showed that cloud computing is one of the new technology trends likely to have a significant impact on the teaching and learning environment. In specific, for example, in moving its E-learning system to a cloud computing platform, Marconi University (Italy) has achieved cost savings and financial flexibility, with a cost reduction of 23% per year [62].

Consumer innovativeness is understood as the trend to willingly include change and try new things, and buy new products more often and more quickly than others [7]. Moreover, most innovativeness studies focus on novelty pick up as the reason for consumers to seek new products [26], it is commonly forgotten that new products also encompass risk which enhances resistance to adoption. E-learning, cloud computing, and, hence, even cloud-based E-learning are no exceptions. While there are a lot of studies of consumer adoption in an online context in general (e.g. Venkatesh et al. [61]), the role of consumer innovativeness in such consumer adoption is rather limited. This study contributes to the literature by the inclusion of innovativeness into Venkatesh et al. [61] model of acceptance and use of information technology.

The purpose of this study is to explore the relevant concepts of E-learning, cloud computing, and to indicate the cloud-based E-learning benefits. Besides that, based on the reviews of literature, customer innovativeness theories, and Unified Theory of Acceptance and Use of Technology (UTAUT) [60, 61], propose the role of consumer innovativeness in E-learning based on cloud computing. In addition, the theoretical model and all hypotheses will be tested. Therefore, the rest of the paper is structured as follows: In the next section 2, the background shows the definition of E-learning and also E-learning 2.0, cloud computing synopsis, and some benefits of cloud-based E-learning. Section 3 includes research model reviews of literature and theoretical framework including hypotheses. Section 4 presents research methods, data, and these analysis results with exploratory factor analysis, reliability analysis (cronbach alpha), confirmatory factor analysis, structural model, analysis of variance and results discussion. Finally conclusions are in section 5.

# 2    Background

## 2.1    E-Learning

E-learning is one of the state-of-the-art educational technologies to facilitate the learning and learners, with the help of software applications and virtual learning environment. There are various names under the umbrella 'E-learning' such as Computer Based Training (CBT), Internet Based Training (IBT), and Web Based Training (WBT). E-learning in generic, with the support of one or more networked computers, helps to transfer the digitized knowledge from the online sources to end user devices such as desktop, laptop, handheld devices [64]. E-learning employs various types of media that deliver text, audio, images, animation, streaming video to E-learners [54].

### E-learning 2.0

E-learning 2.0 is a type of computer-supported collaborative learning system that developed with the emergence of web 2.0 [22, 33]. From an E-learning 2.0

perspective, conventional E-learning systems were based on instructional packets which were delivered to students using assignments. The assignments were evaluated by the teacher. In contrast, the new E-learning places increased emphasis on social learning and social network such as blog, wiki, podcast, virtual network [12, 48].

## 2.2    Cloud Computing

Cloud computing is one of the popular word-tech used all over the information technology world. The term cloud computing is actually derived from the way the Internet puts itself onto network diagrams [46]. Cloud computing is such a type of computing where users do not have to spend any money to build and maintain information technology infrastructure. When users need to use computing resources like application software, users just borrow that facility from a third party organization and access that service via Internet. In return, users pay the service provider as users use the computing power. In short, in the cloud environment, users do not need to buy any hardware and software to run their business applications, thus it helps users minimize their investment on hardware resources and information technology maintenance team [63]. Many companies such as Microsoft, IBM, HP, Dell, VMware are investing on virtualization platforms. They were not only investing to bring easier access of their applications to their customers, but also building their power of next generation cloud technology [47].

Based on the different virtual levels, cloud computing is typically divided into three types according to the packaging of computing resources in different abstraction layers, (1) Infrastructure as a Service (IaaS): refers to taking the servers, storage systems, network communications equipment and other computing resources as a standardized service via the network. (2) Platform as a Service (PaaS): known as the middle of clouds service, provides a mapping that contains system distribution, web server and programming environment, can be used for various stages of software development, testing, and deployment. (3) Software as a Service (SaaS): is the highest-level of cloud computing applications, and also the layer end-user-oriented. It usually refers to the development of software examples and application processes based on the specific infrastructure [52].

## 2.3    Cloud-Based E-Learning

Cloud-based E-learning (CBEL) is the subdivision of cloud computing on educational field for E-learning systems; it is the future for E-learning technology and infrastructure. CBEL has all the supplies such as hardware and software resources to enhance the traditional E-learning infrastructure. Once the educational materials for E-learning systems are virtualized in cloud servers, these materials are ready for use to students and other educational businesses in the form of rent base from cloud vendors [63]. According to Laisheng et al. [35], CBEL architecture is mainly divided into 5 layers, namely hardware resource layer, software resource layer, resource management layer, server layer, and business application layer. Pocatilu et al. [46]; Bhruthari et al. [10]; Jain and Chawla [31] showed that E-learning systems can benefit by cloud computing, (1) Infrastructure: use an E-learning solution on the provider infrastructure. (2) Platform: use and develop an E-learning solution based on the provider development interface. (3) Services: use the E-learning solution given by the provider. In details, according to Zheng and Jingxia [68], CBEL services can be divided into four types as described in Table 1.

**Table 1.** Types of content and cloud computing services

|   | Content | Cloud |
|---|---------|-------|
| 1 | Standard data, audio, video, data, images, text... | *IaaS* |
| 2 | Data can be converted into standard data content. | *SaaS* |
| 3 | Web-based proprietary data, player embedded in web pages... | *SaaS* |
| 4 | Private defined data, player needs to download manually... | *PaaS* |

Source: Zheng and Jingxia [68].

## Cloud-Based E-Learning Benefits

- Anytime and anywhere access: E-learners can access data from anywhere if Internet access is available, as it is stored in the cloud not at any memory chip. This means the user has no need to follow as data follows the user [13].
- Lower costs: E-learners need not have high end configured computers to run the E-learning applications, they can run the applications from the cloud through their computer, mobile phones, laptop... having minimum configuration with Internet connectivity. Since the data is created and accessed in the cloud, the user has no need to spend more money for large memory for data storage in local machines. Organizations also need to pay per use, it is cheaper and need to pay only for the space they need [6].
- Improved performance: CBEL applications have most of the applications and processes in the cloud, client machines do not create problems on performance when they are working [63].
- Instant software updates: CBEL application runs with the cloud computing, the softwares are automatically updated in the cloud. So always E-learners get updates instantly [63].
- Improved document format compatibility: some file formats and fonts do not open properly in some computers or mobile phones, it does not have to worry about those problems. As the CBEL applications open the file from cloud [63].
- Student benefits: students get more advantages through CBEL, they can take online courses, attend the online exams, get feedback about the courses from instructors, and send their projects and assignments through online to their teachers [46].
- Teacher benefits: E-teachers get numerous benefits over traditional teachers, they are able to prepare online tests for students, deal and create better content resources for students through content management, assess the tests, homework, projects taken by students, send the feedback, and communicate with students through online forums [63].

Generally, cloud computing and E-learning are the inevitable trend of information systems and technologies. CBEL is the subdivision of cloud computing on educational field for E-learning systems. There are a lot of benefits of CBEL, that is not only for E-learners, E-teachers or educational organization, and also for service providers. However, there are not many studies on the adoption of CBEL systems. Moreover, while there are many papers on information system adoption as well as customer innovativeness, the innovativeness and adoption in the same model seems to be rare in the literature.

## 3 Research Model

### 3.1 Literature Review

New technology adoption has been examined extensively in information systems research. Theory of Reasoned Action (TRA) was researched in psychosocial perspective in order to identify elements of the trend-conscious behavior [5, 24]. Theory of Planned Behavior (TPB) was constructed by Ajzen [1, 2, 3] from the original TRA theory and added perceived behavioral control element. TPB endorses the researcher to study the influence that consumer innovativeness has on their sensibility with respect to social effects when deciding to use online system [17]. Technology Acceptance Model (TAM) based on the theoretical foundation of the TRA to establish relationships between variables to explain human behavior regarding acceptance of information systems [18, 19]. The most extended of TAM, namely TAM2 [58], TAM3 [59], can be best understood by exploring the determinants to perceived usefulness and perceived ease of use. Innovation Diffusion Theory (IDT) explained the process of technological innovation that is accepted by users [49]. The research on innovation adoption had directed its attention towards understanding whether there might be an orientation to adopt innovations [39]. Goldsmith and Hofacker [26] had extensively investigated the psychological construct of innovativeness that has been defined as the extent to which the consumers adopt innovations.

Unified Theory of Acceptance and Use of Technology (UTAUT) had been built by Venkatesh et al. [60] to explain intention and use behavior of information system users. UTAUT model was developed through theoretical models as TRA [5, 24], TPB [1, 2, 3]; TAM [18, 19], integrated mode of TPB and TAM [55], IDT [43], Motivation Model (MM) [20], Model of PC Utilization (MPCU) [56] and Social Cognitive Theory (SCT) [16, 29]. UTAUT was formulated with 4 core constructs of intention and use as performance expectancy, effort expectancy, social influence and facilitating condition. Venkatesh et al. [61] next adopted an approach that complements the original constructs in UTAUT by inclusion of the factors such as hedonic motivation, price value and habit to propose a theoretical extension which is called UTAUT2. Then, demographic variables such as age, gender and experience, which are part of the original UTAUT, are also included in UTAUT2.

Whereas there are many works about E-learning based on cloud computing (e.g. Zaharescu [66]; Bhruthari et al. [10]; Masud and Huang [40]; Viswanath et al. [63]; Zheng and Jingxia [68]; Jain and Chawla [31]), also about E-learning acceptance and use (e.g. Sun et al. [53]; Will and Allan [65]; Soud and Fisal [51]; Lin et al. [38]...), little is known on the adoption model in the context of the cloud computing (except, e.g. Nguyen et al. [57]). In the meanwhile, although innovativeness has investigated many in, for example, Goldsmith and Foxall [25] and Goldsmith and Hofacker [26]; Citrin et al. [15]; Im et al. [30]; Crespo and Del Bosque [17]; Marcati et al. [39]; Aldas-Manzano et al. [7], its influence on consumer innovativeness adoption in a cybernetic context seems to be unclear. In short, most of the relevant studies have not shown the relationships between consumer innovativeness and CBEL Intention, and also CBEL usage.

## 3.2    Theoretical Framework

The adoption model of CBEL depends on technical elements of information systems or technologies, and characteristics of demographic that introduces the technology and the response of individuals. The theoretical underpinnings of the TRA [5, 24]; TBP [1, 2, 3] are widely used alongside TAM [18, 19] (or TAM2 [58], TAM3 [59]) in the researches on information systems. The initial UTAUT [60] and UTAUT2 [61] refinements have garnered extensive empirical support and provide a robust framework that is well aligned with the adoption of CBEL context was studied (e.g. Nguyen et al. [57]). Specially, in this study, the role of consumer innovativeness is considered in the acceptance and use of E-learning based on cloud computing.

Based on the literature review of E-learning, cloud computing and CBEL; customer innovativeness theories [25, 26]; and Unified Theory of Acceptance and Use of Technology (UTAUT) [60, 61], the model of innovativeness and adoption of E-learning based on cloud computing is built with the following are theoretically supported hypotheses that explore relationships in the model.

*Performance Expectancy (PE)* means that an individual believes that using the system will help them to attain gains in job performance. The five constructs from the different models that pertain to performance expectancy are perceived usefulness in TAM [18, 19], extrinsic motivation in MM [20], job-fit in MPCU [56], relative advantage in IDT [43, 49], and outcome expectations in SCT [16, 29]. The learner believed that the E-learning system was helpful to their performance and the individual learner would be more satisfied with the E-learning [65]. Thus, it hypothesizes that:

*Hypothesis H1a: PE has a positive effect on CBEL intention (CEI).*

*Effort Expectancy (EE)* illustrates that the degree of ease associated with the use of the system. Three constructs from the existing models capture the concept of effort expectancy as perceived ease of use in TAM [18, 19], complexity in MPCU [56], and ease of use in IDT [43, 49]. The effort expectancy of an E-L system would influence users in their deciding whether or not to use the system [65]. Thus, it hypothesizes that:

*Hypothesis H1b: EE has a positive effect on CEI.*

*Social Influence (SI)* is defined as the degree to which an individual perceives that important others believe people should use the new system. Social influence as a direct determinant of behavioral intention is represented as the subjective norm in TAM

[18, 19], social elements in MPCU [56], and image in IDT [43, 49]. According to Venkatesh et al. [60], the role of social influence in technology acceptance decisions is complex and subject to a wide range of contingent influences. Thus, under CBEL, it hypothesizes that:

*Hypothesis H1c: SI has a positive effect on CEI.*

*Facilitating Condition (FC)* is the degree to which a person believes that an organizational and technical infrastructure exists to support the use of the information system. This definition captures concepts embodied by three different constructs on perceived behavioral control in TAM [18, 19], facilitating condition in MPCU [56], and compatibility in IDT [43, 49]. Venkatesh [57] found support for full mediation of the effect of facilitating condition on intention and use. Thus, under CBEL, it hypothesizes that:

*Hypothesis H1d: FC has a positive effect on CEI.*
*Hypothesis H4: FC has a positive effect on CBEL usage (CEU).*

*Price Value (PV)* is defined as a consumer cognitive tradeoff between the perceived benefits of the applications and the monetary cost of using them [21]. The monetary cost and price is usually conceptualized together with the quality of products or services to determine the perceived value of products or services [67]. According to Venkatesh et al. [61], the price value is positive when the benefits of using a technology are perceived to be greater than the monetary cost, and such price value has a positive impact on intention. Thus, under CBEL, it hypothesizes that:

*Hypothesis H1f: PV has a positive effect on CEI.*

*Hedonic Motivation (HM)* has been the fun or pleasure derived from using a technology, and it has been shown to play an important role in determining technology acceptance and use [11]. In information system research, such hedonic motivation has been found to influence the technology acceptance and use directly [28]. According to Childers et al. [14]; Brown and Venkatesh [11], in the consumer context, hedonic motivation has also been found to be an important determinant of technology acceptance and use. Thus, under CBEL, it hypothesizes that:

*Hypothesis H1g: HM has a positive effect on CEI.*

*Habit (HA)* has been defined as the extent to which people tend to perform behaviors automatically because of learning and equate habit with automaticity [34, 37]. Ajzen and Fishbein [4] noted that feedback from previous experiences influence various beliefs and consequently, future behavioral performance. According to Venkatesh et al. [61], the role of habit in technology use has delineated different underlying processes by which habit influences technology use. Thus, under CBEL, it hypothesizes that:

*Hypothesis H1g: Habit has a positive effect on CEI.*
*Hypothesis H5: Habit has a positive effect on CEU.*

*Innovativeness (IN)* has received in depth empirical attention within the diffusion of innovation framework [49]. There are many studies that have used different techniques to define or to measure consumer innovativeness. The two main types of innovativeness have arisen, called general innovativeness (GI) and domain-specific innovativeness (SI) [30, 42]: GI to follow to the openness and creativity of individuals, to their readiness to follow new ways [41], and SI relates to the predisposition to be among the firsts to adopt innovations in a specific domain [26]. According to Citrin et al. [15], increases in innovativeness result in increases in consumer adoption of the online system. Thus, under CBEL, it hypothesizes that:

*Hypothesis H2: Innovativeness has a positive effect on CEI.*
*Hypothesis H3: Innovativeness has a positive effect on CEU.*

*CBEL Intention (CEI)*, consistent with the underlying theory for all of the intention models are reviewed in studies such as Sheppard et al. [50]; Venkatesh et al. [60, 61] for literature review of the intention-behavior relationship, so that behavioral intention has a significant positive influence on technology use. Thus, under CBEL, it hypothesizes that:

*Hypothesis H6: CEI has a positive effect on CEU.*

*Demographic (DE)*, including age, gender, and experience suggested as part of UTAUT2 [61], were included in the analysis. One more characteristic, education, is also added into the research model. Thus, it hypothesizes that:

*Hypothesis H7: Independent and dependent elements are influenced by DE.*

# 4     Research Results

## 4.1     Data Collection and Descriptive Statistics

In order to test the model and all hypotheses which were proposed, information was collected using a structured survey with a set of all scales referring to the different variables identified in the model (see Table 2). According to the literature review, customer innovativeness theories, and the extended Unified Theory of Acceptance and Use of Technology (UTAUT2), data was collected by a survey using convenient sampling. The questionnaires were delivered using Google Docs, E-mail, E-learning forums, and hard copies to respondents who have used or intend to use cloud-based E-learning in Vietnam. A total of 320 respondents was obtained, 282 was finally usable (38 invalid respondents). All scales were scored on a 5-point Likert scale anchored with strongly disagree (1) to strongly agree (5), with 29 indicators. The data were then analyzed by Structural Equation Modeling (SEM) techniques with the application of SPSS and AMOS.

The descriptive statistics are conducted for indicators relating to the users who have used cloud-based E-learning. *Gender*: there are approximately 64% male and 36% female, it is uneven. *Age*: as regards the 19 - 23 age group, 24 - 30 group, and older-30 group, the former is by far the highest at nearly 50%, followed by the latter at 27% and 21% respectively. *Education*: there are nearly 70% of E-learners in university degree, about 24% of E-learners in graduate degree and percentage of the other is lower. *Experience*: although roughly 60% of the people who are good at computing, only 1% people are bad at computing, 39% average experience in computer using. Thus, most of people have experienced in computing. *Cloud computing*: similarities exist between Google Drive and Mediafire, where about 32% respondents use CBEL. 20% use Dropbox, 13% use Sky Drive...

## 4.2    Exploratory and Confirmatory Factor Analysis

Firstly, after eliminating two items (*IN2* and *FC4*) of innovativeness (*IN*) and facilitating condition elements (*FC*) in reliability analysis (Cronbach alpha), because the value of correlation-item of *IN* and *FC* factors < 0.60, the Cronbach alpha of constructs ranges from 0.685 to 0.849. Secondly, eliminating three items (*IN6*, *FC3*, and *PV1*) of innovativeness (*IN*), facilitating condition (*FC*), and price value (*PV*) elements in the 1st Exploratory Factor Analysis (EFA) are due to these factor loading < 0.50. Then, the 2nd EFA, the EFA factor loading of all items range from 0.598 to 0.987. Finally, CFA are conducted to assess and refine the measurement scales. The results of EFA and CFA are shown in Table 2.

**Table 2.** All variables of the model in factor analysis

| Observed variables | | | Factor loading | |
|---|---|---|---|---|
| | | | EFA | CFA |
| | | Cronbach alpha = 0.830;  AVE = 0.675 | | |
| | PE3 | CBEL useful in job | 0.837 | 0.987 |
| PE | PE2 | Using CBEL enables to accomplish tasks quickly | 0.823 | 0.972 |
| | PE1 | Using CBEL increases productivity | 0.795 | 0.635 |
| | PE4 | Increase chances of getting a raise | 0.781 | 0.594 |
| | | Cronbach alpha = 0.784;  AVE = 0.629 | | |
| | EE3 | Learning how to use CBEL is easy | 0.840 | 0.903 |
| EE | EE2 | Interaction with CBEL is clear and understandable | 0.786 | 0.784 |
| | EE4 | Finding CBEL easy to use | 0.775 | 0.753 |
| | EE1 | It is easy to become skillful at using CBE-L | 0.772 | 0.654 |
| | | Cronbach alpha = 0.740;  AVE = 0.542 | | |
| SI | SI1 | People are important to think that should use CBEL | 0.795 | 0.759 |
| | SI2 | People influence behavior think that should use CBEL | 0.793 | 0.695 |
| | SI3 | People whose opinions that value prefer use CBEL | 0.650 | 0.601 |
| | | Cronbach alpha = 0.685;  AVE = 0.613 | | |
| FC | FC1 | The resources necessary to use CBEL | 0.913 | 0.896 |
| | FC3 | Knowledge necessary to use CBEL | 0.724 | 0.715 |
| | | Cronbach alpha = 0.784;  AVE = 0.620 | | |
| PV | PV3 | CBEL is a good value for the money | 0.857 | 0.721 |
| | PV2 | At the current price, CBEL provides a good value | 0.849 | 0.668 |

**Table 2.** (*continued*)

| | | | | |
|---|---|---|---|---|
| | | Cronbach alpha = 0.807; AVE = 0.582 | | |
| HM | HM1 | Using CBEL is fun | 0.826 | 0.785 |
| | HM3 | Using CBEL is enjoyable | 0.766 | 0.783 |
| | HM2 | Using CBEL is entertaining | 0.732 | 0.717 |
| | | Cronbach alpha = 0.804; AVE = 0.579 | | |
| HA | HA2 | Using CBEL has become a habit | 0.826 | 0.909 |
| | HA3 | Addicted to use CBEL | 0.766 | 0.752 |
| | HA1 | Must use CBEL | 0.732 | 0.598 |
| | | Cronbach alpha = 0.849; AVE = 0.586 | | |
| IN | IN1 | Very cautious in trying CBEL | 0.837 | 0.853 |
| | IN5 | Know CBEL before other people do | 0.787 | 0.756 |
| | IN4 | The last one in circle tries to use CBEL | 0.778 | 0.745 |
| | IN3 | Own few CBEL courses | 0.766 | 0.668 |
| | | Cronbach alpha = 0.822; AVE = 0.589 | | |
| CEI | CEI2 | Intend to use CBEL in the future | 0.862 | 0.801 |
| | CEI3 | Will try to use CBEL in daily life | 0.858 | 0.750 |
| | CEI1 | Will plan to use CBEL frequently | 0.857 | 0.743 |
| | | Cronbach alpha = 0.805; AVE = 0.664 | | |
| CEU | CEU1 | How long have you been using CBEL | 0.919 | 0.815 |
| | CEU2 | Every month, how many times have you used CBEL | 0.902 | 0.814 |

AVE: Average Variance Extracted.

The results of the CFA on the overall measurement model yields the following measures: Chi-square $(\chi^2)$/dF = 1.986; p = 0.000; TLI = 0.906; CFI = 0.924; RMSEA = 0.058. The CFA factor loadings of all items range from 0.598 to 0.987. The Average Variance Extracted (AVE) of constructs range from 0.542 to 0.675 (> 0.50) which are good scales (see Table 2). Therefore, the measurement scales for all constructs are satisfactory.

### 4.3   Structural Model

The estimation of structural model was then conducted using ML (Maximum Likelihood) estimation. The indexes for the model showed adequate fit with Chi-square $(\chi^2)$/dF = 1.612; p = 0.000; TLI = 0.952; CFI = 0.962; RMSEA = 0.046. The standardized path coefficients presented in Table 3: Support the positive effect of *PE* on *CEI* with $\gamma = 0.101$ (p = 0.037), that supports H1a. *SI* and *HM* have strongly positive effect on *CEI* with $\gamma = 0.204$ (p = 0.028) and 0.523 (p < 0.001), which in turn H1c and H1f are supported. Support the positive effect of *HA* on *CEI* and *CEU* with $\gamma = 0.189$ (p < 0.001) and 0.079 (p = 0.048), which support H1g and H5. *IN* has a positive effect on

*CEU* with γ = 0.137 (p = 0.023), that supports H3, However, neither the path from *EE, FC,* and *PV* to *CEI,* nor from *FC* and *IN* to *CEU* are non-significant at p = 0.05. Therefore, H1b, H1d, H1e, H2, and H4 are rejected. And the results strongly support H6 by showing an affecting of *CEI* on *CEU* with γ = 0.840 (p < 0.001).

**Table 3.** Analysis results of relationships

|  | H | Relationships | | | Estimate | SE | p-value | Result |
|---|---|---|---|---|---|---|---|---|
| 01 | H1a | PE | → | CEI | 0.101 | 0.045 | 0.037 | *Supported* |
| 02 | H1b | EE | → | CEI | 0.011 | 0.091 | 0.331 | *Rejected* |
| 03 | H1c | SI | → | CEI | 0.204 | 0.083 | 0.028 | *Supported* |
| 04 | H1d | FC | → | CEI | 0.023 | 0.096 | 0.172 | *Rejected* |
| 05 | H1e | PV | → | CEI | 0.043 | 0.238 | 0.589 | *Rejected* |
| 06 | H1f | HM | → | CEI | 0.523 | 0.080 | *** | *Supported* |
| 07 | H1g | HA | → | CEI | 0.189 | 0.050 | *** | *Supported* |
| 08 | H2 | IN | → | CEI | 0.010 | 0.069 | 0.878 | *Rejected* |
| 09 | H3 | IN | → | CEU | 0.137 | 0.060 | 0.023 | *Supported* |
| 10 | H4 | FC | → | CEU | 0.061 | 0.098 | 0.120 | *Rejected* |
| 11 | H5 | HA | → | CEU | 0.079 | 0.039 | 0.048 | *Supported* |
| 12 | H6 | CEI | → | CEU | 0.840 | 0.081 | *** | *Supported* |

SE: Standard Error; *** p < 0.001.

ANOVA analysis is carried out to analyze if there are any differences in the relationship between *PE, EE, SI, FC, PV, HM, HA, IN, CEI,* and *CEU* can be attributed to the demographic variables, namely age, gender, education and experience. The results show that the relationships between independent and dependent variables differ by age (5 factors: *PE, FC, SI, HM,* and *HA*), gender (7 factors: *PE, EE, SI, FC, HA, IN,* and *CEI*), education (5 factors: *PE, EE, SI, HM,* and *HA*), and experience (4 factors: *EE, FC, HM,* and *IN*) are significant with p < 0.05. The results of ANOVA analysis are shown in Table 4. Although there are not differences in *PV* and *CEU* with demographic variables, most of the variables differ. Thus, H7 is supported. Generally, eight out of thirteen hypotheses are supported in this research.

**Table 4.** ANOVA analysis follow age, gender, education and experience

| Demographic | PE | EE | SI | FC | PV | HM | HA | IN | CEI | CEU | Note |
|---|---|---|---|---|---|---|---|---|---|---|---|
| Age | x** | – | x* | x** | – | x* | x* | – | – | – | *5 factors* |
| Gender | x** | x*** | x* | x* | – | – | x* | x*** | x* | – | *7 factors* |
| Education | x* | x* | x** | – | – | x** | x* | – | – | – | *5 factors* |
| Experience | – | x* | – | x*** | – | x* | – | x* | – | – | *4 factors* |

x: particular differences; * p < 0.05; ** p < 0.01; *** p < 0.001.

The results show that when innovativeness is included, performance expectancy, social influence, hedonic motivation, and habit are able to explain both cloud-based E-learning intention about 62% ($R^2$ = 0.621) and cloud-based E-learning use nearly 79% ($R^2$ = 0.785). The findings are also comparable to the baseline model of UTAUT [60] and of UTAUT2 [61] that explained roughly 56% and 40% (UTAUT); 74% and 52% (UTAUT2) of the variance in behavioral intention and technology use respectively. Therefore, the integration of innovativeness with UTAUT predictors in the context of cloud-based E-learning as a new technology is theoretically significant and empirically validated. In details, Fig. 1 illustrates the research model for acceptance and use of cloud-based E-learning, including the presentation of all paths of the model and also all hypotheses (non-significant paths appear as dashed arrows).

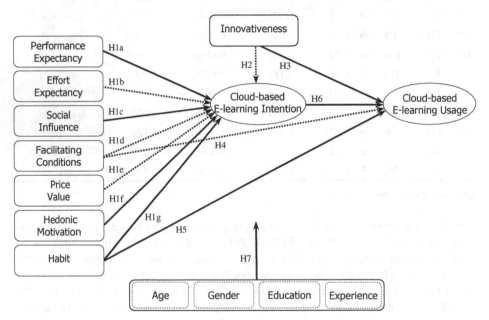

**Fig. 1.** The acceptance and use of cloud-based E-learning model

## 5     Conclusions

With the theoretical exploration of integration of consumer innovativeness with the UTAUT2 antecedents into the same model, the paper proposes a model of E-learning adoption that explains the factors of influence on the consumer intention and use of cloud-based E-learning systems. The model was empirically tested and basically supported. In specific, the determinants of performance expectancy, social influence, hedonic motivation, and habit are positively related to intention to use E-learning that has in turn a positive effect on the use of E-learning. Moreover, innovativeness, and habit directly influence cloud-based E-learning usage. In addition, the paper also shows that the demographic characteristics (age, gender, education, and experience) moderate the effects of UTAUT2 predictors as well as innovativeness on intention to use and use behavior of E-learning. Finally, with the significant support of the combined impact of innovativeness with UTAUT2 determinants, this study continues to contribute to the body of knowledge exploring the predictors of consumer adoption of information systems in general.

In the future study, the authors will work out for the combined effect of the elements and also expand the research scope and object, adjust scales, add more variables to the research model, use random sampling, and propose the recommendations that will help learning strategies of E-learners, cloud-based E-learning implementation of educational organization, and service providers more successful.

# References

1. Ajzen, I.: Behavioral Control, Self-Efficacy, Locus of Control and the Theory of Planned Behavior. Journal of Applied Social Psychology 32, 665–683 (2002)
2. Ajzen, I.: From Intentions to Action: A theory of Planned Pehavior, pp. 11–39. Springer, Heidelberg (1985)
3. Ajzen, I.: The Theory of Planned Behavior. Organization Behavior and Human Decision Process 50, 179–211 (1991)
4. Ajzen, I., Fishbein, M.: The Influence of Attitudes on Behavior In: Albarracin, D., John son, B.T., Zanna, M.P. (eds.) The Handbook of Attitudes, pp. 173–221. Erlbaum, Mahwah, NJ (2005)
5. Ajzen, I., Fishbein, M.: Understanding attitudes and predicting social behavior. Prentice Hall, Englewood cliffs (1980)
6. Al-Jumeily, D., Williams, D., Hussain, A.J., Griffiths, P.: Can We Truly Learn from A Cloud Or Is it just a lot of thunder? In: Developments in E-systems Engineering, pp. 131– 139. IEEE (2010)
7. Aldas-Manzano, J., Lassala-Navarre, C., Ruiz-Mafe, C., Sanz-Blas, S.: The role of consumer innovativeness and perceived risk in online banking usage. International Journal of Bank Marketing 27(1), 53–75 (2009)
8. Ambient Insight: Worldwide Market for Self-paced eLearning Products and Services: 2011-2016 Forecast and Analysis. Ambient Insight Report (2013), http://www.ambientinsight.com
9. Behrend, T.S., Wiebe, E.N., London, J.E., Johnson, E.C.: Cloud computing adoption and usage in community colleges. Behaviour & Information Technology 30(2), 231–240 (2011)
10. Bhruthari, G.P., Sanil, S.N., Prajakta, P.D.: Appliance of Cloud Computing on E-Learning. International Journal of Computer Science and Management Research, 276–281 (2012)
11. Brown, S.A., Venkatesh, V.: Model of Adoption of Technology in the Household: A Baseline Model Test and Ext. Incorporating Household Life Cycle. MIS Quarterly 29(4), 399–426 (2005)
12. Brown, J.S., Adler, R.P.: Open education, the long tail, and learning 2.0. Educause Review 43(1), 16–20 (2008)
13. Certifyme: Announces E-learning Statistics for 2013 (2013), http://www.certifyme.net
14. Childers, T.L., Carr, C.L., Peck, J., Carson, S.: Hedonic and Utilitarian Motivations for Online Retail Shopping Behavior. Journal of Retailing 77(4), 511–535 (2001)

15. Citrin, A.V., Sprott, D.E., Silverman, S.N., Stem Jr., D.E.: Adoption of internet shopping: the role of consumer innovativeness. Industrial Management & Data Systems 100(7), 294–300 (2000)
16. Compeau, D.R., Higgins, C.A.: Computer self-efficacy: Development of a measure and initial test. MIS Quarterly 19(2), 189–211 (1995)
17. Crespo, A.H., Del Bosque, I.R.: The effect of innovativeness on the adoption of B2C e-commerce: A model based on the Theory of Planned Behaviour. Computers in Human Behavior 24(6), 2830–2847 (2009)
18. Davis, F.D.: Perceived usefulness, perceived ease of use and user acceptance of Information Technology. MIS Quaterly 13(3), 319–340 (1989)
19. Davis, F.D.: User acceptance of information technology: System characteristics, user perceptions and behavioral impacts. International Journal of Man-Machine 38, 475–487 (1993)
20. Davis, F.D., Bagozzi, R.P., Warshaw, P.R.: Extrinsic and Intrinsic Motivation to Use Computers in the Workplace. Journal of Applied Social Psychology 22(14), 1111–1132 (1992)
21. Dodds, W.B., Monroe, K.B., Grewal, D.: Effects of Price, Brand and Store Information for Buyers. Journal of Marketing Research 28(3), 307–319 (1991)
22. Downes, S.: E-Learning 2.0. International Review of Research in Open and Distance Learning 6(2) (2005)
23. Ercan, T.: Effective use of cloud computing in educational institutions. Procedia-Social and Behavioral Sciences 2(2), 938–942 (2010)
24. Fishbein, M., Ajzen, I.: Belief, attitude, intention and behavior: An introduction to theory and research. Addision-Wesley (1975)
25. Goldsmith, R.E., Foxall, G.R.: The measurement of innovativeness. In: The International Handbook on Innovation, pp. 321–330 (2003)
26. Goldsmith, R.E., Hofacker, C.F.: Measuring consumer innovativeness. Journal of the Academy of Marketing Science 19(3), 209–221 (1991)
27. Harnal, M.S., Bagga, M.D.: Cloud Computing: An Overview. International Journal 3(7) (2013)
28. Heijden, V.D.: User Acceptance of Hedonic Information Systems. MIS Quarterly 28(4), 695–704 (2004)
29. Hill, T., Smith, N.D., Mann, M.F.: Role of efficacy expectations in predicting the decision to use advanced technologies: The case of computers. Journal of Applied Psychology 72(2), 307–313 (1987)
30. Im, S., Bayus, B.L., Mason, C.H.: An empirical study of innate consumer innovativeness, personal characteristics, and new-product adoption behavior. Journal of the Academy of Marketing Science 31(1), 61–73 (2003)
31. Jain, A., Chawla, S.: E-Learning in the Cloud. International Journal of Latest Research in Science and Technology 2(1), 478–481 (2013)
32. Kamel, S.: The role of virtual organizations in post-graduate education in Egypt: The case of the regional IT institute. In: Tan, F.B. (ed.) Courses on Global IT Applications and Management: Success and Pitfalls, pp. 203–224. Idea Group Publishing, Hershey (2002)
33. Karrer, T.: Understanding E-learning 2.0. Learning Circuits (2007)
34. Kim, S.S., Malhotra, N.K., Narasimhan, S.: Two competing perspectives on automatic use: A Theoretical and Empirical Comparison. Information Systems Research 16(4), 418–432 (2005)

35. Laisheng, X., Zhengxia, W.: Cloud Computing a New Business Paradigm for E-learning. In: International Conference on Measuring Technology and Mechatronics Automation, pp. 716–719 (2011)
36. Leonardo, R.O., Adriano, J.M., Gabriela, V.P., Rafael, V.: Adoption analysis of cloud computing services. African Journal of Business Management 7(24), 2362–2374 (2013)
37. Limayem, M., Hirt, S.G., Cheung, C.M.K.: How Habit Limits the Predictive Power of Intentions: The Case of IS Continuance. MIS Quarterly 31(4), 705–737 (2007)
38. Lin, P.C., Lu, S.C., Liu, S.K.: Towards an Education Behavioral Intention Model for E-Learning Systems: an Extension of UTAUT. Journal of Theoretical and Applied Information Technology 47(3), 1120–1127 (2013)
39. Marcati, A., Guido, G., Peluso, A.M.: The role of SME entrepreneurs innovativeness and personality in the adoption of innovations. Research Policy 37(9), 1579–1590 (2008)
40. Masud, A.H., Huang, X.: An E-learning System Architecture based on Cloud Computing. World Academy of Science, Engineering and Technology 62, 71–76 (2012)
41. Midgley, D.F., Dowling, G.R.: Innovativeness: the concept and its measurement. Journal of Consumer Research 4, 229–242 (1978)
42. Midgley, D.F., Dowling, G.R.: A longitudinal study of product form innovation: the interaction between predispositions and social messages. Journal of Consumer Research 19, 611–625 (1993)
43. Moore, G.C., Benbasat, I.: Development of an instrument to measure the perception of adopting an information technology innovation. Information Systems Research 2(3), 192–222 (1991)
44. Muhambe, T.M., Daniel, O.O.: Post adoption evaluation model for cloud computing services utilization in universities in Kenya. International Journal of Management & Information Technology 5(3), 615–628 (2013)
45. Nguyen, T.D., Nguyen, D.T., Cao, T.H.: Acceptance and use of information system: E-learning based on cloud computing in Vietnam. In: Linawati, Mahendra, M.S., Neuhold, E.J., Tjoa, A.M., You, I. (eds.) ICT-EurAsia 2014. LNCS, vol. 8407, pp. 139–149. Springer, Heidelberg (2014)
46. Pocatilu, P., Alecu, F., Vetrici, M.: Cloud Computing Benefits for E-learning Solutions. Economics of Knowledge 2(1), 9–14 (2010)
47. Popovic, K., Hocenski, Z.: Cloud computing security issues and challenges. In: Proceedings of the 33rd International Convention, pp. 344–349. IEEE (2010)
48. Redecker, C., Ala-Mutka, K., Bacigalupo, M., Ferrari, A., Punie, Y.: Learning 2.0: The impact of Web 2.0 innovations on education and training in Europe. European Commission-Joint Research Center-Institute for Prospective Technological Studies, Seville (2009)
49. Rogers, E.M.: Diffusion of innovations. Free Press, New York (1995)
50. Sheppard, B.H., Hartwick, J., Warshaw, P.R.: The Theory of Reasoned Action: A Meta-Analysis of Past Research with Recommendations for Modifications and Future Research. Journal of Consumer Research 15(3), 325–343 (1988)
51. Soud, A., Fisal, A.R.: Factors that determine continuance intention to use e-learning system: an empirical investigation. In: International Conference on Telecommunication Technology and Applications, vol. 5, pp. 241–246. IACSIT Press, Singapore (2011)
52. Sun Microsystem: Cloud Computing Guide. Sun Microsystems Inc. (2009)
53. Sun, P., Tsai, R., Finger, G., Chen, Y., Yeh, D.: What drives a successful e-Learning? An empirical investigation of the critical factors influencing learner satisfaction. Computers & Education 50, 1183–1202 (2008)
54. Tavangarian, D., Leypold, M.E., Nolting, K., Roser, M., Voigt, D.: Is e-Learning the solution for individual learning? Electronic Journal of e-Learning 2(2), 273–280 (2004)

55. Taylor, S., Todd, P.: Understanding Information Technology Usage: A Test of Competing Models. Information Systems Research 6(2), 144–176 (1995)
56. Thompson, R., Higgins, R., Howell, L.: Personal computing: Toward a conceptual model of utilization. MIS Quarterly 15(1), 125–143 (1991)
57. Venkatesh, V.: Determinants of Perceived Ease of Use: Integrating Perceived Behavioral Control, Computer Anxiety and Enjoyment into the Technology Acceptance Model. Information Systems Research 11(4), 342–365 (2000)
58. Venkatesh, V., Davis, F.D.: A Theoretical Extension of the Technology Acceptance Model: Four Longitudinal Field Studies. Management Science 46(2), 186–204 (2000)
59. Venkatesh, V., Bala, H.: Technology acceptance model 3 and a research agenda on interventions. Decision Sciences 39(2), 273–315 (2008)
60. Venkatesh, V., Morris, M.G., Davis, G.B., Davis, F.D.: User acceptance of information technology: Toward a unified view. MIS Quarterly 27(3), 425–478 (2003)
61. Venkatesh, V., Thong, Y.L.J., Xin, X.: Consumer Acceptance and Use of Information Technology: Extending the Unified Theory of Acceptance and Use of Technology. MIS Quarterly 36(1), 157–178 (2012)
62. Venkatraman Archana: Italian university reduces costs by 23% with cloud platform (2013), http://www.computerweekly.com
63. Viswanath, K., Kusuma, S., Gupta, S.K.: Cloud Computing Issues and Benefits Modern Education. Global Journal of Computer Science and Technology Cloud & Distributed 12(10), 1–7 (2012)
64. Welsh, E.T., Wanberg, C.R., Brown, K.G., Simmering, M.J.: E‑learning: emerging uses, empirical results and future directions. International Journal of Training and Development 7(4), 245–258 (2003)
65. Will, M., Allan, Y.: E-learning system Acceptance and usage pattern, pp. 201–216. Technology Acceptance in Education: Research and Issue (2011)
66. Zaharescu, E.: Enhanced Virtual E-Learning Environments Using Cloud Computing Architectures. International Journal of Computer Science Research and Application 2(1), 31–41 (2012)
67. Zeithaml, V.A.: Consumer Perceptions of Price, Quality, and Value: A Means-End Model and Synthesis of Evidence. Journal of Marketing 52(3), 2–22 (1988)
68. Zheng, H., Jingxia, V.: Integrating E-Learning System Based on Cloud Computing. In: IEEE International Conference on Granular Computing (2012)

# Retracted: Explicit Untainting to Reduce Shadow Memory Usage in Dynamic Taint Analysis

Young Hyun Choi[1], Min-Woo Park[1],
Jung-Ho Eom[2], and Tai-Myoung Chung[1],*

[1] Dept. of Electrical and Computer Engineering, Sungkyunkwan University
300 Cheoncheon-dong, Jangan-gu, Suwon-si, Gyeonggi-do, 440-746, Republic of Korea
{yhchoi,mwpark,tmchung}@imtl.skku.ac.kr
[2] Military Studies, Daejeon University
62 Daehakro, Dong-gu, Daejeon, Republic of Korea
eomhun@gmail.com

**Abstract.** As the growth of computing technologies and smart service, the dimension for importance of security of a system has been increased dramatically. Many researches for solving threats of software vulnerabilities have been proposed in worldwide. Ordinary program testing method for finding defects in software can be categorized into black-box testing and white-box testing. In Black-box testing, the tester does not need to tasks recognition of the internal structure of program, whereas in white-box testing, the tester checks to tasks recognition of internal structure of program. Taint analysis is an efficient black-box testing method for finding exploited crashes by tracking external input to the program. However, taint analysis method is too heavy and slow to provide for commercial analysis program, because of large amount of computation and shadow memory usage. Recent, many experimental approaches to weight down and to speed up the analysis process, but it were lacking in commercial use. In this paper, we propose a method to reduce shadow memory usage by selectively not trace the definite untainting memories. Our evaluation result shows that we can reduce number of taint operation by significant amount.

**Keywords:** Security, Dynamic Taint Analysis, Data Flow Analysis.

## 1 Introduction

Malicious software propagates throughout computer systems and networks leaking information and harming. Adversaries use software vulnerabilities of the system to do malicious actions. According to the annual report from Panda Security, total number of newly introduced malware in 2012 reached 27 million[1]. Moreover, Adversaries traded critical vulnerability of software directly with intruders. This means to find vulnerabilities of software is more important to defend the computer systems and networks[2].

---

* Corresponding author.

B. Murgante et al. (Eds.): ICCSA 2014, Part V, LNCS 8583, pp. 175–184, 2014.
© Springer International Publishing Switzerland 2014

General program testing techniques are classified into two main categories; Black-box testing and white-box testing. In Black-box testing, the tester does not need to tasks recognization of the internal structure of program, whereas in white-box testing, the tester checks to tasks recognization of internal structure of program. Most commercial programs do not release their source code, so black-box testing is usually valid on the binary analysis programs. Taint analysis traces the external input data to the program which helps the analysis to focus on crashes. It is find the higher probability of exploit. Recently, some framework of taint analysis tools are proposed in the academic, but these frameworks are not appropriate for works with commercial analysis program[7,8].

In this paper, we proposed a method to reduce shadow memory usage in dynamic taint analysis. We only trace and watch information of memories and registers that have higher probability of exploit. Therefore we do not assign space in shadow memory for memories and registers that are not being tracked. These help to more faster analysis[2].

The rest of this paper is organized as follows. In section 2 we explain the concept of taint analysis by using an example intermediate language presented in the paper. Section 3 discuss our proposal of reducing overhead in dynamic taint analysis and the we evaluate our schemes with test case in section 4. Finally we conclude this paper and present future works in section 5.

## 2    Taint Analysis

Taint analysis is a kind of information flow analysis. Tainted data has a considerable effect on the other data in the system. Taint analysis is a information flow tracing mechanism which tracks incoming data from the user input throughout whole process. Because of data marking 'tainted(tagged)' proceed from the assumption which user input is untrusted[11]. The purpose of taint analysis is to track influence of tainted data with the taint policy[4]. A taint analysis determines the way of the influence of the tainted information along the execution of program [3,5,12,13]. There are a lot of researches being done on developing analysis frameworks to efficiently analyze programs [6,7,8]. Taint analysis consists of three main factors; taint introduction, taint propagation and taint checking [3]. A detail about each factor should be as follows subsections.

Taint analysis can be divided into three main factors, which are taint introduction, taint propagation and taint checking [5]. To explain taint analysis using formal methods, we present a assemply-like intermediate language in Fig. 1 and its operational semantics in Fig. 2. Operational semantic is written in $[\frac{premise}{state_1 \rightarrow state_2}$ condition $]$ format. When premise is satisfied, program at $state_1$ will be changed to $state_2$. Functions with double brackets are semantic functions which returns semantic value of the given syntax and examples of them are shown in Fig. 3. s[ $x \mapsto v$ ] means x is substituted by v in state s and $\mu$ is replaced by $x_2$. For example, operational semantics of unary operation means that when expression e is evaluated into v in initial state s then the result of *unope* is $U[[v]]$ [2].

We can taint sensitive data we want to protect and examine if those data are sent to external servers. In Table 2. we present a basic taint policy for propagating taint information. Each function is called by appropriate operations as stated in the operational semantics in Fig. 4 [2].

| | | |
|---|---|---|
| *Program* | ::= | *stmt*∗ |
| *stmt s* | ::= | if *b* then $s_1$ else $s_2$ \| goto *exp* \| while *b* do *s* \| *type* store(*exp*, *exp*) |
| *exp e* | ::= | *type* load(*exp*) \| *exp* binop *exp* \| unop *exp* \| *v* |
| *boolean b* | ::= | true \| false \| *exp* = *exp* \| *exp* ≠ *exp* \| *exp* ≤ *exp* \| *exp* < *exp* \| *exp* ≥ *exp* \| *exp* > *exp* \| |
| *binop* | ::= | binary operation |
| *unop* | ::= | unary operation |
| *value v* | ::= | *type* immediate value |
| *type* | ::= | DWORD \| WORD \| BYTE |

**Fig. 1.** Example of Intermediate Language

**Table 1.** Basic Taint Policy

| Functions | Policy Check |
|---|---|
| $P_{input}()$, $P_{binck}()$, $P_{memck}()$ | True |
| $P_{const}()$ | False |
| $P_{unop}(t)$, $P_{assign}(t)$ | t |
| $P_{binop}(t_1, t_2)$ | $t_1 \vee t_2$ |
| $P_{mem}(t_{address})$ | $t_{address}$ |
| $P_{cond}(t)$ | ¬t |
| $P_{goto}(t)$ | ¬t |

# 3 Reducing Shadow Memory Usage in Dynamic Taint Analaysis

In this section, we discuss overhead issues present in dynamic taint analysis and propose a scheme to reduce them for efficient and light-weight use. These rules are similar and based on Min's method[2].

## 3.1 Trace: Memories and Registers

Taint analysis traces the taint input that external input to the program by keeping a shadow memory of taint information structures which are incessantly updated while instructions are executed on program running time. Taint analysis

$$\textbf{INPUT} \frac{v \text{ is input value}}{< get\_input(),\ s > \to s[get\_input() \mapsto v]}$$

$$\textbf{UNOP} \frac{< e\,,s > \to s[e \mapsto v], < unop\ v,\ s[e \mapsto v] > \to s[e \mapsto v][unop\ v \mapsto U[\![v]\!]]}{< unop\ e, s > \to s[e \mapsto v][unop\ v \mapsto U[\![v]\!]]}$$

$$\textbf{BINOP} \frac{\begin{array}{c}< e_1,\ s > \to s[e_1 \mapsto v_1], < e_2, s[e_1 \mapsto v_1] > \to s[e_1 \mapsto v_1][e_2 \mapsto v_2] \to s', \\ < v_1\ binop\ v_2, s' > \to s'[v_1\ binop\ v_2 \mapsto Bin[\![v_1, v_2]\!]]\end{array}}{< e_1\ binop\ e_2,\ s > \to s[e_1 \mapsto v_1][e_2 \mapsto v_2][v_1\ binop\ v_2 \mapsto Bin[\![v_1, v_2]\!]]}$$

$$\textbf{LOAD} \frac{< e,\ s > \to s[e \mapsto v], < load\ v, s[e \mapsto v] > \to s[e \mapsto v][load\ v \mapsto \mu[\![v]\!]]}{< load\ e, s > \to s[e \mapsto v][load\ v \mapsto \mu[\![v]\!]]}$$

$$\textbf{STORE} \frac{\begin{array}{c}< e_1,\ s > \to s[e_1 \mapsto v_1],\ < e_2, s[e_1 \mapsto v_1] > \to s[e_1 \mapsto v][e_2 \mapsto v_2] \to s', \\ < store(v_1, v_2), s' > \to < \mu[\![v_1]\!] \leftarrow v_2, s' > \to s''\end{array}}{< store(e_1, e_2),\ s > \to s''}$$

$$\textbf{GOTO} \frac{< e,\ s > \to s[e \mapsto v], < goto\ v, s[e \mapsto v] > \to < \quad v, [\quad v] > \to s'}{< goto\ e,\ s > \to s'}$$

$$\textbf{TCOND} \frac{< s_1, s > \to s'}{< if\ b\ then\ s_1\ else\ s_2, s > \to s'} \quad \text{if } B[\![b]\!] \text{ is true}$$

$$\textbf{FCOND} \frac{< s_2, s > \to s'}{< if\ b\ then\ s_1\ else\ s_2, s > \to s'} \quad \text{if } B[\![b]\!] \text{ is false}$$

$$\textbf{TWHILE} \frac{< s_1, s > \to s', < while\ b\ do\ s_1, s' > \to s''}{< while\ b\ s_1, s > \to s''} \quad \text{if } B[\![b]\!] \text{ is true}$$

$$\textbf{FWHILE} \frac{}{< while\ b\ do\ s_1, s > \to s} \quad \text{if } B[\![b]\!] \text{ is false}$$

$$\textbf{CONST} \frac{}{< \quad, s > \to s[v \mapsto \quad[\![v]\!]]}$$

**Fig. 2.** Operational Semantics for Example Intermediate Language

| |
|---|
| U[ ] : Returns result of the unary operation |
| Bin[ ] : Returns result of the binary operation |
| B [ ] : Returns true or false of the boolean expression |
| $\mu$[ ] : Maps value of memory in the address given as parameter |

**Fig. 3.** Semantic Functions

$$\text{T-Input} \frac{v \text{ is input value}}{< get\_input(),\ s > \mapsto s[get\_input() \mapsto (v, P_{input}())]}$$

$$\text{T-Unop} \frac{\begin{array}{c}< e, s > \to s[e \mapsto (v, t)], \\ < unop\ v,\ s[e \mapsto (v, t)] > \to s[e \mapsto (v, t)][unop\ v \mapsto (U[\![v]\!], P_{unop}(t))]]\end{array}}{< unop\ e, s > \to s[e \mapsto (v, t)][unop\ v \mapsto (U[\![v]\!], P_{unop}(t))]}$$

$$\text{T-Binop} \frac{\begin{array}{c}< e_1,\ s > \to s[e_1 \mapsto (v_1, t_1)], \\ < e_2, s[e_1 \mapsto (v_1, t_1)] > \to s[e_1 \mapsto (v_1, t_1)][e_2 \mapsto (v_2, t_2)] \to s', \\ < v_1\ binop\ v_2, s' > \to s'[v_1\ binop\ v_2 \mapsto (Bin[\![v_1, v_2]\!], P_{binop}(t_1, t_2))\end{array}}{< e_1\ binop\ e_2,\ s > \to s'[v_1\ binop\ v_2 \mapsto (Bin[\![v_1, v_2]\!], P_{binop}(t_1, t_2))]}$$

$$\text{T-Load} \frac{\begin{array}{c}< e,\ s > \to s[e \mapsto (v, t)], \\ < load\ v, s[e \mapsto (v, t)] > \to s[e \mapsto (v, t)][load\ v \mapsto (\mu[\![v]\!], P_{mem}(t))]\end{array}}{< load\ e, s > \to s[e \mapsto v][load\ v \mapsto (\mu[\![v]\!], P_{mem}(t))]}$$

$$\text{T-Store} \frac{\begin{array}{c}< e_1,\ s > \to s[e_1 \mapsto (v_1, t_1)], \\ < e_2, s[e_1 \mapsto (v_1, t_1)] > \to s[e_1 \mapsto (v_1, t_1)][e_2 \mapsto (v_2, t_2)] \to s', \\ < store(v_1, v_2), s' > \to < \mu[\![v_1]\!] \leftarrow (v_2, P_{assign}(t_2)), s' > \to s''\end{array}}{< store(e_1, e_2),\ s > \to}$$

$$\text{T-Goto} \frac{\begin{array}{c}< e,\ s > \to s[e \mapsto (v, t)], \\ < goto\ v,\ s[e \mapsto (v, t)] > \to < pc \leftarrow (v, P_{goto}()), s[e \mapsto (v, t)] > \to s'\end{array}}{< goto\ e,\ s > \to}$$

$$\text{T-Tcond} \frac{< s_1, s > \to s'}{< if\ b\ then\ s_1\ else\ s_2,\ s > \to s'} \quad \text{if } B[\![b]\!] \text{ is true}$$

$$\text{T-Fcond} \frac{< s_2, s > \to s'}{< if\ b\ then\ els\ e\ s_2, s > \to s'} \quad \text{if } B[\![b]\!] \text{ is false}$$

$$\text{T-TWhile} \frac{< s_1, s > \to s',\ while\ b\ do\ s_1, s' > \to s''}{< while\ b\ do\ s_1, s > \to s''} \quad \text{if } B[\![b]\!] \text{ is true}$$

$$\text{T-FWhile} \frac{}{while\ b\ do\ s_1, s > \to s} \quad \text{if } B[\![b]\!] \text{ is false}$$

$$\text{T-Const} \frac{}{< v,\ s > \to s[v \mapsto (N[\![v]\!], P_{const}())]}$$

**Fig. 4.** Operational Semantics with Taint Policy

is effective on detecting control hijacking vulnerabilities in an executable. For example, if the program counter register stores a tainted value, then control flow of the program can be hijacked by an external input injected by adversary. Thus, the problem is up these minimum size of the memory or register for the hijacking of the control flow. In the case of 32-bit machines, if program counter 4 bytes are hijacked, then adversary can be controlling whole of the 4GB memory space in anywhere. If continuity of tainted data are broken, then the possibility of hijacking the control flow or any adversary's attacks also decreases as well. Fig. 5 describes the memory range that the tainted program counter register can point

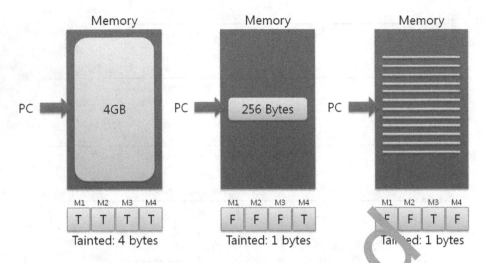

**Fig. 5.** Range of Memory: Changes of Taint Ratio

to in the taint percentage perspective. The specific byte is tainted then marked 'T' and untainted then marked 'F'. When all 4 bytes of the program counter is tainted, it can be overwritten to point at anywhere in the 4GB memory space. However when only 1 byte is tainted, memory address that program counter can point to becomes very limited. Therefore if an adversary has injected shellcode in the process memory, it is less likely that program counter will be overwritten to point to the shellcode for execution. This means in general, lower percentage of tainted bytes has lower possibility of successful exploitation.

## 3.2 Explicit Untainting Memory and Register

We propose a heuristic method of explicit untainting memories and registers that have low percentage of tainted bytes, to make the dynamic taint analysis is more efficient. When the instruction has tainted memory or register as operands, then explicit untainting is checked and grown up. When the instruction is executed, dynamic taint analyzer brings taint information of the operands and updates with the specific operational semantic of the instruction. If the tainted-byte percentage of the output is below the boundary or predefined threshold, it is stop to trace from that point of execution; this process called explicit untainting. The percentage of tainted bytes is calculated by [ total number of tainted bytes / size of the access type ]. Taint policy of our proposal is stated in Table 3.2.

**Binary and Unary Operations.** Unary operations are completed to decision on a single operand such as a memory address or a register. Accordingly, the status of taint information does not change on same taint analysis. Binary operations is a consequentially similar with unary operations. Because, each taint operations is traced and updated the taint information on executes time.

**Boolean Expressions.** Similar to unary operations, the result of evaluating boolean expressions does not have effect on the taint information because the expression itself does not update any memory or a register. Therefore our approach does not do anything different same as Min's method [2].

```
DWORD store(mem1, get_input())
DWORD store(mem2, get_input())

DWORD store(mem2, load(mem2) AND 0x000000FF)

...

DWORD store(reg1, load(mem1) + load(mem2))
goto reg1
```

**Fig. 6.** Example Code for Explicit Untainting

**Branch Operations.** Fig. 6 shows an example intermediate language statements for explicit untainting. Assume that predefined threshold is 25 percent same as Min's method. First, two lines get 4-byte user inputs and store them in memory addresses specified by mem1, mem2 which become tainted by 4 bytes. Then, the next instruction does bitwise AND operation of the value stored in memory address mem2 and the hex value 0x000000FF leaving only lower 1 byte of the memory tainted. Assuming the predefined taint threshold is 25 percent, we mark memory address mem2 as untainted instead of updating the taint information. Finally, target address of goto instruction is sum of values stored in mem1 and mem2 which becomes tainted by 4 bytes because value stored in mem1 is tainted by 4 bytes. Therefore even if we did not track the memory address mem2 along following instructions, target of the goto instruction is still tainted. Note that in this condition, control flow can be hijacked because destination address of the goto instruction is depend on the user input (Using the Min's example and model)[2].

**String Operations.** When user input type is string, then copy to memory 4 bytes on each step. But, when system reads or loads the string, then access to memory 1 bytes by instructions. String operations is over the threshold on process, tainted string bytes will continue to be tainted, because the percentage of tainted bytes is TRUE (100 percent).

### 3.3   Advance Proposed

Min's method takes threshold of 25 percent and reduces usages. Our proposed scheme adopted $P_{binck}()$, $P_{memck}()$ and policy check of $P_{cond}(t)$ is $\neg t$ and $P_{goto}(t)$ is $\neg t$. Alterations of taint policy is bring changes. Fig. 7 explains the situation changes.

**Table 2.** Proposed Taint Policy

| Functions | Policy Check |
|---|---|
| $P_{input}()$, $P_{binck}()$, $P_{memck}()$ | True |
| $P_{const}()$ | False |
| $P_{unop}(t)$, $P_{assign}(t)$ | t |
| $P_{binop}(t_1, t_2)$ | $threshold \geq false\ \ if(percent\ of\ tainted\ bytes)$ |
| | $threshold \leq false\ \ if(percent\ of\ tainted\ bytes)$ |
| $P_{mem}(t_{address})$ | $t_{address}$ |
| $P_{cond}(t)$ | $\neg t$ |
| $P_{goto}(t)$ | $\neg t$ |

Memory location mem1 and mem2 store user inputs which are tainted. If after several instruction phase, memory location mem1 and mem2 are tainted by only 1 bythe with different offset. In the Min's method, memory address mem1 and mem2 are explicitly untainted since they are tainted by 1 byte, resulting reg1 as untainte. However, proposed taint policy, threshold sets 25 percent. But, mem1 and mem2 of 1 byte of tainted data has possibility of tainted effectiveness of warning and explicit untainting. So, we proposed conditional decision scheme. At fig.5 shows memory location with M1, M2, M3, M4. When execute process, taint analyzer combine their tainted data make results are continuous location is T, then we consider this event can be effectiveness of warning, to except from explicit untainting. In Min's method has goto instruction with reg1 as operand therefore, does not seem to be hijackable causing a false negative. But, our scheme has goto instruction with reg1 as operand therefore, increasing probability of hijackable causing a effectiveness of warning is up.

```
DWORD store(mem1, get_input())
DWORD store(mem2, get_input())

DWORD store(mem1, load(mem1) AND 0x0000FF00)
DWORD store(mem2, load(mem2) AND 0x000000FF)

DWORD store(reg1, load(mem1) + load(mem2))
...
goto reg1
```

**Fig. 7.** False Negative Example

## 4   Evaluation

In this section, we evaluate our proposal by tracing an example program and examining number of memories and registers that are untainted. Fig. 8 is the source code of the traced sample program for evaluation. It is a simple program that has two local character buffers to save user input strings. There is no boundary checking for the buffers, therefore it can cause a stack buffer overflow. The

```
#include <stdio.h>

int main(void)
{
      char user[100];
      char pw[100];
      char* pointer=NULL;

      gets(user);
      gets(pw);

      printf("login credentials");
      printf(user);

      if(!strncmp(user,"smba",4) && !strncmp(pw,"pass",4))
          pointer = "Login success";

      printf("Result: %s",pointer);
}
```

**Fig. 8.** Traced Sample Program code for Evaluation

**Table 3.** Summary of Analysis Results

| Description | Min's results | Our results |
|---|---|---|
| Total number of traced instructions | 533,134 | 533,134 |
| Number of tainted instructions | 6,632 | 6,632 |
| Number of explicitly untainted instructions | 4,032 | 5,112 |
| Effectiveness of warning | 1,224 | 1,776 |

threshold is assumed to be 25 percent, which is as same as evaluation threshold range of Min's paper.

We inserted a long input string that triggers the overflow vulnerability in the program when trying to return to the caller of the main function and traced the execution. Out of 533,134 instructions traced from the execution, 6,632 instructions had at least one tainted operand. Out of 6,632 tainted instructions, 1,263 instructions had two operands with number of tainted bytes below the threshold and 2,760 instructions had one tainted operand below the threshold and one untainted operand. As a result, 5,112 instructions can be explicitly untainted. 6,632 to 5,112; the shadow memory is decreased by 1,520. Around 23 percent of tainted instructions can be eliminated. This value is lower than Min's method, but our scheme's effectiveness of warning is higher about 1.5 times.

## 5  Conclusion

In this paper, we explained about the concept of dynamic taint analysis and its application to the field of computer security and networks. Dynamic taint analysis is good for vulnerability detection and malware analysis with black-box testing. But, dynamic taint analysis method is too heavy and slow to provide for analysis program. So, we proposed a method to reduce shadow memory usage by selectively not trace the definite untainting memories for light-weight analysis. Our evaluation results show that by explicit untainting we accomplish the performance improvement.

**Acknowledgments.** This research was funded by the MSIP(Ministry of Science, ICT & Future Planning), Korea in the ICT R&D Program 2014[2014044072003, Development of Cyber Quarantine System using SDN Techniques].

# References

1. 2012 Annual Report PandaLabs,
   http://press.pandasecurity.com/press-room/reports
2. Min, J.-W., Choi, Y.-H., Eom, J.-H., Chung, T.-M.: Explicit Untainting to Reduce Shadow Memory Usage and Access Frequency in Taint Analysis. In: Murgante, B., Misra, S., Carlini, M., Torre, C.M., Nguyen, H.-Q., Taniar, D., Apduhan, B.O., Gervasi, O. (eds.) ICCSA 2013, Part III. LNCS, vol. 7973, pp. 175–186. Springer, Heidelberg (2013)
3. Kang, M., McCamant, S., Poosankam, P., Song, D.: DTA++: Dynamic taint analysis with targeted control-flow propagation. In: 18th Annual Network and Distributed System Security Symposium (2011)
4. Miller, C., et al.: Crash Analysis with Bitblaze. Blackhat, USA (2010)
5. Schwartz, E.J., Avgerinos, T., Brumley, D.: All You Ever Wanted to Know About Dynamic Taint Analysis and Forward Symbolic Execution (but might have been afraid to ask). In: IEEE Symposium on Security and Privacy (2010)
6. Brumley, D., Jager, I., Avgerinos, T., Schwartz, E.J.: BAP: A Binary Analysis Platform. In: Gopalakrishnan, G., Qadeer, S. (eds.) CAV 2011. LNCS, vol. 6806, pp. 463–469. Springer, Heidelberg (2011)
7. Song, D., et al.: BitBlaze: A New Approach to Computer Security via Binary Analysis. In: Sekar, R., Pujari, A.K. (eds.) ICISS 2008. LNCS, vol. 5352, pp. 1–25. Springer, Heidelberg (2008)
8. Clause, J., Li, W., Orso, A.: Dytan: A Generic Dynamic Taint Analysis Framework. In: Proceedings of the 2007 International Symposium on Software Testing and Analysis. ACM (2007)
9. Avgerinos, T., Cha, S.K., Hao, B.L.T., Brumley, D.: AEG: Automatic Exploit Generation. In: Proceedings of the Network and Distributed System Security Symposium (2011)
10. Miller, C., et al.: Crash Analysis with BitBlaze. Blackhat, USA (2010)
11. Choi, Y.-H., Chung, T.-M.: A Framework for Dynamic Taint Analysis of Binary Executable File. In: Proc. ICISA 2013, Pattaya, pp. 374–375 (2013)
12. Scholten, M.: Taint Analysis in Practice, pp. 1–29. Vrije Universiteit Amsterdam, Amsterdam (2007)
13. Newsome, J., Song, D.: Dynamic Taint Analysis for Automatic Detection, Analysis, and Signature Generation of Exploits on Commodity Software. Technical report, School of Computer Science Carnegie Mellon University (2004)

# Using the NIST Reference Model for Refining Logical Architectures[*]

António Pereira[1], Ricardo J. Machado[2], José Eduardo Fernandes[3], Juliana Teixeira[1], Nuno Santos[1], and Ana Lima[1]

[1] CCG - Centro de Computação Gráfica, Guimarães, Portugal
{mpereira,nuno.santos,ana.lima}@ccg.pt,
juliana.teixeira@research.ccg.pt
[2] Centro ALGORITMI, Engineering School of University of Minho, Guimarães, Portugal
rmac@dsi.uminho.pt
[3] Polytechnic Institute of Bragança, School of Technology
and Management Dept. of Informatics and Communications, Bragança, Portugal
jef@ipb.pt

**Abstract.** The emergence of the Internet as a ubiquitous means of communication fostered the growth of new business and service models based on Cloud Computing. Information and Communication Technology companies use reference models to define their Cloud Computing strategies. NIST Cloud Computing Reference Architecture is one of these reference models that assist in the design of business, services, and architecture models. This paper aims to present the use of NIST reference architecture in the design of Cloud Computing architectures by employing a method that enables the application of the reference architecture to the refinement of logical architectures.

**Keywords:** Cloud Computing, Logical Architectures, Reference Model, NIST, Model, Requirements.

## 1 Introduction

The evolution and the increasing availability of Internet enabled the growth of new business and service models for ICT Industry, including models for *Cloud Computing* [1]. Adopting a *Cloud Computing* strategy may have impacts that span beyond just the technology architecture, influencing business and organizational strategies. Organizations look for tools, processes, and best practices to guide them with decisions around a *Cloud Computing* strategy, migration, and implementation. In this context, emerge the reference models, which are frameworks providing a set of best practices to support those decisions for particular problem domain. The use of

---

[*] This work has been supported by the Project AAL4ALL (QREN 13852), co-financed by the Fund through "COMPETE - Programa Operacional Factores de Competitividade", and by FCT – Fundação para a Ciência e Tecnologia within the Project Scope: PEst-OE/EEI/UI0319/2014.

B. Murgante et al. (Eds.): ICCSA 2014, Part V, LNCS 8583, pp. 185–199, 2014.

reference models for *Cloud Computing* environments enables the alignment of business and service architectures. Reference models enable the specification of the main activities and functions for *Cloud Computing* contexts. They address the concerns of the key stakeholders by defining the architecture capabilities and roadmap aligned with the business goals and architecture vision [3].

ICT Industry uses the NIST (U.S. National Institute of Standards and Technology) *Cloud Computing Reference Architecture* (CCRA) [4, 5] to support the decision making process of the design and specification of *Cloud Computing* architectures, models, solutions, and services.

At the same time as the concept of *Cloud Computing* is known, it appears, at the application level, the *Ambient Assisted Living* (AAL) domain with opportunities for developing new products and services that can be available on the Internet [6]. AAL technologies allow enhancing the lives of elderly and dependent people by offering them the ability to carry out routine tasks by themselves. Such technologies use ICT as a core component to generate highly dynamic systems and applications [7]. These systems and applications must assist ubiquitously and adapt to an individual's daily context-aware needs.

*Cloud Computing* promotes the seamless integration of systems and devices in the users daily lives through device and location autonomy [7]. It is necessary to adopt new strategies, new models, and mechanisms that enable the development of consistent, interoperable, and standardized solutions across the *Cloud Computing* ecosystem. This will increase the interest and the penetration rate of new products and services into the *Cloud Computing* market.

The purpose of this paper is to present the application of the NIST CCRA in refining and supporting the design of the *Cloud Computing* architecture for a demonstration case in the AAL domain. The demonstration case is in the context of the AAL4ALL project [8], which aims the development of an ecosystem of AAL products and services. In this context we apply the NIST CCRA to the AAL4ALL *Logical Architecture* [9]. We use a method to verify, identify, and elicit requirements related with *Cloud Computing* functions and activities that must be assured by the AAL4ALL systems, in order to enable the availability of AAL cloud-based services.

The remainder of this paper is organized as follows: section 2, provides an overview of main concepts about reference models, logical architectures, process architectures and *Cloud Computing*, NIST CCRA and some related projects; section 3, describes a method to apply the NIST CCRA to logical architectures; section 4, presents the application of the method to a real case in the AAL domain, namely in the AAL4ALL [8] project; section 5, presents the conclusions of the work and future developments.

## 2    Background and Related Work

This section presents a brief look at the concepts regarding logical architectures, reference models, including NIST, and a few studies in the application of NIST CCRA in the design of information systems and *Cloud Computing* architectures. For simplicity of writing, the remainder of this document simply uses the term *Cloud* to express *Cloud Computing*.

Nowadays, the size and complexity of information systems, together with critical time-to-market needs, demand new software engineering approaches to the design of software architectures. One of these approaches, is the use of reference models that allows to systematically reuse knowledge and components when developing a concrete reference architecture [10]. As defined by [2], a reference architecture is an abstract framework for understanding significant relationships among entities of some environment and for the development of consistent standards or specifications supporting that environment. A reference model is not directly tied to any standards, technologies, or other concrete implementation details, but it does seek to provide common semantics that can be unambiguously used across and between different implementations.

A logical architecture can be considered a view of a system composed of a set of problem-specific abstractions supporting functional requirements [11, 12]. The logical architecture acts as a common abstraction of the system, providing a representation to facilitate its comprehension by all stakeholders regardless their background. The requirements for process-level can be also represented in a logical architecture [11]. The process-level architecture represents the fundamental organization of service development, service creation, and service distribution in the relevant enterprise context. Process architecture is an arrangement of the activities and their interfaces in the process. It takes into account some non-functional requirements, such as performance and availability, and it can be represented with components, connectors, systems, components and connectors, ports, roles, representations and rep-maps, as well as *Architectural Elements* (ΛEs) [11].

Taking into account the scope of this work, we focus on the NIST CCRA (Fig.1) [5], which is the most known and used *Cloud* reference architecture by the ICT industry.

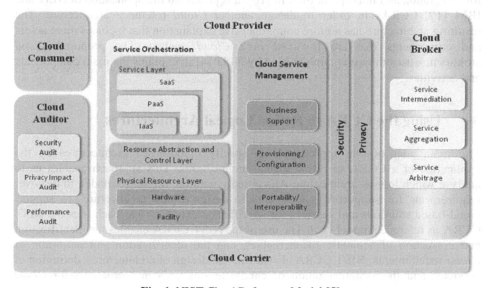

**Fig. 1.** NIST Cloud Reference Model [5]

NIST defines *Cloud* as a computing model that allows ubiquitous and on-demand access to a set of configurable computing resources (i.e., communications networks, servers, storage, applications or services), available on the network, and that can be rapidly provisioned and updated with minimal management effort or service provider interaction. Five main features characterize a *Cloud* [1]: (1) on-demand self-service; (2) broad network access; (3) resource pooling and rapid elasticity; (4) three possible service models (*Software-as-a-Service* - SaaS, *Platform-as-a-Service* - PaaS, and *Infrastructure-as-a-Service* - IaaS); (5) four possible deployment models (*Private, Public Cloud, Hybrid Cloud*, *Community Cloud*).

NIST developed a logical extension of their *Cloud* definition by the development of the NIST CCRA (Fig.1). NIST CCRA is a generic high-level conceptual model that constitutes an effective tool for discussing the requirements, structure, and operation of the *Cloud*. It defines a set of actors, activities, and functions that support the development of *Cloud* architectures. It also enables the analysis of standards for security, interoperability, and portability.

Our analysis identified some approaches that use the NIST CCRA to define *Cloud* functions and activities for other contexts. Other researches use NIST CCRA to analyse, improve, and design *Cloud* architectures. Few works [11, 12, 13] apply this model to define cloud-based infrastructures and services.

Research [13, 14] performed at University of Amsterdam developed the *InterCloud* [15] Architecture Framework (ICAF) [13, 14]. ICAF intends to address problems with multi-domain heterogeneous applications, integration, and interoperability, including the integration with legacy ICT infrastructures services. It also facilitates the interoperability and management of inter-provider *Cloud* infrastructures federation. ICAF uses NIST CCRA and defines additional functionalities that are required by heterogeneous multi-provider *InterCloud* services for integration and interoperability. In [16], it is presented the application of NIST and IBM [17] CCRAs in order to design another *Cloud* reference architecture. The resulting architecture has a more comprehensive explanation that includes more actors and components that are involved in a *Cloud* context. The proposed *Cloud* architecture uses the same method used by NIST [4] to explains the main components and activities.

## 3    Using the CCRA to Refine Logical Architectures

The main goal of our approach is to specify cloud-based architectures by the application of a CCRA to refine logical architectures. This approach enables us to analyse which logical architecture AEs comply with the reference model, to discover requirements for supporting *Cloud*, and to design *Cloud* architectures to support cloud-based solutions and services for specific contexts. Our approach uses the NIST CCRA as reference model [11]. The use of the NIST CCRA enables the identification and correction of semantic incoherencies, and the identification and definition of *Cloud* requirements. NIST CCRA also allows the design of architectures, definition of services, and the systematization of candidate standards and protocols for *Cloud* architectures.

Fig.2 depicts the method of how to use the NIST CCRA in conjunction with logical architectures in order to analyse the suitability of the logical architecture with the reference model.

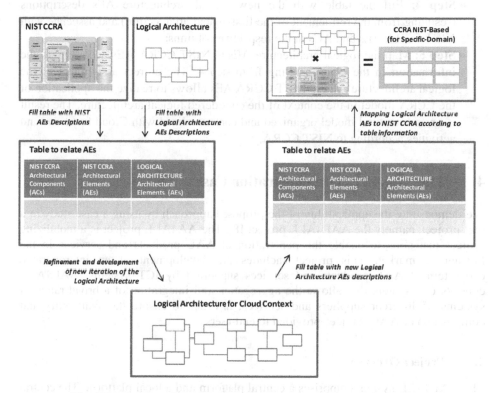

**Fig. 2.** Method to refine logical architectures according to NIST CCRA

The following steps synthesize the application of the method:

- **Step 1:** Select the NIST CCRA *Architectural Components* (ACs) to analyse in logical architecture. Fill the table with their NIST CCRA descriptions (column labelled with *NIST CCRA Architectural Components*) and respective AEs (column labelled with *NIST CCRA Architectural Elements*);
- **Step 2**: Analyse all logical architecture AEs and identify similar semantics with NIST CCRA AEs selected in step 1. Through analysis and comparison of logical architecture AEs descriptions with NIST CCRA AEs descriptions, fill the table with logical architecture AEs related to NIST CCRA AEs. This analysis enables to complement the requirements elicitation efforts related with Cloud contexts;
- **Step 3:** Refine and develop a new iteration of the logical architecture. This implies the redefinition of requirements and logical architecture AEs descriptions according to information in the table. Table information represents the matching of logical architecture AEs descriptions with NIST CCRA AEs

descriptions. The alignment of logical architecture with NIST CCRA is assured by the development of a new iteration of the logical architecture, which results in a cloud-based logical architecture;

- **Step 4:** Fill the table with the new logical architecture AEs descriptions (resulting from the refinement of the first logical architecture) and associate with NIST CCRA AEs and ACs in the respective columns;
- **Step 5:** Map the logical architecture AEs to NIST CCRA AEs according to the information in the table resulting from step 4. The correct assignment of the logical architecture AEs to NIST CCRA AEs allows to realize the application of the CCRA model in the context of the considered logical architecture. The result is an architectural model organized and contextualized with Cloud functions and activities according to NIST CCRA.

## 4    The AAL4ALL Demonstration Case

We demonstrate the applicability of the proposed approach by using a case study in a real project, namely the AAL4ALL project [8]. The AA4ALL project is a mobilizing project that aims to enable the penetration of AAL products and services in the Portuguese market. This project includes the development of an interoperable ecosystem of AAL products and services supported by ICT systems. AAL4ALL consists of a system that allows the aggregation and integration of a broad range of systems of different suppliers and services, in order to ensure the availability and composition of AAL services provided to end-users.

### 4.1    Project Overview

The AAL4ALL system comprises a central platform and a local platform. The central platform aggregates, orchestrates, and processes the AAL services, making them available for use on a *Cloud* model. The local platform aggregates the local systems that support the local services. Fig.3 depicts the elements of the AAL4ALL system.

1. AAL4ALL *Central Platform*: This platform ensures central integration of external systems (i.e., healthcare systems from hospitals, transport management systems, social networking systems, etc.) and local systems (systems installed in the user environment). The composition, processing, and availability of AAL services are performed in a Cloud system that provides AAL4ALL services.
2. AAL4ALL *Local Platform*: This platform, viewed as a local gateway, interconnects the local systems (i.e., sensors, actuators, desktops, cameras, etc.) that capture information from the user environment and user health information. This platform is interconnected with the AAL4ALL *Central Platform* and provides the AAL local services.
3. *External Platforms* (external Clouds): These platforms relate to systems from third party entities that provide services to the AAL AAL4ALL ecosystem. Their systems need to interact with AAL4ALL *Central Platform*.

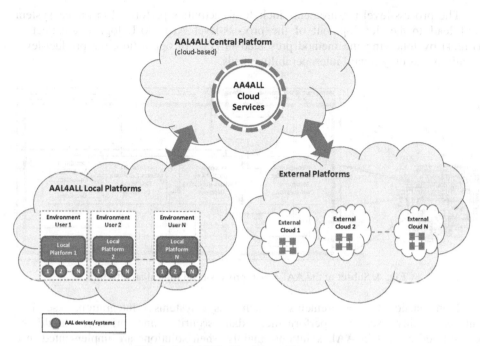

**Fig. 3.** AAL4ALL Cloud-based Ecosystem

## 4.2    AAL4ALL Logical Architecture

The AAL4ALL requirements are represented by use cases that describe the AAL4ALL functionalities that monitor the users routines and installed systems. The initial request for the AAL4ALL project requirements resulted in mixed and confusing sets of misaligned information. The discussions inside the project consortium relative to (1) the multi-domains to be covered, (2) the technologies, solutions and devices to be adopted, as well as (3) the uncertainty relative to interoperability issues (among others), resulted in a lack of consensus for the product-level requirements definition. Adopting a process-level perspective allows eliciting requirements in multi-domain ecosystems, as well as dealing with interoperability issues [9].

The rationale for the design of the models proposed in our approach, in the case of the AAL4ALL project, is based on specifying processes that intent to: (1) execute in a cloud-based software solution; (2) deal with several AAL multi-domains; (3) support interoperability between solutions and devices. The AAL4ALL project aims at developing a unified AAL ecosystem by using a single platform for integrating and orchestrating the products, services, and devices. Therefore, the product-level requirements should reflect the integration of legacy systems, instead of a typical approach that elicits product-level requirements reflecting applications to be developed and their functionalities.

The process-level requirements include all activities performed in the ecosystem and lead to the development of the process-level ALL4ALL logical architecture (Fig.4) by following the method presented in [11]. In opposition, the product-level requirements only regard interoperability needs.

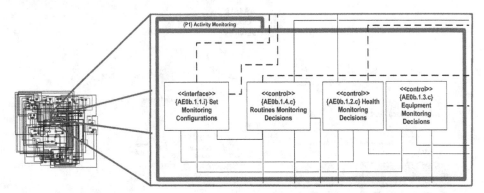

**Fig. 4.** Subset of the AAL4ALL process-level logical architecture

Non-functional requirements, such as systems integration, services interoperability, services performance, data security, and privacy, should be considered critical in AAL solutions, mainly when solutions are implemented in a Cloud-based architecture. The AAL4ALL platforms need to comply with those requirements as recommended by reference architectures.

Reference architectures and standards deal with non-functional requirements and provide best practices oriented to specific domains. A reference architecture can be also used to support non-functional requirements, since it is a framework that organizes system concepts taking into account the application domain characteristics or cross-cutting concerns [9]. Requirements elicitation should take into account domain reference architectures and standards.

## 4.3　Using the CCRA to Refine the AAL4ALL Architecture

The NIST CCRA does not represent the system architecture for a specific *Cloud* system. Rather, it is a tool for describing, discussing, and supporting the development of a system-specific architecture using a common reference framework for *Cloud*. Our approach uses the NIST CCRA to analyse and to elicit *Cloud* requirements, and to adapt the AAL4ALL *Logical Architecture* to a *Cloud* environment. It enables us to evaluate, identify, and include specific *Cloud* requirements for the AAL4ALL architecture.　For simplicity of writing, the remainder of this document simply uses only the term AAL4ALL AE to express AAL4ALL *Logical Architecture* AE. The following paragraphs detail the application of the method explained in section 3.

### Step 1: Select NIST CCRA AEs

We selected the components from the NIST CCRA for which we want to analyse the respective covering in the logical architecture. The selected components, depicted by Fig.5, are: (1) *Software-as-a-Service* (SaaS) Component, from *Service Layer*;

(2) *Business Support Component*; (3) *Provisioning and Configuration Component*; (4) *Portability and Interoperability Component*, from *Cloud Service Management Layer*; (5) *Security and Privacy Component*, (6) SaaS (*Software-as-a-Service*).

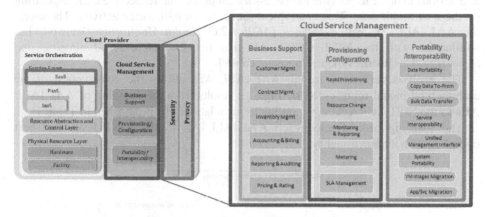

**Fig. 5.** Selection of NIST Components for Analysis

## Step 2: Analyse AAL4ALL AEs

We use AAL4ALL architecture (Fig.4) artifacts to analyse its consistency with NIST CCRA and to adapt it to *Cloud* environment. The analysis look for the intersection of AAL4ALL AEs with NIST CCRA AEs in order to find similar semantics in their AEs descriptions. It uses a tabular representation to relate similar AEs.

Table 1 depicts an example of the way how similar AAL4ALL AEs, and NIST AEs are related. This table contain information about NIST CCRA ACs, NIST CCRA AEs, and AAL4ALL AEs. It illustrates the result of the matching between the AAL4ALL AE with reference *{AE0b.1.2.c} Health Monitoring Decisions* and the NIST CCRA AE with reference *Monitoring and Reporting*.

**Table 1.** AAL4ALL AEs and NIST CCRA AEs related

| NIST CCRA Architectural Components | NIST CCRA Architectural Elements (AE) | AAL4ALL Architectural Elements (AE) |
|---|---|---|
| **Provision and Configuration** | **Monitoring and Reporting** Discovering and monitoring virtual resources, monitoring cloud operations and events and generating performance reports. | **{AE0b.1.2.c} Health Monitoring Decisions** Makes decisions on how the measured information from {AE0b.1.2.i} is used within the AAL4ALL Central Platform. The information can be used by the platform for preventing abnormalities in user's wellbeing while he is at home (routines, sport exercises, during sleep, etc) or to follow measured values through times. |

According to the NIST CCRA AE *Monitoring and Reporting* description, the AAL4ALL AE with reference *{AE0b.1.2.c} Health Monitoring Decisions* corresponds to monitoring operations, event monitoring, reporting activities, and service performance. The AAL4ALL *Central Platform* uses the monitored information (*i.e.*, vital signs) to identify abnormal situations related with the health of the users. If an event occurs, the AAL4ALL *Central Platform* triggers a set of alerts and actions in order to provide the necessary support to the respective user, depending on the user health state. This is a typical scenario of a monitoring activity. Therefore, the AAL4ALL AE with reference *{AE0b.1.2.c} Health Monitoring Decisions* has similarities with the NIST CCRA AE that has the reference *Monitoring and Reporting*. This justifies their relation in Table 1.

The analysis performed for all AAL4ALL AEs allows the construction of the architectural model represented by Fig.6. It results from the mapping of AAL4ALL AEs to the NIST CCRA AEs, according to the relationships established in Table 1.

The mapping task affected each AAL4ALL AE to the NIST layer that best complied with its functionality.

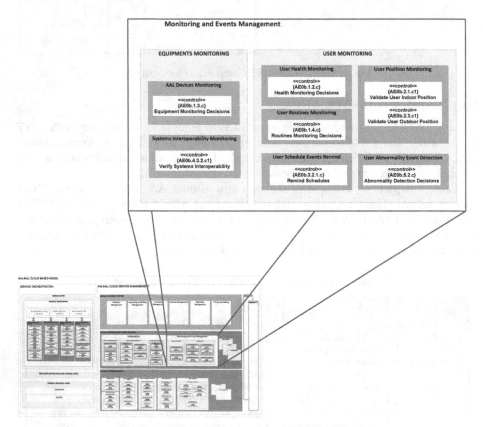

**Fig. 6.** AAL4ALL AEs mapped to NIST CCRA AEs

## Step 3: Refine and development of new Logical Architecture

The results of the analysis enabled us to evaluate and to identify specific *Cloud* requirements for the elicitation phase. The mapping task allowed to find some limitations [9] on the AAL4ALL architecture relative to the NIST CCRA, such as:

(1) the lack of AEs related to business support; (2) the lack of AEs related to security and privacy; (3) AEs with semantics that are not fully compatible with the NIST CCRA.

The analysis performed enabled to complement the requirements elicitation efforts and to redefine the AAL4ALL architecture, giving rise to a new version of the cloud-based AAL4ALL architecture (Fig.7). These results justified the execution of a new iteration on the definition of the AAL4ALL architecture [9].

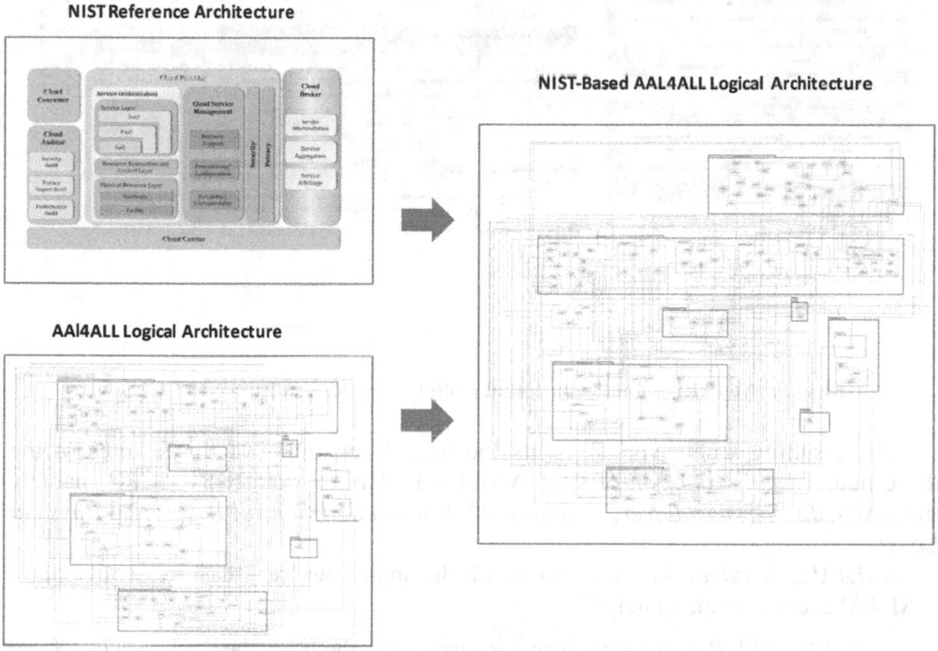

**Fig. 7.** NIST based AAL4ALL Logical Architecture

## Step 4: Fill the table with new logical architecture AE descriptions

In this step we filled the table (similar to the example in Table 1) with AE descriptions of the new AAL4ALL cloud-based, and associated the NIST CCRA AEs and ACs in the respective columns; this facilitated the mapping of AAL4ALL AEs to NIST CCRA AEs.

## Step 5: Mapping AAL4ALL AEs to NIST CCRA AEs

The correct assignment of AAL4ALL AEs to NIST CCRA AEs results in the CCRA model for the AAL4ALL project, viewed as a conceptual architecture. AAL4ALL CCRA is organized and contextualized with *Cloud* functions and activities according to the NIST CCRA. Fig.8 depicts the cloud-based AAL4ALL conceptual architecture.

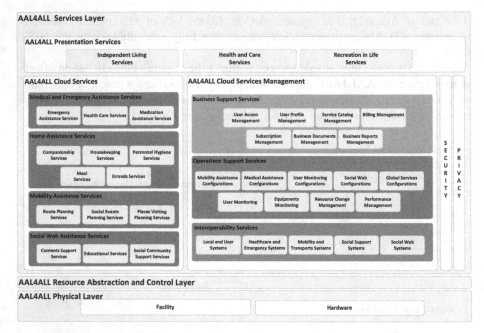

**Fig. 8.** AAL4ALL Cloud Conceptual Architecture derived from NIST CCRA

The resulting architecture presents the logical separation of the features of the three main *Cloud* services layers for AAL4ALL, derived from NIST CCRA, namely the *AAL4ALL Services Layer*, *Resource Abstraction and Control Layer*, and *Physical Layer*.

*AAL4ALL Services Layer* consists of the main services that are available to AAL4ALL ecosystem, namely:

1. *AAL4ALL Presentation Services* - expose to the users the AAL services from AAL4ALL *Central Platform*, providing access to application functionalities organized in an intuitive way and made available to end users; they cover the main AAL services areas, namely: (1) *Independent Living Services*; (2) *Health and Care Services*; and (3) *Recreation in Life* Services.

2. *AAL4ALL Cloud Services*: present the specific domain AAL services divided by the categories: (1) *Assistance and Emergency Medical Services*; (2) *Home Assistance Services*; (3) *Mobility Assistance Services*; (4) *Social Assistance Web Services*.

3. *AAL4ALL Cloud Services Management*: includes all functionalities related to services provided by AAL4ALL *Central Platform* at management and operations of AAL services subscribed by AAL4ALL services consumers. This sub-layer is structured in the following service categories:

   - *Business Support Services:* Composed by services that support the business development and business management activities that enable the AAL services providers to manage the business activities related to the AAL services offer (such as customer management, billing, services payment, and contract management). These services ensure the customer life-cycle management in the AAL4ALL ecosystem.

- *Operations Support Services*: Composed by services that support the AAL4ALL service operations related to customer services provisioning and operation, services and equipment setup and maintenance, software and firmware upgrades, and service and equipment monitoring, among others.

- *Interoperability Services*: composed by services that ensure the services interoperability and systems integration.

4. *Security and Privacy Layers* are important aspects in cloud-based solutions that must address security requirements in their architectures across all layers of architecture, from the application layer to the physical layer. AAL4ALL security services include non-functional requirements such as authentication, authorization, availability, confidentiality, integrity, incident management, and information security monitoring. Privacy services ensure the mechanisms for collecting user information securely, properly, and consistently.

*AAL4ALL Resource Abstraction and Control Layer* ensures the union of the underlying physical resources and software resources, allowing the resource pooling and dynamic resource allocation as well the access control to *Cloud* services.

*AAL4ALL Physical Layer* includes all the features of the physical layer (hardware), such as computing (CPU and memory), network (routers, firewall, network interfaces, data storage components (hard disks), among other elements at the physical infrastructure. It also includes resources at the installation level (facility), such as cooling systems, and communications, among others.

# 5    Conclusions and Future Work

This paper demonstrates our approach of how to use the NIST CCRA to support the design of *Cloud* architectures that enable the design of cloud-based solutions and services for specific contexts, through the refinement of logical architectures.

We present a demonstration case that shows the use of NIST CCRA to refine the design of the AAL4ALL architecture to support the specification of cloud-based services, through the association of the AAL4ALL AEs to corresponding NIST CCRA AEs with similar semantics. The comparison of AEs descriptions enables the identification of compatible AEs. A tabular representation associates each AAL4ALL AE to a related NIST CCRA AE. The result of the analysis allowed complementing the efforts of requirements elicitation and the alignment of AAL4ALL architecture with *Cloud* requirements according to NIST CCRA.

The association of the AAL4ALL AEs into NIST CCRA AEs enabled to apply the CCRA to the AAL4ALL project and to specify the main layers of AAL4ALL services derived from NIST CCRA, namely the AAL4ALL Services Layer, AAL4ALL Resource Abstraction Layer, and AAL4ALL Control Layer and Physical Layer.

Our approach can be used by others to: (i) analyze the coverage of logical architectures with cloud computing issues; (ii) elicit cloud computing requirements;

(iii) define terminology and semantic for cloud contexts; and (iv) specify cloud services and cloud architectures.

As future work, we plan to study how to use the NIST CCRA with other reference models, guidelines, and maturity models in order to have more robustness architectures for *Cloud* contexts.

# References

1. Mell, P., Grance, T.: The NIST Definition of Cloud Computing. NIST Special Publication 800-145, NIST (2011)
2. MacKenzie, C., Laskey, K., Brown, P., Metz, K.: OASIS: Reference Model for Service Oriented Architecture 1.0. OASIS Open (2006)
3. Anbarasu, A.: Cloud Reference Architecture. White paper, Oracle (2012)
4. Bohn, R., Messina, J., Liu, F., Tong, J., Mao, J.: NIST Cloud Computing Reference Architecture. In: 2011 IEEE World Congress on Services (SERVICES), pp. 594–596. IEEE Computer Society (2011)
5. Liu, F., Tong, J., Mao, J., Bohn, R., Messina, J., Badger, M., Leaf, D.: NIST Cloud Computing Reference Architecture. NIST Special Publication 500-292, NIST (2011)
6. Van Den Broek, G., Cavallo, F., Wehrmann, C. (eds.): AALIANCE Ambient Assisted Living Roadmap. IOS Press, Amsterdam (2010)
7. Lee, K., Lunney, T., Curran, K., Santos, J.: Proactive Context-Awareness in Ambient Assisted Living, http://scisweb.ulster.ac.uk/~kevin/ICADIpap.pdf
8. AL4ALL Project - Ambient Assisted Living for All, http://www.aal4all.org/
9. Santos, N., Teixeira, J., Pereira, A., Ferreira, N., Lima, A., Simões, R., Machado, R.J.: A Demonstration Case on the Derivation of Process-Level Logical Architectures for Ambient Assisted Living Ecosystems. In: Garcia, N., Rodrigues, J., Dias, M.S., Elias, D. (eds.) Ambient Assisted Living Book, Taylor and Francis/CRC Press, USA (accepted for publication)
10. Martínez-Fernández, S., Ayala, C., Franch, X., Marques, H., Ameller, D.: A Framework for Software Reference Architecture Analysis and Review, http://www.essi.upc.edu/~dameller/wp-content/papercite-data/pdf/martinez-fernandez2013.pdf
11. Ferreira, N., Santos, N., Machado, R.J., Gašević, D.: Derivation of Process-Oriented Logical Architectures: An Elicitation Approach for Cloud Design. In: Dieste, O., Jedlitschka, A., Juristo, N. (eds.) PROFES 2012. LNCS, vol. 7343, pp. 44–58. Springer, Heidelberg (2012)
12. Kang, S., Choi, Y.: Designing Logical Architectures of Software Systems. In: Sixth Int. Conf. Softw. Eng. Artif. Intell. Netw. Parallel/Distributed Comput. First ACIS Int. Work. Self-Assembling Wirel. Networks, pp. 330–337 (2005)
13. Demchenko, Y., Makkes, M.X., Strijkers, R., De Laat, C.: Intercloud Architecture for interoperability and integration. In: IEEE 4th International Conference on Cloud Computing Technology and Science (CloudCom), pp. 666–674. IEEE Computer Society (2012)
14. Demchenko, Y., Ngo, C., Makkes, M.X., Strijkers, R., De Laat, C.: Defining inter-cloud architecture for interoperability and integration. In: Zimmermann, W., Lee, Y.W., Demchenko, Y. (eds.) Third International Conference on Cloud Computing, GRIDs, and Virtualization, pp. 174–180. IARIA, Wilmington (2012)

15. Jrad, F., Tao, J., Streit, A.: SLA based Service Brokering in Intercloud Environment. In: Proceedings of 2nd International Conference on Cloud Computing (CLOSER 2012). SciTePress (2012)

16. Amanatullah, Y., Lim, C., Ipung, H., Juliandri, A.: Toward cloud computing reference architecture: Cloud service management perspective. In: 2013 International Conference on ICT for Smart Society (ICISS), pp. 1–4 (2013)

17. IBM:Getting cloud computing right. White paper. IBM Corporation (2011)

# Experimental Evaluation of Conceptual Modelling through Mind Maps and Model Driven Engineering

Fernando Wanderley[1], Denis Silveira[2], João Araujo[1],
Ana Moreira[1], and Eduardo Guerra[3]

[1] CITI - Faculdade de Ciências e Tecnologia – Universidade Nova de Lisboa, Lisboa, Portugal
{f.wanderley,joao.araujo,amm}@fct.unl.pt
[2] PROPAD – Universidade Federal de Pernambuco, Pernambuco, Brazil
dsilveira@ufpe.br
[3] INPE – Instituto Nacional de Pesquisas Espaciais, Brazil
eduardo.guerra@inpe.br

**Abstract.** Recent research studies report evidence that many systems' requirements are not fully understood, being difficult to elicit and produce accurate conceptual models more efficiently. We have been investigating how creative models and model-driven development can contribute to overcome this difficulty. Creative models such as mind maps offer effective cognitive support to rapidly produce conceptual models that are closer to stakeholders' expectations. This paper reports on the results of a controlled experiment carried out in academia and industry using a creative and agile modeling approach. This approach uses mind maps to generate conceptual models through transformations, using model-driven engineering techniques. The empirical evidence discussed in this paper shows a significant gain in time spent to build conceptual models.

**Keywords:** Mind Map Modelling, Requirements Engineering, Conceptual Modelling, Experimental Software Engineering.

## 1 Introduction

Conceptual Models are needed to achieve a common understanding of the system domain [1]. Therefore, according to Standish Group [2], companies spend nearly $1.8 trillion on software, hardware, and telecommunications equipment, although **only 35 per cent** of software projects are categorized as successful in satisfying user system requirements. That research has suggested that system failures are due to lack of clear and specific information requirements.

Agile development techniques rely on customer collaboration to define and prioritize software requirements [ref Agile Manifesto]. In this context it is important to create a common vision of the business domain between developers and customers, defining a shared vocabulary among them. Unfortunately, UML diagrams are too technical for most of the customer understand and contribute. In this context, it is important to have a tool that enables a collaborative conceptual modelling between developers and customers.

B. Murgante et al. (Eds.): ICCSA 2014, Part V, LNCS 8583, pp. 200–214, 2014.

We have been investigating how mind maps [3, 4] can be used in conceptual modelling to facilitate communication among domain experts and developers, and how the model-driven engineering (MDE) techniques can help to create conceptual models from mind maps. The resulting approach described in [3, 4] shows the advantages of mind maps and uses MDE techniques to automatically transform requirements expressed in mind maps into conceptual models, such as UML class diagrams.

This paper evaluates this approach to find empirical evidence of the value of mind maps in requirements elicitation and specification. This evaluation focuses on measuring the approach's effort, or *the time spent for construction of conceptual models when creative models are used*. The evaluation was carried out as a controlled experiment involving senior, middle and junior software designers from industry and academia.

The qualities evaluated are time of construction (which is related to the effectiveness of domain knowledge comprehension and communication) and usability. The set of metrics (for usability and effort) of the mind maps analyzed in this study was defined based on the Goal-Question-Metric (GQM) process. GQM [5] is a goal-oriented approach for measuring software systems based on three levels of abstraction.

The results were encouraging as the time spent to produce conceptual models with the help of mind maps were always lower than that obtained to produce those models without mind maps.

The remainder of this paper is structured as follows. Section 2 gives a background on conceptual modelling and their relation with the requirements engineering elicitation and specification activities. Also, it illustrates the approach with an example from industry. Section 3 reports the complete experiment design relating the activities, tasks, artifacts and individuals involved. Section 4 describes the experiment operation. Section 5 reports and analyses the experiment results. Finally, Section 6 discusses some related work, while Section 7 provides some conclusions and points out directions for future work.

## 2    Background

**Conceptual Modelling.** The formal basis of conceptual modelling languages is logic and the first-order logic (FOL) language is sufficient for the specification of most conceptual schemas. However, in many projects the use of logical languages is impractical, and specialized languages are more suitable [1]. One such language is the Unified Modeling Language (UML). Conceptual modeling must precede system design and some requirements engineering tasks should precede conceptual modelling [1]. Table 1 shows the relationship between requirements elicitation and specification, and the essentials of conceptual modelling, where mind maps are proposed for elicitation and class diagrams are proposed for the specification.

**Table 1.** The relation between RE and CM

| Requirements Engineering | Conceptual Modelling |
|---|---|
| Requirements Elicitation (*with Mind Maps*). | A conceptual model of the existing and/or desired domain may be created to achieve a common understanding of the domain(s) [3, 4]. |
| Requirements Specification (*with UML Class Diagrams*). | The conceptual model of an information system is the specification of its functional requirements. This conceptual model specifies all the essential properties (objects and relationships) that the system must have and, together with the specification of the non-functional requirements, corresponds to the system specification. |

Customers are supposed to participate on the requirements elicitation, where a conceptual model is created to share this business model. The process with customer and developers collaboration is also important to create a sharing knowledge about the domain. The requirements specification is more technical and is created based on the knowledge obtained from the first activity. Consequently, it can involve only the development team.

**From Mind Map to Conceptual Model.** Figure 1 shows a creative agile requirements modeling approach that uses MDE techniques to transform mind maps into conceptual model. Given the simplicity of mind maps, the communication between domain experts and the development team is facilitated given the abstraction level (and simplicity) of mind maps. Mind maps are used as a cognitive and systematic tool to support requirements engineers in their elicitation and analysis tasks [6, 7, 8]. In the context of this work, our approach generates conceptual models expressed as UML class diagrams. This model is obtained through transformations, by applying model-driven techniques. The generation process includes the definition and implementation of metamodels for mind maps and class diagrams, and a set of transformation rules written in MOFScript (http://www.eclipse.org/gmt/mofscript) that use those metamodels to generate the class diagrams.

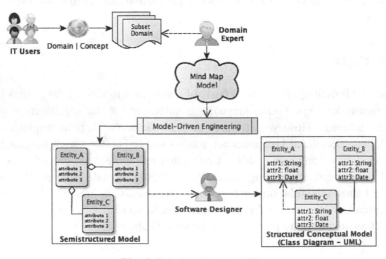

**Fig. 1.** Solution Proposed [2]

The approach consists of three steps: (i) *Domain Analysis*, where mind maps are created to represent the main concepts of the domain (or domain subsets); (ii) *Conceptual Models Derivation* from mind maps, through MDE techniques including a set of transformation rules; (iii) *Refining the Conceptual Model*, where additional model elements (e.g. attributes' types) that were not captured by the transformation are included by the software designer. In this paper we focus on of the evaluation of this approach for adopting mind maps in conceptual modeling. Such evaluation is performed as a semi-controlled experiment, involving senior, middle and junior professionals from industry and academia.

### 2.1 Application to an Industrial Case

To clarify the transformation method, we use Audiobus, a case study from industry and developed by Mobciti (http://www.mobciti.com/). The Mobiciti is a startup company from Brazil that deals with innovation products which link informative, cultural and advertising contents for passengers using public transports. Audiobus is an intelligent system that uses information-geographic systems, which broadcasts digital surround sound for urban transports; customize messages and contents exclusively developed for the surrounding area, provide personalized advertising in each vehicle (see Figure 2).

The next paragraphs illustrate the three main steps of our approach using Audiobus. Note that the concepts identified in the Domain analysis are captured by the mind map of Figure 3.

**Domain Analysis.** Audiobus distributes advertising and interaction contents through a mobile device (called *Device_Android* in Figure 3) installed in the bus as a content receiver and service provider integrated with the bus audio system. Through this mobile device, using HTTP protocol, the system monitors (*Monitor*) essential dynamics information (*DynamicData*) like temperature, battery status and bus location. It is also possible to verify the operation status (*Oper_Status*), audio volume (which can be changed remotely – *RemoteCommand*) and the broadcasting history already transmitted. The broadcasted content has time, duration and are related with geo-referenced information (*DataGeoref*) through coordinates and coverage area (*BeamInput*) and *BeamOutput*).

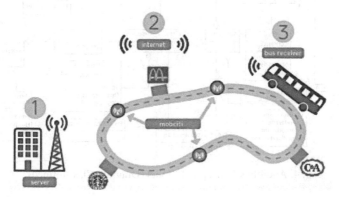

**Fig. 2.** Context of the Audiobus Information System (from Mobciti)

This broadcast content can be music, advertising or a warning (that can be issued in urgency cases for example). This propagation (*BroadcastingType*) can be schedule, georeferenced (dedicated to certain regions), by Bluetooth or instantly. To represent the systems' domain (based on the domain description), the domain expert together with the requirements engineer model the mind map shown in Figure 3.

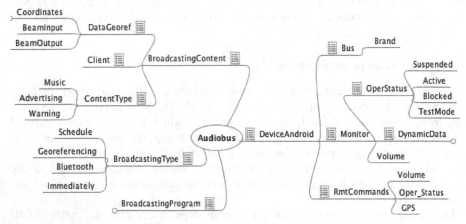

**Fig. 3.** Mind Map of Audiobus System

**Deriving Conceptual Models**. The XML format of the mind map, initially specified by the domain experts and requirements engineers, is the input to the MindDomain tool [2], which transforms mind maps into semi-structured conceptual models. Figure 4 shows the refined generated model, where multiplicities, attributes' types, association and composition relationships are added to the original model.

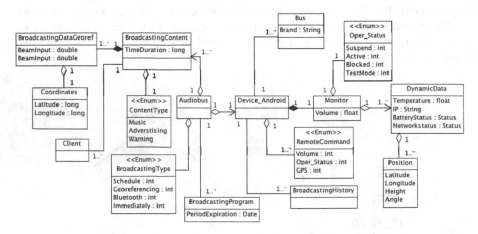

**Fig. 4.** Audiobus Conceptual Model (Diagram Class refined)

**Refining the Conceptual Model**. The transformation gives a first version of the conceptual model, as mind maps contain less information than the required by class diagrams (e.g., multiplicity, attributes' types, other kinds of relationships). Figure 5 depicts a more refined class diagram, where some of the aggregation relationships were changed, and also types and multiplicities were added by the software designer.

In this approach there is a synergy between the domain expert and the requirements engineer (or software designers) as the communication between them is facilitated by the use of mind maps.

# 3    Definition of the Experimental Evaluation

According to Travassos et al. [9], there are different types of experiments. The most appropriate experimentation will depend on the objectives of study, the properties of the software process used during experimentation or the expected final results. The method used in this experimental study is quantitative analysis (collection of periods of time) thus classifying the experiment as a controlled experiment, performed *in-vitro* under laboratory conditions.

The set of metrics (usability and effort) of the mind maps analyzed in this study was defined based on the Goal-Question-Metric (GQM) process. GQM [5] is a goal-oriented approach for measuring software systems based on three levels of abstraction. This approach begins by identifying the measurement goals (conceptual level), taking into account the organization, purpose, quality attributes, views and environment. Each goal is then refined into several questions (operational level), which characterize how a particular goal can be achieved. Metrics are then identified (quantitative level). They provide quantitative information to answer questions from the previous level.

## 3.1    Definition of the Objectives

The two main objectives of the experiment are: *i*) to evaluate the effort (in periods of time) to build conceptual models expressed as UML class diagrams with the cognitive support of a mind map; ii) to evaluate qualitatively the effectiveness of a mind map in understanding the domain analyzed. Based on the structure proposed by [10], Table 2 shows the structure and the objectives defined for this experiment.

**Table 2.** Structure and Objectives of the Experiment

| Structure | Objectives |
|---|---|
| **Objective** of the Study | To analyze and check the **time** taken to produce conceptual models from mind maps, and compare to the **time taken** to obtain those models without using mind maps. |
| **Purpose** | To check the **effort** (in terms of saving time) and **usability** of producing this models with the adoption of mind maps in this process. |
| **Focus** on Quality | Focus on the **time taken** and the **quality** of building the models, like usability and simplicity. |
| **Perspective** | From the software designers' point of view. |
| **Context** | In the context of groups of (conceptual and software) designers from industry and academia. |

## 3.2    Planning

Planning an experiment requires describing what **activities**, **tools**, **artifacts** and **roles** are needed. In other words, it is in planning the experiment that the process for conducting it will be presented. This planning process consists of five steps (Figure 5) described next.

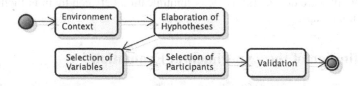

**Fig. 5.** Planning activities of the Experiment process

**Environment Context.** The context of this experiment is about analyzing the time during which the subjects – requirements engineers from academia and industry – develop conceptual models when adopting mind maps. It also analyzes the effectiveness of mind maps based on collecting opinions about the usability and effort of this method by means of a questionnaire.

**Elaboration of Hypotheses.** Having established the objectives and metrics for the analysis and study of the experiment, the hypotheses serve as base premises for analyzing the data collected. Here we describe the set of criteria that addresses the issues raised when defining the experiment. If the hypothesis is rejected, it is possible to draw conclusions based on testing hypotheses previously formulated. Based on the classification and definition of the experiment, null and alternative hypotheses are formalized. The null hypothesis states that there are no real underlying trends or patterns in configuring the experience. The alternative hypotheses are the assumptions from the negation of the null hypothesis, thus configuring elements of interest in the analysis (see Table 3).

**Table 3.** Null and Alternative Hypotheses

| Null Hypotheses (Ho) |
|---|
| (TMCm = TMCs)<br>The time spent for conceptual modeling with a mind map to be equal to the time spent for conceptual modeling without a mind map. |
| (MMCm = MMCs)<br>The average time for conceptual modeling with a mind map to be equal to the average time for conceptual modeling without a mind map. |
| (MdMCm = MdMCs)<br>The median time for conceptual modeling with a mind map to be equal to the median time for conceptual modeling without a mind map. |
| **Alternative Hypotheses (H1)** |
| **Times-**  TMCm< TMCs |
| **Average-** MMCm< MMCs |
| **Median -** MdMCm<MdMCs |

**Selection of Variables**. According to the definition of an experiment, the variables involved are classified as dependent or independent. Independent variables are those over which we can have some level of control. They are the input variables of the experiment. The dependent variables are the ones resulting from the experiment, or the results produced. Table 4 presents the independent variables.

**Table 4.** Independent Variables

| Independent Variables | Control |
|---|---|
| **Different levels of knowledge** | Holding a training event on conceptual modeling to leverage the knowledge. |
| **Lack of knowledge about the tools used** | Conducting practical exercises using the tools and under tuition. |
| **Infrastructure of the machines which performed the experiment** | 6 (six) *notebooks* were provided and equipped with the similar configurations. |
| **Publication of erroneous times** | A chronometric time tool was designed which includes the "pause" function in case the subject has doubts. |

The dependent variables are related to the effective time of modeling requirements artifacts (the conceptual model of information), which are the results of conducting the experiment. Note that the model generated by the subjects are verified by the tutor carrying out the experiment so as to assess whether the model produced is close to, or is adherent to, the template of the model. This is to avoid subjects completing their experiments but producing models that are inconsistent with the expected results.

**Selection of Subjects.** The chosen subjects included undergraduate and graduate students, as well as software designers, developers and software architects working in industry. The subjects were identified and categorized according to the following criteria: (i) **education level** (Specialization, MSc and PhD), (ii) **function** (Developer, Software Designer, Software Architect and Systems Engineer), (iii) **length of experience** (up to 1 year,1-3 years, 3-5 years, 5-8 years and more than 10 years), (iv) **knowledge of modeling** (none, basic– academic knowledge; intermediate– partial use in industry; advanced –complete mastery of use in industry) for conceptual modelling. Having established the criteria, the subjects were distributed as shown in Table 5.

**Table 5.** Distribution of the Subjects

| Category | Context | Criterion | Nr. |
|---|---|---|---|
| Junior | Academia | *Education Level*: Undergraduate - initial. *Experience*: up to 1 year.*Conceptual model*: No model. | 3 |
| Middle | Academia | *Education Level*: Undergraduate – intermediate. *Experience*: 1-3 years. *Conceptual model*: Basic or intermediate. | 3 |
| Senior | Academia | *Education Level*: Undergraduate - final. *Experience*: 1 to 3 years. *Conceptual Model*: Intermediate. | 3 |
| Junior | Industry | *Education Level*: Training: Specialization. *Experience*: 3 to 5 years. *Conceptual Model*: Intermediate | 2 |
| Middle | Industry | *Level of Training*: Specialization. *Experience*: 5 to 8 years. *Conceptual model*: Intermediate. | 4 |
| Senior | Industry | *Education Level*: MSc/ PhD. *Experience*: over 10 years. *Conceptual model*: Advanced. | 3 |
| | | **TOTAL** | **18** |

**Validation.** This phase makes it possible to verify whether the planning was really adhering to the objectives and hypotheses.

### 3.3    Description of the Instrumentation

This section describes the instruments used for carrying out the experiment. The instruments in an experiment are related to the case studies (conducted in the training and implementation phases), and tools to build the models and for the measurements. This section also describes the infrastructure used for conducting the experiment.

**Case Studies.** The case studies are contextualized examples that were used during the training and execution phases of the experiment.

*Training Context.* During the training phase, we used a system for managing the schedule of a taxi company. The system must manage the information related to vehicles, drivers, journeys and clients. It was a simple example used for the subjects to get familiarised with the approach.

*Experiment Context.* For the analysis of the experiment, the selected case for the conceptual modeling activity was based on a global car rental company. This is a subset of the application specified by the Business Rules Group where the full specification can be found at http://www.businessrulesgroup.org/home-brg.shtml.

**Tools.** Table 6 describes the set of tools used to support the implementation of the experiment.

**Table 6.** Tools used in the experiment

| Tool | Description | Finality |
|------|-------------|----------|
| **ArgoUML** | CASE tool for modeling diagrams in UML. | Used for designing the conceptual model (Group A – without mind maps) and refining the conceptual model transformed from the mind map (Group B). |
| **Mind-Domain** | Tool to transform mind map into conceptual models. | Used by Group B (with mind maps) to transform the mind map into a conceptual model. |
| **Test-Watcher** | Tool for managing time of the experiment. | Used by the groups to mark the start time of building the models, to pause (in case of questions — limited to three (3) times) and to record when finished. |
| **Freemind** | Tool to design mind maps. | Used by Group B to create the mind maps. |

## 4    Operation of the Experiment

The operation phase can be seen as a systematic process consisting of three core activities: **preparation, execution** and **validation** of data. During *preparation* the subjects are identified and grouped to perform the experiment, as shown in Table 7. During *execution* the subjects perform the operationalized tasks in the experiment. Finally, during *validation* the results are analyzed.

**Preparation Phase.** This phase started with the training event entitled *"Modeling Software Requirements"*, aiming at introducing the model concepts used and the importance of the conceptual models in the context of Requirements Engineering. The total time spent on training was 1h40min and it was held one week prior to conducting the experiment. During the training phase all subjects received the same descriptions of the system, where they were instructed on how to build conceptual UML models, as well as the construction of mind maps. Upon completion of training, the subjects were divided evenly, according to their experience to obtain balanced groups (see Table 7).

Table 7. Distribution of the Subjects in Groups

| Group A (Without a Mind Map) | | | Group B (With a Mind Map) | | |
|---|---|---|---|---|---|
| Experience | Category | Number. | Experience | Category | Number |
| Junior | Academia | 1 | Junior | Academia | 2 |
| Middle | Academia | 1 | Middle | Academia | 2 |
| Senior | Academia | 2 | Senior | Academia | 1 |
| Junior | Industry | 1 | Junior | Industry | 1 |
| Middle | Industry | 2 | Middle | Industry | 2 |
| Senior | Industry | 1 | Senior | Industry | 2 |
| TOTAL | | 8 | TOTAL | | 10 |

**Execution.** All subjects received individual tutoring and assistance to help with the preparatory steps and production of the models. The process of executing the experiment begins with receiving the descriptions of the domain for each case study, which are abstractions of the conceptual models. It is noteworthy that the texts describing the domains had a "guide" to help the subject to produce each model. The group that used the mind maps received a questionnaire for feedback so as to assess the use of the technique.

**Validation.** Having finalized the production of each model, the tutor of the experiment examines the consistency of the artifacts generated by each subject, using the following checklist: (i) whether the times were measured correctly; (ii) in the spreadsheet which collects the times, if the bias sign was used (i.e., more than 3 interruptions); (iii) if the models were produced "correctly"; (iv) if the subject was in Group B (those using Mind Maps), guarantee s/he fills in the survey questionnaire.

**Threats of Validity.** To avoid errors when collecting data, a tool was developed to extract the data from .CSV files and import it into the Excel tables that served as input artifacts for statistical analyzes. All results were dealt with statistically and guided by the alternative hypotheses described in Table 2. An existing threat to this experimental process is the extent to which the final model is mature and consistent in relation to the domain of the system. However, we know that analyzing the consistency of a model is not a task that is simple to systematize or control. Another important observation is the number of subjects ($n = 18$), which is considered relatively low. Finally, the dimension and the complexity of the case studies can also be a threat.

## 5    Experiment Results

With the objective of analyzing a set of data statistically based on a sample, it is important first of all to perform an exploratory and descriptive analysis of the data collected, i.e., to analyze the distribution of these data. For example, when there are values that are discrepant compared to the average of their Groups, this indicates that there are inconsistent data, known as outliers. The statistical graph used for the analysis proposed is the boxplot. A boxplot is a graph which enables the distribution of a set of data to be represented, based on some of their descriptive parameters, namely: the median ($q_2$), the lower quartile ($q_1$), the upper quartile ($q_3$) and interquartile interval (IQR = $q_3 - q_1$). These descriptive parameters are shown in Table 8; they were used to construct the boxplot in Figure 6. To simplify the representation, in the statistical graphs, each group analysed was labelled according to the method. Group A is the group that did not used mind map and Group B is the group that used mind map.

**Table 8.** Descriptive Parameters of the Boxplot

|  | Group A (without Mind Maps) | Group B (with Mind Maps) |
|---|---|---|
| Min (not outlier) | 913 | 630 |
| 1st Quartile | 1084.25 | 790.25 |
| Median | 1272.5 | 1071.5 |
| 3rd Quartile | 1557.25 | 1324.75 |
| Max (not outlier) | 2127 | 1784 |
| Lim Inf Outlier (Q1-1,5*IIQ) | 374.75 | -11.5 |
| Interval Inter Quartile - IIQ (Q3-Q1) | 473 | 534.5 |
| Lim Sup Outlier (Q3+1,5*IIQ) | 2266.75 | 2126.5 |
| Min (Q1-min not outlier) | 171.25 | 160.25 |
| 1st Quartile | 1084.25 | 790.25 |
| Median | 188.25 | 281.25 |
| 3rd Quartile | 284.75 | 253.25 |
| Max | 569.75 | 459.25 |

Figure 6 shows that the average time to produce conceptual models with mind map is better than without mind maps. The figure shows that the improvement observed in the experiment is between 100 and 200 seconds. We still need more experiments to consolidate our results, but this one serves as a good indication of the use of mind maps with MDE to efficiently generate conceptual models.

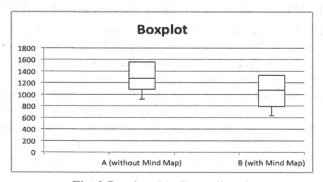

**Fig. 6.** Boxplot of the Data collected

## 5.1 Performance of the Groups

Regarding the hypotheses, the variable of central interest in this experiment is time, where a comparison of the times and mean times between method A (without the use of mind maps) and method B (using mind maps), relating to the design of the conceptual model is carried out. An important point concerns the category relationship (junior, middle and senior) with the context (industry or academia), as shown in Table 8. For statistical relevance, the context classification makes no sense because the sample is very small. For example, in Group A, there is only one (1) junior subject in the academia context, which has nil value when one is dealing with a valid sample. Therefore, as the differences between the subjects are related to time experience, which does not vary much between subjects, all subjects were grouped according to their category (junior, middle and senior) regardless of the industrial or academic context.

**Group A – Without Mind Maps.** The mean times of this group (shown in Table 9 and Figure 7) shows a "break" in the statistical trend regarding the expected behaviour based on the category (experience) of the subjects. This can be understood given that one subject (**#ID = 1**) behaved as if he had more difficulties than the others in delivering his model, thus raising the mean of the group's category (middle), when the experiment was being carried out. This fact is proven because analysis of the subjects' data shows that there were three interruptions and times were lengthy.

Although this situation may have caused unexpected behaviour, the best times of the Group were from subjects of the *senior* group (times of 913 and 1013 seconds) and the worst time (except for this subject quoted above) was equal to 1630 seconds (the junior subject). This finding was not set aside for counting and analysis of the experiment because there were no outliers, as shown in the boxplot at the beginning of this section. The standard deviation of Group A is 403.73 seconds.

**Table 9.** Times of Group A

| #Id | Met. | Time(*sec*) | Cat. |
|-----|------|-------------|------|
| 1   | A    | 2127        | Mid. |
| 2   | A    | 1533        | Mid. |
| 17  | A    | 1013        | Sen. |
| 4   | A    | 1108        | Jr.  |
| 15  | A    | 1630        | Jr.  |
| 10  | A    | 1432        | Mid. |
| 9   | A    | 1113        | Sen. |
| 18  | A    | 913         | Sen. |

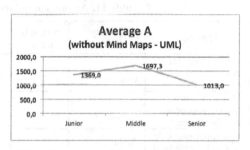

**Fig. 7.** Average of Group A (by Category)

**Group B – With Mind Maps for the conceptual model.** As presented in Table 10 and Figure 8, the times collected from the subjects of Group B (with a mind map) for the conceptual model show an expected behaviour in relation to the statistical tendency of the subjects (experience) correlated to their categories. The two best times (630 and 703 seconds) are related to subjects from the senior group. Only one result, 751 seconds, stands out, relating to a middle subject that was the best in the group. But this was an exception, as the others in the middle category follow the normal trend. The three worst times (1325, 1339 and 1784) are from the subjects of the junior group.

**Table 10.** Times of Group B

| #Id | Met. | Time (*sec*) | Cat. |
|-----|------|--------------|------|
| 16  | B    | 1339         | Jr.  |
| 11  | B    | 938          | Mid. |
| 6   | B    | 1324         | Mid. |
| 7   | B    | 1205         | Mid. |
| 13  | B    | 751          | Mid. |
| 14  | B    | 1325         | Jr.  |
| 5   | B    | 908          | Sen. |
| 12  | B    | 630          | Sen. |
| 3   | B    | 1784         | Jr.  |
| 8   | B    | 703          | Sen. |

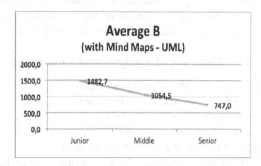

**Fig. 8.** Average of Group B by Category

By looking at Table 10, it can be seen that subject (#ID = 3) had a time of 1784 seconds, therefore a difference of 445 seconds to the penultimate subject whose time was 1339 seconds (also *junior*). This behaviour is similar to the same event that occurred with subject (#ID = 1) from group A. By analyzing the data from this subject, a larger number (equal to 3) of interruptions in relation to his group was noticed.

## 5.2     Qualitative Results

Additionally, to confirm the improvement of the modeling time with or without mind maps, this experiment also allowed us to perform a qualitative analysis of characteristics such as usability, cognition (understanding of the domain) and simplicity (by the use of agile modeling).

These characteristics were analysed from the point of view of the subjects who undertook the modelling with the support of mind maps. They responded a questionnaire whose questions and answers are depicted in Table 11.

**Table 11.** Answers about the Qualitative Analysis

| Question | Quality attribute | Strongly Agree | Agree | Unde-cided | Disagree | Strongly disagree |
|---|---|---|---|---|---|---|
| Can a mind map be considered a model. easy to use? | Usability | 90% | 10% | - | - | - |
| Did mind maps help in understand-ing the domain? | Cognition | 70% | 29% | 1% | - | - |
| Did mind map help in reaching the final conceptual model? | Agile Modeling | 40% | 50% | 10% | - | - |
| Was there a performance improvement? | Performance | 20% | 70% | 10% | | |
| Can the mind map model be used in a context of software development? | Efficiency and Consistency | 70% | 30% | - | | - |

Concerning usability most of the subjects strongly agree that the mind maps are easy to use. Also, they (strongly) agree that mind maps can help in understanding the domain. About 50% agree that mind map help in reaching the final conceptual model. Most of them (70%) agree that there was a performance improvement. Finally, the majority strongly agrees that mind map model be used in a context of software development.

## 5.3     Experiment Analysis and Conclusions

By analysing the experiment data it is possible to conclude that the conceptual modelling by using mind maps in general requires less effort than the same modelling without using it. The time difference is not meaningful in terms of a productivity gain, but it can be considered evidence that it results in a better understanding and comprehension of the software business domain.

This information can be corroborated with the qualitative evaluation, in which most of the experiment participants agree or strongly agree that the mind map is a tool easy to use, and that it helps to reach a domain understanding and a final conceptual model. This is evidence that domain experts can easily use it without previous knowledge on modelling tools for software development.

A curious fact happened in the results was about the participants classified as "junior". Their measured times in performing the tasks did not follow a pattern and because of that it is hard to extract a conclusion from it. For instance, in the groups without mind maps, since there are only two junior participants with very different times, it is hard to classify some of than as a "point outside the curb". Further studies with more participants should be performed for additional conclusions about that.

## 6    Related Work

Czarnecki et al. [11] reports that as domain modeling are performed in the early stage of a design, it is more appropriate that "soft tools" (higher abstraction) for conceptual modelling such as mind maps be used.

Vasconcelos et al. [12] propose GO-MDD, which claims that goal models and MDD can be integrated to fulfil the requirements of a software process maturity model to support the application of GORE methodologies(e.g. i*) in real industry scenarios. Compared to ours, their approach focuses on a different set of metrics (presented in [13], using the GQM approach) as their goal was to support the evaluation of i* models to generate MDD models.

Franch and Grau [14] propose an approach for defining metrics in i* models, in order to analyze the quality of models, as well as to compare alternative models according to some properties. They use a catalogue of patterns for defining metrics. In a related work, Franch proposes a method to better guide the analyst throughout the metrics definition process, over i* models [15]. The method is applied to evaluate business process performance.

España et al. [16] presents a controlled experiment aiming to compare the performance of subjects applying a text-based conceptual model derivation with the performance of subjects applying the communication-based derivation technique. Our work has a similar goal, nevertheless we provide a model with a higher level of abstraction and easy to use by the domain expert. Another contribution is that our controlled experiment was executed with subjects from both academy and industry.

## 7    Conclusions

Current requirements models can be of great help for the software developer, but not so easy to be understood by an end-user or a domain expert not so familiar with such models. This shared comprehension is crucial to customer and developer collaboration in agile methodologies. Creative models, such as mind maps, seem to bridge this gap. But do they? In this paper we evaluate a model-driven approach that uses mind maps to automatically generate conceptual and feature models. This evaluation is focused on measuring the effort of the approach, i.e., if it is worth, concerning the time spent, adopting creative models to help obtain the models mentioned above.

The experiment was realized with professionals from industry and academia. The results were encouraging as the time spent to produce conceptual models with the help of mind maps were always lower than obtaining those models without using mind maps.

This study opens several possibilities in terms of future research, for example to evaluate the use of creative requirements models together with goal-oriented or behavioural models in a model-driven context. Also, other quality attributes could be evaluated in more depth like the usability and reliability.

## References

[1] Olivé, A.: Conceptual Modeling of Information Systems. Springer (2007)
[2] The Standish Group, 2009 Second quarter research report. The Standish Group International, Inc. (2009)
[3] Wanderley, F., Silveira, D.: A Framework to Diminish the Gap between the Business Specialist and the Software Designer. In: QUATIC 2012, Lisbon, Portugal. IEEE Computer Society (2012)
[4] Wanderley, F., Silveira, D., Araujo, J., Lencastre, M.: Generating Feature Model from Creative Requirements using Model Driven Design. In: 16th International Software Product Line Conference, vol. 2, pp. 18–25. ACM Digital Library, New York (2012) ISBN 978-1-4503-1095-6
[5] Basili, V.R., Caldiera, G., Rombach, H.D.: The Goal Question Metric Approach. In: Marciniak, J.J. (ed.) Encyclopedia of Software Engineering, pp. 528–532. John Wiley & Sons, Inc. (1994)
[6] Mahmud, I.: Mind-mapping: An Effective Technique to Facilitate Requirements Engineering in Agile Software Development. In: Proceedings of 14th International Conference on Computer and Information Technology (ICCIT 2011), pp. 22–24 (December 2011)
[7] Rathore, A.: Requirements Management, user stories, mind-maps and story-trees (2007), http://epistemologic.com/2007/04/08/requirements-management-user-stories-mind-maps-and-story-trees/ (accessed in April 2012)
[8] Jaafar, J.: Collaborative Mind map Tool to Facilitate Requirement Engineering (RE). In: Proceedings of the 3rd International Conference on Computing and Informatics, ICOCI 2011, Bandung, Indonesia (2009)
[9] Travassos, G., Gurov, D., Amaral, E.: Introdução a Engenharia de Software Experimental. Relatório Técnico: RT-ES-590/02, COPPE/UFRJ, Rio de Janeiro (2002)
[10] Wohlin, C., Runeson, P., Host, M., Ohlsson, M., Regnell, B., Wesslen, B.A.: Experimentation in software engineering: an introduction. Kluwer, USA (2000)
[11] Czarnecki, K., Gruenbacher, P., Rabiser, R., Schmid, K., Wasowski, A.: Cool Features and Tough Decisions: A Comparison of Variability Modeling Approaches. In: VaMoS 2012, Leipzig, Germany (2012)
[12] de Vasconcelos, A.M.L., Giachetti, G., Marín, B., Pastor, O.: Towards a CMMI-Compliant Goal-Oriented Software Process through Model-Driven Development. In: Johannesson, P., Krogstie, J., Opdahl, A.L. (eds.) PoEM 2011. LNBIP, vol. 92, pp. 253–267. Springer, Heidelberg (2011), doi:10.1007/978-3-642-24849-8_19.
[13] Giachetti, G., Alencar, F., Xavier, F., Pastor, O.: Applying i* Metrics for the Integration of Goal-Oriented Modeling into MDD Processes. UniversitatPolitècnica de Catalunya, Barcelona, Spain (2010)
[14] Franch, X., Grau, G.: Towards a Catalogue of Patterns for Defining Metrics over i* Models. In: Bellahsène, Z., Léonard, M. (eds.) CAiSE 2008. LNCS, vol. 5074, pp. 197–212. Springer, Heidelberg (2008)
[15] Franch, X.: A Method for the Definition of Metrics over i* Models. In: van Eck, P., Gordijn, J., Wieringa, R. (eds.) CAiSE 2009. LNCS, vol. 5565, pp. 201–215. Springer, Heidelberg (2009)
[16] España, S., Ruiz, M., Gonzáles, A.: Systematic derivation of conceptual models from requirements models: a controlled experiment. In: on Research Challenges in Information Science (RCIS) (2012)

# The Use of Metaheuristics to Software Project Scheduling Problem

Broderick Crawford[1,2], Ricardo Soto[1,3], Franklin Johnson[1,4],
Sanjay Misra[5], and Fernando Paredes[6]

[1] Pontifícia Universidad Católica de Valparaíso, Chile
FirstName.Name@ucv.cl
[2] Universidad Finis Terrae, Chile
[3] Universidad Autónoma de Chile, Chile
[4] Universidad de Playa Ancha, Chile
franklin.johnson@upla.cl
[5] Covenant University, OTA, Nigeria
sanjay.misra@covenantuniversity.edu.ng
[6] Universidad Diego Portales, Santiago, Chile
fernando.paredes@udp.cl

**Abstract.** This paper provides an overview of Software Project Scheduling problem as a combinatorial optimization problem. Since its inception by Alba, there have been multiple models to solve this problem. Metaheuristics provide high-level strategies capable of solving these problems efficiently. A set of metaheuristics used to solve this problem is presented, showing the resolution structure and its application. Among these we can find Simulated Annealing, Variable Neighborhood Search, Genetic Algorithms, and Ant Colony Optimization.

**Keywords:** Metaheuristcs, Software Project Scheduling, Optimization

## 1 Introduction

The Software Project Scheduling [1] is a problem which is trying to determine the correct assignment of tasks to developers, to meet the objectives in a given period of time, this is a combinatorial optimization problem (COP) [2], because for the same project may be a large number of possible and feasible plans, each with different cost and duration. The software project management is a complex activity, which should be efficiently performed by the project leader [3,4]. To properly develop a software project, multiple techniques and methodologies are used. These methodologies allow the administration of both, work teams and the management of processes [5]. In research as proposed by Chicano [6], different metaheuristics are used to tackle software project scheduling as a mono-objective problem by automatically generating test cases, which is an incentive to implement various metaheuristics.

The complexity of combinatorial optimization problems is given by the structure of the problem and consequently the way to explore the universe of possible

B. Murgante et al. (Eds.): ICCSA 2014, Part V, LNCS 8583, pp. 215–226, 2014.
© Springer International Publishing Switzerland 2014

solutions [7]. This universe called search space, can become too wide in a discrete space, so the selection of a suitable solution, is a process that requires the use of appropriate strategies.

To solve the combinatorial optimization problems different techniques have been proposed, which can be classified as complete and incomplete techniques [7]. Complete techniques seek to explore the whole search space, so find a global optimum. Some of these techniques are: Linear Programming (LP), Integer Programming (IP) and Branch and Bound among others. These techniques can be impractical for complex cases.

While on the other hand incomplete techniques seek to explore only the most promising areas of the search space, so find a local optimum, but at a lower cost. While incomplete techniques seek to explore only the most promising areas of the search space, so find a local optimum, but at a lower cost. These techniques can be heuristic methods, where the problem is solved through experience based techniques. when a high-level technique derived from other general or natural technique is used, we call metaheuristic [8].

The metaheuristics use various methods to explore the search space in order to achieve a balance between diversification and intensification. The metaheuristics quickly seek the most promising regions in the global search space, and secondly avoid wasting resources in regions that do not contain high-quality solutions.

In this paper a review of the main metaheuristics used to solve the Software Project Scheduling Problem is presented. Furthermore, we show how it has been applied and modeling so as to be solved by these heuristics.

This paper is organized as follows. In Section 2, we describe the Software Project Scheduling Problem. Section 3 presents metaheuristics to solve SPSP. The resolution with Genetic Algorithm, Ant Colony Optimization, Simulated Annealing, Variable Neighborhood Search, and Memetic Algorithms. Finally, the concluding remarks of investigation made in the paper are presented in Section 4.

## 2    The Software Project Scheduling Problem

The Software Project Scheduling Problem (SPSP) was originally proposed by Alba and Chicano in [1] as Project Scheduling Problem (PSP), this is a specific Project Scheduling Problem associated to software project management. This problem is directly related to PSP and more distantly with Resource-Constrained Project Scheduling (RCPS) [9] .

SPSP is to define which tasks are carried out by the software engineers in a software project, and when these tasks are done. To do this asignament we must consider the skills of software engineers. Accordingly the SPSP have three important resources: the skills, are the abilities needed to carry out a task, the tasks, are the activities to complete the project, and the employees which are software engineers.

## 2.1   The Skills

The skills required in a software project for example are: leadership, analytical ability, design expert, advanced developer, among others. Employees have some or all of these skills. And the tasks require these skills. To define the set of skills, we use $SKL = \{skl_1, ..., skl_{|Skl|}\}$, where $|Skl|$ defines the maximum number of skills.

## 2.2   The Tasks

All projects are composed of a set of tasks which must be performed in a specific order. Some of these tasks can be analysis, design, development, and these sub tasks. Alba proposes the use of a graph called task-precedence-graph (TPG) to represent the order of execution of these tasks. This is a non-cyclic directed graph denoted as $G(Vt, Ed)$. Where $Vt = \{t_1, t_2, ..., t_{|Tsk|}\}$ is the set of tasks and $Ed$ is the edge, for example $(t_i, t_j) \in Ed$, means the task $tsk_i$ is a predecessor of task $tsk_j$. The tasks are denoted by $Tsk = \{tsk_1, ..., tsk_{|Tsk|}\}$, where the number of task is $|Tsk|$. The tasks have some attributes as:

$tsk_j^{sk}$ is the group of skills needed to complete the task $j$, where $tsk_j^{sk} \in Skl$.

$tsk_j^{eff}$ represents the effort needed to finish the task $j$.

## 2.3   The Employees

Employees must be assigned to tasks, so that employees have the necessary skills required to complete the task. $EMP = \{emp_1, ..., emp_{|Emp|}\}$ is the group of employees, where $|Emp|$ is the maximum number of employees. The employees have some attributes as:

$emp_i^{sk}$ is the group of skills of employee $i$. $emp_i^{sk} \subseteq Skl$.

$emp_i^{maxd}$ is the maximum of dedication of an employee to the project. $emp_i^{maxd} \in [0, 1]$, when a employee has $emp_i^{maxd} = 1$ work all day in the project, if the employee has $emp_i^{maxd} = 0.5$ work 50 percent of his work day in this project.

$emp_i^{rem}$ is the monthly remuneration of employee $i$. This is an important value, since the cost of the project is obtained from these values.

## 2.4   The Model of SPSP

The objective function may correspond to minimize the cost or duration of the project, or increase the developed quality. Under a multi-objective approach is to minimize both functions simultaneously. From a mono-objective approach, the project manager select the target, he want to minimize in order to allow create schedules according to their needs [10]. In this review, a mono-objective approach is presented.

A solution to SPSP is represented by a matrix. $M = [Emp \times Tsk]$. The size $|Emp| \times |Tsk|$ represents the dimensions determined by employees and tasks. The values of the matrix are real numbers, $m_{ij} \in [0, 1]$, which represent the degree of dedication of employee $i$ to task $j$. When the dedication is $m_{ij} = 0$, the employee $i$ is not assigned to task $j$, if $m_{ij} = 1$, the employee $i$ works all day in task $j$, and if $m_{ij} = 0.75$ the employee $i$ works 75 percent of his work day in the task $j$. An example of a posible solution with 6 tasks and 3 employees can be seen in Fig. 1

| M(ij) | $t_1$ | $t_2$ | $t_3$ | $t_4$ | $t_5$ | $t_6$ |
|-------|------|------|------|------|------|------|
| $e_1$ | 1.00 | 0.50 | 0.00 | 0.25 | 0.00 | 0.00 |
| $e_2$ | 0.25 | 0.50 | 1.00 | 0.50 | 0.00 | 0.25 |
| $e_3$ | 0.50 | 1.00 | 0.00 | 0.50 | 1.00 | 0.25 |

**Fig. 1.** A possible solution for matrix $M$

The generated solutions are feasible if they meet three constraints as follows.

Constraint 1: All tasks are assigned at least one employee as is presented in Eq. 1.

$$\sum_{i=1}^{|Emp|} m_{ij} > 0 \ \forall j \in \{1, ..., Tsk\} \tag{1}$$

Constraint 2: The employees assigned to the task $j$ have all the necessary skills to carry out the task, it is presented in Eq. 2.

$$tsk_j^{sk} \subseteq \bigcup_{i|m_{ij} > 0} emp_i^{sk} \ \forall j \in \{1, ..., Tsk\} \tag{2}$$

Constraint 3: No employee should exceed his maximum dedication to the project. That means, there is no overwork, that is $pj^{overw} = 0$.

When we have a solution, we must evaluate its feasibility, and then the quality of this solution. To obtain the quality firstly we must to calculate the duration of all tasks and cost of the project. We compute the length time for each task according the following formula:

$$tsk_j^{len} = \frac{tsk_j^{eff}}{\sum_{i=1}^{|Emp|} m_{ij}} \tag{3}$$

We define the time of initilization as $tsk_j^{init}$ and the end-time as $tsk_j^{term}$. To calculate it, we use the TGP. For tasks without precedence the time of initialization $tsk_j^{init} = 0$. On the other hand for the tasks with precedence, we must calculate the end-time for all previous tasks. In this case $tsk_j^{init} = max\{tsk_l^{term} \| (tsk_l, tsk_j) \in Emp\}$, the end-time for any kind of task is $tsk_j^{term} = tsk_j^{init} + tsk_j^{len}$.

When we have the $tsk_j^{init}$, $tsk_j^{term}$, and the duration $tsk_j^{len}$ for task $j$, we can construct a Gantt chart. To obtain the project duration $pj^{len}$, we only use the end-time of the last task. We can represent as $pj^{len} = max\{tsk_l^{term} \| \forall l \neq j(tsk_j, tsk_l)\}$. Then we compute the cost associate to a task as $tsk_j^{cos}$, that values is used to obtain the cost of the whole project $pj^{cos}$ using the following formulas:

$$tsk_j^{cos} = \sum_{i=1}^{|Emp|} emp_i^{rem} m_{ij} tsk_j^{len} \tag{4}$$

$$pj^{cos} = \sum_{j=1}^{|Tsk|} tsk_j^{cos} \tag{5}$$

According to the third constraint, we have to consider the overtime work for each employee and the total overtime work. We define the overtime work as $emp_i^{overw}$ as all work the employee $i$ less $emp_i^{maxd}$ at particular time. To compute the project overwork $pj^{overw}$, we use the Eq 6.

$$pj^{overw} = \sum_{i=1}^{|Emp|} emp_i^{overw} \tag{6}$$

To measure and compare the quality of a solution, it is necessary to create a fitness function that minimizes the cost $pj^{cos}$ and duration $pj^{len}$ of the whole project. This means minimizing the fitness function, to this end we define the fitness function as follows:

$$f(x) = \left(w^{cos} pj^{cos} + w^{len} pj^{len}\right). \tag{7}$$

Where $w^{cos}$ and $w^{len}$ are fixed parameters to equate the differences between costs and duration. That fitness function may vary depending on the resolution models proposed by different authors adapted for running their own heuristics.

## 3 Metaheuristics to Solve SPSP

The metaheuristics are high level strategies that allow adapting the search methods to find at least one good enough solution for a particular problem [7]. The metaheuristics use various methods to explore the search space in order to achieve a balance between diversification and intensification. The metaheuristics quickly seek the most promising regions in the global search space, and secondly avoid wasting resources in regions that do not contain high-quality solutions.

There are several ways of classifying metaheuristics, according to their behavior and search strategy. For example, one of the best known classification criteria considers metaheuristics based on trajectory and population [11].

The metaheuristics based on trajectory, starting from a solution and go exploring your neighborhood, and then they update the current solution, forming a trajectory. The algorithms in this category are extensions of simple local search methods to those who is added a mechanism to escape from local minima, so usually the search ends when a maximum number of iterations is reached, there is an acceptable quality solution or stagnation of the process is detected. In this category are Simulated Annealing (SA), Greedy randomized adaptive search procedure (GRASP), variable neighborhood search (VNS), among others [7].

Moreover, the population based metaheuristics are characterized by working with a set of solutions, known as population generated from different types of operators by iteration. In this category are Genetic Algorithms [12], Memetic Algorithms [13], Ant Colony Optimization [14], and Particle Swarm Optimization [15].

According to the above, we present below some metaheuristics used by different authors to solve the software project scheduling problem.

## 3.1    Genetics Algorithms

We started this study with the genetic algorithm metaheuristic [12]. Genetic algorithm is a stochastic search method based on populations. The Genetic Algorithms simulates the behavior of natural evolution, this means making mutations, selection, crossover and inheritance. Populations may be many or few individuals, each individual may be a solution, then the solutions are evaluated and the best individuals are selected for reproduction. Subsequently the crossing of selected individuals is performed, generating new solutions, which are applied to change operators by introducing a mutation; this process helps to explore the solution space. The mechanism applied in a crossover depends on the representation and the problem. It is important to create an appropriate fitness function that allows selecting the best individuals and / or penalizing the worst.

The first work on genetic algorithms to solve SPSP was presented by Alba and Chicano [1], where the behavior of the genetic algorithms is adapted to SPSP. To accomplish this, a new representation and fitness function is proposed.

*Representation and Fitness function:* For representation Alba assumes that employees are not working overtime and maximum dedication is 1. To represent the solution with GA, a binary string chromosome is used.

To represent the possible values in the matrix are needed only 3 bits, because eight values are possible in the matrix. The matrix X is represented into a chromosome which represents the solution. The length of this chromosome is $|E| \cdot |T| \cdot 3$ that is presented in Fig 2.

This representation is used in conjunction with a fitness function that evaluates the cost and penalty of a solution. This can be seen in Eq.8.

$$f(x) = \begin{cases} 1/q & feasible \\ 1/(q+p) & otherwise \end{cases} \qquad (8)$$

Where $q = w^{cos}pj^{cos} + w^{len}pj^{len}$ is the cost of the solution and $pj = undt + reqsk + pj^{overw}$ is a penalty when a solution is not feasible. This function considers the overwork project $pj^{overw}$, the number of tasks without employee assigned

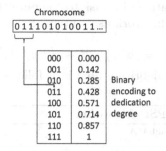

**Fig. 2.** Representation with GA

---

**Algorithm 1.** Pseudocode of a solution using GA
___

```
 1: initialize values
 2: create initial solutions
 3: Evaluate the solution
 4: while condition do
 5:    select solutions
 6:    Recombine
 7:    Evaluate the solution
 8:    Replace
 9: end while
10: Show solution
```
___

*undt*, and the number of skills that are still required to complete the project *reqsk*.

To clarify the proposal by Alba, we present the pseudocode of the proposed solution with genetic algorithms as follows.

## 3.2   Ant Colony Optimization

Ant Colony Optimization (ACO) was created and developed by Dorigo [16]. This metaheuristics take his bases from the behavior of the real ant colonies, and as the ants are able to minimize their routes between the nest and food source.

The first work on Ant Colony Optimization to solve SPSP was presented by Xiao [17], He used a particular type of ACO algorithm called Ant Colony System (ACS), in which all ants are used to perform the update of pheromone values.

*Representation and Fitness function:* Xiao proposes the use of a construction graph on employees and their possible dedications for each project task. He propose a use a matriz $MA = [Emp \times ver]$, where *ver* is the density of nodes, that defines the all possible dedication of employee to a task.

In his work defines two pheromone update rules, a global pheromone update rules and a local pheromone update rules. The global pheromone update rules use the cost, duration and overwork of whole project to compute the quality of a solution to deposit the pheromone. Moreover the local pheromone update rules use only the cost and duration of a task to compute the quality of a solution to deposit the pheromone. This work use a fitness function presented in Eq. 9.

$$f(x) = \left(w^{cos}pj^{cos} + w^{len}pj^{len}\right)^{-1}. \tag{9}$$

The pseudocode of the algorithm used is as follows

---

**Algorithm 2.** ACS for SPSP Algorithm

---
1: initialize parameters $\alpha$, $\beta$, and $MA$
2: initialize solution matrix
3: **repeat**
4:   **repeat**
5:     **repeat**
6:       $t = t + 1$
7:       the ant select a column
8:     **until** $t = |Tsk|$
9:     $a = a + 1$
10:     compute feasible and fitness
11:     update pheromone values(local)
12:   **until** $a = M$
13:   update pheromone values(global)
14:   $g = g + 1$
15: **until** $g = G$
16: output the optimal solution

---

Xiao conduces a series of test with six different heuristics, and he demonstrate that the best heuristic used is based on allocated dedication of an employee to the tasks. Then the ACS algorithm is compared with Genetic algorithm approach on 30 instances. The results show that the best fitness were obtained with ACS.

Recently a new proposal to resolve SPSP with ACO has been proposed by Crawford [18]. In his paper proposes to use another variation of the ACO algorithm called Max-Min Ant System (MMAS) [19] and Hypercube(HC) framework [20].

The solution by MMAS only the best global ant is used for pheromone update, also establishes a new pheromone updating rule based on hypercube framework. In its proposal uses the same matrix representation used by Xiao [17], but provides two new ways to determine the heuristic information.

The first, uses of workload assigned to workers, to not overload an employee already assigned too many tasks. And the second, uses the remuneration of employees, so that is most beneficial select a workers with lower remuneration to reduce total project costs. The pseudocode used is as presented in Algorithm 3.

This proposal can be seen as a more efficient way to solve de SPSP, by implementing a more robust representation a solutions. This proposal introduces in his futures work a concept of autonomous search to improve the performance of the algorithm.

### 3.3 Simulated Annealing, Variable Neighborhood Search, and Memetic Algorithms

This proposed was made by Parra [21]. In his work compare Simulated Annealing (SA), Variable Neighborhood Search (VNS), and Genetic Algorithm (GA), to

**Algorithm 3.** MMAS with HC for SPSP Algorithm

```
 1: repeat
 2:     for g = 1 to Generations do
 3:         for a = 1 to Ants do
 4:             for t = 1 to |Tsk| do
 5:                 ant selects employee dedication
 6:             end for
 7:             compute the feasibility and fitness
 8:         end for
 9:         store the best
10:         update pheromone with HC
11:     end for
12: until condition is complete
```

solve the SPSP under a mono-objective approach. These methods are presented in the following paragraphs and then a explanation of this study is presented.

*Simulated Annealing:* This algorithm is based on the behavior of steel in the annealing process, where initially the particles move very quickly and a disorderly way, with decreasing temperature, the movements are more structured and therefore more ordered [22].

The algorithm starts with a random initial solution, and then a change mechanism is applied to create neighboring solutions. If the result of the neighboring solution is better, it becomes the current solution. This process is repeated until there are improvements in the neighborhood of the current solution: local minimum.

The main disadvantage is that it can be trapped in a cycle. To do this, we could accept some neighbors that are worse than the current solution, which occurs with a given by the acceptance probability function. To better understand the behavior of the algorithm pseudo code is presented below.

**Algorithm 4.** Simulated Annealing Algorithm

```
 1: Define high temperature T
 2: Define cooling schedule T(it)
 3: Define an energy function S
 4: Define initial current_model
 5: repeat
 6:     new_model=random
 7:     calculate P(new_model)
 8:     generate random number
 9:     accept with defined rule
10:     update T
11: until not converged
```

*Variable Neighborhood Search:* It is a metaheuristic in which dynamic changes are made to the structure of the neighbor solutions, with respect to the current solution, specifically in its size [23].

In this way it seeks to diversify exploration in the search space, initially considering solutions first level neighborhoods, then the second and so on, up to the maximum possible level of neighborhoods. Therefore, with the change intends to formulate a mechanism to escape from local minimum, allowing intensify a neighborhood which contains one better than the current solution, and then diversify the compared solutions from other neighborhoods at different levels, i.e. different size.

In this work the author specifically worked with reduced VNS. In this algorithm the activity is similar as basic VNS, except that no local search procedure is used, this algorithm only explores randomly different neighborhoods. This algorithm can find good quality and feasible solutions faster than standard local search algorithm. The pseudocode is applied as follows.

---

**Algorithm 5.** Reduced VNS Algorithm

1: Select $N_k$
2: Find an initial solution $x$
3: $k = 1$
4: **repeat**
5:     $x' = RandomSolution(N_k(x))$
6:     **if** $f(x') < f(x)$ **then**
7:         $x = x'$
8:         $k = 1$
9:     **else**
10:         $k = k + 1$
11:     **end if**
12: **until** $k = k_{max}$
13: output the optimal solution

---

*Memetic Algorithms:* The Memetic Algorithms (MA) were originally proposed by Moscato [24]. They are population-based methods that are combined with local search schemes with genetic operators such as crossover and mutation. The Memetic Algorithm has shown to be an effective and rapid technique for various problems, and achieves an effective balance between the process of intensification and diversification. Furthermore, the Memetic Algorithm are considered hybrid metaheuristic. In general, hybrid metaheuristics are proposals focused on creating high-performance algorithms exploiting and combining the advantages of pure and individual strategies.

In this particular case, a hybrid algorithm is constructed by using a generational replacement genetic algorithm, with local search algorithm through experimental analysis. The algorithm was implemented using the Lamarckism process, working with a probability P and a maximum amount valuation LS. The values of these variables were experimentally established by the author.

These three metaheuristics were implemented and tested using real instances supplied by a software development company. the results were compared according to the time and duration of the project individually, given the mono-objective approach with which the problem was tackled.

These results demonstrated that the best fitness was obtained with MA by using the basis of genetic algorithms and local searches. For larger instances was harder to find better solutions. Other good results were obtained using VNS, but of lower quality than MA. The results with SA shows worse results than VNS and MA.

# 4    Conclusion

This paper provides an overview of metaheuristics used to solve the SPSP. In this work the reasons because metaheuristics offer better solutions than other complete techniques to complex problems are presented. Solutions based on a mono-objective and multi-objective approach are presented. We present solutions for Software Project Scheduling Problem using: Genetic Algorithms, Ant Colony System, Max-Min Ant System, Simulated Annealing, Variable Neihbour Search, and Memetic Algorithms. A comparative analysis of the proposals based on ACO and GA was not done, because the authors working with different instances, which prevent to be objective. We present a problem that is important to adapt to real cases, having a wide applicability in the area of software development and project management.

# References

1. Alba, E., Chicano, J.F.: Software project management with gas. Information Sciences, 2380–2401 (2007)
2. Hromkovic, J.: Algorithmics for Hard Problems: Introduction to Combinatorial Optimization, Randomization, Approximation, and Heuristics. Springer, Heidelberg (2010)
3. Pham, Q.T., Nguyen, A.V., Misra, S.: Apply agile method for improving the efficiency of software development project at vng company. In: Murgante, B., Misra, S., Carlini, M., Torre, C.M., Nguyen, H.-Q., Taniar, D., Apduhan, B.O., Gervasi, O. (eds.) ICCSA 2013, Part II. LNCS, vol. 7972, pp. 427–442. Springer, Heidelberg (2013)
4. Cafer, F., Misra, S.: Effective project leadership in computer science and engineering. In: Gervasi, O., Taniar, D., Murgante, B., Laganà, A., Mun, Y., Gavrilova, M.L. (eds.) ICCSA 2009, Part II. LNCS, vol. 5593, pp. 59–69. Springer, Heidelberg (2009)
5. Mishra, A., Misra, S.: People management in software industry: the key to success. ACM SIGSOFT Software Engineering Notes 35(6), 1–4 (2010)
6. Alba, E.: Análisis y diseño de algoritmos genéticos paralelos distribuidos. PhD thesis (June 1999)
7. Talbi, E.G.: Metaheuristics: From Design to Implementation. Wiley Publishing (2009)
8. Bianchi, L., Dorigo, M., Gambardella, L.M., Gutjahr, W.J.: A survey on metaheuristics for stochastic combinatorial optimization 8(2), 239–287 (2009)
9. Ozdamar, L., Ulusoy, G.: A survey on the resource-constrained project scheduling problem. IIE Transactions 27(5), 574–586 (1995)
10. Chicano, J.F.: Metaheurísticas e Ingeniería del Software. PhD thesis (February 2007)
11. Blum, C., Roli, A.: Metaheuristics in combinatorial optimization: Overview and conceptual comparison. ACM Comput. Surv. 35(3), 268–308 (2003)
12. Chang, C.K., Christensen, M.J., Zhang, T.: Genetic algorithms for project management. Ann. Softw. Eng. 11(1), 107–139 (2001)
13. Hart, W., Krasnogor, N., Smith, J. (eds.): Recent Advances in Memetic Algorithms (2004)

14. Dorigo, M., Stutzle, T.: Ant Colony Optimization. MIT Press, USA (2004)
15. Rada-Vilela, J., Zhang, M., Seah, W.: A performance study on synchronicity and neighborhood size in particle swarm optimization. Soft Computing 17(6), 1019–1030 (2013)
16. Dorigo, M., Gambardella, L.M.: Ant colony system: A cooperative learning approach to the traveling salesman problem. IEEE Transactions on Evolutionary Computation (1997)
17. Xiao, J., Ao, X.T., Tang, Y.: Solving software project scheduling problems with ant colony optimization. Computers and Operations Research 40(1), 33–46 (2013)
18. Crawford, B., Soto, R., Johnson, F., Monfroy, E.: Ants can schedule software projects. In: Stephanidis, C. (ed.) Posters, Part I, HCII 2013. CCIS, vol. 373, pp. 635–639. Springer, Heidelberg (2013)
19. Stutzle, T., Hoos, H.H.: Maxmin ant system. Future Generation Computer Systems 16(8), 889–914 (2000)
20. Johnson, F., Crawford, B., Palma, W.: Hypercube framework for aco applied to timetabling. In: Bramer, M. (ed.) Artificial Intelligence in Theory and Practice. IFIP AICT, vol. 217, pp. 237–246. Springer, Heidelberg (2006)
21. Parra, N., Carolina, D., Salazar, A., Edgar, J.: Metaheuristics to solve the software project scheduling problem. In: CLEI, pp. 1–10 (2012)
22. Mika, M., Waligra, G., Wglarz, J.: Simulated annealing and tabu search for multi-mode resource-constrained project scheduling with positive discounted cash flows and different payment models. European Journal of Operational Research 164(3), 639–668 (2005)
23. Hansen, P., Mladenovi, N., Hansen, P., Mladenovi, N., Gerad, L.C.D.: Variable neighborhood search: Methods and recent applications. In: Proceedings of MIC 1999, pp. 275–280 (1999)
24. Moscato, P.: On evolution, search, optimization, genetic algorithms and martial arts - towards memetic algorithms (1989)

# Multiattribute Based Machine Learning Models for Severity Prediction in Cross Project Context

Meera Sharma[1], Madhu Kumari[2], R.K. Singh[3], and V.B. Singh[4]

[1] Department of Computer Science,
University of Delhi, Delhi, India
[2] Delhi College of Arts & Commerce,
University of Delhi, Delhi, India
[3] Department of Information Technology,
Indira Gandhi Delhi Technical University for Women, Delhi, India
[4] Delhi College of Arts & Commerce, University of Delhi, Delhi, India
{meerakaushik,mesra.madhu,rksingh988,vbsinghdcacdu}@gmail.com

**Abstract.** The severity level of a reported bug is an important attribute. It describes the impact of a bug on functionality of the software. In the available literature, machine learning techniques based prediction models have been proposed to assess the severity level of a bug. These prediction models have been developed by using summary of a reported bug i.e. the description of a bug reported by a user. This work has been also extended in cross project context to help the projects whose historical data is not available. Till now, the literature reveals that bug triager assess the severity level based on only the summary report of a bug but we feel that the severity level of a bug may change its value during the course of fixing and moreover, the severity level is not only characterized by the summary of bug report but also by other attributes namely priority, number of comments, number of dependents, number of duplicates, complexity, summary weight and cc list. In this paper, we have developed prediction models for determining the severity level of a reported bug based on these attributes in cross project context. For empirical validation, we considered 15,859 bug reports of Firefox, Thunderbird, Seamonkey, Boot2Gecko, Add-on SDK, Bugzilla, Webtools and addons.mozilla.org products of Mozilla open source project to develop the classification models based on Support Vector Machine (SVM), Naïve Bayes (NB) and K-Nearest Neighbors (KNN).

# 1    Introduction

The dependence on software is almost inescapable in every sphere of human activities. The developing countries are also getting an edge in the usages of software and especially open source software. During the last four decades, various researchers have developed methods and tools to improve the quality of software. Recently, due to availability of various software repositories namely source code, bugs, attributes of bugs, source code changes, developer communication and mailing list, new research areas in software engineering have been emerged like mining software repositories,

B. Murgante et al. (Eds.): ICCSA 2014, Part V, LNCS 8583, pp. 227–241, 2014.
© Springer International Publishing Switzerland 2014

empirical software engineering and machine learning based software engineering. Various machine learning based prediction models have been developed and is currently being used to improve the quality of software in terms of choosing right developer to fix the bugs, predicting bug fix time, predicting the attributes of a bug namely severity and priority, and bugs lying dormant in the software [1-6]. A large number of bug tracking systems have been developed which help in recording and solving the bugs faced by users. A bug is characterized by many attributes, and severity is one of them. The degree of impact of a bug on the functionality of the software is known as its severity. It is typically measured according to different levels from 1(blocker) to 7(trivial). The summary attribute of a bug report consists of the brief description about the bug. Studies have been carried to predict the bugs either in sever or non-sever category and in different levels using different machine learning techniques [1, 2, and 3]. Recently, an effort has been made to conduct an empirical comparison of different machine learning techniques namely Support Vector Machine, Probability based Naïve Bayes, Decision tree based J48, RIPPER and Random Forest in predicting the bug severity of open and closed source projects [4 and 5]. In all these works, the classifiers have been developed based on the past data of the same project. But, in many cases where the project is new and for which the historical data is not available, it is difficult to predict the severity level of bugs reported in those projects. In this scenario, cross project study can help. An attempt has been made to predict the severity levels of bugs in cross project context on a limited data set using SVM [11].

A software bug report is characterized by the following attributes shown in table 1[12].

In this paper, we have made an attempt to predict the severity level of a reported bug by using multiple attributes namely priority, bug fix time, number of comments, number of bugs on which it is dependent, number of duplicates for it, number of members in cc list, summary weight and complexity of bug in cross project context. Summary weight and complexity are two new attributes which we have derived and discussed in section 2 of the paper. To study and investigate the severity prediction and identification of training candidates in cross project context, we have proposed the following research question:

*Research Question 1:* What is the predictive level that can be achieved by multi attribute based cross project severity prediction models?

In answer of this question, we found that across all the classifiers, performance measures namely accuracy, precision, recall and f-measure lie in the range of 37.34 to 91.63%, 94.99 to 100%, 44.88 to 97.86% and 61.18 to 95.99% respectively for multiattribute based cross project severity prediction.

The rest of the paper is organized as follows. Section 2 of the paper describes the datasets and prediction models. Results have been presented in section 3. Section 4 presents the related work. Threats to validity have been discussed in section 5 and finally the paper is concluded in section 6.

Table 1. Bug Attributes description

| Attribute | Short description |
| --- | --- |
| Severity | This indicates how severe the problem is. e.g. trivial, critical, etc. |
| Bug Id | The unique numeric id of a bug. |
| Priority | This field describes the importance and order in which a bug should be fixed compared to other bugs. P1 is considered the highest and P5 is the lowest. |
| Resolution | The resolution field indicates what happened to this bug. e.g. FIXED |
| Status | The Status field indicates the current state of a bug. e.g. NEW, RESOLVED |
| Number of Comments | Bugs have comments added to them by users. #comments made to a bug report. |
| Create Date | When the bug was filed. |
| Dependencies | If this bug cannot be fixed unless other bugs are fixed (depends on), or this bug stops other bugs being fixed (blocks), their numbers are recorded here. |
| Summary | A one-sentence summary of the problem. |
| Date of Close | When the bug was closed. |
| Keywords | The administrator can define keywords which you can use to tag and categorize bugs e.g. the Mozilla project has keywords like crash and regression. |
| Version | The version field defines the version of the software the bug was found in. |
| CC List | A list of people who get mail when the bug changes. #people in CC list. |
| Platform and OS | These indicate the computing environment where the bug was found. |
| Number of Attachments | Number of attachments for a bug. |
| Bug Fix Time | Last Resolved time-Opened time. Time to fix a bug. |

## 2    Description of Data Sets and Prediction Models

In this paper, an empirical experiment has been conducted on 15,859 bug reports of the Mozilla open source software products namely Firefox, Thunderbird, Seamonkey, Boot2Gecko, Add-on SDK, Bugzilla, Webtools and Addons.mozilla.org. We collected bug reports for resolution "fixed" and status "verified", "resolved" and "closed" because only these types of bug reports contain the meaningful information

**Table 2.** Number of bug reports in each product

| Product | Number of bugs | Observation period |
|---|---|---|
| Firefox | 2712 | Apr. 2001-Aug. 2013 |
| Thunderbird | 115 | Apr. 2000-Mar. 2013 |
| Seamonkey | 6012 | Apr. 1998-July 2013 |
| Boot2Gecko | 1932 | Feb. 2012-Aug. 2013 |
| Add-on SDK | 616 | May 2009-Aug. 2013 |
| Webtools | 366 | Oct. 1998-Aug. 2013 |
| Bugzilla | 964 | Sept. 1994-June 2013 |
| Addons.mozilla.org | 3142 | Oct. 2003-Aug. 2013 |

for building and training the models. The collected bug reports from Bugzilla have also been compared and validated against general change data (i.e. CVS or SVN records). Table 2 shows the data collection in the observed period.

We have used 9 quantified bug attributes namely severity, priority, bug fix time, number of comments, number of bugs on which it is dependent, number of duplicates for it, number of members in cc list, summary weight and complexity of the bug. Severity, Priority, Depends on, Duplicate count and Bug complexity are categorical in nature and rests of the attributes are continuous.

Bug fix time has been calculated as Last_resolved time-Opened time for a bug. A large variation in bug fix time can affect the result. That's why we have considered bugs with fix time less than or equal to 99 days (as maximum number of bugs had fix time in this range). Depending on bug fix time we derived complexity of the bug by defining three levels for fix time. We observed that the bugs are categorized in three levels depending upon their bug fix time and three categories of bug fix time are 0-32 days level 1, 33-65 days level 2 and 66-99 days level 3. Here, we define the complexity of bug from bug fixer point of view. There is a need to quantify summary attribute. This has been done with the help of a two-step process: (1) text mining was applied to get the weight of individual terms of summary by using information gain criteria with the help of a process developed in RapidMiner tool (2) weights of distinct terms in a summary are added to get the weight of the bug summary. There are very less number of reports with number of dependent bugs as most of these bugs are independent. If the bug is independent, the parameter is set to 0 otherwise 1. Similar rule we have followed for duplicate count attribute.

We have carried out following steps for our study:

1. **Data Extraction**
   a. Download the bug reports of different products of Mozilla open source software from the CVS repository: https://bugzilla.mozilla.org/
   b. Save bug reports in excel format.

2. **Data Pre-processing**
   a. Set fix time = Last_resolved time-Opened_ time. Filter the bug reports for fix time less than equal to 99 days.
   b. Set all non-zero values to 1 for depends on and duplicate count attributes.

## 3.  Data Preparation

a.  Derive bug complexity by categorizing bugs in three categories of bug fix time 0-32, 33-65 and 66-99 days.

b.  Calculate the summary weight by using information gain criteria in RapidMiner tool.

## 4.  Modeling

a.  Build three models based on SVM, NB and KNN classifiers.

b.  Train the model by using eight attributes namely priority, bug fix time, number of comments, number of bugs on which it is dependent, number of duplicates for it, number of members in cc list, summary weight and complexity of the bug to predict the bug severity.

c.  Test the model for another dataset.

## 5.  Testing and Validation

a.  Assess the performance of prediction models.

We conducted our study by using Statistica and RapidMiner software tools.

# 3    Results and Discussion

We have applied 3 models based on SVM, NB and KNN classifiers in Statistica software to predict severity in cross project context. In all tables '–' shows that no experiment has been done for that particular combination of testing and training dataset because both training and testing datasets are same.

**Table 3.** Cross project severity prediction accuracy of SVM

| Training Dataset | Testing Dataset (Accuracy in %) | | | | | | | |
|---|---|---|---|---|---|---|---|---|
| | Seamonkey | Thuderbird | Firefox | Boot2Gecko | Bugzilla | AddOnSDK | Addons.mozilla.org | Webtools |
| Seamonkey | - | 59.13 | 76.21 | 53.10 | 34.12 | 46.91 | 61.13 | 63.38 |
| Thuderbird | 62.09 | - | 82.78 | 83.69 | 37.34 | 91.23 | 80.42 | 73.77 |
| Firefox | 91.23 | 64.34 | - | 83.69 | 37.34 | 91.23 | 80.42 | 73.77 |
| Boot2Gecko | 62.09 | 64.91 | 82.77 | - | 37.38 | 91.23 | 80.42 | 73.97 |
| Bugzilla | 58.88 | 53.04 | 60.28 | 45.23 | | 48.86 | 49.26 | 58.19 |
| AddOnSDK | 62.09 | 64.34 | 82.78 | 83.69 | 37.34 | | 80.42 | 73.77 |
| Addons.mozilla.org | 62.09 | 64.34 | 82.78 | 83.69 | 37.34 | 91.23 | | 73.77 |
| Webtools | 62.09 | 64.34 | 82.78 | 83.69 | 37.34 | 91.23 | 80.42 | - |

Table 3 shows cross project severity prediction accuracy of SVM based model for different training data sets. Across all datasets Firefox is the best training candidate for Seamonkey, Boot2Gecko, AddOnSDK and Addons.mozilla.org datasets. Boot2Gecko is the best training candidate for Thunderbird, Bugzilla and Webtools datasets. We observed that in SVM for different training datasets we are getting same accuracy.

**Table 4.** Cross project severity prediction accuracy of NB

| Training Dataset | Testing Dataset (Accuracy in %) | | | | | | | |
|---|---|---|---|---|---|---|---|---|
| | Seamonkey | Thuderbird | Firefox | Boot2Gecko | Bugzilla | AddOnSDK | Addons.mozilla.org | Webtools |
| Seamonkey | - | 45.22 | 78.69 | 45.91 | 29.98 | 37.5 | 46.44 | 51.37 |
| Thuderbird | 62.06 | - | 82.71 | 83.02 | 37.24 | 89.12 | 78.58 | 73.5 |
| Firefox | 63.07 | 59.13 | - | 77.74 | 35.48 | 68.51 | 70.81 | 69.13 |
| Boot2Gecko | 63.15 | 65.18 | 83.89 | - | 45.01 | 91.63 | 82.46 | 78.26 |
| Bugzilla | 59.03 | 47.83 | 62.72 | 77.75 | - | 84.42 | 77.75 | 71.86 |
| AddOnSDK | - | - | - | - | - | - | - | - |
| Addons.mozilla.org | 60.58 | 58.26 | 73.19 | 59.42 | 37.55 | 88.8 | - | 72.95 |
| Webtools | 61.04 | 63.48 | 78.47 | 74.95 | 38.07 | 90.1 | 79.98 | - |

Table 4 shows cross project severity prediction accuracy of NB based model. Across all datasets Boot2Gecko is the best training candidate for Seamonkey, Thunderbird, Firefox, Bugzilla, Webtools, AddOnSDK and Addons.mozilla.org datasets. Thunderbird is the best training candidate for Boot2Gecko dataset. For AddOnSDK dataset, training model was giving a zero standard deviation for the conditional distribution of the independent variable CC Count. So, we were not able to run the experiment for this dataset as training candidate.

**Table 5.** Cross project severity prediction accuracy of KNN

| Training Dataset | Testing Dataset (Accuracy in %) | | | | | | | |
|---|---|---|---|---|---|---|---|---|
| | Seamonkey | Thuderbird | Firefox | Boot2Gecko | Bugzilla | AddOnSDK | Addons.mozilla.org | Webtools |
| Seamonkey | - | 44.35 | 66.74 | 53.31 | 34.02 | 72.4 | 67.95 | 66.39 |
| Thuderbird | 50.7 | - | 62.5 | 56.99 | 33.51 | 71.59 | 62.92 | 57.38 |
| Firefox | 61.33 | 60 | - | 76.6 | 37.03 | 89.77 | 79.44 | 73.22 |
| Boot2Gecko | 61.91 | 63.72 | 81.37 | - | 43.77 | 90.97 | 0.82 | 77.68 |
| Bugzilla | 36.64 | 36.52 | 38.27 | 28.99 | - | 44.64 | 44.14 | 39.62 |
| AddOnSDK | 61.82 | 64.35 | 81.27 | 79.76 | 37.45 | - | 80.17 | 72.68 |
| Addons.mozilla.org | 59.51 | 57.39 | 73.93 | 68.94 | 36.51 | 86.2 | - | 72.4 |
| Webtools | 56.95 | 52.17 | 69.28 | 63.56 | 36.41 | 79.06 | 71.16 | - |

Table 5 shows cross project severity prediction accuracy of KNN based model.

Across all datasets Boot2Gecko is the best training candidate for Seamonkey, Firefox, Bugzilla, AddOnSDK and Webtools datasets. AddOnSDK is the best training candidate for Thunderbird, Boot2Gecko and Addons.mozilla.org datasets.

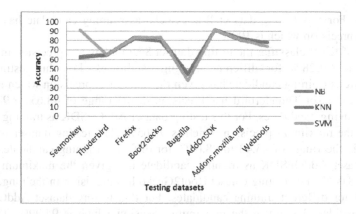

**Fig. 1.** Cross project accuracy (%) for best training candidates in NB, KNN and SVM

Figure 1 shows the maximum cross project accuracy of all testing datasets for the best training candidates in NB, KNN and SVM.

In order to evaluate different performance measures namely precision, recall and f-measure of different classifiers, we have considered severity level 4 only. Due to less number of reports for other severity levels compared to severity level 4, we are getting low performance for these severity levels. This is one of the problems in multi-class prediction where we have imbalance data sets. For empirical validation, we have considered the performance of different machine learning techniques only for severity level 4.

In case of NB classifier, the testing dataset Seamonkey has precision in the range of 92.69 to 100% for different training candidates. For this testing dataset Boot2Gecko as training candidate has given the maximum precision which is 100%. The testing dataset Thunderbird has precision in the range of 59.46 to 98.65% for different training candidates. For this testing dataset Boot2Gecko as training candidate has given the maximum precision which is 98.65%. The testing dataset Firefox has precision in the range of 73.85 to 99.91% for different training candidates. For this testing dataset Thunderbird as training candidate has given the maximum precision which is 99.91%. The testing dataset Boot2Gecko has precision in the range of 49.04 to 99.2% for different training candidates. For this testing dataset Thunderbird as training candidate has given the maximum precision which is 99.2%. The testing dataset Bugzilla has precision in the range of 60.83 to 100% for different training candidates. For this testing dataset Boot2Gecko as training candidate has given the maximum precision which is 100%. The testing dataset AddOnSDK has precision in the range of 39.32 to 98.93% for different training candidates. For this testing dataset Boot2Gecko as training candidate has given the maximum precision which is 98.93%. The testing dataset Addons.mozilla.org has precision in the range of 55.4 to 99.96% for different training candidates. For this testing dataset Boot2Gecko as training candidate has given the maximum precision which is 99.96%. The testing dataset Webtools has precision in the range of 66.3 to 100% for different training candidates. For this testing dataset Boot2Gecko as training candidate has given the maximum precision which is 100%.

In case of KNN classifier the testing dataset Seamonkey has precision in the range of 50.66 to 99.28% for different training candidates. For this testing dataset AddOnSDK as training candidate has given the maximum precision which is 99.28%. The testing dataset Thunderbird has precision in the range of 48.65 to 98.65% for different training candidates. For this testing dataset AddOnSDK as training candidate has given the maximum precision which is 98.65%. The testing dataset Firefox has precision in the range of 43.43 to 98.04% for different training candidates. For this testing dataset AddOnSDK as training candidate has given the maximum precision which is 98.04%. The testing dataset Boot2Gecko has precision in the range of 31.54 to 94.99% for different training candidates. For this testing dataset AddOnSDK as training candidate has given the maximum precision which is 94.99%. The testing dataset Bugzilla has precision in the range of 78.33 to 99.17% for different training candidates. For this testing dataset AddOnSDK as training candidate has given the maximum precision which is 99.17%. The testing dataset AddOnSDK has precision in the range of 47.69 to 98.4% for different training candidates. For this testing dataset Boot2Gecko as training candidate has given the maximum precision which is 98.4%. The testing dataset Addons.mozilla.org has precision in the range of 52.28 to 99.68% for different training candidates. For this testing dataset AddOnSDK as training candidate has given the maximum precision which is 99.68%. The testing dataset Webtools has precision in the range of 50 to 99.26% for different training candidates. For this testing dataset Firefox as training candidate has given the maximum precision which is 99.26%.

In case of SVM classifier the testing dataset Seamonkey has precision in the range of 87.38 to 100% for different training candidates. For this testing dataset all datasets except Bugzilla as training candidates have given the maximum precision which is 100%. The testing dataset Thunderbird has precision in the range of 70.27 to 100% for different training candidates. For this testing dataset all datasets except Bugzilla and Seamonkey as training candidates have given the maximum precision which is 100%. The testing dataset Firefox has precision in the range of 70.47 to 100% for different training candidates. For this testing dataset all datasets except Bugzilla and Seamonkey as training candidates have given the maximum precision which is 100%. The testing dataset Boot2Gecko has precision in the range of 50.34 to 100% for different training candidates. For this testing dataset all datasets except Bugzilla and Seamonkey as training candidates have given the maximum precision which is 100%. The testing dataset Bugzilla has precision in the range of 77.5 to 100% for different training candidates. For this testing dataset all datasets except Seamonkey as training candidates has given the maximum precision which is 100%. The testing dataset AddOnSDK has precision in the range of 48.75 to 100% for different training candidates. For this testing dataset all datasets except Bugzilla and Seamonkey as training candidates have given the maximum precision which is 100%. The testing dataset Addons.mozilla.org has precision in the range of 57.42 to 100% for different training candidates. For this testing dataset all datasets except Bugzilla and Seamonkey as training candidates have given the maximum precision which is 100%.

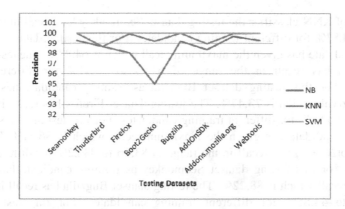

**Fig. 2.** Cross project precision (%) for best training candidates in NB, KNN and SVM

The testing dataset Webtools has precision in the range of 74.07 to 100% for different training candidates. For this testing dataset all datasets except Bugzilla and Seamonkey as training candidates have given the maximum precision which is 100%.

Figure 2 shows the maximum cross project precision of all testing datasets for the best training candidates in NB, KNN and SVM.

In case of NB classifier, the testing dataset Seamonkey has recall in the range of 62.25 to 64.53% for different training candidates. For this testing dataset Bugzilla as training candidate has given the maximum recall which is 64.53%. The testing dataset Thunderbird has recall in the range of 65.31 to 86.27% for different training candidates. For this testing dataset Seamonkey as training candidate has given the maximum recall which is 86.27%. The testing dataset Firefox has recall in the range of 82.89 to 85.95% for different training candidates. For this testing dataset Bugzilla as training candidate has given the maximum recall which is 85.95%. The testing dataset Boot2Gecko has recall in the range of 81.28 to 91.25% for different training candidates. For this testing dataset Seamonkey as training candidate has given the maximum recall which is 91.25%. The testing dataset Bugzilla has recall in the range of 37.86 to 45% to for different training candidates. For this testing dataset Boot2Gecko as training candidate has given the maximum recall which is 45%. The testing dataset AddOnSDK has recall in the range of 91.28 to 97.79% for different training candidates. For this testing dataset Seamonkey as training candidate has given the maximum recall which is 97.79%. The testing dataset Addons.mozilla.org has recall in the range of 80.54 to 86.69% for different training candidates. For this testing dataset Seamonkey as training candidate has given the maximum recall which is 86.69%. The testing dataset Webtools has recall in the range of 73.76 to 82.87% for different training candidates. For this testing dataset Seamonkey as training candidate has given the maximum recall which is 82.87%.

In case of KNN classifier the testing dataset Seamonkey has recall in the range of 62.17 to 64.87% for different training candidates. For this testing dataset Bugzilla as training candidate has given the maximum recall which is 64.87%. The testing dataset Thunderbird has recall in the range of 63.64 to 73.47% for different training candidates. For this testing dataset Bugzilla as training candidate has given the maximum recall which is 73.47%. The testing dataset Firefox has recall in the range of 82.99 to 86.51% for different training candidates. For this testing dataset Bugzilla as training candidate has given the maximum recall which is 86.51%. The testing dataset Boot2Gecko has recall in the range of 83.46 to 88.22% for different training candidates. For this testing dataset Seamonkey as training candidate has given the maximum recall which is 88.22%. The testing dataset Bugzilla has recall in the range of 37.42 to 44.88% for different training candidates. For this testing dataset Boot2Gecko as training candidate has given the maximum recall which is 44.88%. The testing dataset AddOnSDK has recall in the range of 89.93 to 93.62% for different training candidates. For this testing dataset Seamonkey as training candidate has given the maximum recall which is 93.62%. The testing dataset Addons.mozilla.org has recall in the range of 80.31 to 82.61% for different training candidates. For this testing dataset Boot2Gecko as training candidate has given the maximum recall which is 82.61%. The testing dataset Webtools has recall in the range of 73.14 to 78.3% for different training candidates. For this testing dataset Boot2Gecko as training candidate has given the maximum recall which is 78.3%.

In case of SVM classifier the testing dataset Seamonkey has recall in the range of 62.09 to 91.23% for different training candidates. For this testing dataset Firefox as training candidate has given the maximum recall which is 91.23%. The testing dataset Thunderbird has recall in the range of 64.35 to 84% for different training candidates. For this testing dataset Seamonkey as training candidate has given the maximum recall which is 84%. The testing dataset Firefox has recall in the range of 82.78 to 86.97% for different training candidates. For this testing dataset Bugzilla as training candidate has given the maximum recall which is 86.97%. The testing dataset Boot2Gecko has recall in the range of 82.42 to 91.16% for different training candidates. For this testing dataset Seamonkey as training candidate has given the maximum recall which is 91.16%. The testing dataset Bugzilla has recall in the range of 37.34 to 44.89% for different training candidates. For this testing dataset Boot2Gecko as training candidate has given the maximum recall which is 44.89%. The testing dataset AddOnSDK has recall in the range of 91.23 to 97.86% for different training candidates. For this testing dataset Seamonkey as training candidate has given the maximum recall which is 97.86%. The testing dataset Addons.mozilla.org has recall in the range of 80.43 to 85.1% for different training candidates. For this testing dataset Bugzilla as training candidate has given the maximum recall which is 85.1%. The testing dataset Webtools has recall in the range of 73.77 to 91.23% for different training candidates. For this testing dataset Thunderbird as training candidate has given the maximum recall which is 91.23%.

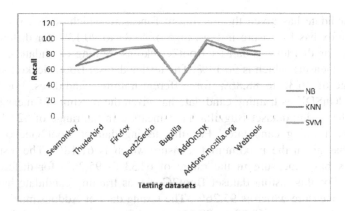

**Fig. 3.** Cross project recall (%) for best training candidates in NB, KNN and SVM

Figure 3 shows the maximum cross project recall of all testing datasets for the best training candidates in NB, KNN and SVM.

In case of NB classifier, the testing dataset Seamonkey has f-measure in the range of 76.09 to 77.8% for different training candidates. For this testing dataset Firefox as training candidate has given the maximum f-measure which is 77.8%. The testing dataset Thunderbird has f-measure in the range of 69.28 to 78.92% for different training candidates. For this testing dataset Boot2Gecko as training candidate has given the maximum f-measure which is 78.92%. The testing dataset Firefox has f-measure in the range of 79.44 to 91.33% for different training candidates. For this testing dataset Boot2Gecko as training candidate has given the maximum f-measure which is 91.33%. The testing dataset Boot2Gecko has f-measure in the range of 63.8 to 90.8% for different training candidates. For this testing dataset Thunderbird as training candidate has given the maximum f-measure which is 90.8%. The testing dataset Bugzilla has f-measure in the range of 50.52 to 62.07% for different training candidates. For this testing dataset Boot2Gecko as training candidate has given the maximum f-measure which is 62.07%. The testing dataset AddOnSDK has f-measure in the range of 56.09 to 95.94% for different training candidates. For this testing dataset Boot2Gecko as training candidate has given the maximum f-measure which is 95.94%. The testing dataset Addons.mozilla.org has f-measure in the range of 67.6 to 90.44% for different training candidates. For this testing dataset Boot2Gecko as training candidate has given the maximum f-measure which is 90.44%. The testing dataset Webtools has f-measure in the range of 73.66 to 87.8% for different training candidates. For this testing dataset Boot2Gecko as training candidate has given the maximum f-measure which is 87.8%.

In case of KNN classifier the testing dataset Seamonkey has f-measure in the range of 56.89 to 77.04% for different training candidates. For this testing dataset Boot2Gecko as training candidate has given the maximum f-measure which is 77.04%. The testing dataset Thunderbird has f-measure in the range of 58.54 to 78.26% for different training candidates. For this testing dataset Boot2Gecko as training candidate has given the maximum f-measure which is 78.26%. The testing dataset Firefox has f-measure in the range of 57.83 to 90.13% for different training candidates. For this testing dataset Boot2Gecko as training candidate has given the maximum f-measure which is 90.13%. The testing dataset Boot2Gecko has f-measure in the range of 46.32 to 88.86% for different training candidates. For this testing dataset AddOnSDK as training candidate has given the maximum f-measure which is 88.86%. The testing dataset Bugzilla has f-measure in the range of 52.61 to 61.18% for different training candidates. For this testing dataset Boot2Gecko as training candidate has given the maximum f-measure which is 61.18%. The testing dataset AddOnSDK has f-measure in the range of 62.33 to 95.26% for different training candidates. For this testing dataset Boot2Gecko as training candidate has given the maximum f-measure which is 95.26%. The testing dataset Addons.mozilla.org has f-measure in the range of 63.89 to 90.12% for different training candidates. For this testing dataset Boot2Gecko as training candidate has given the maximum f-measure which is 90.12%. The testing dataset Webtools has f-measure in the range of 59.73 to 87.4% for different training candidates. For this testing dataset Boot2Gecko as training candidate has given the maximum f-measure which is 87.4%.

In case of SVM classifier the testing dataset Seamonkey has f-measure in the range of 76.61 to 95.42% for different training candidates. For this testing dataset Firefox as training candidate has given the maximum f-measure which is 95.42%. The testing dataset Thunderbird has f-measure in the range of 71.72 to 84.56% for different training candidates. For this testing dataset Seamonkey as training candidate has given the maximum f-measure which is 84.56%. The testing dataset Firefox has f-measure in the range of 77.85 to 91.46% for different training candidates. For this testing dataset Boot2Gecko as training candidate has given the maximum f-measure which is 91.46%. The testing dataset Boot2Gecko has f-measure in the range of 64.58 to 91.12% for different training candidates. For this testing dataset AddOnSDK as training candidate has given the maximum f-measure which is 91.12%. The testing dataset Bugzilla has f-measure in the range of 54.38 to 61.96% for different training candidates. For this testing dataset Boot2Gecko as training candidate has given the maximum f-measure which is 61.96%. The testing dataset AddOnSDK has f-measure in the range of 65.08 to 95.99% for different training candidates. For this testing dataset Boot2Gecko as training candidate has given the maximum f-measure which is 95.99%. The testing dataset Addons.mozilla.org has f-measure in the range of 68.57 to 90.41% for different training candidates. For this testing dataset Boot2Gecko as training candidate has given the maximum f-measure which is 90.41%. The testing dataset Webtools has f-measure in the range of 76.78 to 95.42% for different training candidates. For this testing dataset Thunderbird as training candidate has given the maximum f-measure which is 95.42%.

Figure 4 shows the maximum cross project f-measure of all testing datasets for the best training candidates in NB, KNN and SVM.

Results show that all the three classifiers gave same pattern of accuracy, recall and f-measure. We get maximum values of precision i.e. 100% for SVM across all datasets in comparison of KNN and NB.

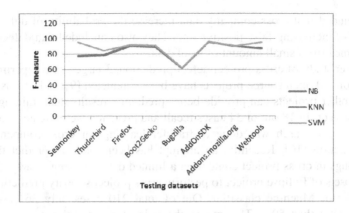

**Fig. 4.** Cross project f-measure (%) for best training candidates in NB, KNN and SVM

## 4    Related Work

An attempt has been made by Chaturvedi and Singh [5] to demonstrate the applicability of machine learning algorithms namely Naïve Bayes, k-Nearest Neighbor, Naïve Bayes Multinomial, Support Vector Machine, J48 and RIPPER to predict the bug severity level bug reports of NASA project. It was found that SVM works with significant accuracy level for open source and NB works with significant accuracy level for closed source projects. Another attempt has been made to predict the severity of reported bug by using summary attribute [1]. Researchers have studied the severity prediction using Naive Bayes classifier and categorized the bugs into severe and non-severe bugs. This study has been further extended to make a comparison with other data mining techniques like Support Vector Machine, Naive Bayes Multinomial and K-Nearest Neighbour for open source projects [3]. Further, a study has been done to determine the severity of reported bug for closed source bug repositories of NASA datasets using NB, MNB, SVM, KNN, J48, and RIPPER [4]. Results show that different machine learning techniques are applicable in determining the bug severity level 2, 3 and 4 with reasonable accuracy and f-measure. Our research has its roots in cross project prediction. Cross project priority prediction have been carried out by researchers and found that the priority prediction in cross project context is satisfactory with more than 70% accuracy and f-measure [6]. In another study authors used 12 real world applications and ran 622 cross-project defect predictions and found that only 21 predictions worked successfully with all precision, recall, and accuracy greater than 75% [7]. They showed that cross-project defect prediction is not symmetric. For example, data of Firefox can predict Internet Explorer defects well with precision 76.47% and recall 81.25% but do not work in opposite direction. They concluded that simply using historical data of projects in the same domain does not lead to good prediction results. They considered a similar problem cross-company defect prediction, using data from other companies to predict defects for local projects and analyzed 10 projects (7 NASA projects from 7 different companies and 3 projects from a Turkish software company SOFTLAB) [8].

They concluded that cross-company data increase the probability of defect detection at the cost of increasing false positive rate. They also concluded that defect predictors can be learned from small amount of local data.

In another study of cross-project defect prediction, 3 large-scale experiments on 34 datasets from 10 open source projects have been conducted [9]. In best cases, training data from other projects can provide better prediction results than training data from the same project. In 18 out of 34 cases recall and precision were greater than 70% and 50%. Another research was done to transfer learning for cross-company software defect prediction [10]. Recently, an attempt has been made to predict the severity levels of bugs in cross project context on a limited data set using SVM [11]. Authors used 7 datasets of Eclipse project to perform cross project severity prediction by using Support Vector Machine classifier. Out of total 210 cases only 36 cases show f-measure greater than 50%. This means the ratio of successful cross-project severity prediction is 17.14% which is very low. This show that selection of training data set should be done more carefully in the context of cross-project severity prediction.

It is clear from the literature that no effort has been made for cross project severity prediction by using bug attributes other than severity. Moreover, we have quantified the reported bug summary based on information gain criteria and derived bug complexity from bug fix time. In this paper, we have automated the prediction of multiclass bug severity and identification of best training candidate in cross project context by using multiple bug attributes.

## 5     Threats to Validity

Following are the factors that affect the validity of our study:

**Construct Validity:** The independent attributes taken in our study are not based on any empirical validation. Bug reports data in the software repository is highly imbalance due to which we are validating prediction results only for severity level 4.

**Internal Validity:** We have considered two derived attributes: complexity and summary weight. Values of these two attributes have not been validated against any predefined values as no predefined values are there in bug repositories.

**External Validity:** We have considered only open source Mozilla products. The study can be extended for other open source and closed source software.

**Reliability:** Statistica and RapidMiner have been used in this paper for model building and testing. The increasing use of these statistical software confirms the reliability of the experiments.

## 6     Conclusion

Bugs lying dormant in the software affect the quality and reliability of the software. A bug is characterized by many attributes. Different classifiers based on machine learning techniques help in predicting and assessing quantified values of these attributes. Earlier, authors have assessed the severity only on the basis of summary of a reported bug. In this paper, we have developed multiattribute based prediction models to assess the severity level of a reported bug in cross project context. The empirical results show that for SVM, KNN and NB classifiers, we get the same pattern of accuracy, recall and f-measure in cross project severity prediction using multiple attributes of a bug. For SVM we get maximum precision i.e. 100 % across all the datasets in comparison of KNN and NB. We found that multiattribute based cross project severity prediction provides accuracy, precision, recall and f-measure in the range of 37.34 to 91.63%, 94.99 to 100%, 44.88 to 97.86% and 61.18 to 95.99% respectively for different classifiers. The proposed work will help the projects whose historical data is not available to predict the severity of bug reports opened in new releases/development cycles. We found that the study to assess the severity level of a bug based on its different attributes will help bug triager in deciding the suitable developer and hence will improve the quality of the software product. In future we will extend our study for more machine learning techniques and datasets of open source and closed source software to empirically validate the study with more confidence.

# References

1. Menzies, T., Marcus, A.: Automated severity assessment of software defect reports. In: IEEE Int. Conf. Software Maintenance, pp. 346–355 (2008)
2. Lamkanfi, A., Demeyer, S., Giger, E., Goethals, B.: Predicting the severity of a reported bug. In: Mining Software Repositories, MSR, pp. 1–10 (2010)
3. Lamkanfi, A., Demeyer, S.Q.D., Verdonck, T.: Comparing mining algorithms for predicting the severity of a reported bug. CSMR, 249–258 (2011)
4. Chaturvedi, K.K., Singh, V.B.: Determining bug severity using machine learning techniques. In: CSI-IEEE Int. Conf. Software Engineering (CONSEG), pp. 378–387 (2012)
5. Chaturvedi, K.K., Singh, V.B.: An empirical Comparison of Machine Learning Techniques in Predicting the Bug Severity of Open and Close Source Projects. Int. J. Open Source Software and Processes 4(2), 32–59 (2013)
6. Sharma, M., Bedi, P., Chaturvedi, K.K., Singh, V.B.: Predicting the Priority of a Reported Bug using Machine Learning Techniques and Cross Project Validation. In: IEEE Int. Conf. Intelligent Systems Design and Applications (ISDA), pp. 27–29 (2012)
7. Zimmermann, T., Nagappan, N., Gall, H., Giger, E.: Cross-project defect prediction: a large scale experiment on data vs. domain vs. process. In: Joint Meeting of the European Software Engineering Conference and the ACM SIGSOFT Symposium on the Foundations of Software Engineering (ESEC/FSE), pp. 91–100 (2009)
8. Turhan, B., Menzies, T., Bener, A.: On the relative value of cross-company and within_company data for defect prediction. Empir. Software Engineering. 14(5), 540–578 (2009)
9. Zhimin, H., Fengdi, S., Ye, Y., Mingshu, L., Qing, W.: An investigation on the feasibility of cross-project defect prediction. Autom Software Engineering 19, 167–199 (2012)
10. Ma, Y., Luo, G., Zeng, X., Chen, A.: Transfer learning for cross-company software defect prediction. Information and Software Technology, Science Direct 54(3), 248–256 (2012)
11. Sharma, M., Chaturvedi, K.K., Singh, V.B.: Severity prediction of bug report in cross project context. In: Int. Conf. Reliability, Infocom Technologies and Optimization (ICRITO), pp. 96–102 (2013)
12. Sharma, M., Kumari, M., Singh, V.B.: Understanding the Meaning of Bug Attributes and Prediction Models. In: 5th IBM Collaborative Academia Research Exchange Workshop, I-CARE, Article No. 15. ACM (2013)

# Self Organizing Networks for 3GPP LTE

Aderemi A. Atayero[*], Oluwadamilola I. Adu, and Adeyemi A. Alatishe

Covenant University, Ota, Nigeria
{atayero,oluwadamilola.adu,
adeyemi.alatishe}@covenantuniversity.edu.ng

**Abstract.** Network elements and their parameters in mobile wireless networks, are largely manually configured. This has been somewhat sufficient; but with the growing data traffic compensated by new and emerging technologies with corresponding larger networks, there is an obvious need to redefine the network operations to achieve optimum performance. A manual configuration approach requires specialized expertise for device deployments, configurations, re-setting network parameters and general management of the network. This process is cost-intensive, time-consuming and prone to errors. Adoption of this approach in the evolved wireless technologies results in poor network performance. Therefore, the introduction of advanced mobile wireless networks has highlighted the need and essence for automation within the network. Self Organizing Networks (SON) developed by 3GPP, using automation, ensures operational efficiency and next generation simplified network management for a mobile wireless network. The introduction of SON in LTE therefore brings about optimum network performance and higher end user Quality of Experience. This paper highlights the SON techniques relevant within an LTE network, a brief description of SON architecture alternatives and then some information on the evolution of SON activities as LTE evolves towards LTE-A.

**Keywords:** LTE, LTE-A, Self-Organizing Networks, Load Balancing.

## 1 Introduction

Over the years there have been series of diverse wireless technology evolutions to meet user and operator demands. One technology or the other has gained preference in one region or the other based on its functionality and/or ease of accessibility. But of all these technological evolutions, 3GPP Long Term Evolution (LTE) has been the most generally and globally accepted of all wireless technologies due to its higher efficiency and flexibility.

Since the introduction of LTE late 2009/early 2010, there have been some testing and deployments which proved the capability of the network, resulting in developments in several domains. User devices like smartphones are being built up to take full advantage of these wireless capabilities proving that it is not just about planning and introducing standards, but about being user-centric in the goals towards making

---

[*] Corresponding author.

B. Murgante et al. (Eds.): ICCSA 2014, Part V, LNCS 8583, pp. 242–254, 2014.

these wireless technologies relevant to the world at large. One of the good problems wireless has is explosive traffic growth which will still increase in the future because even more devices are going to be connected for monitoring activities, communicating, measuring and many others.

Asides the elements visibly present in LTE architecture, there are technologies that make it bigger and better and link it to LTE-Advanced (LTE-A). All of the functionalities of LTE including the "add-on" technologies create a complex and large network, which if not properly managed, defeats the benefits the technology is to provide. Therefore, considering the growing size of the network and the need to keep all operations ubiquitous, seamless and simple, intelligent ways to manage the network's activities is very necessary.

Self Organizing Networks (SON) developed by 3GPP, using automation, ensures operational efficiency and next generation simplified network management for a mobile wireless network. The introduction of SON in LTE therefore brings about optimum performance within the network with very little human intervention.

The drivers for SON are:

- The number and complexities of networks, nodes, elements and parameters
- Existence of multi-technology, multi-vendor and multi-layer operations within the network
- Traffic growth and capacity management
- Consistent quality and service availability
- The need for knowledge-based and interactive networks

The ability to successfully manage all these, leads ultimately to significant operational and capital expenditures (OPEX and CAPEX) savings and excellent performance and user experience. While CAPEX and OPEX savings are derived from the planning and deployment phases of an LTE network, excellent performance/experience and even further OPEX savings are obtained within the optimization and healing activities of the network.

This paper highlights the SON techniques relevant within an LTE network, a brief description of SON architecture alternatives and then some information on the evolution of SON activities as LTE evolves towards LTE-A.

## 2    Self Organizing Networks (SON)

Considering the exploding data traffic and network growth, and the need to keep operations ubiquitous, seamless and simple, the concerns to be dealt with within the LTE network include: plugging/un-plugging a node; optimizing parameters for handoffs and interference management; filling capacity/coverage holes; load imbalance; fault detections and provisioning; energy consumption; inactive or under-active nodes. Therefore, the automation required in the network needs to: a) perform functions which are large-numbered fast, repetitive and complex; b) optimize, detect, report and resolve bottlenecks within the network, and c) perform real-time operations. SON categorizes all automation functions into three: Self Configuration, Self Optimization and Self Healing.

## 2.1    Self Configuration

The essence of self configuration is to achieve a "Plug and Play" system. This process involves three key operations: set-up, authentication and radio configuration. The current self configuration procedures for LTE presents three automated processes: self configuration of eNB, Automatic Neighbour Relations (ANR) and automatic configuration of Physical Cell ID (PCI).

**Self Configuration of eNB.** This is relevant to a new eNB trying to connect to the network. It is a case where the eNB is not yet in relation to the neighbour cells, but to the network management subsystem and the association of the new eNB with the serving gateway (S-GW). It is the basic set-up and initial radio configuration. The stepwise algorithm for self configuration of the eNB is outlined:

1. The eNB is plugged in/powered up.
2. It has an established transport connectivity until the radio frequency transmission is turned on.
3. An IP address is allocated to it by the DHCP/DNS server.
4. The information about the self configuration subsystem of the Operation and Management (O & M) is given to the eNB.
5. A gateway is configured so that it connects to the network. Since a gateway has been connected on the other side to the internet, therefore, the eNB should be able to exchange IP packets with the other internet nodes.
6. The new eNB provides its own information to that self configuration subsystem so that it can get authenticated and identified.
7. Based on these, the necessary software and information for configuration (radio configuration) are downloaded.
8. After the download, the eNB is configured based on the transport and radio configuration downloaded.
9. It then connects to the Operation Administration Management (OAM) for any other management functions and data-ongoing connection.
10. The S1 and X2 interfaces are established.

**Automatic Neighbour Relations (ANR).** ANR is an automated way of adding/deleting neighbour cells. ANR relies on user equipment (UE) to detect unknown cells and report them to eNBs. Its operation can be summarized into: measurements, detection, reporting, decision (add/delete cell) and updating. The step-by-step ANR procedure is outlined:

1. During measurements, the UE detects PCI from an unknown cell.
2. The UE reports the unknown PCI to the serving eNB via Radio Resource Controller (RRC) reconfiguration message.
3. The serving eNB requests the UE to report the E-UTRAN Cell Global ID (ECGI) of the target eNB. The eNB is able to detect devices faster that way.
4. The UE reports ECGI by reading the broadcast channel (BCCH) channel.
5. Based on the ECGI, the serving eNB retrieves the IP address from the Mobility Management Entity (MME) to further set-up the X2 interface, since an

initial X2 interface set-up would have happened during the target eNB's self configuration.

6. Function is extended to inter-RAT and inter-frequency cases with suitable messaging.

**ANR with Operation Administration & Management (OAM) Support.** ANR with OAM support is a more centralized system of operation. The OAM is the management system of the network. ANR procedures with OAM support are outlined:

1. The new eNB registers with OAM and downloads the neighbour information table which includes the PCI, ECGI and IP addresses of the neighbouring eNBs.
2. The neighbours update their own tables with the new eNB information.
3. The UE reports the unknown PCI to the serving eNB.
4. The eNB sets-up the X2 interface using the neighbour information table formed previously.

**Automatic Configuration of Physical Cell Identification (PCI).** A Physical Cell Identifier (PCI) is an important configuration parameter of a cell. Every cell has a PCI which is in the Synchronization Channel (SCH), for synchronization with the UE on the downlink. Unlike the global identifiers which are unique throughout the network, there are only 504 PCIs in the E-UTRAN; therefore, the PCIs are redundant throughout the network and are only unique within a defined region. PCI configuration must satisfy two rules [1]:

• The PCI of one cell must not be same with its neighbour cells.
• The PCIs of the neighbour cells must not be same.

In today's algorithms for automatic PCI assignments, conflicts may occur in the way they are allocated. Therefore, to achieve the aim of SON, work is currently being done to ensure automatic configuration of PCIs become a part of the standardized configuration.

## 2.2    Self Optimization

Self configuration alone is not sufficient to guarantee effective management of the end-to-end network, the need for knowledge-based end-to-end monitoring is also very crucial. After configurations, automated processes/algorithms should be able to regularly compare the current system status parameters to the target parameters and execute corrective actions when required. This process ensures optimum performance at all times. This process is known as Self Optimization. There are several self optimization processes in the standard today, which will be outlined beginning from the most popular.

**Load Balancing.** This is one of the earliest self optimization strategies. It is relevant in cases where a heavily loaded cell and a lightly loaded one are neighbours. It uses an automated function/algorithm to avoid cell overloading and consequent performance degradation. These algorithms adjust parameters of elements intelligently to prevent further issues (like ping pong handovers) while trying to balance the load. Load balancing therefore seeks to achieve the following:

- Balance of load between neighbour cells.
- Improvement in system capacity as a result of regularization of cell congestions.
- Efficient and effective management of the network for optimum performance.

In existing networks, parameters are manually adjusted to obtain a high level of network operational performance. In LTE the concept of self-optimizing networks (SON) is introduced, where the parameter tuning is done automatically based on measurements. A challenge is to deliver additional performance gain further improving network efficiency. The use of load balancing (LB), which belongs to the group of suggested SON functions for LTE network operations, is meant to deliver this extra gain in terms of network performance. For LB this is achieved by adjusting the network control parameters in such a way that overloaded cells can offload the excess traffic to low-loaded adjacent cells, whenever available. In a live network high load fluctuations occurs and they are usually accounted for by over-dimensioning the network during planning phase. A SON enabled network, where the proposed SON algorithm monitors the network and reacts to these peaks in load, can achieve better performance by distributing the load among neighboring cells [2].

When the loads among cells are not balanced, the block probabilities of heavily loaded cells may be higher, while their neighboring cells may have resources not fully utilized. In this case load balancing can be conducted to alleviate and even avoid this problem. There has been a lot of research done on load balancing, which can be classified into two categories: block probability-triggered load balancing [2-4], and utility-based load balancing [5-7].

In the first category, the overhead is low because the load balancing is triggered only when the block probability is larger than a certain threshold. However, the block probability is not minimized, since load balancing can be done before block happening to reduce it. For the second category, i.e., utility-based load balancing schemes, the performance is better because the load balance and throughput are considered in both cell selection and handover phases. However, their overheads are heavy, because the load of each cell has to be exchanged instantaneously [8].

Post third generation (3G) broadband mobile telecommunications such as HSPA+, LTE, and LTE-Advanced use several iterative methods for load balancing but these methods usually require precision, rigor and these carry a high computational cost. In [9,10], the use of soft computing which capitalizes on uncertainty, approximation and imprecision, a new trend in load balancing was designed and simulated using an open source system level simulator developed by Vienna University of Technology, The load indicators used by these iterative methods were used to develop two neural encoded fuzzy models for load balancing. The key performance indicators used are number of satisfied/unsatisfied users, load distribution index and virtual load.

Although load balancing algorithms are uniquely developed and implemented by individual vendors, all the algorithms go by the same steps to accomplishment:

1.  Load is measured for each cell covered by its own eNB and this measurement information is shared among the eNBs of the neighbouring cells over the X2 interface.
2.  The vendor-unique algorithm is applied to check if it is necessary to redistribute load among neighbouring cells.
3.  The necessary parameters are adjusted for effective and efficient load balancing.

For successful load balancing optimization, it is necessary to examine what the load is exactly; this determines the algorithm to be implemented. These algorithms differ from vendor to vendor.

However, the roadmap for efficient/effective load balancing in LTE is targeted towards a system with accurate measuring method and first-hand information of the traffic requirement and radio conditions of each user. Therefore, together with the scheduler and X2 interfaces to other eNBs, the algorithm makes accurate decisions for load-based handover. This kind of handover solution, can give the target cell sufficient information to avoid immediate handover (or ping-pong handover) back to the source cell based on normal handover thresholds [11].

**Mobility Robustness / Handover Optimization (MRO).** Handover coordination is very necessary in ensuring seamless mobility for user devices within a wireless network. In 2G/3G systems, setting handover parameters is a manual and time-consuming task and sometimes too costly to update after initial deployment [12]. Mobility Robustness Optimization (MRO) automates this process to dynamically improve handover operations within the network, provide enhanced end-user experience and improved network capacity.

To achieve this aim, the question to be critically answered is "What triggers handover?" Therefore, 3GPP categorize handover failures into:

-   Failures due to too late handover triggering
-   Failures due to too early handover triggering
-   Failures due to handover to a wrong cell

Also, unwanted handovers may occur subsequent to connection set-up, when cell-reselection parameters are not in agreement with the handover parameters.

Therefore, the MRO algorithm is aimed at detecting and minimizing these failures as well as reducing inefficient use of network resources caused by unnecessary handovers and also reducing handovers subsequent to connection set-up.

As specified by 3GPP, enabling MRO requires that: a) the relevant mobility robustness parameters should be automatically configurable by the eNB SON entities; b) OAM should be able to configure a valid range of values for these parameters; and c) the eNB should pick a value from within this configured range, using vendor-specific algorithms for handover parameter optimization.

For efficient/effective MRO, there must be linkage to policies to ensure other parameters/QoE is not affected. This implies that all parameter modifications must align with other similar interacting SON algorithms (such as Load Balancing). Therefore, there is a need for communication between SON algorithms to resolve probable conflicts and ensure stability.

**Coverage and Capacity Optimization.** Coverage and Capacity Optimization (CCO) is a self optimization technique used in managing wireless networks according to coverage and capacity. CCO measures the health of the network and compares with performance target and policies as defined by individual operators. It has been identified by 3GPP as a crucial optimization area in which the SON algorithm determines the optimum antenna configuration and RF parameters (such as UL power control parameters) for the cells that serve a particular area and for a defined traffic situation, after the cells have been deployed.

For successful implementation of CCO SON algorithms, there is need to take into serious consideration, the difference between coverage optimization and capacity optimization. Coverage optimization involves identifying a "hole" in the network and then adjusting parameters of the neighbouring cells to cover the hole. However, increasing cell coverage affects spectral efficiency negatively due to declining signal power, which results in lesser capacity. It is therefore not possible to optimize coverage and capacity at the same time, but a careful balance and management of the trade-offs between the two will achieve the optimization aim [12].

Adapting to network changes (such as addition/removal of eNBs and change in user distribution) manually is costly and time consuming. Hence, the CCO algorithms operate endlessly, gathering measurements and executing actions if needed. CCO is a slow process in which decisions are made based on long-run statistics.

Below is a list of functions the CCO algorithm is to perform as identified by 3GPP; but 3GPP does not specify how to perform these functions but are operator-defined:

- E-UTRAN coverage holes with 2G/3G coverage.
- E-UTRAN coverage holes without any other coverage.
- E-UTRAN coverage holes with isolated island coverage.
- E-UTRAN coverage holes with overlapping sectors.

**Random Access Channel (RACH) Optimization.** RACH configuration within a network has major effects on the user experience and the general network performance. RACH configuration is a major determinant for call setup delays, hand-over delays and uplink synchronized state data resuming delays. Consequently, the RACH configuration significantly affects call setup success rate and hand-over success rate. This configuration is done in order to attain a desired balance in the allocation of radio resources between services and the random accesses while avoiding extreme interference and eventual degradation of system capacity. Low preamble detection probability and limited coverage also result from a poorly configured RACH. The automation of RACH configuration contributes to excellent performance with little/no human intervention; such that the algorithm monitors the current conditions (e.g. change in RACH load, uplink interference), and adjusts the relevant parameters as necessary. RACH parameter optimization provides the following benefits to the network:

- Short call setup delays resulting in high call setup rates
- Short data resuming delays from UL unsynchronized state
- Short handover delays resulting in high handover success rate

More generally, RACH optimization provides reduced connection time, higher throughput, and better cell coverage and system capacity. All the UE and eNB measurements are provided to the SON entity, which resides in the eNB. An eNB exchanges information over the X2 interface with its neighbours for the purpose of RACH optimization. The PRACH Configuration is exchanged via the X2 setup and eNB configuration update procedures. An eNB may also need to communicate with the O&M in order to perform RACH optimization.

**Inter-Cell Interference Coordination (ICIC).** Mutual interference may occur between the cells in an LTE network. Interference unattended to leads to signal quality degradation. Inter-cell interference in LTE is coordinated based on the Physical Resource Block (PRB). It involves coordinating the utilization of the available PRBs in the associated cells by introducing restrictions and prioritization, leading to significantly improved Signal to Interference Ratio (SIR) and the associated throughput. This can be accomplished by adopting ICIC RRM (Radio Resource Management) mechanisms through signalling of Overload Indicator (OI), High Interference Indicator (HII), or downlink transmitter power indicator.

Multi-layer heterogeneous network layout including small cell base stations are considered to be the key to further enhancements of the spectral efficiency achieved in mobile communication networks. It has been recognized that inter-cell interference has become the limiting factor when trying to achieve not only high average user satisfaction, but also a high degree of satisfaction for as many users as possible. Therefore, inter-cell interference coordination (ICIC) lies in the focus of researchers defining next generation mobile communication standards, such as LTE-A [15].

The servicing operator for each cell carries out interference coordination, by configuring the ICIC associated parameters such as reporting thresholds/periods and prioritized resources. The ICIC SON algorithm is responsible for the automatic setting and updating of these parameters.

The ICIC SON algorithm work commenced in Release 9 but was not completed here. It is targeted at self configuration and self optimization of the control parameters of ICIC RRM strategies for uplink and downlink. To achieve interference coordination, the SON algorithm leverages on exchange of messages between eNBs in different cells through the X2 interface. The SON algorithm enables automatic configuration/adaptation with respect to cell topology, it requires little human intervention and leads to optimized capacity in terms of satisfied users.

**Energy Saving.** Mobile network operators are very keen on finding network energy saving solutions to minimize power consumption in telecommunication networks as much as possible. This will lead to reduced OPEX (since energy consumption is a major part of an operator's OPEX) and enable sustainable development on the long-run. Energy saving is very crucial today, especially with the increasing deployment of mobile radio network devices to cope with the growing user capacity.

OPEX due to energy consumption within a network can be significantly controlled by: a) the design of low-powered network elements; b) temporarily powering off unused capacity; and c) working on the power amplifiers, since they consume majority of the available energy in a wireless network.

The normal practice is the use of modems to put the relevant network elements in stand-by mode. These modems have a separate management system. To achieve an automated system of saving energy, the network elements should be able to remotely default into stand-by mode using the minimum power possible when its capacity is not needed, and also switch-off stand-by mode remotely when needed, without affecting user experience.

The energy saving solutions in the E-UTRAN, which are being worked on by 3GPP, to be used as the basis for standardization and further works are: Inter-RAT energy savings; Intra-eNB energy savings; and Inter-eNB energy savings 3GPP has also stipulated the following conditions under which any energy saving solutions should operate, since energy savings should ideally not result in service degradation or network incompetence:

- User accessibility should be uncompromised when a cell switches to energy saving mode.
- Backward compatibility and the ability to provide energy savings for Rel-10 network deployment that serves several legacy UEs should be met.
- The solutions should not impact the physical layer.
- The solutions should not impact the UE power consumption negatively.

### 2.3  Self Healing

Like self optimization, self healing also handles network maintenance. The management of faults and consequent corrections demands a lot of human input; however, with the exponential growth of the network, addition of devices, there is an inevitable need for automation. Self healing involves automatic detection and localization of failures and the application of the necessary algorithms to restore system functionality.

**Cell Outage Detection and Compensation.** The equipment usually detects faults in itself automatically. But in a situation where the detection system itself is faulty and has therefore failed to notify the OAM, such unidentified faults of the eNBs are referred to as sleeping cells [13]. Cell Outage Detection and Compensation automatically handles these eNB failures by combining several individual mechanisms to determine if an outage has occurred, and then compensating for the failures after soft recovery techniques fail to restore normal service. The automated detection mechanism ensures the operator knows about the fault before the end user. The SON compensation system temporarily mitigates the problem.

### 2.4  SON Architecture Alternatives

The SON architecture defines the location of SON within the network. When implemented at a high level in the network (OAM), it is called Network Management System (NMS); while implementation at lower levels (network elements) like the eNBs is called Element Management System (EMS). For self-configuration techniques of SON, a self configuration subsystem is created in the OAM which handles the self configuration process. For self optimization, the subsystem can be created in the

OAM or the eNB or both. Therefore, depending on the location of SON algorithms, SON architecture may be described as being centralized, distributed or hybrid (a combination of centralized and distributed).

**Centralized SON.** This is an example of the Network Management System (NMS) where the algorithms are created and executed in the OAM as shown in Figure 1. In this type of SON architecture, the algorithms are present in just a few locations thereby making it simple and easy to implement. However, the OAM systems are currently vendor-specific, resulting in optimization limitations across different vendors, hence, defeating the aim of simplicity. In order to benefit maximally from centralized SON, there are several works going on to standardize the Northbound Interface, which is the link between the NMS and the EMS. Implementation of centralized SON is eNB self configuration, where the algorithm is created and executed in the OAM.

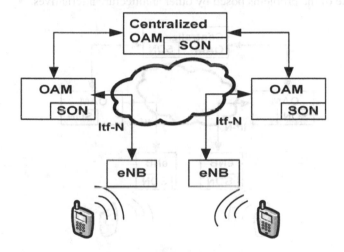

**Fig. 1.** Centralized SON Architecture

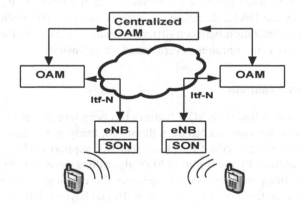

**Fig. 2.** Distributed SON Architecture

**Distributed SON.** An example of the EMS in which the algorithms are deployed and executed at the eNBs is distributed SON. Therefore the SON automated processes may be said to be present in many locations at the lower level of the architecture as seen in Figure 2. Due to the magnitude of deployment to be carried out caused by a large number of eNBs, the distributed SON cannot support complex optimization algorithms.

In order to fully benefit from this architecture type, work is being done towards extending the X2 interface (interface between the eNBs). However, distributed SON offers quick optimization/ deployment when concerned with one/two eNBs. An example of this is in ANR and load balancing optimizations.

**Hybrid SON.** An architecture in which the optimization algorithms are executed in both OAM and the eNBs is called Hybrid SON. The hybrid SON (illustrated in Figure 3) solves some of the problems posed by other architecture alternatives.

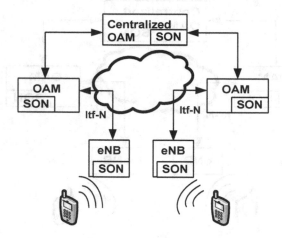

**Fig. 3.** Hybrid SON Architecture

The simpler optimization processes are executed at the eNBs while the complex ones are handled by the OAM; therefore, it supports various optimization algorithms and also supports optimization between different vendors. However, the hybrid SON is deployment intensive and requires several interface extensions.

## 2.5    LTE – SON Evolution

3GPP standardization in line with SON features has been targeted at favouring multi-vendor network environments. As has been discussed, many works are on-going within 3GPP to define generic standard interfaces that will support exchange of common information to be utilized by the different SON algorithms developed by each vendor. The SON specifications are being developed over the existing 3GPP network management architecture defined over Releases 8, 9, 10 and beyond. Outlined hereafter, is the evolution of SON activities as LTE evolved [14].

**Release 8** marked the first LTE network standardization; therefore, the SON features here focused on processes involved with initial equipment installation and integration. Release 8 SON activities include:

- eNB Self Configuration: This involves Automatic Software Download and dynamic configuration of X2 and S1 interfaces.
- Automatic Neighbour Relation (ANR)
- Framework for PCI selection
- Support for Mobility Load Balancing

**Release 9** marked enhancements on Release 8 LTE network; therefore, SON techniques in Release 9 focused on optimization operations of already deployed networks. Release 9 SON activities include:

- Automatic Radio Network Configuration Data Preparation
- Self optimization management
- Load Balancing Optimization
- Mobility Robustness/Handover optimization (MRO)
- Random Access Channel (RACH) Optimization
- Coverage and Capacity optimization (CCO)
- Inter-Cell Interference Coordination (ICIC)

**Release 10** SON in LTE activities include enhancements to existing use cases and definition of new use cases as follows:

- Self optimization management continuation: CCO and RACH
- Self healing management: Cell Outage Detection and Compensation
- OAM aspects of Energy saving in Radio Networks
- LTE self optimizing networks enhancements
- Enhanced Inter-Cell Interference Coordination (eICIC)
- Minimization of Drive Testing

**Release 11** SON activities include:

- UTRAN SON management: ANR
- LTE SON coordination management
- Inter-RAT Energy saving management
- Further self optimizing networks enhancements: MRO, support for Energy saving.

**Release 12** SON activities include:

- Enhanced Network-Management-Centralized CCO
- Multi-vendor plug and play eNB connection to the network.

The 3GPP SON standardization is a work in progress and is expected to cover all focus areas of wireless technology evolution, as it relates to network management, optimization and troubleshooting in multi-tech, multi-cell, multi-actor and heterogeneous networks.

# 3  Conclusion

The evolving mobile internet requires broadband mobile wireless networks with excellent performance hinged on efficient and effective network management. The introduction and growth of LTE-SON establishes the possibility of such management by automating the configuration and optimization of wireless networks. This in turn lowers capital and operational costs, enhances network flexibility and improves overall network performance.

# References

[1] Feng, S., Seidel, E.: Self-Organizing networks (SON) in 3GPP Long Term Evolution. Nomor Research, Munich (2008)

[2] Tonguz, O.K., Yanmaz, E.: The mathematical theory of dynamic load balancing in cellular networks. IEEE Trans. on Mobile Computing 7(12), 1504–1518 (2008)

[3] Eklundh, B.: Channel utilization and blocking probability in a cellular mobile telephone system with directed retry. IEEE Trans. Comm. 34(3), 329–337 (1986)

[4] Jiang, H., Rappaport, S.S.: CBWL: A new channel assignment and sharing method for cellular communication systems. IEEE Trans. Vehicular Technology 43(4), 313–322 (1994)

[5] Das, S., Viswanathan, H., Rittenhouse, G.: Dynamic load balancing through coordinated scheduling in packet data systems. In: IEEE Proc. INFOCOM (2003)

[6] Bu, T., Li, L., Ramjee, R.: Generalized proportional fair scheduling in third generation wireless data networks. In: IEEE Proc. INFOCOM (April 2006)

[7] Son, K., Chong, S., Veciana, G.: Dynamic association for load balancing and interference avoidance in multi-cell networks. IEEE Trans. on Wireless Communications 8(7), 3566–3576 (2009)

[8] Alatishe, A.A., Ike, D.: A review of Load Balancing Techniques in 3GPP LTE. International Journal of Computer Science Engineering 2(4), 112–116 (2013)

[9] Atayero, A.A., Luka, M.K., Alatishe, A.A.: Neural Encoded Fuzzy Models for Load Balancing in 3GPP LTE. International Journal of Applied Information Systems 3(7) (September 2012)

[10] Atayero, A.A., Luka, M.K.: Integrated Models for Information Communication Systems and Network: Design and Development. In: ANFIS Modelling of Dynamic Load Balancing in LTE. IGI-Global Publishers, USA (2012) ISBN13: 978-14666-220-1

[11] 3G Americas, The benefits of SON in LTE: Self optimizing and self organizing networks. Whitepaper (December 2009)

[12] 3GPP TS 36.902, Evolved Universal Terrestrial Radio Access Network (E-UTRAN); Self-Configuring and Self-Optimizing Network (SON) Use Cases and Solutions (May 2011)

[13] NEC Corporation, Self Organizing Network: NEC's proposal for next-generation radio network management. Whitepaper (February 2009)

[14] 4G Americas, Self-optimizing networks in 3GPP Release 11: The benefits of SON in LTE. Whitepaper (October 2013)

[15] Pauli, V., Seidel, E.: Inter-Cell Interference Coordination for LTE-A. Nomor Research GmbH, Munich (2011)

# A Two-Way Loop Algorithm for Exploiting Instruction-Level Parallelism in Memory System

Sanjay Misra[1], Abraham Ayegba Alfa[2], Sunday Olamide Adewale[3],
Michael Abogunde Akogbe[2], and Mikail Olayemi Olaniyi[2]

[1] Covenant University, OTA, Nigeria
[2] Federal University of Technology, Minna, Nigeria
[3] Federal University of Technology, Akure, Nigeria
Sanjay.misra@covenantuniversity.edu.ng,
abrahamsalfa@gmail.com, adewale@futa.edu.ng,
{michael.akogbe,mikail.olaniyi}@futminna.edu.ng

**Abstract.** There is ever increasing need for the use of computer memory and processing elements in computations. Multiple and complex instructions processing require to be carried out almost concurrently and in parallel that exhibit interleaves and inherent dependencies. Loop architectures such as unrolling loop architecture do not allow for branch/conditional instructions processing (or execution). Two-Way Loop (TWL) technique exploits instruction-level parallelism (ILP) using TWL algorithm to transform basic block loops to parallel ILP architecture to allow parallel instructions processes and executions. This paper presents TWL for concurrent executions of straight forward and branch/conditional instructions. Further evaluation of TWL algorithm is carried out in this paper.

**Keywords:** Branch/conditional, loops, ILP, multiple issues, parallelism.

## 1 Introduction

Pipelining enables infinite number of instructions to be executed concurrently (or at the same time), though in distinct pipeline stages at a particular moment [1]. Pipelining is used to overlap the execution of instructions and to attain better performance. This prospective overlap for several sets of instructions is known as instruction-level parallelism (ILP) [1]; reason being that these instructions can be executed in parallel. There are two very distinct approaches for exploiting ILP: hardware and software technology [1], [2].

The factors limiting exploiting of ILP in compilers/processing elements continue to widen because of complexity in size and nature of instructions constructs. Some of the short coming observed with existing techniques include: majority support multiple issues of straight instructions, prediction techniques are required to transform conditional/branch instructions to straight instructions, which often cause stall of memory/compiler system [3].

In BB architecture, there is only one entry and exit points in loop body (that is inflexible to accept interrupts for new instructions scheduling until the entry or exit is

B. Murgante et al. (Eds.): ICCSA 2014, Part V, LNCS 8583, pp. 255–264, 2014.
© Springer International Publishing Switzerland 2014

encountered) [3]. There is basically little or no overlap among instructions in basic block architecture. The opportunities to speed up program execution through parallelism exploitation became feasible; that is executing parts of the program at the same time as against increasing sequential execution speeds. Unrolling loop technique is an attempt to achieve that by replicating original loop body into several other sub-loops for multiple issues of instructions for concurrent processing and execution.

The remainder of the paper is organized as follows. The next section introduces unique features and problems in existing techniques for exploiting ILP in memory system. In section 3, challenges are identified. Following that, a new technique is formed and results related to ILP exploitations are discussed in section 4 and, finally, conclusions are drawn in section 5.

## 2     Related Works

### 2.1     Pipeline Parallelism

Flynn; Smith and Weiss [4], [2] introduced pipelining into the architecture of CPUs to permit instruction execution in minimum time possible. Pipeline parallelism technique involves partitioning of each instruction set into several other segments that required different hardware resources for purpose of completing execution within a cycle. A CPI (cycle per instruction) equals to one can be attained given any scenario to be true [4]. Multiple issues architecture takes advantage of parallelism inherent in programs of application such as sequential stream of codes, compare and control data dependencies found in instruction, identifying sets of independent instructions to be issued concurrently without altering the correctness of program [5].

### 2.2     Loop Architectures

Data and control dependencies form the basis for execution of a cyclic code containing conditions (or branch). Loop dependencies analysis continues to widen in complexity especially when every statement can be executed severally giving rise to two (or more) separate iteration executions (i.e. loop-carried dependencies; as a result of dependencies existing between statements) in the same loop body [6]. There exists greater responsibility to identify which paths are most repeated, that every path of the program for real time application exhibit constraints in time, overall time of execution must be minimized [5].

*Unrolling of Loop:* This technique involves many enlarged basic blocks for the purpose of exploiting parallelism by avoiding branch instructions. Enlarging basic block means repetitive operations in form of loops using an algorithm to achieve an efficient and small scale code. Unrolling loop algorithm repeats a loop body for a number of times, if the bound variables are unchanged and defined at compile time [8], [9]. The loop unrolling architecture and code are illustrated in Figure 1. Unrolling loop resolves no control issue for instruction at the start of every iteration (runs check on the body of the loop for entry point) [9], [8] and [5].

## 2.3     Execution Cycle of Instructions

The execution of a single basic block of instructions on compiler is partitioned into a series of independent operations referred to as execution cycle of instruction. Parathasarathy [7] gives description of five basic steps for executing a compiler instruction using memory operand is as follows:

*Fetch:* The control unit fetches the instruction from the instruction queue and the instruction pointer (IP) is incremented.

*Decode:* The function of instruction is decoded by the control unit to ascertain what the instruction does. Input operands of instruction are pushed to the arithmetic logic unit (ALU) and signals are transmitted to the ALU specifying the operation to be carried out.

*Fetch operands:* If the instruction requires an input operand stored on memory, the control unit takes advantage of a *Read operation* to recall the operand and copy it into internal registers. Internal registers are not visible to program of the user.

*Execute:* The instruction is executed by the ALU using the named registers and internal registers as operands and sends the result to named registers and/or memory. The status flags rendering information about the state of processor is updated by the ALU.

*Store output operand:* If the output operand is located in memory, the control unit takes advantage of a write operation to store the data.

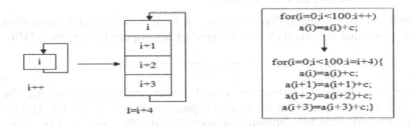

**Fig. 1.** Loop unrolling architecture and code. *Source- [5]*

# 3     Two-Way Loop Technique

This section discusses the TWL algorithm, its implementation and its mathematical modeling.

## 3.1     Two-Way Loop Algorithm

Two-Way Loop algorithm supports multiple issues/concurrent instructions executions of straight and branch paths of loops. It modifies unrolling of loop technique by severally enlarging basic block for parallelism exploitation by allowing multiple branch instructions executions.

I.     Identify conditional branch instructions //across several loop unrolling
II.    Transform instructions in Step I into predicate defining instructions // instructions that set a specific value known as a predicate
III.   Instructions belonging to straight and branch constructs are then modified into predicate instructions // both of them execute according to the value of the predicate
IV.    Fetch and execute predicated instructions irrespective of the value of their predicate// across several loops unrolling
V.     Instructions retirement phase
VI.    If predicate value = TRUE // continue to the next and last pipeline stage
VII.   If predicate value = FALSE // nullified: results produced do not need to be written back and hence lost


### 3.2    Evaluation

*Time of Execution:* The impact of ILP can be measured by the speedup in execution time (that is speedup of ILP) is defined by Equation 1

$$ILP\ Speedup = \frac{T0}{T1}\ .$$  (1)

where,

T0 = execution time of pipelining technique
T1 = execution time of TWL technique

*Performance:* According to Flynn Benchmark, Execution time = total time required to run program (that is wall-clock time for product development and testing) [10].

$$Performance = \frac{1}{(execution\ time)} \leq 1\ ,$$  (2)

*Utilization*: Is number of instructions issued/number completed per second. The mean time that a request speeds in the system exposes more ILP. Cantrell [11] develops a benchmark and formula to compute number of instructions executed ($\mu$) if the mean time of execution is $T$ seconds, is given by: $\mu = \frac{1}{T}$

$$Utilization: \rho = \lambda T = \frac{\lambda}{\mu} = 1\ .$$  (3)

The mean waiting time (i.e. no parallelism is present in program), Tw = ∞
$\lambda = rate\ of\ issue\ of\ instructions.$


### 3.3    Mathematical Model for Two-Way Loop Algorithm

These mathematical notations show relationships between loops and instructions in a two-way loop.

I.    *Loops /iterations constructs:* Unrolling is replicating the body of a loop multiple times to reduce loop iteration overhead.

```
for h = [ 0...Q]
X(h) = Y(h) + Z(h)
```

Unrolled 3 times, gives;
```
for h = [0...Q] by 3
X(h) = Y(h) + Z(h)
X(h+1) = Y(h+1) + Z(h+1)
X(h+2) = Y(h+2) + Z(h+2)
X(h+3) = Y(h+3) + Z(h+3)
```

II.  *Unrolling construct:* Unrolling loop can be transformed from original basic-block architecture into several deep inner loop bound for any given loop of the form:

```
for h = [ 0 ... Q]
X(p) = f(p)
```

The target machine/memory system usually has its vector length and several inner deep loop nests, where the inner loop has a constant loop bound without loop dependencies

```
for h = [0 ...Q] by 32
for k =[ 0 ...32]
X(h) = f(h)
```

III.  *Interleave/overlap constructs:* To overlap (or interchange) the order of nested loops in the instructions basic-block transformation. Estimated by:

```
for h = [0...Q]
for p = [ 0...Q]
A(h,p) = f(h,p)
```

Then, an interchange of the h and p loops gives:
```
for p = [0...Q]
for h = [0...Q]
X(h,p) = f(h,p)
```

This interchange code proceeds for column-major order computation (preferably if X is stored in column-major order). To improve memory system and increase parallelism exploited for processing elements by transforming loops of basic-block architecture to several nested loops and performs interchange of loops order inwards to bring about overlapping of processes. Consider the multiple-nested loops of a matrix multiplication:

```
for h = [ 0...R]
for p = [ 0...Q]
for k = [0...k]
Z(h,p) += X(h,k) * Y(k,p)
```

Applying tiling technique, all three loops gives:

```
for hh = [0...R] by Y
for pp = [0...Q] by Y
for kk =  [0...k] by Y
for h = hh...hh + Y
for p = pp...pp + Y
for k = kk...kk + Y
for Z(h,p) + =  X(h,k) * Y(k,p)
```

Suppose that: Y(h) is not compiled after this code executes, it need not be stored explicitly and memory space is saved.

IV. *Branch and conditional constructs*: When instructions execute in loop with conditions or branches, it can be expressed as:

```
for h = [ 0...Q]
for p = [ 0...R]
for k =  [0...K]
If  k = h
Y(h,k) = f(h,k)
Else k ≠ h; i.e. (k = p)
Z(p,k) = g(p,k)
```

# 4     Results

Experimental execution time for TWL technique against pipelining technique is presented in Table 1.

**Table 1.** The Mean Time of Executions for the Simulation Test

| Execution Time | Loops fields | | | |
|---|---|---|---|---|
| | l1 | l2 | l3 | l4 |
| T0 (sec) | 851 | 205 | 337 | 514 |
| T1 (sec) | 502 | 306 | 282 | 291 |

Table 1 gives the mean time of executions carried out in TWL technique test for 50 students (several instructions set), 4 sub-loops forms performed in parallel and concurrently with the pipelining technique (statistical sample error of ± 0.05) for students' record.

$$\text{Total execution time } T_{et} = \sum T_{e0} + \sum T_{e1}$$

where,

$T_{e0}$ = execution time for pipelining technique

$T_{e1}$ = execution time for TWL (new) technique

$\sum T_{e0} = 851 + 205 + 337 + 514 = [1] = 1907$

$\sum T_{e0} = 502 + 306 + 282 + 291 = [2] = 1381$

$\therefore T_{et} = \sum T_{e0} + \sum T_{e1} = [1] + [2]$

$\qquad = 1907 + 1381$

$\qquad = 3288$

Percentages of time of execution of pipelining and TWL techniques are given by:

$$Percentage \ of \ T_{e0} = \frac{\sum T_{e0}}{\sum T_{et}} \times \frac{100}{1}$$

$$= \frac{1907}{3288} \times \frac{100}{1} = 0.5799878 \times 100 = 579987 = 58\% \ (Approx.)$$

and,

$$Percentage \ of \ T_{e1} = \frac{\sum T_{e1}}{\sum T_{et}} \times \frac{100}{1}$$

$$= \frac{1381}{3288} \times \frac{100}{1} = 0.4200122 \times 100 = 4200122 = 42\%$$

The performances ($P_0$) of pipelining technique and ($P_1$) of new technique, are given by:

$$P_0 = \frac{1}{T_{e0}} = \frac{1}{1907} = 5.2438 \times 10^{-4} \ (Approx.)$$

$$P_1 = \frac{1}{T_{e1}} = \frac{1}{1381} = 7.2411 \times 10^{-4} \ (Approx.)$$

Units are (CPS) Cycle per Second. The performance ($P_1$) is better than ($P_0$) because the mean execution time is low for the TWL technique.

From equation 1, computing the speedup ($n$):

$$n = \frac{T_{e0}}{T_{e1}} = \frac{1907}{1381} = 1.381 \ (Approx.)$$

This implies the TWL technique is 1.381 times faster than the pipelining technique.

*Scale of Graphs:* 100 (or 0.01) units are used to represent the quantities measured on Y- axes. While, 1 unit is used to capture the values of magnitude of loops ($l$) on X-axes as shown in Figures 2 and 3.

The difference in executions time between the pipelining technique ($T_{e0}$) and TWL technique ($T_{e1}$) is illustrated in Figure 2.

*Remarks:* The time of execution of 42% for TWL makes it more preferable to 58% for pipelining technique. The time taken to complete parallel executions in TWL is relatively same as time taken to complete one loop process/execution in the pipelining technique. Time spent for the pipelining technique is much due to sequential scheduling of the loops, while that of TWL technique is low, several loops are executed concurrently due to support for multiple issues and instruction level parallelism.

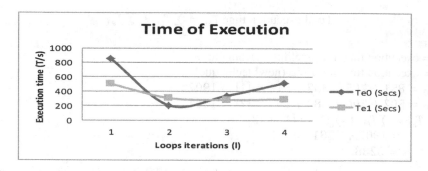

**Fig. 2.** A graph showing the difference between execution time of TWL technique and pipelining technique

The difference in the rate of loops execution per cycle (frequency) can be determine for the pipelining technique ($F_0$) and TWL technique ($F_1$) is shown in Figure 3.

*Remarks:* The frequency of pipelining technique ($F_0$) is very low because of lack of support for the ILP (i. e. more loops for less frequencies). While, the frequency of TWL ($F_1$) is relatively stable because of presence of multiple issues, and multiple processes/executions are carried out at the same time.

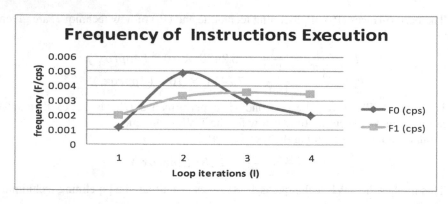

**Fig. 3.** A graph of frequency of TWL technique compared to pipelining technique

*Capacity:* Performance can be perceived as throughput and utilization. Throughput is total work done per unit time (measured as number of requests completed per second). While, utilization is number of instructions issued and completed successfully per second.

Number of instructions executed ($\mu$), if the mean time of execution is $T_e$ seconds (per day), is given by:

$$\mu = \frac{1}{T_{e0}} = 0.00052438 \times 3600 \times 24 = 0.3020451$$

$$\mu = \frac{1}{T_{e1}} = 0.00072411 \times 3600 \times 24 = 62.563104$$

and, $\lambda = no\ of\ completed\ loops = 4$

From Equation 4, utilization for:

pipelining technique, $\rho = \lambda T_{e0}$

$$= \frac{\lambda}{\mu} = \frac{4}{0.3020451} = 13.243056$$

and, TWL, $\rho = \lambda T_{e1}$

$$= \frac{\lambda}{\mu} = \frac{4}{62.563104} = 0.06393545$$

$\therefore$ This means that $\rho$ for TWL is less than 1 according to Cantrell's benchmark, which suggests more instructions and parallel processes can be carried-out in a day.

## 5    Conclusion

The paper presented a machine design using TWL algorithm for exploiting ILP, decreases time of execution over the existing technique by two, the utilization of 0.068 gives rise to increased capabilities to processing elements/compiler to issue multiple instructions at the same time, perform parallel executions, transforms basic block of instructions to several dependent and independent instructions as against pipelining technique.

## References

1. Hennessy, J., Patterson, D.A.: Computer Architecture, 4th edn., pp. 2–104. Morgan Kaufmann Publishers Elsevier, San Francisco (2007)
2. Smith, J.E., Weiss, J.: PowerPC 601 and Alpha 21064: A tale of two RISCs. IEEE Journal of Computer 27(6), 46–58 (1994)
3. Jack, W.D., Sanjay, J.: Improving Instruction-Level Parallelism by Loop Unrolling and Dynamic Memory Disambiguation. An M.Sc. Thesis of Department of Computer Science, Thornton Hall, University of Virginia, Charlottesville, USA, pp. 1–8 (1995)
4. Flynn, M.J.: Computer Architecture: Pipelined and Parallel Processor Design, 1st edn., pp. 34–55. Jones and Bartlett Publishers, Inc., USA (1995) ISBN: 0867202041
5. Pozzi, L.: Compilation Techniques for Exploiting Instruction Level Parallelism, A Survey. Department of Electrical and Information, University of Milan, Milan. Italy Technical Report 20133, pp. 1–3 (2010)
6. Bacon, D.F., Graham, S.L., Sharp, O.J.: Complier Transformations for High Performance Computing. Journal of ACM Computing Surveys, 345–420 (1994)
7. Rau, B.R., Fisher, J.A.: Instruction-Level Parallel Processing: History Overview and Perspective. The Journal of Supercomputing 7(7), 9–50 (1993)
8. Pepijn, W.: Simdization Transformation Strategies - Polyhedral Transformations and Cost Estimation. An M.Sc Thesis, Department of Computer/Electrical Engineering, Delft University of Technology, Delft, Netherlands, pp. 1–77 (2012)

9. Vijay, S.P., Sarita, A.: Code Transformations to Improve Memory Parallelism. In: Proceedings of the 32nd Annual ACM/IEEE International Symposium on Microarchitecture, pp. 147–155. IEEE Computer Society, Haifa (1999)
10. Cantrell, C.D.: Computer System Performance Measurement. In: Unpublished Note Prepared for Lecture CE/EE 4304, Erik Jonsson School of Engineering and Computer Science, pp. 1–71. The University of Texas, Dallas (2012), http://www.utdallas.edu/~cantrell/ee4304/perf.pdf
11. Marcos, R.D.A., David, R.K.: Runtime Predictability of Loops. In: Proceedings of the Fourth Annual IEEE International Workshop on Workload Characterization, I.C., Ed., Austin, Texas, USA, pp. 91–98 (2001)

# Investigation of SPM Approaches
# for Academic IT – Projects

Varsha Karandikar, Ankit Mehra, and Shaligram Prajapat

International Institute of Professional Studies, Devi Ahilya University,
Indore, India
{vjk183,ankit.mehra92,shaligram.prajapat}@gmail.com

**Abstract.** In IT world, millions of software projects have been initiated, being developed and deployed every year. Almost billion dollars are invested for successful and useful software's. But many of these projects are not able to satisfy user's need. Many of them are not being delivered as per the allotted budget and schedule due to lack of project management standards. Especially in CS and IT education system of central India, academic projects are not evaluated on these grounds. In this paper, we focus on the study of quality and success of academic projects, and investigating the various approaches of project monitoring and management for developing usable software. The paper puts deep insight towards existing approaches in practice and opens scope for analysis of academic CS projects in Indore districts near future.

**Keywords:** SPM (Software Project Management), CS, IT.

## 1 Introduction

In the world of professional education, success of technical education depends on the quality of projects done by students/ candidates. Due to enormous number of rise in intake capacity, very limited numbers of mentors are there to track candidates (in 2nd and 3rd tier technical institutions). Due to heterogeneous platforms, environments, streams and interests, managing and controlling them by very few faculty members is hard problem.

As depicted in fig 1(a), one can infer that 7.14% institutes are using autonomous project tracking tools and techniques to monitor academic projects. This leads to a scope of possibilities of proposed studies as demonstrated from study results in Fig 1(a).

B. Murgante et al. (Eds.): ICCSA 2014, Part V, LNCS 8583, pp. 265–282, 2014.

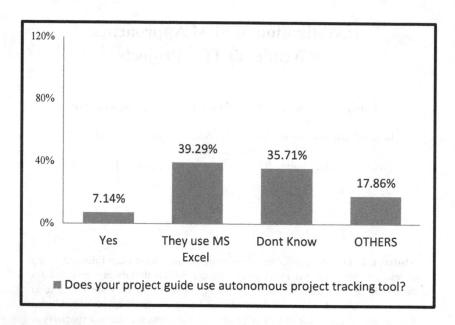

**Fig. 1a.** Showing use of project management tool in institutions

The above fact leads to following observations:

a)  Some projects may not finish before deadlines.

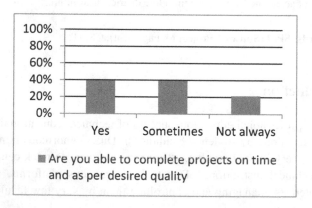

**Fig. 1b.** Students view regarding completion of their projects

b)  Projects may finish but proper documentation is not created as a by-product, in successive phases.
c)  Due to academic nature, constant guidance and motivation may not be provided, that leads to failure of project.
d)  Students depend on other resources like web, past projects, help from seniors, friends, external references.

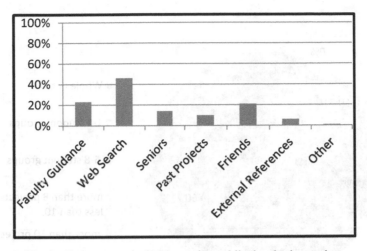

**Fig. 1c.** Resources consulted by student while developing projects

e) Due to lack of sufficient and proper guidance, sometimes potential projects are not well developed.
f) Projects are considered in the same manner as other subjects in the regular college curriculum and less attention is paid towards it by the administration.
g) Students are not aware about the proper procedure and the deliverables to be produced beforehand.

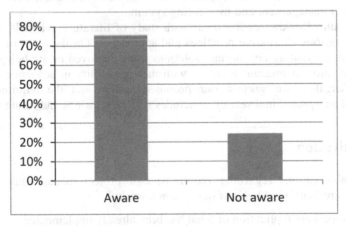

**Fig. 1(d).** Showing awareness of project development process

h) Due to more student-mentor ratio, not every student gets same time for discussion with mentors, and sometimes problems related to projects doesn't get solved.
i) Internal progress reporting is not conducted in a scheduled manner and it is difficult for a mentor to keep track of every project developed under his/her guidance.

These problems were the main issues being raised during the interviews and survey with students and mentors.

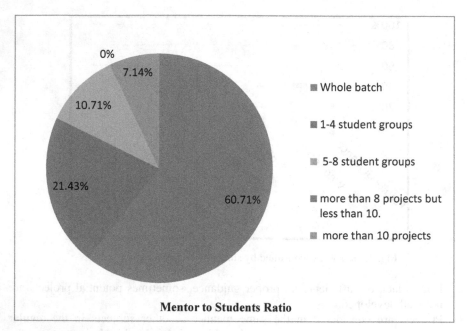

**Fig. 1e.** Survey among students' shows that there are more students per guide

Subsequent sections deals with related studies and available alternative system for student project management and monitoring system.

The Literature Review section deals with study of literature available in problem domain and various alternative practices and approaches. On the basis of this study, we emphasize on finding appropriate solutions to be deployed or are being deployed in some institution at present; together with their feasibility in academic IT project domain. After that, we suggest our proposed system and the requirements are gathered and analyzed. Final section concludes with conclusions and future work.

## 2    Motivation

Following observations regarding requirement of project management are worth mentioning here that forms basis of our research work:

1) Most projects are replication of what has been already implemented.
2) Project management methodology/ framework (PMBOK, CMM) are not followed.
3) Stakeholders/potential users and scope of the system are not properly identified. Most projects are limited to problems which have already been solved.
4) No room is kept for improving the current solution or   trying out innovative idea by using IT for solving problems of society
5) More focus on learning technology stack while design and analysis phases are ignored which is the reason behind successful projects.
6) Time management is not proper and so deadlines get missed at times or the quality of product being developed is suffered.

7) While developing projects in teams, students lack the required guidance for dividing the modules to team members and mostly, the work is carried out by one or more talented and knowledgeable person. This approach hampers the learning of other members who do not collaborate.

8) Projects are not developed with a view of cost-benefit analysis, requirement of user training, and infrastructure requirements. The projects being developed are just for the sake of grades in the academic curriculum.

Above points motivate us to seek out alternative frameworks for quality project management in IT education sectors.

## 3    Review of Literature

Very few open literatures are available in SPM for CS/IT branches. Following are standard and relevant perceptions:

In[1], Jalote highlighted that failure of any project  may be due to unclear objectives, bad planning, new technology, lack of awareness of project management methodology, and   insufficient team members. And at least 3 of the 5 reasons mentioned above are related to project management. Stakeholders of SPM are not aware or need to put efforts on these parameters while monitoring.

In [2], Pressman described that, "planning, monitoring, controlling of people, process and events form preliminary components for delivering operational and implementable products". And in the academic projects monitoring, some or more attributes are missing and overridden.

In [3], the author Jones has analyzed large number of software ranging from system software, information systems, outsourced projects to even military applications. The analysis revealed that unsuccessful projects had one or more common problems including poor project planning, poor measurements, poor milestone tracking, poor cost estimating, poor change control and poor quality control. The academic projects developed in institutions also fail due to one or more of the reasons cited above.

In [4,] emphasis is given on the fact that many research papers on project management are not able to address the engineering and management disciplines of software projects for industry. They have attempted to illuminate this section by providing comprehensive guidance in project management methodologies and frameworks currently being used in the industry.  Similar efforts are required in the field of academics for SPM.

In [5], many project development methodologies (RAD, Incremental, Prototyping) have been compared on some parameters but emphasis given to agile methodology which is less lenient and provides success-oriented approach in today's perspective. That is to say, that there exists a plethora of project management approaches in the current scenario and each approach has been helpful in certain set of situations. This implies that all these approaches require introspection for their utilization in various projects. With this view, issues for choosing modern project management have been discussed.

In [6], the authors emphasized that PERT and CPM are used to plot tasks and dependencies and for determining the probabilities of tasks done, while Agile project management uses a realistic approach by plotting graphs between actual work parameters in a relative manner. Use of PERT and CPM charts enables a software manager to take decision about schedule, human resource, thereby leading the project to provide expected outcomes.

In PERT and CPM, the graphs become complex, for large projects. In software projects, the dependencies represented by PERT/CPM graphs does not make much sense. CPM method does not pay importance to randomness inherent in planning software systems.

In [7], the author laid emphasis on the agile technology which is widely used in many developing projects in today's scenario. This methodology incorporates best features from its previous methodologies and applies it in iterative manner for fulfillment of objectives. This methodology is flexible, and provides more freedom. It allows development teams to adapt to requirements

With its use in many industries, some points in opposition have also risen. Agile project management is flexible, but cannot solve all problems. That is, it can be applied to a subset of problem domain. In this methodology, manager holds a central role.

In [8] the author carried on an empirical study on project management practices by conducting survey among 995 project managers. Their study indicated CMM, PRINCE2, were used by less number of project managers. More emphasis was on standard practices. Gantt chart was most used scheduling tool. The study also indicated most project managers found these techniques inadequate for complex projects, and expertise was desired in their application.

In [10], the authors analyzed PRINCE2® which is a workable methodology and can be applied within modern business environment. This methodology may result in stifling/hampering the creativity and innovation of the manager.

In [11], the authors emphasize RAD methodology. RAD tools and methodology retains their existence for its high caliber of speedy development cycle, gui interface, encompassing support for building reusable and extensible systems. Migration to the RAD environment was difficult for developers having long experiences in traditional systems. The knowledge of how to utilize RAD in an organization requires careful evaluation.

In the same manner, our work involves extracting the best practices being used in industry and reconfiguring them as per the needs of the educational institutions of Indore.

Frequent papers are available for industry projects but not for academic projects because there is no standard process for project management especially for CS/IT and also there is a shortage of skilled faculty with SPM mindsets.

Our work involves a study of current practices of project management in institutions and its impact on students as well as mentors. It also determines the effect on quality of projects developed by students during their curriculum and the process followed by them.

We analyze the existing /alternate system being used in many organizations on the basis of a survey carried among a number of mentors and students.

On the basis of survey, we accumulated some of the key points and deficiencies of current system as indicated by students and mentors during the survey. Using the analyzed facts and information, we propose some essential elements and best practices that could be followed in an institute to enhance the project management process and ease out the work of mentors and student.

# 4    Requirement Analysis

Different systems and processes are being followed by mentors while managing projects in an organization. Since the number of projects to be managed are much more, the mentor follow some policies from whole project management process i.e., it involves steps from project initiation to project submission and then further evaluation. Some measures/deadlines (final submission) for assessment are fixed by the institution while mentor is free to follow any policy during guidance. To explore in depth, we performed following requirement analysis phases.

**Interview and Associated Findings**

At this stage, we interviewed in- person with faculty members of an institute who have coordinated in academic projects as a mentor. We asked them several open-ended as well as close-ended questions and recorded their responses. The questions asked were:

**a) What are your views regarding project management process?**

The process initiate with the announcement of Project work to be done along with the required deliverables and deadlines. Examples and samples are provided whenever required. Continuous evaluation is done during the period and final grade is calculated with effects of this evaluation along with the final demo which is assessed by an external examiner.

**b) Views on internal assessment or final demonstration?**

Almost 74% students pointed out that evaluation should be combination of internal examination as well as performance in final viva.

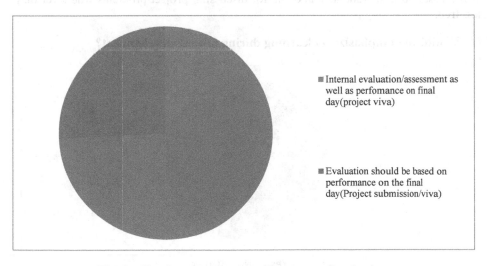

**Fig. 2a.** Showing responses regarding pattern of evaluation

**c) If they focus on internal assessment, what are the means of internal evaluation?**

As per the interview responses, 75% people pointed out that internal evaluation should take place in the form of document submission, presentations, working software demo.

**d) Do you conduct regular/periodic sessions for discussing the problems faced by students?**

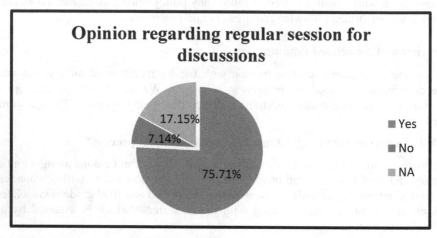

**Fig. 2b.** Regular sessions by mentors for discussing problems related to project

Mentors were asked whether they conduct regular sessions for solving problems related to project. Among them, 75.71% mentors told they conduct regular sessions, while rest told that students can come for discussing project problems whenever they are free.

**e) Would you emphasize on learning during project development?**

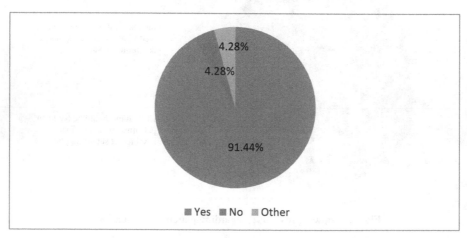

**Fig. 2c.** Illustrating mentor's view on evaluation parameters

In this survey, we asked mentors on which parameters do your evaluate projects developed by students in a curriculum. As seen in responses, most of the mentors considered knowledge gained during project phases and efforts put down on projects as the most significant elements for evaluation.

**f)  Are there any other criteria for the above?**

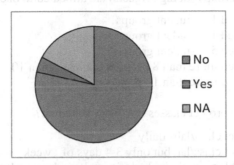

**Fig. 2d.** Response for other criteria's of evaluation

Mentors were asked whether there exists other criterion of evaluation not included in above question. Most of the mentors responded with 'No', i.e., no other criterion exists.

**g)  How do you track students' progress?**
Through periodic interactions and meeting between guide and students by the means of Demos, Presentations, Deliverables and similar measures.

**h)  Do you provide an introductory session to prospective students?**

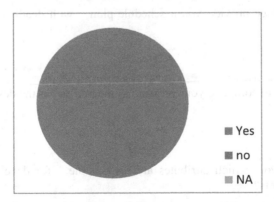

**Fig. 2e.** Showing responses of mentors regarding introductory session

Mentors were asked whether they conduct an introductory session for students to make them aware of project development process. As well as, they provide information about schedule plan, along with deliverables to be submitted. Evident from graph, all mentors provide an introductory session before commencement of projects.

This finished the analysis of interviews conducted among project guides. In this, it was indicated that mentor do follow standard approaches at their end, but this has not resulted in desired outcomes i.e. successful and usable software.

## b) Questionnaire

A survey was carried among students community regarding current system of academic project in institutions. The survey was conducted using online services Google Forms and Survey monkey. The questions asked were:

1. How many student groups (doing projects) are allocated to one project guide?
a) Each guide is allotted 1-4 student groups
b) Each guide is allotted 1-4 student groups
c) Each guide is allotted 5-8 student groups
d) Each guide is allotted more than 8 projects but less  than 10.
e) Each guide is allotted more than 10 projects

2. Do you have regular project classes in college schedule?

a) Yes, we have project class/labs daily
b) Yes, we have project class/lab but only 3-4 days of  week
c) No, we discuss problems with guide during college hours when they are free
3. How do you receive notifications about submission dates, presentation and reporting details?

a) Via Sms from Class representative
b) From friends
c) From Project Guide
d) Via Social media
e) Any of the above
f) Others(Please specify)

4. Does your college provide you a schedule plan  with deadlines before starting project?

a) Yes
b) No

5. Do you manage to complete your project in time and as per the desired quality?
a) Yes
b) Not always
c) Sometimes

6. According to you, which attributes are responsible  for developing successful academic projects?

Please select appropriate label for each attribute.

| Attributes | Strongly Agree | Agree | Neutral | Disagree | Strongly Disagree |
|---|---|---|---|---|---|
| Proper Planning | | | | | |
| Constant motivation and guidance from project guide | | | | | |

| Following strict timeline in each phase of project development | | | | |
|---|---|---|---|---|
| Developing standard documentation (as per IEEE format) | | | | |
| Help from external references (if any) | | | | |
| More emphasis on projects and practical tasks in curriculum | | | | |
| Fair share of time available for discussing project problems with guide | | | | |

## Major Findings

The survey was conducted to know students' opinion about current system. We received almost 70 responses for statistical analysis. The responses received were analyzed using pictorial graphs.

The graphs are given below:

a) Students were asked whether they had scheduled project labs in their curriculum. (See fig 2(f))

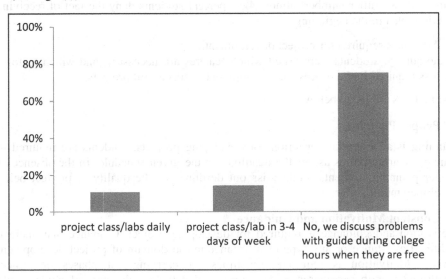

**Fig. 2f.** Responses of students showing frequency of project classes

Responses indicate that project labs are not conducted as per schedule for discussing projects. Almost 80% students told that they discuss problems with mentors whenever they are free.

b) Students were asked whether they are provided with a schedule plan with other required details for project. This information should be provided before commencement of project, so that it would aid in building the mindset of project development among students.

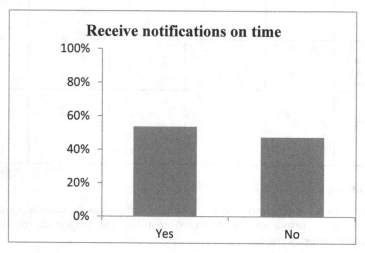

**Fig. 2g.** Opinion about pre-announcement of schedule

As per the responses, around 48% students indicated the availability of schedule plan while a similar number; almost 42% percent students deny the fact of receiving schedule plan in the beginning.

c) Attributes required for project development.

In the survey, students were asked which features are necessary, and which factors are less required in the process of development of successful projects.

The features are listed below:

### i) Proper Planning

Planning is an important criterion for developing projects. Students are required to manage their activities as per the deadlines in the given schedule. In the absence of proper planning, students might miss out deadlines or the quality of product being developed might suffer.

### ii) Constant Motivation and Guidance

Motivation and guidance is required for developing project. Guidance may include technical assistance or any type of assistance in the domain of project development phases. Motivation is required sometimes for carrying out phases of project development with perseverance. This implies that the involvement of guide would result in developing successful and usable product along with proper documentation.

### iii)  Developing Standard Document

Documentation is considered very important while developing software projects. In the case of failure of developed product, documents would be great help in recognizing what went wrong and suggesting solutions for improvement. Developing standardized documents is more important since then only it would be able to define your system well and would be recognized a measure of judging your system. Apart from this, documentation has other benefits as well.

### iv)  Help from External References

Assistance from external reference is an important parameter since the developer is amateur and could not know about every aspect of project development, be it coding, analysis or design. Hence, some professional help is always required.

### v)   More Emphasis of Projects and Practicals

More emphasis on project and practical subjects is required to make students habitual for developing projects as well as preparing their mindsets of project management.

As depicted by graph(fig 2(h)), students have provided a fair review for all features indicating that there are some flaws in current system and these features should be incorporated in project management practices and approaches.

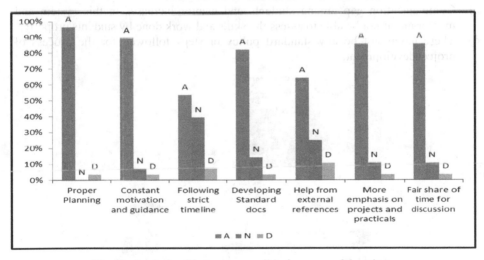

**Fig. 2h.** Analysis of factors responsible for successful projects

A-Agree
N-Neutral
D-Disagree

## 5    Alternative System

On the basis of interviews and survey, we came out with two procedures of project management being adopted currently in institutions.

**a. Centralized System of Project Management**

In this system, a single mentor is allocated for a batch of around 40-50 students. As per the study carried among mentors and faculties, the academic project activity involves the following steps:

  i) Students are first asked for their project titles and list of all students is submitted to the mentor by class representative (CR). The mentors maintain the list on paper.
 ii) Then, after some defined interval students are asked to submit the deliverables like SRS, SDD, and all. Students continue to work on the projects on their own and by taking help from external references like friends, seniors.
iii) Mentor might conduct a presentation/demo after a fixed duration of time for reporting the current status of project and stating problems/obstructions.
 iv) The final submission date is announced by the administration department and the students are asked to submit their report and present a demo towards external as well as internal examiner.

**Limitations**

a) Mentors might not be able to track all projects properly due to academic/organization's obligations.
b) It is difficult for a mentor to devote his/her time for so many projects, which may result in project failure or some lacuna in learning.
c) Communication gap among students and mentor increases in this system and mentor might not be able to assess the skills and work done by students properly.
d) There might not be any standard policy or steps followed for the process of project development.

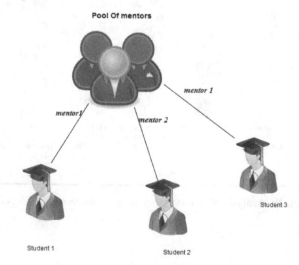

**Fig. 3.** Showing allocation of mentors to students

**b. Distributed Project Monitoring**

In this system, more than one project guide might be allocated for a batch of students. Each mentor is allocated two or three projects group. Each project group may contain one, two, or more than two students.

We suggest an algorithm for Distributed project monitoring:

**Data Structures Used**

S'= Set of student groups formed for developing projects.

L= Set of standard learning projects.

M= Set of mentors being allotted to project.

B=size of batch

O= Set of projects brought by students and needed to be verified and approved by mentors before beginning with development phases.

P= Set of project which are to be developed by students.

AM= Set of all Project mentors along with their allotted student groups

o represents operation performed on set

L'= Grade sheet of whole batch B.

**Algorithm**

1. Assume: $B = \{a1, a2, \ldots, an\}$
   Consider a set $S' = \{S_i: i=1,2,\ldots,n\}$, $S_i \subset B$ and $S_i \cap S_j = \emptyset$ and $S_1 \cup S_2 \ldots \cup S_{n=} B$

2. Consider a set L of standard projects and a set O of suggested projects

3. M is the set of mentors. $M = \{m_i\}$ where i= 1, 2, …., 12

4. We assume an ordered pair $(S_i, P_i)$ where $S_i \in S'$ and $P_i \in O$ if SRS of O is approved by $m_i$ else $P_i \in L$

5. Let cardinality of (S) =n
   And cardinality of (M) = m

$\Rightarrow$ n/ m = p=Cardinality of set $AM_i$

$$\bigcup_{l=1}^{p} {}_{J=1}^{m} AM_{ij}$$

//at this stage allocation of mentors, projects has been done and tracking should be started)

6. Consider set $T' = \{T_1, T_2, \ldots, T_n\}$
   where $T_i = \{t_1, t_2, t_3, \ldots, t_n\}$ such that $\exists\ T_j$ for every $S_j$

7. For every $S_j$
   while($T_j \neq \emptyset$ )
   {
     if(Sj o $t_i$ == true)
     {
        $T_j = T_j - t_i$;
        i++;
     }
     if($T_j = \emptyset$)
     {
        $L' = L' + (S_j, grade)$
     }
   }

where ($S_j$ o $t_i$ == true) should be verified by respective mentor(m)

In this algorithm, we have given tracking of phases, with respect to student group and tracking schedule T, defined by each mentor in O(mn) + O(mp) time.

The key issues of this system are summarized as following:

### a) Unbiased Allocation of projects and project groups

In this system, pool of project mentors is allocated for a batch. Two or three projects are developed under the guidance of each mentor. The project team and projects titles might be explicitly allotted to the system or students are given the freedom for choosing the partner as well as project. Whichever policy is adopted, it should be in the interest of student community and it should lead to increase the quality of projects developed.

If the system is decision maker, then there are two approaches for this:

i)  The project title is provided by system, the students have the freedom to decide their group themselves.

ii)  The project group is decided by mentor and the group can decide the topic themselves.

iii)  Both the project group and project title are allocated by system. An important criterion for allocation would be to consider similarity of skill sets, interests and experience. Example, A software involving database should be allocated to a mentor with expertise in databases.

### b) Communication

Since there will be few students under different mentors, the students are responsible for maintaining communication with their respective mentors. The class representative or batch facilitator will not convey the timings or deadlines. A separate communication channel needed to be maintained between student and mentors.

### c) Planning

Every mentor will be following their own schedule and would issue deadlines for report, document as per their convenience. The students will be required to follow instructions of their concerned mentor. Otherwise, a standard schedule plan needed to be prepared with the consent of all mentors.

### d) Coordination among mentors

All the mentors will be guiding same batch of student, but different student groups. So, it would be necessary to main some sort of coordination among mentors regarding some issues like final dates of viva, how the students would be communicated, acceptance of other issues.

### e) Fair Evaluation guidelines

There would be a standard procedure for evaluation for unbiased grading. Otherwise, student might feel cheated for system.

### f) Tracking and monitoring the progress of students

Since there are more than one mentor for batch, each mentor at individual level needs to ensure that the project groups allotted to him/her are working on projects, their doubts are getting solved, deliverables are submitted on time.

**g) Adequate discussion time devoted to all projects, so that learning can be enhanced**

Yes, fair share of time must be made available to all project groups so that each group may be getting the motivation, guidance and their problems might be solved more readily.

We propose two more systems of Project management which are also capable of providing efficient project management in order to overcome the problems of schedule, team management, communication and all such issues. The proposed systems are described as follows:

## A) Peer to Peer Monitoring Process

This scheme involves only peer and no supervisor. Students review each other projects and provide feedback for improvement. For academic project management in institutions, this system can be implemented by assigning senior students as mentors for junior students.

### Pros

- Senior students as mentors might be able to solve      problems more readily since they might have experienced the same problem in their time.
- Senior students can guide the junior students      which would also result in improving their leadership and management skills. It would be a two way relationship.

### Cons

- Since this system allows interaction only among students, one possible outcome would be students would be less focused for their projects due to lack of motivation.
- In the case where senior students are not well experienced, this might result in serious consequences like failure of project.

## B) Hybrid Approach

In this approach, both peer to peer and centralized system will be applied for managing academic projects. A student group will be monitored by both senior students as well as a mentor. Such system is already operational in online course offerings by Coursers, edx and similar services. Here, peer monitoring is carried out by teaching assistants (TAs) and tutoring is done by teachers. So, in this system, the burden of teacher is lessened and the teaching assistants get a practical learning as well as coaching experience. This approach can also be applied for monitoring academics projects with seniors and Project mentors.

### Pros

- This system results in better productivity since TA's are continuous monitoring project activities.
- The mentor's work is lessened by peer mentors and his/her job remains to solve the expert problems.

### Cons

- Coordination problems may arise in the absence of a common interface.
- Sometimes, students might get perplexed with difference in views of mentor and peers.

# 6      Conclusion and Further Work

In this paper, we have presented the state of the art to the modern practices of software Project management, which are in use today.In future, it can be extended with further study to analyze the impacts of reasons for software failure stated by Jalote[1] on the quality of academic CS projects. Another issue is to involve an investigation on finding which are the attributes that affect the process of Software Project Management and can be further improved for much better results. The process would end with effective framework and powerful tool as an assistant of student, mentor, evaluation system, and society.   Motivation in the form of ranking and recommendations should be given to the work with the platform/tool on which student has worked or is very much aware, together with learning tools. In this way, he/she would focus on other stages of project management.

# References

1. Jalote, P.: Software Project Management. Pearson education (2010)
2. Pressman, R.: Software Engineering- A practitioner's Approach 6e, 628 (2005)
3. Jones, C.: Software Project Management Practices: Failure Versus Success (October 2004)
4. Hewagamage, C., Hewagamage, K.P.: Redesigned Framework and Approach for IT Project Management. International Journal of Software Engineering and Its Application 5(3) (July 2011)
5. Kennedy, D.M.: Software Development Teams In Higher Education: An Educator's View. In: Ascilite 1998 (1998)
6. White, D., Fortune, J.: Current practice in project management: An empirical study. International Journal of Project Management 20(1), 1–11 (2002)
7. O'Sheedy, D.G.: A Study of Agile Project Management Methods used for IT Implementation Projects in Small and Medium-sized Enterprises. Southern Cross University, ePublications@SCU, Thesis 2012 (2012)
8. Attarzadeh, I., Ow, S.H.: Modern project management: Essential kills and techniques. Communications of the IBIMA 2 (2008)
9. Martin, R.C.: PERT, CPM, and Agile Project Management (October 5, 2003), http://www.codejournal.com/public/cj/previews/PERTCPMAGILE.pdf
10. Ruth Court, FTC Kaplan. An Introduction to the PRINCE2 project methodology. CIMA (April 2006)
11. Agarwal, R., Prasad, J., Tanniru, M., Lynch, J.: Risks of Rapid Application Development. Communications of ACM.© 2000 ACM 0002-0782/00/1100

# An Approach for Security Patterns
# Application in Component Based Models

Rahma Bouaziz[1,2], Slim Kallel[2], and Bernard Coulette[1]

[1] IRIT, University of Toulouse
Toulouse, France
{Rahma.bouaziz,Bernard.Coulette}@irit.fr
[2] ReDCAD, University of Sfax,
Sfax, Tunisia
slim.kallel@fsegs.rnu.tn

**Abstract.** Since applications have become increasingly complex and because the design of secure systems necessitates security expertise, security patterns are now widely used as guidelines proposed by security experts in order to solve a recurring security problem. In order to encourage application designers to take advantage from security solutions proposed by security patterns, we think that it is necessary to provide an appropriate mechanism to implement those patterns. We propose a full security pattern integration methodology from the earliest phases of software development until the generation of the application code. The proposed solution uses the UML component model as an application domain of security patterns and bases on the use of UML profiles and model transformations with the ATL language. For the generation of code and for keeping the separation between the functional code of the component based application and security solution, we use the aspect paradigm. An illustration of the proposed approach is provided using the Role Based Access Control (RBAC) pattern. A case study of GPS system is also provided to demonstrate the application of the proposed approach.

**Keywords:** component based approach, Security pattern, RBAC, UML profile, Model Driven Development.

## 1 Introduction

Over the last two decades the usage of components has brought significant help in development of complex systems. There have been a number of component-based models developed for this purpose and some of them have already become well-established and used in different industrial domains. Therefore, security represents one of the most important aspects to deal with in the development of such systems. There by, security failure can have serious consequences and imply direct impact on the environment or on the users.

It has been argued that security concerns should be addressed at every phase of the development process [1]. Several approaches and methodologies have been proposed

B. Murgante et al. (Eds.): ICCSA 2014, Part V, LNCS 8583, pp. 283–296, 2014.

to help non-expert in security to implement secure software systems. One of them is Security Patterns [2], which is defined as an adaptation of design patterns to security. Security engineering with patterns is currently a very active area of research since security patterns capture the experience of experts in order to solve a security problem in a more structured and reusable way. A security pattern presents a well-understood generic solution to a particular recurring security problem that arises in specific contexts. Moreover, the advantages of security patterns are that they provide novice security developers with security guidelines so that they will be able to solve security problems in a more systematic and structured way.

However, many security patterns have been proposed without guiding developers for their concrete application. Still, there is very little work concerning the integration of security patterns in software development process [3]. Even more there is no clear, well-documented and accepted process dealing with their full integration from the earliest phases of software development until the generation of the application code especially when security concerns should be addressed at every stage of the development process [1]. That is what we are aiming to present in this paper.

Based on model driven development, we propose an approach dedicated to non-experts in security to help them to model and implement secure software systems. The developer produces the application source code semi-automatically by making a series of transformations. Various model transformations, e.g., model merging and model marking can be used. We leverage on this idea to propose a new security pattern integration process. The main point is that security patterns will be considered as models and especially we are aiming to propose security patterns as UML profiles.

The remainder of this paper is organized as follows. A brief presentation of basic concepts needed throughout the paper is given in Section 2. Section 3 presents the motivations. Section 4 details our approach for implementing secure systems. In Section 5 we illustrate our approach using the RBAC pattern. Section 6 presents the code generation process. A use case of GPS system is presented in Section 7. In section 8 we present a state of the art of work dealing with security in software development and security patterns. Section 9 ends the paper with a conclusion and an outlook.

## 2     Preliminaries

This section presents the key concepts required for the paper to be introduced. We briefly present some concepts related to Component-based Software Engineering (CBSE) and Model Driven Development (MDD).

### 2.1     Component Based Software Engineering

With CBSE [4] applications are developed with the reuse of prefabricated software parts called software components. These components can be validated by an expert and then reused in the context of a software development process. The component approach is mainly based on the principle of reuse by assembling components that greatly simplifies the understanding and development of a system.

For many years, component-based software engineering has essentially focused on providing methods and techniques to support the development of software functionalities in an efficient way. Yet, for certain types of applications such as dependable, real-time or embedded systems, other factors (i.e. reliability, safety, security, ...) are as important as the functionality itself [5].

## 2.2   Model Driven Development

The idea promoted by MDD is to use models systematically at different phases of system's development lifecycle. In other words, models provide input and output at all stages of system development until the final system itself is generated. The advantage of an MDE process is that it clearly defines each step to be taken, forcing the developers to follow the defined methodology. Models provide an integrated view of the system, increase the portability of the solution and facilitate the maintenance of the system. At the same time it promotes reuse since domain models can be reused for a variety of applications. In MDD, everything is considered as a model or a model element; hence, it is important to use a well-defined modeling language, such as UML [6], to describe each model precisely.

# 3   Motivations

The importance of security in actual systems is growing since most attacks on software systems are based on vulnerabilities caused by software that has been poorly designed and developed [7]. That's the reason why systems engineers need proven and generic security expert solutions that can be applied to security problems in order to be able to reduce the number of successful attacks against these systems. Security patterns area convenient way of satisfying this need. Applying and defining security patterns for developing secure software systems are currently a very active area of research [8]. However, some limitations are obvious.

First, most of existing approaches [9][10] focus on the definition and the application of security patterns in design level without providing any mechanism for implementing these patterns. Other approaches [11][3] propose concrete implementation of these patterns by providing middleware services that ensure the pattern functionalities. There is little work concerning the full integration of security patterns from the earliest phases of software development and providing automatic generation of the secure application code [12].

Second, there is currently no comprehensive methodology to assist system developers (non-expert in security) in security pattern integration. There is no guidance on how such security patterns can be integrated into current software component or model based system development methods. The wrong integration of the security pattern may cause system malfunction.

Third, numerous patterns currently exist for the construction of security mechanisms, but the lack of automated tools that support their integration in software systems development gives that security patterns are often neglected at the design level and do not constitute an intuitive solution that can be used by software designers.

Fourth, the code that applies security patterns is generally not well-modularized, as it is tangled with the code implementing each component's core functionality and scattered across the implementation of different components.

In this paper, we provide a security pattern integration process with tool support in order to encourage developers to take advantages from security solutions proposed as security patterns.

## 4    The Security Pattern Integration Approach

We propose an approach for modeling and implementing secure component-based applications. Our main objective is to provide an end-to-end process and it support tool to automatically applying security patterns at the different phases of development process.

We propose the framework represented in Fig. 1 that aims to decouple the application domain expertise (functional concerns) from the security expertise (security concerns) that are both needed to build a secure application. This framework consists in two levels:

Design level. We provide in this level concepts to easily design secure component model by automatically integrating security pattern into UML2 Component diagrams. Artifacts in this level are:

(A)   Component Metamodel that defines primitives and basic concepts to model component-based applications. In this approach, we use the UML2 metamodel.

(B)   Security Patterns that contain the specification of specific security solutions. We have chosen the set of security patterns defined in [14].

(C)   Component Security Profiles that extend the UML component metamodel by new security concepts, which depend on the defined pattern concepts. We define an UML profile for each family of security pattern (e.g. Acces control, fault tolérence, ...). .

(D)   Component Model represents the functional application model, which is conforms to the component metamodel.

(E)   Security Pattern Application Rules (SPARs) allow to **semi-automatically** integrating the security patterns to the component model using the defined UML profiles. For each security pattern, we define a set of rules which are implemented as model to model transformations.

(F)   Secure Component Model is the result of applying SPARs rules to the component Model.

Implementation level. We provide a mechanism to implement, in a modular way, the generated secure component model using aspect oriented programming.

(G)   Functional components code is the code of the application without any security property. Such code can be automatically generated (model-to-text transformation) or be a result of set of refinements of the component model. An intermediate model can be also defined as a plate-form specific model.

(H)   Aspects code is automatically generated from the secure component model. We propose a set transformation rules to automatically generate AspectJ aspects from the secure component model..

(I)   Secure application code is the result of waving aspect code within the functional application code using an aspect weaver.

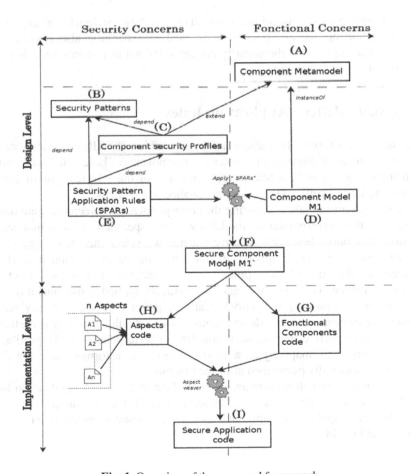

**Fig. 1.** Overview of the proposed framework

We propose a three steps process for implementing secure component-based applications.

The first step corresponds to the design of the functional application code. The designer may use the Papyrus suite tool [15], for example, to specify his application as a UML 2 component diagram. The resulting model should not containany security concepts. In this phase, the designer can also use any UML profile that support specific component models like CCM, EJB or Fractal.

The second step corresponds to the design and the integration of asecurity concern into the component model. The designer selects the security pattern to be applied according to the confronted problem. The software designer should determine which design pattern will be applied according to his context and which policy type to enforce in which situation. He can apply more then one security pattern in the same model. The designer should just apply the security patterns rules implemented using ATL (Atlas transformation language) [16]. The result of this transformation is a new component model that supports all the security concepts defined in the selected pattern.

The third step consists insemi-automatically generating the functional application code that implements the component model, and in generating also a skeleton of aspect code that implements the security concepts defined as patterns. Details of this step are out of scope in this paper.

## 5     Security Pattern Application Rules

We define a set of security pattern application rules (SPARs) to automate the integration of security patterns into software components. These rules are deduced through the relationships between security concepts of the selected pattern and the corresponding UML profile. These rules are applied in two steps:

The first step corresponds to ensuring the correspondence between the main pattern concepts and the correspondence model elements (specified as a component, a connection or a port). For each security pattern, we select the main concept that should be applied by the designer (i.e., the name of the application artifact that corresponds to the role of the applied security pattern). The definition of this correspondence depends on security patterns previously applied to the same model.

The second step ensures the automatically mapping the other security pattern concepts to the corresponding model elements. Concretely, this mapping is performed by applying the respective stereotypes defined in the corresponding UML profile. We have implemented this mapping as a model-to-model transformations using ATL. This step is automatically performed after the first one.

In the following, we will present an example of Security Pattern Application Rules taking the RBAC (Role-based Access Control) pattern [21] as an example of security patterns. We also applied our approach and the proposed rules on other security patterns defined in [14].

Role-Based Access Control Pattern. RBAC is an authorization mechanism in which access decisions to an object are based on roles that users hold within an organization. The rights to execute a set of operations are grouped in roles and users are assigned to one or more roles. This pattern allows the designer to rigorously define an efficient access control model at the design phase. Fig. 2 presents the basic RBAC pattern.

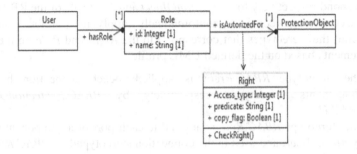

**Fig. 2.** Basic RBAC Pattern

RBAC profile.As presented above, we defined an UML profile - the standard UML lightweight extension mechanism- for this security pattern. In previous work [22] we have proposed a process of generating an UML profile that extend UML 2 component model with security pattern concepts. Fig. 3 summarizes the RBAC profile, which contains a set of stereotypes, tagged values, and OCL [17] constraints previously presented in the RBAC pattern:

– The stereotypes *user* and *ProtectionObject* extend the UML *Component* metaclasses.
– The stereotype *Role* extends the *Port* metaclass.
– The stereotype *Right* extends the *Connector* metaclass.

Due to lack of space, we do not present the OCL constraints in this paper.

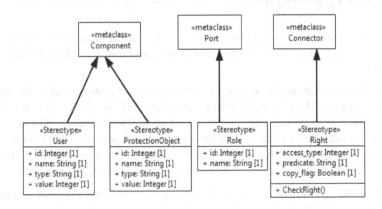

**Fig. 3.** RBAC Profile

RBAC Application rules description. In this section, we detail the proposed security pattern application rules for RBAC pattern. As presented in Section 4, application of these rules includes two steps.

In the first step, the designer should just identify in his component model the user of his secure system as well as the objects (i.e. resources) to be protected. Concretely, the stereotypes *RBAC.ProtectionObject* and *RBAC.User* will be added to components that correspond respectively to *ProtectionObject* and *User* role of the RBAC pattern.

The second step consists in automatically applying (without any designer intervention) the stereotypes that correspond to the *Right* and *Role* on the respective model elements based on the defined UML profile.

- The stereotype *RBAC.Right* is applied to each connection between two components stereotyped respectively by *RBAC.ProtectionObject* and *RBAC.User*.
- The stereotype *RBAC.Role* is applied to each port of a component stereotyped by *RBAC.User*, and used in the connection stereotyped by *RBAC.Right*.

The defined rules (i.e. the application of the defined stereotypes) are implemented using ATL (Atlas Transformation Language). Listing 1 shows an excerpt of the ATL code corresponding to the first step.

```
rule Component2secureComponent {
from notsecure : MM!"uml::Component"
to secure : MM1!"uml::Component"(
name<- notsecure.name ..... )
do
{secure.applyStereotype(thisModule.getStereotype('RBAC.ProtectionObject'));}
}
```

**Listing.1.** Part of the RBAC pattern application rules

# 6    Code Generation

We propose a code generation process for generating the functional application code and the security code using aspect oriented programming [18]. This process is made of two steps.

**Functional Code Generation.** Several approaches and commercial tools support the generation of code skeleton in different technologies (EJB, .NET, C++, etc.) from the UML component diagram, based on a set of predefined libraries. The designer can also generate the corresponding code based on a MDA approach [21], which requires the automatic transformation of the model into a platform specific model. The corresponding code can be easily generated using model-to-text generator.

**Security Code Generation.** For each security pattern, we propose a template to generate AspectJ code and the required helper Java classes. We generate a skeleton of aspect code, which should be completed by the developer, according to the functional application code generated during the first step. *Pointcuts* intercept the call of critical methods, while *advices* ensure the functionalities of the patterns. We generate different types of advice (around, before, and after) depending on the security pattern.

For example, the generated aspect code for RBAC policy is defined as following. The generated pointcut intercepts the call of all methods performed by the system users. We generate an around advice, which verifies before the execution of each intercept method, that the caller (user) has the specified role (i.e. has the access) to execute this method. Due to space limitation, we do not detail the aspect template for RBAC pattern.

## 7    Case Study

In this section, we present how we applied our approach to GPS system (Global Position system).The Basic GPS consists of three parts: the *space segment*, the *control segment* and the *user segment*. In what follows, we focus on the space and user segments. Indeed, in these two parts we have identified requirements for access control to services offered by components. In this example, we consider mainly the management of access control to various services offered by phone operators especially downloading geographic maps in real-time and managing secure access to satellites.

The basic GPS system described above works as follows:

(1) The GPS Terminal receives continuously the signal of Satellite as well as that of SecureSatellite. The SecureSatellite is active if the user has access rights to it.

(2) The GPS Terminal sends a request to download map to the Phone Operator.

(3) The Phone Operator allows the user, depending on its access rights, to download the requested geographic map.

With regard to the Basic GPS system, we have identified two use cases: downloading geographic maps and access to secure satellites

The UML component model of the basic GPS system is shown in Fig. 4. It contains five components:

- Satellite enables to emit permanently a navigation message containing all the necessary data for the receiver to perform all the navigation calculations.
- SecureSatellite emits secure signals
- GPSTerminal receives the message transmitted by the Satellite and must have access rights to the signals sent by SecureSatellite. It requires the map downloading service from the Operator component.
- Operator (Phone operator) offers the service to download maps. It requires maps requested by members from MapDataBase.
- MapDataBase offers the possibility to the Operator to have the map downloaded by the applicant.

RABC pattern is applied as a solution of the access problem in this system.

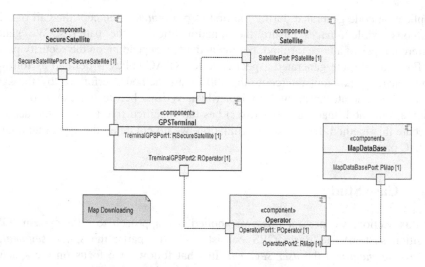

**Fig. 4.** Component Diagram of Basic GPS system

After GPS application is described, the developer specifies "User ='GPSTerminal', 'Operator'" and "Protection Object='SecureSatellite', 'Operator', 'MapDataBase'" then he ATL file (part of it is presented in Listing.1) in which the RBAC pattern is described corresponding to the parameter enteredby the developer and applies the RBAC pattern. Stereotypes are added automatically to indicate that the RBAC pattern is applied. For example, a stereotype is added to indicate that the component GPS terminal corresponds to the *User* role and to indicate that the *Right* role corresponds to the connection between each user and projection object component in the diagram. Fig.5 shows the component diagram of the basic GPS system after the RBAC pattern has been applied.

## 8    Related Work

In this section, we present works addressing general approaches for secure system development and approaches which target security pattern integration.

Many studies have already been done on modeling security in UML. The works presented in [22] describes an extension UMLsec of UML that enables to express security relevant information within the diagrams in a system specification. UMLsec is defined as a UML profile. In [23] the authors present a modeling language for the model-driven development of secure, distributed systems based on UML. Their approach targets role-based access control with additional support for specifying authorization constraints. SecureUML is a modeling language that defines a vocabulary for annotating UML-based models with information relevant for access control. These two approaches are not in competition, but they complement each other by providing different points of view on the secure information system.

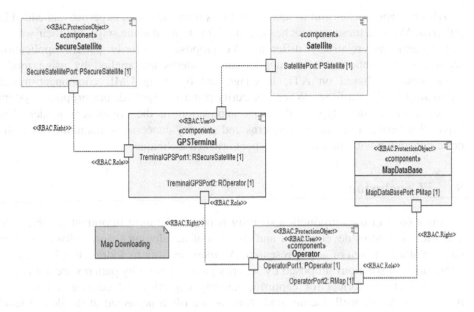

**Fig. 5.** Component Diagram of Basic GPS systemafter SPARs application

Concerning patterns instantiation and application, several proposals have already been made. The works presented in [24] present an approach that uses a formal design pattern representation and a design pattern instantiation technique of automatic generation of component wrapper from design pattern. Additionally, several approaches introduce their own tool-based support for the pattern instantiation. In [25] the authors propose an approach for representing and applying design patterns. Authors in [26] provide a simple UML profile for design pattern which allows the explicit representation of design patterns in UML models with a workable model transformation approach and a transformation process. Another method described in [27] presents a methodology for the creation of automated transformations that can apply a design pattern to an existing program. In [28] the authors propose a method supporting design pattern application in software project. The elaborated method is based on semantics defined via UML profile and transformationof models.

Aiming to provide software engineers with guidelines regarding the application of security patterns in systems development lifecycle, several methodologies have been proposed in the literature. In [29] authors provide an analysis of the most important works related to security patterns. They discuss their applicability for the analysis and design of secure architectures in real and complex environments.

In [30] authors propose a security pattern integration technique dealing with model transformation using ATL. Moreover, authors in [11] use Petri nets to first model security patterns on an abstract level. A methodology for integrating security patterns into all stages of the software development lifecycle is proposed in [31]. Other methodologies present the use of the aspect oriented software design approach to model security patterns as aspects and weave them into the functional model [32][33].

All these approaches aim to take advantages from the use of expertise embedded in patterns. Many of these approaches use similar actions at some stage, but their way to use patterns are relatively different. We propose a security pattern application technique in component-based applications. Patterns are applied by using model transformations based on ATL language and by using UML component model extensions (UML profiles). When a security pattern is applied, our proposed system leaves a mark (stereotypes) about its application in the component model. The separation between functional concerns and security concerns is maintained until the code generation with the use of the aspect oriented technique.

# 9    Conclusion

As mentioned in the introduction security is one of the most important properties to consider in systems development and the goal of developing secure software system has remained an area of active research. A promising way to address this issue is the application of solutions proposed by security patterns. Security patterns are defined in a well-structured format for capturing security expertise and communicating it to novices. Although well documented, patterns are often neglected at the design level and do not constitute an intuitive solution that can be used by software designers. This can be the result of the need of integration process to let designers apply those pattern's solutions in practical situations and to work with patterns at higher levels of abstraction. Towards this direction, we have proposed, in this paper, a methodology enabling semi-automatic application of security patterns. It includes a method for describing security pattern transformation rules. Our methodology covers the most used security patterns. We are currently developing a tool supporting the overall approach. A preliminary work has been done in [34] using the Role Based Access Control (RBAC) Pattern. We aim to catalog security patterns in the form of a repository and to integrate them in an MDE process to build secure component based software applications.We also planto consider dependencies between security patterns.

# References

1. Devanbu, P.T., Stubblebine, S.: Software Engineering for Security: a Roadmap. In: Proceedings of the Conference of the Future of Software Engineering (2000)
2. Yoder, J., Barcalow, J.: Architectural patterns for enabling application security. In: Fourth Conference on Patterns Languages of Programs (1997)
3. Diego, S., Advisors, R., Maña, A., Yagüe, M.I.: Integration of Security Patterns in Software Models based on Semantic Descriptions
4. Szyperski, C.: Component Software: Beyond Object-Oriented Programming. Addison-Wesley Longman Publishing Co., Boston (2002)
5. Sentilles, S.: Towards efficient component based software development of distributed embedded systems (2009)
6. OMG: Omg unified modeling language specification,
   http://www.omg.org/spec/UML/2.0/

7. Halkidis, S.T., Tsantalis, N., Chatzigeorgiou, A., Stephanides, G.: Architectural Risk Analysis of Software Systems Based on Security Patterns. IEEE Transactions on Dependable and Secure Computing 5, 129–142 (2008)
8. Schumacher, M.: Security Engineering with Patterns. LNCS, vol. 2754. Springer, Heidelberg (2003)
9. Haralambos Mouratidis, P.G.: Security Patterns for Agent Systems
10. Rudolph, A.F., Gurgens, S., Rudolph, C.: Towards a Generic Process for Security Pattern Integration (Info). In: 20th International Workshop on Database and Expert Systems Application, pp. 171–175 (2009)
11. Horvath, V., Dörges, T.: From security patterns to implementation using petri nets. In: Proceedings of the Fourth International Workshop on Software Engineering for Secure Systems, SESS 2008, pp. 17–24. ACM Press, New York (2008)
12. Diego Ray, A.M., Integration, M.I.Y.: of Security Patterns in Software Models based on Semantic Descriptions. Presented at the
13. Bouaziz, R., Hamid, B., Desnos, N.: Towards a better integration of patterns in secure component-based systems design. In: Murgante, B., Gervasi, O., Iglesias, A., Taniar, D., Apduhan, B.O. (eds.) ICCSA 2011, Part V. LNCS, vol. 6786, pp. 607–621. Springer, Heidelberg (2011)
14. Schumacher, M., Fernandez-Buglioni, E., Hybertson, D., Buschmann, F., Sommerlad, P.: Security Patterns: Integrating Security and Systems Engineering. Wiley Series in Software Design Patterns. Wiley (2006)
15. Papyrus UML, http://www.papyrusuml.org/scripts/home/publigen/content/templates/show.asp?L=EN&P=55&vTicker=alleza&ITEMID=3
16. ATL, http://www.eclipse.org/atl/
17. OCL 2.3.1, http://www.omg.org/spec/OCL/2.3.1/
18. Aspect-oriented software development website, http://aosd.net
19. IBM: rational rose website, http://www-01.ibm.com/software/awdtools/developer/rose/
20. Entreprise Architect website, http://www.sparxsystems.com/
21. OMG: MDA Specifications, http://www.omg.org/mda/specs.htm
22. Jürjens, J.: UMLsec: Extending UML for secure systems development. In: Jézéquel, J.-M., Hussmann, H., Cook, S. (eds.) UML 2002. LNCS, vol. 2460, pp. 412–425. Springer, Heidelberg (2002)
23. Lodderstedt, T., Basin, D., Doser, J.: SecureUML: A UML-Based Modeling Language for Model-Driven Security. In: Jézéquel, J.-M., Hussmann, H., Cook, S. (eds.) UML 2002. LNCS, vol. 2460, pp. 426–441. Springer, Heidelberg (2002)
24. Yau, S.S.: Integration in component-based software development using design patterns. In: Proceedings 24th Annual International Computer Software and Applications Conference, pp. 369–374. IEEE Comput. Soc. (2000)
25. El Boussaidi, G., Mili, H.: A model-driven framework for representing and applying design patterns. In: 31st Annual International Computer Software and Applications Conference, COMPSAC 2007, vol. 1, pp. 97–100. IEEE (2007)
26. Wang, X.-B., Wu, Q.-Y., Wang, H.-M., Shi, D.-X.: Research and Implementation of Design Pattern-Oriented Model Transformation. In: 2007 International Multi-Conference on Computing in the Global Information Technology (ICCGI 2007), pp. 24–24. IEEE (2007)

27. Cinnéide, Ó., Nixon, M., Automated, P.: software evolution towards design patterns. In: Proceedings of the 4th International Workshop on Principles of Software Evolution, IWPSE 2001, p. 162. ACM Press, New York (2002)
28. Kajsa, P., Majtás, L.: Design patterns instantiation based on semantics and model transformations. In: van Leeuwen, J., Muscholl, A., Peleg, D., Pokorný, J., Rumpe, B. (eds.) SOFSEM 2010. LNCS, vol. 5901, pp. 540–551. Springer, Heidelberg (2010)
29. Ortiz, R., Moral-García, S., Moral-Rubio, S., Vela, B., Garzás, J., Fernández-Medina, E.: Applicability of security patterns. In: Meersman, R., Dillon, T.S., Herrero, P. (eds.) OTM 2010, Part I. LNCS, vol. 6426, pp. 672–684. Springer, Heidelberg (2010)
30. Yu, Y., Kaiya, H., Washizaki, H., Xiong, Y., Hu, Z., Yoshioka, N.: Enforcing a security pattern in stakeholder goal models. In: Proceedings of the 4th ACM Workshop on Quality of Protection, QoP 2008, p. 9. ACM Press, New York (2008)
31. Fernandez, E.B., Larrondo-Petrie, M.M., Sorgente, T., Vanhilst, M.: A Methodology to Develop Secure Systems Using Patterns. Integrating Security and Software Engineering 5, 107–126 (2006)
32. Georg, G., Ray, I., France, R.: Using Aspects to Design a Secure System. In: Proc. of the Eighth IEEE International Conference on Engineering of Complex Computer Systems (ICECCS 2002), pp. 117–126. ACM Press, Greenbelt (2002)
33. Indrakshi Ray, R.F.: An aspect-based approach to modeling access control concerns
34. Bouaziz, R., Coulette, B.: Secure Component Based Applications Through Security Patterns. In: International Conference on Computational Science and Engineering, CSE 2012 (2012)
35. Khoury, P.E., Mokhtari, A., Coquery, E., Hacid, M.-S.: An Ontological Interface for Software Developers to Select Security Patterns. In: DEXA Workshops, pp. 297–301 (2008)
36. Asnar, Y., Massacci, F., Saïdane, A., Riccucci, C., Felici, M., Tedeschi, A., El Khoury, P., Li, K., Seguran, M., Zannone, N.: Organizational Patterns for Security and Dependability: From Design to Application. IJSSE 2(3), 1–22 (2011)

# Evaluation of Web Session Cluster Quality Based on Access-Time Dissimilarity and Evolutionary Algorithms

Veer Sain Dixit[1,*], Shveta Kundra Bhatia[2], and V.B. Singh[3]

[1] Computer Science Department,
Atma Ram Sanatan Dharma College, University of Delhi,
New Delhi, India
veersaindixit@rediffmail.com
[2] Computer Science Department, Research Scholar, University of Delhi,
New Delhi, India
shvetakundra@gmail.com
[3] Computer Science Department,
Delhi College of Arts and Commerce, University of Delhi,
New Delhi, India
singh_vb@rediffmail.com

**Abstract.** Web session cluster refinement is one of the major research issues for the improvement of cluster quality in recent days. The motive of refinement using Evolutionary Algorithms is quite obvious because in any clustering algorithm the obtained clusters shall have some data items that are inappropriately clustered, hence, never giving us well separated and cohesive clusters. Hence the quality of clusters is improved using refinement techniques. Initial clusters are formed using K-Means clustering algorithm which suffers from local minima problem. The refinement on clusters is performed on the basis of access and time features (Modified Knockout Refinement Algorithm) which is a distance based dissimilarity, Genetic Algorithm (GA), Particle Swarm Optimization (PSO) and a combination of MKRA with GA and MKRA with PSO. Results are evaluated on five synthetic datasets and three real datasets. Further, it is shown experimentally that effectiveness of combining MKRA with evolutionary techniques produces better quality clusters.

**Keywords:** Genetic Algorithm (GA), Particle Swarm Optimization (PSO), Davies Bouldin Index (DB), Cohesion (COH), Separation (SEP).

## 1    Introduction

The exponential growth of the information on the internet has prompted the need for developing tools to serve preferences of individual users browsing the internet. The web log data can be suitably used to extract usage behavior. The log data needs to be

---

* Corresponding author.

B. Murgante et al. (Eds.): ICCSA 2014, Part V, LNCS 8583, pp. 297–310, 2014.

pre-processed for the formation of the session file, where a session is a set of requests made by the end users during a single visit to the site in a certain time frame. The behavior of the users can be extracted using various techniques such as Association Rules, Sequential Patterns, Usage Clusters, Page Clusters and User Classifications [1].

In this research, our main focus is on web session clustering which a problem of grouping web sessions with similar behavior is resulting in the maximization of similarity within a group and minimizing the similarity of objects in two different clusters. Here our intent is to refine clusters for quality improvement. There are several proposed clustering algorithms showing their inability to handle large data sets effectively. The convergent K-Means algorithm has popularly been used to cluster large data sets because of its linear time complexity, space complexity and order independence for a given initial seed. The K-Means algorithm is used to perform initial clustering on large as well as small datasets and the variation in quality of clusters is analyzed. We propose to eliminate the factor of heterogeneity by using optimization algorithms such as Genetic algorithm, Particle swarm optimization. The process of refinement process leads to formation of clusters that are well separated from each other and are compact as compared to original clusters. Evolutionary Algorithms are used for refinement to overcome the problem of local minima of K-Means as evolutionary algorithms find global optimal solutions but with a limitation of high execution time and difficulty of managing control parameters due to which they have been typically tested on small data sets. The elimination of sessions from the clusters has improved the similarity within the cluster while a few sessions still contribute to the heterogeneity of the cluster. To solve the above issue we combine MKRA with GA and MKRA with PSO.

Using the quality measures of Cohesion, Separation and Davies Bouldin index of obtained clusters as the evaluation measures [2, 3] we compare our results. The results show that our method of evaluating dissimilarity is effective and the combination of distance based dissimilarity and evolutionary techniques results in better quality clusters.

The rest of the paper is organized as follows: In section 2, there is an illustration of the related work. In Section 3, the problem is formulated and derived. Experimental results are evaluated in Section 4. Section 5 concludes the paper.

## 2     Related Work

A large number of real world applications use clustering for grouping objects as a solution to mine information for problem solving [4]. Clustering of web data is the process of grouping web data into clusters so that similar objects are in a single group and dissimilar objects are in different groups [5, 6]. The goal is to meet user preferences and organize the data into groups so as to increase web information accessibility, understand users' navigation behavior, improving information retrieval and content delivery on the Web. Web data clustering can be link based [7, 8] and session-based [9, 10]. Link based clustering groups together generated sessions on the basis of access patterns in sessions and session based clustering on the viewing of web pages. The records of user's activities within a Web site are stored in a log file. Each record in the log file contains the client's IP address, the date and time the request is received, the requested object and some additional information such as protocol of request, size of the object etc. A good survey on various clustering algorithms can be found in [11].

Web session clustering has been worked upon by researchers to solve the problems faced by web based applications. The BIRCH algorithm was developed by Fu [14] that generalized the session in attribute oriented induction with respect to a data structure called page hierarchy-partial ordering. Xie and Phoha [12] suggested web user clustering from access log using the belief function. Shahabi and Kashani [13] suggested path similarity measures used to group the sessions. Gonzalcs, Mabu, Taboada & Hirasawa [15] improved the clustering results by applying Genetic Algorithms through user feedback. Oyanagi [16] presents an application of matrix clustering to Web usage data. Castellano [17] proposed a relational fuzzy clustering algorithm to discover profiles and extracting real user preferences.   Along with the above clustering techniques based on different features K-means is one of the most widely used clustering algorithms. Due to certain constraints of the k-means algorithm research has been performed for its improvement but with modifications in its pre-conditions. Bentley [18] suggested kd-trees to improve triangle inequality to enhance K-Means. Bradley and Fayyad [19] presented a procedure for computing a refined starting condition that is based on a technique for estimating the modes of a distribution. A Kernel-Means algorithm was established by Scholkops [20] maps data points from the input space to a feature space through a non linear transformation minimizing the clustering error in feature space. Dhillon [21] suggested refining clusters in high dimensional text data by combining the first variation principal and spherical K-Means. Elkan [22] suggested the use of triangle inequality to accelerate K-Means. Kanungo [23] and Pelleg [24] worked towards the acceleration of the K-Means algorithm.

Once clusters have been formed Multilevel refinement schemes were used for refining and improving the clusters produced by hierarchical agglomerative clustering by Karypis [25]. Refinement of web usage data clustering from K-Means with Genetic Algorithm was suggested by Sujatha and Iyakutty [26]. PSO based data clustering was initiated by Merewe and Engelbrecht [27] who used PSO with k-means clustering technique. Xiao, Dow, Eberhart, Miled and Oppelt [28] used Self Organizing Map along with PSO to improve the efficiency of the clustering process. Omran, Salman and Engelgrecht [29] proposed a hybrid of PSO and k-means where clustering is done using PSO and refinement using k-means. [15] Gave a framework for web mining using an evolutionary algorithm named Genetic Relation Algorithm (GRA) that performs additional searching for documents according to user's interests. An introduction to Genetic Algorithms is given by Mitchell [30]. Asllani and Lari [31] used genetic algorithm for dynamic and multiple criteria web-site optimizations. K-Means was combined with Particle Swarm Optimization [32] to enhance data clustering where as PSO reaches an optimal solution the clustering process switches to K-Means for faster and accurate processing. K-Means was combined with the Genetic Algorithm (GA) in [33] where K-Means was used as a search operator instead of crossover.

In the previous work modifications are performed in the clustering techniques to improve quality of clusters. In this research, limitations of the K-Means algorithm have been taken care of to form clusters taking into account the improvement of the quality of clusters by refining them after the formation of clusters. Refinement is based on dissimilarities between sessions using features of access and time along with GA and PSO.

# 3    Refinement Techniques

Mining of Web Usage data has been used for identification of user's preferences on the internet. Web session clustering, is a process of grouping web sessions into clusters so that homogeneous sessions are in the same cluster and heterogeneous sessions lie in different clusters. Before clustering, data needs to be cleaned and processed for session identification. The pre-processing task builds a session file where each session is a collection of requests made by a user during his visit based on the unique IP addresses and session time-outs. Clustering of user sessions help in exhibiting similar browsing behaviors and more than one group of pages related to sessions having similar interests.

Consider a set of N session's defining the Page View Matrix (PVM) i.e.

$$S = \{S_1, S_2, S_3 \ldots S_N\} \tag{1}$$

Which are to be clustered where each

$$S_i = \{w_1, w_2, w_3 \ldots w_x\} \tag{2}$$

With x being the number of web pages.

Where

$$w_j = \begin{cases} 1, \text{ if the web page j has been viewed} & \text{where } j=1\ldots x \\ 0, \text{ if the web page j has not been viewed} \end{cases} \tag{3}$$

The sessions are to be grouped into clusters (groups)

$$G = \{G_1, G_2, G_3 \ldots G_K\} \tag{4}$$

Where K is the number of clusters such that

$$G_1 \cup G_2 \cup G_3 \ldots \ldots \cup G_K = S,\ G_i \neq \emptyset,\ G_i \cap G_j = \emptyset \text{ for } 1 \leq i \neq j \leq K \tag{5}$$

As per the above formulation the session set defined in equation (1) is to be clustered as given in equation (4) satisfying the criteria described in equation (5). Session clusters in our work are generated using the K- Means algorithm. It is one of the most widely used clustering algorithms as the run time for the algorithm is linear in nature and is easy to implement, however the algorithm is dependent on the initial partitioning and the data ordering along with a known value of K. The algorithm partitions the data points into clusters, so as to minimize the sum of the distances as given in equation (6) between the data points and the centroid of the clusters.

$$J = \sum_{i=1}^{K} \sum_{x_j \in P_i} \left\| x_j - c_i \right\|^2 \tag{6}$$

The authors define quality of a cluster in terms of cohesion as given in equation (7) and separation as given in equation (8). The obtained clusters have some measure of dissimilarity within, which motivates the author for refinement of clusters to attain improved quality clusters measured using the Davies Bouldin (DB) Index as given in equation (9).

$$Cohesion(COH) = \max(\ diam_{\ i,j=1......m}(x_i, x_j)) \tag{7}$$

Where $x_i$, $x_j \in C$, i,j= 1.....m.

$$Separation(S) = \min(\ distance(x_l, x_j)) \tag{8}$$

Where $x_i \in C_1$, $x_j \in C_2$. i=1.....m, j= 1.....n.

$$DB = 1/K \sum_{i,j-1}^{K} \max_{i \neq j} \left[ \frac{diam(C_i) + diam(C_j)}{d(C_i, C_j)} \right] \tag{9}$$

Cohesion measures how close in terms of maximum distance the elements are in a cluster by minimizing the distance between two farthest points in a cluster and separation ensuring that clusters are far enough from each other. DB index defines the ratio of the sum of within-cluster scatter to between cluster separations.Hence the authors intent is to find clusters that minimize cohesion and maximize separation with an overall low Davies Bouldin Index.

In this paper, the proposed MKRA [34] algorithm is compared with evolutionary algorithms.

## 3.1    Genetic Algorithm

A genetic algorithm (GA) is an iterative search technique that imitates the process of evolution in nature. This technique is used to generate solutions that are useful for optimization and search problems. Genetic algorithms generate solutions to optimization problems using techniques such as inheritance, mutation, selection, and crossover. The algorithm involves a population, each individual is represented by a finite string of symbols, known as the genome, encoding possible solutions in a given problem space. The algorithm leads to selection of genomes on the basis of a fitness function. Genetic algorithms are applied in areas, which are very huge to be searched exhaustively.

Genetic Algorithm starts with an initial population of particles (sessions in our application) generated at random that can have a binary, character based, real valued, tree or any other representation. New populations are generated decoding the active population and examining them on the basis of a fitness function. Individuals having a good fitness function are selected from the previous generation. An individual with high-fitness stands a better chance of reproducing, while low-fitness ones are more likely to be eliminated. The new individuals are found in the search space by genetically inspired operators crossover and mutation. Crossover is performed with a crossover probability or crossover rate between two selected individuals, called parents. By interchanging parts of their genomes two new individuals are formed, called offspring. Sub strings are exchanged after a randomly selected crossover point. This operator tends to enable the evolutionary process to move towards promising regions of the search space. The mutation operator is implemented to prevent premature convergence to local optima. The termination condition of genetic algorithms may be specified as some fixed maximum number of generations or till the point an acceptable fitness level is reached. The parameters that control the GA are the size of the population, rate of crossover and the probability of mutation. The above parameters can significantly affect the performance of the algorithm. In this paper, it is considered that the evolutionary methodology helps in selection of good sessions from the clusters improving the quality from the original ones.

In our application Sessions are extracted according to their probability as long as the fitness improves leading to a refined cluster with best fit sessions.

```
Procedure GENETIC ALGORITHM REFINEMENT
INPUT: Set of particles (Original Cluster)
PROCESS:
Step 1: Generate an initial population consisting of population size individuals.
        Each attribute is switched on with probability p_initialize
Step 2: For all individuals in the population
Step 3: Perform mutation
Step 4: Choose two individuals from the population and perform crossover with
        probability p_crossover. The type of crossover can be selected by crossover
        type.
Step 5: Perform selection, map all individuals to sections on a roulette wheel whose
        size is proportional to the individual's fitness and draw population size
        individuals at random according to their probability.
Step 6: As long as the fitness improves, go to step 3.
Output: Set of selected particles (Refined Cluster)
```

**Fig. 1.** Algorithm for Genetic Algorithm Refinement

## 3.2    Particle Swarm Optimization

Particle swarm optimization is a global optimization algorithm given by Kennedy, Eberhart and Shi for solving problems in which a best solution can be represented. Particle swarm optimization (PSO) is an attractive heuristic search method due to its easy implementation, stable convergence characteristics and good computational efficiency. In PSO based refinement we formulated sessions as particles for the particle swarm optimization algorithm. The sessions taken as particles move from their initial positions guided by various components and assigns weights to each of the sessions. Only those sessions are selected for generating new clusters that have a value greater than equal to the average of the weights of all the sessions in the cluster. It can be applied to a large number of optimization problems as it is very effective and quick. The technique takes care of the problem of local minima and hence can be used for refinement to improve cluster quality.

```
Procedure PARTICLE SWARM OPTIMIZATION REFINEMENT
INPUT: Set of particles (Original Cluster)
PROCESS:
Step 1: For each particle calculate fitness value.
Step 2: If the fitness value is better than the best fitness value (pBest) in history then
        set current value as the new pBest.
Step 3: Choose the particle with the best fitness value of all the particles as the gBest
        For each session.
Step 4: Calculate particle velocity
        Update particle position
Step 5: While maximum iterations or minimum error criteria is not attained.
        Assign weights to each session based on the above calculated fitness values.
        Select only those particles that have a weight greater than the average of the
        weights of all particles in the population.
OUTPUT: Set of selected particles (Refined Cluster)
```

**Fig. 2.** Algorithm for Particle Swarm Optimization Refinement

## 3.3    Combination of MKRA and GA

In order to improve our refinement process we use a combination of MKRA and GA where the selected sessions from MKRA are given as an input to GA. Considering the efficiency of Genetic Algorithm we cascade it with the MKRA algorithm combining the searching abilities of both techniques in one algorithm. In this implementation the sessions are selected based on the fitness function of the genetic algorithm.

```
Procedure MKRA-GA REFINEMENT
INPUT: Set of Original Clusters (OC)

PROCESS:
Step 1: Select a cluster.
    Step 2: Take transpose of the Page View Matrix (PVM)of the cluster
    Step 3: Refinement using MKRA leading to formation of RC₁
    Step 4: Refinement of RC₁ using GA leading to formation of RC₂
    Step 5: Repeat for every cluster.

OUTPUT: Set of Refined Clusters (RC)
```

**Fig. 3.** Algorithm for combination of MKRA and GA Refinement

## 3.4    Combination of MKRA and PSO

A combination of MKRA and PSO has also been implemented for refinement in a cascaded fashion. The purpose is to obtain groups having more similarity by eliminating sessions dissimilar to the rest in the same cluster to reduce the search space for improving web site structure and recommendation of web pages. Original and refined cluster qualities are discussed and analyzed by using different performance measures.

```
Procedure MKRA-PSO
INPUT: Set of Original Clusters (OC)

PROCESS:
Step 1: Select a cluster.
   Step 2: Take transpose of the Page View Matrix (PVM)of the cluster
   Step 3: Refinement using MKRA leading to formation of RC₁
   Step 4: Refinement of RC₁ using PSO leading to formation of RC₂
   Step 5: Repeat for every cluster.

OUTPUT: Set of Refined Clusters (RC)
```

**Fig. 4.** Algorithm for combination of MKRA and PSO Refinement

# 4     Experiments, Results and Discussion

Data used in the experiment is taken from three different real log files from the internet traffic archive containing information about all web requests to the websites. The datasets used are as follows:

The EPA trace contains all HTTP requests of a day to the EPA server located at Research Triangle Park, NC. The logs are an ASCII file with one line per request, with the following columns: host making the request, date, request given in quotes, HTTP reply code and bytes in the reply. The logs were collected from 23:53:25 on Tuesday, August 29 1995 through 23:53:07 on Wednesday, August 30 1995, a total of 24 hours. There were 47,748 total requests.

The NASA dataset HTTP requests of one month to the NASA Kennedy Space Center server in Florida were recorded. The logs are an ASCII file with one line per request, with the following attributes: host making the request, Timestamp of the request, request given in quotes, HTTP reply code and bytes in the reply. The log was collected from 00:00:00 July 1, 1995 through 23:59:59 July 31, 1995, a total of 31 days. . The raw web log file used for the experiment contained 131031 web requests.

The SDSC log data contains a day's trace to the SDSC WWW server located at the San Diego Supercomputer Center in San Diego, California. The logs are an ASCII file with one line per request, having the following entries: host making the request, timestamp of the request, filename of the requested item, operation performed by the server, remainder of the transaction log is either the HTTP request, or information associated with a failed operation. The logs were collected from 00:00:00 through 23:59:41 on Tuesday, August 22 1995, a total of 24 hours. There were 28,338 requests.

The synthetic data sets have been extracted from the pre-processed NASA web log. The pre-processed log file is divided into five parts of different sizes to study and evaluate the performance of refinement techniques on small and large datasets.

The data sets used for the experiments are listed as follows:

**Table 1.** Data Sets used

| Real Data Sets | Number of Sessions |
|---|---|
| EPA | 47748 |
| NASA | 131031 |
| SDSC | 28338 |
| **Synthetic Data Sets** | |
| DS1 | 500 |
| DS2 | 1500 |
| DS3 | 2500 |
| DS4 | 5000 |
| DS5 | 6000 |

The comparison of results of refinement using the K-Means and MKRA algorithm on all the data sets is as follows:

**Table 2.** Comparison of Davies Bouldin Index for various refinement techniques

| Dataset | Number of Clusters | K-Means | MKRA |
|---|---|---|---|
| DS1 | 10 | 2.8446 | 2.494 |
| | 15 | 2.669 | 2.2824 |
| | 20 | 3.094 | 2.2857 |
| DS2 | 10 | 3.1907 | 2.7935 |
| | 15 | 3.068 | 2.7711 |
| | 20 | 3.4351 | 3.1555 |
| DS3 | 10 | 3.4398 | 3.117 |
| | 15 | 3.2849 | 2.9897 |
| | 20 | 3.6325 | 3.4795 |
| DS4 | 10 | 2.9493 | 2.5793 |
| | 15 | 2.9532 | 2.6894 |
| | 20 | 3.2713 | 3.0654 |
| DS5 | 10 | 2.7734 | 2.4382 |
| | 15 | 2.7154 | 2.3815 |
| | 20 | 3.0535 | 2.8103 |
| NASA | 10 | 2.9919 | 2.7208 |
| | 15 | 2.6317 | 2.2784 |
| | 20 | 2.4674 | 2.1814 |
| EPA | 10 | 2.3483 | 2.0747 |
| | 15 | 2.0666 | 1.56 |
| | 20 | 2.1649 | 1.7166 |
| SDSC | 10 | 2.2521 | 1.5968 |
| | 15 | 1.8799 | 1.4362 |
| | 20 | 1.9557 | 1.2934 |

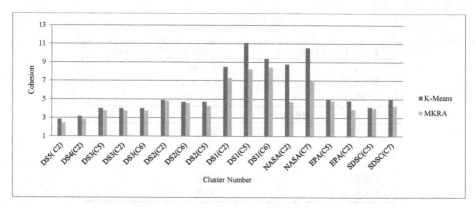

**Fig. 5.** Comparison of Cohesion for Original Clusters and Refined Clusters

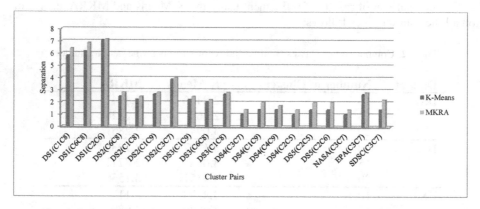

**Fig. 6.** Comparison of Separation between cluster pairs before and after refinement

From table 2, the results show that MKRA refinement technique gives clusters with improved quality as compared to the K-Means clustering technique because of the lower DB Index values. Figure 5 shows the improvement of intra cluster cohesion of the refined clusters as the distance between two farthest points in a cluster has reduced. Figure 6 shows that elimination of heterogeneous components from the clusters results in making the clusters well separated from each other by increasing the inter cluster distances. From the above analysis it is clear that MKRA provides us with good quality clusters.

Hence, now comparing MKRA with the evolutionary refinement techniques on the K clusters obtained from NASA, EPA and SDSC datasets using K-Means where value of K is 10, 15 and 20 we get the following results. Table 3 shows the DB index values found for the original clusters (K=10, 15 and 20) of the three data sets and the refined clusters obtained using MKRA, GA, PSO, MKRA-GA and a combination of MKRA-PSO. As lower values of DB index signify better cluster quality it is observed that for the NASA dataset (K=10) MKRA shows quality improvement of 10% than original, GA, PSO; for (K=15) a 20 - 30% improvement in quality is observed using MKRA as compared to GA, PSO; for (K=20) a 10 - 20% improvement in quality is observed using MKRA as compared to GA, PSO. A combination of MKRA-GA for K=10 shows quality improvement of 5% than MKRA; for (K=15) a 10 - 15% improvement in quality is observed using MKRA-GA as compared to GA, PSO and MKRA; for (K=20) a 10 - 15% improvement in quality is observed using MKRA-GA as compared to GA, PSO and MKRA. A combination of MKRA-PSO for K=10 shows quality improvement of 10% than MKRA; for (K=15) a 10 - 15% improvement in quality is observed using MKRA-PSO as compared to GA, PSO and MKRA; for (K=20) a 5% improvement in quality is observed using MKRA-PSO as compared to GA, PSO and MKRA.

Figure 7, 8 and 9 show the variation in the Davies Bouldin Index for clusters using the various refinement techniques. For a clustering solution, if DB Index values are low it represents better cluster formation. It is evident from the figures that a combination of MKRA and PSO is the best refinement technique for all the three data sets. Figure 10 shows the execution time of the refinement techniques on each cluster for K=10. The MKRA algorithm takes minimum time for refinement as compared to other techniques when applied individually. The execution time for a combination of techniques for refinement is higher but with improved quality clusters.

**Table 3.** Davies Bouldin Index values for 10, 15 and 20 clusters for all refinement techniques

|      | Number of Clusters | MKRA | GA | PSO | MKRA-GA | MKRA-PSO |
|------|------|------|------|------|------|------|
| NASA | 10 | 2.7208 | 2.9952 | 2.9627 | 2.7165 | 2.7021 |
|      | 15 | 2.2784 | 2.55 | 2.5391 | 2.1045 | 2.2534 |
|      | 20 | 2.1814 | 2.3273 | 2.3723 | 2.0188 | 2.1028 |
| EPA  | 10 | 2.0747 | 2.1629 | 2.1477 | 1.9932 | 1.7699 |
|      | 15 | 1.56 | 1.6532 | 1.9255 | 1.5102 | 1.3109 |
|      | 20 | 1.7166 | 1.5693 | 2.0195 | 1.4398 | 1.4011 |
| SDSC | 10 | 1.5968 | 2.2124 | 2.0951 | 1.5334 | 1.4532 |
|      | 15 | 1.4362 | 1.7932 | 1.8 | 1.3786 | 1.3245 |
|      | 20 | 1.2934 | 1.8543 | 1.8732 | 1.2366 | 1.2101 |

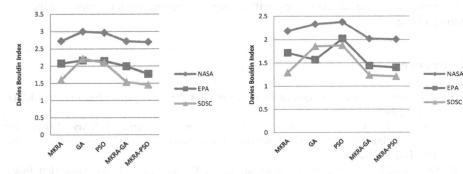

**Fig. 7.** Comparison of DB index for 10 Clusters

**Fig. 9.** Comparison of DB index for 20 Clusters

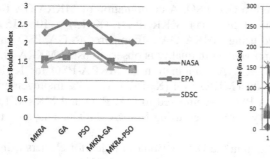
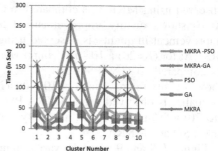

**Fig. 8.** Comparison of DB index for 15 Clusters

**Fig. 10.** Comparison of execution time for 10 Clusters for NASA dataset

## 5    Conclusion and Future Work

This paper discusses the problem of web usage clusters having large number of web sessions leading to poor quality of clusters. Evaluations were performed on three original web log files of NASA, EPA and SDSC and five synthetic data sets where MKRA was compared with GA, PSO, MKRA-GA and MKRA-PSO refinement techniques. It is evident from the observations that the quality of refined clusters using GA, PSO is better than the MKRA algorithm for the not very large EPA and SDSC datasets whereas MKRA generates good quality clusters after refinement for the huge NASA dataset as compared to GA, PSO. It is evident from the observations that combination of MKRA-GA and MKRA-PSO results in significant improvement of cluster quality as compared to MKRA, GA, PSO refinement techniques on the three real datasets irrespective of their sizes. The MKRA algorithm takes minimum time for refinement as compared to other techniques when applied individually. The combination of techniques gives clusters with improved quality but with a higher execution time.

In future the author proposes to compare results of K-Means with a soft clustering technique with varying distance measures. Further cascading of the GA and PSO for refinement shall be evaluated along with a proposed recommender system for the best quality clusters.

## References

[1] Mobasher, Discovery of aggregate usage profiles for web personalization. WebKDD, Boston (2009)
[2] Deborah, L., Baskaran, R., Kannan, A.: A Survey on Internal Validity Measure for Cluster Validation. International Journal of Computer Science & Engineering Survey (IJCSES) 1(2) (2010)
[3] Sanghoun, O., Chang, W.A., Moongu, J.: An Evolutionary Cluster Validation Index (2008)

[4] Nock, R., Nielsen, F.: On Weighting Clustering. IEEE Transactions and Pattern Analysis and Machine Intelligence 28(8), 1223–1235 (2006)

[5] Baldi, P., Frasconi, P., Smyth, P.: Modeling the Internet and the Web. Wiley (2003)

[6] Chakrabarti, S.: Mining the Web. Morgan Kaufmann Publishers (2003)

[7] Banerjee, A., Ghosh, J.: Click stream clustering using weighted longest common subsequences. In: Proceedings of the Web Mining Workshop at the 1st SIAM Conference on Data Mining (2001)

[8] Cadez, I.V., Heckerman, D., Meek, C., Smyth, P., White, S.: Model-based clustering and visualization of navigation patterns on a Web site. Data Mining and Knowledge Discovery 7(4), 399–424 (2003)

[9] Eiron, N., McCurley, K.: Untangling compound documents on the Web. In: Proceedings of the Fourteenth ACM Conference on Hypertext and Hypermedia (2003)

[10] Flake, G., Lawrence, S., Giles, C.L., Coetzee, F.: Self-organization and identification of Web Communities. IEEE Computer 35(3) (2002)

[11] Berkhin, P.: Survey of clustering data mining techniques. Springer, Heidelberg (2006)

[12] Xie, Y., Phoha, V.V.: Web user clustering from access log using belief function. In: Proceedings of the First International Conference on Knowledge Capture (K-CAP 2001), pp. 202–208. ACM Press (2001)

[13] Shahabi, C., Banaei-Kashani, F.: A framework for efficient and anonymous web usage mining based on client-side tracking. In: Kohavi, R., Masand, B., Spiliopoulou, M., Srivastava, J. (eds.) WebKDD 2001. LNCS (LNAI), vol. 2356, pp. 113–144. Springer, Heidelberg (2002)

[14] Fu, Y., Sandhu, K., Shih, M.: Clustering of Web users based on access patterns. Proceedings of WEBKDD (1999)

[15] Gonzales, E., Mabu, S., Taboada, K., Hirasawa, K.: Web Mining using Genetic Relation Algorithm. In: SICE Annual Conference, pp. 1622–1627 (2010)

[16] Oyanagi, S., Kubota, K., Nakase, A.: Application of matrix clustering to web log analysis and access prediction. In: Third International Workshop on Mining Web Log Data Across All Customers Touch Points, EBKDD 2001 (2001)

[17] Castellano, G., Fanelli, A.M., Mencar, C., Torsello, M.: Similarity based Fuzzy clustering for user profiling. In: Proceedings of International Conference on Web Intelligence and Intelligent Agent Technology. IEEE/WIC/ACM (2007)

[18] Bentley, J.: Multidimensional Binary Search Trees Used for Associative Searching. ACM 18(9), 509–517 (1975)

[19] Bradley, P.S., Fayyad, U., Reina, C.: Scaling Clustering Algorithms to Large Databases. In: 4th International Conference on Knowledge Discovery and Data Mining, KDD 1998. AAAI Press (August 1998)

[20] Scholkopf, B., Smola, J., Muller, R.: Technical Report: Nonlinear component analysis as a kernel eigen value problem. Neural Comput. 10(5), 1299–1319 (1998)

[21] Dhillon, I.S., Fan, J., Guan, Y.: Efficient clustering of very large document collections. In: Data Mining for Scientific and Engineering Applications, pp. 357–381. Kluwer Academic Publishers (2001)

[22] Elkan, C.: Using the Triangle Inequality to Accelerate k-Means. In: Proceedings of the Twentieth International Conference on Machine Learning (ICML 2003), pp. 609–616 (2003)

[23] Kanungo, T., Mount, D.M., Netanyahu, N., Piatko, C., Silverman, R., Wu, A.: An efficient kmeans clustering algorithm: Analysis and implementation. IEEE Trans. Pattern Analysis and Machine Intelligence 24(7), 881–892 (2002)

[24] Pelleg, D., Moore, A.: Accelerating exact kmeans algorithm with geometric reasoning. In: Proceedings of the Fifth ACM SIGKDD International Conference on Knowledge Discovery and Data Mining, New York, pp. 727–734 (1999)

[25] Karypis, G., Han, E., Kumar, V.: Multilevel Refinement for Hierarchical Clustering. Department of Computer Science & Engineering Army HPC Research Center (1999)

[26] Sujatha, N., Iyakutty, K.: Refinement of Web usage Data Clustering from K-means with Genetic Algorithm. European Journal of Scientific Research 42(3), 478–490 (2010) ISSN 1450-216X

[27] Merwe, V.D., Engelbrecht, A.: Data clustering using particle swarm optimization. In: The 2003 Congress on Evolutionary Computation, CEC 2003, vol. 1, pp. 215–220. IEEExplore (2003)

[28] Xiao, X., Dow, E.R., Eberhart, R., Miled, Z., Oppelt, R.: Gene Clustering using Self-Organizing Maps and Particle Swarm Optimization. In: Guo, M. (ed.) ISPA 2003. LNCS, vol. 2745, pp. 154–160. Springer, Heidelberg (2003)

[29] Omran, M., Salman, A., Engelbrecht, A.: Dynamic clustering using particle swarm optimization with application in image segmentation. Pattern Analysis and Applications, 332–344 (2006)

[30] Mitchell, M.: An Introduction to Genetic Algorithms, ch. 1-6, pp. 1–203. MIT Press (1998)

[31] Arben, A., Alireza, L.: Using genetic algorithm for dynamic and multiple criteria web-site optimizations. European Journal of Operational Research, 1767–1777 (2007)

[32] Ahmadyfard, A., Modares, H.: Combining PSO and K-Means to Enhance Data Clustering. In: International Symposium on Telecommunications. Published by IEEE (2008)

[33] Krishna, K., Murty, M.N.: Genetic K-Means Algorithm. IEEE Transactions Published in: Systems, Man, and Cybernetics, Part B: Cybernetics 29(3) (1999)

[34] Dixit, V.S.: Refinement of Clusters Based on Dissimilarity Measures. International Journal of Multidisciplinary Research and Advances in Engineering (IJMRAE) 6(1) (January 2014) (accepted to be published)

# Inferring User Interface Patterns from Execution Traces of Web Applications

Miguel Nabuco[1], Ana C.R. Paiva[1,2], and João Pascoal Faria[1,2]

[1] Department of Informatics Engineering
[2] INESC-TEC
Faculty of Engineering of University of Porto
Porto, Portugal

**Abstract.** This paper presents a dynamic reverse engineering approach to extract User Interface (UI) Patterns from existent Web Applications. Firstly, information related to user interaction is saved, in particular: user actions and parameters; the HTML source pages; and the URLs. Secondly, the collected information is analysed in order to calculate several metrics (e.g., the differences between subsequent HTML pages). Thirdly, the existent UI Patterns are inferred from the overall information calculated based on a set of heuristic rules. The overall reverse engineering approach is evaluated with some experiments over several public Web Applications.

## 1 Introduction

Web Applications are no longer solely used to display information. They have been used more and more to handle tasks that before could only be performed by desktop applications [11]. Also, due to this era of globalization, there is a constant need for communication and interaction with different locations around the world, and Web Applications play a key role in this process.

However, opposite to desktop and mobile applications, Web Applications suffer from a lack of standards and conventions [6]. For example, distinct Web sites may use different tags for the same HTML elements, the same functionality may be implemented with dissimilar code, and the same task may be handled in diverse ways. This characteristic has a negative impact over Web testing because it inhibits the reuse of test code.

Despite this diversity in Web Applications, there are some elements that developers frequently use to provide the users a sense of comfort and easiness, such as User Interface (UI) Patterns [27]. UI Patterns are recurring solutions that solve common UI design problems. They provide an easy way to perform certain tasks, such as searching for information (Search pattern) or authenticating to access private data (Login pattern). When a user sees an instance of a UI Pattern, he can easily infer how the UI is supposed to be used.

Despite a common behavior, UI Patterns may have slightly different implementations along the Web. However, it is possible to define generic and reusable test strategies to test them after a configuration process to adapt the tests

B. Murgante et al. (Eds.): ICCSA 2014, Part V, LNCS 8583, pp. 311–326, 2014.

to those different possible applications [16]. That is the main idea behind the
Pattern-Based GUI Testing (PBGT) project in which context this research work
is developed. In the PBGT approach, the user builds a test model containing
instantiations of the aforementioned test strategies (UI Test Patterns) for testing
occurrences of UI Patterns on Web Applications.

The goal of the work described in this paper is to develop a reverse engineering
(RE) process to automatically identify the presence of UI Patterns on existent
Web Applications, and facilitate the building of test models in the context of
PBGT. Such test models need to be completed and configured manually in order
to generate test cases for testing the Web Application.

The rest of the paper is structured as follows. Section 2 presents an overview
of the PBGT approach, setting the context for this work. Section 3 describes the
UI patterns that our approach is able to detect. Section 4 describes the reverse
engineering process and heuristics. Section 5 demonstrates the effective use of
the approach in several existent Web Applications. Section 6 addresses related
work. Section 7 provides the conclusions, sums up the positive points and the
limitations of this approach, and points out the future work.

## 2    PBGT Overview

Pattern Based GUI Testing (PBGT) is a testing approach that aims to increase
the level of abstraction of test models and to promote reuse. PBGT [15] has the
following components (see also Figure 1):

- **PARADIGM** — A domain specific language (DSL) for building GUI test
  models based on UI patterns;
- **PARADIGM-ME** — A modeling and testing environment to support the
  building of test models;
- **PARADIGM-TG** — A test case generation tool that builds test cases
  from PARADIGM models;
- **PARADIGM-TE** — A test case execution tool to execute the tests, analyze the coverage and create reports;
- **PARADIGM-RE** — A dynamic reverse engineering approach and tool to
  extract UI Patterns from existent Web Applications without access to their
  source code. This tool can also generate test models to use in PARADIGM-
  ME.

The focus of this paper is on this extracting process (PARADIGM-RE). In
Figure 1, the activities (rounded corner rectangles) with the human figure mean
that they are not fully automatic requiring manual intervention. The activities
with the cog mean that part (or all) of that activity is automatic. The numbers
within the activities define their sequencing.

The definition of the PARADIGM DSL has started by analyzing a set of
existing UI Patterns [27]. This DSL contains a set of UI Test Patterns that
specify generic test strategies to test those UI Patterns along their different
implementations. Besides UI Test Patterns, the DSL has also a set of connectors
that allows defining the sequencing of test strategies execution.

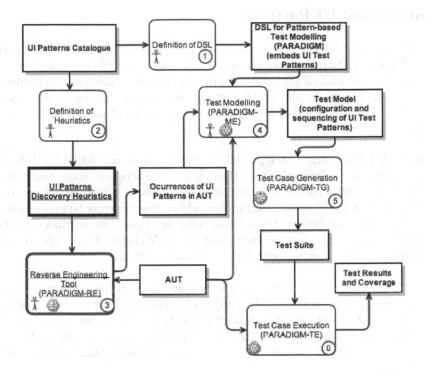

**Fig. 1.** PBGT overview

In order to diminish the modelling effort, the PARADIGM-RE tries to infer the existing UI Patterns within an Application Under Test (AUT) based on a set of heuristics. PARADIGM-RE will also automatically create a PARADIGM model, with the appropriate UI Test Patterns to test the identified UI Patterns. This model consists of a XML file that contains all the information about the UI Patterns found: their definition and configurations. The PARADIGM model generated by the RE tool does not contain the connectors between the UI Patterns, only the UI Patterns and their input data.

Afterwards, the extracted model may be completed manually in PARADIGM-ME. In particular, the connectors need to be added to define the sequence of execution. Afterwards, the model needs to be configured. During this configuration process, the tester provides preconditions (defining when the corresponding test strategy can be executed) [14] and checks to verify if a particular element is behaving as expected. Then, PARADIGM-TG can generate test cases based on the model, creating a test suite. PARADIGM-TE will then execute the test cases over the AUT to find errors and generate test coverage reports to assess the quality of the test suite.

# 3   Supported UI Patterns

The UI Patterns that the RE approach is able to detect are described in the sequel. These UI Patterns were chosen because they are the ones used by the modeling environment (PARADIGM-ME). To provide maximum compatibility between these two tools, the UI Pattern set chosen was the same. Also, most of the Web Applications can be almost entirely modeled using this set of UI Patterns.

- **Login UI Pattern.** The login UI pattern is commonly found in Web Applications, especially in the ones that restrict the access to some functionalities/data. Besides two input fields (usually password is a cyphered text box, meaning that the characters input are only visible as symbols) and a submit button, sometimes this pattern has a "remember me" button that saves the authentication data for the next visit to the Website. The authentication process has two possible outcomes: valid and invalid.

- **Search UI Pattern.** The search UI pattern consists of one or more input fields where the user inserts the subjects he wants to look for; and a submit button to start the search, sending the data to the server, retrieving the results and showing them to the user. When the search succeeds, it shows the results; when it fails, it may show a message explaining that no results were found.

- **Sort UI Pattern.** The sort UI pattern sorts a list of data, for example price, name and relevance, according to some defined criteria, e.g., ascending or descending. For example, in a Web store a user can sort a list of a specific type of products according to their price in order to identify the cheapest one.

- **Master Detail UI Pattern.**
  The master detail UI pattern is present in a Web page when selecting an element from a set results in filtering/updating another related set accordingly. As an example, consider a Web Application that displays the most sold audio CD in a certain week. There is an option, for example, a dropdown box that lets the user choose the week he wants to analyze (master). By changing the master selection, the list of the most sold CDs (detail) changes accordingly. Usually, besides the detail update, the majority of the Web page remains the same.

- **Menu UI Pattern.** The menu UI pattern is very common in Web pages. It defines a tree structure with several options, for instance, to access some options/functionalities of a Web site.

- **Input UI Pattern.** An input UI pattern is simply an input field (usually a text box) where a user can insert data.

## 4    Reverse Engineering Approach

The Web does not have specific design standards and the UI patterns can have slightly different implementations along the Web, so the definition of a global approach that can infer them is not a trivial activity. As depicted in Figure 2, the reverse engineering approach presented in this paper comprises the following components that are executed sequentially:

- **Extractor** — extracts the following information (using Selenium [24], a tool for automating interaction with Web Applications) from a user interaction with a Web Application:
  - the source HTML code of the Web pages visited;
  - the actions performed by the user (e.g., a click, a text input);
  - the URL of each page visited.
- **Analyser** — analyses the information gathered before and computes some metrics (e.g., differences between the source code of two subsequent Web pages);
- **Inferrer** — infers the UI Patterns based on heuristic rules defined over metrics and data calculated during the analysis phase (e.g., percentage of HTML changes between two subsequent visited pages).

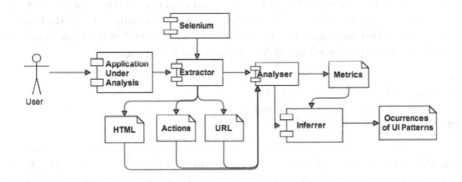

**Fig. 2.** Reverse Engineering Process

These components are described in more detail in the sequel.

### A - Extractor

During the extraction phase, the PARADIGM-RE tool gathers the information needed to infer the UI patterns. This section describes such information and how it is extracted.

The information extracted comprises the following:

- **Source code of HTML pages loaded** — the tool records the HTML source code of the Web pages loaded during user interaction; we denote by **HTML$_i$** the source code of the $i$-th page loaded, with $1 \leq i \leq N$, where $N$ is the total number of pages loaded;
- **URLs** — the tool records the URLs of the Web pages loaded; we denote by **URL$_i$** the URL of the $i$-th page loaded, with $1 \leq i \leq N$;
- **User actions (UA)** — the tool records in a HTML table, using Selenium [12], all the actions performed by the user during interaction; we denote by **UA$_k$** the $k$-*th* user action recorded, with $1 \leq k \leq M$, where M is the total number of user actions recorded.
  Each recorded user action **UA$_k$** has the following information:
  - **Command$_k$** — is the action that the user performed. Can be either"*type*", "*click*", "*clickAndWait*" (e.g., when the user clicks in a Web page element that loads a new page afterwards), "*typeAndWait*" (typing that loads different pages dinamically) and "*select*" (e.g., when the user clicks an option from a HTML dropdown list).
  - **TargetId$_k$** — is the id of the UI element object of the interaction.
  - **Value$_k$** — is related to "*type*" and "*typeAndWait*" actions and saves the string typed by the user in the text box. On "*select*", value is the option selected by the user.

We also denote by **Page$_k$** the index (between 1 and N) of the page where **UA$_j$** (between 1 and M) occurred. It is worth noting that the only actions that cause load of a new Web page are the "*clickAndWait*" and "*typeAndWait*" actions. An example of an action saved using Selenium is the following:

```
<tr>
<td>type</td>  <!- -Command- ->
<td>name=artist</td>  <!- -Target - ->
<td>u2</td>  <!- -Value - ->
</tr>
```

In this action, the user typed "u2" (the value) in the text box with the *id* "name=artist" (target). An example of the information saved during a user interaction with a Web Application can be seen in Table 1.

The RE tool is implemented in Java. The extraction process is performed according to the following workflow:

1. When the RE tool starts, it launches the Firefox Web browser and navigates to the specified URL of the AUT, given as a parameter.
2. The RE tool saves the initial HTML source page as $HTML_1.txt$.
3. The user starts Selenium IDE within Firefox.
4. The RE tool saves the user actions performed. After a user action of the kind "*clickAndWait*" or "*typeAndWait*", the RE tool saves the HTML of the new loaded page.
5. Step 4 is repeated until the user presses the "Escape" key on the keyboard, signaling the end of the extraction process, i.e., the end of an execution trace.

**Table 1.** Example Execution Trace

| UA index | Page index | Command | TargetId | Value |
|----------|------------|---------|----------|-------|
| - | - | open | /account/login.php | - |
| 1 | 1 | Type | name=formloginname | pbgt |
| 2 | 1 | Type | name=formpw | pbgtpass |
| 3 | 1 | Click | id=rememberusernamepwd | - |
| 4 | 1 | clickAndWait | name=login | - |
| 5 | 2 | clickAndWait | css=button[type="button"] | - |
| 6 | 3 | Click | link=Me | - |
| 7=M | 3=N | clickAndWait | link=Log Out | - |

## B - Analyser

In order to infer the UI patterns, the analyser calculates the following metrics and derived data, based on the information previously extracted:

- **HTMLDiff** — For every pair of consecutive HTML source files, $HTML_i$ and $HTML_{i+1}$, it calculates the difference between them, denoted $HTMLDiff_{i+1}$, containing the lines of HTML code updated, as determined by the *java-diff-utils* library [13].
- **RatioTotal** — For each $HTMLDiff_i$ file, it computes the ratio between its size and the average size over all the $HTMLDiff$ files, denoted and calculated as follows:

$$RatioTotal_i = \frac{size(HTMLDiff_i)}{\frac{\sum_{j=2}^{N} size(HTMLDiff_j)}{N-1}}, 2 \leq i \leq N \qquad (1)$$

- **RatioPrevious** — For each pair of consecutive HTML source files, it computes the difference between their sizes, calculated according to the formula:

$$RatioPrevious_{i+1} = S_{i+1} - S_i \qquad (2)$$

$$S_i = size(HTML_i) \qquad (3)$$

$$S_{i+1} = size(HTML_{i+1}) \qquad (4)$$

- **URL Keywords** — URLs can have keywords that may be useful to help inferring some UI patterns. We denote by **keywords(URL$_i$)** the set of keywords contained in $URL_i$.
- **Textboxes ids and types** — Some textboxes may have valuable information in their names or ids (found in the HTML page source), which may lead to the detection of a UI Pattern. We denote by **TargetId$_k$** and **Type$_k$** the *id* and *type* of user action $UA_k$.

- **Number of occurrences of certain strings in HTML pages** — The number of times certain strings appear in a HTML page can help in pattern inference. We denote by **count(s, HTML$_i$)** the number of occurrences of string $s$ in $HTML_i$.

After all the calculations, the UI inferring process starts.

## C - Inferrer

Occurrences of UI patterns in the pages visited are inferred from the information gathered in the previous phases, based on a set of heuristics defined specifically for each UI pattern. The heuristics defined for a UI pattern P are combined with different weights to produce a final score $S(P)$ that indicates the confidence of the presence of that UI pattern.

The heuristics weights (in Tables 2, 3, 4, 5) were defined based on a training set of Web Applications. For each application inide that set, we run the RE-Tool and tune the value of the weights for the existent UI patterns. The final values are the ones that inferr more patterns within those Web Applications. This training process is not completely automatic because the user needs to provide information about when a specific UI pattern is present.

The weights are assigned so that a score $S(P) \geq 1$ indicates a sufficiently high confidence to assume that the UI pattern occurs. So the score is calculated using the formula:

$$S(P) = \sum_{i}^{N} W(i)V(i),$$ (5)

In this formula, $i$ is the heuristic being evaluated, $N$ is the total number of heuristics for that specific UI pattern, $W(i)$ is the weight of heuristic $i$ and $V(i)$ indicates if the heuristic $i$ holds (value 1) or not (value 0).

There are also pre-conditions associated with some UI Patterns that must hold in order to procede with the score calculation.

## D - UI Patterns Heuristics

In the sequel we present the heuristics and corresponding weights defined for each UI Pattern.

### Login

The score calculation process starts just when there are two "*type*" commands with different ids in the same page. This is the pre-condition. After that, the RE tool looks for common keywords, such as "login" (heuristic **A**), "user" and "pass" (heuristic **B**), in their *ids*. Most often, the input fields (usually textboxes) to receive passwords transform the characters entered in cyphered tokens, which is coded in HTML assigning "password" to the "*type*" of the input field (heuristic

**C**) as shown next.

*<input id="appassword" name="password" type="password">*

The URL is also parsed to find "login" (or "logon") keywords that, although not found frequently in the URL, increase the degree of confidence in the presence of a login UI pattern (heuristic **D**).

Table 2 summarizes the heuristics defined for finding occurrences of the Login UI Pattern, including the mandatory pre-condition. Some of the conditions presented below are derived from the UAs.

**Pre-condition:** Two "type" commands $UA_j$ and $UA_k$ with $TargetId_i \neq TargetId_k$ occurring in page $i$. (Without loss of generality, we assume that $UA_j$ refers to the username and $UA_k$ refers to the password.)

**Table 2.** Login pattern heuristics

| ID | Heuristic | Weight |
|----|-----------|--------|
| A | "login" is-substring-of $TargetId_j$ | 0.8 |
| B | "user" is-substring-of $TargetId_j \wedge$ "pass" is-substring-of $TargetId_k$ | 0.8 |
| C | "password" is a substring of $Type_k$ | 0.6 |
| D | {"login","logon"} $\cap$ keywords$(URL_i) \neq \emptyset$ | 0.8 |

## Search

The process to look for Search UI Pattern starts only if the execution trace (e.g., Table 1) has one "*type*" command preceding a "*clickAndWait*" command. So, the inferring process applied to identify a Search UI pattern starts by analyzing the input field where the user inserted the content that he wants to search for. In several Websites, such field is a text box named '*search*", or contains the word '*search*" in its *id* or *title* (heuristic **E**). For instance, in the Amazon Website, the search box id contains the keyword search, as it can be seen in the example below with the information retrieved by Selenium from the user interaction.

*<input type='text' id='twotabsearchtextbox' title='Search For' value=" name='field-keywords' autocomplete='off'>*

When performing a Search operation, most of the times the URL may have additional interesting information, such as, the word "search" (heuristic **F**) and the content to search for (heuristic **G**). For example, in the Australian Charts Website, if a search for "daft punk" is performed, the URL will be:

`www.australian-charts.com/search.asp?search=daft+punk&cat=s`

To complete the Search pattern inference, the RE tool counts the number of times that the content to look for appears in the resulting Web page. If the total is higher than a specific number, there is more confidence that a Search UI pattern was found (heuristic **H**). The overall heuristics used to find occurrences of the Search UI pattern are summarized in Table 3.

**Pre-condition**: A *"type"* command $UA_k$ precedes a *'clickAndWait'* command in page $i$.

**Table 3.** Search pattern heuristics

| ID | Heuristic | Weight |
|----|-----------|--------|
| E | "search" is-substring-of $TargetId_k$ | 0.8 |
| F | "search" $\in$ keywords($URL_{i+1}$) | 0.7 |
| G | $Value_k \in$ keywords($URL_{i+1}$) | 0.6 |
| H | count($Value_k, HTML_{i+1}$) > 10 | 0.8 |

### Sort

To infer the presence of a Sort UI pattern, some heuristics were defined based on the *RatioPrevious*, *RatioTotal* and *URL keywords* calculated before. Considering that most of the times, performing a sort operation, does not change much the HTML source code of the current page, it is a good indicator of the presence of a Sort UI pattern when:

- *RatioPrevious* is low ($< 5$ KB) meaning that there are few differences between the HTML source code of the current page and the previous one (heuristic **K**);
- *RatioTotal* is low ($< 0.15$) indicating that differences between the HTML source code of current page and the previous one is considerably less that the average of differences between all pairs of consecutive pages of the execution trace (heuristic **L**).

In addition, it is very common to find some keywords within the corresponding URL when a Web page has a Sort UI pattern, such as *asc* (ascending), *desc* (descending), *order* (heuristic **J**), *sort* and *sortType* (heuristic **I**). The heuristics defined to find occurrences of the sort UI pattern are summarized in Table 4.

### Master Detail

As it is explained in the previous section, the two HTML captures (before – $HTML_i$ – and after – $HTML_{i+1}$ – changing a master value) are very similar, containing very few changes. There are two simple metrics defined on top of this difference. The RatioPrevious value (heuristic **M**), along with the RatioTotal value (heuristic **N**), will allow the algorithm to know the amount of information changed between the two pages and its correlation with the rest of the pages in that Web Application. The other metric is the number of lines in the HTMLDiff

**Table 4.** Sort pattern heuristics

| ID | Heuristic | Weight |
|----|-----------|--------|
| I  | {"sort","sortType"} $\cap$ keywords($URL_{i+1}$) $\neq \emptyset$ | 0.8 |
| J  | {"asc","desc", "order"} $\cap$ keywords($URL_{i+1}$) $\neq \emptyset$ | 0.6 |
| K  | $RatioPrevious_{i+1} < 5$ KB | 0.4 |
| L  | $RatioTotal_{i+1} < 0.15$ | 0.4 |

**Table 5.** Master Detail pattern heuristics

| ID | Heuristic | Weight |
|----|-----------|--------|
| M  | $RatioPrevious_{i+1} < 5$ KB | 0.5 |
| N  | $RatioTotal_{i+1} < 0.1$ | 0.5 |
| O  | number of lines($HTMLDiff_{i+1}$) $< 6$ | 0.5 |

file (heuristic **O**). The heuristics used for finding occurrences of this pattern are summarized in Table 5.

We tested these heuristics on the Web Applications training set and tuned the values that would determine or not the presence of a Master Detail pattern. For the case of RatioPrevious we set the value to 5 KB meaning that when the difference between two consecutive pages is less than 5 KB value there is more probability of having found a Master Detail. For the case of RatioTotal, we assumed value 0.1. For the case of the number of lines, when this value is lower than 6 it means that there is more probablity of inferring a Master Detail.

## Input

In each page, an occurrence of the Input pattern is inferred for each object ID that is the target of at least one "type" command in that page, and is not involved in a Search or Login pattern in the same page (as search string, username or password).

## Menu

The ambiguity of the definition of menu in a Web Application makes its recognition very difficult. A menu can be anywhere on the page, with or without subsections, and be spread horizontally or vertically. It is usually on the top, bottom, left or right part of the page. There are no specific HTML elements to construct a menu, and so the Website developers are free to implement a menu in the way they desire. However, a good number of them use a specific menu type, called "Site navigation" [28]. Its structure is similar to the following example:

```
<div><ul>
<li><strong>Home</strong></li>
<li><a href="about.html">About Us</a></li>
<li><a href="clients.html">Our Clients</a></li>
```

*<li><a href="products.html">Our Products</a></li>*
*<li><a href="contact.html">Contact Us</a></li>*
*</ul></div>*

Usually the *href* attributes in a menu are set to local pages. From all the HTML page sources, the algorithm tries to find common elements among 80% of them (since there can be pages in a Web Application that contain no Menu pattern), and sees if, from the common elements, there is a general structure like the one described above. There can be either zero or one instance of a Menu UI pattern per execution trace.

## 5   Evaluation

The RE tool was initially experimented iteratively over a number of Web Applications, a training set, with the goal of refining and fine-tuning the heuristics and weights used to maximize the capacity to find UI patterns.

If during the RE process the tool detects the same UI Pattern instance several times, the tool is capable of ignoring it or add more configurations to the one already detected.

After the training phase, the RE tool was used to detect UI Patterns in several public known and widely used Web Applications, in an evaluation set. The Web Applications chosen for this evaluation set were different than the ones used to refine the heuristics in the training set, to prevent biased results.

This time, the purpose was to evaluate the RE tool, i.e., determine which UI patterns' occurrences the tool was able to detect in each application execution trace (ET) and compare them to the patterns that really exist in such trace. The results are presented in tables that show the number of instances of each UI pattern that exist in the ET, the ones that the tool correctly found and the ones that the tool mistakenly found (false positives).

Five applications were chosen from the top 15 most popular Websites [9]: Amazon, YouTube, Facebook, Wikipedia and Yahoo. These Web Applications were chosen because they contain instances of the UI patterns the approach proposes to detect. Also, they are well designed and contain very few known bugs.

The numerical results were combined and can be found in Table 6.

**Table 6.** Evaluation set results

| Pattern | Present in ET | Correctly found | False positives |
|---|---|---|---|
| Login | 9 | 6 | 0 |
| Search | 24 | 21 | 0 |
| Sort | 11 | 4 | 0 |
| Input | 28 | 26 | 3 |
| Menu | 3 | 1 | 0 |
| Master Detail | 24 | 9 | 9 |
| **Total** | **99** | **67(67.7%)** | **12** |

Overall, the success rate of the tool was quite satisfactory. The tool found 67 of 99 pattern occurrences (67.7%).

Despite some miscounts, it identified most of the patterns. The Menu and the Master Details patterns were the most problematic. This is due to the ambiguity of the Master Detail definition, meaning that a Master Detail can be presented in many ways, making it difficult to specify precise heuristics. The Search and Login patterns were almost always found correctly due to their heuristics being more strict and specific. The three Input false positives were in fact text boxes belonging to the three Search patterns that were not found.

# 6   Related Work

Reverse engineering is "*the process of analysing the subject system to identify the system components and interrelationships and to create representations of the system in another form or at a higher level of abstraction*" [5]. In reverse engineering, the subject system is not changed. There are different methods of applying reverse engineering to a system: the dynamic method, in which the data are retrieved from the system at run time without access to the source code [20], the static method, which obtains the data from the system source code [30], and the hybrid method, which combines the two previous methods.

There are several ways to obtain information from application execution traces [25] [2]. ReGUI [17] is a tool that reduces the effort of obtaining models of the structure and behaviour of Graphical User Interfaces (GUI) [18]. Another approach for behaviour model extraction combined static and dynamic information and was used in model checking [8]. EvoTrace [10] uses execution traces to trace the evolution of a software system. Amalfitano [1] presented a technique for testing Rich Internet Applications (RIAs) that generates test cases from the applications execution traces.

Several tools have been created to obtain information from Web Applications. Re-Web [23] obtains dynamically information from Web server logs that helps to find structural and navigational problems in Web Applications. WARE [7] is a static analyzer that creates UML diagrams from the Web Application source code, Crawljax [24] is a tool that obtains graphical sitemaps, by automatically crawling through a Web Application and Selenium is an open-source capture-replay tool that saves the users interaction in HTML files.

Concerning UI Patterns, there are different sources that enumerate the most popular/frequently used ones [21] [22] [26] [29]. Although certain pattern layouts may be different from different sources or implementations, they all present a common behaviour and functionality. As an example, the Master Detail pattern may be presented graphically in several ways but its functionality stays the same: selecting a master option will filter and show different results for the detail area.

In the area of inferring patterns from Web Applications, it was discovered that, despite the fact that there are approaches to extract information from the Web, besides our previous work applying Inductive Logic Programming (ILP) to execution traces to infer UI Patterns [19] [16], none of the approaches found deal with UI pattern detection. They mostly deal with other sets of patterns, more related to the area of Web mining. The purpose of these patterns is to find relations between different data or to find the same data in different formats. Brin [3] presents an approach to extract relations and patterns for the same data spread through many different formats. Chang [4] proposes a similar method to discover patterns, by extracting structured data from semi-structured Web documents.

## 7    Conclusions and Future work

This paper proposed a dynamic reverse engineering method to identify recurrent behavior present in Web Applications, by extracting information from an execution trace and afterwards inferring the existing UI Patterns and their configurations from that information. The result is then exported into a PARADIGM model to be completed in PARADIGM-ME. Then, the model can be used to generate test cases that are performed on the Web Application.

The evaluation of the overall approach was conducted in several worldwide used Web Applications. The result was quite satisfactory, as the RE tool found most of the occurrences of UI patterns present in each application as well as their exact location (in which page they were found).

Despite the satisfactory results obtained, the work still needs some improvement. The features planned for future versions of the RE tool include the support for a larger set of UI patterns and the definition of more precise heuristics based on other web elements (such as manipulating the HTML DOM tree).

**Acknowledgments.** This work is financed by the ERDF — European Regional Development Fund through the COMPETE Programme (operational programme for competitiveness) and by National Funds through the FCT — Fundação para a Ciência e a Tecnologia (Portuguese Foundation for Science and Technology) within project FCOMP-01-0124-FEDER-020554.

## References

1. Amalfitano, D.: Rich Internet Application Testing Using Execution Trace Data. In: Third International Conference on Software Testing, Verification, and Validation Workshops, pp. 274–283 (2010)
2. Andjelkovic, I., Artho, C.: Trace Server: A tool for storing, querying and analyzing execution traces. In: JPF Workshop (2011)
3. Brin, S.: Extracting patterns and relations from the world wide web. In: Atzeni, P., Mendelzon, A.O., Mecca, G. (eds.) WebDB 1998. LNCS, vol. 1590, pp. 172–183. Springer, Heidelberg (1999)
4. Chang, C.H., Hsu, C.N., Lui, S.C.: Automatic information extraction from semi-structured Web-pages by pattern discovery. Decision Support Systems Journal 35, 129–147 (2003)
5. Chikofsky, E.J., Cross, J.H.: Reverse engineering and design recovery: a taxonomy. IEEE Software Journal 7(1), 13–17 (1990)

6. Constantine, L.L., Lockwood, L.A.D.: Usage-centered engineering for Web applications. IEEE Software Journal 19(2), 42–50 (2002)
7. Di Penta, M.: Integrating static and dynamic analysis to improve the comprehension of existing Web applications. In: Proceedings of 7th IEEE International Symposium on Web Site Evolution, pp. 87–94 (2005)
8. Duarte, L.M., Kramer, J., Uchitel, S.: Model extraction using context information. In: Wang, J., Whittle, J., Harel, D., Reggio, G. (eds.) MoDELS 2006. LNCS, vol. 4199, pp. 380–394. Springer, Heidelberg (2006)
9. eBizMba. Top 15 most popular Websites,
   http://www.ebizmba.com/articles/most-popular-Websites
   (accessed June 2013)
10. Fischer, M., Oberleitner, J., Gall, H., Gschwind, T.: System evolution tracking through execution trace analysis. In: CSMR 2005, pp. 112–121 (2005)
11. Garrett, J.J.: Ajax: a new approach to Web applications (2006),
    http://www.adaptivepath.com/publications/essays/archives/000385.php
12. Gheorghiu, G.: A look at Selenium. Better Software (October 2005)
13. Java-diff-utils. The DiffUtils library for computing diffs, applying patches, generating side-by-side view in Java, http://code.google.com/p/java-diff-utils/ (accessed June 2013)
14. Monteiro, T., Paiva, A.: Pattern based gui testing modeling environment. In: Fourth International Workshop on TESTing Techniques & Experimentation Benchmarks for Event-Driven Software, TESTBEDS (2013)
15. Moreira, R., Paiva, A., Memon, A.: A Pattern-Based Approach for GUI Modeling and Testing. In: Proceedings of the 24th Annual International Symposium on Software Reliability Engineering (ISSRE 2013) (2013)
16. Morgado, I.C., Paiva, A.C.R., Faria, J.P., Camacho, R.: GUI reverse engineering with machine learning. In: 2012 First International Workshop on Realizing Artificial Intelligence Synergies in Software Engineering (RAISE), pp. 27–31 (June 2012)
17. Morgado, I.C., Paiva, A., Faria, J.P.: Dynamic Reverse Engineering of Graphical User Interfaces. International Journal on Advances in Software 5, 223–235 (2012)
18. Morgado, I.C., Paiva, A., Faria, J.P.: Reverse Engineering of Graphical User Interfaces. In: The Sixth International Conference on Software Engineering Advances, Barcelona, pp. 293–298 (2011)
19. Nabuco, M., Paiva, A., Camacho, R., Faria, J.: Inferring UI patterns with inductive logic programming. In: 8th Iberian Conference on Informations Systems and Technologies (CISTI) (2013)
20. Grilo, A.M.P., Paiva, A.C.R., Faria, J.P.: Reverse Engineering of GUI Models for Testing. In: 5th Iberian Conference on Information Systems and Technologies (CISTI), Santiago de Compostela, Spain (2010)
21. Neil, T.: 12 standard screen patterns,
    http://designingWebinterfaces.com/
    designing-Web-interfaces-12-screen-patterns (accessed June 2013)
22. Pattenry. UI design patterns and library builder, http://paternry.com (accessed June 2013)
23. Ricca, F., Tonella, P.: Understanding and restructuring Web sites with ReWeb. IEEE Multimedia 8, 40–51 (2001)
24. Roest, D.: Automated Regression Testing of Ajax Web Applications. Faculty EEMCS, Delft University of Technology Msc thesis (2010)

25. Steven, J., Ch, P., Fleck, B., Podgurski, A.: jRapture: A capture/replay tool for observation-based testing. In: Proceedings of the International Symposium on Software Testing and Analysis, pp. 158–167. ACM Press (2000)
26. Toxboe, A.: UI patterns - user interface design pattern library, http://ui-patterns.com (accessed June 2013)
27. Welie, M., Gerrit, C., Eliens, A.: Patterns as tools for user interface design. In: Workshop on Tools for Working With Guidelines, Biarritz, France (2000)
28. W3C Wiki. Creating multiple pages with navigation menus, http://www.w3.org/wiki/Creating_multiple_pages_with_navigation_menus (accessed June 2013)
29. van Welie, M.: Interaction design pattern library (2013), http://www.welie.com/patterns/ (accessed June 2013)
30. Silva, J.C., Silva, C., Gonçalo, R.D., Saraiva, J., Campos, J.C.: The GUISurfer tool: towards a language independent approach to reverse engineering GUI code. In: Proceedings of the 2nd ACM SIGCHI Symposium on Engineering Interactive Computing Systems (2010)

# Interactive Test Case Design Algorithm

Fedor Strok[1,2]

[1] Yandex, 16 Leo Tolstoy St. Moscow, Russia
[2] School of Applied Mathematics and Information Science
National Research University Higher School of Economics
Bol. Trekhsvyatitelskii 3, Moscow, Russia
fdrstrok@yandex-team.ru,fdr.strok@gmail.com

**Abstract.** Pairwise testing has become a popular approach for test case generation. In contrast to manual enumeration, test case generation is fully automated. However, how test scenarios were obtained is hidden from the user. Attribute exploration is a technique that unites best practices, providing semi-supervised procedure to explore test-cases.

**Keywords:** formal concept analysis, pairwise testing, attribute exploration.

## 1 Introduction

One of the biggest challenges in software testing is choosing the proper strategy to cover all possible cases. One of the most widely used practices is to define main parameters that influence the output and check their possible value combinations. The naive way is just to check for all possible combinations of such parameters, and it immediately leads to exponential complexity. Even in the case of 6 boolean parameters quality assurance expert is going to check 64 combinations.

As experience shows, tools and methodologies are very useful for real-world test tasks [4,7,8]. While it is well-covered for the case of white-box testing, there is lack of tools for black-box testing. Since it is crucial to consider test time execution and support, it is import to reduce the number of tests preserving the coverage of the system under testing [10].

First alternative is a manual test generation. A software test engineer considers all possible cases, 'written down' in some order, which seems logical to him or her. The main risk is to lose some information behind the cases. Actually, it could be rather exhausting to cover all possibilities for a large number of parameters.

A popular alternative way is pairwise testing [2], or its generalization, n-wise testing. We define parameters and domains, and pass them like a model into a black-box algorithm [1], that gives us a set of test-cases, which satisfies special condition (for each pair of input parameters all possible discrete combinations of those parameters are tested). In the general case it can produce different cases in different runs, while it can be fixed by passing a random seed to it. The main advantage of the approach is its insensitivity to the number of parameters. However, it could be rather computationally-hard task.

B. Murgante et al. (Eds.): ICCSA 2014, Part V, LNCS 8583, pp. 327–336, 2014.

Usually, there are dependencies between input parameters. The natural form to express such dependencies in mathematical terms is implication - a statement in the form: 'if ..., then ...'. The 'if'-part is called premise, and the 'then' is called conclusion. Consideration of parameter's interdependence could decrease the complexity of result cases in terms of their quantity, by excluding some of possible combinations.

Our approach is focused on implications. We use the algorithms of Formal Concept Analysis to provide software engineers with a tool, that helps to explore the domain in a semi-automatic way. It guarantees sound and complete description if the expert gives valid answers to the system.

The rest of the paper is organized as follows: in Section 2 we introduce basic notions of Formal Concept Analysis. In Section 3 we focus on the procedure of attribute exploration. Section 4 provides example of attribute exploration in the field of positive integers. We make conclusions in Section 5.

## 2    Formal Concept Analysis

Formal Concept Analysis [6] is a technique allowing to derive a formal ontology from a collection of objects and attributes. It was introduced in 1984 by Rudolf Wille. It relies on lattice and order theories [3]. Numerous applications are found in the field of machine learning, data mining, text mining and biology.

**Definition 1.** *A formal context $K$ is a triple $K := (G, M, I)$, where $G$ denotes a set of objects, $M$ is a set of attributes, and $I \subseteq G \times M$ is a binary relation between $G$ and $M$.*

It can be interpreted in the following way: for objects in $G$ there exists a description in terms of attributes in $M$, and relation $I$ reflects that an object has an attribute: $(g, m) \in I \iff$ object $g$ possesses $m$.

An example of a formal context is provided below:

Objects:

**1** – equilateral triangle
**2** – right triange,
**3** – rectangle,
**4** – square,

Attributes:

**a** – 3 vertices,
**b** – 4 vertices,
**c** – has a right angle,
**d** – all sides are equal

For a given context 2 mappings are considered:

$\varphi: 2^G \to 2^M$ $\varphi(A) \overset{\text{def}}{=} \{m \in M \mid gIm \text{ for all } g \in A\}$.

$\psi: 2^M \to 2^G$ $\psi(B) \overset{\text{def}}{=} \{g \in G \mid gIm \text{ for all } m \in B\}$.

For all $A_1, A_2 \subseteq G$, $B_1, B_2 \subseteq M$

1. $A_1 \subseteq A_2 \Rightarrow \varphi(A_2) \subseteq \varphi(A_1)$
2. $B_1 \subseteq B_2 \Rightarrow \psi(B_2) \subseteq \psi(B_1)$
3. $A_1 \subseteq \psi\varphi(A_1)$ $B_1 \subseteq \varphi\psi(B_1)$

**Definition 2.** *Mappings $\varphi$ and $\psi$, satisfying properties 1-3 above, define a Galois connection between $(2^G, \subseteq)$ and $(2^M, \subseteq)$, which means:*
$\varphi(A) \subseteq B \Leftrightarrow \psi(B) \subseteq A$

According to tradition, notation $(\cdot)'$ is used instead of $\varphi$ and $\psi$. $(\cdot)''$ stands both for $\varphi \circ \psi$ and $\psi \circ \varphi$ (depending on the argument). For arbitrary $A \subseteq G$, $B \subseteq M$

$$A' \overset{\text{def}}{=} \{m \in M \mid gIm \text{ for all } g \in A\}, \quad B' \overset{\text{def}}{=} \{g \in G \mid gIm \text{ for all } m \in B\}.$$

**Definition 3.** (Formal) concept *is a pair* $(A, B)$:
$A \subseteq G$, $B \subseteq M$, $A' = B$, $B' = A$.

In the example with geometric figures a pair $(\{3, 4\}, \{b, c\})$ is a formal concept. For a formal context $(G, M, I)$, $A, A_1, A_2 \subseteq G$ - set of objects, $B \subseteq M$ - set of attributes, the following statements hold for operation $(\cdot)'$:

1. $A_1 \subseteq A_2 \Rightarrow A_2' \subseteq A_1'$
2. $A_1 \subseteq A_2 \Rightarrow A_1'' \subseteq A_2''$
3. $A \subseteq A''$
4. $A''' = A'$ and $A'''' = A''$
5. $(A_1 \cup A_2)' = A_1' \cap A_2'$
6. $A \subseteq B' \Leftrightarrow B \subseteq A' \Leftrightarrow A \times B \subseteq I$

**Definition 4.** *Closure operator on set $G$ is a mapping $\gamma: \mathcal{P}(G) \to \mathcal{P}(G)$, which maps every $X \subseteq G$ to closure $\gamma X \subseteq G$, under the following conditions:*

1. $\gamma\gamma X = \gamma X$ *(idempotence)*
2. $X \subseteq \gamma X$ *(extensivity)*
3. $X \subseteq Y \Rightarrow \gamma X \subseteq \gamma Y$ *(monotonicity)*

**Definition 5.** Implication $A \to B$, where $A, B \subseteq M$, takes place if $A' \subseteq B'$, in other words if each object having $A$ also has all attributes from $B$.

Implications comply with Armstrong axioms:

$$\frac{}{X \to X} \tag{1}$$

$$\frac{X \to Y}{X \cup Z \to Y} \tag{2}$$

$$\frac{X \to Y, Y \cup Z \to W}{X \cup Z \to W} \tag{3}$$

# 3    Attribute Exploration

Attribute exploration is a well known algorithm within Formal Concept Analysis [5]. It is applicable to spheres of knowledge of arbitrary type [9,11]. The main idea is to explore the object domain in a semi-automatic manner. It means that an expert is required, but his duty is to answer specific questions about possible dependencies in the area. Questions are provided in the form of implications, asking whether they are true or false. If the answer is true, the implication is added to the base of knowledge. In the case of answer false, the expert is asked to provide a counterexample violating proposed dependency.

In other words, exploration algorithm wants to explore all possible combinations of a given attribute set. It is typical that objects in this field of knowledge are too difficult to enumerate them. So the algorithm starts with a set of examples. Then it computes canonical base of implications for the provided formal context. Then a domain expert is asked if the computed implications are valid in general. If it is true, then existing context represents all possible combinations in the domain. Otherwise, there exists a counterexample in the domain, which should be added to the context, and then canonical base should be calculated again.

The general strategy is quite intuitive: we start exploring the domain with some knowledge of typical examples and dependencies. To extend knowledge database we either add a rule, or provide another example that violates the currently studied dependency. The main advantage of this approach is that it is done algorithmically.

---

**Algorithm 1.** NEXT CLOSURE($A$, $M$, $\mathcal{L}$)

---

**Input:** Closure operator $X \mapsto \mathcal{L}(X)$ on attribute set $M$ and subset $A \subseteq M$.
**Output:** lectically next closed itemset $A$.
    **for all** $m \in M$ in reverse order **do**
        **if** $m \in A$ **then**
            $A := A \setminus \{m\}$
        **else**
            $B := \mathcal{L}(A \cup \{m\})$
            **if** $B \setminus A$ does not contain elements $< m$ **then**
                **return** $B$
    **return** $\perp$

---

# 4    Practical Examples

## 4.1    Numbers - Attribute Exploration

Let us consider the domain of natural numbers [12]. As a set of possible attributes we can consider the following: even (2*n), odd (2*n+1), divisible_by_three (3*n), prime (has no positive divisors other than 1 and itself), factorial (is a factorial

---

**Algorithm 2.** ATTRIBUTE EXPLORATION

---

**Input:** A subcontext $(E, M, J = I \cap E \times M)$ of $(G, M, I)$, possibly empty.
**Input:** Interactive: confirm that $A = B''$ in a formal context $(G, M, I)$, $M$ finite, or
  give an object showing that $A \neq B''$.
**Output:** The canonical base $\mathcal{L}$ of $(G, M, I)$ and a possibly enlarged subcontext
  $(E, M, J = I \cap E \times M)$ with the same canonical base.
$\mathcal{L} := \emptyset$
$A := \emptyset$
**while** $A \neq M$ **do**
    **while** $A \neq A^{JJ}$ **do**
        **if** $A^{JJ} = A^{II}$ **then**
            $\mathcal{L} := \mathcal{L} \cup \{A \to A^{JJ}\}$
            exit while
        **else**
            extend $E$ by some object $g \in A^I \setminus A^{JJI}$
    $A := NextClosure(A, M, \mathcal{L})$
**return** $\mathcal{L}, (E, M, J)$

---

of a positive number). We can use the implementation from [12]. We start from an empty set of objects. The canonical base for such context is $\emptyset \to M$. So we get a question:
  $=> even, factorial, divided\_by\_three, odd, prime$
Is the following implication valid?

Obviously, not all numbers obtain all attributes. At least we can consider number 2, which is $even, factorial, prime$. We add 2 to our context and base is recalculated.

| G \ M | even | factorial | divided_by_three | odd | prime |
|---|---|---|---|---|---|
| 2 | × | × | | | × |

  $=> even, factorial, prime$
Is the following implication valid?
Now we can think of number 5, which is $prime, odd$

| G \ M | even | factorial | divided_by_three | odd | prime |
|---|---|---|---|---|---|
| 2 | × | × | | | × |
| 5 | | | | × | × |

$=> prime$
Is the following implication valid?
Now we are about to either say that all numbers are prime, or provide a non-prime number, e.g. 6

| G \ M | even | factorial | divided_by_three | odd | prime |
|---|---|---|---|---|---|
| 2 | × | × | | | × |
| 5 | | | | × | × |
| 6 | × | × | × | | |

*factorial* => *even*

Is the following implication valid?

Now we have both 2 and 6, which are simultaneously even and factorial. There is a counterexample, we should find a number, which is factorial, but not even, which is 1.

| G \ M | even | factorial | divided_by_three | odd | prime |
|-------|------|-----------|------------------|-----|-------|
| 2 | × | × | | | × |
| 5 | | | | × | × |
| 6 | × | × | × | | |
| 1 | | × | | × | × |

*odd* => *prime*

Is the following implication valid?

That does not hold for number 9.

| G \ M | even | factorial | divided_by_three | odd | prime |
|-------|------|-----------|------------------|-----|-------|
| 2 | × | × | | | × |
| 5 | | | | × | × |
| 6 | × | × | × | | |
| 1 | | × | | × | × |
| 9 | | | × | × | |

*factorial, odd* => *prime*

Is the following implication valid?

We have the only number which is factorial and odd - 1, and it is prime. *factorial, divided_by_three* => *even*

Is the following implication valid?

Now we have to remember what implication is. The only case when implication does not hold is when premise is true, and conclusion is false. The least factorial which is divided_by_three is 6, which is already even. *prime, divided_by_three* => *even, factorial, odd*

Is the following implication valid?

We have number 3, which is just odd.

| G \ M | even | factorial | divided_by_three | odd | prime |
|-------|------|-----------|------------------|-----|-------|
| 2 | × | × | | | × |
| 5 | | | | × | × |
| 6 | × | × | × | | |
| 1 | | × | | × | × |
| 9 | | | × | × | |
| 3 | | | × | × | × |

*prime, divided_by_three* => *odd*

Is the following implication valid?

The only prime, which is divided_by_three is three itself, so it is true. *even* => *factorial*

Is the following implication valid?

Not all even numbers are factorials, e.g. 8.

| G \ M | even | factorial | divided_by_three | odd | prime |
|---|---|---|---|---|---|
| 2 | × | × | | | × |
| 5 | | | | × | × |
| 6 | × | × | × | | |
| 1 | | × | | × | × |
| 9 | | | × | × | |
| 3 | | | × | × | × |
| 8 | × | | | | |

$even, odd => factorial, prime, divided\_by\_three$
Is the following implication valid?
We do not have numbers which are both even and odd. $even, divided\_by\_three =>$ $factorial$
Is the following implication valid?
We have number 12, which is even and divided_by_three, but it is not a factorial.

| G \ M | even | factorial | divided_by_three | odd | prime |
|---|---|---|---|---|---|
| 2 | × | × | | | × |
| 5 | | | | × | × |
| 6 | × | × | × | | |
| 1 | | × | | × | × |
| 9 | | | × | × | |
| 3 | | | × | × | × |
| 8 | × | | | | |
| 12 | × | | × | | |

$even, prime => factorial$
Is the following implication valid?
The only even prime number is 2. And the exploration process is over.
  The final context:

| G \ M | even | factorial | divided_by_three | odd | prime |
|---|---|---|---|---|---|
| 1 | | × | | × | × |
| 2 | × | × | | | × |
| 3 | | | × | × | × |
| 5 | | | | × | × |
| 6 | × | × | × | | |
| 8 | × | | | | |
| 9 | | | × | × | |
| 12 | × | | × | | |

The set of implications:

- $factorial, odd \rightarrow prime$
- $factorial, divided\_by\_three \rightarrow even$
- $prime, divided\_by\_three \nrightarrow odd$
- $even, odd \rightarrow factorial, prime, divided\_by\_three$
- $even, prime \rightarrow factorial$

## 4.2   Numbers - Model-Based

Let us consider the same problem in terms of pairwise testing. We formulate models in terms of PICT, a pairwise testing tool by Microsoft [2]. The initial model looks in a very simple way - we just have to determine the factors:

- Even: 1, 0
- Factorial: 1, 0
- Divs3: 1, 0
- Odd: 1, 0
- Prime: 1, 0

The results of pairwise generation are provided below:

| Even | Factorial | Divs3 | Odd | Prime |
|------|-----------|-------|-----|-------|
| 1 | 1 | 0 | 0 | 1 |
| 0 | 0 | 1 | 1 | 0 |
| 1 | 0 | 1 | 1 | 1 |
| 0 | 1 | 0 | 1 | 0 |
| 1 | 0 | 1 | 0 | 0 |
| 0 | 1 | 1 | 0 | 1 |
| 0 | 0 | 0 | 0 | 0 |

To get results as in the previous section, we have to modify the input model in the following way:

1. #Parameters:
   - Even: 1, 0
   - Factorial: 1, 0
   - Divs3: 1, 0
   - Odd: 1, 0
   - Prime: 1, 0
   - $Result: 12, 9, 6, 3, 2, 1
2. #Implications:
   - IF [Even] = 1 THEN [Odd] = 0 ELSE [Odd] = 1;
   - IF [Odd] = 1 AND [Factorial] = 1 THEN [$Result] = 1;
   - IF [Even] = 1 AND [Prime] = 1 THEN [$Result] = 2;
   - IF [Divs3] = 1 AND [Prime] = 1 THEN [$Result] = 3;
   - IF [Divs3] = 1 AND [Even] = 1 THEN [$Result] IN 6, 12;
   - IF [Divs3] = 1 AND [Odd] = 1 AND [Prime] = 0 THEN [$Result] = 9;
   - IF [Even] = 1 AND [Factorial] = 1 AND [Divs3] = 0 THEN [$Result] = 2;
3. #Data-specific dependencies
   - IF [$Result] = 1 THEN [Even] = 0 AND [Factorial] = 1 AND [Divs3] = 0 AND [Odd] = 1 AND [Prime] = 0;
   - IF [$Result] = 2 THEN [Even] = 1 AND [Factorial] = 1 AND [Divs3] = 0 AND [Odd] = 0 AND [Prime] = 1;

- IF [$Result] = 3 THEN [Even] = 0 AND [Factorial] = 0 AND [Divs3] = 1 AND [Odd] = 1 AND [Prime] = 1;
- IF [$Result] = 6 THEN [Even] = 1 AND [Factorial] = 1 AND [Divs3] = 1 AND [Odd] = 0 AND [Prime] = 0;
- IF [$Result] = 9 THEN [Even] = 0 AND [Factorial] = 0 AND [Divs3] = 1 AND [Odd] = 1 AND [Prime] = 0;
- IF [$Result] = 12 THEN [Even] = 1 AND [Factorial] = 0 AND [Divs3] = 1 AND [Odd] = 0 AND [Prime] = 0;

And for such a model description PICT outputs the following set of cases:

| Even | Factorial | Divs3 | Odd | Prime | $Result |
|------|-----------|-------|-----|-------|---------|
| 0 | 0 | 1 | 1 | 1 | 3 |
| 1 | 1 | 1 | 0 | 0 | 6 |
| 0 | 0 | 1 | 1 | 0 | 9 |
| 1 | 0 | 1 | 0 | 0 | 12 |
| 1 | 1 | 0 | 0 | 1 | 2 |
| 0 | 1 | 0 | 1 | 0 | 1 |

## 5   Conclusion

Formal Concept Analysis provides useful technique for the problem of test case design. It unites best practices of manual development and automatic generation. It provides sound and complete description of the investigated domain, based on expert knowledge. The output of the system consists of two main parts: the description of typical objects in the domain, and interdependence between parameters in terms of implications.

An important advantage of proposed technique is extensibility. If we add a new attribute, we can just copy the previous examples into new formal context, assuming that new attribute is absent for all objects and proceed with the procedure of attribute exploration. It holds even for the beginning of procedure. We can start with non-empty set of objects and implications simultaneously.

Described algorithm could be used as standalone solution for test case design, as well as, tool to get existing dependencies in the domain. Obtained implications could be valuable in pairwise testing to adjust the model description.

However, we should admit that current approach is limited in terms of attribute description. For now, it is highly dependent on the boolean nature of attributes. One of the main directions of future work is to work with attributes of general form.

## References

1. Bach, J., Shroeder, P.: Pairwise Testing A Best Practice That Isnt. In: Proceedings of the 22nd Pacific Northwest Software Quality Conference (2004)
2. Czerwonka, J.: Pairwise Testing in Real World: Practical Extensions to Test Case Generators. In: Proceedings of the 24th Pacific Northwest Software Quality Conference (2006)

3. Davey, B., Priestly, H.: Introduction to Lattices and Order, 2nd edn. Cambridge Mathematical Textbooks. Cambridge University Press (2002)
4. DeMillo, R.A., McCracken, W.M., Martin, R.J., Passafiume, J.F.: Software Testing and Evaluation. Benjamin/Cummings Publishing Company, Menlo Park (1987)
5. Ganter, B.: Attribute exploration with background knowledge. Theoretical Computer Science 217(2), 215–233 (1999)
6. Ganter, B., Wille, R.: Formal Concept Analysis: Mathematical Foundations. Springer (1999)
7. Graham, D.R. (ed.): Computer-Aided Software Testing: The CAST Report. Unicom Seminars Ltd., Middlesex (1991)
8. Grochtmann, M., Wegener, J., Grimm, K.: Test case design using classification trees and the classification-tree editor CTE. In: Proceedings of Quality Week, vol. 95, p. 30 (May 1995)
9. Obiedkov, S., Kourie, D.G., Eloff, J.H.: Building access control models with attribute exploration. Computers and Security 28(1), 2–7 (2009)
10. Offutt, A.J., Pan, J., Voas, J.M.: Procedures for reducing the size of coverage-based test sets. In: Proc. Twelfth Int'l. Conf. Testing Computer Softw. (1995)
11. Revenko, A., Kuznetsov, S.O.: Attribute Exploration of Properties of Functions on Ordered Sets. In: CLA, pp. 313–324 (2010)
12. https://github.com/ae-hse/fca/

# Recognizing Antipatterns and Analyzing Their Effects on Software Maintainability

Dénes Bán and Rudolf Ferenc

University of Szeged, Department of Software Engineering
Árpád tér 2. H-6720 Szeged, Hungary
{zealot,ferenc}@inf.u-szeged.hu

**Abstract.** Similarly to design patterns and their inherent extra information about the structure and design of a system, antipatterns – or bad code smells – can also greatly influence the quality of software. Although the belief that they negatively impact maintainability is widely accepted, there are still relatively few objective results that would support this theory.

In this paper we show our approach of detecting antipatterns in source code by structural analysis and use the results to reveal connections among antipatterns, number of bugs, and maintainability. We studied 228 open-source Java based systems and extracted bug-related information for 34 of them from the PROMISE database. For estimating the maintainability, we used the ColumbusQM probabilistic quality model.

We found that there is a statistically significant, 0.55 Spearman correlation between the number of bugs and the number of antipatterns. Moreover, there is an even stronger, -0.62 reverse Spearman correlation between the number of antipatterns and code maintainability. We also found that even these few implemented antipatterns could nearly match the machine learning based bug-predicting power of 50 class level source code metrics.

Although the presented analysis is not conclusive by far, these first results suggest that antipatterns really do decrease code quality and can highlight spots that require closer attention.

**Keywords:** Antipatterns, Software maintainability, Empirical validation, OO design, ISO/IEC 25010, SQuaRE.

## 1 Introduction

Antipatterns can be most simply thought of as the opposites of the more well-known design patterns [6]. While design patterns represent "best practice" solutions to common design problems in a given context, antipatterns describe a commonly occurring solution to a problem that generates decidedly negative consequences [3]. Also an important distinction is that antipatterns have a refactoring solution to the represented problem, which preserves the behavior of the code, but improves some of its internal qualities [5]. The widespread belief is that the more antipatterns a software contains, the worse its quality is.

B. Murgante et al. (Eds.): ICCSA 2014, Part V, LNCS 8583, pp. 337–352, 2014.
© Springer International Publishing Switzerland 2014

Some research even suggests that antipatterns are symptoms of more abstract design flaws [10,17]. However, there is little empirical evidence that antipatterns really decrease code quality.

We try to reveal the effect of antipatterns by investigating its impact on maintainability and its connection to bugs. For the purpose of quality assessment, we chose our ColumbusQM probabilistic quality model [2], which ultimately produces one number per system describing how "good" that system is. The antipattern-related information came from our own, structural analysis based extractor tool and source code metrics were computed using the Columbus Code-Analyzer reverse engineering tool [4]. We compiled the types of data described above for a total of 228 open-source Java systems, 34 of which have corresponding class level bug numbers from the open-access PROMISE [14] database. With all this information we try to answer the following questions:

**Research Question 1.** *What kind of relation exists between antipatterns and the number of known bugs?*

**Research Question 2.** *What kind of relation exists between antipatterns and the maintainability of the software?*

**Research Question 3.** *Can antipatterns be used to predict future software faults?*

We obtained some promising results showing that antipatterns indeed negatively correlate with maintainability according to our quality model. Moreover, antipatterns correlate positively with the number of known bugs and also seem to be good attributes for bug prediction. However, these results are only a small step towards the empirical validation of this subject.

The rest of our paper is structured as follows. In Section 2 we present our approach for extracting antipatterns and analyzing their relationship with bugs and maintainability. Next, Section 3 summarizes the achieved empirical results. Section 4 lists the possible threats to the validity of our work. Then, in Section 5 we highlight the related work. Finally, we conclude the paper in Section 6.

## 2   Approach

For analyzing the relationship between antipatterns, bugs, and maintainability we calculated the following measures for the subject systems:

- an absolute measure of maintainability per system (we used our ColumbusQM probabilistic quality model [2] to get this value).
- the total number of antipatterns per system.
- the total number of bugs per system.

For the third research question, we could compile an finer grained set of data – since the system-based quality attribute is not needed here:

- the total number of antipatterns related to each class in every subject system.

- the total number of bugs related to every class in every subject system.
- every class level metric for each class in every subject system.

The metric values were extracted by the Columbus tool [4], the bug number information comes from the PROMISE open bug database [14] and the pattern related metrics are calculated by our own tool described in Subsection 2.2.

## 2.1 Used Metrics

We used the following source code metrics for antipattern recognition – chosen because of the interpretation of antipatterns described in Section 2.2:

- **AD** (**A**PI **D**ocumentation): ratio of the number of documented public members of a class or package to the number of all of its public members.
- **CBO** (**C**oupling **B**etween **O**bjects): The CBO metric for a class means the number of directly used different classes by the class.
- **CC** (**C**lone **C**overage): ratio of code covered by code duplications in the source code element to the size of the source code element, expressed in terms of the number of syntactic entities (statements, expressions, etc.).
- **CD** (**C**omment **D**ensity): ratio of the comment lines of the source code element (CLOC) to the sum of its comment and logical lines of code (CLOC+LLOC).
- **CLOC** (**C**omment **L**ines **O**f **C**ode): number of comment and documentation code lines of the source code element; however, its nested, anonymous or local classes are not included.
- **LLOC** (**L**ogical **L**ines **O**f **C**ode): number of code lines of the source code element, without the empty and comment lines; its nested, anonymous or local classes are not included.
- **McCC** (**McC**abe's **C**yclomatic **C**omplexity): complexity of the method expressed as the number of independent control flow paths in it.
- **NA** (**N**umber of **A**ttributes): number of attributes in the source code element, including the inherited ones; however, the attributes of its nested, anonymous or local classes (or subpackages) are not included.
- **NII** (**N**umber of **I**ncoming **I**nvocations): number of other methods and attribute initializations which directly call the method (or methods of a class).
- **NLE** (**N**esting **L**evel **E**lse-If): complexity expressed as the depth of the maximum "embeddedness" of the conditional and iteration block scopes in a method (or the maximum of these for the container class), where in the if-else-if construct only the first if instruction is considered.
- **NOA** (**N**umber **O**f **A**ncestors): number of classes, interfaces, enums and annotations from which the class is directly or indirectly inherited.
- **NOS** (**N**umber **O**f **S**tatements): number of statements in the source code element; however, the statements of its nested, anonymous or local classes are not included.
- **RFC** (**R**esponse set **F**or **C**lass): number of local (i.e. not inherited) methods in the class (NLM) plus the number of directly invoked other methods by its methods or attribute initializations (NOI).

- **TLOC** (**T**otal **L**ines **O**f **C**ode): number of code lines of the source code element, including empty and comment lines, as well as its nested, anonymous or local classes.
- **TNLM** (**T**otal **N**umber of **L**ocal **M**ethods): number of local (i.e. not inherited) methods in the class, including the local methods of its nested, anonymous or local classes.
- **Warning P1, P2 or P3**: number of different coding rule violations reported by the PMD analyzer tool[1], categorized into three priority levels.
- **WMC** (**W**eighted **M**ethods per **C**lass): The WMC metric for a class is the total of the McCC metrics of its local methods.

## 2.2   Mining Antipatterns

The whole process of analyzing the subject source files and extracting antipatterns is shown in Figure 1.

First, we convert the source code – through a language specific format and a linking stage – to the LIM model (**L**anguage **I**ndependent **M**odel), a part of the Columbus framework. It represents the information obtained from the static analysis of code in a more abstract, graph-like format. It has different types of nodes that correspond to e.g. classes, methods, attributes, etc. while different edges represent the connections between these.

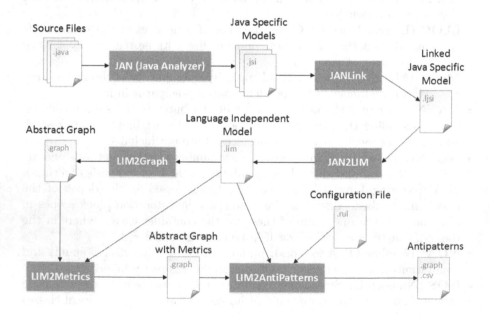

**Fig. 1.** The stages of the analysis

---

[1] http://pmd.sourceforge.net

From this data, the LIM2Metrics tool can compute various kinds of source code metrics – e.g. logical lines of code or number of statements in the *size* category, comment lines of code or API documentation in the *documentation* category and also different complexity, cohesion and inheritance metrics (see also Section 2.1). As these values are not a part of the LIM model, the output is pinned to an even more abstract graph – fittingly named "graph". This format literally only has "Nodes" and "Edges" but nodes can have dynamically attached and dynamically typed attributes. Since the LIM model is "strict" – e.g. it can only have statically typed attributes defined in advance – the "graph" format is more suitable as a target for the results of metrics, pattern matches, etc.

Each antipattern implementation can define one or more externally configurable parameters, mostly used for easily changeable metric thresholds. These come from an XML-style rule file – called RUL – that can handle multiple configurations and even inheritance. It can also contain language-aware descriptions and warning messages that will be attached to the affected graph nodes.

After all these preparations, our tool can be run. It is basically a single new class built around the Visitor design pattern [6] which is appropriate as it is a new operation defined for an existing data structure and this data structure does not need to be changed to accommodate the modification. It "visits" the LIM model and uses its structural information and the computed metrics from its corresponding graph nodes to identify antipatterns. It currently recognizes the 9 types of antipatterns listed below. We chose to implement these 9 antipatterns because they appeared to be the most widespread in the literature and, as such, the most universally regarded as a real negative factor. They are described in greater detail by Fowler and Beck [5], here we just provide a short informal definition and explain how we interpreted them in the context of our LIM model. The parameters of the recognition are denoted with a starting \$ sign and can be configured in the RUL file mentioned above. The referenced object-oriented source code metrics can be found in Subsection 2.1.

- **Feature Envy** (FE): A class is said to be envious of another class if it is more concerned with the attributes of that other class than those of its own. Interpreted as a method that accesses at least \$MinAccess attributes and at least \$MinForeign% of those belong to another class.
- **Lazy Class** (LC): A lazy class is one that does not do "much", only delegates its requests to other connected classes – i.e. a non-complex class with numerous connections. Interpreted as a class whose CBO metric is at least \$MinCBO but its WMC metric is no more than \$MaxWMC.
- **Large Class Code** (LCC): Simply put, a class that is "too big" – i.e. it probably encapsulates not just one concept or does too much. Interpreted as a class whose LLOC metric is at least \$MinLLOC.
- **Large Class Data** (LCD): A class that encapsulates too many attributes, some of which might be extracted – along with the methods that more closely correspond to them – into smaller classes and be a part of the original class through aggregation or association. Interpreted as a class whose NA metric is at least \$MinNA.

- **Long Function** (LF): Similarly to LCC, if a method is too long, it probably has parts that could – should – be separated into their own logical entities, thereby making the whole system more comprehensible. Interpreted as a method where either one of the LLOC, NOS or McCC metrics exceed $MinLLOC, $MinNOS or $MinMcCC, respectively. The NOS threshold is necessary because if every statement of the long function is spread across only a few lines then the LLOC metric will not be high. The McCC threshold ensures that even if the length of the function does not exceed its limits we still get a warning if it is too complex for its relatively smaller size.
- **Long Parameter List** (LPL): The long parameter list is one of the most recognized and accepted "bad code smells" in code. Interpreted as a function (or method) whose number of parameters is at least $MinParams.
- **Refused Bequest** (RB): If a class refuses to use its inherited members – especially if they are marked "protected," by which the parent expresses that descendants *should* most likely use it – then it is a sign that inheritance might not be the appropriate method of implementation reuse. Interpreted as a class that inherits at least one protected member that is not accessed by any locally defined method or attribute. Note that the missing parameter is intentional because we decided that one such member is severe enough.
- **Shotgun Surgery** (SHS): Following the "Locality of Change" principle, if a method needs to be modified then it should not cause the need for many other – especially remote – modifications, otherwise one of those can easily be missed leading to bugs. Interpreted as a method whose NII metric is at least $MinNII.
- **Temporary Field** (TF): If an attribute only "makes sense" to a small percent of the container class then it – and its closely related methods – should be decoupled. Interpreted as an attribute that is only referenced by at most $RefMax% of the members of its container class.

### 2.3 ColumbusQM Software Quality Model

Our probabilistic software quality model [2] is based on the quality characteristics defined by the ISO/IEC 25010 standard [8]. The computation of the high level quality characteristics is based on a directed acyclic graph whose nodes correspond to quality properties that can either be internal (low-level) or external (high-level). Internal quality properties characterize the software product from an internal (developer) view and are usually estimated by using source code metrics. External quality properties characterize the software product from an external (end user) view and are usually aggregated somehow by using internal and other external quality properties. The edges of the graph represent dependencies between an internal and an external or two external properties. In addition to the external nodes defined by the standard (black nodes) we introduced new ones (light gray nodes) and even kept those that were contained only in the old ISO/IEC 9126 standard (dark gray nodes). The aim is to evaluate all the external quality properties by performing an aggregation along the edges of the graph, called Attribute Dependency Graph (ADG). We calculate

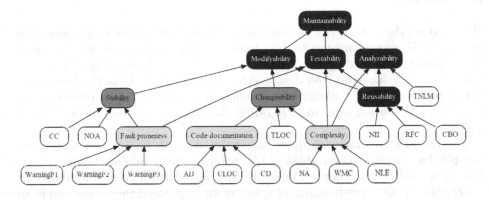

**Fig. 2.** The quality model used to calculate maintainability

a so called "goodness value" (from the [0,1] interval) to each node in the ADG that expresses how good or bad (1 is the best) is the system regarding that quality attribute. We used the ADG presented in Figure 2 – which is a further developed version of the ADG published by Bakota et al. [2] – for assessing the maintainability of the selected subject systems. The informal definition of the referenced low-level metrics are described in Subsection 2.1.

### 2.4 PROMISE

PROMISE [14] is an open-access bug database whose purpose is to help software quality related research and make the referenced experiments repeatable. It contains the code of numerous open-source applications or frameworks and their corresponding bug-related data on the class level – i.e. not just system aggregates. We extracted this class level bug information for 34 systems from it that will be used for answering our third research question in Section 3.

### 2.5 Machine Learning

From the class level table of data we wanted to find out if the numbers of different antipatterns have any underlying structure that could help in identifying which system classes have bugs. This is exactly the type of problem machine learning is concerned with.

Machine learning is a branch of artificial intelligence that tries to predict the workings of the mechanism that generated its input data – its training set – to be able to correctly label unclassified instances later. As empirically they perform best in similar cases, we chose decision trees as the method of analysis, specifically J48, an open-source implementation of C4.5 [15]. We preformed the actual learning with Weka [7].

# 3   Results

We analyzed the 228 subject systems and calculated the measures introduced in Section 2. These systems are all Java based and open-source – so we can access their source codes and analyze them – but their purposes are very diverse – ranging from web servers and database interfaces to IDEs, issue trackers, other static analyzers, build tools and many more. Note that the 34 systems that also have bug information from the PROMISE database are a subset of the original 228.

For the first two research questions concerned with finding correlation, we compiled a system level database of the maintainability values, numbers of antipatterns and the numbers of bugs. As the bug-related data was available for every class in the corresponding 34 systems, we aggregated these values to fit in on a per system basis. The resulting dataset can be seen in Table 1, sorted in ascending order by the number of bugs.

After this, we performed correlation analysis on the collected data. Since we did not expect the relationship between the inspected values to be linear – only monotone –, we used the Spearman correlation. The Spearman correlation is in fact a "traditional" Pearson correlation, only it is carried out on the ordered ranks of the values, not the values themselves. This shows how much the two data sets "move together." The extent of this matching movement is somewhat masked by the ranking – which can be viewed as a kind of data loss – but this is not that important as we are more interested in the *existence* of this relation rather than its type.

When we only considered the systems that had related bug information, we found that the total number of bugs and the total number of antipatterns per system has a Spearman correlation of **0.55** with a p-value below 0.001. This can be regarded as a significant relation that answers our **first research question** by suggesting that the more antipatterns there are in the source code of a software system, the more bugs it will likely contain. Intuitively this is to be expected but with this empirical experiment we are a step closer to being able to treat this belief as a fact. The discovered relation is, of course, not a one-to-one compliance but it illustrates the negative effect of antipatterns on the source code well.

If we disregard the bug information but expand our search to all the 228 analyzed systems, we can inspect the connection between the number of antipatterns and the computed maintainability values. Here we found that there is an even stronger, **-0.62** reverse Spearman correlation, also with a p-value less than 0.001. Based on this observation, we can also answer our **second research question**: the more antipatterns a system contains the less maintainable it will be, meaning that it will most likely cost more time and resources to execute any changes. This result corresponds to the definitions of antipatterns and maintainability quite well as they suggest that antipatterns – the common solutions to design problems that seem beneficial but cause more harm than good on the long

**Table 1.** The extracted metrics of the analyzed systems

| Name | Nr. of Antipatterns | Maintainability | Nr. of Bugs |
|---|---|---|---|
| jedit-4.3 | 2,351 | 0.47240 | 12 |
| camel-1.0 | 685 | 0.62129 | 14 |
| forrest-0.7 | 53 | 0.73364 | 15 |
| ivy-1.4 | 709 | 0.49465 | 18 |
| pbeans-2 | 105 | 0.48909 | 19 |
| synapse-1.0 | 398 | 0.58202 | 21 |
| ant-1.3 | 933 | 0.51566 | 33 |
| ant-1.5 | 2,069 | 0.41340 | 35 |
| poi-2.0 | 2,025 | 0.36309 | 39 |
| ant-1.4 | 1,218 | 0.45270 | 47 |
| ivy-2.0 | 1,260 | 0.44374 | 56 |
| log4j-1.0 | 224 | 0.59301 | 61 |
| log4j-1.1 | 341 | 0.56738 | 86 |
| Lucene | 3,090 | 0.47288 | 97 |
| synapse-1.1 | 717 | 0.56659 | 99 |
| jedit-4.2 | 1,899 | 0.46826 | 106 |
| tomcat-1 | 5,765 | 0.30972 | 114 |
| xerces-1.2 | 2,520 | 0.13329 | 115 |
| synapse-1.2 | 934 | 0.55554 | 145 |
| xalan-2.4 | 3,718 | 0.15994 | 156 |
| ant-1.6 | 2,821 | 0.38825 | 184 |
| velocity-1.6 | 614 | 0.43804 | 190 |
| xerces-1.3 | 2,670 | 0.13600 | 193 |
| jedit-4.1 | 1,440 | 0.47400 | 217 |
| jedit-4.0 | 1,207 | 0.47388 | 226 |
| lucene-2.0 | 1,580 | 0.47880 | 268 |
| camel-1.4 | 2,136 | 0.62682 | 335 |
| ant-1.7 | 3,549 | 0.38340 | 338 |
| Mylyn | 5,378 | 0.65841 | 340 |
| PDE_UI | 7,523 | 0.53104 | 341 |
| jedit-3.2 | 1,017 | 0.47523 | 382 |
| camel-1.6 | 2,804 | 0.62796 | 500 |
| camel-1.2 | 1,280 | 0.63005 | 522 |
| xalan-2.6 | 11,115 | 0.22621 | 625 |

run – really do lower the expected maintainability – a value representing how easily a piece of software can be understood and modified, i.e. the "long run" harm.

This relation is visualized in Figure 3. The analyzed systems are sorted in the descending order of their contained antipatterns and the trend line of the maintainability value clearly shows improvement.

To answer our third and final research question, we could compile an even finer grained set of data. Here we retained the original, class level form of bug-related data and extracted class level source code metrics. We also needed to transform the antipattern information to correspond to classes so instead of aggregating

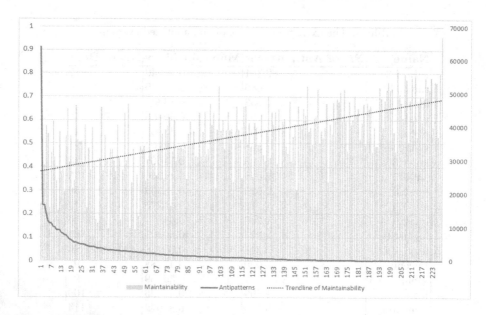

**Fig. 3.** The trend of maintainability in case of decreasing antipatterns

them to a system level we kept class antipatterns unmodified while collecting
method and attribute based antipatterns to their closest parent classes. Part of
this data set is shown in Table 2.

The resulting class level data was largely biased because – as it is to be
expected – more than 80% of the classes did not contain any bugs. We handled
this problem by subsampling the bugless classes in order to create a normally
distributed starting point. Subsampling means that in order to make the results
of the machine learning experiment more representative, we used only part of
the data related to those classes that did not contain bugs. We created a *sub*set
by randomly *sampling* – hence the name – from the bugless classes so that their
number becomes equal to the "buggy" classes. This way the learning algorithm
could not achieve a precision higher than 50% just by choosing one class over
the other. We then applied the J48 decision tree – mentioned in Section 2 – in
three different configurations:

- using only the different antipattern numbers as predictors
- using only the object-oriented metrics extracted by the Columbus tool as
  predictors
- using the attributes of both categories

Every actual learning experiment was performed with a ten-fold cross validation. We tried to calibrate the built decision trees to have around 50 leaf nodes
and around 100 nodes in total. This is an approximately good compromise between under- and overlearning the training data. The results are summarized in
Table 3.

**Table 2.** Part of the compiled class level data set

| System | Name | Bugs | LLOC | TNOS | ... | ALL | FE | LC | LCC | LCD | LF | LPL | RB | SHS | TF |
|---|---|---|---|---|---|---|---|---|---|---|---|---|---|---|---|
| ant-1.3 | org.apache.tools.ant.AntClassLoader | 2 | 230 | 124 | ... | 4 | 0 | 0 | 0 | 0 | 1 | 0 | 0 | 0 | 3 |
| ant-1.3 | org.apache.tools.ant.BuildEvent | 0 | 51 | 21 | ... | 0 | 0 | 0 | 0 | 0 | 0 | 0 | 0 | 0 | 0 |
| ant-1.3 | org.apache.tools.ant.BuildException | 0 | 63 | 27 | ... | 6 | 0 | 0 | 0 | 0 | 0 | 0 | 0 | 6 | 0 |
| ant-1.3 | org.apache.tools.ant.DefaultLogger | 2 | 74 | 30 | ... | 7 | 0 | 0 | 0 | 0 | 0 | 0 | 6 | 0 | 1 |
| ant-1.3 | org.apache.tools.ant.DesirableFilter | 0 | 20 | 11 | ... | 0 | 0 | 0 | 0 | 0 | 0 | 0 | 0 | 0 | 0 |
| ant-1.3 | org.apache.tools.ant.DirectoryScanner | 0 | 472 | 353 | ... | 25 | 0 | 0 | 0 | 0 | 3 | 0 | 19 | 1 | 2 |
| ant-1.3 | org.apache.tools.ant.IntrospectionHelper | 2 | 229 | 142 | ... | 2 | 0 | 0 | 0 | 0 | 1 | 0 | 0 | 0 | 1 |
| ant-1.3 | org.apache.tools.ant.Location | 0 | 29 | 13 | ... | 1 | 0 | 0 | 0 | 0 | 0 | 0 | 0 | 0 | 1 |
| ant-1.3 | org.apache.tools.ant.Main | 1 | 364 | 254 | ... | 3 | 0 | 0 | 0 | 0 | 2 | 0 | 0 | 0 | 1 |
| ant-1.3 | org.apache.tools.ant.NoBannerLogger | 0 | 21 | 8 | ... | 0 | 0 | 0 | 0 | 0 | 0 | 0 | 0 | 0 | 0 |
| ant-1.3 | org.apache.tools.ant.PathTokenizer | 0 | 37 | 16 | ... | 0 | 0 | 0 | 0 | 0 | 0 | 0 | 0 | 9 | 25 |
| ant-1.3 | org.apache.tools.ant.Project | 1 | 710 | 406 | ... | 37 | 0 | 0 | 1 | 0 | 2 | 0 | 0 | 0 | 1 |
| ant-1.3 | org.apache.tools.ant.ProjectHelper | 3 | 151 | 267 | ... | 2 | 0 | 0 | 0 | 0 | 1 | 0 | 0 | 0 | 0 |
| ant-1.3 | org.apache.tools.ant.RuntimeConfigurable | 2 | 45 | 18 | ... | 0 | 0 | 0 | 0 | 0 | 0 | 0 | 0 | 0 | 0 |
| ant-1.3 | org.apache.tools.ant.Target | 1 | 103 | 42 | ... | 1 | 0 | 0 | 0 | 0 | 0 | 0 | 0 | 1 | 0 |
| ant-1.3 | org.apache.tools.ant.Task | 0 | 64 | 19 | ... | 15 | 0 | 0 | 0 | 0 | 0 | 0 | 8 | 6 | 1 |
| ant-1.3 | org.apache.tools.ant.TaskAdapter | 0 | 25 | 13 | ... | 0 | 0 | 0 | 0 | 0 | 0 | 0 | 0 | 0 | 0 |
| ant-1.3 | org.apache.tools.ant.taskdefs.Ant | 0 | 133 | 87 | ... | 0 | 0 | 0 | 0 | 0 | 0 | 0 | 0 | 0 | 0 |
| ant-1.3 | org.apache.tools.ant.taskdefs.AntStructure | 0 | 179 | 134 | ... | 1 | 0 | 0 | 0 | 0 | 1 | 0 | 0 | 0 | 0 |
| ant-1.3 | org.apache.tools.ant.taskdefs.Available | 0 | 101 | 46 | ... | 0 | 0 | 0 | 0 | 0 | 0 | 0 | 0 | 0 | 0 |

Table 3. The results of the machine learning experiments

| Method | TP Rate | FP Rate | Precision | Recall | F-Measure |
|---|---|---|---|---|---|
| Antipatterns | 0.658 | 0.342 | 0.670 | 0.658 | 0.653 |
| Metrics | 0.711 | 0.289 | 0.712 | 0.711 | 0.711 |
| Both | 0.712 | 0.288 | 0.712 | 0.712 | 0.712 |

These results clearly show that although antipatterns are – for now – inferior to OO metrics in this field, even a few patterns concerned with single entities only can approximate their bug predicting powers quite well. We note that it is expected that the "Both" category does not improve upon the "Metrics" category because in most cases – as of yet – the implemented antipatterns can be viewed as predefined thresholds on certain metrics. With this we answered our **third research question** as antipatterns can already be considered valuable bug predictors and with more implemented patterns – spanning multiple code entities – and heavier use of the contextual structural information they might even overtake them.

## 4 Threats to Validity

Similarly to most works, our approach also has some threats to validity. First of all, when dealing with recognizing antipatterns, or any patterns, the accuracy of mining is always a question. To make sure that we really extract the patterns we want, we created small, targeted source code tests that checked the structural requirements and metric thresholds of each pattern. To be also sure that we want to match the patterns we *should* – i.e. those that are most likely to really have a bad effect on the quality of the software –, we only implemented antipatterns that are well-known and well-documented in the literature. This way the only remaining threat factor is the interpretation of those patterns to the LIM model.

Then there is the concern of choosing the correct thresholds for certain metrics. Although they are easily configurable – even before every new inspection –, in order to have results we could correlate and analyze, we had to use some specific thresholds. These were approximated by expert opinions taking into consideration the minimum, maximum and average values of the corresponding metrics. This method can be further improved by implementing statistics based dynamic thresholds which is among our future goals.

Another threat to validity is using our previously published quality model for calculating maintainability values. Although we have done many empirical validations of our probabilistic quality model in previous works, we cannot state that the used maintainability model is perfect. Moreover, as the ISO/IEC 25010 standard does not define the low-level metrics, the results can vary depending on the quality model's settings (chosen metrics and weights given by professionals). It is also very important to have a source code metrics repository with a large enough number of systems to get an objective absolute measure for maintainability. These factors are possible threats to validity, but our results and continuous empirical validation of the model proves its applicability and usefulness.

Our results also depend on the assumption that the bug-related values we extracted from the PROMISE database are correct. If they are inaccurate that means that our correlation results with the number of bugs are inaccurate too. But as many other works make use of these data, we consider the possibility of this negligible.

Finally, we have to face the threat that our data is biased or that the results are coincidental. We tried to combat these factors by using different kinds of subject systems, balancing our training data to a normal class distribution before the machine learning procedure and considering only statistically significant correlations.

## 5 Related Work

The most closely related research to our current work was done by Marinescu. In his publication in 2001 [12], he emphasized that the search for given types of flaws should be systematic, repeatable, scalable and language-independent. First, he defined a unified format for describing antipatterns and then a methodology for evaluating those. He showed this method in action using the GodClass and DataClass antipatterns and argued that it could similarly be done for any other pattern. To automate this process, he used his own TableGen software to analyze C++ source code – analogous to the Columbus tool in our case, – save its output to a database and extract information using standard queries.

In one of his works from 2004 [13], he was more concerned with automation and made declaring new antipatterns easier with "detection strategies." In these, one can define different filters for source metrics – limit or interval, absolute or relative – even with statistical methods that set the appropriate value by analyzing all values first and computing their average, mean, etc., to find outliers. Finally, these intermediate result sets can be joined by standard set operations like union, intersection, or difference. When manually checking the results he defined a "loose precision" value next to the usual "strict" one that did not label a match as a false positive if it *was* indeed faulty but not because of the searched pattern. His conclusion is an empirically 70% accurate tool that can be considered successful.

These "detection strategies" were extended with historical information by Rapu et al. [16]. They achieved this by running the above described analysis on not only the current version of their subject software system – Jun 3D graphical framework – but on every fifth from the start. This way they could extract two more metrics: persistence that means how much of its "lifetime" was a given code element faulty, and stability that means how many times the code element changed during its life. The logic behind this was that e.g. a GodClass antipattern is dangerous only if it is not persistent – the error is due to changes, not part of the original design – or not stable – when it really disturbs the evolution of the system. With this method they managed to halve the candidates in need of manual checking in the case of the above example.

Trifu and Marinescu went further by assuming that these antipatterns are just symptoms of larger, more abstract faults [17]. They proposed to group several antipatterns – that may even occur together often – and supplemented them with contextual information to form "design flaws." Their main goal was to make antipattern recognition – and pattern recognition in general – more of a well-defined engineering task rather than a form of art.

Our work is similar to the ones mentioned above in that we also detect antipatterns by employing source code metrics and static analysis. But, in addition, we inspect the correlations of these patterns to the maintainability values of the subject systems and also consider bug-related information to more objectively prove the common belief that they are indeed connected.

In another approach, Khomh et al. [9] concentrated more on statistical methods and asked whether there is a connection between antipatterns and change-proneness. They used their own extractor tool "DECOR" and null-hypotheses to answer this question and concluded that antipatterns really increase change-proneness. We similarly use statistical methods – beside machine learning – but instead of change-proneness we concentrate on maintainability and bug information.

Lozano et al. [10] overviewed a broader sweep of related works and urged researchers to standardize their efforts. Apart from the individual harms antipatterns may cause, they aimed to find out that from exactly when in the life cycle of a software can an antipattern be considered "bad" and – not unlike [17] – whether these antipatterns should be raised to a higher abstraction level. In contrast to the historical information, we focus on objective metric results to shed light on the effect of antipatterns.

Also another approach by Mäntylä et al. [11] is to use questionnaires to reveal the subjective side of software maintenance. The different opinions of the participating developers could mostly be explained by demographic analysis and their roles in the company but there was a surprising difference compared to the metric based results. We, on the other hand, make these objective, metric based results our priority.

Although we used literature suggestion and expert opinion based metric thresholds for this empirical analysis, our work could be repeated – and possibly improved – by using the data-driven, robust and pragmatic metric threshold derivation method described by Alves et al. [1]. They analyze, weigh and aggregate the different metrics of a large number of benchmark systems in order to statistically evaluate them and extract metric thresholds that represent the best or worst $X\%$ of all the source code. This can help in pinpointing the most problematic – but still manageably few – parts of a system.

# 6    Conclusions

In this paper we presented an empirical analysis of exploring the connection between antipatterns, the number of known bugs and software maintainability. First, we briefly explained how we implemented a tool that can match antipatterns on a language independent model of a system. We then analyzed more than

200 open-source Java systems, extracted their object-oriented metrics and antipatterns, calculated their corresponding maintainability values using our probabilistic quality model and even collected class level bug information for 34 of them. By correlation analysis and machine learning methods we were able to draw interesting conclusions.

On a system level scale, we found that in the case of the 34 systems that also had bug-related information there is a significant positive correlation between the number of bugs and the number of antipatterns. Also, if we disregarded the bug data but expanded our search to all 228 analyzed systems to concentrate on maintainability, the result was an even stronger negative correlation between the number of antipatterns and maintainability. This further supports what one would intuitively think considering the definitions of antipatterns, bugs, and quality.

Another interesting result is that the mentioned 9 antipatterns in themselves can quite closely match the bug predicting power of more than 50 class level object-oriented metrics. Although they – as of yet – are inferior, with further patterns that would span over source code elements and rely more heavily on the available structural information, this method has the potential to overtake simple metrics in fault prediction.

As with all similar works, ours also has some threats to its validity but we feel that it is a valuable step towards empirically validating that antipatterns really do hurt software maintainability and can highlight points in the source code that require closer attention.

**Acknowledgments.** This research was supported by the European Union and the State of Hungary, co-financed by the European Social Fund in the framework of TÁMOP-4.2.4.A/2-11/1-2012-0001 "National Excellence Program", and by the Hungarian national grant GOP-1.1.1-11-2011-0006.

# References

1. Alves, T.L., Ypma, C., Visser, J.: Deriving metric thresholds from benchmark data. In: 2010 IEEE International Conference on Software Maintenance (ICSM), pp. 1–10. IEEE (2010)
2. Bakota, T., Hegedűs, P., Körtvélyesi, P., Ferenc, R., Gyimóthy, T.: A probabilistic software quality model. In: 2011 27th IEEE International Conference on Software Maintenance (ICSM), pp. 243–252 (2011)
3. Brown, W.J., Malveau, R.C., McCormick III, I.H.W., Mowbray, T.J.: AntiPatterns: Refactoring Software, Architectures, and Projects in Crisis. John Wiley & Sons, Inc., New York (1998)
4. Ferenc, R., Beszédes, Á., Tarkiainen, M., Gyimóthy, T.: Columbus-reverse engineering tool and schema for c++. In: Proceedings of the International Conference on Software Maintenance, pp. 172–181. IEEE (2002)
5. Fowler, M., Beck, K.: Refactoring: Improving the Design of Existing Code. Addison-Wesley object technology series. Addison-Wesley (1999)

6. Gamma, E., Helm, R., Johnson, R., Vlissides, J.: Design Patterns: Elements of Reusable Object-oriented Software. Addison-Wesley Longman Publishing Co., Inc., Boston (1995)
7. Hall, M., Frank, E., Holmes, G., Pfahringer, B., Reutemann, P., Witten, I.H.: The weka data mining software: An update. SIGKDD Explor. Newsl. 11(1) (November 2009)
8. ISO/IEC: ISO/IEC 25000:2005. Software Engineering – Software product Quality Requirements and Evaluation (SQuaRE) – Guide to SQuaRE. ISO/IEC (2005)
9. Khomh, F., Di Penta, M., Guéhéneuc, Y.G.: An exploratory study of the impact of code smells on software change-proneness. In: 16th Working Conference on Reverse Engineering, WCRE 2009, pp. 75–84 (2009)
10. Lozano, A., Wermelinger, M., Nuseibeh, B.: Assessing the impact of bad smells using historical information. In: Ninth International Workshop on Principles of Software Evolution: In Conjunction with the 6th ESEC/FSE Joint Meeting, IW-PSE 2007, pp. 31–34. ACM (2007)
11. Mäntylä, M., Vanhanen, J., Lassenius, C.: Bad smells - humans as code critics. In: Proceedings of the 20th IEEE International Conference on Software Maintenance, pp. 399–408 (2004)
12. Marinescu, R.: Detecting design flaws via metrics in object-oriented systems. In: Proceedings of TOOLS, pp. 173–182. IEEE Computer Society (2001)
13. Marinescu, R.: Detection strategies: Metrics-based rules for detecting design flaws. In: Proc. IEEE International Conference on Software Maintenance (2004)
14. Menzies, T., Caglayan, B., He, Z., Kocaguneli, E., Krall, J., Peters, F., Turhan, B.: The promise repository of empirical software engineering data (June 2012), http://promisedata.googlecode.com
15. Quinlan, J.R.: C4.5: Programs for Machine Learning. Morgan Kaufmann Publishers Inc., San Francisco (1993)
16. Rapu, D., Ducasse, S., Girba, T., Marinescu, R.: Using history information to improve design flaws detection. In: Proceedings of the Eighth European Conference on Software Maintenance and Reengineering, CSMR 2004, pp. 223–232 (2004)
17. Trifu, A., Marinescu, R.: Diagnosing design problems in object oriented systems. In: Proceedings of the 12th Working Conference on Reverse Engineering, WCRE 2005, pp. 155–164. IEEE Computer Society (2005)

# The Impact of Version Control Operations on the Quality Change of the Source Code

Csaba Faragó, Péter Hegedűs, and Rudolf Ferenc

University of Szeged Department of Software Engineering
Árpád tér 2. H-6720 Szeged, Hungary
{farago,hpeter,ferenc}@inf.u-szeged.hu

**Abstract.** The number of software systems under development and maintenance is rapidly increasing. The quality of a system's source code tends to decrease during its lifetime which is a problem because maintaining low quality code consumes a big portion of the available efforts. In this research we investigated one aspect of code change, the version control commit operations (add, update, delete). We studied the impact of these operations on the maintainability of the code. We calculated the ISO/IEC 9126 quality attributes for thousands of revisions of an industrial and three open-source software systems. We also collected the cardinality of each version control operation type for every investigated revision. Based on these data, we identified that operation Add has a rather positive, while operation Update has a rather negative effect on the quality. On the other hand, for operation Delete we could not find a clear connection to quality change.

**Keywords:** Software Maintainability, Software Erosion, Source Code Version Control, ISO/IEC 9126, Case Study.

## 1 Introduction

Software quality plays a crucial role in modern development projects. There is an ever-increasing amount of software systems in maintenance phase, and it is a well-known fact that software systems are eroding [14], meaning that in general their quality is continuously decreasing due to the ever-ongoing modifications in their source code, unless explicit efforts are spent on improvements [3]. Our aim is to check the connection between the developers' interactions and the quality change, in order to identify which patterns typically increase, and which decrease code maintainability.

Our longer term plan is to discover as much of these patterns as possible. In the beginning we focus only on the data found in the version control systems. Later we plan to include other available data, especially micro interactions performed within the IDE during development, and data found in issue tracking systems. Based on the results, we can hopefully formulate advices to software developers on how to avoid the maintainability decrease. Furthermore, this could also help to better allocate efforts spent on increasing quality.

In this research we focus on the *version control operations*. Specifically, we check the effects of file additions, updates and deletions on the maintainability

B. Murgante et al. (Eds.): ICCSA 2014, Part V, LNCS 8583, pp. 353–369, 2014.

of the source code. We checked how the higher number or higher proportion of a version control operation within a commit typically affects maintainability. Basically, we assumed that file additions have positive impact, as they introduce new, clean, reasoned code. Initially we expected the same result also for file deletions, as this operation is typically used during code refactoring, which is an explicit step towards better maintainability. On the other hand, we expected that file updates tend to decrease maintainability.

To summarize our goals, we formulated the following research questions:

**RQ1:** *Does the amount of file additions, updates and deletions within a commit impact the maintainability of the source code?*

**RQ2:** *Are there any differences between checks considering the absolute number of operations (Add, Update, Delete) and checks investigating the relative proportion of the same operation within commits?*

The paper is organized as follows. Section 2 introduces works that are related to ours. Then, in Section 3 we present the methodology used to test the underlying relationship between version control operations and maintainability changes. Section 4 discusses the results of the performed statistical tests and summarizes our findings. In Section 5 we list the possible threats to the validity of the results, while Section 6 concludes the paper.

## 2    Related Work

As a particular software quality related activity, refactoring is a widely researched field. Lots of works build models for predicting refactorings based on version control history analysis [18–20]. Moser et al. developed an algorithm for distinguishing commits resulted by refactorings from those of other types of changes [13]. Peters and Zaidman investigated the lifespan of code smells and the refactoring behavior of developers by mining the software repository of seven open-source systems [15]. The results of their study indicate that engineers are aware of code smells, but are not really concerned by their impact, given the low refactoring activity.

There are works which focus on the effect of software processes on product quality [11]. Hindle et al. deal with understanding the rationale behind large commits. They contrast large commits against small commits and show that large commits are more perfective, while small commits are more corrective [8]. Bachmann and Bernstein explore among others if the process quality, as measured by the process data, has an influence on the product quality. They showed that product quality – measured by number of bugs reported – is affected by process data quality measures [4].

Another group of papers focus on estimating some properties of maintenance activities (e.g. the effort needed to make changes in the source code, comprehension of maintenance activities, complexity of modification) [7, 12, 21]. Tóth et al. showed that the cumulative effort to implement a series of changes is larger than the effort that would be needed to make the same modification in only one step [21]. The work of Gall et al. focuses on detecting logical couplings from CVS release history data. They argue that the dependencies and interrelations between classes and modules that can be extracted from version control operations affect the maintainability of object-oriented systems [6]. Fluri et al. examine the co-evolution of code and comments as a vital part of code comprehension. They found that newly added code – despite its growth rate – barely gets commented; class and method declarations are commented most frequently, but e.g. method calls are far less; and that 97% of comment changes are done in the same revision as the associated source code change [5]. Unlike these works, we do not use the version control data to predict refactorings or software quality attributes, but to directly analyze the effect of the way version control operations are performed on software maintainability.

Atkins et al. use the version control data to evaluate the impact of software tools on software maintainability [1]. They explore how to quantify the effects of a software tool once it has been deployed in a development environment and present an effort-analysis method that derives tool usage statistics and developer actions from a project's change history (version control system). We also try to evaluate the maintainability changes through version control data; however, we investigate the general effect of version control operations regardless of tool usage.

Pratap et al. in their work [16] present a fuzzy logic approach for estimating the maintainability, while in this research we use a probabilistic quality model.

## 3   Methodology

In this section we summarize the kind of data we collected and how we elaborated on them to gain the results.

### 3.1   Version Control Operations

We take the version control operations as predictors of source code maintainability. In this work we investigated the number of the version control operation types: we only checked how many Adds, Updates and Deletes existed in the examined commit. These three numbers of every commit formed the predictor input of the analysis. E.g., if a certain commit contains 2 file additions, 5 file updates and 1 file deletion, then the input related to that commit would be (2, 5, 1). The fourth version control operation – Rename – was not considered, because there were hardly any commits containing this operation.

As the used quality model handles Java files only, we removed the non-Java source related statistics. E.g., if a commit contained 3 updates, 2 of Java files and one of an XML file, then we simply treated this as a commit of 2 updates.

### 3.2   The Applied Quality Model

The dependent variable of the research was the quality of the source code. This was estimated by the ColumbusQM probabilistic software quality model [2], which is based on the ISO/IEC 9126 standard [10].

The model calculates a composite measure of source code metrics like logical lines of code, complexity, number of coding rule violations etc. The calculation is based on expert weights and a statistical aggregation algorithm that uses a so-called benchmark as the basis of the qualification. The resulting maintainability value is expressed by a real number between 0.0 and 1.0. The higher number indicates better maintainability. See the work of Bakota et al. [2] for further details about the model.

### 3.3  Maintainability Change

The system's maintainability change can be calculated as the difference of the maintainability values of the current revision and the previous one. However, a simple subtraction is not sufficient for two reasons:

- The quality model provides the quality value based on a distribution function. The absolute difference between e.g. 0.58 and 0.54 is not the same as between 0.98 and 0.94. The latter difference is bigger as improving a software with already high quality is harder than improving a medium quality system.
- The same amount of maintainability change (e.g. committing 10 serious coding rule violations into the source code) has a much bigger effect on a small system than on a large one.

To overcome these shortcomings, we applied the following transformations:

- We used the quantile function of the standard normal distribution to calculate the original absolute value from the goodness value. This is feasible because the goodness value is derived from a probability function with normal distribution. We performed this transformation with the `qnorm()` R function [17]. The transformed values served as the basis of the subtractions.
- We multiplied the results of the subtractions (the maintainability value differences) by the current size of the system, more specifically, the current total logical lines of code (TLLOC, number of non-comment non-empty lines of code).

Figure 1 illustrates why the quantile conversion is necessary. The same difference on the y axis is not the same after quantile conversion (x axis), as expected.

We defined the quality change of the first commit to be 0.0.

### 3.4  Two-Sample Wilcoxon Rank Tests

For investigating **RQ1** and **RQ2**, two subsets of the commits were defined in several ways detailed below. The partitioning was performed based only on the version control operations in each case. After the partitioning we examined the maintainability changes of the commits belonging to these subsets. To check if the differences are significant or not, we used the two-sample Wilcoxon rank test (also known as Mann-Whitney U test) [9]. The Wilcoxon rank test is a so-called paired difference test, which checks if the population mean ranks differ in two data sets. Unlike mean this is not sensitive to the extreme values.

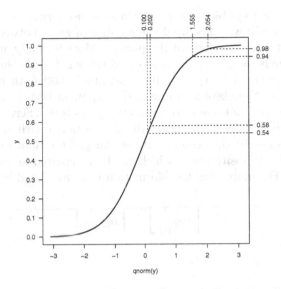

**Fig. 1.** Illustration why quantile conversion is necessary

The tests were performed by the `wilcox.test()` function in R [17]. The result of the test is practically the p-value, which tells us the probability of the result being at least as extreme as the actual one, provided that the null-hypothesis is true. In every case the null-hypothesis was that there is no difference between the distribution of the maintainability change values in the two commit sets. The alternative hypothesis was the following: the elements (maintainability differences) in one subset are less or greater than those in the other subsets.

Instead of executing a two direction Wilcoxon rank test (which would consider only the absolute magnitude of the difference, and not the direction – i.e. which one is greater), we executed the one direction test twice: first considering that the values in the first subset are less than those in the second, and in the second case we checked the opposite direction. We chose this approach as we needed the direction as well (we were not satisfied with the answer that the values are different in one subset compared the other, we also wanted to know which of them are less and which are greater).

As we performed the test twice each time, two p-values resulted. Let us denote them with $p_1$ (in case of values in the first set are less than those in the second one) and $p_2$ (the opposite direction). E.g., in case of a concrete division it turned out that the p-value is 0.0046 being numbers in one subset greater than those in the other, which also means that the p-value of having smaller values in the first set is 0.9954. Please note that the sum of these p-values are always 1.0, i.e. 100%. From the two p-values we consider the better one, noting the direction this result was executed with. Therefore the result is practically always exactly twice as good as it would be in case of a two direction test. E.g., in case of comparing two exactly same datasets, the resulting p-value is 0.5. This is considered when analyzing the results.

In order to be able to publish the results in a concise format, we introduced an approximate approach: we calculated the number of zeros between the decimal point and the first non-zero digit of the p-value. More formally, if the canonical form of the p-value is $(a \cdot 10^b)$, the transformed value is the absolute value of the exponent minus one (i.e. $|b| - 1$). E.g., if the p-value is 0.0046, then the canonical form is $4.6 \cdot 10^{-3}$, so the absolute value of the exponent is 3, minus 1 yields 2.

Please note that at least one of the two exponents is 0. Therefore for an even more compact interpretation, the non-null value is taken with appropriate sign (positive if the values in the second dataset are greater than in the first one, and negative in the opposite case), which can be calculated as the difference of the two p-values. Formally, this transformation was calculated by the following function:

$$f = \left\lfloor log \frac{1}{p_1} \right\rfloor - \left\lfloor log \frac{1}{p_2} \right\rfloor \tag{1}$$

### 3.5  Divisions

This section describes how the two subsets of the whole commit set were defined. All of the below mentioned partitions were performed for every version control operation type (Add, Update and Delete).

First we define the notion of *main dataset* which can be one of the following:

- The whole dataset, including all the revisions.
- The subset of the commits where the examined commit operation type occurs at least once.
- The subset of the commits where all the commit operations are of the same type.

We partitioned the main dataset into two parts (*first dataset* and *second dataset*) in the following ways:

- Divide the main dataset into two, based on the median of the absolute number of the examined operations. The greater values go into the first dataset, the second dataset is the complementary of the first one considering the main dataset.
- Divide the main dataset into two based on the median of the proportion of the examined operations, with similar division.
- Take the main dataset as the first dataset, and the second dataset as its complementary considering the whole dataset. This division can be defined only if the main dataset is not the whole dataset.

After eliminating those combinations which are not relevant, we ended up with seven combinations for dataset division per commit operation type. All of these are illustrated with the example of operation Add and the assumption that the presence of this operation has positive impact on the maintainability.

**DIV1:** *Take all commits, divide them into two based on the absolute median of the examined operation.* It checks if commits containing high number of operation Add have better effect on maintainability than those containing low number of operation Add.

**DIV2:** *Take all commits, divide them into two based on the relative median of the examined operation.* It checks if the commits in which the proportion of operation Add is high have better effect on maintainability compared to those where the proportion of operation Add is low. To illustrate the difference between DIV1 and DIV2 consider a commit containing 100 operations, 10 of them are Addition (the absolute number is high but the proportion is low) and a commit containing 3 operations, 2 of them are Additions (the absolute number is low, but the proportion is high).

**DIV3:** *The first subset consists of those commits which contain at least one of the examined operations, and the second one consists of the commits without the examined operation.* It checks if commits containing file addition have better effect on the maintainability than those containing no file additions at all.

**DIV4:** *Considering only those commits where at least one examined operation exists, divide them into two based on the absolute median of the examined operation.* This is similar to DIV1 with the exception that those commits which does not contain any Add operation are not considered. This kind of division is especially useful for operation Add, as this operation is relatively rare compared to file modification, therefore this provides a finer grained comparison.

**DIV5:** *Considering only those commits where at least one examined operation exists, divide them into two based on the relative median of the examined operation.* Similar to DIV2; see the previous explanation.

**DIV6:** *The first subset consists of those commits which contain the examined operation only, and the second one consists of the commits with at least one another type of operation.* This checks if commits containing file additions exclusively have better effect on the maintainability compared to those containing at least one non-addition operation. This division is also especially useful in case of file updates.

**DIV7:** *Considering only those commits where all the operations are of the examined type, divide them into two based on the absolute median of the examined operation.* This division is used to find out if it is true that commits which contain more file additions result better maintainability compared to those containing less number of additions. It is especially useful in case of file updates, as most of the commits contain exclusively that operation.

Please note that 2 of the theoretically possible 9 divisions were eliminated because they always yield trivial divisions (100% - 0%):

- All commits and its complementary. The complementary of all commits is always empty.

– Relative median division of commits containing the examined operation only. In these cases the proportion of the examined operation is always 100%, therefore one of the 2 datasets would be empty.

Table 1 illustrates these divisions.

**Table 1.** Divisions

|  | Complementary | Absolute Median | Relative Median |
|---|---|---|---|
| **All Commits** | - | DIV1 | DIV2 |
| **Operation Exists** | DIV3 | DIV4 | DIV5 |
| **Operation Exclusive** | DIV6 | DIV7 | - |

The tests were executed on all of these combinations during the experiment. In case of median divisions, if the median was ambiguous, both cases were tested (checking into which subset these elements should be added), and the better division (the more balanced division) is taken. In order to present the result in concise format the aforementioned exponent values were calculated for every possible combination and they were summed per system and operation. The mathematical background behind the addition is based on the exponents. If the p-values are independent, then the root probability is the product of the original probabilities, in which case the exponent of the resulting value would be approximately the sum of the exponents.

As a result we get a matrix with the version control operations in the rows and analyzed systems in the columns, and an integer value in each cell. We stress that this is only an approximation, fist of all, because the divisions are not independent. However, it is adequate for a quick overview, and for comparing the results of different systems. We also drill down in one case to illustrate how the calculated numbers were aggregated.

## 3.6   Random Checks

To validate the results, a random analysis was performed as well in the following way. We kept the original source control operations data and the values of the quality changes. But we permuted randomly the order of the revisions they were originally assigned to, just like a pack of cards (using the `sample()` R function). We performed randomization several times, permuting the already permuted series and executed the same analysis with the randomized data as with the original to assess the significance of our actual results.

The expected values of the exponents in random case can be derived from the diagram in Figure 2: 80% having 0, $9 - 9\%$ having -1 and 1, $0.9 - 0.9\%$ having -2 and 2, $0.09 - 0.09\%$ having -3 and 3, etc. With other words, the probability of the absolute value of the random exponents being at least 1 is 20%, 2 is 2%, 3 is 0.2%, etc.

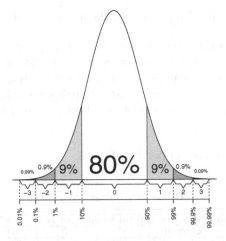

**Fig. 2.** Illustrating the calculated exponent values

As we have 3*7=21 tests per project all together, statistically 21*0.2=4.2 of them would be a non-null value, and 0.42 of them having an absolute value of at least 2. Therefore we set the acceptance criterion for the test that the absolute value of the exponents to be at most 2, which corresponds to the p-value 0.02. As we checked 4 projects (see below), statistically this means that 1 or 2 of the 4*21=84 cases would be false significant.

The expected absolute value in random case is about 1. Based on a check we found that the absolute value is at least 1 in about 66% of the cases, at least 2 in about 24%, at least 3 in about 7%, at least 4 in about 1.7%, at least 5 in about 0.35% of the cases, and so on. Based on this we accept the absolute values 4 and higher as significant.

## 4   Discussion

### 4.1   Examined Software Systems

The analysis was performed on the source code of 4 software systems. One of them was an industrial one, of which we had all the information from the very first commit. The others were open-source ones. Unfortunately, we did not find any open-source project of which we had all the commits from the beginning of the development in the same version control system. In order to gain as adequate results as possible, we considered only those projects for which we had at least 1,000 commits affecting at least one Java file. Furthermore, the too small code increase could also have significant bias, therefore we considered only those systems where the ratio of the maximal logical lines of code (typically the size of the system after the last available commit) and the minimal one (which was typically the size of the initial commit) was at least 3. We found 3 such systems which met these requirements.

Therefore, all together we performed the analysis on the following 4 systems:

- **Ant** – a command line tool for building Java applications[1]
- **Gremon** – a greenhouse work-flow monitoring system.[2] It was developed by a local company between June 2011 and March 2012.
- **Struts 2** – a framework for creating enterprise-ready java web applications.[3]
- **Tomcat** – an implementation of the Java Servlet and Java Server Pages technologies.[4]

Table 2 shows the basic properties of them.

**Table 2.** Analyzed systems

| Name | Min. | Max. | Total | Java | Total number of | | | Rev. with 1+ | | | Rev. with only | | |
|---|---|---|---|---|---|---|---|---|---|---|---|---|---|
| | TLLOC[5] | | Commits | | A | U | D | A | U | D | A | U | D |
| Ant | 2,887 | 106,413 | 6,118 | 6,102 | 1,062 | 20,000 | 204 | 488 | 5,878 | 55 | 196 | 5,585 | 19 |
| Gremon | 23 | 55,282 | 1,653 | 1,158 | 1,071 | 4,034 | 230 | 304 | 1,101 | 89 | 42 | 829 | 8 |
| Struts 2 | 39,871 | 152,081 | 2,132 | 1,452 | 1,273 | 4,734 | 308 | 219 | 1,386 | 94 | 41 | 1,201 | 12 |
| Tomcat | 13,387 | 46,606 | 1,330 | 1,292 | 797 | 3,807 | 485 | 104 | 1,236 | 77 | 32 | 1,141 | 23 |

## 4.2   Summarized Results of the Wilcoxon Tests

The results of the methodology introduced in Section 3.4 are shown in Table 3. The absolute number reflects the magnitude of the impact, while the sign gives the direction (maintainability increase or decrease). Figure 3 illustrates the same results as follows: the upper light gray bars represent the file additions, the lower darker gray bars the file updates, and the black vertical lines the file deletions. The file additions are all located on the positive part, the file updates on the negative, and deletions are hectic, with lower absolute values.

**Table 3.** Sum of the exponents

| | Gremon | Ant | Struts 2 | Tomcat |
|---|---|---|---|---|
| Add | 5 | 62 | 20 | 14 |
| Update | -11 | -29 | -11 | -3 |
| Delete | 4 | -12 | -6 | 1 |

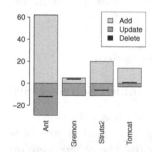

**Fig. 3.** Exponents with bars

These results cannot be interpreted on their own, they only provide a rough overview. The divisions are not independent; furthermore, in some cases the exactly same divisions are checked several times.

[1] http://ant.apache.org
[2] http://www.gremonsystems.com
[3] http://struts.apache.org/2.x
[4] http://tomcat.apache.org
[5] Total Logical Lines Of Code – Number of non-comment and non-empty lines of code

## 4.3   Wilcoxon Tests Details

For details on the above numbers consider Table 4. The sum of the rows are also shown, which helps us drawing the attention on the most promising results. We recall the probabilities in random case (see Subsection 3.6), to illustrate the magnitude of the numbers in the first 3 columns.

**Table 4.** Exponent details

| Operation | Division | Gremon | Ant | Struts 2 | Tomcat | $\sum$ |
|-----------|----------|--------|-----|----------|--------|--------|
| Add       | DIV1     | 1      | 15  | 6        | 4      | 26     |
|           | DIV2     | 1      | 15  | 6        | 4      | 26     |
|           | DIV3     | 1      | 15  | 6        | 4      | 26     |
|           | DIV4     | 0      | 7   | 1        | 1      | 9      |
|           | DIV5     | 1      | 1   | 0        | 0      | 2      |
|           | DIV6     | 1      | 6   | 1        | 1      | 9      |
|           | DIV7     | 0      | 3   | 0        | 0      | 3      |
| Update    | DIV1     | -2     | 2   | 0        | 0      | 0      |
|           | DIV2     | -2     | -11 | -3       | -1     | -17    |
|           | DIV3     | -2     | -4  | 0        | -1     | -7     |
|           | DIV4     | -1     | 2   | 0        | 1      | 2      |
|           | DIV5     | -1     | -8  | -4       | -1     | -14    |
|           | DIV6     | -2     | -11 | -3       | -1     | -17    |
|           | DIV7     | -1     | 0   | -1       | 0      | -2     |
| Delete    | DIV1     | 0      | -1  | 0        | 0      | -1     |
|           | DIV2     | 0      | -1  | 0        | 0      | -1     |
|           | DIV3     | 0      | -1  | 0        | 0      | -1     |
|           | DIV4     | 0      | 0   | -1       | 0      | -1     |
|           | DIV5     | 1      | -2  | -3       | 0      | -4     |
|           | DIV6     | 3      | -5  | -2       | 0      | -4     |
|           | DIV7     | NA     | -2  | 0        | 1      | -1     |

The diagrams in Figure 4 illustrate the results of the Wilcoxon rank tests visually, where the values found in the summary column of Table 4 are illustrated. High absolute length of a bar means high significance within the project. Comparison is also interesting between the projects: high absolute lengths on the same place are considered as a strong result.

In case of operation Add (left bars, light gray) all of the bars are non-negative for every system. The bars related to DIV1, DIV2 and DIV3 are the tallest, and in 3 out of the 4 cases the bars for DIV4, DIV5, DIV6 and DIV7 are similar.

The bars for operation Update (middle bars, dark gray) are a bit more hectic; in general we can say that the height of most of the bars are negative. Furthermore, in case of DIV2, DIV5 and DIV6 we have long negative bars.

The hectic results of operation Delete (right bars, black) are also illustrated. Now let us check the results in the tables.

**Addition.** The results in the first 3 divisions (DIV1, DIV2, and DIV3) are in all cases the same, because addition exists in less than half of the commits (see the definitions in Section 3). The overall result of the Wilcoxon test (26) is very high for these divisions, the highest absolute value in the table. On 3 out of the 4 projects the test yielded significant result (exponent $\geq 2$). This definitely means that *commits containing additions have better effect on the maintainability compared to those containing no file additions.*

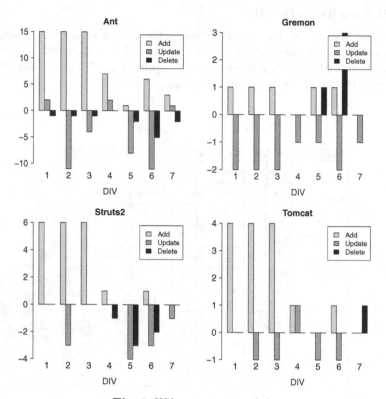

**Fig. 4.** Wilcoxon test result bars

The result of DIV4 is also relatively high (9), however, this is caused by a high value for one project, with less support of the others. This means *that for commits containing an addition, in some cases the higher absolute number of addition results better maintainability, comparing with the commits of lower number of additions.* It is interesting that this is not the case if the proportion of additions in taken (DIV5 with result of 2): *higher proportion of file addition does not result in significantly better maintainability.*

DIV6 checks if commits containing exclusively file additions have significantly different effect on maintainability compared to those containing other operations as well. The overall result (9) is remarkably high; however, the result is modulated by the fact that in case of 3 projects the connection is weak. The reason could be the low number of commits containing file additions only (the p-value is affected also by the number of elements: if the same result is supported by a higher number of elements, the p-value is lower). The result of DIV7 tells that if all file operations are file additions, then the connection between the number of operations and the maintainability is weak.

**Update.** Unlike in case of file addition, file update results of the first 3 divisions are quite different. Based on DIV1 (final result is 0) there is no difference between the effects of the commits with low and high absolute number of Update operations on maintainability. However, it contains a contradiction: besides having 2 zeroes, the result contains a +2 (meaning that the high number of Updates significantly improves the maintainability) and a -2 (meaning it significantly decreases). We have not found the reason of this contradiction, possibly further investigations of other data is necessary.

On the other hand, DIV2 provides a very significant result (-17), and this is supported by every analyzed system with varying degree. This suggests that the proportion of operation Update really matters from maintainability point of view. In general, the higher the proportion of the operation Update within a commit, the worse its effect on the maintainability. DIV7 resulted in exactly the same values, because the divisions were the same: the commit either contains exclusively update or contains other operations as well.

Based on DIV3 we can say that the mere existence of operation Update has a negative effect on maintainability (comparing with those not containing any Update). This was significant in 3 out of the 4 systems, with no contradiction on the 4th. DIV4 is similar to DIV1 (absolute median division of those commits which contain at least one Update) with similar low significance and small contradiction. DIV5 is similar to DIV2 (relative median division of those commits which contain at least one Update) with similar but lower exponents.

The result of DIV6 (-17) is significant, which is supported by most of the checked systems. This means that in general, the presence of operation Update has a negative effect on the maintainability. DIV7 is similar to DIV1 or DIV4 (absolute median division of commits containing exclusively Update), and the result is not significant at all.

The Update operation has negative effect on maintainability, but the way how it appears (alone or together with other operation) really matters. It seems that the presence of other operations suppresses the effect of the update.

**Delete.** The effect of the operation Delete seems a bit contradictory. In case of Ant and Struts 2 we have non-positive results only. In case of Gremon and Tomcat the values are non-negative but with lower absolute values than the others. The NA means not available; in case of Gremon there were not enough commits which contained exclusively Delete operations and we could not perform a division based on the number of operations so that both sets would contain enough number of elements to be able to compare. There were 8 such commits, and we executed the test only if at least 5 elements in both subset existed.

The highest absolute values can be found in case of DIV6, meaning that there is a significant difference in the effect on the maintainability between commits containing exclusively Delete operations compared to those containing other operations as well, but the values are very contradictory. In 2 out of the 4 cases the results suggest that deletion significantly decreases the maintainability. This is a bit strange because it suggests that it is more likely that the more maintainable code is removed than those harder to maintain. Deletion could typically occur in case of refactoring, and we would expect that the hard-to-maintain code is removed and better-to-maintain code appears instead, but it seems that this is not the case.

On the other hand, in case of Gremon just the opposite is true with relatively high confidence. We have not found any explanation to this contradiction, and we cannot be certain that the reason is the fact that Gremon is an industrial software, implemented by paid programmers, while the others are not, implemented by volunteers, or something entirely different.

### 4.4     Answers to the Research Questions

**RQ1:** *Does the amount of file additions, updates and deletions within a commit impact the maintainability of the source code?*

Consider Table 3. For interpretation of the magnitude of these values please consider the random probabilities in Section 3.6. In case of operation Add all the values are positive, and all of them can be considered to be significant. Therefore, we can state that operation Add has positive impact on the maintainability. In case of operation Update all the values are negative. The absolute value of one of them (-3 in case of Tomcat) is relatively low which would not be convincing in itself. However, along with the others we can state that operation Update has negative impact on the maintainability. In case of operation Delete we have 2 positive and 2 negative results, containing low and high absolute values as well. Considering these data only we cannot formulate a valid statement for this operation.

**RQ2:** *Are there any differences between checks considering the absolute number of operations (Add, Update, Delete) and checks investigating the relative proportion of the same operation within commits?*

Consider Table 4. The values found in DIV1 (absolute median) should be compared with DIV2 (relative median) and those in DIV4 with DIV5. In case of operation Add there is no difference between DIV1 and DIV2. In case of comparing DIV4 with DIV5 we find that the values in the sum column are 9 and 2, respectively. This seems to be a good result at first glance; however, this is caused by only one value, and all the other 7 values are not significant. Therefore, based on these values only we cannot formulate anything for operation Add. In case of operation Update the relative median (DIV2 and DIV5) results in significantly lower values than those of absolute median (DIV1 and DIV4). Therefore, we can state that in case of Update the high proportion of the operation causes the maintainability decrease, rather than the absolute number of it. In case of operation Delete we again cannot formulate any statement.

## 5     Threats to Validity

The facts which might potentially threaten some of the validity of the results are the following.

The data are not fully complete, which is true for many of the researches of course; ours is not an exception either. The commit data for the Gremon project is complete: all the commits are available from the very beginning. On the other hand, large amount of data in case of open-source projects are missing. E.g., in case of every project the initial commit contained a large amount of development which came from another version control system. To alleviate this problem we chose projects having plenty enough code enhancements during the development. In some cases a lot of development was done in another branch, and this appears as a huge merge in the examined branch. This could also have significant bias.

The results of the different systems show similar tendency in most of the cases; however, in some cases the results are diverging. This is especially true for operation Delete. We have not found the reason of these divergences.

There is no definition to the quality of the source code expressed by a number. The ColumbusQM tool used by us is one of several approaches, with its own advantages and drawbacks. We find this model as a good, well founded one, but we are aware that it is not perfect: it is being improved continuously. We treat this as an external threat and hope that more and more precise information about the software quality will support the results with higher confidence instead of threatening it.

# 6    Conclusions and Future Work

This research is part of a longer term study, aimed to identify the patterns of the developers' behavior which causes significant impact on the source code quality.

We studied the impact of version control commit operations on the maintainability change. Only the bare number of operations was considered, nothing else. We received some interesting answers, which are the following.

We found that file additions have positive, or at least better impact on maintainability, compared to the effect of those commits containing no or small number of additions.

On the other hand, the research showed that file updates have significant negative effect on the maintainability.

We found no clear connection between file deletions and maintainability. Based on this research the net effect of this operation is rather negative than positive, which contradicts with our initial assumption.

We identified the similarities and the differences between the high number and the high proportion of existence of a version control operation within commits. Later on this fact might also be an important part of the formula which will hopefully explain the influence of the developer's interactions on the source code quality.

Answers to these any maybe many other questions might help improving the knowledge where exactly the software erosion decreases.

**Acknowledgments.** This research was supported by the Hungarian national grant GOP-1.1.1-11-2011-0006, and the European Union and the State of Hungary, co-financed by the European Social Fund in the framework of TÁMOP 4.2.4. A/2-11-1-2012-0001 „National Excellence Program".

# References

1. Atkins, D.L., Ball, T., Graves, T.L., Mockus, A.: Using Version Control Data to Evaluate the Impact of Software Tools: A Case Study of the Version Editor. IEEE Transactions on Software Engineering 28(7), 625–637 (2002)
2. Bakota, T., Hegedűs, P., Körtvélyesi, P., Ferenc, R., Gyimóthy, T.: A Probabilistic Software Quality Model. In: Proceedings of the 27th IEEE International Conference on Software Maintenance (ICSM 2011), pp. 368–377. IEEE Computer Society, Williamsburg (2011)
3. Bakota, T., Hegedus, P., Ladányi, G., Körtvélyesi, P., Ferenc, R., Gyimóthy, T.: A Cost Model Based on Software Maintainability. In: Proceedings of the 28th IEEE International Conference on Software Maintenance (ICSM 2012), pp. 316–325. IEEE Computer Society, Riva del Garda (2012)
4. Bernstein, A., Bachmann, A.: When Process Data Quality Affects the Number of Bugs: Correlations in Software Engineering Datasets. In: Proceedings of the 7th IEEE Working Conference on Mining Software Repositories, MSR 2010, pp. 62–71 (2010)
5. Fluri, B., Wursch, M., Gall, H.C.: Do Code and Comments Co-evolve? On the Relation between Source Code and Comment Changes. In: 14th Working Conference on Reverse Engineering, WCRE 2007, pp. 70–79. IEEE (2007)
6. Gall, H., Jazayeri, M., Krajewski, J.: CVS Release History Data for Detecting Logical Couplings. In: Proceedings of the Sixth International Workshop on Principles of Software Evolution, pp. 13–23. IEEE (2003)
7. Hayes, J.H., Patel, S.C., Zhao, L.: A Metrics-Based Software Maintenance Effort Model. In: Proceedings of the Eighth Euromicro Working Conference on Software Maintenance and Reengineering (CSMR 2004), pp. 254–260. IEEE Computer Society, Washington, DC (2004)
8. Hindle, A., German, D.M., Holt, R.: What Do Large Commits Tell Us?: a Taxonomical Study of Large Commits. In: Proceedings of the 2008 International Working Conference on Mining Software Repositories, MSR 2008, pp. 99–108. ACM, New York (2008)
9. Hollander, M., Wolfe, D.A.: Nonparametric Statistical Methods, 2nd edn. Wiley-Interscience (January 1999)
10. ISO/IEC: ISO/IEC 9126. Software Engineering – Product quality 6.5. ISO/IEC (2001)
11. Koch, S., Neumann, C.: Exploring the Effects of Process Characteristics on Product Quality in Open Source Software Development. Journal of Database Management 19(2), 31 (2008)
12. Mockus, A., Weiss, D.M., Zhang, P.: Understanding and Predicting Effort in Software Projects. In: Proceedings of the 25th International Conference on Software Engineering (ICSE 2003), pp. 274–284. IEEE Computer Society, Washington, DC (2003)
13. Moser, R., Pedrycz, W., Sillitti, A., Succi, G.: A Model to Identify Refactoring Effort during Maintenance by Mining Source Code Repositories. In: Jedlitschka, A., Salo, O. (eds.) PROFES 2008. LNCS, vol. 5089, pp. 360–370. Springer, Heidelberg (2008)
14. Parnas, D.L.: Software Aging. In: Proceedings of the 16th International Conference on Software Engineering, ICSE 1994, pp. 279–287. IEEE Computer Society Press, Los Alamitos (1994)

15. Peters, R., Zaidman, A.: Evaluating the Lifespan of Code Smells using Software Repository Mining. In: Proceedings of the 2012 16th European Conference on Software Maintenance and Reengineering, CSMR 2012, pp. 411–416. IEEE Computer Society, Washington, DC (2012)
16. Pratap, A., Chaudhary, R., Yadav, K.: Estimation of software maintainability using fuzzy logic technique. In: 2014 International Conference on Issues and Challenges in Intelligent Computing Techniques (ICICT), pp. 486–492 (February 2014)
17. R Core Team: R: A Language and Environment for Statistical Computing. R Foundation for Statistical Computing, Vienna, Austria (2013), http://www.R-project.org/
18. Ratzinger, J., Sigmund, T., Vorburger, P., Gall, H.: Mining Software Evolution to Predict Refactoring. In: Proceedings of the First International Symposium on Empirical Software Engineering and Measurement, ESEM 2007, pp. 354–363. IEEE Computer Society, Washington, DC (2007)
19. Schofield, C., Tansey, B., Xing, Z., Stroulia, E.: Digging the Development Dust for Refactorings. In: Proceedings of the 14th IEEE International Conference on Program Comprehension, ICPC 2006, pp. 23–34. IEEE Computer Society, Washington, DC (2006)
20. Stroggylos, K., Spinellis, D.: Refactoring–Does It Improve Software Quality? In: Fifth International Workshop on Software Quality, WoSQ 2007: ICSE Workshops 2007, p. 10. IEEE (2007)
21. Tóth, G., Végh, Á.Z., Beszédes, Á., Schrettner, L., Gergely, T., Gyimóthy, T.: Adjusting Effort Estimation Using Micro Productivity Profiles. In: Proceedings of the 12th Symposium on Programming Languages and Software Tools (SPLST 2011), pp. 207–218 (October 2011)

# A Structured Approach for Eliciting, Modeling, and Using Quality-Related Domain Knowledge

Azadeh Alebrahim, Maritta Heisel, and Rene Meis

Paluno – The Ruhr Institute for Software Technology, Germany
{firstname.lastname}@paluno.uni-due.de

**Abstract.** In requirements engineering, properties of the environment and assumptions about it, called *domain knowledge*, need to be captured in addition to exploring the requirements. Despite the recognition of the significance of capturing and using the required domain knowledge, it might be missing, left implicit, or be captured inadequately during the software development. This results in an incorrect specification. Moreover, the software might fail to achieve its quality objectives because of ignored required constraints and assumptions. In order to analyze software quality properly, we propose a structured approach for eliciting, modeling, and using domain knowledge. We investigate what kind of quality-related domain knowledge is required for the early phases of quality-driven software development and how such domain knowledge can be systematically elicited and explicitly modeled to be used for the analysis of quality requirements. Our method aims at improving the quality of the requirements engineering process by facilitating the capturing and using of implicit domain knowledge.

**Keywords:** Quality requirements, domain knowledge, problem frames, knowledge management, requirements engineering.

## 1 Introduction

The system-to-be comprises the software to be built and its surrounding environment structured as a collection of domains such as people, devices, and existing software [1]. The environment represents the part of the real world into which the software will be integrated. Hence, in requirements engineering, properties of the domains of the environment and assumptions about them, called *domain knowledge*, need to be captured in addition to exploring the requirements [2,3]. Note that we do not mean *application domain* under the term *domain*, but entities in the environment that are relevant.

Despite the recognition of the significance of capturing the required domain knowledge, it might be missing, left implicit, or be captured inadequately during the software development process [1]. Domain knowledge is often undocumented and tacit in the minds of the people involved in the process of software development [4]. The common ad-hoc nature of gaining domain knowledge is error-prone. Hooks and Farry [5] report on a project where 49% of requirements errors were due to incorrect domain knowledge. Capturing inadequate assumptions about the environment of the flight guidance software led to the crash of a Boeing 757 in Colombia in December 1995 [6].

Several requirements engineering methods exist, e.g., for security. Fabian et al. [7] conclude in their survey about these methods that it is not yet state of the art to consider

B. Murgante et al. (Eds.): ICCSA 2014, Part V, LNCS 8583, pp. 370–386, 2014.

domain knowledge. The software development process involves knowledge-intensive activities [8]. It is an open research question of *how* to elicit domain knowledge as part of the software development process correctly for effective requirement engineering [9]. Lamsweerde [1] and Jackson [10] underline the importance of eliciting domain knowledge in addition to the elicitation of requirements to obtain correct specifications. However, there is sparse support in capturing and modeling domain knowledge.

In this paper, we propose a method for capturing implicit and quality-relevant domain knowledge, and making it explicit for reuse in a systematic manner during software development. Our approach consists of a meta-process and an object-process which are structured in the steps eliciting, modeling, and using domain knowledge. Both processes are independent from any specific tool or notation. This facilitates the integration of the processes into requirements analysis and design processes. The meta-process is applied for a given software quality together with a quality analysis method only once to define how to elicit, model, and use the relevant domain knowledge for the specific software quality and the given analysis method. Results of previous applications of the meta-process for the same software quality together with a different analysis method can be reused. The object-process is applied for a given software project. The domain knowledge is elicited, modeled, and used using the principles that are output of the meta-process for the software quality and quality analysis method under consideration.

We illustrate the application of the meta-process using three quality analysis methods that were already developed for eliciting, modeling and using quality-relevant domain knowledge. These methods are the Quality Requirements Optimization (QuaRO[1]) method [11], which analyzes and detects interactions between security and performance requirements based on pairwise comparisons, the Problem-Oriented Performance Requirements Analysis (POPeRA) method [12], which identifies and analyzes potential performance problems, and the Problem-based Privacy Analysis (ProPAn) method [13,14], which identifies privacy threats on the requirements analysis level. We will illustrate the object-process using a smart grid scenario as given application and our three methods as output of the meta-process.

The benefit of our method lies in improving the quality of the requirements engineering process. This is achieved by providing a systematic method that facilitates the capturing and modeling of implicit domain knowledge as reusable artifacts.

In the following, Sect. 2 introduces the smart grid scenario and Sect. 3 the background of our work. Sections 4 and 5 describe the meta- and object-process, which are our main contributions. Sect. 6 discusses related work, and Sect. 7 concludes.

## 2    Introducing the Smart Grid Application

In this section, we introduce the real-life case study "smart grids" adapted from the NESSoS project[2]. To use energy in an optimal way, smart grids make it possible to couple the generation, distribution, storage, and consumption of energy. Smart grids use

---

[1] The QuaRO method is a comprehensive method for optimizing requirements according to stakeholders' goals. In this paper, we only focus on the part concerning requirements interaction detection.

[2] http://www.nessos-project.eu/

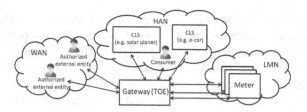

**Fig. 1.** The context of a smart grid system based on [15]

**Table 1.** An excerpt of relevant terms for the smart grid

| | |
|---|---|
| **Gateway** | represents the central communication unit in a *smart metering system*. It is responsible for collecting, processing, storing, and communicating *meter data*. |
| **Meter data** | refers to readings measured by the meter regarding consumption or production of a certain commodity. |
| **Meter** | represents the device that measures the consumption or production of a certain commodity and sends it to the gateway. |
| **Authorized external entity** | could be a human or IT unit that communicates with the gateway from outside the gateway boundaries through a *WAN*. The roles defined as external entities that interact with the gateway and the meter are, for example, *consumer, grid Operator, supplier, gateway operator, a gateway administrator*. |
| **WAN** | (Wide Area Network) provides the communication network that connects the gateway to the outside world. |
| **Consumer** | refers to the end user or producer of commodities (electricity, gas, water, or heat). |
| **CLS** | (Controllable Local Systems) are systems containing IT-components in the Home Area Network (HAN) of the consumer that do not belong to the Smart Metering System but may use the Gateway for dedicated communication purposes. |

information and communication technology (ICT), which allows for financial, informational, and electrical transactions. For the smart grid, different quality requirements have to be taken into account. Detailed information about the energy consumption of the consumers can reveal privacy-sensitive data about the persons staying in a house. Hence, we are concerned with privacy issues. A smart grid involves a wide range of data that should be treated in a secure way. Additionally, introducing new data interfaces to the grid (smart meters, collectors, and other smart devices) provides new entry points for attackers. Therefore, special attention should be paid to security concerns. The number of smart devices to be managed has a deep impact on the performance of the whole system. This makes performance of smart grids an important issue. Figure 1 shows the context of a smart grid system based on a protection profile that was issued by the Bundesamt für Sicherheit in der Informationstechnik [15]. First, define some terms specific to the smart grid domain taken from the protection profile represented in Table 1.

Due to space limitations, we focus in this paper on the functional requirement *"The smart meter gateway shall submit processed meter data to authorized external entities. (RQ4)"*, security requirements *"Integrity (RQ10)/Confidentiality (RQ11)/Authenticity (RQ12) of data transferred in the WAN shall be protected"*, performance requirement *"The time to retrieve meter data from the smart meter and publish it through WAN shall be less than 5 seconds. (RQ24)"*, and privacy requirement *"Privacy of the consumer data shall be protected while the data is transferred in and from the smart metering system. (RQ17)"*. We derived these requirements from the protection profile [15].

# 3    Problem-Oriented Requirements Engineering

This section outlines basic concepts of the problem frames approach proposed by Michael Jackson [10]. We illustrate our process using quality analysis methods that are based on problem frames as requirements engineering method. Note that our proposed process can also be applied using any other requirements engineering approach.

Requirements analysis with problem frames proceeds as follows: to understand the problem, the environment in which the *machine* (i.e., software to be built) will operate must be described first. To this end, we set up a *context diagram* consisting of *machines*, *domains* and *interfaces*. Domains represent parts of the environment which are relevant for the problem at hand. Then, the problem is decomposed into simple subproblems that fit to a *problem frame*. Problem frames are patterns used to understand, describe, and analyze software development problems. An instantiated problem frame is a *problem diagram* which basically consists of a *submachine* of the machine given in the context diagram, relevant domains, interfaces between them, and a *requirement*. The task is to construct a *(sub-)machine* that improves the behavior of the environment (in which it is integrated) in accordance with the requirement.

We describe problem frames using UML class diagrams [16], extended by a specific UML profile for problem frames (UML4PF) proposed by Hatebur and Heisel [17]. A class with the stereotype ≪machine≫ represents the software to be developed. Jackson distinguishes the domain types biddable domains (represented by the stereotype ≪BiddableDomain≫) that are usually people, causal domains (≪CausalDomain≫) that comply with some physical laws, and lexical domains (≪LexicalDomain≫) that are data representations. To describe the problem context, a *connection domain* (≪ConnectionDomain≫) between two other domains may be necessary. Connection domains establish a connection between other domains by means of technical devices. Figure 2 shows the problem diagram for the functional requirement *RQ4*. It describes that smart meter gateway submits meter data to an authorized external entity. The submachine *SubmitMD* is one part of the smart meter gateway. It sends the *MeterData* through the causal domain *WAN* to the biddable domain *AuthorizedExternalEntity*. When we state a requirement we want to change something in the world with the machine to be developed. Therefore, each requirement expressed by the stereotype ≪requirement≫ constrains at least one domain. This is expressed by a dependency from the requirement to a domain with the stereotype ≪constrains≫. A requirement may refer to several domains in the environment of the machine. This is expressed by a dependency from the requirement to these domains with the stereotype ≪refersTo≫. The requirement *RQ4* constrains the domain *WAN*, and it refers to the domains *MeterData* and *AuthorizedExternalEntity*.

In the original problem frames approach, the focus is on functional requirements. We extended the UML-based problem frames approach by providing a way to attach quality requirements to problem diagrams [18]. We represent quality requirements as annotations in problem diagrams. Since UML lacks notations to specify and model quality requirements, we use specific UML profiles to add annotations to the UML models. We use a UML profile for dependability [17] to annotate problem diagrams with security requirements. For example, we apply the stereotypes ≪integrity≫, ≪confidentiality≫, and ≪authenticity≫ to represent integrity, confidential-

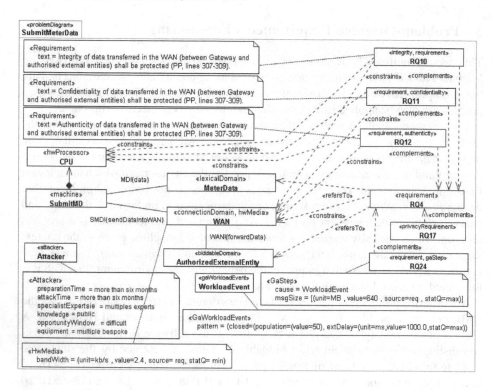

**Fig. 2.** Problem Diagram for submitting meter data to external entities

ity, and authenticity requirements as it is illustrated in Figure 2. To annotate privacy requirements, we use the privacy profile [13] that enables us to state that the privacy of a stakeholder shall be preserved against a counterstakeholder using the stereotype ≪privacyRequirement≫. To provide support for annotating problem descriptions with performance requirements, we use the UML profile MARTE (Modeling and Analysis of Real-time and Embedded Systems) [19]. We annotate each performance requirement with the stereotype ≪gaStep≫ to express a response time requirement. In the problem diagram for submitting meter readings (Figure 2), the functional requirement *RQ4* is complemented by the following quality requirements: *RQ10* (integrity), *RQ11* (confidentiality), *RQ12* (authenticity), *RQ17* (privacy), and *RQ24* (performance).

## 4   Structured Meta-Process for Eliciting, Modeling, and Using Domain Knowledge

This section describes the meta-process composed of three steps for eliciting, modeling, and using domain knowledge for a specific software quality shown in Figure 3. The starting point is the domain expertise that exists for the software quality at hand. The meta-process is conduted for a specific software quality optionally together with a quality analysis method. Once we have elicited and modeled domain knowledge for a

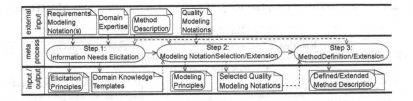

**Fig. 3.** Meta-process for eliciting, modeling, and using domain knowledge

specific software quality in steps one and two, we use it in step three by extending the given quality analysis method or defining a new one that uses the elicited and modeled domain knowledge for analyzing quality requirements. The modularity of the meta-process allows us to reuse the outputs from previous applications of the meta-process for a meta-process that considers the same software quality. We consider three methods *QuaRO* [11], *POPeRA* [12], and *ProPAn* [13], as mentioned in Section 1. The artifacts in the top of Figure 3 represent the external inputs for the steps of the method. Those in the bottom of the figure represent the output of the steps providing input for further steps and/or for the object-process (see Section 5). The rounded rectangles in the input and output notes indicate that the artifact describes a procedure that later can be executed. In the following, we describe each step of the meta-process followed by its application to the software qualities performance, security, and privacy.

**Step 1: Information Needs Elicitation.** This step is concerned with eliciting the relevant information that has to be collected when dealing with specific software qualities. The source of information is the *domain expertise*. This expertise can stem from *"developers of the software product, software engineering textbooks and other types of documentation, external information which is available to the public (e.g. IEEE standards), existing documents in the organization, formal or informal (guidelines, books, procedures, manuals), results of empirical work (controlled experiments, case studies), and published experience reports (e.g. lessons learnt)"* [20]. Optionally, a *requirements modeling notation* that is used in the application, and a *method description* of a quality analysis method that shall be extended could be further inputs if existing and needed.

We extract the information needs for the software quality from the domain expertise with respect to the optionally given method description. We document these information needs as structured templates called *Domain Knowledge Templates*. These templates have later to be instantiated in the first step of the object-process. In addition to the domain knowledge templates, the output of step 1 provides guidance how to elicit relevant domain knowledge systematically. We call such guidance *Elicitation Principles*. An optionally given requirements modeling notation can be used to be referred to in the elicitation principles. Domain knowledge templates represent *what* domain knowledge has to be elicited, and elicitation principles represent *how* this domain knowledge has to be elicited. Elicitation principles are used as an input for the first step of the object-process and describe how this step has to be carried out.

In the following, we show applications of the meta-process for the software qualities performance, security, and privacy. In all shown applications, we use the UML-based problem frames (see Section 3) as the requirements modeling notation.

**Table 2.** Domain knowledge template for performance and mapping to the MARTE profile

| Quality: Performance | | | | |
|---|---|---|---|---|
| Domain Knowledge Template | | | Mapping to MARTE | |
| Domain Knowledge Description | Possible Values | | Value | Property |
| For each Problem Diagram | | | | |
| Number of concurrent users | Natural | | | GaWorkloadEvent. pattern. population |
| Arrival pattern | ArrivalPattern | | | GaWorkloadEvent. pattern |
| Data size | DataSize (bit, Byte, KB, ...) | | | GaStep. msgSize |
| For each Causal Domain | | | | |
| Memory | capacity | DataSize (bit, Byte, KB, ...) | | HwMemory. memorySize |
| | latency | Duration (s, ms, min,hr, day) | | HwMemory. timing |
| Network | bandwidth | DataRate (b/s, Kb/s, Mb/s) | | HwMedia. bandWidth |
| | latency | Duration (s, ms, min,hr, day) | | HwMedia. packetTime |
| CPU | speed | Frequency (Hz, kHz, MHz, GHz) | | HwProcessor. frequency |
| | Number of cores | Natural | | HwProcessor. nbCores |

*Applying step 1 for performance.* To elicit the domain knowledge that performance analysts require to analyze performance for early software development phases (POPeRA and QuaRO methods), we make use of the *Domain Expertise* presented by Bass et al. [21,22]. Performance is concerned with the *workload* of the system and the available *resources* to process the workload [21]. The workload is described by triggers of the system, representing requests from outside or inside the system. Workload exhibits the characteristics of the system use. It includes the number of concurrent users and their arrival pattern. The arrival pattern can be periodic (e.g. every 10 milliseconds), stochastic (according to a probabilistic distribution), or sporadically (not to capture by periodic or stochastic characterization) [22]. Processing the requests requires resources. Each resource has to be described by its type in the system, such as CPU, memory, and network, its utilization, and its capacity, such as the transmission speed for a network.

The developed *Domain Knowledge Template* for performance is shown in Table 2. The columns "Domain Knowledge Description" and "Possible Values" show the domain knowledge to be elicited for performance and its possible values. The column "Value" has to be filled out in the first step of the object-process. Once we have captured the information needs for performance analysis as domain knowledge templates, we have to give guidance how to elicit them (*Elicitation Principles*). The first part of the domain knowledge template for performance contains information relevant for each problem diagram that contains a performance requirement. The second part of the template shows the domain knowledge to be elicited for each causal domain that is part of a problem diagram with a performance requirement. We iterate over these causal domains in the requirement models (lexical and machine domains are special types of causal domains). For each domain, we have to check if it represents or contains any hardware device that the system is executed on or any resource that can be consumed by the corresponding performance requirement. If this is the case, the resource has to be specified as indicated in the domain knowledge template. For each resource type (CPU, network, memory), we have to state if the resource is already existing in the requirement models or it is not modeled yet.

*Applying step 1 for security.* To guarantee security, we need domain knowledge about the type of possible attackers that influence the restrictiveness of a security

**Table 3.** Domain knowledge template for security and mapping to the dependability profile

| Quality: Security | | | |
|---|---|---|---|
| **Domain Knowledge Template** | | | **Mapping to profile** |
| **Domain Knowledge Description** | **Possible Values** | **Value** | **Property    (Dependability profile)** |
| Preparation time | one day, one week, two weeks, … | | Attacker.preparationTime |
| Attack time | one day, one week, two weeks, … | | Attacker.attackTime |
| Specialist expertise | laymen, proficient, expert, … | | Attacker.specialistExpertise |
| Knowledge of the TOE | public, restricted, sensitive, critical | | Attacker.knowledge |
| Window of opportunity | unnecessary/unlimited, easy, … | | Attacker.opportunity |
| IT hardware/software or other equipment | standard, specialized, bespoke, … | | Attacker.equipment |

requirement. Different types of attackers can be considered. For example, a software attacker targets at manipulating the software, whereas a network attacker aims at manipulating the network traffic. To describe the attacker we use the properties as described by the Common Methodology for Information Technology Security Evaluation (CEM) [23] (*Domain Expertise*) for vulnerability assessment of the TOE (target of evaluation i.e., system-to-be). The properties to be considered (according to CEM) are given in the *Domain Knowledge Template* shown in the first column of Table 3.

Now, we describe the *Elicitation Principles* that support us in capturing domain knowledge. One attacker should be identified for each modeled security requirement. It should be checked if such an attacker for each security requirement exists. If not, we have to identify a suitable attacker according to the related security requirement. The domain knowledge template has to be instantiated for each attacker.

*Applying step 1 for privacy.* One limitation of the Problem-based Privacy Analysis method (ProPAn) [13] (*Method Description*) is that it can only reason about stakeholders and counterstakeholders that are part of the requirement model. But often stakeholders of personal information are not directly interacting with a software and hence maybe overlooked by the requirements engineer. Thus, there is the need to elicit privacy relevant domain knowledge, namely, the indirect (counter)stakeholders of the domains of the context diagram. Therefore, we performed the meta-process (see Figure 3) for the ProPAn-method (method description). Due to the fact that ProPAn is based on the problem frames approach, we have it as a requirements modeling notation in the meta-process. Based on the stakeholder analysis literature [24,25] (*Domain Expertise*), we developed domain knowledge templates in the form of extensible questionnaires [14]. The questionnaires shall help the requirements engineer in cooperation with domain experts to identify indirect (counter)stakeholders and indirect or implicit relationships between domains. An indirect stakeholder is a biddable domain that is not already part of the context diagram and from whom personal information is possibly stored or processed in the system-to-be. An indirect counterstakeholder is also a biddable domain that is not already part of the context diagram and may gain information about entities through a domain of the context diagram. Table 4 contains the questionnaire for causal and lexical domains. The *Elicitation Principle* means that one has to answer for each domain of the requirement model the questions of the corresponding questionnaire.

**Table 4.** Domain knowledge elicitation questionnaire for causal and lexical domains

| No. | Question |
|---|---|
| 1 | **Elicitation of Counterstakeholders** |
| 1.1 | Is there a competitor that also uses the domain? |
| 1.2 | Could the domain be attacked by a hacker? |
| 1.3 | Does the domain provide information to legislators or law enforcement agencies? |
| 1.4 | Is the domain also used in other systems? State possible counterstakeholders that have access to the domain in these systems. |
| 2 | **Elicitation of Stakeholders** |
| 2.1 | Is the domain also used in other systems? State possible stakeholders of these systems from whom information is accessible through the domain. |
| 2.2 | Is initially personal information of stakeholders stored in the domain? |
| 2.3 | Does the domain store or process personal information of stakeholders directly, indirectly, or implicitly connected to it? |

**Step 2: Modeling Notation Selection/Extension.** The aim of this step is to select a suitable notation for modeling quality-relevant domain knowledge in a way that it can be used for the requirements analysis and integrates into the optionally given requirements modeling notation. According to the elicited domain knowledge from the previous step, we investigate whether there are existing *Quality Modeling Notations* for the software quality that are sufficient for integrating the domain knowledge into the existing requirement models. In such a case, we select an appropriate notation. Otherwise, we have to extend existing notations with required artifacts or define a new one. The *Selected Quality Modeling* notation will be applied to the requirement models in the object-process in order to support the modeling of domain knowledge. In addition, we obtain *Modeling Principles* as output of this step. They provide guidance for the translation of the domain knowledge elicited in the domain knowledge templates of the previous step into the selected quality modeling notation.

*Applying step 2 for performance* We selected the UML profile for Modeling and Analysis of Real-Time and Embedded systems (MARTE) [19] adopted by OMG consortium for modeling performance-related domain knowledge (*Selected Quality Modeling Notation*). It extends the UML modeling language to support modeling of performance and real-time concepts. They make use of the stereotypes from the MARTE profile to express the domain knowledge that was elicited in the first step of our method. Expressing domain knowledge as stereotypes helps us to integrate domain knowledge in the existing requirement models. In the case that we are concerned with a hidden resource, the hidden resource has to be modeled explicitly as a causal domain. It additionally has to be annotated with a performance relevant stereotype from the MARTE profile representing the kind of resource it provides. The column "Mapping to MARTE" in Table 2 shows how the domain knowledge elicited in step one can be mapped to the MARTE stereotypes and attributes (*Modeling Principles*).

*Applying step 2 for security.* We chose the dependability profile (*Selected Quality Modeling Notation*) proposed by Hatebur and Heisel [17] for modeling security-related domain knowledge identified in the previous step. We make use of the stereotype ≪attacker≫ and its attributes to express the attackers and their characteristics. Each identified attacker has to be modeled explicitly as a biddable domain. The stereotype ≪attacker≫ has to be applied to it. The attacker is then assigned to the

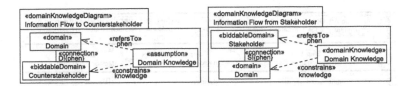

**Fig. 4.** Domain knowledge patterns for the modeling of privacy-relevant domain knowledge

corresponding security requirements which are also provided by the dependability profile. The column "Mapping to profile" in Table 3 shows how the elicited security-specific domain knowledge can be integrated in the requirement models using the stereotypes and attributes from the dependability profile (*Modeling Principles*).

*Applying step 2 for privacy.* For the domain knowledge extension of ProPAn as presented in [14], we do not need to select or extend a *Quality Modeling Notation*. We are able to represent the possible information flows stemming from indirect stakeholders and possible access relationships of counterstakeholders using domain knowledge diagrams, which are already provided by the UML4PF profile. To assist the modeling process of the privacy-relevant domain knowledge, we identified two *"domain knowledge frames"* (shown in Figure 4) for the two types of questions in the questionnaire. These can easily be instantiated using the answers of the questionnaires. The Domain is instantiated with the domain for which the questionnaire was answered and the (Counter)Stakeholder with the elicited (counter)stakeholder (*Modeling Principles*).

**Step 3: Method Definition/Extension.** This step aims at defining or extending a method for quality requirements analysis. The methods QuaRO and POPeRA are examples for newly defined methods, where quality relevant domain knowledge has to be considered from the beginning of the analysis process. The ProPAn method is an example for extending an existing method with quality-relevant domain knowledge. In case of extending an existing method, the *Method Description* has to be considered as input. Additionally, we take the selected quality modeling notations into account for defining a new method or extending an existing one.

*Applying step 3 for performance.* We defined the POPeRA method [12] (*Defined Method Description*) for detecting and analyzing potential performance problems. The method first identifies performance-critical resources using the modeled performance-relevant domain knowledge. Next, it identifies problem diagrams, where the inbound requests exceed the processing capacity of the performance-critical resource because of a high workload. These resources represent potential bottlenecks.

*Applying step 3 for security and performance.* The QuaRO method (*Defined Method Description*) uses the structure of problem diagrams to identify the domains, where quality requirements might interact. When the state of a domain can be changed by one or more sub-machines at the same time, their related quality requirements might be in conflict. Modeling domain knowledge regarding security and performance allows us to detect additional domains, where security and performance might conflict. Resources modeled by domain knowledge represent such conflicting domains. The reason

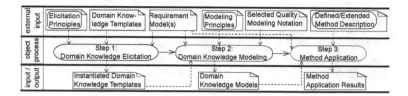

**Fig. 5.** Object-process for eliciting, modeling, and using domain knowledge

is that the achievement of security requirements requires additional resources affecting the achievement of performance requirements negatively. Modeling the attacker and its characteristics determines the strength of the security mechanism to be selected, which affects the resource usage. Please note that the modularity of the proposed meta-process allowed us to reuse the results of step 1 and step 2 of the meta-process for performance for the definition of the QuaRO method.

*Applying step 3 for privacy.* The domain knowledge diagrams created by instantiation of the domain knowledge patterns are integrated into the graph generation algorithms of ProPAn. Analogously to the graph generation rules based on problem diagrams, there is also possibly an information flow from a referred to each constrained domain in domain knowledge diagrams. Additionally, counterstakeholders that are constrained in a domain knowledge diagram may be able to access information from the referred domains (*Extended Method Description*). Thus, more possible privacy threats can be detected as in the original ProPAn method.

## 5  Structured Object-Process for Eliciting, Modeling, and Using Domain Knowledge

In this section, we describe the object-process composed of three steps for eliciting, modeling, and using domain knowledge for selected software qualities and a specific software application. We use the outputs of the meta-process (see Section 4) as inputs of the corresponding steps of the object-process to analyze the respective quality properties. Figure 5 illustrates the steps of the object-process.

In the following, we describe each step of the object-process followed by its application to the software qualities performance, security, and privacy and the concrete software application smart grid that we introduced in Section 2.

**Step 1: Domain Knowledge Elicitation.** To elicit quality-relevant domain knowledge for a specific software application to be developed, we instantiate *Domain Knowledge Templates* (output of the first step of the meta-process). For the instantiation, we make use of the *Elicitation Principles* (also output of the first step of the meta-process) and the given *Requirement Model* of the specific software application. The elicitation principles provide guidance for the instantiation of the domain knowledge templates for the given requirement models. As output, we obtain *Instantiated Domain Knowledge Templates* that serve as input for the modeling step (step 2).

*Applying step 1 for performance.* For each performance requirement, we instantiate the *Domain Knowledge Template* for performance (see Table 2) according to the information contained in the existing documents for the smart grid application [15,26]. We exemplify the instantiation of the template for the performance requirement *RQ24*, which complements the functional requirement *RQ4*.

According to the elicitation principles (see Section 4), we have to iterate over the causal domains that are part of a problem diagram containing the respective performance requirement to identify relevant resources. In Figure 2, the causal domain *WAN* represents a performance-specific resource, namely a network resource. The machine domain *SubmitMD* contains the hidden resource *CPU*, which is not modeled yet. To fill the properties for the column "value" in Table 2, we need additional information that is missing in the Protection Profile [15] and Open Meter [26] documents. Hence, we looked for the necessary domain knowledge in the existing literature such as [27]. Based on this search, we assume that there are almost 50 electricity providers [3] in Germany that receive meter readings from the gateway. As the number of concurrent users is not further specified in the documents under consideration, we take the worst case, which is 50 concurrent users as "value" for the *number of concurrent users* and closed as "value" for the *arrival pattern*. Data size of meter readings to be transmitted to the gateway is relevant for the property *data size*. It can be between 1 KB and 16 MB [27]. It varies according to the period of time, in which meter data has to be sent to authorized external entities. It amounts to 1 KB by immediate sending of meter data after reading and 16 MB by sending meter data every two hours. This would be between 40 KB and 640 MB for 40 smart meters. Hence, the "value" is 640 MB. According to the documents from the Open Meter project, for the external communication a Power Line Communication (PLC) can be used. In this case, the minimum speed must be 2.4 Kbps for a reliable communication ("value" for the property *bandWidth*). The rest of properties is either unknown or irrelevant for the requirement RQ24.

*Applying step 1 for security* For eliciting security-relevant domain knowledge, we have to instantiate the *Domain Knowledge Template* for each identified attacker once (*Elicitation Principle*). We identified three network attackers for three security requirements *RQ10*, *RQ11*, and *RQ12*. The reason is that the meter data to be transmitted through the network *WAN* can be manipulated by a network attacker (see Figure 2). There is no information in the Protection Profile [15] about the attacker that the system must be protected against. Therefore, we assume that the system must be protected against the strongest attacker. Hence, we select for each property in the domain knowledge template for security the strongest one to obtain values for the column "Value". By doing this, we obtain instances of the domain knowledge template shown in Table 3.

*Applying step 1 for privacy.* We answered the questionnaires (*Domain Knowledge Templates*) for all domains of the context diagram of the smart grid (*Elicitation Principle*) and identified various indirect stakeholders and counterstakeholders. For example, due to questions 1.2 and 1.4 we found out that controllable local systems (CLS) could be attacked by *hackers* or provide personal information to *malicious producers of CLS*, e.g. usage profiles of e-cars.

---

[3] http://www.strom-pfadfinder.de/stromanbieter/

**Step 2: Domain Knowledge Modeling.** In this step, we model the domain knowledge that we elicited in the previous step. For modeling domain knowledge and integrating it in the existing requirement models, we make use of the *Instantiated Domain Knowledge Templates*. By means of *Modeling Principles*, we annotate the *Requirement Models* with elicited domain knowledge. We use the *Selected Quality Modeling Notation* for annotating requirement models. As a result, we obtain *Domain Knowledge Models* which ideally are integrated into the existing requirement model.

*Applying step 2 for performance.* We use the MARTE stereotypes ≪hwMedia≫, ≪gaStep≫, ≪hwProcessor≫, and ≪gaWorkloadEvent≫ (*Selected Quality Modeling Notation*) for modeling the performance-specific domain knowledge captured in the *Instantiated Domain Knowledge Template*. This is done according to the *Modeling Principles* given by the mapping shown in Table 2. The modeled domain knowledge is shown in Figure 2.

*Applying step 2 for security.* In this step, we model the network attacker and its characteristics according to the *Instantiated Domain Knowledge Template*, if it is not modeled yet. We model the network attacker explicitly as a biddable domain for the confidentiality requirement *RQ11* (*Modeling Principles*). Then, we apply the stereotype ≪attacker≫ from the dependability profile which is the *Selected Quality Modeling Notation* selected in step 2 of the meta-process. We assign the attributes of the stereotype ≪attacker≫ using mapping provided by Table 3 (*Modeling Principles*). The attacker and its properties are shown in Figure 2.

*Applying step 2 for privacy.* Because of ProPAn's tool support[4], the elicited indirect (counter)stakeholders (*Instantiated Domain Knowledge Templates*) of the smart grid scenario are automatically modeled by instantiating the domain knowledge patterns according to the answers of the questionnaires (*Modeling Principles*).

**Step 3: Method Application.** The third step is concerned with applying a specific quality analysis method (*Defined/ Extended Method Description*), we defined or extended in the third step of the meta-process. The given *Requirement Models* and the *Domain Knowledge Models* obtained from the previous step are used as inputs.

*Applying step 3 for performance.* By applying the POPeRA method (*Defined/ Extended Method Description*) we identified *CPU* as a performance-critical resource (see Figure 2). Such resources are modeled as domain knowledge in the problem diagrams (*Domain Knowledge Models*) where the software might fail to achieve the performance requirements. Then, using the identified performance-critical resource *CPU*, we analyzed whether the processing capacity of *CPU* suffices to satisfy the performance requirement *RQ24* and other requirements that have to be achieved using this resource with regard to the existing workload (modeled as domain knowledge). We identified *CPU* as potential bottleneck.

*Applying step 3 for security and performance.* By applying the QuaRO method (*Defined/ Extended Method Description*), we identified potential interactions among

---

[4] Available at http://www.uni-due.de/swe/propan.shtml

security and performance requirements. Performance requirement *RQ24* might be in conflict with security requirements *RQ10*, *RQ11*, and *RQ12* (see Figure 2).

*Applying step 3 for privacy.* As mentioned in the application of step 2 for privacy, we identified hackers and malicious producers of CLS as indirect counterstakeholders of the CLS (*Domain Knowledge Models*) and analyzed the privacy threats that possibly exist in the smart grid scenario for the stakeholder customer and counterstakeholder hacker (*Defined/Extended Method Description*). The analysis shows that there is possibly a privacy threat originating from a hacker or a malicious producer via the HAN using a CLS. This privacy threat is not covered by an assumption or threat in the protection profile [15]. Hence, our extended method now finds relevant privacy threats that previously have been overlooked.

# 6  Related Work

There exist only few approaches dealing with capturing and representing knowledge needed for a successful consideration of software qualities in software development.

Zave and Jackson [2] identify four areas in which the foundation of the requirements engineering discipline is weak. One of these areas is domain knowledge. Among others, the authors emphasize the importance of capturing domain knowledge for the satisfaction of requirements. However, they do not provide a structured way or specific notations to model domain knowledge, and only consider functional requirements.

According to Probst [28], the goal of knowledge management (KM) is the improvement of processes and capabilities by utilizing knowledge resources such as skills, experience, routines, and technologies. The author proposes a KM model that structures the KM process as activities identification, acquisition, development, distribution, preservation, and use of knowledge, called building blocks of KM. The steps of our method can be easily mapped to these building blocks. Knowledge identification identifies which knowledge and expertise exists. This is a prerequisite for conducting our method. It leads to identify the need for capturing, modeling, and using domain knowledge. Knowledge acquisition is concerned with obtaining knowledge from involved stakeholders, domain experts, or using documents. This activity corresponds to the step *information needs elicitation* in our meta-process. Knowledge development aims at producing new knowledge. It can be related to the step *domain knowledge elicitation* in the object-process. The objective of knowledge distribution is to make the knowledge available and usable. This activity corresponds to the step *modeling notation selection* in the meta-process. Knowledge preservation avoids the loss of gained expertise by preserving the knowledge after it has been developed. This building block can be mapped to the step *domain knowledge modeling* which stores the captured domain knowledge in requirement models. Consequently, the knowledge has to be deployed in the production process (knowledge use). This is achieved in our method in the steps *method definition* and *method application*. The mapping of the steps of our method to the KM building blocks shows that we followed successfully the concepts involved in the field of KM.

There exist several approaches for the elicitation of domain knowledge in the field of domain engineering [29,30]. These approaches focus on the development of reusable software and therefore also on the analysis of the application domain. During the domain analysis phase, domain knowledge is systematically collected and documented. In the field of domain engineering the term "domain" corresponds to the term "system" in Jackson's terminology. In this paper, we collect and document domain knowledge in a more fine-grained way which allows us analyzing software quality requirements.

Peng et al. [31] present a method for the analysis of non-functional requirements based on a feature model. This method elicits the domain knowledge before the analysis of non-functional requirements. In contrast, we suggest to elicit the required domain knowledge for a specific software quality. We think that our method leads to a more complete and targeted elicitation of domain knowledge.

In the NFR framework [32], knowledge about the type of NFR and the domain has to be acquired before using the framework. This knowledge is captured to understand the characteristics of the application domain and to obtain NFR-related information to be used for identifying the important NFR softgoals. Examples of such domain knowledge are organizational priorities or providing terminologies for different types of NFRs. This kind of domain knowledge differs from ours, as it is used as initial information to identify the goals and requirements. The knowledge we capture and model is more fine-grained and is required in addition to the quality requirements. Moreover, we provide a systematic method for capturing and modeling domain knowledge, whereas the NFR framework does not provide any guidelines on how to acquire such knowledge.

## 7   Conclusions

For an adequate consideration of software quality during requirements analysis, we have to identify and to take into account the quality-specific domain knowledge. By means of three different requirement analysis methods, we have pointed out the need for eliciting, modeling, and using domain knowledge. Hence, to avoid requirements errors due to incorrect domain knowledge, domain knowledge should be considered with the same emphasis as requirements during requirements analysis.

In this paper, we proposed a structured method consisting of a meta-process and an object-process for eliciting, modeling, and using quality-specific domain knowledge at the requirements analysis level.

The meta-process is quality-dependent. It therefore has to be carried out once for each kind of quality requirement to be considered. To facilitate the reuse of captured and modeled domain knowledge, we provide individual templates and guidelines suitable for each kind of quality requirement. These templates and guidelines are reusable if the same quality shall be considered, but in a different notation or for a different analysis method. Additionally, all outputs of the meta-process can be reused by instantiation of the object-process and applying it to a concrete software application.

We instantiated the first two steps of the meta-process for three kinds of quality requirements, namely performance, security, and privacy. Then, we showed how the elicited and modeled domain knowledge can be used in the three methods POPeRA, QuaRo, and ProPAn in the third step of the meta-process. We instantiated the object-process with the corresponding outputs of the meta-process to apply our methods POPe-RA, QuaRo, and ProPAn on the concrete software application smart grid.

Our approach is independent from any specific tool or notation. Hence, it can easily be integrated into existing requirement analysis methods. Our proposed method helps requirements engineers to develop processes for the consideration of quality requirements in a structured way and independently of the tools or notations they use.

As future work, we plan to develop further quality analysis methods using the proposed meta-process. Furthermore, we want to investigate to which extent the outputs of meta-processes carried out for different modeling notations differ and how they are related. In addition, we strive for empirically validating our method to determine the effort spent for exceuting the method and to further improve the elicitation templates.

# References

1. Lamsweerde, A.: Requirements Engineering: From System Goals to UML Models to Software Specifications. Wiley (2009)
2. Zave, P., Jackson, M.: Four dark corners of requirements engineering. ACM Trans. Softw. Eng. Methodol. 6, 1–30 (1997)
3. van Lamsweerde, A.: Reasoning about alternative requirements options. In: Borgida, A.T., Chaudhri, V.K., Giorgini, P., Yu, E.S. (eds.) Conceptual Modeling: Foundations and Applications. LNCS, vol. 5600, pp. 380–397. Springer, Heidelberg (2009)
4. Prieto-Díaz, R.: Domain analysis: an introduction. SIGSOFT Softw. Eng. Notes 15(2), 47–54 (1990)
5. Hooks, I.F., Farry, K.A.: Customer-centered Products: Creating Successful Products Through Smart Requirements Management. AMACOM (2001)
6. Modugno, F., Leveson, N., Reese, J., Partridge, K., Sandys, S.: Integrated safety analysis of requirements specifications. In: Requirements Engineering, pp. 65–78 (1997)
7. Fabian, B., Gürses, S., Heisel, M., Santen, T., Schmidt, H.: A comparison of security requirements engineering methods. Requirements Engineering – Special Issue on Security Requirements Engineering 15, 7–40 (2010)
8. Robillard, P.N.: The Role of Knowledge in Software Development. Commun. ACM 42, 87–92 (1999)
9. Niknafs, A., Berry, D.M.: The impct of domain knowledge on the effectiveness of requirements idea generation during requirements elicitation. In: Proc. of the 20th IEEE Int. RE Conf., pp. 181–190 (2012)
10. Jackson, M.: Problem Frames. Analyzing and structuring software development problems. Addison-Wesley (2001)
11. Alebrahim, A., Choppy, C., Faßbender, S., Heisel, M.: Optimizing functional and quality requirements according to stakeholders' goals. In: Mistrik, I. (ed.) Relating System Quality and Software Architecture, pp. 75–120. Elsevier (2014)
12. Alebrahim, A., Heisel, M.: A problem-oriented method for performance requirements engineering using performance analysis patterns. In: FGCS (submitted, 2014)
13. Beckers, K., Faßbender, S., Heisel, M., Meis, R.: A problem-based approach for computer-aided privacy threat identification. In: Preneel, B., Ikonomou, D. (eds.) APF 2012. LNCS, vol. 8319, pp. 1–16. Springer, Heidelberg (2014)
14. Meis, R.: Problem-Based Consideration of Privacy-Relevant Domain Knowledge. In: Hansen, M., Hoepman, J.-H., Leenes, R., Whitehouse, D. (eds.) Privacy and Identity 2013. IFIP AICT, vol. 421, pp. 150–164. Springer, Heidelberg (2014)
15. Kreutzmann, H., Vollmer, S., Tekampe, N., Abromeit, A.: Protection profile for the gateway of a smart metering system. Technical report, BSI (2011)

16. UML Revision Task Force: OMG Unified Modeling Language (UML), Superstructure (2009), http://www.omg.org/spec/UML/2.3/Superstructure/PDF
17. Hatebur, D., Heisel, M.: A UML profile for requirements analysis of dependable software. In: Schoitsch, E. (ed.) SAFECOMP 2010. LNCS, vol. 6351, pp. 317–331. Springer, Heidelberg (2010)
18. Alebrahim, A., Hatebur, D., Heisel, M.: Towards systematic integration of quality requirements into software architecture. In: Crnkovic, I., Gruhn, V., Book, M. (eds.) ECSA 2011. LNCS, vol. 6903, pp. 17–25. Springer, Heidelberg (2011)
19. UML Revision Task Force: UML Profile for MARTE: Modeling and Analysis of Real-Time Embedded Systems (2011), http://www.omg.org/spec/MARTE/1.0/PDF
20. Land, L., Aurum, A., Handzic, M.: Capturing implicit software engineering knowledge. In: Proceedings of the 2001 Australian Software Engineering Conference, pp. 108–114 (2001)
21. Bass, L., Klein, M., Bachmann, F.: Quality attributes design primitives. Technical report, Software Engineering Institute (2000)
22. Bass, L., Clemens, P., Kazman, R.: Software architecture in practice. Addison-Wesley (2003)
23. International Organization for Standardization (ISO) and International Electrotechnical Commission (IEC): Common Evaluation Methodology 3.1. ISO/IEC 15408 (2009)
24. Sharp, H., Finkelstein, A., Galal, G.: Stakeholder Identification in the Requirements Engineering Process. In: DEXA Workshop, pp. 387–391 (1999)
25. Alexander, I.F., Robertson, S.: Understanding Project Sociology by Modeling Stakeholders. IEEE Software 21(1), 23–27 (2004)
26. Remero, G., Tarruell, F., Mauri, G., Pajot, A., Alberdi, G., Arzberger, M., Denda, R., Giubbini, P., Rodrguez, C., Miranda, E., Galeote, I., Morgaz, M., Larumbe, I., Navarro, E., Lassche, R., Haas, J., Steen, A., Cornelissen, P., Radtke, G., Martnez, C., Orcajada, K.H., Wiedemann, T.: D1.1 Requ. of AMI. Technical report, OPEN meter proj. (2009)
27. Deconinck, G.: An evaluation of two-way communication means for advanced metering in Flanders (Belgium). In: Instrumentation and Measurement Technology Conference Proceedings (IMTC), pp. 900–905 (2008)
28. Probst, G.J.B.: Practical Knowledge Management: A Model that Works. Prism (1998)
29. Kang, K.C., Cohen, S.G., Hess, J.A., Novak, W.E., Peterson, A.S.: Feature-Oriented Domain Analysis (FODA) Feasibility Study. Technical report, Carnegie-Mellon University Software Engineering Institute (November 1990)
30. Frakes, W., Prieto-;Diaz, R., Fox, C.: DARE: Domain analysis and reuse environment. Annals of Software Engineering 5(1), 125–141 (1998)
31. Peng, X., Lee, S., Zhao, W.: Feature-Oriented Nonfunctional Requirement Analysis for Software Product Line. Journal of Computer Science and Technology 24(2) (2009)
32. Chung, L., Nixon, B., Yu, E., Mylopoulos, J.: Non-functional Requirements in Software Engineering. Kluwer Academic Publishers (2000)

# Debugger-Based Record Replay and Dynamic Analysis for In-Vehicle Infotainment

Hanno Eichelberger[1,2], Jürgen Ruf[1], Thomas Kropf[1],
Thomas Greiner[2], and Wolfgang Rosenstiel[1]

[1] University of Tübingen, Germany
{forename.surname}@uni-tuebingen.de
[2] Hochschule Pforzheim University, Germany
{forename.surname}@hs-pforzheim.de

**Abstract.** In the first operation tests of In-Vehicle Infotainment systems remaining failures which were not detected by other testing techniques can often be observed. These failures have to be reconstructed in the laboratory in order to be able to debug them. Record and replay concepts enable the automatic reconstruction of execution sequences, but often they disturb the normal execution, require comprehensive instrumentation or can not be implemented for every platform. Our approach considers these criteria by implementing record and replay with a symbolic debugger. Manually locating the root causes of failures during replay is time-consuming. Therefore, we apply dynamic analyses; they can detect anomalies and state changes during a run. The paper presents a collection of analyses which can not be smoothly executed online during operation, but do not cause drawbacks at replay.

**Keywords:** record replay, dynamic analysis, in-vehicle infotainment.

## 1 Introduction

Nowadays embedded systems and embedded software are used in different areas. Important domains are vehicles and cars. Modern cars contain hundreds of millions of lines of source code. Big parts of this source code are used for In-Vehicle Infotainment (IVI) systems. These systems assist the driver, e.g. by providing route information through a navigation system. They run on highly specialized embedded hardware (commonly x86- or ARM-architectures).

Lack of quality, mainly failures, in Infotainment software can have dangerous and very uncomfortable effects. In the following practical examples of such failures are described. The navigation software of the iPhone iOS6 has led several persons in Australia by mistake into the Muarry-Sunset National park. After several hours of walking and driving in dangerous terrain the drivers could find back into areas with mobile connection. Such failure can be caused by wrong route calculations or wrong route drawings. It has been observed that some navigation systems crash on specific GPS locations. Thereby, the user can miss information to get to the correct street. This effect can be caused by additional case-specific operations in the GPS processing.

B. Murgante et al. (Eds.): ICCSA 2014, Part V, LNCS 8583, pp. 387–401, 2014.
© Springer International Publishing Switzerland 2014

A lot of failures can be detected during unit testing by applying common verification techniques: static code analysis or model-based testing. Unfortunately, embedded software architectures can be complex and hard to verify with static code analysis, e.g. static pointer analyses of C code are difficult and can be better verified during runtime. Model-based testing strongly depends on the granularity of the specification. It can omit to test specific cases, when they were not specified by the architects. Thereby, the final integrated system can still contain failures. These failures are very time-consuming to patch, even though the product has to be placed on the market as early as possible.

When failures occur during the first operation tests, these failures have to be reconstructed several times in order to debug them. For example if a navigation system fails on a specific route, the GPS input has to be reconstructed in order to simulate and trigger the failure and debug it. For preventing effortful manual reconstruction, the failure sequence can be recorded once the failure occurs and replayed for debugging (so called Replay Debugging). Recording should not disturb the normal operation and the replayed run should be equal to the original run. Most Replay Debugging approaches are concepted for specific types of platforms or modify the source code of the program (causing side effects). There is no platform independent solution which does not instrument the source code.

When a failure can be replayed, it is still difficult to locate the root cause of the failure in the source code. The static defect in the source code is called fault. Manually detecting these faults is very time-consuming. The fault can be located many operation steps before the failure occurs. For specific failures it is not even possible to locate the failure, because only a not-expected behavior of the system can be observed and no exception is thrown.

Dynamic analysis techniques enable the automated diagnosis of runs. Detected potential error states which are the effect of faults might give hints for debugging the faulty behavior. This dynamic analysis can compare the source coverage of runs, check the execution time of operations or follow the change of variables and pointers. The automated analysis of runs can give hints for manually finding and debugging faults and the resulting failures. Thereby, the debugging process of replayed failures can be accelerated. Running the analysis in parallel during operation slows down the normal execution and disturbs it.

**Contributions.** The following list provides contributions for solving the mentioned challenges in order to achieve high quality embedded software efficiently.

- Reconstruct failures efficiently, not disturbing the application with recording.
- Accelerate the location of root causes of not-expected behaviors. Therefore, implement dynamic analyses efficiently.
- Avoid side effects through changing the source code of the software, by implementing instrumentation-free record/replay and dynamic analysis.
- Provide a platform-independent solution, especially for embedded systems.

## 2    Method

In order to efficiently debug failures occurred during operation in embedded software we propose the following method (see Figure 1).

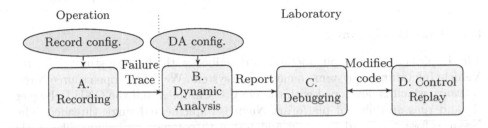

**Fig. 1.** Overview for debugger-based record/replay and dynamic analysis

**A. Recording:** First the incoming events to the software are captured during operation. Recording mechanisms are implemented with a symbolic debugger in order to avoid instrumentation and to achieve platform compatibility. Therefore, the debugger is controlled with a script. The developer can decide which events are relevant and have to be captured. Thus, the recording can be kept lean. The events are recorded into a trace file. When a faulty behavior of the system is observed by the user, the execution can be stopped and the trace file can be stored as failure trace. It is possible that the system results in an unexpected behavior instead of an exception.

**B. Dynamic Analysis:** The failure sequence can be loaded and deterministically replayed in the laboratory. The replay mechanisms are implemented with a symbolic debugger as well. The software is executed with the debug interface on the same hardware as during operation in order to arrange the same system behavior as during the original run. During replay the failure occurs again based on the deterministic replay. Manually debugging the complete execution sequence or even several processing paths of events is very time-consuming. Therefore, we apply dynamic analyses during replay in order to automatically detect potential anomalies or relevant state changes. These information can give a hint on the fault. Analyses performed online during operation disturb the normal execution, but do not cause drawbacks during replay, because no interactions with the user or external components are required for the execution of the replay.

**C. Debugging:** Based on the report by the dynamic analyses, the developer can manually fix the fault. This process is integrated in the debugger-based framework. The step results in a patched program.

**D. Control Replay:** The modified program can be tested with a control replay. It is executed with the recorded event failure sequence in order to observe whether the faulty behavior occurs again or not. If the faulty behavior still occurs the developer can jump back to C. or B..

# 3   Implementation

In this section the implementation of our method for debugger-based replay and dynamic analysis is demonstrated by an example software. The approach can easily be transferred to other applications which run on different platforms.

## 3.1   Case Study Navit

The implementation of our method is described with the case study of an In-Vehicle Infotainment system - a navigation system. We use the open source Navit software which is developed with the GNU C Compiler and the GNU Debugger [1] and runs on different platforms. Navit is implemented single-threaded. The Navit software is hard to verify and test with common techniques like static code analysis or model-based testing. It bases on a plugin framework and each component is implemented as a plugin (see architecture of Navit in Figure 2). Each plugin communicates with other plugins through dynamically solved callback references. The references are implemented as general-type pointers and it is only possible to follow the invocation flow by exploring the control flow. Therefore, it is difficult for static code analyses to detect dependencies between plugins and invocation paths which can cause runtime errors. Testing the system based on the specification model (e.g., UML) is difficult as well. The input space of GPS, user input and map loading is big and can not be covered by exploring the input space. The modeling and generation of specific driving behavior, e.g. driving two times in a wrong direction, requires the manual implementation of complex input data generators. Finally, it is still possible, that the specification contains logical errors which were not intended by the architects.

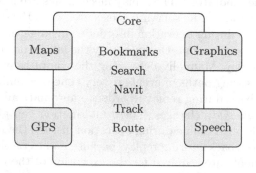

**Fig. 2.** Navit plugin architecture (Source: Navit - A car navigation system [2])

Our method is applied after static code analysis and model-based testing. These techniques can omit failures which can only be detected in the final integrated system during operation. Our method can be applied to record and debug these remaining failures efficiently.

## 3.2    Record Replay GPS and User Input

In this section the implemented record/replay mechanisms with a symbolic debugger are described. The GNU Debugger is used as base tooling. The concept is described based on the Navit event processing but can be easily ported to other applications. The event processing of Navit bases on a single-threaded event processing framework. We minimally improved the source code of Navit by implementing the access to devices through polling instead of processing event callbacks. In our implementation a timer is invoked every 10 ms. This timer triggers the GPS plugin each second and the Graphics plugin every 10 ms for polling devices states. The invocation of the GPS and Graphics plugins are described in Figure 3.

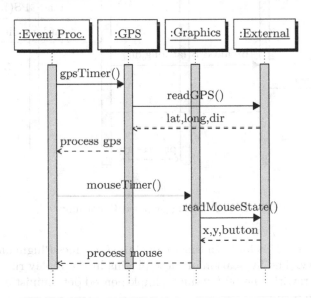

**Fig. 3.** Navit GPS and Mouse device polling

The event processing timer calls the GPS plugin in order to access the current GPS state. This state is accessed by calling an external GPS library or server. The GPS plugin receives the GPS state and invokes the routine for processing the GPS coordinates. Afterwards, the event processing timer calls the next device in sequence, e.g. the mouse. The mouse is accessed with the Graphics plugin, which reads the current mouse state with an external library and forwards it to the mouse event processing routine.

Implementing the event processing and the access of devices this way has the advantage that the workload for device inputs is kept constant (the workload differs only for different route calculations). All asynchronous calls can be wrapped to polling processing. In the following the record/replay for polling event processing is described for the Navit example.

Figure 4 describes the recording process of GPS events into a log file with the GDB. The processing starts with the triggering of the GPS plugin by the event processing timer. It reads the current GPS status with an external library. Afterwards, the GDB breaks the execution and stores the current GPS data into a log file. For each breaking of the GPS a log entry is written. After logging, the GPS data is processed as during a normal run.

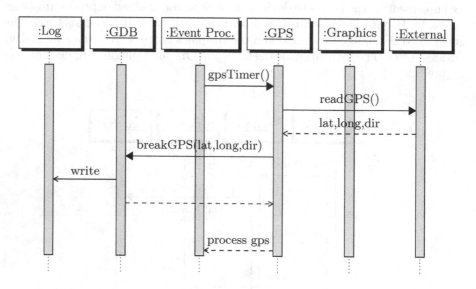

**Fig. 4.** Debugger-based Recording

The log file contains for each timer call of the Graphics Plugin and of the GPS plugin an entry. The invocation of these plugins in the replay run has the same sequence. Therewith, the replay can be implemented deterministically, achieving always the same run.

During replay the execution of the callback timer proceeds like during the original run, but is controlled by the GDB (see Figure 5). It breaks before the accessing of the external GPS server and skips this call with a debugger *jump* command. Afterwards, the GPS data from the original run is set (or injected) as current GPS data with the GDB. These data are forwarded to the GPS processing routine after the GDB break ends. Record/replay for mouse or other devices is similar.

The breakpoints are configured by the user through a script. Therewith, the record/replay can be specified for any kind of software. The breakpoints are defined as hardware-breakpoints and few overhead for breaking is required (see section 4). If the GDB is not supported on the used platform, any other programmable debugger can be used as well. For our experiments the controlling of the debugger was implemented with the GDB Python interface.

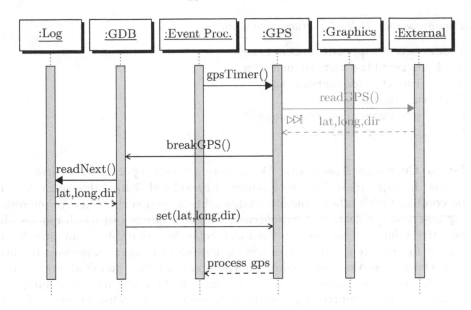

**Fig. 5.** Debugger-based Replay

## 3.3 Dynamic Analysis

During replay each intermediate program state can be accessed with the debugger. It can step over relevant code on operation-granularity and can process state checks. This way provides the possibility to implement fine-granular dynamic analyses. We developed several dynamic analyses which can support the developer for debugging. This paper regards the event processing of the GPS events. A *processing run* considers each operation for processing of a GPS event. In our method only the last 5, 10 or 15 GPS event processing runs of a recorded execution sequence are analysed (see section 4). This is commonly the place where the failure occurs. In the following different dynamic analyses are listed.

**Delta Coverage.** The delta debugging considers the difference between good and a bad runs [10]. The delta coverage algorithm (see following listing) tests which lines of code are executed more in one processing run than in the processing run before. Therefore, the last executed line is added as string representation (filename, line number) to a list *currentrun* (line 1) for every step in a processing run. If *currentline* not occurred in *prevrun* it is added to the report (line 2-3). After one processing run is finished (see line 4-6) *currentrun* is copied to *prevrun*. In this way, it can be checked where in the source code a run differs from a previous run. The difference can contain logic which causes a failure. Similarly to coverage changes, the changes of global or local variable values can be considered.

*Pseudocode for Delta Coverage algorithm*

```
1 currentrun.append(currentline)
2 if currentline not in prevrun:
3     report.add(currentline)
4 if runfinished:
5     prevrun=currentrun.copy()
6     currentrun.clear()
```

**Delta Coverage Assertion.** Delta Coverage can report a lot of irrelevant deltas. In order to shrink the amount of considered deltas, this concept can be combined with other dynamic analyses, like assertion checking. An interesting assertion predicate is $foreach(var : localvars, var >= 0)$ which reports all negative values for variables in the local scope. Negative values can often be a reason for bugs (e.g. array-accessing or drawing of objects on screen). In this way, checking all variables against an assertion can detect potential error-states without specifying value ranges for each variable. Our algorithm (see following listing) checks at source code locations where coverage deltas occurred if any local variable holds a negative value (line 3-6).

*Pseudocode for Delta Coverage algorithm combined with negative value checking*

```
1 currentrun.append(currentline)
2 ...
3 if currentline not in prevrun:
4     for var,val in localvars:
5         if isfloat(val) and float(val) < 0:
6             report.add(var,val,currentline)
```

Similarly, when a line occurs in a previous run, but not in the current run, the line can be marked as critical in a first replay. In a second replay of the execution sequence assertions can be checked for this case as well.

**Timing.** When the system reacts with low performance (e.g. when selecting a specific route destination) it is interesting which source lines occurred often and acquired a lot of time. Therefore, the approximate time needed for the execution of a specific source line can be summed up. In a hash set for each line an entry is added. When a source line is passed, the required time is added up. At the end of each processing run the lines which required most time are written to the report.

**Pointer Changes.** It may occur that pointers change the target they were pointing to, e.g. when pointers are incorrectly modified by external black-box libraries. In order to detect such pointer changes the dynamic analysis can check the target of each global pointer for each operation of a processing run. In Navit global pointers are used for callback routines; this causes exceptions when they are switched to a wrong target. Furthermore, changing pointers which point to arrays with specific size can cause segmentation faults at next usage.

## 3.4    Practical Example

The Navit software contains a practical logic bug in the drawing routine of the vehicle class visualized in the following listing. This bug is difficult to detect with other testing techniques, because of the reasons described in section 3.1.

*Pseudocode Navit vehicle draw method bug*

```
1 void vehicle_draw(struct vehicle *this ,..., int angle,...)
2 {
3  if (angle < 0)
4     angle+=360;
5  draw_it(...,angle,...);
```

The condition in line 4 checks whether the angle of the direction of the vehicle is smaller than zero. When it is smaller, 360 is added to the angle. For minus values bigger than $-360$ the algorithm works well. But it is possible, that the GPS external library returns angle values smaller than $-360$ in special cases, e.g. $-380$. The presented routine would not handle these values correctly and just add 360 resulting in a minus value as well, e.g. $-20$. Unfortunately, the next drawing processing steps of the vehicle cursor can not handle negative values. A negative value for the direction smaller than $-360$ results in a faulty behavior: the cursor for the vehicle is not drawn on the map anymore. In this case the faulty behavior is not linked to a source line, because no exception is thrown. The correct implementation of the code line 4 would be $angle = mod(angle, 360)$.

Recording sequences of GPS coordinates containing several directions with positive values and one value smaller than $-360$, the faulty behavior of the cursor not being drawn on the map can be recorded. During replay we applied *Delta Coverage* analysis on 10 GPS processing runs. It detected 1093 coverage deltas between the runs. One of the deltas was detected between a run with a positive angle and the one with the negative angle. It pointed to the line 4 of the code above.

Manually searching for the relevant deltas is time-consuming. In our approach several dynamic analyses are combined in order to shrink the amount of possible error sources and detect exactly the fault line. In our case a potential error source for the navigation system is given by negative values for variables, because GPS values are directly translated into map coordinates and the direction should be transformed as well. Therefore, we applied the *Delta Coverage Assertion* to our example in order to list negative values at source code locations where coverage deltas occurred. This way line 4 in source code above was detected (beside one other detected line as potential fault candidate). Therewith, we could locate the root cause of the wrong behavior, by checking the intersection for a specific type of deltas (coverage deltas) and variable states (negative values). Other kinds of bugs can be detected with different combinations of dynamic analyses.

# 4    Evaluation

In this section four aspects of our approach are evaluated: performance, ease of implementation, avoidance of instrumentation and platform independence.

## 4.1    Performance

We measured the performance of record and replay with the GDB and compared it to the state of the art record/replay tool Rebranch [7] which provides debugging support during replay as well, but only records the control flow. The measurements are presented in Figure 6.

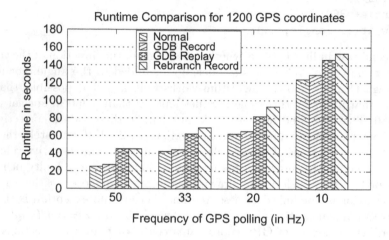

**Fig. 6.** Record and Replay performance overhead

The measurements consider a route with 1200 GPS coordinates, which were driven by a simulator car. These coordinates are read from a file by a Mockup GPS server. The GPS frequency was tested at 50Hz, 33Hz, 20Hz and 10Hz, in parallel, the mouse input was captured at 100Hz. The mouse input was read directly from the device. It can be observed, that the recording run requires only few overhead in comparison with a normal run (without recording). In our measurements recording requires on average 3 seconds overhead. This results in an average overhead of 1.05X. The replaying overhead is bigger, because some GDB operations for setting variable values require more overhead, than for accessing variables. Certainly, the replay performance is not as important as the recording performance, because replay in the laboratory does not require user interaction. The overhead for Rebranch recording was measured at on average 1.52X. In comparison to Rebranch recording, the GDB recording approach was on average 23 seconds faster, which resulted in better system reaction and smoother visualization.

We also measured the overhead for dynamic analyses. The analyses' overhead for stepping over each code operation with a debugger is quite high, especially for a lot of operations which require few computational effort. These analyses can not be executed online during the operation of the software. Therefore, it is better to execute them during replay. The measurements for the dynamic analyses are presented in Figure 7. It presents the runtime for the replay of 15, 10 and 5 GPS coordinates with a frequency of 1Hz and the overhead for dynamic analyses of GPS processing during replay. For dynamic analyses the GDB single steps over ~12000-23000 operations for each GPS processing run controlled by a python script. The overhead for dynamic analysis for *Delta Coverage*, *Delta Coverage Assertion* and *Timing* was always under 22X. For *Pointer Change* analysis we measured 55-60X, because all global pointers (amount of 26 pointers) have to be followed. These performance overheads are common for dynamic analyses.

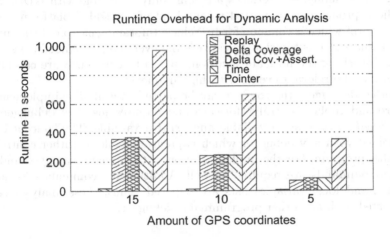

**Fig. 7.** Dynamic analyses performance overhead

The measurements were executed on an Intel Core Duo CPU 3.00 GHz 8 GB RAM with Ubuntu Linux version 12.04.

## 4.2   The Ease of Implementation

For the implementation of record and replay only few source lines are required. The recording script for GPS and mouse events of Navit only requires <50 lines. The replay script requires <45 lines. Dynamic analyses can be developed easily as well. Each analyses algorithm is integrated into a replay script. The following numbers of lines are required for dynamic analysis scripts additional to the replay code: Delta Coverage<20 lines, Delta Coverage Assertion<40 lines, Timing<15 lines, Pointer Changes<30 lines. It is obvious that the implementation of such analysis scripts is manageable.

### 4.3  Avoidance of Instrumentation

For our approach no instrumentation of the source code is required. The source code can be kept unmodified and just has to be compiled with debug information. Therewith, the processing routines of GPS and mouse devices do not have to be instrumented for recording and for replay. Dynamic analyses require access to the complete program state during the whole run. For achieving that, the source code has to be instrumented after each statement. Using a debugger avoids the instrumentation of each source line and thereby avoids side effects.

### 4.4  Platform Independence

The experiments presented in this paper are implemented with the GDB on an x86 platform. Most embedded platforms provide a debugger interface (e.g. JTAG). These debugger interfaces are commonly accessible with GDB. Therewith, the approach can be ported to any kind of embedded platform, e.g. the ARM platform which is common for In-Vehicle Infotainment, too. If the supplied debugging interface is too slow to capture the events of the embedded software without disturbing the normal execution, special tracing hardware can be used for recording the relevant events with high frequency.

For other debuggers the concept can be ported as well. The implementation of record and replay is straightforward and requires just few debugger commands: *break*, *jump*, *set* a variable and *print* a variable. The concept can be implemented with any debugger which is programmable or rather controllable by another program. For the implementation of the dynamic analyses only few debugger commands are required as well. Additionally, commands for accessing global and local variables lists are required. Thus, these analyses can be implemented with any other programmable debugger.

## 5   Related Work

In this section approaches from the state of the art in the area of record/replay and dynamic analysis are listed.

### 5.1  Record Replay

Pinplay [3] is a framework which bases on the Pin dynamic instrumentation system. Pinplay records the execution of a software on instruction level. Thereby, the performance is very slow (sometimes over 100X) and not appropriate for multimedia applications. We tested it with a small OpenGL example and it was too slow to control the scene by mouse. Pin translates the source code into an intermediate language for instrumentation. It only works for specific platforms.

Scribes [4] implements record/replay by recording system calls and signals deterministically with synchronization points. The use of synchronization points enables the record/replay on multicore-platforms. The performance overhead of

Scribes is very low; the authors measured at 1.15X for open office. For Scribes it is required to deploy a specific Linux kernel build which supports the recording. We replayed the execution of a painting program and during the replay it was not visualized and was therewith not observable. The integration of a symbolic debugger is not mentioned by the authors.

Lee et al. [5] present an approach for the record and replay of multi-threaded embedded software. It records calls to operating system libraries as well as inter-process communication. The implemented tool replaces some platform-specific system functions by wrapper functions during compilation time. The recording overhead was measured by the authors under 1.09X for console applications, but it was not evaluated for multimedia applications. The GNU Debugger is integrated into the tool for debugging during replay. The authors do not present automated debugging techniques during replay.

VMWare in version 6 and 7 [6] supported replay debugging. Using virtualization the complete system state can be check pointed and external inputs from devices can be captured in the virtualization layer. Thereby, the performance overhead for recording can be limited to few percent additional to the virtualization overhead. Unfortunately, virtualization is not supported by every platform, especially not by embedded platforms.

Rebranch [7] records the control-flow of programs and replays it in the GDB for debugging. Rebranch loads the program with a special loader implemented for Linux which avoids the instrumentation of the program. The overhead for recording Navit is on average 1.52X for Rebranch. The Navit GUI wasn't visualized during replay with Rebranch.

RERAN [8] implements record and replay for Android operating systems. It realizes event capturing by using specific Android SDK tools. Replay is implemented with a platform specific replay agent. The concept is not transferable to other embedded systems.

## 5.2  Dynamic Analysis

The book *Why Programs Fail* [9,10] explains how program executions can be recorded and how failure root causes can be localized during replay. It describes several concepts for dynamic analyses including delta debugging, anomaly detection and failure slicing. Some of these concepts are considered in our work as well, e.g. for our delta coverage approach. In contrast to the concepts in the book, in the failures considered by our approach it is not obvious at which line in the source code the failure occurred. Thus, several approaches like slicing can not be applied. The authors applied the approaches mainly to web applications but not to multimedia embedded systems.

The WiDS Checker [11] implements record/replay and dynamic assertion checking during replay for distributed systems. Similar to our approach, it only records the relevant events exchanged between the distributed nodes. It builds a happened-before relation graph in order to deterministically replay the execution on a simulator. The user can define assertions which are checked during replay for each received event. It does not support assertion checking on operation level and is just applicable to the WiDS platform.

The Friday tool [12] applies a symbolic debugger during the replay of the distributed execution and uses breakpoints and watchpoints for runtime monitoring. It only applies the concept for specific errors for distributed systems.

The ARVE (Aspect-oriented Runtime Verification Environment) uses the GNU Debugger for dynamic analyses in order to achieve platform compatibility as well. It triggers the software under test based on the specification, but does not implement any record and replay capabilities.

Jangali [14] implements record/replay and several dynamic analyses only for JavaScript. It bases on the instrumentation of the source code with shadow values. Hereby, the performance overhead for recording is very high, on average 26X. The overhead for dynamic analyses is on average 30X. Similar to our approach, dynamic analyses are implemented with few source lines using the framework.

DrDebug [15] uses Pinplay for record/replay and implements slicing for bugs. As mentioned before, Pinplay is too slow for multimedia applications. Slicing can mainly be applied to failures for which the exception source line is obvious.

In conclusion, there is no approach for record/replay with dynamic analyses which considers all the four pillars: performance, ease of implementation, avoidance of instrumentation and platform-independence. Our approach implements these pillars using a symbolic debugger as base.

## 6   Conclusion

This paper presents an approach for efficiently detecting root causes of observed faulty behavior of embedded software. It implements record and replay of failure input sequences of devices using a symbolic debugger. During replay different dynamic analyses are executed in order to detect anomalies and state changes in the program execution which are potential candidates for failure root causes. The performance overhead for recording incoming events is very few - in our measurements on average 1.05X. The record/replay can be adapted to the type of program in order to avoid time-consuming recording (e.g., recording on instruction level). Dynamic analyses can be implemented easily using debugger scripts. In the presented examples less than 40 lines of code are required for each presented analysis. Using a debugger, instrumentation of the program's source code can be avoided. Debuggers are similarly implemented on different platforms and for our presented approach only few debugger operations are required. In future we will extend record/replay for multi-threaded software. We plan to test the approach on different types of embedded hardware. We also want to extend the approach with Runtime Verification analysis during replay [16].

**Acknowledgement.** This work is funded by the State of Baden-Wuerttemberg, Germany, Ministry of Science, Research and Arts within the scope of the Cooperative Research Training Group.

# References

1. GDB: The GNU Project Debugger, http://www.sourceware.org/gdb
2. Navit - Car Navigation System, http://www.navit-project.org
3. Patil, H., Pereira, C., Stallcup, M., Lueck, G., Cownie, J.: PinPlay: a Framework for Deterministic Replay and Reproducible Analysis of Parallel Programs. In: 8th IEEE/ACM International Symposium on Code Generation and Optimization, pp. 2–11. ACM (2010)
4. Laadan, O., Viennot, N., Nieh, J.: Transparent, Lightweight Application Execution Replay on Commodity Multiprocessor Operating Systems. In: 2010 ACM SIG-METRICS International Conference on Measurement and Modeling of Computer Systems, pp. 155–166. ACM (2010)
5. Lee, Y.-H., Song, W.Y., Girme, R., Zaveri, S., Chen, Y.: Replay Debugging for Multi-Threaded Embedded Software. In: 2010 IEEE International Conference on Embedded and Ubiquitous Computing, pp. 15–22. IEEE (2010)
6. Replay Debugging on Linux. VMWare Technical Note (2009), https://www.vmware.com/pdf/ws7_replay_linux_technote.pdf
7. Wang, N., Han, J., Fang, J.: A Transparent Control-Flow based Approach to Record-Replay Non-Deterministic Bugs. In: 7th International Conference on Networking, Architecture and Storage, pp. 189–198. IEEE (2012)
8. Gomez, L., Neamtiu, I., Azim, T., Millstein, T.: RERAN: Timing- and Touch-Sensitive Record and Replay for Android. In: Proceedings of the 35th International Conference on Software Engineering, pp. 72–81. IEEE (2013)
9. Zeller, A.: Isolating Cause-Effect Chains from Computer Programs. In: 10th ACM SIGSOFT Symposium on Foundations of Software Engineering, pp. 1–10. ACM (2002)
10. Zeller, A.: Why Programs Fail: a Guide to Systematic Debugging, 2nd edn. Morgan Kaufmann Publishers (2009)
11. Liu, X., Lin, W., Pan, A., Zhang, Z.: WiDS Checker. In: 4th USENIX Conference on Networked Systems Design & Implementation, pp. 257–270. USENIX (2007)
12. Geels, D., Altekar, G., Maniatis, P., Roscoe, T., Stoica, I.: Friday: Global Comprehension for Distributed Replay. In: 4th USENIX Conference on Networked Systems Design & Implementation, pp. 285–298. USENIX (2007)
13. Shin, H., Endoh, Y., Kataoka, Y.: ARVE: Aspect-Oriented Runtime Verification Environment. In: Sokolsky, O., Taşıran, S. (eds.) RV 2007. LNCS, vol. 4839, pp. 87–96. Springer, Heidelberg (2007)
14. Sen, K., Kalasapur, S., Brutch, T., Gibbs, S.: Jalangi: A Selective Record-Replay and Dynamic Analysis Framework for Javascript. In: 9th Joint Meeting on Foundations of Software Engineering, pp. 488–498. ACM (2013)
15. Wang, Y., Patil, H., Pereira, C., Lueck, G., Gupta, R., Neamtiu, I.: DrDebug: Deterministic Replay based Cyclic Debugging with Dynamic Slicing. In: 12th IEEE/ACM International Symposium on Code Generation and Optimization, pp. 98–108. ACM (2014)
16. Eichelberger, H., Kropf, T., Greiner, T.: Rosenstiel. W.: Runtime Verification Driven Debugging of Replayed Errors. In: Proceedings of the Ph.D. Workshop of ICTSS 2013, pp. 1–4 (2013), http://ictss.sabanciuniv.edu/phd-workshop

# Service Layer for IDE Integration of C/C++ Preprocessor Related Analysis

Richárd Dévai[1], László Vidács[2], Rudolf Ferenc[1], and Tibor Gyimóthy[1]

[1] Department of Software Engineering, University of Szeged, Hungary
Devai.Richard@stud.u-szeged.hu, {ferenc,gyimothy}@inf.u-szeged.hu
[2] MTA-SZTE Research Group on Artificial Intelligence, Hungary
lac@inf.u-szeged.hu

**Abstract.** Software development in C/C++ languages is tightly coupled with preprocessor directives. While the use of preprocessor constructs cannot be avoided, current IDE support for developers can still be improved. Early feedback from IDEs about misused macros or conditional compilation has positive effects on developer productivity and code quality as well. In this paper we introduce a service layer for the Visual Studio to make detailed preprocessor information accessible for any type of IDE extensions. The service layer is built upon our previous work on the analysis of directives. We wrap the analyzer tool and provide its functionality through an API. We present the public interface of the service and demonstrate the provided services through small plug-ins implemented using various extension mechanisms. These plug-ins work together to aid the daily work of developers in several ways. We provide (1) an editor extension through the Managed Extensibility Framework which provides macro highlighting within the source code editor; (2) detailed information about actual macro substitutions and an alternative code view to show the results of macro calls; (3) a managed package for discovering the intermediate steps of macro replacements through a macro explorer. The purpose of this work is twofold: we present an additional layer designed to aid the work of tool developers; second, we provide directly usable IDE components to express its potentials.

## 1  Introduction

Preprocessor directives – like macros and conditional compilation – constitute an integral part of the source code of C/C++ software, especially when applications are built for several target architectures, or in case of software product lines where several parallel configurations exist in the code [11]. An empirical study on open source applications shows that preprocessor directives make up a relatively high 8.4% of source code lines on average [3]. Although the preprocessor is useful for forward engineering and development, it behaves as an obstacle in case of program understanding and reverse engineering tasks. The fundamental problem about preprocessing from a program comprehension point of view is that the compiler gets the preprocessed code and not the original source code

B. Murgante et al. (Eds.): ICCSA 2014, Part V, LNCS 8583, pp. 402–417, 2014.

that the developer sees. In many cases the two codes are markedly different. These differences influence the code quality in case of directive-intensive programs. Heavy use of directives is usually considered harmful [17] and it results in weak code quality and maintainability. A large amount of work addressed the elimination of directive use, with only partial results. The first point where the software developer is facing problems with macros is when a runtime error occurs at a source code line which contains macros only. The usual debugger stops at the line in question, but there is no information on what is the real code that the compiler used. Besides constant-like macros, many times the macro name is replaced by whole C/C++ loops or complex expressions spreading across several lines – all hidden from the developer. Furthermore, several macros have multiple definitions depending on conditional directives, which fact makes it hard to find the actual definition manually. These labor-intensive activities can increase the overall effort spent on development or maintenance tasks.

Widespread integrated development environments today, like the Visual Studio for C++ [14], give fairly limited support for the developer. As the preprocessor language is independent from the C/C++ language, the analysis of directives requires a separate analyzer, extra risk and effort for tool developers. Although the benefits of such extension are clear, out of the box solutions are usually not shipped with IDEs, and the developers are still forced to do workarounds to investigate macro calls. In our recent work [25] we introduced a Visual Studio Add-In that utilizes preprocessor related analysis and presents macro folding information together with a static view on macro calls. This paper builds on previous results and takes into account two observations: first, macro folding required to alter the source code, which is not acceptable by developers; and second, dynamic views are much more usable in concrete scenarios than static views introduced previously. In this work we intend to keep the workplace of the developer clean, while adding small pieces of information in a targeted and dynamic way. In line with the philosophy of flexible development environments and extensibility mechanisms, we provide a service layer for directive related information. This means that we use a wrapper for our preprocessor analyzer tools and provide an API for IDE extension developers to access macro calls, definitions, etc. Visual Studio extensions introduced in this paper also use this service layer to demonstrate its functionality.

In this paper we present the following main contributions:

1. Service layer for preprocessor related analysis
2. IDE extensions built up on service layer
   - Macro and conditional directive highlights in the code editor
   - Side-by-side view to reveal the results of macro expansions
   - Macro definitions view to follow macro expansion internals

The paper is organized as follows: we briefly mention current IDE capabilities and propose additional views of the source code as a motivating example in the next section. Section 3 introduces extension mechanisms used in recent versions of Visual Studio, while in Section 4 we present our service layer to enable access for preprocessor related analysis results. We present new source code views as a demonstration of the usability of the service layer using various extensibility mechanisms in Section 5. Related work is discussed in Section 6, while conclusions and future plans are outlined in Section 7.

## 2  Motivation

The source of program understanding problems is that the compiler gets the preprocessed code and not the original source code that the programmer sees. Let us consider a compiler error message pointing to line 49 in Figure 1:

```
46
47 ⊟void TiXmlString::reserve (size_type cap)
48   {
49     FILT_WND(clear);
50     if (cap > capacity())
51     {
         T:V... IC... ... . ....
```

**Fig. 1.** Example C++ code in Visual Studio code editor

To resolve the problem the developer has to look for the macro name in the code to find out the replaced text, which was actually compiled. The Visual Studio provides several ways for code search, one may search among files in the project, or even select to find classes or other program entities. Unfortunately the preprocessor language is a pure textual language and is unrelated to C or C++. Using the *Find references* feature one can find the #define directive of the FILT_WND macro. Yet, there can be several results, because the same macro name can be defined in several ways using conditional compilation. This enables to have different replacement texts for each platform or configuration. The next question is: Which one of the definitions is finally used? In recent versions the Visual Studio helps with conditional highlights, but even after finding the right definition, usually the search continues, because the macro definition contains further macro calls to look for. This is a tedious and time consuming task.

```
46
47    void TiXmlString::reserve (size_type cap)
48    {
49      if (security_level > 2) { clearWindowW(log) };
50      if (cap > capacity())
51      {
        T:V... IC:... . .
```

**Fig. 2.** Preprocessed code view using IDE extension

A better solution could be to alter the project configuration to produce the preprocessed file as well during the building process. One can compare the original and preprocessed code using .i files. However in this case the reason for the compiler error can be seen, but the concrete macro definition remains hidden, where the error could be corrected.

Our contributions include Visual Studio plug-ins to aid developers overcome these problems – like macro highlights in Figure 2. Furthermore, we provide a service layer to integrate preprocessor information into Visual Studio in a usable way for plug-in developers.

## 3   Visual Studio Extension Frameworks

In this section we first outline the history of plug-in possibilities of Visual Studio, and then briefly compare the most recent solutions in terms of their usability. The history of Visual Studio extension mechanisms goes back to the very first release. We distinguish four main types of extension mechanisms which were supported in recent releases:

- *Visual Basic for Application Macros*
- *Visual Studio Extensibility* and *Automation Add-Ins*
- *Managed Package Framework*
- *Editor Extension Point* components

The type of extension we should use is strongly depends on the objective we would like to achieve. From our point of view majority of interest lies packages, but our service layer can be consumed by any type of extensions. In this section the main historical changes of extensions are briefly introduced and at last we will give some direction about their common usage. We do not discuss *VBA Macros* as they were removed with the 2012 release.

To give a brief overview of changes in Visual Studio releases on extension frameworks we refer to Table 1. As shown in the table, development of *Add-Ins* stopped in the release in 2012, but generally prior versions brought only slight improvements of its API (we present Development Tools Environment (DTE) versions in the table). At the beginning *Add-In* APIs were poorly documented, but it was compensated by web sources – some driven also by Microsoft specialists – like *MZ-Tools*[1]. At the beginning, these resources approached from the Visual Basic side, while the C# support and the low number number of examples could be told more insufficient for beginners. There are also slight semantic differences between the usage and behavior of APIs for Add-Ins because of lingual differences, and one has to understand architectural flavors of the IDE to touch it the right place and with proper technique. After the opening of *Visual Studio Code Gallery* the support was improved by the increased number of examples. Search options and categorisation enhanced the proper use of different extension types. Open source projects both on *CodePlex*[2] and *GitHub*[3] meant a great help for developers in discovery of *Add-In* API's capabilities.

---

[1] MZ-Tools Add-In resources:
   http://www.mztools.com/resources_vsnet_addins.aspx
[2] CodePlex: https://www.codeplex.com
[3] GitHub: https://github.com

**Table 1.** Visual Studio extension type changes in recent releases

| Release | 2005 | 2008 | 2010 | 2012 | 2013 |
|---------|------|------|------|------|------|
| Add-In | DTE80 | DTE90 | DTE100 | | Deprecated |
| MPF | Release | | WPF API | | |
| EEP | | | Release | | |

For a deeper level of integration *MPF* can come into the picture. At first *MPF* was supposed to be used by Microsoft partners for commercial purposes. Only Microsoft's VSIP partners were able to distribute packages until the Visual Studio release in 2010. Hence its use was not spread as widely as in case of Add-In's. Users of MPF could rely only experienced partnership developer's blogs, but the Add-Ins remained first choice because of their simplicity. In 2010 Packages became freely available, but Add-Ins are still more supported by the community compared to them.

The newest and suppletory extension type for Visual Studio is *Editor Extension* as shown in Figure 1. Editor extensions are based on a lightweight plug-in technology called Managed Extensibility Framework which was also introduced in 2010. Their most important advantage is their simplicity. They can be used as editor enhancements to highlight code by formatting tokens through language classifiers, to add tags for background highlight, and to decorate code in the editor with custom UI elements using adornments.

In terms of API capabilities, packages enable access to deeper integration level, and besides they provide custom and consumable services through public APIs. Their main purpose is to extend Visual Studio with so called *Language Services*, hence they can be used to build up complex services as well. On the other hand, their use requires higher level of expertise. Packages are the base building blocks of the IDE and they even make it possible to reuse the IDE on other purposes with Shell Isolated packages like the *MS SQL Management Studio*. Partially the services provided for *MPF* components also available from *Add-Ins*, but they are designed to use a simple and lightweight API, so this is not a common method of their usage. The connection type of these two extensions are different at many points. It is advised by Microsoft from the release of 2013 to migrate *Add-Ins* to *MPF*. Except the simplified event system (also used for connection of Add-Ins) packages can use all API features provided for Add-Ins.

With the appearance of *Editor Extensions* one can access and easily extend the editor's presentation layer showing additional information to the programer in a lightweight method. Editor extensions owned pretty high interest as part of the reworked and *Windows Presentation Foundation* based IDE released in 2010. They give supplementary support for code highlighting and other feedback abilities just inside the editor window was depend on *Language Services* before, and can't be accessed through separated, lightweight API.

To summarize extensions by their most common usage, *Add-Ins* are the best choice whether we would like to extend the IDE with a standalone application as easy as possible. Add-Ins providing services directly to the user without a registered API, supported with controllers and tool windows to ease usability. An Add-In can use services of packages, and can manage commands registered either by itself or by other extensions. Even nowadays Add-In is a popular choice to create a simple extension component, however for complex applications it is more common to use MPF (as we also did). Editor extensions are mostly essential to give text decoration/highlight support, since their usage is limited.

## 4  Service Layer to Access Macro Expansions

The primary goal of our service layer is to extend macro-related capabilities of the Visual Studio IDE and to enable easier access to detailed preprocessor analysis. Our aim was to ease of its access from as many types of Visual Studio extensions as possible. While services registered by packages can be accessed form all type of extensions, we decided to prefer this technique over others. In the following we present an overview on the context and structure of the service, while its use is demonstrated in Section 5.

### 4.1  Service Layer Architecture and Context Outline

The Visual Studio extension structure is outlined in Figure 3. The central part of the figure is the Service Provider component of the IDE. It is responsible for dispatching services to various types of plug-ins. The *PPService* component represents our published layer of services and internal processes behind it. This

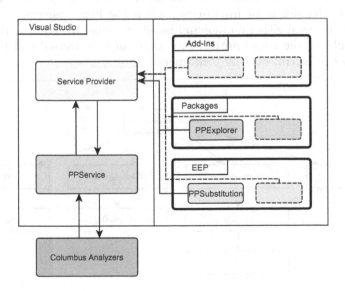

**Fig. 3.** Context and higher level architecture of the service layer

component registers our service into the Visual Studio's service provider and also uses provided services. The core functionality of the *PPService* component is implemented by *Columbus Analyzers*. This component contains the *Preprocessor Analyzer* (CANPP) and builds our internal representation of the compilation unit (see 4.3 for more details). The service component depends on various data extractors like the Preprocessor ASG extractor as well.

The service is based upon different plug-in interfaces of Visual Studio that provided for MPF and Extensibility Framework. On the right hand side of the figure three actual types of plug-ins are represented. These plug-ins use our services through the provider component. This part of the figure presents a sample package and an editor extension we implemented to demonstrate the capabilities of the service. Note that our service is joining the service provider through an MPF interface, so the *PPService* component could also belong to the set of *Packages* on the right hand side of the picture.

### 4.2    Service Layer API

The service component hides the analysis process of directives and maintains a central repository for preprocessor details of the whole solution. The internal analysis can be triggered by some user action in the IDE (eg.: the build of a project). The component raises events about recent actions (eg.: about finished analysis), and API consumers are notified about these events. In case of course code changes new results invalidates previous data set of the actual project. The repository is designed for a later support of versioning mechanisms to track differences between analyzed versions. The current mechanism is connected to builds, which means that directive evaluation does not takes place during typing as the syntax highlights or intellisense, which use fuzzy parsers. Analysis results can be accessed either on demand based on actual file name and positions within the file, or the whole project structure can be traversed and the needed information can be collected by consumers.

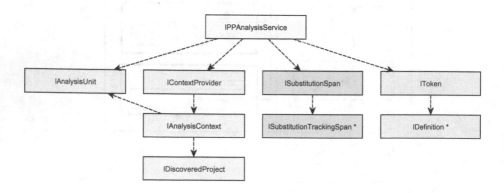

**Fig. 4.** Interface hierarchy of the service layer

In the rest of this section we give a brief description of our preprocessing data access interfaces presented in Figure 4. Interfaces are accessible through the *IPPAnalysisService*, which is registered by the package into the service provider. This interface is the base of data sharing, placed at the top of Figure 4. IContextProvider can be used to arrange analysis data to projects. Data can be extracted from *IDiscoveredProject* and an exact analysis round is presented by *IAnalysisConext*. The data for different source files are handled by *IAnalysisUnit*.

We would like to give the chance to a consumer to access simpler data without extracting details through complex traversals on our internal representation. Therefore results are accessible through different kinds of perspectives: a simpler and easy to use interface handling whole replacement texts, and a detailed interface where each token can be accessed.

Providing most basic type of data *ISubstitutionSpan* and its tracking version are available. We can access space information about all kind of macro names that are subjects of substitution process during macro processing. This interface gives information about macro calls and their substituted counter parts in code and preprocessed code file. Tracking version of the interface is ready to be used in Editor Extensions to create highlights on code files in IDE. These spans are also classified by different type of macro name areas like normal macro names and conditional subtypes as clauses evaluated to true or false, etc.

On next level the service provides data in tree structures. This interface group reflects our detailed internal representation (introduced in the next section below). These interfaces represent text tokens and their associated definitions used during macro substitution progress in a simplified structure. *IToken* interface gives information about the text, its file position, the substitution related child which is an other *IToken* calculated by the preprocessor during macro substitution. In addition, contains information about the used definition which lead to the child, finally information about whether the actual token is function or it is within the condition of a conditional directive. Definitions are presented through the *IDefinition*interface, which provides name and placement information. It also implements an interface called *IVsCodeDefViewContext* which makes the user able to feed the definition into the service of Visual Studio's Code Definition Window. Traverse of the token tree is also aided by several extension methods, although the actual analysis data is available easily through *IPPAnalysisService* by file name as well.

To present the usage of our interfaces we give a brief sample from the Macro Explorer managed package written in C# in Listing 1. A picture of this view in Visual Studio can be seen in Figure 8.

```
ISubstitutionSpan actSpan =
  _spans.FirstOrDefault(
    span => span.LineOriginal == line &&
      (span.ColumnOriginal <= column &&
       span.ColumnOriginal +
       span.LengthOriginal >= column));

if (actSpan != null) {
```

```
IToken actToken =
  _tokens.FirstOrDefault(
    token => token.Line - 1 == actSpan.LineOriginal &&
    token.Column - 1 == actSpan.ColumnOriginal);

if (actToken != null) {
  explorerTree.Items.Clear();
  TreeViewItem root = AddItem(null, actToken);
  if (actToken.UsedDefinition != null) {
    jumpCodeDefinitionToDefinion(
          actToken.UsedDefinition);
  }
  explorerTree.Items.Add(root);
}
}
```

**Listing 1.** Interface consuming code sample

In the example we search for a target span by line and column information. Whether we find the required span then we select the joining token from the actual token collection by the head column and line number of selected span. Than we clear the tree view shown by the Macro Explorer. Next we traverse the token tree and build the represented hierarchy with *TreeViewItem*s. After that we give the definition of the root to the service of the Code Definition Window and at last we add the *TreeViewItem* created for the root to the tree view.

## 4.3   Internal Representation of Directives

Detailed analysis is done by the analyzer tool of the Columbus Framework. In our previous work we defined a schema (metamodel) for the preprocessor [22]. The Columbus Schema for C/C++ Preprocessing describes the original source code, the final preprocessed code and all transformation steps in between. Schema instances represent preprocessor constructs of concrete programs. We analyze one configuration at a time (dynamic instances.) Our representation contains

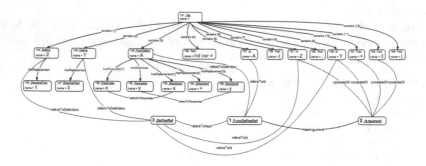

**Fig. 5.** Sample schema instance of a function-like macro

all kinds of preprocessor construct, however in current work we use the macro-related part of the schema. Schema instances are produced by a tool, which can be smoothly incorporated into build processes, as it behaves as a usual preprocessor as well. Using schema instances macro expansions can be tracked at all places of programs, e.g. in conditional expressions as well, in a step-by-step way. The output of the tool can be written out in XML format. Figure 5 presents the dynamic schema instance of a function-like macro example. Macro definitions are denoted with (Func)Define nodes. Definitions contain replacement text and may contain parameters. Macro calls are linked to active macro definitions via (Func)DefineRef reference nodes, which also mark actual arguments of function-like macros. Detailed macro call information, including all internal steps, can be extracted by traversing the instance graph along these references. These references take into account conditional compilation and concatenation and stringize operators used within macro replacement texts. For further information we refer to our previous work [22].

## 5   Applications of the Service Layer

In this section we present how the service layer is used to implement various mechanisms to support program understanding of preprocessor-related program parts. In our previous work we implemented a folding mechanism integrated into the code editor window [25]. The folded and unfolded states in the source code window correspond well to macro names (folded) and replacement texts (unfolded). Macro folds can also be nested in a way to show nested macro calls and even argument substitutions in case of function-like macros. The mechanism is appropriate for presenting the whole macro expansion process in a step by step manner. However due to practical reasons we needed to revise this concept. To implement folding in code edit window the actual source code must be modified in two ways: (1) folding markers (▼, ▲, ► and ◄) are inserted and (2) macros replacement text is replaced directly within the editor. To keep the code editor clean from unwanted modifications, a different approach is used in this paper. Code highlights are used to mark macros in the source code and a parallel window is shown to check the results of the macro replacement. In addition, the process of macro replacements can be followed using the macro explorer.

Figure 6 shows a Visual Studio screen, where all components of our tool set are present: (1) code editor window; (2) side-by-side view for parallel observation of the original code and the result of preprocesing; (3) macro definition explorer; and (4) macro definitions view.

### 5.1   Macro Highlights and Side-by-Side View

As mentioned above, the motivation for highlighting code is to avoid changing the code just for presenting macro information. The aim is to give hands-on information around the code editor, but leave it untouched as much as possible. The only change in the outlook of the code editor is that macro calls are highlighted (see the light green code in the left hand side of Figure 7). Code highlight

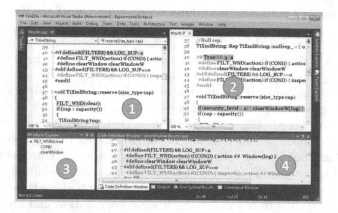

**Fig. 6.** Macro related extensions built on the service layer in Visual Studio

positions can be obtained using the service layer API. The natural question that follows highlight is to check what is the final value of the macros after preprocessing. This is a central problem for a developer in a situation as outlined in our motivating example. The compiler in fact compiles the replaced text, which is not visible in a usual code editor. In case of an error message pointing at the position of the macro call, the developer needs to perform code search for the corresponding macro definition(s) to see what went wrong. Our service layer also can generate a copy of the source code for a user request where Macro names are being substituted. However in this format Macro calls are being replaced with the final macro replacement texts, but comments are held on to keep the code near to the origin version as possible. The *Macro Explorer* window is designed to be placed next to the code editor, thus parallel view of the original source code and the corresponding preprocessed code can be observed. Replaced code is highlighted using different color than in code editor as can be seen in Figure 7.

## 5.2 Macro Explorer and Definitions View

Macro highlights and the side by side view provide static information on macro replacements. Although observing the final preprocessed code is mostly sufficient, macro bodies may contain further macro calls, in many cases even 10-20 macros take part in a full macro replacement process. In these cases one may have to investigate internal steps of macro replacement. Searching manually for each macro definition is a time consuming task. Several macros have even more than one definitions surrounded by #if conditions. These conditions depend on macros as well, hence selecting the right one from several definitions is not straightforward. The macro explorer view and the code definition view is prepared for these situations. The macro explorer is a managed package and provides a hierarchical view of macro definitions in the order they take part in the macro replacement (see Figure 8).

```
40  ⊟#if defined(FILTERS) && LOG_SUP>2
41  |  #define FILT_WND(action) if (COND) { action ## Window(log) }
42  |  #define clearWindow clearWindowW
43  ⊟#elif defined(FILTERS) && LOG_SUP==0
44  |  #define FILT_WND(action) if (COND) { inspect(1); action ## Window() }
45     #endif
46
47  ⊟void TiXmlString::reserve (size_type cap)
48     {
49        FILT_WND(clear);
50        if (can > canacity())
```
110 %   ▾

```
tinystr.if  ✕
39
40  #if True && 3>2
41     #define FILT_WND(action) if (COND) { action ## Window(log) }
42     #define clearWindow clearWindowW
43  #elif defined(FILTERS) && LOG_SUP==0
44     #define FILT_WND(action) if (COND) { inspect(1); action ## Window() }
45  #endif
46
47  void TiXmlString::reserve (size_type cap)
48  {
49     if (security_level > 2) { clearWindowW(log) };
50     if (cap > capacity())
```

**Fig. 7.** Side-by-side view and macro highlight example in Visual Studio (two side windows shown below each other)

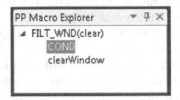

**Fig. 8.** Macro definitions explorer view in Visual Studio

Selecting any of the definitions in the explorer activates the code definition view to position to the selected macro definition in the source code (in the bottom part of the IDE ). Note that definitions are usually placed in separate headers, not in the same file as the edited source code, where the macro is originally called. Several headers also take part in the include hierarchy, and the actual definition may be contained by them. The positioning also takes into account the conditional directives as can be seen in Figure 9.

Last but not least, the concatenating operator may cause strange constructs which are hard to find manually and also possibly result in coding problems and weak code quality. Using the concatenating operator (## in the replacement text of a macro, the macro parameter can be concatenated to a fixed string and the resulting token may produce a new, hidden macro call. These type of calls are

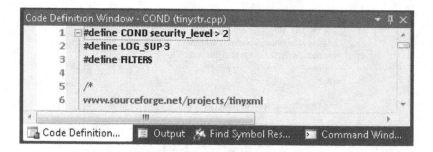

**Fig. 9.** Code definition window example in Visual Studio

rare, but could not be found by code search, as the macro name is not present in the code in its final form. The macro explorer helps developers overcome these situations as well, although our tool is currently a research prototype. Analysis of the Mozilla Firefox revealed 24 such concatenations which resulted in new macro calls, and these calls are used 337 times, which is not negligible.

## 6   Related Work

Preprocessor directives are still widely used as no real size program with configurations exist without them. Ernst, Badros and Notkin [3] analyzed 26 commonly used Unix software packages and found that preprocessor directives made up the relatively high 8.4% of lines on average.

To overcome the preprocessor as a barrier in program understanding, researchers tackled problems of various areas. Analysis and visualization of include directives is a research topic from the early years, while in a recent work Spinellis [19] proposes a solution for the automatic removal of unnecessary includes, based on computed dependencies of program elements. Dealing with software configurations is a well studied topic as well. Krone and Snelting [7] proposed concept lattices to aid reengineering configurations. Latendresse [9,10] proposed a symbolic evaluation algorithm for finding the conditions required for a particular source line to get through the conditional compilation. CViMe and C-CLR tools are Eclipse plugins, which collect and present configuration-controller macros [16]. Sutton and Maletic implemented analyzer tools on the top of the srcML infrastructure to reveal portability issues based on include files and configuration macros [20]. In the work of Garrido the analysis of preprocessor constructs was integrated into the C refactoring tool, where she implemented a configuration independent solution [5,6]. The Refactoring Browser by Vittek [26] carries out automated modifications on a C source code. An interesting idea in this work is that of handling macros as special include files (the macro body is included), but handling of ## operators is not solved in some cases. To handle the problem of configurations, this tool relies on user input. Livadas and Small developed a preprocessor inserting special lines into the preprocessed file to support the source code highlighting methods of the Ghinsu program slicing tool [12].

Those working on C or C++ analyzers are confronted by the problem of pre-processor directives. Therefore, a lot of effort has been made to avoid their usage. Mennie and Clarke proposed a method to transform some macros and condition-als into C/C++ code [13]. Spinellis tackled the problem of global renaming of variables, preprocessor-aware solutions have been implemented in the CScout tool [18]. Saebjoernsen et al. [15] propose a mapping between the C language and the preprocessor to find inconsistent macro usage. The preprocessor-problem occurs also in the context of aspect mining and aspect-refactoring. Adams et al. worked on the problem of aspect refactoring, and also how to refactor various conditional compilation usage patterns into aspects [1]. In our previous work, we defined the macro dependency graph (MDG) for dependence based slicing of pre-processor macros [24]. Using the MDG C++ slices were extended with macro slices and better precision is achieved in case of more than 75% of backward slices [23]. Despite the wide range of initiations, current software development tools lack of support for the developers. In a recent paper Feigenspan et al. investigate the use of coloring techniques depending on the preprocessor con-ditionals in the FeatureCommander tool [4]. Folding is an interactive extension of the textual view of the source code. The idea of folding in the context of preprocessing is presented by Kullbach and Riediger [8] and we applied the idea in our previous work. The folding mechanism was successfully employed within the GUPRO program understanding environment [2]. In this work we targeted the IDE and decided not to touch the code but use highlighting and parallel code window instead. The Understand for C++ reverse engineering tool pro-vides cross references between the use and definition of software entities [21]. This includes the step-by-step tracing of macro calls in both directions as well. The tool is appropriate for tracking back the uses of a give macro definition but the information is imprecise in certain situations like macro calls generated by ## operators. A similar solution to macro folding is implemented in the Emacs editor. In C-mode, the M-x c-macro-expand command in Emacs will run the C preprocessor on the actual region and display the results in another buffer. This is similar to unfolding a macro. Besides the folding mechanism shown in our previous paper, we provide parallel code window and a more intuitive view for stepwise investigation of macro definitions taking part in the expansion process.

# 7    Conclusions

Advanced integrated development environments influence the daily work of de-velopers, thus having positive effect on productivity. IDEs also provide extension mechanisms to include plug-ins. Plug-ins enrich the environment to support sev-eral aspects of coding and help the developer maintain high code quality. In this paper our aim was to integrate preprocessor related analysis in Visual Studio. Built on our previous works on detailed macro analysis and IDE Add-Ins, we extend the capabilities of the Visual Studio with a service layer to provide ac-cess for other plug-ins to the internals of preprocessing. We presented the high level and low level architecture of this service and demonstrated its usability. We implemented packages and an editor extension to show the interoperability of mechanisms, and showed how these plug-ins can aid the daily work of developers. Using our services and tools one can overcome obstacles in program understanding caused by preprocessor conditionals, dependent macro definitions and concatenated macro parameters.

Our future plans include a usability study with the help of developers to identify directions of enhancements. We already identified rooms for development in integrating folding in code view, improving performance and implementing synchronized side-by-side view windows.

**Acknowledgments.** This research was supported by the Hungarian national grant GOP-1.1.1-11-2011-0006.

# References

1. Adams, B., De Meuter, W., Tromp, H., Hassan, A.E.: Can we refactor conditional compilation into aspects? In: AOSD 2009: Proceedings of the 8th ACM International Conference on Aspect-Oriented Software Development, pp. 243–254. ACM, New York (2009)
2. Ebert, J., Kullbach, B., Riediger, V., Winter, A.: GUPRO - Generic Understanding of Programs. In: Mens, T., Schürr, A., Taentzer, G. (eds.) Electronic Notes in Theoretical Computer Science, vol. 72, Elsevier (2002)
3. Ernst, M.D., Badros, G.J., Notkin, D.: An empirical analysis of C preprocessor use. IEEE Transactions on Software Engineering 28(12) (December 2002)
4. Feigenspan, J., Kästner, C., Apel, S., Liebig, J., Schulze, M., Dachselt, R., Papendieck, M., Leich, T., Saake, G.: Do background colors improve program comprehension in the #ifdef hell? Empirical Software Engineering 18(4), 699–745 (2013)
5. Garrido, A., Johnson, R.: Analyzing multiple configurations of a c program. In: Proceedings of the 21st International Conference on Software Maintenance (ICSM 2005), pp. 379–388. IEEE Computer Society (2005)
6. Garrido, A., Johnson, R.: Embracing the c preprocessor during refactoring. Journal of Software: Evolution and Process 25(12), 1285–1304 (2013)
7. Krone, M., Snelting, G.: On the inference of configuration structures from source code. In: Proceedings of ICSE 1994, 16th International Conference on Software Engineering, pp. 49–57. IEEE Computer Society (1994)
8. Kullbach, B., Riediger, V.: Folding: An Approach to Enable Program Understanding of Preprocessed Languages. In: Proceedings of the 8th Working Conference on Reverse Engineering (WCRE 2001), pp. 3–12. IEEE Computer Society (2001)
9. Latendresse, M.: Fast symbolic evaluation of C/C++ preprocessing using conditional values. In: Proceedings of the 7th European Conference on Software Maintenance and Reengineering (CSMR 2003), pp. 170–179. IEEE Computer Society (March 2003)
10. Latendresse, M.: Rewrite systems for symbolic evaluation of c-like preprocessing. In: Proceedings of the 8th European Conference on Software Maintenance and Reengineering (CSMR 2004), pp. 165–173. IEEE Computer Society (March 2004)

11. Liebig, J., Apel, S., Lengauer, C., Kästner, C., Schulze, M.: An analysis of the variability in forty preprocessor-based software product lines. In: Proceedings of the 32Nd ACM/IEEE International Conference on Software Engineering, ICSE 2010, vol. 1, pp. 105–114. ACM, New York (2010)

12. Livadas, P., Small, D.: Understanding code containing preprocessor constructs. In: Proceedings of IWPC 1994, Third IEEE Workshop on Program Comprehension, pp. 89–97 (November 1994)

13. Mennie, C.A., Clarke, C.L.A.: Giving meaning to macros. In: Proceedings of IWPC 2004, pp. 79–88. IEEE Computer Society (2004)

14. Microsoft Visual Studio (2014), http://www.microsoft.com/visualstudio/

15. Saebjoernsen, A., Jiang, L., Quinlan, D.J., Su, Z.: Static validation of c preprocessor macros. In: AOSD 2009: Proceedings of the 8th ACM International Conference on Aspect-Oriented Software Development, pp. 149–160. IEEE Computer Society (2009)

16. Singh, N., Gibbs, C., Coady, Y.: C-clr: a tool for navigating highly configurable system software. In: ACP4IS 2007: Proceedings of the 6th Workshop on Aspects, Components, and Patterns for Infrastructure Software, p. 9. ACM, New York (2007)

17. Spencer, H., Collyer, G.: #ifdef considered harmful, or portability experience with C News. In: USENIX Summer Technical Conference, pp. 185–197 (June 1992)

18. Spinellis, D.: A refactoring browser for c. In: ECOOP 2008 International Workshop on Advanced Software Development Tools and Techniques (WASDeTT) (2008)

19. Spinellis, D.: Optimizing header file include directives. Journal of Software Maintenance and Evolution: Research and Practice 22 (2010)

20. Sutton, A., Maletic, J.I.: How we manage portability and configuration with the c preprocessor. In: Proceedings of the 23rd International Conference on Software Maintenance (ICSM 2007), pp. 275–284 (2007)

21. Understand for C++ homepage (2009), http://www.scitools.com

22. Vidács, L., Beszédes, A., Ferenc, R.: Columbus Schema for C/C++ Preprocessing. In: Proceedings of CSMR 2004 (8th European Conference on Software Maintenance and Reengineering), pp. 75–84. IEEE Computer Society (March 2004)

23. Vidács, L., Beszédes, A., Ferenc, R.: Macro impact analysis using macro slicing. In: Proceedings of ICSOFT 2007, The 2nd International Conference on Software and Data Technologies, pp. 230–235 (July 2007)

24. Vidács, L., Beszédes, Á., Gyimóthy, T.: Combining Preprocessor Slicing with C/C++ Language Slicing. Science of Computer Programming 74(7), 399–413 (2009)

25. Vidács, L., Dévai, R., Ferenc, R., Gyimóthy, T.: Developer Support for Understanding Preprocessor Macro Expansions. In: Kim, T.-h., Ramos, C., Kim, H.-k., Kiumi, A., Mohammed, S., Ślęzak, D. (eds.) ASEA/DRBC 2012. CCIS, vol. 340, pp. 121–130. Springer, Heidelberg (2012)

26. Vittek, M.: Refactoring browser with preprocessor. In: Proceedings of the Seventh European Conference on Software Maintenance and Reengineering (CSMR 2003), Benevento, Italy, pp. 101–110 (March 2003)

# A State-Based Testing Method for Detecting Aspect Composition Faults

Fábio Fagundes Silveira[1], Adilson Marques da Cunha[2], and Maria Lúcia Lisbôa[3]

[1] Federal University of São Paulo – UNIFESP, São José dos Campos, Brazil
fsilveira@unifesp.br
[2] Aeronautics Institute of Technology – ITA, São José dos Campos, Brazil
cunha@ita.br
[3] Federal University of Rio Grande do Sul – UFRGS, Porto Alegre, Brazil
llisboa@inf.ufrgs.br

**Abstract.** Aspect-Oriented Software Development is a contemporary technique of software development that aims to improve the separation of concerns issues faced by traditional approaches. It improves the modularity of crosscutting concerns into units called *aspects*. However, this feature raises concerns about the quality of aspect-oriented programs (AOP). Existing functional testing approaches do not directly investigate the aspect composition problem, its resultant interactions and representation on a dynamic model. This paper describes a state-based testing method for AOP that targets class–aspect and aspect–aspect faults. To support the developed method, we introduce a model to represent the dynamic behavior of aspects interactions, a strategy to derive testing sequences, and a testing tool. Results of our assessment show that the approach is capable of detecting faults based on fault-models available in the literature. Furthermore, it is able to reveal another source of faults on AOPs, the aspect composition fault.

**Keywords:** aspect-oriented programming, aspect composition fault.

## 1 Introduction

Aspect-Oriented Software Development (AOSD) [1,2] is a contemporary technique of software development that aims to improve the separation of concerns issues faced by traditional approaches such as Object-Oriented and Procedural programming. Aspect-Oriented (AO) development improves the modularity of certain crosscutting concerns into units called *aspects*. Besides separation of concerns, AOSD also enables reuse, and aspects composition is one the keys to achieve it.

Weaving aspects into OO programs can modify not only their structures, but also relationships among their components. The main functionalities of programs can be combined with the ones delivered by aspects, which in turn can be merged among themselves.

Whenever a new software engineering technique emerge, it is necessary to promote investigations to verify the reusability of methods, techniques, strategies and testing tools. AOSD can be considered as an evolution of the Object-Oriented (OO) paradigm, thus it needs specific testing approaches. Testing of AO programs has gained considerable attention from researchers over the years. There have been significant advances in

B. Murgante et al. (Eds.): ICCSA 2014, Part V, LNCS 8583, pp. 418–433, 2014.

testing AO programs (AOP), regarding structural [3,4,5], functional [6,7,8] and mutation testing techniques [9,7,10]. Regarding functional-based testing technique, existing approaches do not directly investigate the aspect composition problem [11], its resultant interactions, and their representation on a dynamic model.

In this paper, we present a state-based testing method for AO programs. It provides class–aspect and, more specifically, aspect–aspect faults detecting capabilities. In addition, to support the developed method, a model to represent dynamic behavior of aspects interactions, an extended strategy to derive testing sequences, and a testing tool prototype are introduced.

To evaluate the method, we have applied it into three different case studies, from distinct domains. The strategy adopted to assess the application of the method consisted in measuring the effectiveness of testing sequences generated by this method through the use of mutation testing [12]. One of these experiments is described here, along with its results analysis and discussions. The results of our assessment show that the method is capable of detecting faults present on the fault-model proposed in [13,14]. Furthermore, it is able to reveal the fifth source of faults on AOPs, introduced by our previous work, the aspect composition fault [11].

The remainder of this paper is organized as follows. Section 2 introduces aspects composition concepts adopted by this work and AO software testing. Section 3 briefly summarizes related works on AO testing and modeling. In Section 4, the major contribution of the paper is presented along with its components and a case study description. Next, in Section 5, the main results and discussion are described. Finally, concluding remarks and future works are given in Section 6.

## 2   Aspects Composition and AOP Testing

Abstractly, the idea of executing together software elements created separately is called *composition*. In this paper as well as on the AOP perspective considered here, the term *aspects composition* refers to relationships among aspects (i.e., generalization, association and dependency) weaved into the target system. Thus, composition has a completely different meaning from weaving and/or combination process. The latter occurs when one or more aspects are weaved into an application without necessarily interacting to each other [11]. Different languages provide several techniques of composition, including subprograms invocation, inheritance and generic instantiation. Aspects composition schemes refer to conceptual models and mechanisms for composition such as inheritance, aggregation and delegation [15].

In this paper, the "⊕" symbol is used to represent a composition schema. For instance, concerns $C_1$ and $C_2$ can be composed both to raise a $C_3$ concern or to represent that these concerns are able to get running together, possibly augmenting a functionality. Thus, it is possible to represent them as: $C_1 \oplus C_2 = C_3$ or $(C_1 \oplus C_2)$. Therefore, there are certain circumstances where a composition is not possible or it is not reached. This situation is so-called composition anomaly [15] and is represented here as $(C_1 \oplus C_2) = \varnothing$.

Whereas many statements on literature emphasize that the adoption of AO Software Development (AOSD) eventually leads to higher quality, AO does not provide correctness by itself. According to Alexander et al. [13], there are four potential faults on a woven program: a) the fault resides in a portion of the core concern, which is not affected by the woven aspect; b) the fault resides in a portion that is specific to the aspect; c) the fault is an emergent property created by interactions among aspects and primary abstractions (e.g., classes); and d) the fault consists of an emergent property, when more than one aspect is woven into a primary abstraction, resulting from aspects woven order, where their precedence are not respected. To the best of our knowledge, there are no approaches on the literature that directly investigate the aspect composition testing.

Interactions among aspects may or may not intentionally take place. If intentionally, the developer aims to exploit the composition characteristics, such as reusability. If not, the developer builds or modifies an aspect that is not part of an earlier composition (i.e., previous aspect composition configuration), but might from now on take part of it.

There are some problems regarding to interaction of two or more different aspects applicable to the same join point: incompatibility and ordering. Composing aspects could result into an incompatible situation, e.g., as they present related properties, leading to redundancy, or if they present inconsistency, where an aspect can invalidate functionality offered by the other. Regarding the ordering problem, aspects can implement temporal restriction properties with each other.

Considering these facts, it is possible to outline some problems that could be verified when testing aspects composition, beyond those ones proposed in [13,14]: a) functionalities of aspects not attained; b) errors in superaspects might be propagated to subaspects; 3) functionalities provided by one or more aspects could affect other functionalities provided by other aspects; 4) temporal restrictions specifications might not be respected when several complex aspect are woven; and 5) an aspect might change the hierarchical structure of a given composition, leading the system to an inconsistent state.

According to these facts aforementioned, the development of testing approaches for testing aspects composition still continues a major issue, enabling to detect faults resulting from interactions among aspects.

## 3    Related Works

A number of researches about software testing on the AO technique shows up important works on this area. Alexander et al. [13] proposed the seminal fault model to AO programs considering six types of faults: $f1$) incorrect strength in pointcut patterns; $f2$) incorrect aspect precedence; $f3$) failure to establish expected postconditions; $f4$) failure to preserve state invariants; $f5$) incorrect focus of control flow; and $f6$) incorrect changes in control dependencies. Bsekken and Alexander [14] showed a fault model for AspectJ pointcut identifying four pointcut fault categories.

Mortensen and Alexander [16] proposed the combination of two traditional testing techniques for AO programs: coverage and mutation testing. The paper provides a set of mutation operators: *Pointcut strengthening* (PCS), *Pointcut weaking* (PCW) and *Precedence changing* (PRC). Anbalagan and Xie [17] presented an framework and a tool that implements two operator mutants defined in [16]. Delamare et al. [9] showed a tool called ajMutator and an approach based on test-driven development concept to for testing AspectJ features. Ferrari et al. [18] introduced a tool called Proteum/Aj, overcoming some limitations faced by previous tools [17,9]. Ferrari et al. [19] presented an exploratory analysis on fault-proponess of AOPs from three real world AO systems.

Xu et al. [20] showed up an approach that combines state models (classes and aspects) and flow graphs (methods and advices) to produce an hybrid testing model called *Aspect Scope State Model* (ASSM). Xu et al. [21] presented a state-based approach to AO programs testing, inspired by the FREE model [22], called *Aspectual State Model* (ASM). An evolution of these approaches are detailed in [23,24]. Liu and Xang [6] introduced a state-based approach for AOP, considering changes in the state-based behaviour introduced by advices in different aspects. Xu et al. [8] presented a Model-based Aspect Checking and Testing framework (MACT), based on finite state models.

Lemos et al. [25] presented the derivation of a control and data flow model for AOPs, based upon the static analysis of the AspectJ object code at the unit level, and more recently at the integration level [4] . A recently report published by the Brazilian community summarizes their contributions to the area of AOSD [26]. It lists five key challenges for testing: a) identifying new potential problems; b) defining new proper underlying models; 3) customizing existing test selection criteria and/or defining new ones; d) providing adequate tool support; and e) experimenting and assessing the approaches. With respect to the item "b", this paper represents a contribution towards new underlying models.

Although consistent and relevant, none of these approaches and works previously mentioned directly investigate the aspect composition problem from the functional-based testing perspective, its resultant interactions, and their representation on a dynamic model.

# 4  The METEORA Method

Testing the aspects composition is a major issue, since these interactions might produce unexpected results, both in affected classes functionality or exactly in the involved aspects. On this scenario, we propose a state-based testing method for AO software called METEORA. This method consists in: i) establishing well defined phases so that the tester could reach a better and more systematic testing activity; ii) making use of an extended model to represent dynamic behavior of aspects interactions; iii) producing and storing artifacts that can take advantage on later stages when applying other testings approaches, as regression ones; and iv) a systematic manner to enable the detection of faults with respect to aspects composition.

As presented in Section 3, Xu et al. [21] have proposed an state-based approach for AO programs based on the FREE model [22]. However, such approach does not provide a way to deal specifically with aspects interactions issues. According to authors, to get AOPs modeled are necessary three elements: base elements (classes), crosscutting elements (aspects) and crosscutting relationships (classes and aspects). Thus, a fourth element was added here: the one which represents relationships among aspects. As a consequence, a new model for representing states of AOPs was developed, called inter-class-aspectual interactions state-based model (MEIICA). This model introduces several extensions to the base one, in order to enable the representation of added, removed and/or changed states by aspects or aspects compositions weaved into classes.

METEORA uses the MEIICA model, which is an extension of ASM and FREE models. This method comprises an hybrid approach, since it allows, by using one of its components, to model interactions between class states (FREE), class–aspect states (ASM) and aspect–aspect states (MEIICA). It can be also considered as an incremental approach, due to reuse of some core concerns' test cases, peculiar to the testing of its classes. METEORA applies integration tests, where interactions aforementioned are verified. It also includes the concept of cluster testing, suggesting that aspects which act as a whole to deliver an specific functionality should be testing together. The adoption of cluster concept consists in an useful alternative to reduce the explosion state problem. Based on this reasoning, tests are applied over an unit under testing (UUT) or a cluster under testing (CUT).

Assertion comprehends another important concept adopted by METEORA to evaluate pre and postconditions of methods, class and aspect invariants, besides class and aspect state invariants. Such an evaluation is made by an oracle, which checks assertions specified by the tester during the test cases execution, determining whether a test case has unveiled or not a fault. The oracle in METEORA is implemented by a testing tool prototype that supports the method.

As part of a functional approach, some of artifacts produced by METEORA are derived from requirements specification or documentation of classes and aspects under testing. METEORA is capable of detecting faults present on the fault-model proposed by Alexander et al. [13] and Bsekken and Alexander [14]. Furthermore, it is able to reveal the fifth source of faults on AOPs, introduced by our previous work – the *aspect composition fault* [11].

A major concern is to distinguish composition fault from precedence fault: the latter one has to do exclusively with temporal restrictions among aspects. However, a composition fault is not necessarily revealed in a stricted way by temporal issues. It can occur in other situations, resulting from the definition of *composition fault*: a process or a incorrect data definition, e.g., an incorrect instruction or statement present into aspects, derived from one of the following situations: $s1$) when an aspect is crosscutted by another aspect; $s2$) when an aspect addresses pointcuts declared by another aspect; or $s3$) when an advice invokes methods declared in another aspect. In addition to that, another situation ($s4$) of composition fault can occurs when aspects in an isolated way correctly deliver their functionalities, but when composed with one another show up a different behavior from their specifications. Notice that even when well-established precedence are met, conditions above may put the system into an erroneous state.

A fundamental issue regarding testing activities is the development of testing tools. These tools automate and allow practical applications of testing criteria. They also help to perform tests on greater and more complex systems. Besides that, testing tools implement methods and techniques and can be used to demonstrate their effectiveness. With this in mind, a testing tool prototype called *KTest* was developed, aiming to support the testing applied by METEORA. This prototype enables: a) the MEIICA interpretation of the application under testing, allowing the CRT generation by using the Round-trip $Path_{Malg}$ algorithm; b) the abstract test case derivation from the CRT; c) the storage of such artifacts in XML files; d) the execution of test input into the application under testing; e) the generation and storage of actual results; and f) the generation of testing reports with assertions evaluation specified in MEIICA.

## 4.1   MEIICA

This extended model provides a way to represent the dynamic behavior of AO programs, focusing on aspect interactions. It is organized into a well defined structure, composed by states, transitions and pseudo-states, besides other information associated with such a structure.

To provide an effective representation of new elements which exist in AO technique, new UML entities have been added or extended by using stereotypes based on textual descriptions. Moreover, this model also comprises colors in order to improve the legibility of diagrams where several types of states and transitions are present. So, it is possible to highlight structural elements that the tester consider more important.

As depicted in Fig. 1, MEIICA is able to represent states added (red, <<asp_ins>>), removed (gray, <<asp_rem>>) and/or changed (green, <<asp_alt>>) by aspects. In a similar way, this model represents transitions among states of classes and aspects (red, <<act>>), among states of the same aspect (green, <<aat>>), among states of different aspects (blue, <<a2at>>), transitions added (yellow, <<instr>>) into or removed (gray, <<remtr>>) from core concerns. Transitions that refer to around advices are drawed in red as well.

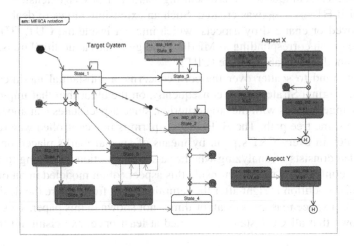

**Fig. 1.** (Color online) Extended model notation – MEIICA

Besides these transitions, a new class of transitions was developed, called *internal* or *immediate* ones. This class comprehends events (methods) triggered automatically, not explicit stimulated (i.e., advices invoke). In MEIICA, these transitions, unnamed, are binded to join points, which have their own graphical representations.

Symbols "○", "●" and "⊗" respectively represent before, around and after advices, where additional behaviors are weaved.

Pseudo-states "Ⓧⁱ" and "Ⓗ" are used to represent labels or return points on the model where an extra behavior has been included: the former indicates a pointer to a state located in another region, and the latter the return to the point (previously join point) where the deviation has taken place. The first pseudo-state aims to simplify the model, allowing a better orthogonal modeling of aspects' states. Thus, it is possible to avoid repetition of states which tend to be scattered all over the model. Consequently, the number of states decreases and such a situation contributes in a positive way to reduce the well-known state explosion problem. The second pseudo-state has a complementary functionality to the first one, depicting the return from an orthogonally modelled region to an specific point in the model, originary of the internal advice call. Precedence among aspects are indicated by numerical labels.

## 4.2   A New Strategy to Derive Test Sequences

The FREE approach [22] uses a strategy called *Round-trip path* to derive test sequences. Basically, it requires the construction of a transition tree from a finite state-machine. This tree contains all test sequences that begin and end with the same state. This strategy consider a tree node terminal when the state it represents is already presented in the same path or it is a final state.

As strategies originally developed to the OO paradigm, in general, can not be directly applied to the AO technique due to crosscutting characteristics, an extension of Binder's strategy called RTP$_{METEORA}$ was created. Such an extension allows to deal with states added, removed or changed by aspects, which interact inside the CUT. RTP$_{METEORA}$ traverses the graph corresponding to MEIICA and generates a modified transition tree, called Combined Reacheability Tree (CRT).

As aspects tend to scatter over the core concerns, a substantial modification was made on the original Strategy and, consequently, on the algorithm that implements it. The main change regards with the stop criterion, since aspect states can appear several times upon a same tree path. The RTP$_{METEORA}$ permits in a controlled way repetitions of aspect states in every CRT's path, by means of special marks placed on the tree. The main idea consists of analyzing whether a state repetition belonging to an aspect is within the context of a transition inside this aspect (often modeled in an orthogonal region) or this repetition is right after the returning point from where the deviation has taken place. For space reasons, this algorithm is not shown in this paper. This extended strategy allows that all CUT states are visited at least once, exercising all transitions once as well.

The CRT has also special transitions, as the ones already presented (internal or immediate). They are named b{tr}, ar{tr} or a{tr}, representing implicit calls to before, around and after advices weaved at the tr transition. Each path from the initial state until a leaf node on the CRT corresponds to a sequence test or an Abstract Test Case (ATC).

Regarding coverage criteria, METEORA makes use of an extended version of *All Round-trip Paths* [22], called All Round-trip Paths$_{METEORA}$. This coverage is achieved when all paths generated by RTP$_{METEORA}$ are exercised. The criteria extended on this work allows to detect not only the same faults of the original one, but also the specific faults related to aspects composition.

Aiming to clarify the concepts pointed out up to here, the main phases of METEORA are described in the next section.

## 4.3   Application Phases of METEORA

*1. Analysis of Classes/Aspects Documentation* – the first step is to make a critical analysis upon classes and aspects documentation, mainly concerned with their dynamic models. This understanding is a crucial point to apply subsequent phases; *2. MEIICA Development* – in this phase, the model, which represents dynamic interactions among states of classes and aspects as well aspects and aspects, is built based on information gained from the previous phase. A subphase might be considered here: validate MEIICA against checklists adapted from [22], to ensure that the model built is a testable one; *3. Combined Reacheability Tree (CRT) Generation* – this phase consists of transcribing a tree from the MEIICA, which contains all transitions sequences (method and advice calls) that begin and end with the same state, with certain redundancy of states and transitions of aspects, i.e., an extended round-trip path. *4. Test Case Sequences Transcription* – test case sequences are transcribed from the CRT generated on the previous phase. Each full branch in the tree becomes an Abstract Test Case (ATC). On this work, a test case is considered abstract when concrete parameters are not assigned to constructors and methods' parameters. This phase is automated by applying an algorithm called KTestGenCTC, a slightly modification from the original depth-first traversal algorithm; *5. Concrete Test Cases Derivation* – in this phase, the tester shall manually specify concrete values to constructors and methods' parameters, in order to satisfy constraints associated with transitions in the CRT. This manually assignment is due to path sensitization problem, which is in general undecidable and must be heuristically solved [22]. On this work, concrete test cases are equivalent to Test Input (TI), but not necessarily all transitions of a given ATC will be into a TI, since there are many internal transitions or explicit invokes in an ATC made by advices; *6. Testing Execution* – once TIs were defined, the application under testing is executed. All concrete test cases are submitted to the application by the testing tool prototype built; *7. Result Analysis* – at last, this phase aims to check whether testing output results are the expected ones, which are met on abstract test cases generated by phase 4. The expected results are obtained by sequences of states, which correspond to events (transitions) execution. This phase produces a testing result report, allowing the tester to analyze, for instance, all states and transitions expected and achieved as well as results of assertions evaluation.

## 4.4   Case Study: An AO Banking Application

METEORA has been applied over three different case studies, belonging to distinct domains. The one chosen to be reported here refers to a Bank Accounts Control System (BACS), substantially adapted from [27]. The BACS aims to provide the management of a variety of bank accounts, such as current, savings, fixed-income investment, to name but a few. Some of its Functional Requirements (FR) and Non-Functional Requirements (NFR) are clearly crosscutting and hence implemented as aspects. The implementation of this system is inserted into a scenario composed by orthogonal aspects, which are directly applied into a target application, as well as composed by aspects compositions, where is possible to verify some types of aspects interactions.

As stated in [27], business rules can be addressed by aspects without needing to make changes into the core concerns. Some of these business rules can be classified as FR.

Thus, among the new FR specified to be part of BACS, it can be cited: a) a new type of bank account which will provide a safety mechanism within an credit limit, called overdraft account. This account allows customers to have a negative balance within a threshold established by the bank institution; b) additional management for overdraft accounts, such as new policies to accommodate extensions/alteration of overdraft limit, based on customers' financial profile or organization policies; and c) new constraints for blocking accounts according to certain financial activities circumstances.

With regard to NFR, BACS must provide persistence (with transactions and synchronism control mechanisms), security[1] and logging. Clearly, these concerns affect the system in a crosscutting way, thus they shall be implemented as aspects.

Two testing scenarios with two different CUT were built to apply METEORA into the BACS. Due to lacking of space, just the scenario which refers to overdraft account and persistence concerns are described here, since the main idea was to test these two composed concerns.

Documentation and requirements specification analysis were performed by applying *Phase 1* of METEORA. Considering that the SpecialAccount class is inserted into the BACS by an aspect, the aspect which implements it (SpecialAccountCtr - $SAC$) was also added to the CUT. Hence, this scenario was created with the CUT composed by the aforementioned class, the SpecialAccountCtr aspect and all aspects which reside in persistence package: the Transaction ($Tr$) abstract aspect and its CCTransaction ($CCTr$) subaspect; and the Synchronization ($Sy$) abstract aspect and its CCSynchronization ($CCSy$) subaspect. This way, the composition notation for this scenario is: $((Tr \oplus CCTr) \oplus (Sy \oplus CCSy) \oplus SAC$.

Even though the original Account class has four states (OPEN, FROZEN, INACTIVE and CLOSED), for simplicity, just two of them were modeled: OPEN (renamed to POSITIVE) and CLOSED. The OVERDRAWN state of SpecialAccountCtr aspect represents accounts with a negative balance (i.e., *balance < 0 and balance >= limit*).

State diagram of the transaction subconcern reports basically two states: COMMITED and ROLLEDBACK. The former denotes activities where transactional operations are effectively performed as an atomic one, i.e., database operations are committed. The latter state represents situations where a transaction failure has occurred, i.e., all database operations must be rolled back.

The synchronism subconcern is comprised by PERSISTENCE_CLEAN and PERSISTENCE_DIRTY states. The first one indicates a situation where attribute values of a persistent object (that represents data in the datastore) have not been changed in the current transaction. The second state represents a persistent object that has been changed in the current transaction, i.e., its persistent state is now different from its memory state.

Fig. 2 depicts the MEIICA of this CUT, created as a result of *Phase 2* application, which represents the composition's behavior previously presented.

In this figure, it is shown that whenever an account object is instantiated, an internal event is automatically fired at the exit point (after advice – $\otimes$ symbol) on the new event, transitioning it to the PERSISTED_CLEAN state. After that it must return, via history symbol ($\text{H}$), to the last join point from where the internal transition has occurred. So, the POSITIVE state is reached. An account can become overdrawn if occurs a debit

---

[1] For simplicity, the security concern here is implemented only with authentication control.

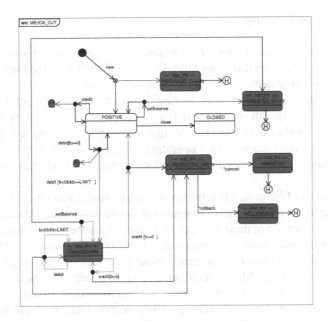

**Fig. 2.** (Color online) MEIICA of the cluster under testing (CUT)

event and the resultant balance is less than zero. Notice that around advice present on the debit event has two different internal transitions: one leads to the P1 state, which represents a label to the PERSISTED_DIRTY, and the other leads to the OVERDRAWN state. As precedence indicates, when a debit occurs, it is intercepted by an around advice (● symbol) and, if its postcondition ($b < 0$ *and* $b >= LIMIT$) is evaluated as true, the PERSISTENT_DIRTY state of synchronization aspect is automatically reached. From this point, a commit or a rollback transition can be fired, transitioning it respectively to COMMITTED or ROLLEDBACK states. In both states a transition to an history symbol returns it to the last join point (on the debit event) and the OVERDRAWN state of SpecialAccountCtr aspect is then entered.

As one can observe, just credit and debit operations are transactional. Thus, commit and rollback events are just fired over them. Moreover, the PERSISTEN_DIRTY state is entered from different types of transitions: red one from POSITIVE state and blue one from OVERDRAWN. Whereas the former indicates a relationship between a class state and an aspect state, the latter indicates a relationship between aspect states. In regard to commit and rollback events, both of them denote relationship between aspect states, since PERSISTENCE_DIRTY belongs to synchronization aspect and both COMMITTED and ROLLEDBACK belong to the transaction one.

By means of using *KTest* tool, *Phases 3* and *4* of METEORA were applied, originating the CRT and thirteen (13) ATCs. Based on manual assignment made by the tester, the *Phase 5* has produced TIs, described in XML format.

The *Phase 6* refers to execute TIs created on the previous phase. Again, the *KTest* has helped and automated this phase. Finally, the *Phase 7* of METEORA's application has generated testing result reports, where some results are discussed in Section 5.

## 5   Results and Discussion

This Section describes three factors from where analysis and discussion of results were based on: i) mutation scores; ii) supportability of METEORA to the fault model proposed in [13]; and iii) detection of aspects composition faults.

The strategy adopted by this work to evaluate the application of METEORA consisted of measuring the effectiveness of testing sequences generated by this method through the use of mutation testing [12]. Mutation testing is based on seeding the source code with faults by applying mutation operators. Resulting modified programs are then referred as mutants. The next step is to determine whether testing identifies these seeded faults.

According to DeMillo et al. [12], mutation testing provides an objective measure of the adequability (effectiveness) level of analyzed test sequences, called score mutation. It is calculated as the ratio of dead mutants over the total number of non-equivalent ones. Equivalent mutants are semantically equivalent to the original program. Thus, they can not be killed by any test case. Equivalence is usually shown by a manual process during the execution of test cases. More specifically, AO equivalent mutants are those where modified pointcuts (mutants) match the same join points as the original pointcut.

The following mutant operators were used for generating mutants from the case studies: Pointcut strengthening (PCS), Pointcut weakening (PCW), and Precedence changing (PRC) [16]; Method Name Change (MNC), Statement Deletion (SSDL), Logical Context Negation (OCNG), and Statement Insertion (SINS), Modifier Method Change (MMC) and Boolean Replacement (CBR) [28].

The first analysis from obtained experiments results regards with the adequability measure of test sequences generated by METEORA and calculated by mutation scores.

Test sequences for the CUT, described in this paper, have presented a high level of adequability. As previously mentioned, this CUT is comprised by aspects responsible for special account and persistence (with transactions and synchronism control mechanisms) and by a class of the core concerns. Table 1 shows the total of mutants generated by operators, besides equivalent and killed mutants, and the mutation score.

After generation of mutants, some of them have remained alive after testing execution. Debugging has revealed that some of mutations generated by PCS and PCW operators were not enough to strengthening or weakening matched join points. Hence,

**Table 1.** Mutation scores for the CUT

| Operator | Generated Mutants | Equivalent Mutants | Killed Mutants | Mutation Score |
|----------|-------------------|--------------------|----------------|----------------|
| PCS      | 15                | 01                 | 14             | 1,000          |
| PCW      | 14                | 02                 | 12             | 1,000          |
| PRC      | 11                | –                  | 11             | 1,000          |
| MNC      | 14                | –                  | 13             | 0,928          |
| EMM      | 13                | 01                 | 11             | 0,916          |
| SSDL     | 13                | –                  | 13             | 1,000          |
| OCNG     | 15                | –                  | 14             | 0,933          |
| TOTAL    | 95                |                    |                |                |

pointcuts were still selecting the same set of join points. This has determined the equivalence of mutants. With regard to PRC operator, although it has inverted the precedence order between two aspects of the CUT, a precedence statement like "thisAspect, *" into another aspect was in fact more significant than the one which has suffered mutation. Again, this has classified the mutant as equivalent one.

The application of METEORA has shown that it supports the fault model proposed in [13]. To enable that each of these faults ($f1$ to $f6$) were detected, a large number of operations involving the aforementioned operators were done. For example, we cite the syntactical changes applied to pointcuts with PCS operator. In this case it is possible: replace wildcards in pointcuts by explicit methods' name; delete special operators (e.g., "+" in AspectJ) to alter the matching in a hierarchical structure of components; or moving wildcards, for instance, from left to right (e.g., *.drop*, *.dro*, *.dr*, ...). Based on these operations, it is possible to argue that such operators are language-dependent.

The $f1$ (incorrect strength in pointcut patterns) was detected due to introduced faults into the code by applying mutations cited before. Regarding $f2$ (incorrect aspect precedence), this fault was detected by using the PRC operator in two ways: the former by inverting the precedence statement and the latter by commenting them. Results obtained have demonstrated that depending on the precedence order assumed by the weaver, not always the absence of this statement can reveal $f2$. Hence, more safety mutations are attained by inverting precedence statements. The third fault, $f3$ (failure to establish expected postconditions) was detected by inserting statements which broke the contract between client and server. That is, aspects declared as *"privileged"* can directly access attributes of crosscutted entities, violating method postconditions in which clients expect to be true. The SINS operator was used in most of these mutations. In a similar way, $f4$ (failure to preserve state invariants) was disclosed. EMM and CBR operators were used to detect $f4$. For instance, it was possible to violate state invariants by negating the boolean values present in certain modifiers methods used by aspects. Notice that two subclasses of $f4$ can be considered here: failure to preserve state invariants of classes and aspects. With respect to $f5$ (incorrect focus of control flow), it was not just detected as commonly presented in the literature, using recursive methods. Non-recursive methods which depend on control flow were used as well. PCW and PCS were used on experiments to mutate pointcuts matched by cflow and cflowbelow statements. The last fault presented on this model, $f6$ (incorrect changes in control dependencies), was revealed by mutations introduced by using the SSDL operator. In all cases, it removes the proceed call, generally present into around advices. An exception of its presence relates to situations where the programmer intends to completely change the method behavior. Therefore, this situation was not identified in none of experiments. By removing this statement, the original control dependencies are fully altered, resulting in different states from the ones expected.

## 5.1 Aspects Composition Faults

The main focus of this paper falls into aspect composition and their consequent interactions. So, more detailed results are discussed here regarding aspects composition faults.

Section 4 has described the four situations from where might arise an aspect composition fault ($s1$ to $s4$). As one can observe, a composition fault can be also detected in case of incorrect precedence order. This way, the occurrence of $f2$ (incorrect aspect precedence) shall be sufficient but not necessary to denote a composition fault. So, during all the experiments we have tried to model and detect such a faults with no correlation to $f2$.

The application of *Phase 7* of METEORA has detected several composition faults. Two of them are described here. In the first example, the **PERSISTENT_DIRTY** state was not reached due to restrictions applied in a pointcut that matches modifiers methods (**set***(..)) and is also addressed by a pointcut of another aspect ($s2$). As a consequence, objects from the hash table (which contain "dirty" objects, i.e., the ones whose values have been changed in the current transaction) were not updated. In short, objects involved in transactional methods execution have not been persisted due the lack of *commit* statement. The notation of this anomalous composition regarding to transaction and synchronism concerns is: $(Tr \oplus Sy) = \varnothing$.

Fig. 3(a) illustrates this situation with a fragment of the testing result report generated by the application of *Phase 7* of METEORA. This test report presents the following data: state where transition comes from and its type (From/tF); expected transition and its type (TrATC / tTrATC); achieved transition and its type (TrTI_OUT/tTrTI_OUT); precondition and its evaluation (Pre/Ev); postcondition and its evaluation (Post/Ev); expected target state and its type (Target/tT); state invariant and its evaluation (InvSta/Ev); and finally, attributes values before (BVlr) and after (AVlr) the transition execution.

A boolean result is assigned to each generated report, which corresponds to the testing result evaluation for a given test case. Notice the bold regions in this figure, where assertions were evaluated as false. Another important fact is the difference between the expected transition (*commit*) and the obtained one (*credit*). The type of transaction ($\ll a2at \gg$ stereotype) denotes an interaction between states of different aspects (synchronism and transaction). Notice yet the type of ar{credit} immediate transition: <<act,a2at>>. This denotes a transition between states of a class and an aspect, but in one of these situations ($s2$) that corresponds to the definition of the aspect composition fault.

(e)

| n | From | TrATC | tTrATC | TrTI_OUT | Pre | Ev | Post | Ev | Target | InvSta | Ev | Inv | Ev |
|---|------|-------|--------|----------|-----|----|------|----|--------|--------|----|-----|----|
| 1 | ALPHA | a{new} | act | a-new | | | state=='PERSISTED_CLEAN' | true | PERSISTENT_CLEAN | state=='PERSISTENT_CLEAN' | true | | |
| 2 | PERSISTENT_CLEAN | new | | new | | | bal>=0&&bal>=limit | true | POSITIVE | bal>=0 | true | id!=0 | true |
| 3 | POSITIVE | ar{credit} | act.a2at | ar-credit | | | state=='PERSISTED_DIRTY | false | PERSISTENT_DIRTY | state=='PERSISTENT_DIRTY' | false | | |
| 4 | PERSISTENT_DIRTY | commit | a2at | credit | | | state=='COMMITED' | false | COMMITED | state=='COMMITED' | false | | |
| 5 | COMMITED | credit | act.bjpt | --- | | | bal>=0&&bal>=limit | false | POSITIVE | bal>=0 | false | d!=0 | false |

(a)

| n | From | TrATC | tTrATC | TrTI_OUT | Pre | Ev | Post | Ev | Target | InvSta | Ev | Inv | Ev |
|---|------|-------|--------|----------|-----|----|------|----|--------|--------|----|-----|----|
| 1 | ALPHA | a{new} | act | a-new | | | state=='PERSISTED_CLEAN' | true | PERSISTENT_CLEAN | state=='PERSISTENT_CLEAN' | true | | |
| 2 | PERSISTENT_CLEAN | new | | new | | | bal>=0&&bal>=limit | true | POSITIVE | bal>=0 | true | id!=0 | true |
| 3 | POSITIVE | ar{debit} | act.a2at | ar-debit | | | state=='PERSISTED_DIRTY' | false | PERSISTENT_DIRTY | state=='PERSISTENT_DIRTY' | false | | |
| 4 | PERSISTENT_DIRTY | commit | a2at | debit | | | state=='COMMITED' | false | COMMITED | state=='COMMITED' | false | | |
| 5 | COMMITED | debit | act.bjpt | ar{credit} | | | bal<0&&bal=limit | false | OVERDRAWN | bal<0&&bal>=limit | false | d!=0 | false |
| 6 | OVERDRAWN | ar{credit} | a2at | commit | | | state=='PERSISTED_DIRTY' | false | PERSISTENT_DIRTY | state=='PERSISTENT_DIRTY' | false | | |
| 7 | PERSISTENT_DIRTY | commit | a2at | credit | | | state=='COMMITED' | false | COMMITED | state=='COMMITED' | false | | |
| 8 | COMMITED | credit | at.bjpt | --- | | | bal>=0&&bal>=limit | false | POSITIVE | bal>=0 | false | d!=0 | false |

**Fig. 3.** Testing result report with composition faults

The non-usage of modifiers methods (set) of Account class by SpecialAccountCtr aspect ($SAC$) corresponds to the second example of composition fault described here.

Thus, persistent objects were not saved in datastore. Pointcuts of the persistence ($Prs$) concern matches these modifiers methods present into the Account class hierarchy to deliver their services. As one can observe, this fault was only detected in the debit method call, which is advised by the around advice of SpecialAccountCtr aspect.

When the credit method was invoked, the commit operation has been normally performed. So, the notation for this fault is: $(Sy \oplus Tr) = Prs$ and $(SAC \oplus Prs) = \varnothing$.

It is important to emphasize that this fault adheres in a rigorous way to the aspect composition fault, since both of concerns (special account and persistence) deliver correctly their services when individually woven into core concerns. Special account concern have properly dealt with situations where accounts became overdrawn, making right analysis over result balances according to thresholds established for customers. Persistence concern has correctly delivered its services with required transaction and synchronism control mechanisms.

Nevertheless, when both of these aspects were weaved (composed) together, debit operations (main focuses of the first aspect) which have resulted in negative balances were not persisted, although they were performed. However, credit operations, besides normally performed, were correctly persisted.

Fig. 3(b) depicts this situation, which corresponds to $s4$ in the aspect composition fault definition. Bold regions highlight violated assertions and obtained and expected transitions (e.g., commit $x$ debit). The two unaligned ellipses in this figure point out that commit operation was performed as expected when credit methods were invoked, even though the missed commit related to the debit method.

It is valuable to notice that in this case, temporal restrictions ($f2$) have not shown any interference on the test result.

# 6    Conclusion and Future Work

The Aspect-Oriented (AO) technique, as a contemporary approach, points out that more studies and experiments should be carefully led to indicate what are the best testing approaches applicable to this way of development.

Aspects tend to modify not only structures of object-oriented programs, but also relationships among their components. Functionalities of core concerns can be combined with the ones delivered by aspects, which in turn, can be merged among themselves. These relationships motivate the appearance of different faults in aspect-oriented programs (AOP).

This paper presented the METEORA, a state-based testing method for AO software. This method provides class–aspect and aspect–aspect faults detecting capabilities. In addition, to support the developed method, a model to represent dynamic behavior of aspects interactions, an extended strategy to derive testing sequences, and a testing tool prototype were developed and briefly described.

To evaluate the proposed method, we used three different case studies. Mutation testing [12] were applied to measuring the effectiveness of testing sequences generated by METEORA. The results of our assessment show that the method is able of revealing faults caused by aspects composition in weaving processes, including the ones proposed in [16,14]. Moreover, the method is able to reveal the fifth source of faults on AOPs, introduced by our previous investigation [11].

Next steps of this research include: i) to apply the mutant operators and the tools proposed in [10,9]; ii) to conduct an exploratory study with the intent of evaluating in a more rigorous way the proposed approach; iii) to extend the method, applying the same concepts adopted by METEORA over dynamic aspects; and iv) to investigate the possibility of developing specific mutant operators to generate aspects with composition faults.

**Acknowledgments.** The authors would like to thank FAPESP, CAPES and CNPq for financial support.

# References

1. Filman, R., Elrad, T., Clarke, S., Akşit, M. (eds.): Aspect-Oriented Software Development. Addison-Wesley, Boston (2005)
2. Kiczales, G., Lamping, J., Menhdhekar, A., et al.: Aspect-oriented programming. In: Akşit, M., Matsuoka, S. (eds.) ECOOP 1997. LNCS, vol. 1241, pp. 220–242. Springer, Heidelberg (1997)
3. Lemos, O., Franchin, I., Masiero, P.: Integration testing of object-oriented and aspect-oriented programs: A structural pairwise approach for java. Science of Comp. Programming 74, 861–878 (2009)
4. Lemos, O., Masiero, P.: A pointcut-based coverage analysis approach for aspect-oriented programs. Info. Sciences 181(13), 2721–2746 (2011)
5. Xiong, L., Li, J.: A structural testing approach for aspect-oriented programs based on data and control flow. In: 2013 4th IEEE International Conference on Software Engineering and Service Science (ICSESS), pp. 85–88 (May 2013)
6. Liu, C.-H., Chang, C.-W.: A state-based testing approach for aspect-oriented programming. J. Inf. Sci. Eng. 24(1), 11–31 (2008)
7. Xu, D., Ding, J.: Prioritizing state-based aspect tests. In: 2010 Third International Conference on Software Testing, Verification and Validation (ICST), pp. 265–274 (April 2010)
8. Xu, D., el Ariss, O., Xu, W., Wang, L.: Testing aspect-oriented programs with finite state machines. Softw. Test., Verif. Reliab. 22(4), 267–293 (2012)
9. Delamare, R., Baudry, B., Le-Traon, Y.: Ajmutator: A tool for the mutation analysis of aspectj pointcut descriptors. In: International Conference on Software Testing, Verification and Validation Workshops, ICSTW 2009, pp. 200–204 (April 2009)
10. Ferrari, F.C., Rashid, A., Maldonado, J.C.: Towards the practical mutation testing of aspectj programs. Science of Computer Programming 78(9), 1639–1662 (2013)
11. Silveira, F.F., da Cunha, A.M., de Resende, A.M.P., Lisbôa, M.L.B.: The testing activity on the aspect-oriented paradigm. In: Proceedings of the 1st Workshop on Testing Aspect-Oriented Programs at AOSD 2005 (2005)
12. DeMillo, R.A., Lipton, R.J., Sayward, F.G.: Hints on test data selection: help for the practicing programmer. J. -Computer 11, 34–41 (1978)
13. Alexander, R.T., Bieman, J.M., Andrews, A.A.: Towards the systematic testing of aspect-oriented programs. Tech. Rep. CS-4-105, Colorado State University, Fort Collins, Colorado (2004)

14. Bsekken, J., Alexander, R.: A candidate fault model for aspectj pointcuts. In: 17th International Symposium on Software Reliability Engineering, ISSRE 2006, pp. 169–178 (November 2006)

15. Bergmans, L., Tekinerdogan, B., Glandrup, M., Aksit, M.: On composing separated concerns: Composability and composition anomalies. In: Workshop on Advanced Separation of Concerns (2000)

16. Mortensen, M., Alexander, R.T.: An approach for adequate testing of aspectJ programs. In: Proceedings of the 1st Workshop on Testing Aspect-Oriented Programs at AOSD 2005 (2005)

17. Anbalagan, P., Xie, T.: Automated generation of pointcut mutants for testing pointcuts in aspectj programs. In: 19th International Symposium on Software Reliability Engineering, ISSRE 2008, pp. 239–248 (November 2008)

18. Ferrari, F.C., Nakagawa, E.Y., Rashid, A., Maldonado, J.C.: Automating the mutation testing of aspect-oriented java programs. In: Proceedings of the 5th Workshop on Automation of Software Test, AST 2010, pp. 51–58. ACM, New York (2010)

19. Ferrari, F., Burrows, R., et al.: An exploratory study of fault-proneness in evolving aspect-oriented programs. In: Proceedings of the 32nd ACM/IEEE International Conference on Software Engineering, ICSE 2010, vol. 1, pp. 65–74. ACM, New York (2010)

20. Xu, W., Xu, D., Goel, V., Nygard, K.: Aspect flow graph for testing aspect-oriented programs. In: Proceedings of the 8th IASTED International Conference on Software Engineering and Applications. ACTA Press (2004)

21. Xu, D., Xu, W., Nygard, K.: A state-based approach to testing aspect-oriented programs. In: SEKE 2005: Proceedings of the 17th International Conference on Software Engineering and Knowledge Engineering, Taipei (July 2005)

22. Binder, R.V.: Testing object-oriented systems: models, patterns, and tools. Addison-Wesley, Boston (2001)

23. Xu, D., Xu, W.: State-based incremental testing of aspect-oriented programs. In: AOSD 2006: Proceedings of the 5th International Conference on Aspect-Oriented Software Development, pp. 180–189. ACM Press, New York (2006)

24. Xu, D., Xu, W., Wong, W.E.: Testing aspect-oriented programs with uml design models. International Journal of Software Engineering and Knowledge Engineering 18(3), 413–437 (2008)

25. Lemos, O., Vincenzi, A., Maldonado, J., Masiero, P.: Control and data flow structural testing criteria for aspect-oriented programs. J. Syst. and Software 80(6), 862–882 (2007)

26. Kulesza, U., Soares, S., Chavez, C., et al.: The crosscutting impact of the AOSD Brazilian research community. Journal of Systems and Software 86, 905–933 (2013)

27. Laddad, R.: AspectJ in Action: Practical Aspect-Oriented Programming. Manning Publications Co., Greenwich (2003)

28. Ma, Y.-S., Kwon, Y.-R., Offutt, J.: Inter-class mutation operators for java. In: Proceedings of the International Symposium on Software Reliability Engineering, ISSRE 2002, p. 352. IEEE Computer Society, Washington, DC (2002)

# Creating a Framework for Quality Decisions
# in Software Projects

Pasqualina Potena, Luis Fernandez-Sanz, Carmen Pages, and Teresa Diez

Computer Science Department, University of Alcalá
28871 Alcalá de Henares, Madrid, Spain
{p.potena,luis.fernandez,carmen.pages,teresa.diez}@uah.es

**Abstract.** This work analyzes the challenges that quality decisions represent to software project managers. Projects' goals are normally determined by the paradigm of the Iron Triangle of project management. Managers need to know which are the effects of a quality assurance (QA) decision on the three axis: which effects in quality they can get but at what cost and which effects may appear in terms of schedule. This decision problem is clearly related to existing disciplines like SBSE, multi-objective optimization and methods for ROI calculation and value-based software engineering. This survey paper critically reviews the contributions of these disciplines to support QA decisions together with basic information from a pilot survey carried out as part of the developments of the Iceberg project funded by EU Programme Marie Curie.

**Keywords:** software quality, QA decisions, influence factors, Iron triangle.

# 1   Introduction

The project Iceberg is a four-years research project funded by EU under the Marie Curie IAPP programme started in February 2013. It works through the cooperation of research and industrial partners from Italy and Spain, namely: University of Alcalá, CINI, Assioma.net and DEISER. Iceberg project addresses the challenge of providing a support for QA decisions to project managers, thinking in questions like "Given a high quality constraint, what is the cost to achieve, measurably, the goal? Is there a way to minimize such cost, standing the required high quality?" "Given a budget constraint (which prevents from performing all the required QA activities) what is the cost for missing quality activities? Missing activities implies bad quality: how does this "bad quality" manifests itself during operation, and how much does it cost?".

A decision involves passing judgment on an issue under consideration and it is commonly understood to be the act of reaching a conclusion or of making up one's mind [35]. Software projects involve making decisions across multiple lifecycle phases that span from requirements through implementation-integration to maintenance phase. Different types of decisions may be made depending on several factors, such as application domain. However, all software projects involve making decisions in different times and occasions from the very beginning: e.g. even deciding launching a

B. Murgante et al. (Eds.): ICCSA 2014, Part V, LNCS 8583, pp. 434–448, 2014.

project to solve a problem or to satisfy a request is a decision. In one extreme, even the act of not making a decision is a decision. For example, if a project manager chooses to ignore a project member's request for more resources or time, the manager is making a decision not to act and must deal with the consequences of this noncommittal decision [35].

All the decisions during project should be responsibility of the project manager in the end as he/she should assume the whole responsibility of the results of the software project and at the same time he/she has the authority to adopt the decisions in the sphere of the project. All the decisions and actions of the project manager should be oriented to satisfy the goals of the different stakeholders involved in a project. As stated in deliverable 2.1 of the Iceberg project [53] and in the Iceberg project proposal [36], main objectives of projects are defined by the classical model of restrictions of the triangle of project management. This model also known as the iron triangle (see Figure 1) was invented by Dr. Martin Barnes in 1969 to show the connection between time, cost and output (correct scope at the correct quality) [2].

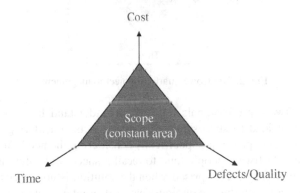

**Fig. 1.** The project management triangle or iron triangle

This classical triangle has been adapted by subsequent refinements adopting the shape of a square with four factors where the defects/quality axis is split into two dimensions: scope and quality (see Figure 2) [35]. A major issue in this direction is that the software quality cannot be analyzed separately, because the project managers must assure the respect to constraints on schedule and costs. A quality decision, for example, can lead to implement static code analysis (e.g. implementing tools, new processes, providing training, etc.) but its impact on project schedule, for example, can cause delays in completion of projects tasks while number of defects might be reduced up to certain extent leading to cost savings: in the end, the project manager needs to know if this is helpful and convenient for the project goals. To illustrate a typical case of the multidimensionality of decisions to be made, we can mention the case of implementation of software inspections in Unisys center in Australia [17]. Systematic collection of results during 8 month trough indicators and metrics allowed evaluation of the decision of implementing inspections on requirements, design and test cases. Effects on the three dimensions were measured:

- Effort involved was 0.53 man-years with a money value of cost of training of 80K$ (50K$ for attendee labor and 30K$ of training courses) plus 116K$ of extra effort. It required additional extension of schedule of 0.25 months.
- Savings in effort due to early detection was 8.2 man-years, reducing 4 months of schedule and given a ROI of 1:15.5 with a money value of 1.8M$, It was possible to see that fixing test cases defects implies 5 hours/each while a design one involved 8 hours and a requirements one 11.6 hours, confirming the escalation of defect costs through the life cycle, finally ending at 24.7 hours for fixing defects detected during testing.

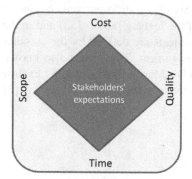

**Fig. 2.** The Iron Square of project management

In general, software professionals tend to not understand how to systematically make decisions that lead to successful software projects considering the inputs and outputs of their decision processes specifically considering the needs and expectations of the stakeholders. Software people tend to recall a successful solution for a particular problem, but then they are surprised when the solution is not successful in solving a similar problem in which the stakeholders have different expectations [35]. The root cause is the existence of inconsistent processes of decision: information goes into a decision process, a person-dependent miracle occurs and a solution or decision comes out. Professionals often consider people to be experts or gurus in their fields if they can make good decisions easily.

One of the few high level models proposed for general decisions in software projects is the one named as PEAK (see Figure 3). It does not address the adoption of a particular decision but a general procedure for managing a decision process.

Briefly PEAK specifies the main inputs to a decision model: (i) the Problem statement providing a clear presentation of what needs to be solved (e.g. Is it profitable for my project including static code analysis?), (ii) the problem's assumptions (freeware tools are available, training of personnel requires an average of 2 days, etc.), and (iii) the decision maker's Experience and Knowledge (e.g., knowledge/experience gained from previous projects). The output mainly comprises: (1) feasible alternative solutions associated with their benefit and costs, and (2) schedule/time-related decisions (e.g., the stakeholders to involve and stakeholders' preferences with respect issues concerning the alternative solutions). The risk assumed with alternative solutions is also output of the decision model. Therefore, alternative solutions have associated, for example: (i) the probability that the solutions will not work as envisioned; (ii) the level of risk that the solutions or decisions will not lead to a successful result; and (iii) the stakeholders' opinion on relations between risks and solutions alternatives.

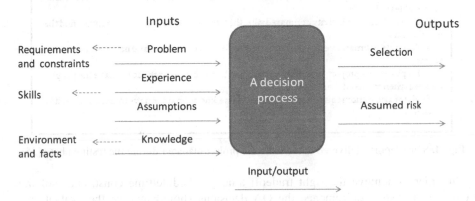

**Fig. 3.** PEAK decision model [35]

Another set of good practices for decisions is provided by one the of the process areas in CMMi 1.2 [10]. DAR (Decision Analysis and Resolution) process area provides guidelines to determine which issues should be subjected to a formal evaluation process and then applying formal evaluation processes to these issues. They are oriented to any type of decisions during projects but specially mentions decisions related to topics linked to medium or high risk, changing work products under configuration management (a technique always included in SQA assets), decisions which would cause schedule delays over a certain percentage or specific amount of time, etc. The recommended practices include "Establish Guidelines for Decision Analysis", "Establish Evaluation Criteria", "Identify Alternative Solutions", "Select Evaluation Methods", "Evaluate Alternatives" and "Select Solutions".

For example, project managers make their decisions during testing related to: (i) what to do if there isn't enough time for thorough testing, and (ii) how to know when it is time to stop testing [25]. The decision making process starts with an analysis of reports delivered to the software project manager during the testing phase (see Figure 4 for examples of reports).

The decision makers might also think about different aspects such as: which functionality is most important to the project's (e.g., the one that has the biggest safety impact, and has the largest financial impact on users) or which the highest risk areas might be according to the developers' opinion. In fact, as mentioned in [25], the decision to stop testing can be difficult. Deadlines or the testing budget may impact on decisions, but it is better to stop testing when reliability, functionality, and quality standards have been met.

Any combination of quality decisions may have a considerable impact on the cost, time and software quality. Therefore, an automated support would be very helpful in each software life-cycle. The decision makers would be able to quantify such impact in order to suggest the best quality decision, which minimizes the costs while satisfying the schedule/time, cost and quality constraints.

1) Defect find-and-close rates by week, normalized against the level of effort (are we finding defects, and can developers keep up with the number found and the ones necessary to fix?)

2) Number of tests planned, run, passed by week (do you know what we have to test, and are we able to do so?)

3) Defect found for activity compared with the total defects found (which activity find the most defects?)

4) Schedule estimates versus the actuals (will we make the dates, and how well do we estimate?)

5) People on the project, planned versus actual by week or month (do we have the people we need when we need them?)

6) Major and minor requirements changes (do we know what we have to do, and does it change?)

**Fig. 4.** Some reports delivered to the software project manager during the testing phase [25]

In order to achieve the right tradeoff among schedule/time constraints, software quality and costs requirements, the QA decisions should involve the evaluation of new alternatives to the current (i) software application level (e.g., by the configuration of software components, the introducing new components into the system, etc.) and (ii) project management level (e.g., the shift in allocations of people). For example, a decision taken for implementing static code analysis may be good for the satisfaction of a certain level of software quality, but at the same time it may require a high cost for implementing (e.g. tools, new processes, training, etc.). A major challenge is then finding the best balance among many different competing and conflicting constraints.

Hence, instead of viewing the project management and the application level as separate control/optimization problems, an explicit coordination between these levels would allow finding the appropriate tradeoff between the, usually conflicting, decisions of the levels. As a consequence, this latter contribution would allow finding the appropriate tradeoff among the, usually conflicting, constraints on cost, time/schedule, software quality.

Our approach attempts to find optimal decisions for each level (i.e., the application and the project management level) and to support explicit cooperation between layers. Our solution allows the verification of the impact of decisions (i.e., the quality decisions made at the application level, and the schedule/time and cost decisions made at the project management level) on system quality.

The paper is structured as follows. Section 2 describes our research challenge. Related work is described in Section 3, and finally Section 4 summarizes our findings and the novelty of our contribution.

## 2      Research Challenges

The main objective of the Iceberg project is the generation of a framework set for quality decision-making based on the cost, schedule/time and quality tradeoff. As a consequence it will provide a set of prototypal frameworks for quality decision-making based on the mentioned parameters. In particular, the frameworks aim to develop and manage software projects with a tight cooperation between the application level and the project management level. Secondary objectives are represented by the following points.

(i) We intend to analyze solutions that concern quality decision-making, cost and schedule/time issues all along the software lifecycle thus understanding, e.g.:

- What approaches have been reported regarding quality decision-making in the single lifecycle phases?
- How the quality decisions are (e.g., adaptation mechanisms typically used in adaptive systems) treated in the scientific literature? What is the relationship between these identified decisions and software defects (or incidents and other concepts)?
- What are the common causes of decision-making (such not satisfying of constraints on reliability)?
- What approaches have been reported dealing with human and organizational factors? For example, how do the approaches deal with the problem of automating and optimizing shift allocations to people in order to meet certain service levels?
- Which are the schedule/time/cost-related properties considered by the existing approaches for quality decision-making? What is the importance of these identified properties? What is the relationship between these identified properties and the software quality?

**Table 1.** Research objectives

| Objective | Description | Verifiable results | Measures of Success |
|-----------|-------------|--------------------|--------------------|
| RO1 | *Optimizing the quality-decision making process with a tight cooperation between the application level and the project management level.* | - Agreement on a list of quality decisions and project management decisions.<br>- Validation of the effectiveness and efficiency of SBSE techniques.<br>- Validation of the effectiveness and efficiency of the cooperation between the application and the project management level through sensitivity analysis (i.e., by introducing some level of uncertainty and measuring its effect on the management process and software quality). | - Preliminary prototype of a SBSE technique for decision making process based on quality, cost, and schedule/time tradeoff.<br>- Preliminary prototype for the project management level.<br>-Preliminary prototype for the integration of the prototypes for the project management and application level. |
| RO2 | *Defining a meta-model that expresses the dependencies among quality decisions, defects issues, cost factor and schedule factor.* | - Agreement on a list of quality decisions (relationship between decisions and software defects)<br>- Agreement of a list of causes of decision-making.<br>-Agreement of a list of schedule/time/cost-related properties | Preliminary prototype that combines quality decisions and project management decisions for the testing phase. |

(ii) In order to determine measures for assessing the three factors (i.e., cost, schedule/time and quality), we intend to conduct an interview-survey in which several representatives of industry will be involved. Our study will be based on a method specified in [44]. We plan to conduct our survey in multiple stages that span from questionnaire preparation through data collection and analysis to validity addressing. This information is essential to validate which options from SOTA (State Of The Art) could be feasible for practitioners according to SOPA (State Of The Practice). Specifically, we stated our research objectives as described in Table 1.

Specifically, these are high level work's goals (i.e., long term objectives) decided for achievement. We will refine these high level goals into more concrete sub goals (i.e., short term objective) until it is possible to objectively measure their fulfillment through the contrast with data and experiences of the industrial partners involved in the project.

## 3    Related Work

Research in software quality has flourished in the past years, in particular in the fields of new tools, techniques, and development methodologies. Their aim is to predict/estimate software quality, or to provide support for the quality decisions that software stakeholders take during the software life-cycle.

Several research efforts have been devoted to the definition of methods and tools able to predict/estimate the quality of software systems. Due to the different amount and type of information available, different prediction/estimation approaches have been introduced in each phase. However, all existing approaches basically provide guidelines: (i) to determine nonfunctional properties (e.g., reliability, availability) for a composite system (see, e.g., survey [3]) by using analytical models [37] and empirical study [26]; (ii) to predict defects in source code (e.g., see survey [30]); (iv) to estimate software defects (e.g., [59]); or (v) to design fault tolerant systems, or analyze fault tolerant architectures (see, [5] for a list of these approaches). Furthermore, other papers deal, for example, with: (1) the comparison of bug prediction approaches [16]; (2) the investigation of the usefulness of elementary design evolution metrics to identify defective classes [45]); (3) the empirical validation of object-oriented metrics on open source software for fault prediction [29]; and (4) the combination of techniques for a reliability target (e.g., the one of operational and debug testing [14]).

Decision-making techniques have been introduced to facilitate the reasoning process all along the software lifecycle. Different techniques have been introduced in order to, for example: (1) use software architecture for documenting and communicating design decisions and architectural solutions [9]; (2) analyze the impact of architectural decisions on system quality (e.g., the Architecture Tradeoff Analysis Method [41]), (3) estimates costs, (short-term and long-term) benefits and uncertainty of architectural design decision (e.g., the CBAM method [40]), (4) derive test plans from requirements [28], or use architectural artifacts (like software architecture specification models, architectural design decisions, architectural documentation) in testing the implementation of a system (e.g., test cases, test plans, coverage measures) and executes code-level test cases to check the implementation [5], and (4) select testing techniques according to the features of the software to test [15].

Optimization techniques have largely been used to automate, for example: (i) the testing process (e.g., the one of the mutation testing [38]); (ii) the search for an optimal architecture design with respect to a (set of) quality attribute(s) (see, survey [1]); or (iii) the adaptation of a software architecture (both its structure and behavior) with non-functional attributes tradeoff (e.g., [54]). Existing approaches basically are based on simple optimization models (e.g., in [54] the adaptation cost is minimized) or multi-objective optimization models (for example, for test case generations [6] or cross-project defect prediction [8]) maximizing a set of objectives (e.g., maximizing data flow coverage and minimizing the size of the test set [52], or minimizing adaptation costs and system probability of failure [50]).

However, decisions are not only made at the application level, but also at the project management level (i.e., schedule/time-related decisions are made). Research efforts have been spent for software development estimation, by using, for example a statistical model for managing selection bias effects [39], or the soft system methodology to establish a benchmark for managing cost overruns in software projects [19]. Other papers have focused on project staffing and scheduling using different approaches such a Mixed-Integer Linear Program (MILP) [33], a hybrid MILP/constraint programming benders decomposition algorithm [47], and a MIP-based approach [24].

Emerging computing application paradigms require systems that are not only reliable, compact and fast, but which also optimize many different competing and conflicting objectives, like response time, throughput and consumption of resources [31]. Any combination of quality decisions may have a considerable impact on cost, time/schedule decisions. Therefore, a major issue in this direction is that decisions in a single system level (i.e., the application or the project management level) cannot be analyzed separately, because they (sometime adversely) affect each other. These evaluations can suffer of large elapsed time when the search space size increases. In such cases, the complete enumeration of possible alternatives results inefficient.

Our approach falls within the Search-Based Software Engineering (SBSE), an approach in which search based optimization algorithms are used to solve problems in software engineering. In the past five years SBSE has proved to be a widely applicable and success-full approach. SBSE techniques have been applied to several problems throughout the software engineering lifecycle, from requirements and project planning to maintenance and reengineering. The SBSE approach results attractive because it provide a suite of adaptive automated and semiautomated solutions in situations typified by large complex problem spaces with multiple competing and conflicting objectives [32]. SBSE helps developers to raise their focus from a human-based search for an (near-) optimal adaptive software system building (and management) to the parameterization of models whose solution is machine-based (i.e. fully automated). Even though interest in the exploration of the SBSEs potential as a means for quality decision making and software project management has also grown rapidly, there are still big research challenges to be addressed.

For example, for quality decisions, SBSE search methodologies been used in order to support: (i) the generation of software test cases [42], and develop testing tools such as AUSTIN for unit testing C programs [46], (ii) the large-scale QoS-aware service compositions [58], and automate the search for an optimal architecture design based on functional and nonfunctional requirements tradeoffs [50], and (iii) the distributed system's allocation of software components to hardware nodes (i.e., deployment architecture) while guaranteeing a specific level of QoS properties [48].

SBSE techniques have also been largely applied in problems in software project management (see, e.g., to help software project managers find the best values for initial team size and schedule estimates for a given project so that cost, time and productivity are optimized [55]). A quite extensive list of these approaches can be found in [23]. As stated in [23], "the use of search-based approaches for project management is still in the development/research stage." Moreover, notwithstanding the increasing interest and diffusion of SBSE practices, the approaches typically do not support automation for this task. As remarked in [23], only the work in [56] (to the best of our knowledge) proposes a tool for project management based on SBSE approach.

As mentioned in [23], research effort has been devoted to project scheduling and resource allocation. However, all these approaches basically provide guidelines to plan projects. Their primary input is represented by information about (i) work packages (e.g., cost, duration, dependencies), and (ii) staff skills. Shortly, as described in [23], they process these input information and produce the results, which consist of an optimal work package ordering and staff allocation. They are guided by a single or multiobjectives fitness function. It is typically minimized, for example: the completion time of the project, or the risks to associate to the development process (e.g., delays in the project completion time, or reduced budgets available).

As outlined in [23], SBSE methodologies have been also applied to build effort estimation models or to enhance the use of other estimation techniques (e.g., genetic programming has been used in [18] to validate the component-based method for software sizing, and a tabu search approach has been adopted in [11] to estimate software development effort). An overview of existing approaches is provided in [23], as well as their advantages/limitations and open challenges are outlined. These approaches could be exploited, for example, (i) to support the choice of a reliable measure to compare different estimation models; or (ii) to investigate prediction uncertainty and risk of inaccurate prediction by means of using sensitivity analysis or multi-objective optimization (they only have been used to obtain exact prediction, i.e., one point estimate for a project).

Some search-based papers for project managements are focused on the problem of process risk (e.g., [27]) and the product risks (e.g., [43]). [1] Finally, the overtime planning is also considered in [22]. Specially, this work introduces a multiobjective decision support approach to help balance project risks and duration against overtime.

---

[1] "Risks to the product concern the possibility that there may be flaws in the product that make it less attractive to customers, while process risks concern the problems that may cause delays in the project completion time, or reduced budgets available forcing compromise."[23].

Even though interest in the exploration of the SBSEs potential as a means for quality decision making and software project management has also grown rapidly, there are still big research challenges to be addressed. Coordination aspects between the application level and the project management level have already been exploited. We can remark the following points.

(i) *"Build-or-buy" decisions in software architecture, system delivery time constraints, and testing have been considered together.* In [13], a framework for supporting "build-or- buy" decisions in software architecture has been introduced. Specifically, the work presents a non-linear cost/quality optimization model based on decision variables indicating the set of architectural components to buy and to build in order to minimize the software cost under reliability and delivery time constraints. The model can be ideally embedded into a Cost Benefit Analysis Method to provide decision support to software architects. Such formulation involves further variables representing the amount of unit testing to be performed on each in-house developed component.

(ii) *Reliability and costs together have been considered in different contexts*, for example to provide guidelines in (1) evaluating the effort spent to test the software, deal with the resource allocation during the test process or quantify the costs of service failure repair/mitigation actions (see, e.g., [34], [4], [12] detailed below), or (2) comparing the costs of defect-detection techniques [57].

In [34] it is formulated a reliability constrained cost minimization problem, where the decision variables represent the component failure intensities. Specifically, in order to represent the dependency of the component cost on the component failure intensity, i.e., the cost to attain certain failure intensity, we exploit three different types of cost functions (i.e. linear, logarithmic exponential, inverse power). This model works after the components have been chosen because its solution provides insights about the failure intensities that the (selected) components have to attain to minimize the system cost.

Resource allocation during the test process in modular software systems is dealt, for example, in [4]. Specifically, this work presents a framework for performing resource allocation (budget and time) during the test process of a software system. The framework exploits a model developed with the goal of finding the maximum reliability of the software system while satisfying a budget limit on the total test cost and minimum reliability of components. The paper assumes that a software system has already been specified, designed and coded.

The work in [12] presents an approach for service selection taking into account costs and reliability requirements. In particular, it defines a set of optimization models that allow quantifying the costs of service failure repair/mitigation actions aimed at keeping the whole system reliability over a given threshold.

(iii) *The Value-Based Software Engineering (VBSE) principles have been adopted.* VBSE is the approach in which models and measures of value are of use for managers, developers and users as they make tradeoff decisions between, for example, quality and cost or functionality and schedule.

*Models for achieving product and process improvement have been introduced.* The goal of these models is to ensure a capable process, i.e., a process that produces a significantly reduced number of exploitable defects (see, the work in [49] that pro-vides advice for those making a business case for building software assurance into software products during software development). Examples of process improvement models include the Software Engineering Institute's Capability Maturity Model Inte-gration (CMMI) framework, along with the retired Capability Maturity Model for Software (Software CMM). These models address process capability by assessing the presence, or absence, of proxies (i.e., essential practices that are generally considered to ensure against defects).

*Research efforts have also been focused in dealing with the automated selection and configuration of methods and tools.* The work in [51] analyzes challenges of manag-ing engineering tool variability in context of engineering project environment confi-gurations. For example, best-practice method support (e.g., the QATAM technique for the evaluation of QA strategies and their tradeoffs in the context of a software process) is discussed. The work also presents a conceptual approach using semantic modeling of project requirements and tool capabilities.

(iv) *How human and organizational factors influence software quality practices and productivity has been investigated.* The work in [21] conducts a survey to understand which is the situation of real testing practice and which factors mainly related to pro-fessionals (attitude, training or similar items) are having a real influence in software quality in terms of the perception of participants.

## 4     Conclusions: What Is the Impact If the Framework Is Successful?

In this section, we present the overall conclusions of our project in the context of findings and novelty of our contribution. To the best of our knowledge, this is the first attempt to combine cost, schedule/time, software quality and project management and application level coordination, for developing, and managing software applications. Addressing this goal has required an effort in surveying existing approaches and con-tributions to determine which line of action is capable of producing the results pur-sued by the Iceberg project.

Results from published limited experiences suggest that optimizing design and management of next generation systems can only be handled effectively by modeling and exploiting an explicit coordination between the, usually conflicting, quality deci-sions and the project management decisions (i.e., schedule/time and cost-related deci-sions). As a consequence, the most promising line for the Iceberg project's goals is the work with SBSE methodologies combined with multi-objective optimization (and other existing decision-making methods) and software quality validation techniques.

This theoretical approach has been also contrasted with the regular practices of software organizations as expressed in preliminary results of a pilot survey to 9 indus-trial partners. The data revealed that organizations are not collecting a wide set of data related to quality, cost/effort, schedule and size (this one required for density meas-ures). They are not implementing too many QA techniques, mainly testing and static code analysis (apart from more or less sophisticated configuration management) so decisions on the application or not of QA methods widely covered in literature are not applicable. Their decisions tried to cover almost all the possible factors (from development and QA methods and processes to human and organizational ones) but available collected data are not enough to provide evidences of their effects on the management indicators (quality, cost and time).

As a consequence, we have to address some  major challenges for combining possibilities of existing approaches and limitations of real environments with a wide perspective, i.e. providing framework for decisions applicable to a broad range of real organizations. We envision that our approach when implemented in a support model will be capable of assisting software designers/maintainers and software project managers during the whole software component lifecycle.

To reach this goal, we will require the contributions from academia and industrial experts in different fields including not only search-based optimization and quality/cost/time assessments but also the experimentation of our approach on real world case studies by considering realistic model parameter values, as well as integration of our frameworks. Subsequently we plan to analyze effort and time necessary to incorporate our solutions into real-world systems which, e.g. already feature built-in adaptation mechanisms (e.g., by exploiting the work in [7]). We also intend to integrate human architecture in this adaptation process (see, for example, the work in [20], that provide support for the decisions that system adaptation manager take upon the effect of software-level changes on human interactions and vice versa).

**Acknowledgments.** This research has been partially supported by European Commission funding under the 7h Framework Programme IAPP Marie Curie program for project ICEBERG no. 324356.

# References

1. Aleti, A., Buhnova, B., Grunske, L., Koziolek, A., Meedeniya, I.: Software Architecture Optimization Methods: A Systematic Literature Review. IEEE Transactions on Software Engineering 39(5), 658–683 (2013)
2. APM. A History of the Association for Project Management 1972-2010. In: Association for Project Management, Buckinghamshire (2010)
3. Balsamo, S., Marco, A.D., Inverardi, P., Simeoni, M.: Model-Based Performance Prediction in Software Development: A Survey. IEEE Trans. Soft. Eng. 30(5), 295–310 (2004)
4. Berman, O., Cutler, M.: Resource allocation during tests for optimally reliable software. Computers & OR 31(11), 1847–1865 (2004)
5. Bertolino, A., Inverardi, P., Muccini, H.: Software architecture-based analysis and testing: a look into achievements and future challenges. Computing 95(8), 633–648 (2013)
6. Buzdalov, M., Buzdalova, A., Petrova, I.: Generation of tests for programming challenge tasks using multi-objective optimization. In: Blum, C., Alba, E. (eds.) GECCO (Companion), pp. 1655–1658. ACM (2013)
7. Camara, J., Correia, P., de Lemos, R., Garlan, D., Gomes, P., Schmerl, B., Ventura, R.: Evolving an adaptive industrial software system to use architecture-based self-adaptation. In: 2013 ICSE Workshop on Software Engineering for Adaptive and Self-Managing Systems (SEAMS), pp. 13–22 (May 2013)

8. Canfora, G., Lucia, A.D., Penta, M.D., Oliveto, R., Panichella, A., Panichella, S.: Multiobjective Cross-Project Defect Prediction. In: ICST, pp. 252–261. IEEE (2013)
9. Clements, P., Bachmann, F., Bass, L., Garlan, D., Ivers, J., Little, R., Merson, P., Nord, R., Stafford, J.: Documenting Software Architectures: Views and Beyond, 2nd edn. Addison Wesley (2011)
10. CMMi Team, CMMI® for Development, Version 1.2 CMMI-DEV, V1.2, CMU/SEI-2006-TR-008, ESC-TR-2006-008, Software Engineering Institute (2006)
11. Corazza, A., Martino, S.D., Ferrucci, F., Gravino, C., Sarro, F., Mendes, E.: Using tabu search to configure support vector regression for effort estimation. Empirical Software Engineering 18(3), 506–546 (2013)
12. Cortellessa, V., Marinelli, F., Mirandola, R., Potena, P.: Quantifying the influence of failure repair/mitigation costs on service-based systems. In: 24th International Symposium on Software Reliability Engineering, ISSRE 2013 (2013) (to appear)
13. Cortellessa, V., Marinelli, F., Potena, P.: An optimization framework for "build-or-buy" decisions in software architecture. Computers & OR 35(10), 3090–3106 (2008)
14. Cotroneo, D., Pietrantuono, R., Russo, S.: Combining operational and debug testing for improving reliability. IEEE Transactions on Reliability 62(2), 408–423 (2013)
15. Cotroneo, D., Pietrantuono, R., Russo, S.: Testing techniques selection based on ODC fault types and software metrics. Journal of Systems and Software 86(6), 1613–1637 (2013)
16. D'Ambros, M., Lanza, M., Robbes, R.: An extensive comparison of bug prediction approaches. In: 7th IEEE Working Conference on Mining Software Repositories (MSR), pp. 31–41
17. Damian, D., Zowghi, D., Vaidyanathasamy, L., Pal, Y.: An Industrial Case Study of Immediate Benefits of Requirements Engineering Process Improvement at the Australian Center for Unisys Software. Empirical Software Engineering 9(1-2), 45–75 (2004)
18. Dolado, J.J.: A Validation of the Component-Based Method for Software Size Estimation. IEEE Trans. Software Eng. 26(10), 1006–1021 (2000)
19. Doloi, H.K.: Understanding stakeholders' perspective of cost estimation in project management. International Journal of Project Management 29(5), 622–636 (2011)
20. Dorn, C., Taylor, R.N.: Coupling Software Architecture and Human Architecture for Collaboration-aware System Adaptation. In: Proceedings of the International Conference on Software Engineering, ICSE 2013, pp. 53–62. IEEE Press (2013)
21. Fernández-Sanz, L., Villalba, M.T., Hilera, J.R., Lacuesta, R.: Factors with negative influence on software testing practice in spain: A survey. In: O'Connor, R.V., Baddoo, N., Cuadrago Gallego, J., Rejas Muslera, R., Smolander, K., Messnarz, R. (eds.) EuroSPI 2009. CCIS, vol. 42, pp. 1–12. Springer, Heidelberg (2009)
22. Ferrucci, F., Harman, M., Ren, J., Sarro, F.: Not going to take this anymore: multiobjective overtime planning for software engineering projects. In: ICSE, pp. 462–471. IEEE / ACM (2013)
23. Ferrucci, F., Harman, M., Sarro, F.: Search-Based Software Project Management. Software Project Management in a Changing World. Springer (to appear, 2014)
24. Firat, M., Hurkens, C.A.J.: An improved MIP-based approach for a multi-skill workforce scheduling problem. J. Scheduling 15(3), 363–380 (2012)
25. Futrell, R.T., Shafer, L.I., Shafer, D.F.: Quality Software Project Management. Prentice Hall PTR, Upper Saddle River (2001)
26. Goseva-Popstojanova, K., Singh, A.D., Mazimdar, S., Li, F.: Empirical Characterization of Session-Based Workload and Reliability for Web Servers. Empirical Software Engineering 11(1), 71–117 (2006)

27. Gueorguiev, S., Harman, M., Antoniol, G.: Software Project Planning for Robustness and Completion Time in the Presence of Uncertainty Using Multi Objective Search Based Software Engineering. In: Proceedings of Conference on Genetic and Evolutionary Computation, GECCO 2009, pp. 1673–1680. ACM (2009)
28. Güldali, B., Funke, H., Sauer, S., Engels, G.: TORC: test plan optimization by requirements clustering. Software Quality Journal 19(4), 771–799 (2011)
29. Gyimothy, T., Ferenc, R., Siket, I.: Empirical validation of object-oriented metrics on open source software for fault prediction. IEEE Transactions on Software Engineering 31(10), 897–910 (2005)
30. Hall, T., Beecham, S., Bowes, D., Gray, D., Counsell, S.: A Systematic Literature Review on Fault Prediction Performance in Software Engineering. IEEE Trans. Software Eng. 38(6), 1276–1304 (2012)
31. Harman, M., Langdon, W.B., Jia, Y., White, D.R., Arcuri, A., Clark, J.A.: The GISMOE Challenge: Constructing the Pareto Program Surface Using Genetic Programming to Find Better Programs (Keynote Paper). In: Proceedings of the IEEE/ACM International Conference on Automated Software Engineering, ASE 2012, pp. 1–14 (2012)
32. Harman, M., Mansouri, S.A., Zhang, Y.: Search-based Software Engineering: Trends, Techniques and Applications. ACM Comput. Surv. 45(1), 11:1–11:61 (2012)
33. Heimerl, C., Kolisch, R.: Scheduling and staffing multiple projects with a multi-skilled workforce. OR Spectrum 32(2), 343–368 (2010)
34. Helander, M.E., Zhao, M., Ohlsson, N.: Planning models for software reliability and cost. IEEE Trans. Software Eng. 24(6), 420–434 (1998)
35. Hoover, C., Rosso-Llopart, M., Taran, G.: Evaluating Project Decisions: Case Studies in Software Engineering. Addison-Wesley Professional (2009)
36. Iceberg consortium. Project proposal. How to estimate costs of poor quality in a Software QA project: a novel approach to support management decisions (2012)
37. Janevski, N., Goseva-Popstojanova, K.: Session Reliability of Web Systems Under Heavy-TailedWorkloads: An Approach based on Design and Analysis of Experiments. IEEE Transactions on Software Engineering 99(preprints), 1 (2013)
38. Jia, Y., Harman, M.: An Analysis and Survey of the Development of Mutation Testing. IEEE Transactions on Software Engineering 37(5), 649–678 (2011)
39. Jørgensen, M.: The influence of selection bias on effort overruns in software development projects. Information & Software Technology 55(9), 1640–1650 (2013)
40. Kazman, R., Asundi, J., Klein, M.: Quantifying the costs and benefits of architectural decisions. In: Proceedings of the 23rd International Conference on Software Engineering, ICSE 2001, pp. 297–306 (May 2001)
41. Kazman, R., Klein, M., Barbacci, M., Longstaff, T., Lipson, H., Carriere, J.: The architecture tradeoff analysis method. In: Proceedings of the Fourth IEEE Intern. Conference on Engineering of Complex Computer Systems, ICECCS 1998, pp. 68–78 (August 1998)
42. Kifetew, F.M., Panichella, A., De Lucia, A., Oliveto, R., Tonella, P.: Orthogonal Exploration of the Search Space in Evolutionary Test Case Generation. In: Proceedings of the Intern. Symposium on Software Testing and Analysis, ISSTA 2013, pp. 257–267. ACM (2013)
43. Kiper, J.D., Feather, M.S., Richardson, J.: Optimizing the V& V Process for Critical Systems. In: Proceedings of the 9th Annual Conference on Genetic and Evolutionary Computation, GECCO 2007, pp. 1139–1139. ACM, New York (2007)
44. Kitchenham, B.: Procedures for Performing Systematic Reviews (TR/SE-0401). Technical report, Keele University (2004)

45. Kpodjedo, S., Ricca, F., Galinier, P., Guéhéneuc, Y.-G., Antoniol, G.: Design evolution metrics for defect prediction in object oriented systems. Empirical Software Engineering 16(1), 141–175 (2011)
46. Lakhotia, K., Harman, M., Gross, H.: Austin: An open source tool for search based software testing of c programs. Information and Software Technology 55(1), 112–125 (2013)
47. Li, H., Womer, K.: Scheduling projects with multi-skilled personnel by a hybrid MILP/CP benders decomposition algorithm. J. Scheduling 12(3), 281–298 (2009)
48. Malek, S., Medvidovic, N., Mikic-Rakic, M.: An Extensible Framework for Improving a Distributed Software System's Deployment Architecture. IEEE Transactions on Software Engineering 38(1), 73–100 (2012)
49. Mead, N., Allen, J., Conklin, W., Drommi, A., Harrison, J., Ingalsbe, J., Rainey, J., Shoemaker, D.: Making the Business Case for Software Assurance (CMU/SEI-2009-SR-001). Technical report. Software Engineering Institute, Carnegie Mellon University, Pittsburgh (2009),
http://resources.sei.cmu.edu/library/assetview.cfm?AssetID=8831
50. Mirandola, R., Potena, P., Scandurra, P.: Adaptation space exploration for service-oriented applications. Science of Computer Programming 80(Part B), 356–384 (2014)
51. Moser, T., Biffl, S., Winkler, D.: Process-driven Feature Modeling for Variability Management of Project Environment Configurations. In: Proceedings of International Conference on Product Focused Software, PROFES 2010, pp. 47–50. ACM (2010)
52. Oster, N., Saglietti, F.: Automatic Test Data Generation by Multi-objective Optimisation. In: Górski, J. (ed.) SAFECOMP 2006. LNCS, vol. 4166, pp. 426–438. Springer, Heidelberg (2006)
53. Pierantuono, R., Russo, S., Battipaglia, I., Gaiani, C., Fernandez, L., Rodriguez, D.: Industrial needs collection and state of the art surveys (2013)
54. Potena, P.: Optimization of adaptation plans for a service-oriented architecture with cost, reliability, availability and performance tradeoff. Journal of Systems and Software 6(3), 624–648 (2013)
55. Rodriguez, D., Ruiz, M., Riquelme, J.C., Harrison, R.: Multiobjective Simulation Opti misation in Software Project Management. In: Proceedings of the 13th Annual Conference on Genetic and Evolutionary Computation, GECCO 2011, pp. 1883–1890. ACM (2011)
56. Stylianou, C., Gerasimou, S., Andreou, A.: A Novel Prototype Tool for Intelligent Software Project Scheduling and Staffing Enhanced with Personality Factors. In: IEEE Intern. Conference on Tools with Artificial Intelligence (ICTAI), vol. 1, pp. 277–284 (2012)
57. Wagner, S.: Towards Software Quality Economics for Defect-Detection Techniques. In: Software Engineering Worksho. 29th Annual IEEE/NASA, pp. 265–274 (April 2005)
58. Ye, Z., Zhou, X., Bouguettaya, A.: Genetic Algorithm Based QoS-Aware Service Compositions in Cloud Computing. In: Yu, J.X., Kim, M.H., Unland, R. (eds.) DASFAA 2011, Part II. LNCS, vol. 6588, pp. 321–334. Springer, Heidelberg (2011)
59. Zeng, H., Rine, D.: Estimation of Software Defects Fix Effort Using Neural Networks. In: COMPSAC Workshops, pp. 20–21. IEEE Computer Society (2004)

# Multi-heuristic Based Algorithm
# for Test Case Prioritization

Michael N. Nawar and Moheb M. Ragheb

Department of Computer Engineering, Cairo University
Cairo, Egypt
{michael.nawar,moheb.ragheb}@eng.cu.edu.eg

**Abstract.** Regression testing is the process of retesting the software after it has been modified and ensuring that there is no new errors have been introduced in the software due to these modifications. As the size of the software projects increases, the regression testing became a very costly process, so the need of detecting the faults in the software project as fast as possible became more and more important. Test case prioritization arranges test cases for execution to increase the probability of early fault detection during the regression testing. In this paper, three simple test case prioritization heuristics are presented, where every heuristic calculates the average number faults found per each test case. The three heuristics are combined together to develop a multi-heuristic based algorithm that arrange test cases based on their priorities using the scores obtained from the three heuristics. The effectiveness of the three heuristics and the multi-heuristic based algorithm are illustrated with the help of APFD (Average Percentage Faults Detected) metric. The main aim of this paper is to show how using simple heuristics for test cases prioritization would help in error early detection during regression testing, and to show how the proposed multi-heuristic based algorithm has significant increase in terms of APFD even if the algorithm is using simple heuristics.

## 1 Introduction

Regression testing is an important activity in the software life cycle, but it can also be very expensive and can account for a large proportion of the software maintenance budget [13]. To reduce the cost of the regression testing, multiple research has been inducted in the area of regression test optimization problem. It have been shown in [15] that the order in which the test cases of a test suite are executed has an influence on the rate at which faults can be detected.

**The Test Case Prioritization Problem.** Given a program $P_v$, its maintained test suite $T_v^m$ , a new version $P_{v+1}$, and a function $f(T)$ from a test suite T to a real number.

Problem is to find an ordered test suite $T_v^o \in Perm(T_v^m)$ such that $\forall T \in Perm(T_v^m)$, $f(T) \leq f(T_v^o)$. Here $Perm(T)$ is the set of all possible orderings of

B. Murgante et al. (Eds.): ICCSA 2014, Part V, LNCS 8583, pp. 449–460, 2014.

$T$, and $f(T)$ is an objective function that assigns to a test suite a relative score that indicates how well it satisfies the goals of prioritization.

Different objective functions can be pursued by testers when prioritizing test cases. They may wish to find bugs faster, cover more parts of code faster, find severe faults sooner, run easier-to-execute test cases earlier, and so on.

Test case prioritization for early fault detection has two main advantages. First, it reduces the cost associated with each bug, since the cost is partially related to the time it is discovered. Early bug detection, will lead to early debugging and hence fast bug fixing. Second, if the regression testing is executed within limited resources, early fault detection ensures that most of the faults are detected.

The main problem in the test case prioritization that no single approach is able to provide a complete solution to this problem. Therefore, in this paper the main contribution is to combine multiple prioritization techniques to get benefit from the advantages of each technique, and to find an improved solution for the test case prioritization problem. The proposed multi-heuristic algorithm do not only uses many sources of information but also provides flexibility in using other heuristics or prioritization techniques.

This paper is organized as follows: Section 2 presents an overview of research areas related to the topic of this paper. Section 3 presents the proposed heuristics and multi-heuristic prioritization methods.Section 4 presents the experimental study and discusses the results. Section 5 contains conclusions and a discussion on the future work.

## 2    Related Work

In this section, research areas related to the topic of this paper are elaborated. First, the state of the art test case prioritization approaches are discussed. Then the set ranking techniques that could be used to arrange the test cases based on the score of each element in the set are discussed. Finally, the optimal parameter searching algorithms that could be used for tuning the performance of the set ranking techniques are presented.

### 2.1    Test Case Prioritization

There exist different approaches for solving the test case prioritization problem. These approaches differ in: objective functions, the type of information exploited, the algorithm used to optimize the targeted objective function.

Code coverage prioritization techniques are methods to prioritize test cases based on coverage criteria, such as requirement coverage, statement coverage, branch coverage, etc. . . . Therefore, the code coverage prioritization techniques could be considered as white box testing techniques.

Miscellaneous approaches have been proposed to solve the test case prioritization problem. In [15], [16], Rothermel et al. investigated several prioritizing techniques, such as total statement (or branch) coverage prioritization and additional statement (or branch) coverage prioritization, that can improve the rate of fault detection. Zhang et al. [21] propose a technique which could incorporate varying test coverage requirements and prioritize accordingly.

Walcott et al. [19] formulate a time-aware test suite prioritization where regression test prioritization is performed in a time constrained execution environment in which testing only occurs for a fixed time period and also the execution time of test cases are known. They use genetic algorithms to find solutions to this optimization problem.

Mirarab [12] proposed a Bayesian network to predict the probability of test cases finding faults using multiple heuristics. Proposed BN-based framework provides flexibility in the utilized measurements, and enables combining different heuristics together based on their rational relations rather than ad-hoc methods.

Some approaches are based on the greedy algorithms. The 2-Optimal (Greedy) Algorithm [14] is K-Optimal Greedy Approach, where K = 2.The K-Optimal approach selects the next K elements that, taken together, consume the largest part of the problem. The 2-Optimal approach has been found to be fast and effective.

Other noncoverage based techniques in the literature include customer requirement-based prioritization [11], history-based test prioritization [10], and cost effective-based prioritization that prioritize test cases based on cost of analysis and cost of prioritization [5], [6].

## 2.2  Set Ranking

Set ranking is the process of ordering the elements of the sets such that an objective function is optimized. The elements in a set could have single score, or multiple scores. The sets where each element has a single score could be ranked using simple techniques like: sorting element based on their scores, using priority matrix [17], using Schulze method [18], etc...

Sets where each element has multiple scores could be ranked using multiple techniques like: (a) Using multi-level sorting for element based on their scores, (b) Creating a representative score for each element in the set using a combination of the element's scores then ranking the set using any single score ranking technique, or (c) Developing an algorithm that combine multiple heuristic in the same time, like Asmuni et al [2] who developed a fuzzy algorithm using multiple heuristic for the construction of timetables.

## 2.3  Optimal Parameter Searching

Optimal parameter searching for complex algorithms is a well known problem. Many widely used algorithm like support vector machines (SVMs) and conditional random field classifiers (CRF) need parameter tuning. The common approach for finding the optimal parameters is to use expert judgment and trial-and-error, but this approach has several problems. Most significantly, this method is impractical when the number of parameters increase. Also, intuition can be misleading, with the result being that some good parameters are never tested. Genetic and evolutionary algorithms are used to search for optimal paramters for several problems. Blome et al [3] used a genetic algorithm to find an optimal configuration for the local region principal components analysis (LRPCA) face recognition baseline algorithm. Akay et al [1] used artificial bee colony algorithm for solving the real-parameter optimization problems. Zhang et al [20] used ant colony optimization algorithm to find optimal parameters of SVM. Other techniques are used for solving the same problem. Givens et al [8] used a generalized linear mixed-effects model (GLMM) and Harzallah et al [9] used a rank-based Friedman Test to analyze the effects of parameters on an LDA+PCA algorithm.

## 3   Proposed Prioritization Methods

In this section, three heuristic test prioritization methods are presented, then the three heuristics are combined together to develop a multi-heuristic based algorithm. The objective is to develop simple and effective test prioritization methods. Therefore, the major tenet while developing these heuristics and the multi-heuristic algorithm is that they should require as little information as possible.

### 3.1   Heuristic H1

The major idea of this heuristic method is that test cases that execute a higher number of modified statements should be given a higher priority than test cases that execute a smaller number of modified statements. The intuition for this heuristic method is based on the premise that a test case that executes a higher number of modified statements has a higher probability of revealing a fault than a test case with a smaller number of executed modified statements. More formally, let

$$H1(t) = C(f) * \frac{time(t)}{max(time(t_f))}$$

where $t$ is the test case that we want to calculate its score, $f$ is the function called by the test case $t$, $C(f)$ is the number of statements changed in the function $f$, $t_f$ is the set of all test cases that test the function $f$, $time(t)$ is the execution time of the test case $t$ last time it has been executed, and the $max(time(t_f))$ is the maximum execution time of the test cases from the set $t_f$ last time they have been executed.

### 3.2   Heuristic H2

The major idea of this heuristic method is that test cases with higher number of impacted statement due to change in code should be given a higher priority than test cases that have a smaller number of impacted statements. The intuition for this heuristic method is based on the premise that a function that has a higher number of impacted statements has a higher probability of containing a fault than a function with a smaller number of impacted statements. And therefore, the test case that executes a higher number of statements in the function has a higher probability of revealing a fault than a test case with a smaller number of executed statements. More formally, let

$$H2(t) = \frac{I(f)}{Length(f)} * \frac{time(t)}{max(time(t_f))}$$

where $t$ is the test case that we want to calculate its score, $f$ is the function called by the test case $t$, $I(f)$ is the number of statements impacted in the function $f$ because of the change in the function, $Length(f)$ is the total number of statements in the function $f$, $t_f$ is the set of all test cases that test the function $f$, $time(t)$ is the execution time of the test case $t$ last time it has been executed, and the $max(time(t_f))$ is the maximum execution time of the test cases from the set $t_f$ last time they have been executed.

The impact $I(f)$ could be calculated using a simple algorithm as follows:

---

**Algorithm 1.** Calculate the Impact

---

**Require:** *function f*
**Ensure:** *I*
1: $I \leftarrow 0$
2: *Queue q* ← *new Queue*
3: **for** *each changed line l in f* **do**
4:     *q.enqueue(l)*
5: **end for**
6: **while not** *q.empty()* **do**
7:     $l \leftarrow queue.dequeue()$
8:     $I \leftarrow I + 1$
9:     **for** *each use u for the def l* **do**
10:         **if** *u not traversed before* **then**
11:             *q.enqueue(u)*
12:         **end if**
13:     **end for**
14: **end while**
15: **return** *I*

---

This algorithm simply assumes that each changed statement in the function is a definition of a variable, and this definition will affect the lines that will use variables assigned this definition. And so on, the newly affected variables, will affect new lines. This algorithm will continue till all affected statements in the function is traversed.

## 3.3  Heuristic H3

The major idea of this heuristic method is that test cases that control a higher number of modified statements and observe their output should be given a higher priority than test cases that control a smaller number of modified statements and observe their output. The intuition for this heuristic method is based on the premise that a test case that control a higher number of modified statements and observer their output has a higher probability of revealing a fault than a test case that control a smaller number of modified statements and observe their output. More formally, let

$$H3(t) = ( \sum_{\forall i \in C(f)} O_i * Co_i) * \frac{time(t)}{max(time(t_f))}$$

where $t$ is the test case that we want to calculate its score, $f$ is the function called by the test case $t$, $C(f)$ is the number of statements changed in the function $f$, $O_i$ is the observability of the statement $i$, $Co_i$ is the controllability of the statement $i$, $t_f$ is the set of all test cases that test the function $f$, $time(t)$ is the execution time of the test case $t$ last time it has been executed, and the $max(time(t_f))$ is the maximum execution time of the test cases from the set $t_f$ last time they have been executed. The controllability of the statement could be expressed as:

$$Co_i = 2^{-\# \ branches \ before \ statement}$$

The observability $O_i$ of a statement $i$ could be calculated using a simple algorithm as follows:

---

**Algorithm 2.** Calculate the Observability

---

**Require:** *function f, statement i*
**Ensure:** $O_i$
 1: **if** *i is a return statement* **then**
 2:    return  $O_i \leftarrow 1$
 3: **end if**
 4: *Queue q $\leftarrow$ new Queue*
 5: *q.enqueue(i, 1)*
 6: **while not** *q.empty()* **do**
 7:    $(l, j) \leftarrow queue.dequeue()$
 8:    **for** *each use u for the def l* **do**
 9:       **if** *u is a return statement* **then**
10:          return  $O_i \leftarrow \frac{1}{j+1}$
11:       **end if**
12:       **if** *u not traversed before* **then**
13:          *q.enqueue(u, j + 1)*
14:       **end if**
15:    **end for**
16: **end while**
17: return  $O_i \leftarrow 0$

---

This algorithm simply assumes that each changed statement in the function is a definition of a variable, and this definition will affect the lines that will use variables assigned this definition. And so on, the newly affected variables, will affect new lines. This algorithm will continue till a return statement is found (i.e. an output that could be observed), or till there is no more statement to traverse.

## 3.4   The Multi-heuristic Algorithm

The proposed multi-heuristic algorithm could be described as follows:

---

**Algorithm 3.** The Multi-heuristic Algorithm

**Require:** $T[1 \ldots n]$ **and** $C[0 \ldots 26]$
**Ensure:** $T'[1 \ldots n]$
1: $Scores[1 \ldots n] \leftarrow 0, 0, \ldots, 0$
2: **for** $i = 1 \ldots n$ **do**
3:     $score \leftarrow 0$
4:     **for** $j = 1 \ldots n$ **and** $j \neq i$ **do**
5:         **if** $H1(T[i]) < H1(T[j])$ **then**
6:             $score \leftarrow score + 1$
7:         **else if** $H1(T[i]) > H1(T[j])$ **then**
8:             $score \leftarrow score + 2$
9:         **end if**
10:        **if** $H2(T[i]) < H2(T[j])$ **then**
11:            $score \leftarrow score + 3$
12:        **else if** $H2(T[i]) > H2(T[j])$ **then**
13:            $score \leftarrow score + 6$
14:        **end if**
15:        **if** $H3(T[i]) < H3(T[j])$ **then**
16:            $score \leftarrow score + 9$
17:        **else if** $H3(T[i]) > H3(T[j])$ **then**
18:            $score \leftarrow score + 18$
19:        **end if**
20:    **end for**
21:    $Scores[i] \leftarrow Scores[i] + C[score]$
22: **end for**
23: $T'[1 \ldots n] \leftarrow SortOnScores(T[1 \ldots n], Scores[1 \ldots n])$
24: **return** $T'$

---

The input to the algorithm is a set of test cases $T[1 \ldots n]$ and a list of coefficients $C[0 \ldots 26]$. In the first line, the overall score of each test case is set to 0. In line 2, each test case $t_i$ is traversed. In line 3, the score variable is set to 0. In line 5, each test case $t_j$, $j \neq i$ is traversed. In the lines 5 to 19, every two test cases $t_i, t_j$ are compared together in terms of $H1$, $H2$, and $H3$ then the score variable is set to a number from 0 to 26 based on the values of the heuristics of the two test cases. In line 21, the overall score of the test case $t_i$ is incremented by the corresponding coefficient to the score. Finally, line 23 sort the test cases based on their overall score and then return the new prioritized test cases in line 24.

**The Algorithm Parameter Tuning.** In this part, the algorithm used for setting the coefficients $C[1 \ldots n]$ of the multi-heuristic based algorithm will be explained. The coefficients are set using a simple algorithm as follows:

---

**Algorithm 4.** Parameter Tuning Algorithm

---
**Require:** $T[1 \ldots n]$ **and** $M$
**Ensure:** $C[0 \ldots 26]$
  1: $Score \leftarrow 0$
  2: **for** $i = 1 \ldots M$ **do**
  3:     $TC[0 \ldots 26] \leftarrow SimulatedAnnealing(T[1 \ldots n])$
  4:     **if** $Score < ScoreFunction(C[0 \ldots 26])$ **then**
  5:       $Score \leftarrow ScoreFunction(C[0 \ldots 26])$
  6:       $C \leftarrow TC$
  7:     **end if**
  8: **end for**
  9: **return** $C$

---

This algorithm takes a set of test cases $T[1 \ldots n]$, and number of trials $M$ to find optimal coefficients $C[0 \ldots 26]$. The algorithm simply check for the solution of the coefficients using simulated annealing method for $M$ times, then finally return the best solution obtained in all the trials. The $ScoreFunction$ used by the algorithm in line 4 and also used by the $SimulatedAnnealing$ algorithm is the APFD function explained in equation (1).

# 4    Empirical Results

This section presents the results of an empirical evaluation aiming to show that multi-heuristic based test-case prioritization results in a noticeable performance improvement. We also want to analyze the newly defined heuristics and multi-heuristics techniques in comparison to well-known techniques.

## 4.1    Evaluation Measurements

In order to quantify the efficiency gains achieved with a certain test-case prioritization, the metric APFD was introduced by Rothermel et al. [15]. This metric is the weighted average percentage of faults detected over the life of a testsuite. The APFD of a test-suite $T$ consisting $n$ test-cases and $m$ mutants is defined as:

$$APFD = 1 - \frac{TF_1 + TF_2 + \cdots + TF_n}{nm} + \frac{1}{nm} \tag{1}$$

Here, $TF_i$ is the first test-case in ordering $T'$ of $T$ which reveals fault i. We use this metric in order to compare the different prioritization techniques.

## 4.2   Experiment Design

The target objects of the experiments need to be Java programs with at least a few versions available. As subject programs, we have used five Java programs from the Software-Artifact Infrastructure Repository (SIR) [7]: ant, xml-security, jmeter, nanoxml, and siena. The ant program is a Java-based build tool, extensively used in industry for building and deploying Java (especially webbased) applications. Jmeter is a Java application for load-testing and performance measurement. Xml-security implements some XML security standards. Nanoxml is a small parser for XML files in Java language. Siena (Scalable Internet Event Notification Architecture) is an Internet-scale event notification middle-ware for distributed event based applications deployed over wide-area networks, responsible for selecting notifications that are of interest to clients (as expressed in client subscriptions) and then delivering those notifications to the clients via access points.

Table 1 summarizes the information of our subject programs. For each program, the table shows: the number of available consecutive versions of the program (Versions), the total lines of code in the most recent version of the program in thousands lines (LOC (K)), the number of test cases available for the most recent version of the program (Tests), and finally the type of the test suite (Type).

**Table 1.** Target Programs and Their Attributes

| Program | Version | LOC(K) | Tests | Type |
|---------|---------|--------|-------|------|
| ant | 11 | 80.5 | 877 | JUnit |
| xmlsecurity | 9 | 16.8 | 78 | JUnit |
| jmeter | 7 | 43.4 | 83 | JUnit |
| nanoxml | 6 | 7.6 | 216 | TSL |
| siena | 8 | 6 | 567 | TSL |

Two of these five programs will be used for the tuning of the parameters of the multi-heuristics algorithm (jmeter and siena), while the remaining three programs will be used in testing the proposed three heuristics and the multi-heuristic algorithm.

These five programs, as released by their developers, do not have faults that could be revealed by their test suites. To generate faulty variations of the objects, we have planted mutation faults into the programs, and then check whether the test suite can detect these changes (kill mutants) or not. We have used the pool of mutation faults have been obtained from the authors of [4], to make the results presented in this work comparable to those of the mentioned study, since their research is focused on regression testing and they are using the same objects under test.

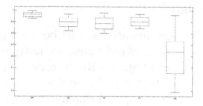

**Fig. 1.** The APFD Metric on Ant Program

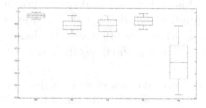

**Fig. 2.** The APFD Metric on Xml-security Program

## 4.3 Evaluation

Following the tradition of previous papers about test-case prioritization we use box-plots to illustrate the results of the APFD analysis. The box-plots illustrate minimum, 25th percentile, median, 75th percentile and maximum for each of the used prioritization methods. Figures 1, 2 and 3 shows the APFD metric of the overall result for all the releases of the ant, sml-security and nanoxml programs respectively. Each figure from 1, 2 and 3 contains five box-plots representing from right to left: the original test cases without ordering APFD, heuristic #1 APFD, heuristic #2 APFD, heuristic #3 APFD, and the multi-heuristic algorithm APFD.

As can be seen in Figures 1, 2 and 3, the APFD metric of the multi-heuristic based algorithm is higher than the three heuristics and the original ordering in all the cases, and the APFD metrics of the three heuristics are always better

**Fig. 3.** The APFD Metric on Nanoxml Program

than the original ordering in all cases. The results of the three heuristics are almost the same in all the cases, but it seems that heuristic #1 is slightly better than heuristics #2 and #3.

# 5    Conclusion

This paper described three heuristics for the test case prioritization for regression testing. Also, it presented the results of an empirical study that investigated their relative effectiveness. The data and analysis indicate that the results from the experiment are promising and suggest that multi-heuristic may improve the effectiveness of test prioritization, without significantly affecting the cost of prioritization.

There are numerous ways to extend this research work. The heuristics used in the multi-heuristic algorithm could be replaced by other heuristics or prioritization algorithm. The parameter tuning algorithm can be solved by other advanced techniques more than the simulated annealing. Also, the experimental study presented in this paper was relatively small to get conclusive statement about the effectiveness of the multi-heuristic algorithm we plan to perform an experimental study on larger systems to have better understanding of the advantages and limitations of multi-heuristic test prioritization.

# References

1. Akay, B., Karaboga, D.: A modified artificial bee colony algorithm for real-parameter optimization. Information Sciences 192, 120–142 (2012)
2. Asmuni, H., et al.: An investigation of fuzzy multiple heuristic orderings in the construction of university examination timetables. Computers & Operations Research 36(4), 981–1001 (2009)
3. Bolme, D.S., Beveridge, J.R., Draper, B.A., Phillips, P.J., Lui, Y.M.: Automatically searching for optimal parameter settings using a genetic algorithm. In: Crowley, J.L., Draper, B.A., Thonnat, M. (eds.) ICVS 2011. LNCS, vol. 6962, pp. 213–222. Springer, Heidelberg (2011)
4. Do, H., Rothermel, G.: On the use of mutation faults in empirical assessments of test case prioritization techniques. IEEE Trans. Softw. Eng., 733–752 (2006)
5. Elbaum, S., Malishevsky, A., Rothermel, G.: Incorporating Varying Test Costs and Fault Severities into Test Case Prioritization. In: Proc. 23rd Intl Conf. Software Eng. (ICSE 2001), pp. 329–338 (2001)
6. Elbaum, S., Malishevsky, A.G., Rothermel, G.: Test Case Prioritization: A Family of Empirical Studies. IEEE Trans. Software Eng. 28(2), 159–182 (2002)
7. Galileo Research Group:Software-artifact infrastructure repository, SIR (2009)
8. Givens, G.H., Beveridge, J.R., Draper, B.A., Bolme, D.S.: Using a generalized linear mixed model to study the configuration space of pca+lda human face recognition algorithm. In: Perales, F.J., Draper, B.A. (eds.) AMDO 2004. LNCS, vol. 3179, pp. 1–11. Springer, Heidelberg (2004)
9. Harzallah, H., Jurie, F., Schmid, C.: Combining efficient object localization and image classification. In: International Conference Computer Vision (2009)

10. Kim, J.M., Porter, A.: A History-Based Test Prioritization Technique for Regression Testing in Resource Constrained Environments. In: Proc. 24th Intl Conf. Software Eng., pp. 119–129 (2002)
11. Leon, D., Podgurski, A.: A Comparison of Coverage-Based and Distribution-Based Techniques for Filtering and Prioritizing Test Cases. In: Proc. Intl Symp. Software Reliability Eng., pp. 442–453 (2003)
12. Mirarab, S.: A Bayesian Framework for Software Regression Testing. Master of applied science thesis. Waterloo, Ontario, Canada (2008)
13. Onoma, K., Tsai, W.-T., Poonawala, M., Suganuma, H.: Regression Testing in an Industrial Environment. Comm. ACM 41(5), 81–86 (1988)
14. Parsa, S., Khalilian, A.: On the Optimization Approach towards Test Suite Minimization. International Journal of Software Engineering and its Applications 4(1), 15–28 (2010)
15. Rothermel, G., et al.: Test case prioritization: An empirical study. In: Proceedings of the IEEE International Conference on Software Maintenance (ICSM 1999). IEEE (1999)
16. Rothermel, G., Untch, R., Chu, C., Harrold, M.J.: Prioritizing Test Cases for Regression Testing. IEEE Trans. Software Eng. 27(10), 929–948 (2001)
17. Scholtes, P.R., Joiner, B.L., Streibel, B.J.: The team handbook. Joiner/Oriel Incorporated (2003)
18. Schulze, M.: A new monotonic, clone-independent, reversal symmetric, and condorcet-consistent single-winner election method. Social Choice and Welfare 36(2), 267–303 (2011)
19. Walcott, K.R., Soffa, M.L., Kapfhammer, G.M., Roos, R.S.: Time aware test suite prioritization. In: ISSTA, pp. 1–12 (2006)
20. Zhang, X.L., Chen, X.F., He, Z.J.: An ACO-based algorithm for parameter optimization of support vector machines. Expert Systems with Applications 37(9), 6618–6628 (2010)
21. Zhang, X., Nie, C., Xu, B., Qu, B.: Test case prioritization based on varying testing requirement priorities and test case costs. In: Proceedings of the IEEE International Conference on Quality Software (QSIC), pp. 15–24 (2007)

# On the Impact of Debugging on Software Reliability Growth Analysis: A Case Study

Marcello Cinque[1], Claudio Gaiani[2], Daniele De Stradis[2],
Antonio Pecchia[1], Roberto Pietrantuono[1], and Stefano Russo[1]

[1] Dipartimento di Ingegneria Elettrica e delle Tecnologie dell'Informazione
Università di Napoli Federico II, Via Claudio 21 - 80125 Napoli, Italy
{macinque,antonio.pecchia,roberto.pietrantuono,sterusso}@unina.it
[2] Assioma.Net, Via G. Spano, 6/11 - 10134 Torino, Italy
{claudio.gaiani,daniele.destradis}@assioma.net

**Abstract.** Reliability is one of the most relevant software quality attributes. The literature offers a variety of mathematical models - namely, software reliability growth models (SRGMs) - to estimate the reliability of a software product at a given time, as well as to predict the reliability that will be achieved as testing activities progress. One of the typical assumptions of SRGMs is the immediate debugging of detected faults. In reality, the impact of the debugging process cannot be neglected at all. This paper reports the results of a real-world case-study in which we analyze the debugging process of a Customer Relationship Management (CRM) system, and study its impact on SRGM-based reliability estimation and prediction.

**Keywords:** Software reliability, SRGM, reliability estimation, debugging, reliability prediction.

## 1 Introduction

Reliability estimation and prediction play a key role for software project schedule and cost/quality control. Over the past three decades, an important field of research has concerned **software reliability growth models** (SRGMs). SRGMs are a wide class of mathematical models conceived to fit inter-failure times from test data, in order to estimate the next time to failure based on the observed trend. They can be used to: *i)* estimate software reliability at a given time, as well as the number of residual faults in the software, *ii)* predict the expected reliability given a fixed budget for testing to spend (e.g., in terms of testing time or testing effort), *iii)* schedule the optimal time to release, given a desired reliability level, *iv)* compare actual and estimated release time in order to identify delays and their causes.

SGRMs make a set of assumptions to meet the mentioned objectives, the most usual ones being: immediate debugging, perfect debugging, dependent inter-failure times, equal probability to find a failure across time units [1]. In the literature, a greater attention is being paid to the immediate debugging assumption, since its impact is more relevant than other factors in real projects. While

B. Murgante et al. (Eds.): ICCSA 2014, Part V, LNCS 8583, pp. 461–475, 2014.

some works introduce modeling approaches to include repair times [17] [18], several empirical studies make it evident that debugging is a complex process to model in real-world projects [19], [20], [21]. Indeed, there are many factors impacting the computation of the actual repair time and the regularity of the debugging process, including the type of defect[1], its priority or severity, and human factors (e.g., skills of people involved in the fixing process). These make such an assumption easy to be violated (especially as complexity and size of a software project increase), with repair times far from being *immediate*. In many cases, the debugging process might even become a bottleneck for project releases, and its impact cannot be neglected at all.

The impact of an irregular and variable debugging process hampers a correct modeling and influences the assessment of release quality estimates and of SRGM-based predictions. This can determine errors in taking decisions on when to stop testing, and in the estimate of the residual defectiveness.

In this paper we analyze 3,392 real-world issues of an industrial case-study[2] collected over two years. On these data, we first *i)* use SRGMs to characterize the software reliability growth under the assumption of immediate debugging; then, *ii)* we characterize out the debugging process; *iii)* finally, we evaluate the impact of the debugging time evolution on reliability estimation and prediction, and thus on release scheduling performed by SRGMs. The study shows that:

1. Collected issues are amenable to be modeled by SRGMs; we applied a set of 7 models to fit data and found the truncated logistic and truncated normal SRGM being the best fitting models. Despite the real data do not fulfill classical SRGM assumptions, such as dependent inter-failure times and equal failure detection probability, the models have shown to be robust. However, since they are built upon *opened* issues, nothing can be said about the non-immediate debugging assumption. Therefore, these models are useful for predictions only provided that the debugging time is negligible.

2. The observed debugging process has a non-negligible time, on average equals to 12.8 days. The statistical characterization highlights a good quality of the process. Issues are closed regularly and in a reasonable time and, as desirable, severe issues are solved faster than minor ones. It also emerged that the queuing time of issues is not negligible (about a third of the overall fixing time), hence it has to be considered to avoid inaccurate results.

3. The non-immediate debugging has an impact on both reliability estimation and prediction, and thus on optimal release time, in a different way. In both cases the impact is dependent on the debugging process quality in terms of debugging time and debugging time variation, but while the impact on reliability estimation is quite insensitive with respect to the testing time dimension, the release schedule prediction error can greatly variate: as

---

[1] Note that the literature on debugging typically uses the term "defect", while in the SRGM field they use the term "fault": in this work we use them synonymously to denote the adjudged cause of an observed failure.

[2] The actual name of the system in not disclosed here due to confidentiality reasons.

testing times proceeds, the optimal release schedule prediction can be affected considerably by the debugging time.

The paper is organized as follows. Section 2 presents related work in the area of SRGMs, while Section 3 introduces available dataset. Section 4 presents the SRGMs modeling; Section 5 discusses the debugging process and its impact on reliability and best schedule estimation; Section 6 concludes the work.

## 2   Related Work

Reliability analysis can be conducted through modeling, by measurements, and by hybrid approaches [5] [6] [7] [23] [24]. Reliability is very tightly related to testing, since it is expected to grow as more time/resources are devoted to testing [8]. In the testing phase, one of the most successful approaches for reliability analysis is SRG modeling. Many SRGMs are available in the literature. A very successful one was proposed by Goel and Okumoto in 1979 [9], describing the failing process by an exponential *mvf* distribution. Other common models were proposed later, including: the generalized version of the exponential model, which uses the Weibull distribution [13]; the S-Shaped model [10], conceived to capture the possible increasing/decreasing behaviour of the failure rate during the testing process; the Gokhale and Trivedi log-logistic model [11], that also follows an increasing/decreasing pattern describing the initial phase of testing as characterized by a slow initial learning phase. More recent models, as those based on the Gompertz SRGM, proposed by Ohishi *et al.* [15], derived from the statistical theory of extreme-value. Several tools have been developed to deal with parametrization and fitting of models (such as SMERFS, SoRel, PISRAT, and CASRE). All these models are based on a set of common assumptions, the most impacting ones being the immediate and perfect debugging.

In the past, some research has defined SRGMs accounting for the perfect debugging assumption, namely by including the imperfect debugging in the model. For instance, there is a class of SRGMs known as infinite-failure models, which, contrarily to the finite-failure models, assume that an infinite number of faults would be detected in infinite testing time [2]. These are meant to capture debugging where faults may be re-introduced. An example is the Musa-Okumoto logarithmic Poisson execution time model [3], the more recent failure-size proportional model proposed in [4], as well as the model by Jain *et al.* [16].

As for the *immediate* repair assumption, some researchers modeled the debugging process through queuing models, considering also the non-immediate debugging time. For instance, the work in [17] uses a queue model for the correction process, while authors in [18] discuss both finite and infinite server queuing models for reliability measurement through SRGMs. On one side there are these models that can take into account the debugging times from a theoretical point of view; on the other side, there are empirical studies that analyze the characteristics of bug fixing process in real projects, which make it evident that debugging is a complex process to model. Thus, parametrizing models can be a non-trivial task., as there are several factors that impact the computation of the actual debugging time. For instance, the study in [19] reports an analysis on 1500 defects revealed in 5 years on an IBM middleware, classifying defects per topic and developer expertise, showing that the time to repair is impacted from these two factors. Zhang et al. [20] found some factors influencing the time lag between the defect assignment to a developer and the actual starting of the repair action, through a case-study on 3 open source software. They found that the assigned severity, the bug description, and the number of methods and changes in the code as impacting factors. The work in [21] reports a study specifically focused on finding bottlenecks in the issue management process, through a case-study of the Apache web-server and the Firefox browser. The main cause of inefficiency is the time lag in which the correction is verified after the repairing to confirm the correct resolution. These are all factors that cause irregularities in the debugging process, and can make the impact of debugging on reliability estimation/prediction more variable and difficult to control. In this work, such impact is studied with reference to one specific real-world case-study.

## 3   Data Source

The target system is a Customer Relationship Management (CRM) software for a multinational company operating in the healthcare sector. The system follows a classical three-layer architecture, with a Frontend, a Backend, and a Database, glued together with an Enterprise Application Integration (EAI) layer. The system provides classical CRM functionalities, such as, sales management, user profiles, agenda, contacts, inventory, procurement of goods, and various reporting tools. Data used in this study are issues collected from the tracker of the target system. In particular, we use data extracted from 3392 issues collected for 30 months, ranging from September 2012 to January 2014. For every issue, several attributes are available, such as:

- *State*: the working state of the issue; the following states are considered (detailed in section 5): new, published, in study, launched, completed, tested, delivered, suspended, and closed;
- *Timestamps*: the dates of every state transition are tracked; the most important timestamps are the ones related to the *new* state (when an issue is opened) and the *closed* state (when an issue is solved); the timestamps related to intermediate states are also useful to analyze the phases of the workflow, such as, how much time an issue is queued waiting to be fixed (e.g., from the *new* to the *in study* state) or how much time is devoted to the fixing itself (e.g., from the *in study* to the *closed* state);
- *Assignees*: the number of resources allocated on the issue;
- *Affected Version*: the version of the system affected by the issue;
- *Severity*: the severity of the issue, classified in blocking, major and minor;
- *Affected Component*: the name of the component (and subcomponent) affected by the issue;
- *Resolution*: final classification of the issue, as *fixed*, *won't fix*, or *not a defect*.

Data have been cleaned to remove inconsistencies and useless issues (for instance, newly opened issues that are canceled without being treated, duplicate issues, and issues that are not a defect). Finally, 3,335 issues have been considered in the study.

## 4  Reliability Analysis through SRGMs

In this Section, we present the analysis conducted on reliability growth *vs.* testing time for our data source. At this stage, we assume an immediate and perfect debugging. Hence, the model is built using the opening time of the issues. We consider the most common class of SRGMs, those describing the testing process as a non-homogeneous Poisson process (NHPP). These are characterized by the parameter of the stochastic process, $\lambda(t)$, indicating the fault detection intensity (known as failure intensity), and by the *mean value function* (*mvf*), $m(t)$, that is the expectation of the cumulative number of faults detected at time $t$ [11]: $N(t)$: $m(t) = E[N(t)]$; $\frac{dm(t)}{dt} = \lambda(t)$. The different SRGMs are described by their *mvf*, that appears in the form $m(t) = aF(t)$, where $a$ is the expected number of total faults exposed in an infinite testing time.

A common belief about SRGMs is that there exists no one single model able to work well with any set of data; the best model needs to be selected for each specific context. For our purpose, we consider the list reported in Table 1, in order to capture the actual behavior of the testing process. It also reports the corresponding expression of the mean value function (*mvf*); the estimated number of faults $a$ is the *mvf*'s first parameter.

To select the best fitting SRGM, we fit the issue data with every SRGM listed in Table 1 by the EM algorithm [14], and then perform a goodness of fit (GoF) test by means of the Kolmogorov-Smirnov (KS) test. Among the SRGMs with KS test satisfied, we use the *Akaike Information Criterion* (AIC) to select the model, taking the SRGM with the lowest AIC value (as in [14]).

**Table 1.** Software Reliability Growth Models

| Model | m(t) function |
|---|---|
| Exponential | $a \cdot (1 - e^{-bt})$ |
| S-shaped | $a \cdot [1 - (1 + gt)e^{-bt}]$ |
| Weibull | $a \cdot (1 - e^{-bt^{\gamma}})$ |
| Log Logistic | $a \cdot \frac{(\lambda t)^{\kappa}}{1+(\lambda t)^{\kappa}}$ |
| Log Normal * | $a \cdot \Phi(\frac{log(t)-\mu}{\sigma})$ |
| Truncated Logistic | $a \cdot \frac{(1-e^{-t/\kappa})}{(1+e^{-(t-\lambda)/\kappa})}$ |
| Truncated Normal * | $a \cdot \frac{\Phi((t-\mu)/\sigma)}{1-\Phi(-\mu/\sigma)}$ |
| * $\Phi$ indicates the normal distribution | |

Figure 1 shows the entire set of the raw data and the model fitting them. We can note a pronounced saturation around the day 100, causing no model satisfying the KS test. For visually capturing the trend, we however reported the model having the lowest AIC value among the SRGMs, which is a truncated logistic one. At a closer look, we noticed that the saturation point around day 100 corresponds to the release of the first major version; data from the day 190 on refer to issues belonging to version 2.0 of the software. Thus, we split the dataset into two groups, according to the release (Figure 2(a)-2(b)).

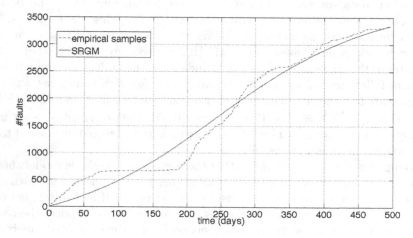

**Fig. 1.** Cumulative number of opened issues

**Table 2.** SRGM Fitting Results

| Version | Current #Faults | Selected SRGM | Current Estimate of #Faults | KS Test true at | Exp. #Faults at $t = \infty$ | Scale param. | Shape param. | AIC |
|---------|-----------------|---------------|------------------------------|------------------|-------------------------------|--------------|--------------|-----|
| 1 | 665 | Trunc. Normal | 663.93 | 90% | 671.38 | 13.00 | 34.78 | -1,491.77 |
| 2 | 2,647 | Trunc. Logistic | 2,640 | 90% | 2,808.22 | 20.95 | 85.54 | -6,834.93 |

Table 2 shows the statistics of the selected models for the two versions. The estimates in this case are very close to actual data, both satisfying the KS test. Such models can provide estimates and predictions in terms of: residual faults at a given time, percentage of detected faults over the total expected ones; failure intensity; reliability. Note that these measures are equivalent to each other, since the expected cumulative number of faults at time $t$ is the $mvf(t)$ function, whose first derivative is the failure intensity function $\lambda(t)$; the latter can be used in the computation of reliability (e.g., [11], [22]). Considering these measures, testers can evaluate, for instance, what is the best time to release.

As for version 1, it is evident that the testing process saturates, detecting less and less faults as the testing proceeds. The process detected more than 99% of

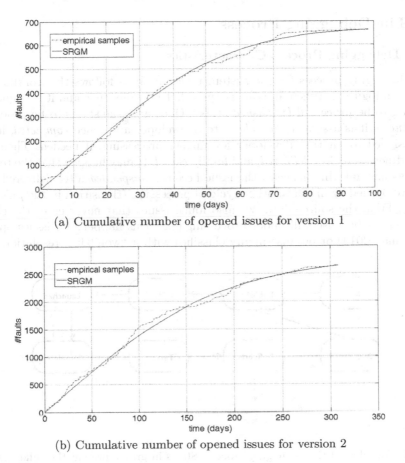

(a) Cumulative number of opened issues for version 1

(b) Cumulative number of opened issues for version 2

**Fig. 2.** Cumulative number of opened issues and fitted SRGMs for both versions

the total expected faults and will take much time to detect residual ones: thus this has been a good time to release. For version 2, testers detected roughly 94% of the total expected faults. If, for instance, they decide to release with the same quality as version 1, i.e., at 99%, the model predicts a testing time of 448 days, thus still 448 - 308 = 140 days of residual testing days. Based on these and similar analyses, decisions can be taken on when to stop testing.

The analysis has been conducted on the opening time of the issues. This means that 99% of quality is assumed to be the 99% of the total estimated faults that have been *opened*, i.e., *detected*: this is the actual released quality only under the assumption that the correction of those faults was immediate. The actual quality is given by the *closed* issues, namely *removed* faults, whose fixing contributes to the actual reliability growth. In the next Section, we first analyze the debugging process in our case study, then we remove the immediate debugging assumption in the SRGMs, and discuss the changes in the reliability analysis.

# 5   The Debugging Process

## 5.1   Debugging Process Characteristics

The debugging process for the system under analysis follows the workflow depicted in Figure 3. When an issue is opened, it becomes *new* and it is enqueued, waiting to be processed (*published* state). Once an issue starts to be processed (*in study*), it is assigned (*launched*) to a developer and, once *completed*, it can be assigned to another developer for further processing, if needed. Then, the amendment is *tested, delivered,* and finally *closed.* It may happen that the testing process may fail. In this case, the issue becomes *suspended* after delivered, and then re-opened again for another cycle of processing (transition in the *published* state). From the source data, we also found issues that never enter the closed state, either because still under processing (e.g., this happens for issues opened in January 2014) or because finally classified with a "won't fix" resolution.

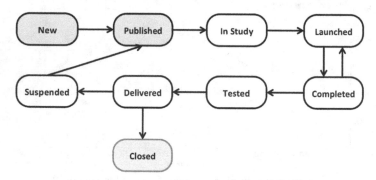

**Fig. 3.** Workflow of the debugging process. States in gray represent the "idle" part of the process, where issues are enqueued waiting to be processed.

Clearly, the overall debug process is far from being immediate: the mean Time To Fix (TTFix) on the overall dataset is equal to 12.8 days - for a single issue, the TTFix being the time to transit from the *new* state to the *closed* state. More in detail, we have evaluated the following statistics from source data:

- $MTTFix$ (in days): the expected value of the TTFix;
- $MED_{TTFix}$ (in days): the median of the TTFix;
- $StdDev_{TTFix}$ (in days): the standard deviation of the TTFix;
- $MQT$ (in days): the Mean Queuing Time, that is, the expected value of the time an issue is enqueued waiting to be processed;
- $MED_{QT}$ (in days): the median of the queuing time;
- $StdDev_{QT}$ (in days): the standard deviation of the queuing time;
- $\#Res$: average number of resources allocated to the issue;
- *Kurtosis*: it indicates the peakedness of the distribution;
- *Skewness*: it measures the asymmetry of the distribution; a positive (negative) value indicates a right-tailed (left-tailed) distribution.

The statistics computed are summarized in Table 3, where issues have been grouped according to their severity (minor, major, and blocking). From the statistics, it can be noted that overall the debug process has a good level of quality. Issues are closed in a reasonable time framework (from 9 to 16 days on average, 5-6 days median), even if very variable (high standard deviation), depending on the complexity of the issue. Moreover, as desirable, blocking issues are solved faster than major issues, that, in turn, are solved faster than minor issues. Another indication of the quality of the process is provided by kurtosis and skewness indicators. The high value of kurtosis denotes that the most of variance is due to few high peaks, as opposed to the undesirable situation of frequent small deviations from the mean time to fix; these high peaks are of course desired in the left side of the distribution, indicating that a lot of defects have a short time to fix: thus a good value of kurtosis does not suffice by itself, but the skewness matters too. In particular, a positive skew is desired, as in our case, indicating that the time to fix distribution is right-tailed, with a lot of defects with low time to fix. An example of distribution of the number of issues according to the TTFix is shown in Figure 4 for minor issues. It can be observed that the distribution is right-tailed with a pronounced peak in the left side. The distribution of major and blocking issues, here omitted, exhibit an even better shape, as expected by looking at their kurtosis and skew values.

**Table 3.** Statistics of the TTFix distributions. Mean values and standard deviations are in days.

| Severity | $MTTFix$ | $MED_{TTFix}$ | $StdDev_{TTFix}$ | $MQT$ | $MED_{QT}$ | $StdDev_{QT}$ | $\#Res$ | $Kurtosis$ | $Skewness$ |
|----------|----------|---------------|------------------|-------|------------|---------------|---------|------------|------------|
| Minor    | 15.88    | 6.16          | 29.29            | 5.67  | 0.89       | 16.39         | 3.69    | 42.81      | 6.29       |
| Major    | 12.69    | 5.76          | 25.28            | 3.42  | 0.81       | 9.98          | 3.75    | 52.26      | 6.85       |
| Blocking | 9.85     | 4.70          | 21.42            | 2.46  | 0.65       | 8.42          | 3.72    | 79.89      | 8.24       |

As for the resources allocated to defects, we note that on average 3.7 resources are allocated on defects, irrespective from their severity. Finally, considering the queuing time, again we can observe that the MQT of blocking issues is shorter than the one of major and minor ones, as desirable. However, this time is not negligible and equals to about one third of the overall debug process on average. Hence, estimations based solely on the measure of the time to process the issue, ignoring the queuing time, could end up with inaccurate results.

## 5.2    Impact of Non-immediate Debug

This Section discusses the impact of debugging on the SRGM-based analysis. Synthetic statistics about the observed debugging process tell that its overall features reflect a good process, with low average and median time to fix, a net shape of the distribution, severities times to fix as expected, and so on. Figure 5(a) and 5(b) report the raw data about the cumulative number of opened

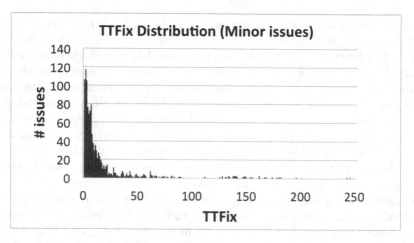

**Fig. 4.** Distribution of the number of minor issues according to the TTFix

(testing process) and closed (debugging process) issues, along with SRGMs fitting them. The graphs show what is the impact of the debugging times on the achieved quality.

The closed issue curve is the one actually contributing to reliability increase (namely, when the fault is actually removed); the opening curve would represent the reliability increase only under immediate repairs. Thus, in the following, we consider the difference between the two curves and their corresponding models in order to infer conclusions about the debugging impact.

Let us define $\Delta_{issues}(t)$ and $\Delta_{time}(F)$, respectively, as: *i)* the difference between the opened and closed issues at time $t$ (i.e., *pending issues* at $t$, which is the vertical distance between the raw data curves), and *ii)* the time required to close a given number of opened issues, $F$ (namely, the *delay* of the debug process compared to testing, which is the horizontal distance between the raw data curves). We also define the differences between the corresponding models as $\rho_{mvf}(t)$ and $\rho_{time}(F)$. These are the differences between the opened and closed issues at time $t$ and at $mvf = F$, respectively, as estimated/predicted by the corresponding SRGMs. The $\Delta$ values are used to: *i)* evaluate the difference between the actually achieved quality (in terms of number of removed faults) and the believed one[3], which is the quality under immediate repair assumption (i.e., the opened issues), as well as *ii)* the difference between the actual time required to close $F$ issues and the believed time (again, under immediate debugging, through the opening curve). This is the impact of assuming immediate debugging on quality/time estimates. On the other hand, the $\rho$ values are used

---

[3] Quality in the following is expressed through the (predicted) number of closed issues (i.e., removed faults) or the (predicted) percentage of closed issues with respect to the total one; for what said previously about the equivalence of this information to failure intensity and thus reliability, "quality estimation" and "quality prediction" are equivalent to "reliability estimation" and "reliability prediction".

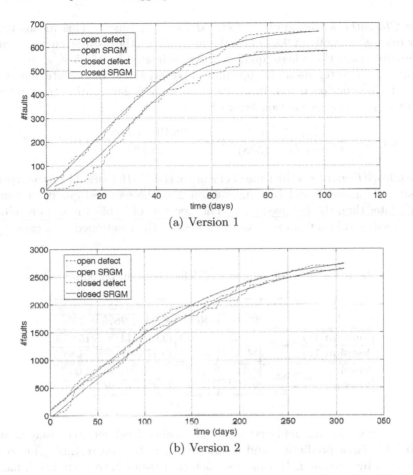

Fig. 5. Cumulative number of opened-closed issues and fitted SRGMs

to assess the same differences on *predicted* values, which are needed to take decisions like "when to stop testing". This is the impact of the immediate debugging assumption on predictions made through SRGMs.

For version 1, 99% of the total estimated issues has already been detected and it has been released, while the version 2 is still at 94% and it is still to be released: thus, we compute on version 1 the $\Delta$ differences on actual data to see the impact on estimated quality/time, whereas, on version 2, we compute the $\rho$ values on future predictions, at percentages greater than the achieved 94%.

Let us first consider version 1. At the last day, 98.19, the total opened issues were 665, namely about the 99% of total estimated ones. The actual quality at that time is given by the closed issues, that are 578 , thus the 86.14 % of the total estimated one, rather than the believed 99%. The error is therefore:

$$\epsilon_{\Delta_{issues}} = \frac{\Delta_{issues}(98.19)}{Closed(98.19)} \cdot 100 = \frac{665 - 578}{578} \cdot 100 = 15.05\% \tag{1}$$

where $Closed(t)$ is the number of closed issues at time $t$. This means that if tester released actually at 99% of total issues and use the opening curve assuming immediate repair, the release quality is overestimated by 15.05 %.

Similarly, if tester used the opening curve assuming immediate repair, the removed 578 issues occurred, in his view, after 63.98 days, rather than at 98.19; thus there is a time estimation error of:

$$\epsilon_{\Delta_{time}} = \frac{\Delta_{time}(578)}{ClosedTime(578)} \cdot 100 = \frac{98.19 - 63.98}{98.19} \cdot 100 = 34.84\% \qquad (2)$$

where $ClosedTime(F)$ is the time of closing of the $F-th$ issus. This is interpreted as: using the immediate debugging assumption, the required quality is reached 34.84% later than the believed time. The first raw of Table 4 reports results for release quality values from 95% to 98%, besides the mentioned 99% case.

**Table 4.** Results on the impact of debugging time on both versions

| Version | Achieved or Predicted release quality | | | | |
|---|---|---|---|---|---|
| | 95% | 96% | 97% | 98% | 99% |
| **Version 1:** $\epsilon_{\Delta_{issues}}$ | 14.15% | 14.59% | 13.61% | 14.06% | 15.05% |
| **Version 1:** $\epsilon_{\Delta_{time}}$ | 14.63% | 16.07% | 20.37% | 31.28% | 34.84% |
| **Version 2:** $\epsilon_{\rho_{mvf}}$ | 0.04% | 0.18% | 0.33% | 0.63% | 1.05% |
| **Version 2:** $\epsilon_{\rho_{time}}$ | 0.31% | 1.18% | 3.58% | 14.89% | $\infty$ |

If the tester has not achieved a desired quality level yet, s/he may want to use SRGMs for a prediction and decide on when to stop testing. This is well represented by version 2. In this case, detected issues have been 2,647, namely the 94.4 % of total estimated ones. We evaluate the impact of debugging time on prediction accuracy supposing that tester wants to release at 95%, 96%, 97%, 98%, and 99% of total faults (namely: 2,662, 2,690, 2,718, 2,746, 2,774 faults). In these cases, if s/he uses the opening curve, the release should be at the days: 321, 339, 363, 396, and 447. However, using the closing curve, the number of removed faults corresponding to the above release days are: 2,661, 2,685, 2,709, 2,729, 2,745. Quality overestimation errors would be caused. For instance, suppose that tester wants to release at 97%. In this case, the quality overestimation error would be:

$$\epsilon_{\rho_{mvf}} = \frac{\rho_{mvf}(363)}{SRGM(Closed(363))} \cdot 100 = \frac{2718 - 2709}{2709} \cdot 100 = 0.33\% \qquad (3)$$

Similarly to the version 1 case, there will also be an error about the time prediction. If tester uses the opening curve assuming immediate debugging to release at 97%, the opened 2718 issues in 363 days will be closed only at day 376, causing an error of[4]:

$$\epsilon_{\rho_{time}} = \frac{\rho_{time}(2718)}{SRGM(ClosedTime(2718)} \cdot 100 = (376 - 363)/363 \cdot 100 = 3.58\% \quad (4)$$

where $SRGM(ClosedTime(F))$ is the predicted time required to close $F$ issues. The second part of Table 4 reports results form 95% to 99% release criteria.

As it may be noticed, the errors on quality overestimation are quite small in version 2, compared to version 1, and are slightly increasing with the desired quality level. The small error denotes a very good debugging process, whose curve is strictly following the opening one. Notwithstanding, it is interesting to note how the error on the time prediction is higher, and increases rapidly for increasing values of the desired release quality, due to the saturation of both curves. From 98% to 99%, it increases up to infinite. This is interpreted as follows: if tester wants to release at 98% of the total estimated faults, and uses the opening curve assuming immediate repair, it would predict a testing time of 14.89% days less than the actually required testing time. If this desired quality goes beyond the 98%, such an error increases abruptly, reaching infinite at 99%. Thus, depending on the desired quality and on debugging process characteristics, this *testing time underestimation error* may be very high and is much more sensitive than the *quality overestimation error*.

In general, such time error always goes to infinite at some point (precisely, at the saturation point of the closing curve); in the practice, it can go to infinite considerably earlier if the debug process is not as close to the testing process as in the version 2 case. For instance, in version 1, for the same type of error (computed on raw data) the infinite occur soon after 578 issues, i.e., at only 86.14 % of the total estimated ones.

To summarize, the worse the debugging process, the greater the error on quality estimation is, and the earlier the time prediction error goes to infinite: but while the *quality estimation/prediction error* is directly related to the number of pending issues quite independently from the release time (e.g., in the same way at 70%, 80%, or 90%), the *time estimation/prediction error* is much more sensitive: at high quality values, the underestimation of the required testing time can be very high, depending on the saturation of the opening and closing curves.

# 6   Conclusion

In this paper we analyzed the impact of the debugging time on reliability estimation and prediction. Characterization of issues found in a real-world industrial

---

[4] Note that, unlike the case of $\Delta_{time}$ value, here the difference is taken between the predicted time to close the number of issues that tester wants to remove and the predicted time to open that number of issues. For $\Delta_{time}$ values, we take the time to close the number of issues actually closed subtracted by the time at which that number of issues was opened (i.e., the "believed" time for achieving that quality).

project indicates that the time taken by the debugging process is not negligible. For example, the debugging time causes an overestimation of the perceived software quality up to 15.5% in our dataset; similarly, it causes the underestimation of the testing time that would be required to obtain a given quality for a software product. In the future we aim to analyze further datasets and issues found in industrial projects in order to achieve a representative characterization of the debugging time and more accurate reliability estimation.

**Acknowledgment.** This work has been supported by the European Commission in the context of the FP7 project "ICEBERG", Marie Curie Industry-Academia Partnerships and Pathways (IAPP) number 324356.

# References

1. Stringfellow, C., Amschler Andrews, A.: An Empirical Method for Selecting Software Reliability Growth Models. Empirical Software Engineering 7(4) (2002)
2. Farr, W.: Software Reliability Modeling Survey. In: Lyu, M.R. (ed.) Handbook of Software Reliability Engineering, pp. 71–117. McGraw-Hill (1996)
3. Musa, J.D., Okumoto, K.: A logarithmic Poisson execution time model for software reliability measurement. In: Proc. 7th Int. Conf. on Software Engineering (ICSE), pp. 230–238 (1984)
4. Zachariah, B., Rattihalli, R.N.: Failure Size Proportional Models and an Analysis of Failure Detection Abilities of Software Testing Strategies. IEEE Trans. on Reliability 56(2) (2007)
5. Dugan, J.B.: Automated Analysis of Phase-Mission Reliability. IEEE Trans. on Reliability 40, 45–52 (1991)
6. Garzia, M.R.: Assessing the Reliability of Windows Servers. In: Proc. of IEEE Dependable Systems and Networks Conference, DSN 2002 (2002)
7. Pietrantuono, R., Russo, S., Trivedi, K.S.: Online Monitoring of Software System Reliability. In: Proc. of the European Dependable Computing Conference (EDCC), pp. 209–218 (2010)
8. Cotroneo, D., Pietrantuono, R., Russo, S.: Combining Operational and Debug Testing for Improving Reliability. IEEE Trans. on Reliability 62(2), 408–423 (2013)
9. Goel, A.L., Okumoto, K.: Time-dependent error-detection rate model for software reliability and other performance measures. IEEE Trans. on Reliability 28(3) (1979)
10. Yamada, S., Ohba, M., Osaki, S.: S-Shaped Reliability Growth Modeling for Software Error Detection. IEEE Trans. on Reliability 32(5) (1983)
11. Gokhale, S.S., Trivedi, K.S.: Log-logistic software reliability growth model. In: Proc. 3rd Int. High-Assurance Systems Engineering Symposium, pp. 34–41 (1998)
12. Yamada, S., Ohtera, H., Narihisa, H.: Software reliability growth models with testing effort. IEEE Trans. on Reliability R-35 (1986)
13. Goel, A.L.: Software Reliability Models: Assumptions, Limitations and Applicability. IEEE Trans. on Software Engineering SE-11(12) (1985)
14. Okamura, H., Watanabe, Y., Dohi, T.: An iterative scheme for maximum likelihood estimation in software reliability modeling. In: Proc. 14th Int. Symposium on Software Reliability Engineering (ISSRE), pp. 246–256 (2003)
15. Ohishi, K., Okamura, H., Dohi, T.: Gompertz software reliability model: Estimation algorithm and empirical validation. Journal of Systems and Software 82(3) (2009)

16. Jain, M., Manjula, T.: Software reliability growth model (SRGM) with imperfect debugging, fault reduction factor and multiple change-point. In: Deep, K., Nagar, A., Pant, M., Bansal, J.C. (eds.) Proceedings of the International Conf. on SocProS 2011. AISC, vol. 131, pp. 1027–1037. Springer, Heidelberg (2012)
17. Musa, J.D., Iannino, A., Okumoto, K.: Software Reliability, Measurement, Prediction and Application. McGraw Hill (1987)
18. Huang, C.-Y., Huang, W.-C.: Software Reliability Analysis and Measurement Using Finite and Infinite Server Queueing Models. IEEE Trans. on Reliability 57(1), 192–203 (2008)
19. Nguyen, T.T., Nguyen, T.N., Duesterwald, E., Klinger, T., Santhanam, P.: Inferring developer expertise through defect analysis. In: 34th International Conference on Software Engineering (ICSE), pp. 1297–1300 (2012)
20. Zhang, F., Khomh, F., Zou, Y., Hassan, A.E.: An empirical study on factors impacting bug fixing time. In: 19th Working Conference on Reverse Engineering, WCRE (2012)
21. Ihara, A., Ohira, M., Matsumoto, K.: An analysis method for improving a bug modification process in open source software development. In: Proc. of the Joint International Annual ERCIM Workshops on Principles of Software Evolution (IWPSE) and Software Evolution (Evol), pp. 135–144 (2009)
22. Pietrantuono, R., Russo, S., Trivedi, K.S.: Software Reliability and Testing Time Allocation: An Architecture-Based Approach. IEEE Trans. on Software Engineering 36(3), 323–337 (2010)
23. Cinque, M., Cotroneo, D., Pecchia, A.: Event Logs for the Analysis of Software Failures: A Rule-Based Approach. IEEE Trans. on Software Engineering 39(6), 806–821 (2013)
24. Frattini, F., Ghosh, R., Cinque, M., Rindos, A., Trivedi, K.S.: Analysis of bugs in Apache Virtual Computing Lab. In: 43rd Annual IEEE/IFIP International Conference on Dependable Systems and Networks, DSN (2013)

# Simplifying Maintenance by Application of Architectural Services

Jaroslav Král[1] and Michal Žemlička[2,3]

[1] Masaryk University, Faculty of Informatics,
Botanická 68a, 602 00 Brno, Czech Republic
kral@fi.muni.cz
[2] Charles University, Faculty of Mathematics and Physics
Malostranské nám. 25, 118 00 Praha 1, Czech Republic
zemlicka@sisal.mff.cuni.cz
[3] Czech Technical University, Faculty of Information Technology,
Thákurova 9, 160 00 Praha 6, Czech Republic
michal.zemlicka@fit.cvut.cz

**Abstract.** Software maintenance is the most difficult and extremely expensive activity of lifecycle of software systems. We show that maintenance cost estimation depends on many factors that are not taken into account by widely used maintenance cost estimation methods. The factors include variants of systems architecture, especially service- and component-oriented ones, variants of software development processes, communication means, and the software artifact maintenance duration and history. We present an analysis of reasons and sources of maintenance effort needs. We show that the maintenance issues and effort can be substantially reduced in systems having a special form of service-oriented architecture — software confederations.

**Keywords:** service-oriented architecture, software confederations, maintenance simplification, maintenance effort reduction.

## 1 Introduction

Software, like other technical products, becomes worn out or obsolete after some time of use. During the worn-out period the usefulness of the system fails whereas the maintenance effort of the systems grows. The maintenance then becomes too expensive or unable to avoid the overall permanent obsolescence of the system. In such a situation the software must be rewritten or retired.

The key aspects affecting software maintenance are:

1. Quality of requirements specification – when the requirements specifications are biased, the maintenance and obsolescence of the product starts already during the development phase.
2. Quality of the system architecture design – the main effects of good system architecture are:

B. Murgante et al. (Eds.): ICCSA 2014, Part V, LNCS 8583, pp. 476–491, 2014.
© Springer International Publishing Switzerland 2014

(a) Code changes done during the maintenance tend to be local. It allows to keep the maintenance effort growth not so fast with time (it is possible to prolong the product lifecycle or exclude the product retirement totally[1]). The architecture simplifies the use of third party products, free software inclusive.

(b) The architecture moreover supports agile development (incremental and iterative with agile procedures).

We believe that the impact of the system architecture and modern development processes on the system maintenance is underestimated.

3. Used development tools and means – modern programming languages and frameworks allow significant reduction of parts that should be rewritten. If the modules are small and autonomous/independent enough, agile development or maintenance methods can be used.

The paper is structured as follows: We discuss the main issues of the maintenance of large systems. Then we describe a variant of service-oriented architecture (SOA, [1,2]) called confederation and show that it can solve the issues.

## 2  Maintenance Traps

Besides the maintenance costs directly spent due to software changes, there are expenses caused by the business losses at user side (process changes, performance/throughput degradation, changes in dependent systems, failures of business processes, etc.). These indirect (additional) maintenance costs can be crucial.

It is not clear how the real maintenance costs are influenced by the use of component architecture. The general use of component and agile techniques (service-oriented ones inclusive) indicates that the proper use of component architecture (SOA inclusive) and agile development process substantially reduces maintenance costs and enhances the maintenance quality. We will discuss the facts supporting this conclusion.

Improper or insufficient project specification causes additional changes or enhancements of the specifications during maintenance. It influences the system in the way that the original logic of the system is modified (typically logic of some parts is different from the logic of the others) and the system is then more difficult to change and maintain.

This issue is important as there is about some 40–50 per cent of projects having problems (they are challenged). It is well known that more than 50 per cent of errors detected during the development of challenged systems are caused by the errors made in early stages of system lifecycle. It follows that such systems are difficult to maintain – mainly due to induced errors.

The crucial traps of software maintenance is the consequence of the following facts:

---

[1] It means that it the system could be retired for other reasons, typically in the case when significant changes in user needs occur or even the agenda supported by the system may get obsolete.

- The systems are getting larger, more complex, and often more open.
- The frequency of the changes of requirements grows during the system life-time.
- The innovations of the systems are getting more frequent.
- The duration of systems development grows and it cannot be substantially shortened[2]. If we therefore innovate a large system so that we completely redevelop it, the duration $D$ is so large that the reconstructed system is obsolete and still error prone before it is released for use (compare the antipattern Reorg Cycle [4]).

A solution is to use agile development philosophy. But it works only if the outputs of mini-development cycles (e.g. sprints in Scrum methodology [5]) are frequently released. It has substantial drawbacks unless the outputs are autonomous software artifacts. We propose a software architecture pattern enabling such a solution. The pattern combines the methods and tools of SOA and component-oriented systems. The maintenance then in fact becomes a low intensity continuous development process using basic principles of agile development [6]. Our proposals are more straightforward and, we believe, more powerful than the methods in [7].

The solutions proposed below can be, besides large enterprises, used in small and medium enterprises (SME) often without any need to use complex proposals from [8,9].

## 3    Known Facts

Software maintenance is a very expensive task. Many studies indicate that maintenance costs of a system are substantially higher than the development costs (two or three times higher [10]). The immense direct maintenance costs are often substantially smaller than the expenses due to the hidden consequences of the maintenance, for example:

- The direct maintenance costs are further growing so the management cancels the system although the system capabilities are still useful but the system tends to be unmaintainable.
- Software maintenance in its present form causes the unreliability of software products. The consequence is that the today software systems can hardly be considered to be standard engineering products (compare warranty disclaimer attached to many software license agreements). Software failures imply important losses.
- The relation between development and maintenance costs [10] implies that hard-to-develop systems are almost unmaintainable.

---

[2] Let $S$ be a metric of system size (e.g. KLOC – thousands of lines of code) and $D$ its development duration. Let the system is developed as a (logical) monolith. Then $D$ must be greater than $kS^b$ where $k$ is a constant and $b$ is about $1/3$, see COCOMO II [3] for details. The area $D < kS^b$ is called inaccessible area.

- The high maintenance effort causes many problems to developers but also to the users. The system with a high maintenance effort is as a rule unstable and requires some additional maintenance effort at the user side (e.g. due to the continuous retraining). It results in dissatisfaction and large additional costs.
- The frequency of failures of software has the typical bath tube shape. During the system life cycle the frequency of system failures initially falls down, then it is quite stable and then it rises again. In the last case as well as in the case of system obsolescence the system must be completely rewritten. The complete rewriting of the system can be very difficult in the case of global systems used at many places and supported by many servers having thousands of users. The complete reconstruction/redevelopment of the software implies a painful data conversion and often interruption of the system services. It means that worn-out systems must be often cancelled. The reasons are that the philosophy of the systems becomes obsolete or (more frequently) needed enhancements become increasingly global (they consist of growing number of local code changes causing the system to be poorly documented). Moreover, the changes tend to influence more and more parts of the system, what makes them more risky, labor consuming, and expensive.

The maintainability problem could be substantially reduced via the application of object-oriented (OO) technologies. OO technology reduces the maintenance problem but all the above points remain valid.

## 4   How to Reduce the Maintenance Effort

There are several ways of reduction of the maintenance effort. Some ways are simple, others are more sophisticated. In fact there are the following obvious strategies:

A  Reduction of the size of a newly written code by the use of:
   - frameworks,
   - legacy systems,
   - (customizable) third-party systems.
B  Reduction of the maintenance effort of the newly written code by an increasing quality of requirements and better development processes.
C  Reduction of system complexity by coarse decomposition of the system.

The strategy A has the following main variants:

A1 *Reduction of the system size.* Many functions of software systems can be often omitted without any loss of usefulness. There are even cases when a service performed manually is preferable to the computerized one. Brilliant examples of it are described by Goldratt in [11,12]. It is not, however, easy to find a good balance between automated and manual functions.

A2 *Reuse and purchasing.* The size of the newly written code can be reduced in the cases when we can reuse our existing software (programs, libraries, applications like legacy systems) or if we can integrate purchased software provided by prestigious vendors (at best complete applications) into our system. Note that the integration of systems smaller than complete application (e.g. classes of OO programs, see [13]), is also possible but less effective. The kernel of the system is as a rule formed by a network of application-like components.

A3 *The most useful first.* The Pareto rule as well as our practical experience indicate that often the substantial part (about 80 per cent) of system usefulness is due a small part of the system (about 20 per cent). In other words: if we develop the most useful parts of the system first, we spent only 1/5 effort to have 4/5 of final usefulness.

It has further advantages. If the 1/5 of code is opened for use, the opened parts are usually well specified and the experience gained during the use of the most useful parts is a strong precondition for the enhancement of the quality of requirements specification of the remaining parts. It is enabled by modern software architectures, especially the component or service-oriented ones. The architectures enable iterative or incremental agile development processes [5,14,8,15,16].

So we should start with the most useful functions. It is often not too difficult to detect them. The development should be incremental, sometimes iterative [16]. The strategy A3 requires good design and coding practices.

If the developed system is large, its development strategy B can be based on the following turns:

B1 *Software prototyping [17].* It is an effective way of improvement of the quality of requirements and therefore of reduction of the need to rewrite the newly developed code.

B2 *Coarse-grained decomposition and user-oriented interface.* It is the application of the consequences of the A2 strategy. The maintenance effort can be reduced if the system is decomposed into quite large components providing complex functionality accessible via commands and/or messages, the syntax and semantics of which is flexible and business oriented. In other words the messages are intuitively in agreement with user knowledge and intuition (and with the knowledge of the developers of the components communicating with the given one)[3]. It has the pleasant consequence that not only implementation details but also the design philosophy (object-oriented or relational-database-oriented interface) is completely hidden. Incremental

---

[3] The authors were members of several teams developing the control software for a flexible manufacturing systems (FMS). The FMS was designed as an island of automation in a factory producing machine tools. It is generally accepted that the islands of automation is not a good solution. It was however not the case due to the fact, that system interface was fully user oriented. The system has been in use for more than twenty years. It is possible that it is still used.

development has many software engineering advantages. If used, it makes the application of the modern principles of development straightforward. Typical is the application of incremental development strategy. The components can be complete applications and should be designed as increments during the incremental development.

The messages should describe "what" to do rather than "how" to achieve it. The messages should also simulate the communication between human beings as much as possible.

B3 *Monitoring the communication between the (large) components of the system.* The maintenance of the system can be reduced if the messages between the components of the system are stored into a log file for a later analysis.

To summarize, we need the tools and software architectures supporting at least:

- Agile development and maintenance.
- Incremental development (components, SOA).
- Prototyping.
- Flexible coarse-grained user-domain-oriented interface of business applications.
- Messaging system with logging and the possibility of agile message routing control.
- Business-domain-oriented communication.
- Easy integration of legacy systems and third party products.

## 4.1   Limits of Code-Oriented Maintenance

It is well known that coding is almost no problem in software development provided that the system vision and requirements specification have satisfactory quality and that the system has a corresponding architecture. The choice of the architecture influences many crucial system properties. For example, RAMS (reliability, assessibility, maintainability, and safety and system capabilities like flexibility of use or the extent of ability in technical and business processes .

As the properties of systems depend on their architectures, it is very difficult, if not impossible, to link system properties with the formal properties of the code. The properties of the systems having modern architectures depends heavily on the policies and rules for human actors how to react on business irregularities. The methods of code maintenance based on formal code analysis – compare maintainability model [18] or [19] – they are useful but they do not solve typical bottleneck of software system maintenance. The underestimation of it could be an important reason of the steadily growing software maintenance costs [20].

The above requirements are quite easily fulfilled in the case that the system has SOA being a peer-to-peer (P2P) network of large autonomous components [21,22] (e.g. information systems). It is preferable to build the systems from loosely coupled components cooperating using coarse-grained message protocols transporting possibly encapsulated business documents. Such communication should be understandable to users. Note, however, that the legibility can be based on surprising attitudes. The interchanged documents can have the form of natural language messages, forms, or even spreadsheets. It can be achieved in systems called *software confederations* [1] or *confederations* for short. Software confederations must be used in many cases[4].

# 5   Software Confederations

We propose a solution of the maintenance issues based on the use of software confederations [15,23,24]. It can be used to solve the problem of permanent obsolescence (Reorg Cycle Antipattern) and other issues appearing in the development, maintenance, and use of large software systems.

A software confederation is from the logical point of view a P2P network integrating relatively autonomous software entities providing basic system capabilities. We call these software entities *application services*. The application services are used as black boxes communicating using coarse-grained messages understandable by users. Such messages often transport business documents.

Such messages have the following substantial advantages:

– It enables user agility during system development and system maintenance.
– It simplifies the development of business intelligence.
– It is good for the solutions of business issues (court trials inclusive)
– It simplifies optimization of business processes and enables their agility.
– It simplifies or even enables application of agile system development.
– It is a precondition for the design of architecture as a service.
– It simplifies and enables higher quality of system maintenance.
– It simplifies the implementation of the recommendation of information hiding what simplifies maintenance. It enables independency of lifecycles of individual services, simplifies error localization, and restricts propagation of code changes.

A very important property of the coarse-grained messages is that they can be inspired by the user knowledge domain. They are a formalization of service-level requirements. It substantially simplifies the setting as well as the fulfillment of service-level agreement (SLA) during system development and maintenance.

Such a solution has many further advantages: It simplifies implementation, deployment, and acceptance of the system. It stimulates positive approach of the users to the system, it lowers their load and lowers the number of errors caused by the users. It simplifies acceptance of the system, preserves reasonable relations on the workplace and to some extent it can have positive influence on the quality of data in the system.

---

[4] The main reason is that in the systems like the information systems of state administration or of an international enterprise and many others the system must be developed via integration of the information systems of organizational subunits (offices, local enterprises, etc.).

The messages are understandable to the users. And as they are based on a long-term experience with long-time used real-world services, they can be very stable. It is a crucial advantage for requirements specification as well as for maintenance.

The application services communicate via a network of organizational artifacts being also the peers of the P2P network. They are used as connectors in sense of component oriented philosophy [25] as they connect other services. They are connectors as services (CaaS). They can be viewed as architecture services as they are in fact architecture building blocks.

Application services behave inside a confederation somewhat like the computers in computer networks. CaaS behaves provides capabilities including the ones similar to the capabilities provided by communication hardware (routers or switches).

## 5.1   Stable Legible Business-Oriented Interfaces

Software confederations are often implemented in the following steps:

- Design and implementation of the kernel of the middleware. This step is usually omitted if Internet services are used.
- Existing applications and newly developed components are connected to the middleware by wrappers providing the access to the full functionality and/or data of the components intended to be public. We shall call such wrappers *basic gates* (BG).
- User interface(s) of the system is (are) provided by specific services – portals – allowing a transparent integrated access to all services of the confederation.

The basic gate of a component $C$ tends to have properties dictated by the implementation philosophy of $C$. For example if a component is developed using the object-oriented philosophy, the language of its basic gate tends to have form of sequences of the method calls. It is not pleasant for the user of the component as well as for the designers of the components collaborating with $C$ due to the fact that if the design philosophy of a component is changed, the language of its basic gate will be also changed. It must be reflected in the collaborating components. It causes painful changes in interfaces even in the case when the functions/services provided by the component for the given user are not changed. It is not optimal and often it is not acceptable at all. In any case it increases maintenance effort.

A very powerful turn how to implement legible interface is to design application services as three-tier structure (connector C, basic gate BG, application A). The capabilities of the connector can be implemented as a classic connector of component-oriented systems in the sense described by Mikic-Rakic and Medvidovic [25] or Plasil and Visnovsky [26]. It is, the triple is in fact a single program (Fig. 1).

A substantially more powerful and flexible solution is to implement the connector C itself as a service (see Fig. 2). We call it *front-end gate* (FEG). It in fact is a connector as a service.

**Fig. 1.** Classical connected application composed from connecting part, wrapper, and connected application

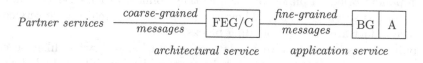

**Fig. 2.** Front-end gate or connector as a service

Front-end gate works as a transformer compiling sequences of fine-grained messages into coarse-grained messages. It in the opposite direction also translates coarse-grained messages into sequences of fine-grained messages.

The standard policy is that the messages for an application service (BG, application) can be send only via the front-end gate and that BG replies using the front-end gate only.

This policy can be in necessary cases generalized so that the application service has more front-end gates.

Different user groups can require different message formats (message languages) and their requirements can vary. The basic gates cannot accept the messages in all the languages as it would make them too complex and in many cases it could make some functions of the components inaccessible. It can be impossible if some components are used as black boxes.

The different groups of users can have different and sometimes contradictory requirements (on message formats inclusive). The solution is a confederation applying a generalized use of front-end gates.

As mentioned above, the maintenance is reduced if the system and/or component interface is user knowledge domain oriented. In other words the language of the messages should be near to the language the user uses in his application/knowledge domain – the user should intuitively understand the messages. The same is true for the designers and developers of the collaborating components.

So we need the interface of the components (services) to be user oriented, but the designers of the components (e.g. data tier) need a design-oriented interface, for example an interface based on SQL.

The user-oriented interfaces have important advantages: technical as well as functional. The technical ones are, e.g., easy prototyping, effective monitoring, simpler testing, information hiding [27]. The user-oriented advantages are flexibility, the user involvement into development and maintenance processes, agile business processes, incremental development of business processes, etc.

## 5.2   Front-End Gates (Services as Connectors)

To summarize the FEG is from the system programmer view a standard peer. Its functions are:

- The transformation of message formats, it is a language transducer.
- It serves also as communication switch.

The construction of FEG can use turns known from the theory and the art of the construction of compiler back-ends. Confederations have typically a four-level architecture:

1. Upper obligatory level – P2P of services.
2. P2P like architecture of particular services. We call these peers *mesocomponents*.
3. Mesocomponents can be assembled from components in the UML [28] sense.
4. UML components are assembled from objects/classes. The unlimited composition of services is a powerful tool enabling to make the structure of very large systems understandable and therefore maintainable. There are indications that it could simplify the run of confederations on clouds.

FEG can be easily modified to enable unlimited composition of services, easy prototyping, generalized routing and agile business processes. Such generalized connectors are designed as white boxes. It enables the agile development of the confederations. The details can be found in [24,16].

The abilities of the confederations can be significantly modified by the changes of the CaaS code. They can be strongly modified by the change of the policy of their use. The power of the policies is often significantly higher than in the more popular variants of SOA not using the CaaS.

# 6   Implementation Issues of Software Confederations

The concept of CaaS is very powerful and flexible. Various modifications of CaaS can be used to redirect or to monitor messages, to change communication protocols, etc. Various CaaS can be viewed as instances of the following general structure.

Application services include various software artefacts being often wrapped application or wrapped portlets. Their interfaces have often no interface fulfilling above conditions.

The drawback of the solution from the Fig. 3 is that the interfaces provided by basic gates are not coarse-grained and user-oriented. The interface provided by the basic gates (BG) is developer oriented and fine grained. It prefers the form of the (remote) procedure call. The pair (BG, application) is used as a black box provided by third-party products. Then there is a very limited possibility to change the BG. The interface provided by BG is not flexible enough and does not hide implementation details.

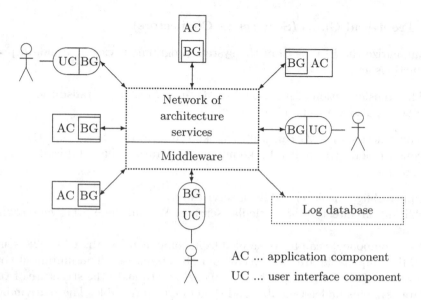

**Fig. 3.** The structure of a simple software confederation

Its advantage is that it has access to all capabilities of the applications. Confederations have from the implementation point of view a two-tier structure. Confederations have, however, from the logical point of view rich and very flexible architecture [29].

- The combination of the changes of the codes of architectural services and policies of their uses provides rich set of composition and patterns of the collaboration of services. For example, there is a very simple way to build unlimited hierarchy of composite services.
- Individual services can have and disclose their internal architecture.
- Architecture services can be easily generalized to be control elements of business processes.

There are several technical possibilities how to implement software confederations. The most straightforward implementation uses Internet services as a message transport tool (middleware). The components are connected to middleware via gates (called basic gates in the sequel). Components are servers in the terms of Internet. The users of the confederation are Internet clients (see Fig. 1). Different user roles can have different user interfaces. It enhances the system and therefore decreases the need of perfective maintenance. We can, however, use other middleware types.

User interface should be easily modifiable in order to be easily maintainable. It can be easily achieved if the component interface is implemented as a tandem consisting of newly developed possibly XSLT [30,31] based application and an Internet browser. The application should be again a peer in the P2P forming

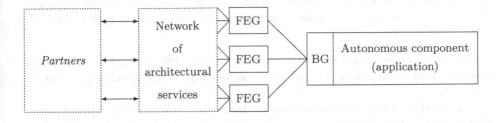

**Fig. 4.** The use of front-end gates

the confederation. The communication between components can be monitored –
the copies of the messages can be stored into a log database.

It is good to design middleware such that the messages are texts – sentences
of a formal language with a possibly complicated syntax. The language can be
defined via XML [32]. A more detailed discussion can be found in [15].

# 7   Maintenance of Software Confederations

Let us now return to the maintenance problem: The rich syntax of the messages
simplifies the control whether the components obey the interface agreements. It
simplifies the system maintenance.

Let us discuss how to implement the above given list of maintenance effort
reducing measurements:

A1 *System size reduction.* If the system is developed incrementally, there is a
   good chance that useless functions will not be developed or they will be
   canceled soon. It can be easily applied if the developed system is a software confederation. We can apply the Pareto 80–20 rule and increase the
   usefulness of the resulting system.
A2 *Reuse* of the existing software, integration of third party products. Software
   confederations were invented to solve this problem. Integration is a quite
   simple task in confederations.
A3 *The most useful first.* It is an inherent property of software confederations
   that they are open, i.e. new components can be added quite easily (compare the properties of Internet). It follows that the incremental development
   can and should start from the most useful components. Note that the modifications of the confederation are simple as they can be implemented via
   adding new elements or via the modification of a small number of components (usually one). Moreover there is a good chance that the component
   modification can be hidden, i.e. it need not influence the component interface. Software confederations simplify software prototyping and enable their
   use for different purposes, for example:
   – A prototype $P$ of a component $C$ not existing yet is developed. The
     prototype can often be implemented as a (part of) user interface.

- The messages are redirected from $C$ to $P$. The copies of the messages are stored in a log database for the analysis of computer traffic – see [33] for details.

Coarse-grained decomposition as well as component or service reuse can be easily implemented. It is also quite simple to support powerful monitoring systems usable for users.

The analysis of system failures starts from the analysis of the log database. Experience indicates that it is a very effective way to localize bugs. The bug detection can be further enhanced if the messages contain some further information simplifying the bug detection and analysis. This tool is especially easy to implement if the messages have XML or even natural language based formats.

One sometimes argues that the formats of messages should be object-oriented. In this case the messages must code method calls. The particular messages must then be quite simple as the methods tend to be simple. So one user message must be transformed into many simple messages and there are too many messages to be analyzed. There can be problems with the system effectivity. This problem can be partially solved if the components in the sense of UML [28] are used. But this solution is still a too low-level and too procedurally-oriented concept. UML-based components are good for the design of autonomous components. It is doubtful whether they are appropriate for the design of the confederations as their interface is not enough user and/or developer friendly, not application domain oriented, not independent on component implementation philosophy.

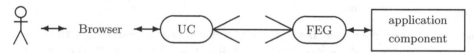

**Fig. 5.** The path from a user to a component providing required functionality

## 8    Conclusions

Software confederations offer many software engineering advantages like openness, reusability, modifiability, maintainability, and others. It is such a powerful technique that we need not use any complex tools like UML [28].

Although the main ideas of the confederation paradigm are very simple, it is not easy for practitioners to use this paradigm as they must give up the idea, that the best way to design large systems is to design them as a logically monolithic ones. If used properly, the confederation attitude simplifies substantially the development as well as the maintenance.

One can argue, that is not enough evidence that it is true. An indirect confirmation can be found in the history of computing. The Y2K problem disclosed that there were many systems written in COBOL that had not been requiring any maintenance for many years. The systems written in COBOL were an analogy of P2P in the batch environment. Such systems were sets of autonomous programs communicating via files. Software confederations have properties making them substantially more powerful than monolithic systems. So we can hope that they will be at least so stable and will require at most so little maintenance effort as the systems written in COBOL.

Software confederations are in fact the only known feasible way to support global management. They are, in fact, engines of global economy. Their importance is, however, underestimated. There are some technical problems to be solved: security issues, transactions, and documentation. These problems, however, do not make the advantages doubtful.

We have shown that maintenance cost estimations are insufficient as they neglect following aspects:

- **Specification quality** where the indicator of worse maintainability is the fact that the problem was challenged and that the specifications had to be changed.
- **Used architecture** – impact of the architecture depends on the use type; typical variants are:
  - monolith,
  - logical monolith developed using iterations,
  - component system developed in the classical sense,
  - SOA built in top-down manner (typical for the use of ITIL or COBIT methodologies).
  - software confederations – SOA built in bottom-up manner from properly wrapped business applications.
- **System age** – the maintenance complexity and costs grow with time.
- **The use of different variants of agile development** what is related to the used architecture.
- **The use of frameworks and other tools** as some of the system maintenance can be done within the framework/tool maintenance. As the frameworks and tools are often used by many projects, the costs for their maintenance ar for individual projects lower.

These challenges concern not only COCOMO II but also – we suppose – all the variants of the Function Points method. The methods in their original form are in principle fully usable only for the project being realized using the waterfall model. It is complicated to use them for estimation of integration projects using services and recommendations of Open Group [9] or OASIS [8]. The usability of standard procedures is then more or less restricted to the development of so called application services providing basic steps of business processes.

Very large systems cannot be practically developed otherwise then by some variant of confederation or another variant of SOA.

A proper use of confederations can substantially reduce maintenance challenges and costs. This fact is not properly taken into account in software practice and the estimation (forecast) of basic project metrics. There are other possible benefits regarding to the quality dimensions reliability, safety, and accessibility.

We will investigate appearing trends to use content management systems and cloud systems to implement service-oriented systems. The new techniques can be applied to confederations as well.

# References

1. Král, J., Žemlička, M.: Electronic government and software confederations. In: Tjoa, A.M., Wagner, R.R. (eds.) Twelfth International Workshop on Database and Experts System Application, pp. 383–387. IEEE Computer Society, Los Alamitos (2001)
2. Erl, T.: Service-Oriented Architecture: Concepts, Technology, and Design. Prentice Hall PTR, Upper Saddle River (2005)
3. COCOMO: COCOMO II (1995), http://sunset.usc.edu/csse/research/COCOMOII/cocomo_main.html
4. Armour, P.: The reorg cycle. Communications of the ACM 46, 19–22 (2003)
5. Cohn, M.: Succeeding With Agile: Software Development Using Scrum. Addison-Wesley Professional (2009)
6. Beck, K., Beedle, M., van Bennekum, A., Cockburn, A., Cunningham, W., Fowler, M., Grenning, J., Highsmith, J., Hunt, A., Jeffries, R., Kern, J., Marick, B., Martin, R.C., Mellor, S., Schwaber, K., Sutherland, J., Thomas, D.: Manifesto for agile software development (2001)
7. Li, Z., Keung, J.: Software cost estimation framework for service-oriented architecture systems using divide-and-conquer approach. In: SOSE, pp. 47–54. IEEE (2010)
8. MacKenzie, C.M., Laskey, K., McCabe, F., Brown, P.F., Metz, R.: Reference model for service-oriented architecture 1.0, OASIS standard (October 12, 2006)
9. Open Group: Draft technical standard SOA reference architecture (2009)
10. Boehm, B.W.: Software Engineering Economics. Prentice Hall (1981)
11. Goldratt, E.M., Cox, J.: The Goal: A Process of Ongoing Improvement, 2nd edn. North River Press, Great Barrington (1992)
12. Goldratt, E.M.: Critical Chain. North River Press, Great Barrington (1997)
13. Finch, L.: So much OO, so little reuse. Dr. Dobb's Journal (1998)
14. Beck, K.: Extreme Programming Explained: Embrace Change. Addison Wesley, Boston (1999)
15. Král, J., Žemlička, M.: Software architecture for evolving environment. In: Kontogiannis, K., Zou, Y., Penta, M.D. (eds.) 13th IEEE International Workshop on Software Technology and Engineering Practice, pp. 49–58. IEEE Computer Society, Los Alamitos (2006)
16. Král, J., Žemlička, M.: Human aspects of machine-to-machine communications and cooperation. The Smart Computing Review 1(2), 104–115 (2011)
17. Sommerville, I.: Software Engineering, 5th edn. 42765 in International Computer Science. Addison-Wesley, Reading (1996)
18. Heitlager, I., Kuipers, T., Visser, J.: A practical model for measuring maintainability. In: Proceedings of the 6th International Conference on the Quality of Information and Communications Technology (QUATIC 2007), pp. 30–39. IEEE Computer Society Press (2007)
19. Vargas, R.T., Nugroho, A., Chaudron, M., Visser, J.: The use of uml class diagrams and code change-proneness. In: Proceedings of the International Workshop on Experiences and Empirical Studies in Software Modelling, EESSMod 2012, 2nd edn. ACM, New York (2012)

20. Omnext: How to save on software maintenance costs, White paper (2010), http://www.omnext.net/downloads/Whitepaper_Omnext.pdf
21. Král, J., Žemlička, M.: Autonomous components. In: Jeffery, K., Hlaváč, V., Wiedermann, J. (eds.) SOFSEM 2000. LNCS, vol. 1963, pp. 375–383. Springer, Heidelberg (2000)
22. Jiao, W.: Using autonomous components to improve runtime qualities of software. IET Software 5, 1–20 (2011)
23. Král, J., Žemlička, M.: Implementation of business processes in service-oriented systems. International Journal of Business Process Integration and Management 3(3), 208–219 (2008)
24. Král, J., Žemlička, M.: Support of service systems by advanced SOA. In: Lytras, M.D., Ruan, D., Tennyson, R.D., Ordonez De Pablos, P., García Peñalvo, F.J., Rusu, L. (eds.) WSKS 2011. CCIS, vol. 278, pp. 78–88. Springer, Heidelberg (2013)
25. Mikic-Rakic, M., Medvidović, N.: Architecture-level support for software component deployment in resource constrained environments. In: Bishop, J.M. (ed.) CD 2002. LNCS, vol. 2370, pp. 31–50. Springer, Heidelberg (2002)
26. Plášil, F., Višňovský, S.: Behavior protocols for software components. IEEE Transactions on Software Engineering 28(11), 1056–1076 (2002)
27. Parnas, D.L.: Designing software for ease of extension and contraction. IEEE Transactions on Software Engineering 5(2), 128–138 (1979)
28. Object Management Group: Unified modeling language (2011)
29. Žemlička, M., Král, J.: Flexible business-oriented service interfaces in information systems. In: Filipe, J., Maciaszek, L. (eds.) Proceedings of Enase 2014 - 9th International Conference on Evaluation of Novel Approaches to Software Engineering. SciTePress (2014)
30. W3 Consortium: XSL transformations (XSLT) (1999), http://www.w3.org/TR/xslt
31. W3 Consortium: XSL transformations (XSLT) version 2.0 (2007), http://www.w3.org/TR/xslt20
32. W3 Consortium: Extensible markup language (XML) 1.0 (5th edn.) (2008), http://www.w3.org/TR/xml/
33. Král, J., Žemlička, M.: Service orientation and the quality indicators for software services. In: Trappl, R. (ed.) Cybernetics and Systems, vol. 2, pp. 434–439. Austrian Society for Cybernetic Studies, Vienna (2004)

# On the Structural Code Clone Detection Problem: A Survey and Software Metric Based Approach

Mustafa Kapdan[1,2], Mehmet Aktas[2], and Melike Yigit[1,3]

[1] Corporate Development and Information Technologies Department, Turkish Airlines, Istanbul, Turkey
mkapdan@thy.com
[2] Computer Engineering Department, Yildiz Technical University, Istanbul, Turkey
aktas@yildiz.edu.tr
[3] Computer Engineering Department, Bahcesehir University, Istanbul, Turkey
melikeyigit@thy.com
Institute of Science

**Abstract.** Unnecessary repeated codes (*clones*) have not been well documented and are difficult to maintain. Code clones may become an important problem in software development cycle and they must be fixed in all occurrences. This condition increases significantly software maintenance costs and required effort/duration for understanding the code. Over the years, many techniques have been proposed in order to minimize or prevent the code cloning problems. The main focus of these techniques is on the detection of clones. In such studies, code cloning is studied under two main categories: simple and structural. Simple clone is defined as the similarity that arises from the repetition of the code snippet in the software. Structural clone is defined as the similarity in software structure (i.e. design patterns and object oriented programming class relations). Simple clone detection techniques fail to determine the reasons of code repetition whether it is due to design or not, as they do not look at the code from a wider perspective for repetitive code snippets. In this study, we survey the existing structural clones approaches. We also introduce an approach that utilizes software quality metrics for detecting the structural code clones.

**Keywords:** Code Clone, Simple Code Clone, Structural Code Clone, Software Metrics, Object Oriented Metrics.

## 1 Introduction

In software development process, reusing an existing code with small changes (or without change) is a situation that is often encountered. This style of coding is simple called as code cloning. If a software system has repetitive code snippets in various levels of its software architecture, then these code pieces are called as structural clones. Software clones are often considered as a sign of lack in

B. Murgante et al. (Eds.): ICCSA 2014, Part V, LNCS 8583, pp. 492–507, 2014.
© Springer International Publishing Switzerland 2014

software quality [1]. To improve the quality of code and produce more effective and useful software, code clone detection is highly important. To this end, over the years; the reasons of the emergence of code clones; size of code clones in software project; and the causes of the code clones have been investigated in different studies in detail. For instance; Roy and Cordy [2], in their literature review, examine some of these issues including reasons of emergence of clones, size of code clones in projects and damages of code clones that are all described in Section 2.

Detection of code clones is very important to avoid their side effects in software systems. To this end, both syntactically or functionally similar code clones should be exposed. Over the years, many detection methods based on different types of clones have been proposed. Simple clone detection techniques vary according to the characteristics and source code representations. This diversity can be examined under 5 different categories: text based, token based, abstract synchronized tree based, program dependent graph based and metric based. These simple clone detection techniques cannot look from a broad perspective into code clone snippets, thus they do not determine the reasons of repetition. Hence, there is a need for structural code clone techniques. The benefits of structural code clone detection can be summarized as follows: identifying similarities in the software architecture between different software systems, increasing re-usability, enhancing program outcomes and function without changing the internal structure. This article examines the structural code clone detection methods in detail. Especially, we focus on software metrics based structural code clone detection methods and review state of art in this area.

The organization of this article is as follows. In Section 2, theoretical background is explained. In Section 3, some terminology regarding code clones are defined. Section 4 gives brief overview of software code clone detection techniques. In Section 5, the proposed metric based structural code clone detection technique is described. Section 6 concludes the paper and presents future work.

## 2    Theoretical Background

In this section, Roy and Cordy's [2] literature review explaining emergence of code clones, size of code clones in software projects and the damages of code clones is summarized in order to clarify theoretical knowledge of code clones.

**Reasons of Emergence of Clones:** Code clone reasons show variety. These reasons can be examined under 5 main categories as follows: development strategy, hesitation in the development process, to overcome limitations of the request, lack of information and accidentally copying [2].

- **Development strategy:** Working on old version of the project for maintenance, merging different software systems, using automatic code generation tools are some examples where the system pushes developer to create code clone deliberately [2].

- **Hesitation in development process:** Newly added codes to an already working software system may cause errors. Developers, who think like that, always hesitate to write new code into a system. Such developers prefer to reuse the existing codes, by copying and pasting of the existing codes in the system. In addition, some of the developers can make code clones deliberately in order to follow path of system architecture, even if the cloned part is not useful [2].
- **To overcome limitations of request:** There may be some reasons that push developers make code clones. These reasons are mechanism of the programming language which does not support reusable code; the lack of time for writing error-prone and reusable code; lack of understanding of system; deadlines for the completion of work; measuring developer performance based on the number of code lines [2].
- **Lack of information:** Some of the inexperienced developers are unable to develop modular, reusable code due to lack of basic software coding knowledge such as object oriented programming, abstraction and interface. Hence, they may prefer to use code clones [2].
- **Accidentally copying:** Developers, who unaware of each other and work on the same code base, can create the code clones. In such cases different developers implement the similar code segments, which may utilize the same third-party code snippets. In turn, this creates code clones [2].

**Size of Code Clones in Projects:** Recent studies show that, industrial software systems commonly contain code clones [3,4,5,6]. According to these studies, significant portion of code clones vary depending on the software domain [7,8]. Baker [3] has showed that between 13% and 20% of software projects have code clones. Lague [9] has showed that between 6% and 8% of software projects has functional code clones. Baxter [10] has presented that 31% of software projects are detected to have code clones. Mayrand [11] has identified between 5% and 20% of industrial software projects that include code clones. Kasper and Godfrey studies [12] have showed that 10% - 15% of code clones have been detected in software systems. All these researches show that code cloning in software projects is a commonly encountered problem.

**Damages of Code Clones:** In large scale software systems, there are many reasons for code clones. In a given system, the presence of code clones does not affect system operation, but they increase the complexity maintenance costs of the system [13]. In addition, code clones may cause many problems as well [3,4,5,6,11,14]. Code problems arising from clones can be summarized in the following items.

- **Increased maintenance costs:** If a bug is found in any cloned code, all of the other clones should be found in order to fix this bug. Recent studies show that maintenance cost is proportion to size of the code clones in software system [3,15]. After software release, cost of the changes vary from 40% and 70% of the total cost [16].

- **Increasing number of possible errors:** Using a slightly modified copied code (or using the code without any modification) have a significant impact on propagation of errors. When a code clone contains a bug, this bug is copied everywhere and the total number of possible errors increases [15].

- **Increased update costs:** When a clone code needs to be modified, this operation must be performed on each clone. This increases the maintenance and updating costs [16].

- **Increasing required resource size:** The code size of the system gets increased due to copy-paste coding style. While this is not a problem for commodity computers, it may become a problem in mobile devices due to their limited resources. For example, this may necessitate a change in the hardware with the software of the mobile device [14].

- **Increasing response time for change request:** In order to meet customer request, new development or changes on existing code should be performed. System should be fully understood before meeting customer requirements to prevent possible new bugs, but the clone codes make it hard to understand. Hence, change requests take larger time in systems that have code clones [5].

- **Poor system architecture:** Code clones may arise from lack of knowledge on the usage of abstraction and object oriented design. In turn, this creates a poor system architecture. In this type of poor system architecture, reusable code cannot be implemented easily, thus maintenance cost increases [3].

- **Increasing probability of errors:** While transferring previously written code into a new system, most of the developers take old code base without analyzing code or without changing its logic flow. Thus, developers may take old code clones (from third-party code bases) into new system. Use of this method is prone to error. This creates errors which are likely to occur [6].

## 3   Code Clone Terminology

General terms of software code clone terminology and descriptions are summarized in Table 1 are given below. This terminology includes Code Snippets, Code Clone, Clone Pair, Clone Class and Clone Types.

**Code Snippet:**   This is used for a small portion of the code, which can also contain comments. It can be any sequential statement or function definition. Code snippet can be identified uniquely from file name of the original file, which contains it. It includes the start line number and the end line number of code segment of the file [2].

**Code Clone:** If one code snippet shows similarities to any other code snippet depending on defined similarity algorithm, this code snippet is called a clone of the other code snippet [2].

**Clone Pair:** If the program consists of two pieces of code that are similar to each other, the pair is called a clone pair [2].

**Clone Class:** The set of code snippets that are similar to the each other is called as Clone Class. In the cluster, each code snippet is similar to the remaining code snippets [2].

**Clone Types:** This can be examined under the 2 categories: simple code clone and structural clone. In simple clone type, code snippets have basic similarities between the two varieties [2]. These similarities are basically analyzed in 4 categories as follows:

- **Type I:** Textually similar code snippets after removing comments and blank lines.
- **Type II:** Code snippets that are not textually similar, but syntactically similar. For example: int a = 3, int b = 3. Although, the expression names (a, b) are not similar each other textually, they functionally similar to each other.
- **Type III:** This type differentiates from Type II by adding or removing expression into the code snippet. One clone part can have more or less expression than its pair.
- **Type IV:** Textually and syntactically, both code snippets are completely different from each other but they return the same results.

Detection of Clone Types mentioned above works on the code snippets. But the similarities between these code snippets can arise from high level similarities in the software architecture. Structural clone is a result of the architectural similarity. It can contain one or more type of Code Clones mentioned above.

**Table 1.** Clone Type Classification

| Clone Type | Simple Clone | | Structural Clone |
|---|---|---|---|
| | **Text Based** | **Functional** | |
| Type I | ✓ | ✗ | ✓ |
| Type II | ✓ | ✗ | ✓ |
| Type II | ✓ | ✗ | ✓ |
| Type IV | ✗ | ✓ | ✓ |

Table 1 summarizes clone types. As shown in the table, structural clone can contain any of 4 simple code clones. Besides, it can detect similarities in the architectural level.

## 4    Literature Overview of Detection Types

In Software Engineering Research Area, extensive research has been done on the code clones. In order to reduce or prevent problems in the code clone, many techniques have been proposed in literature. The focus of this research is on the code clone detection. Therefore, the literature overview has also been focused on detection of clones.

Clone detection tools can find out text based or functional similarities between source files that constitute software project. Their principle is based on looking for code snippets which has high rate similarities with other parts. But the main issue here, comparing code snippets with all other parts is really high-cost process. In order to reduce this computation cost, some techniques have been developed. Figure 1 shows the general work flow concept of the these techniques.

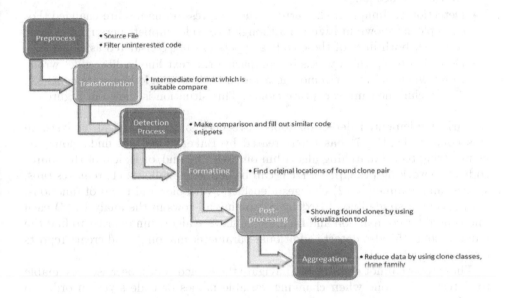

**Fig. 1.** General work flow of detection techniques

In the literature, there are multiple code clone detection tools. Here, both academic and commercial products are available. These tools have a logic flow with 2 phases: data conversion phase and identifying similarities phase. Data conversion phase: The program's source code is converted into intermediate data format. By using such data format, similarity algorithms work easily and efficiently. Similar code snippets are determined at identifying similarities phase. The intermediate data format affects comparison algorithm and its result. Thus, this data format plays an important role in the classification of detection tools. Depending on comparison level and detection algorithm, these tools are able to detect different types of clones. For this reason, various techniques, which are used by detection tools, can be summarized as follows.

**Text-Based Techniques:** Many published detection tools are only text-based. In this approach, the program's source code is assumed as ordered text lines. While comparing any of the two code snippets with each other, the same text is tried to be caught. The maximum number of similar text lines is indicated as clone. In this technique, the source data is not subjected to too much conversion. A direct comparison algorithm is mostly used in the source code. However, in some approaches such as comments and white space removal operations are carried out for small-scale transformation. These approaches usually compare the code line by line. Such detection methods have led to some mistakes. Some of these errors are listed below [2].

- Relocation in the line of code style leads to errors. For example, as shown in Figure 2, every two pieces of code are not considered to be identical, but they are clones [2].
- Detection technique makes errors when expression names are changed. For example, as shown in Figure 3 although two code snippets perform the same function, both lines of these code snippets are not considered as clones [2].
- Detection tool, which works by comparing the text line by line, gives wrong result when adding or removing a parenthesis to single line statements [2]. This technique cannot capture clones. This situation is shown in Figure 4.

Early implemented detection tools have been based on lexical analysis. In this context, DUP [17] has been created by Baker. This tool finds clones by using string based matching algorithm on each line and/or lexical of the source code. The working principle of DUP can be defined as follows: (1) removes tabs, spaces, and comments, (2) changes identifiers, variables and types of functions, (3) provides a single line of text by combining all rows in the analysis, (4) each line is used for comparison and then use suffix tree algorithm in order to find the longest clone, (5) also detects additional parameter mappings and create reports to show mappings that are found.

These tools cannot explore and navigate the copied codes, because it is unable to detect code clone when changing variable names or code style. In order to avoid problems arising from changes in the comment or blank lines, it works by comparing words of string rather than each character of string.

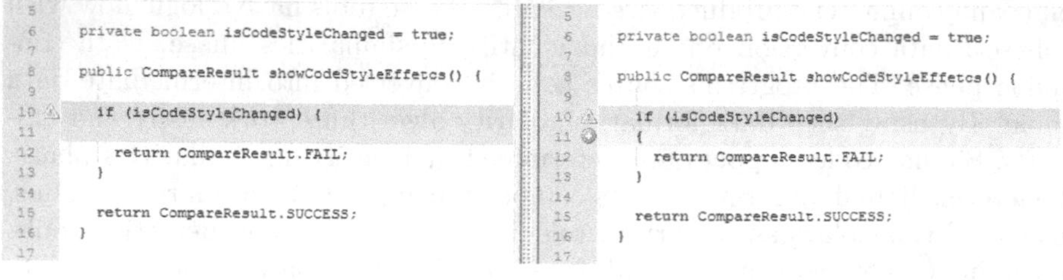

**Fig. 2.** Error arising from code style

```
  5                                                        5
  6    private boolean isVariableNameChanged = true;       6    private boolean isVariableNameChanged = true;
  7                                                        7
  8    public CompareResult showEffetcsOfChangingVariableNames() {   8    public CompareResult showEffetcsOfChangingVariableNames() {
  9                                                        9
 10    VariableName changedName = new VariableName();      10    VariableName originalName = new VariableName();
 11                                                       11
 12    isVariableNameChanged = changedName.isMyNameChanged();   12    isVariableNameChanged = originalName.isMyNameChanged();
 13                                                       13
 14    if (isVariableNameChanged)                         14    if (isVariableNameChanged)
 15    {                                                  15    {
 16       return CompareResult.FAIL;                      16       return CompareResult.FAIL;
 17    }                                                  17    }
 18                                                       18
 19    return CompareResult.SUCCESS;                      19    return CompareResult.SUCCESS;
 20    }                                                  20    }
```

**Fig. 3.** Error arising from changing variable name

```
  5    private boolean isSingleLineStatementExist = true;    5    private boolean isSingleLineStatementExist = true;
  6                                                        6
  7    private boolean isCurlyBracesChanged = true;        7    private boolean isCurlyBracesChanged = true;
  8                                                        8
  9    public CompareResult showEffetcsOfCurlyBraces() {   9    public CompareResult showEffetcsOfCurlyBraces() {
 10                                                       10
 11    if (isSingleLineStatementExist)                    11    if (isSingleLineStatementExist)
 12    {                                                  12
 13    isCurlyBracesChanged = Statement.COVERED_BY_CURLY_BRACES;   13    isCurlyBracesChanged = Statement.UNCOVERED_BY_CURLY_BRACES;
 14    }                                                  14
 15                                                       15
 16    if (isCurlyBracesChanged) {                        16    if (isCurlyBracesChanged) {
 17       return CompareResult.FAIL;                      17       return CompareResult.FAIL;
 18    }                                                  18    }
 19                                                       19
 20    return CompareResult.SUCCESS;                      20    return CompareResult.SUCCESS;
 21    }                                                  21    }
```

**Fig. 4.** Error arising from adding or removing to single line statements

Ducase and colleagues identified another text-based approach to find clones [18,19]. This approach is not based on parsing and therefore can easily adapt to any language. This method reads the source code creates arrays of lines, removes spaces and comments and then detects matches by using dynamic model matching algorithm based on string. It gives clone pairs' line numbers and deletes potential code clone rows. However, the computational complexity of this method is $O(n\hat{2})$ and it is too costly.

**Token-Based Techniques:** In this approach, all the source code is translated into a sequence of tokens. Then these lines of token arrays are scanned to find out code clones. Cloned sub-ordered tokens are found as an output of the scanning process. Compared with text-based system, this technique eliminates text-based techniques errors as a result of coding format.

The most advanced token-based technique is the CCFinder tool. It has been created by Kamiya and colleagues [5].The working principle of CCFinder can be defined as follows: (1) parse source file into tokens by using lexical analysis and (2) combine the tokens into single line token array. CCFinder uses this token array by adding new tokens or removing old ones depending on requirement. And type of each identifier variable is replaced with a special token. By using this replacement with special token, effect of changing variable names can be eliminated. Each parsed token is added into suffix tree algorithm and then this suffix tree is used to find out clone pair and clone classes.

There is many other token based implementation exist in clone detection field. CP-Miner [7,20] use special token as CCFinder but it differentiate from CCFinder by its methodology. CP-Miner use data mining techniques in order to find out code clones. DUP which is created by Baker [3,17,21] is also similar CCFinder but it does not use transformation from source code to special tokens. It adds lines of source code into suffix tree algorithm and then makes computation.

**Tree-Based Techniques:**    The source codes are transformed into abstract syntax tree (AST) depending on the programming language. Tree based techniques looks for cloned sub-tree by using matching algorithms on parsed trees. The cloned sub-trees can be shown as clone pair or clone classes.

One of the most important abstract syntax tree techniques called as CloneDR that have been made by Baxter and colleagues [10]. The working principle of CloneDR as follows: (1) creates annotated parse tree by using CloneDR compiler, (2) use metric based hash functions in matching algorithm in order to detect clones in sub-trees. Similar trees are grouped into source code clones. Another tree based clone detection tool, Bauhaus ccdiml, has been proposed by [22]. The differences between ccdiml and CloneDR can be summarized as followings: (1) ccdiml does not use metric based hash function in order to find out code clones, (2) CloneDR can work simultaneously with name changing however, ccdmil cannot, (3) abstract syntax tree is used as a base for matching algorithms by CloneDR, but ccdmil use it as an intermediate language.

**Program Dependency Graph Techniques:**    The program flow information is also included in data, which is extracted from source codes. In this way, apart from all the other techniques, one step further, the program also contains information on business logic. After obtaining the program graph, sub-graph clones are found by using isomorphic sub-graph matching algorithms. PDG-DUP is created by Komondoor and Horwitz [23,24]. This is one of the dependent graph approaches. This tool finds structural sub-graphs by using its program slicing method. Komondoor and Horwitz also identifies clones without breaking the semantics of the original code and offers an approach to group together. Another program slicing-based approach is created by Gallagher and Lucas [25]. The aim of their work is that using program slicing method on all variables on the system to make a decision about parsed fragment whether it is clone or not. However, they are not sure about the result of analysis, thus they only explain the advantages and disadvantages of their methodology. Chen [26], has presented a dependent diagram technique for compression programs by taking into account the flow of data and code syntax.

**Metric-Based Techniques:**    Metric-based approaches collects different metrics from code snippets and creates a vector. It compares the metric vectors rather than directly comparing the code segments. Software metrics to identify similar code clone detection technique has been proposed by different studies. In these studies, firstly, a software metric cluster which is named as fingerprint function is extracted. By using these fingerprint functions, for one or more lexical units (e.g. class or function) are calculated. Then, code clones are found by comparing fingerprints of these lexical units. Patenaude and colleagues [27] created metric based approach for code clones. Their used metrics can be summarized as follows: (1) number of call from inside a method, (2) number of arguments, (3) McCabe cyclomethicone complexity measure, (4) number of non-local variables, and (5) number of local variables. These metrics were defined for the Java programming language. Kanika and colleagues [28] presents a metric calculation tool, which called MCD Finder for Java. MCD Finder metric calculation uses the Java byte code in programs (rather than to convert the source code and make metric calculations). Using byte code provides a platform-independent approach.

**Structural Clones:** Structural analysis of clone includes syntactic and logical analysis of codes. Therefore, the code in the software, the software's design templates are used in the analysis of structural clones. Initially it draws design templates, outlines the structure of the system. Developers may have to modify the current structure due to changing requests on requirements. Structural clone detection depends on domain analysis directly. For detection of different structural clones, different techniques are required. In some cases, this requires expert developers to interpret the development, thus a fully automated detection tool creation is hard. But still, structural clone detection tools give more information to the user in the architectural sense rather than simple clone detection tools. Some tools have been developed for the detection of structural clones. One of these tools is Clone Miner which is created by Basit and colleagues [29]. The working principle of this tool can be defined as follows: (1) detect simple clones, (2) make an analysis for structural code clone by grouping this detected simple clone as class level or file level. De Lucia and colleagues [30] use a technique of structural clones to perform work on web pages clones. In this study, a threshold value of any given web page is determined and then compares the page with others. If other page similar more than this threshold, these two pages are alleged to be clones of each other. Here, the Levenshtein distance algorithm are used to measure the similarity of the pages.

De Lucia and colleagues [30] made another approach that detects specific structural code clone on web pages. In this approach, they use hyperlink, which supply connection between web pages, in order to detect structural code clone. This approach use graph-based template matching algorithms.

Marcus and Maletic [31] propose a clone detection approach that has a different perspective than others. In the clone detection approaches that are examined so far identifiers are shown result of similarity analysis. In this study, the structural clone analysis are made with reverse logic flow (in source code identifier names) are made on the basis of the structural clone analysis. When identifier names are similar to each other of two different structural clone systems, this approach works but the change of identifier names will fail to identify structural clones in this work.

Design templates that are similar to the structure of the class and that provides application-level abstraction structure are called as micro pattern [32]. This micro-structural clone analysis can be made out of design templates. Shi and Olsson [33] created the Pinot tool that does the clone analysis based on design templates in the source code. Downside of this approach, search templates in the system are known. However, detection of structural clones should make calls on unknown templates. All of these techniques have been summarized in Table 2 in chronological order.

**Table 2.** Detection tools classification

| Detection Tool | Year | Clone Type | Detection Approach |
|---|---|---|---|
| Baker, Dup [17] | 1992 | Simple | Text-Based / Token-Based |
| Baxter and colleagues, CloneDR. [9] | 1998 | Simple | AST Based |
| Patenaude and colleagues [27] | 1999 | Simple | Metric Based |
| Marcus and Maletic [31] | 2001 | Simple/Structural | Metric Based /Token Based |
| Komondoor and Horwitz, PDG-DUP [23] | 2001 | Simple | Program Dependency Based |
| Kamiya and colleagues, CCFinder [5] | 2002 | Simple | Token Based |
| Gallagher and Lucas,[25] | 2003 | Simple | Program Dependency Based |
| Ducasse and colleagues [18] | 2004 | Simple | Metric Based |
| De Lucia and colleagues,[30] | 2004 | Structural | Program Dependency Graph |
| Basit and colleagues, Clone Miner [29] | 2005 | Simple/Structural | All simple techniques plus data mining |
| Li and colleagues, CP-Miner [7] | 2006 | Simple | Token Based |
| Bauhaus, ccdiml [22] | 2006 | Simple | AST Based |
| Shi and Olsson, Pinot [33] | 2006 | Structural | All simple techniques |
| Kanika and colleagues, MCD Finder [28] | 2013 | Simple | Metric Based |

# 5   Metric Usage in Structural Clone Analyzing

Frequently existence of simple clones in software system is a sign that there can be high level of similarities (i.e. similarities in the software architecture). We list the reasons why a software design pattern may be used by coping and pasting [7].

- Using the same analysis pattern [3],
- Using the same design patterns [34],
- Architecture level of the components used in the design of the similarity,
- To solve similar problem, developers use the same architecture [10]

After searching literature, we notice that metric based approaches never used in order to identify structural code clones, thus this subject is an open issue. Structural clones, based on software metrics measured values should be able to detected. Analyzing this issue for finding structural clones according to measured

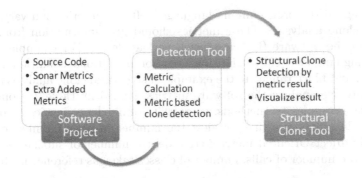

**Fig. 5.** The work flow of proposed method

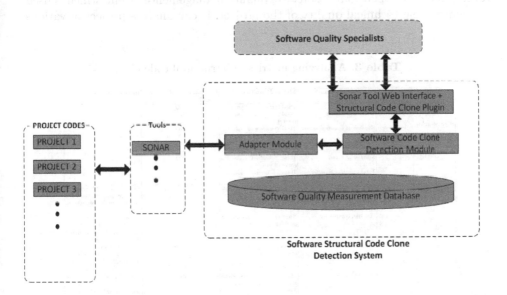

**Fig. 6.** General Architecture of proposed methodology

software metrics' values is the main subject of this paper. In this respect, Figure 5 illustrates workflow of our methodology.

This article focuses on the identification of the structural code clone using the software metrics. The list of software metrics are shown in Table 3 are identified to be used in the determination of the structural code clones. In this study, we use the open-source Sonar Software Quality measurement system and build a plug-in for structural clone detection. The proposed system architecture is given in Figure 6. As shown in this architectural design, our goal is that using different software quality metrics measurement tools to gather software metric measurements, which will lead us to do structural clone analysis.

Depending on the programming language, software metrics can vary for the structural clone analysis . These metrics should give information from different levels of the software (i.e. class level, package level). For example, in Java programming language, class definitions, method definitions, conditional or loop statements, and block range are the example of structural clone analysis metrics. In this study, we analyze the software metrics defined in Table 3: Complexity metrics (class complexity, methods, complexity of file dependencies); Code Size Metrics (code number of lines, review the number of rows, number of blank lines); and Object Oriented metrics (tree depth, number of subclasses derived for a class, the number of calls, number of classes taken as reference in the class) [18].

In the proposed system, we plan to use Sonar [35] quality metric measurement tool. Sonar is an open source software quality management system, which is used to measure the technical quality of the project. It can analyze project in various

**Table 3.** Analyzing metrics for structural code clone

| Metric Category | Metric Name | Metric Definition | Is suitable for Structural clone? | Intended Purpose |
|---|---|---|---|---|
| Size Metrics | Line number | Total line count of project | NO | - |
| | Comment number | Total comment count of project | NO | - |
| | Number of commented code line | Total commented code line count of project | NO | - |
| | Number of uncommented line | Total uncommented code line count of project | NO | - |
| | Number of token | Total tokenized code line count of project | NO | - |
| | Total identifier count | Total tokenized identifer count of project | NO | - |
| Complexity Metrics | Class complexity | Average complexity of class | YES | Same complexity value of classes,in project, which can be structurally similar to each other. |
| | Function complexity | Number of statement is method | YES | Same complexity value of,functions in project, which can be structurally similar to each other. |
| | File complexity | Average complexity of files of project | YES | Same complexity value of files in project, which can be structurally similar to each other. |
| | File Dependency | The number of files which cause dependency between files. | YES | Same dependency value of packages in project, which can be structurally similar to each other. |
| Object Oriented Metrics | Depth in tree | The number of depth of class at inheritance level | YES | Same depth value of classes in project, which can be structurally similar to each other. |
| | Number of use other class | Total number of other class which is used in the class. | YES | Same used number class value of classes in project,which can be structurally similar to each other. |
| | Number of child | Total number of child of the class | YES | Same number of child value of classes in project,which can be structurally similar to each other. |
| | Number of used by other class | Total number of used times of class by others | YES | Same used times value of class by other classes in project,which can be structurally similar to each other. |
| | Number of arguments | Total number of arguments which is passed to method | YES | Same argument number value of method in project,which can be structurally similar to each other. |
| | Number of returning parameter | Total number of arguments which is returning by method | YES | Same returning argument number value of method in project,which can be structurally similar to each other. |
| | Number of calling other methods | Total number of calling method in the specific method | YES | Same number of calling other method value of method in project, which can be structurally similar to each other. |

levels from method to whole project. The Sonar tool can identify the Type I and Type II clones, and also partly Type III simple clones. Sonar supports more than 100 metrics. We use some of these metrics explained in Table 3 in our study for detecting structural clones and classify these metrics under three main categories that are complexity, design, documentation and size.

Our target structure will use metric based approach in order to identify structural clone because the metric based approach is the most effective approach among simple clone for memory usage. Preliminary studies of this work has been discussed in [36]. For this purpose, new metrics shown in Figure 7 are added to Sonar which is required metrics for structural clone detection. These metrics are divided into two sub-categories that are relational and self identifying metrics. In addition, each metric has its own impact factor for using in comparison. Then, target system will use output of these metric values in order to compare two or more projects.

| Project-Related Metrics | Variable-Related Metrics | Method-Related Metrics |
|---|---|---|
| +numberOfDefaultClasses : int = 0 | +numberOfTotalLocalMethodVariables : int = 0 | +numberOfImplementedInterfaces : int = 0 |
| +numberOfInterfaces : int = 0 | +numberOfTotalClassVariables : int = 0 | +numberOfPrivateMethods : int = 0 |
| +numberOfDefaultAccessInterfaces : int = 0 | +numberOfPrivateClassVariables : int = 0 | +numberOfProtectedMethods : int = 0 |
| +numberOfAbstractClasses : int = 0 | +numberOfDefaultClassVariables : int = 0 | +numberOfDefaultMethods : int = 0 |
| +numberOfDefaultAbstractClasses : int = 0 | +numberOfProtectedClassVariables : int = 0 | +numberOfTotalOverridenMethods : int = 0 |
| +numberOfFinalClasses : int = 0 | +numberOfPublicClassVariables : int = 0 | +numberOfProtectedOverridenMethods : int = 0 |
| +numberOfDefaultFinalClasses : int = 0 | +numberOfTotalStaticVariables : int = 0 | +numberOfDefaultOverridenMethods : int = 0 |
| +numberOfEnums : int = 0 | +numberOfPrivateStaticVariables : int = 0 | +numberOfPublicOverridenMethods : int = 0 |
| +numberOfDefaultEnums : int = 0 | +numberOfProtectedStaticVariables : int = 0 | +numberOfTotalAbstractMethods : int = 0 |
| +getImpactFactorOfAttribute() : int | +numberOfDefaultStaticVariables : int = 0 | +numberOfPrivateAbstractMethods : int = 0 |
| | +numberOfPublicStaticVariables : int = 0 | +numberOfDefaultAbstractMethods : int = 0 |
| | +numberOfTotalFinalVariables : int = 0 | +numberOfPublicAbstractMethods : int = 0 |
| | +numberOfPrivateFinalVariables : int = 0 | +numberOfStaticMethods : int = 0 |
| | +numberOfProtectedFinalVariables : int = 0 | +numberOfFinalMethods : int = 0 |
| | +numberOfDefaultFinalVariables : int = 0 | +numberOfFinalStaticMethods : int = 0 |
| | +numberOfPublicFinalVariables : int = 0 | +numberOfVoidReturnMethods : int = 0 |
| | +numberOfTotalFinalStaticMethods : int = 0 | +numberOfBooleanReturnMethods : int = 0 |
| | +numberOfPrivateFinalStaticMethods : int = 0 | +numberOfCharReturnMethods : int = 0 |
| | +numberOfProtectedFinalStaticMethods : int = 0 | +numberOfByteReturnMethods : int = 0 |
| | +numberOfDefaultTotalFinalStaticMethods : int = 0 | +numberOfShortReturnMethods : int = 0 |
| | +numberOfPublicFinalStaticMethods : int = 0 | +numberOfIntegerReturnMethods : int = 0 |
| | +getImpactFactorOfAttribute() : int | +numberOfFloatReturnMethods : int = 0 |
| | | +numberOfLongReturnMethods : int = 0 |
| | | +numberOfDoubleReturnMethods : int = 0 |
| | | +numberOfObjectReturnMethods : int = 0 |
| | | +getImpactFactorOfAttribute() : int |

**Fig. 7.** Newly Added Metrics to Sonar

# 6   Conclusion and Future Plans

Clone detection is one of the active research areas in the literature of software quality. Many studies have been performed to remove clones from systems. In addition, protection of software systems from clones has been studied in the developments life cycle. In this study, clone types, their detection mechanisms and related tools are reviewed. Then structural code clone detection method is proposed by combining the structural code clone detection techniques and metric-based code clone detection techniques. This proposed method helps the users, who use clone detection technique, to choose right clone detection tools. Development and evaluation of this proposed methodology are planned as a future work. As a result of this development, we plan on building a plug-in that can be integrated with Sonar tool to be used for structural clone detection.

# References

1. Fowler, M.: Refactoring: improving the design of existing code. Addison-Wesley Professional (1999)
2. Roy, C.K., Cordy, J.R.: A survey on software clone detection research. Technical report, Citeseer (2007)
3. Baker, B.S.: On finding duplication and near-duplication in large software systems. In: Proceedings of 2nd Working Conference on Reverse Engineering, pp. 86–95. IEEE (1995)
4. Casazza, G., Antoniol, G., Villano, U., Merlo, E., Di Penta, M.: Identifying clones in the linux kernel. In: Proceedings of the First IEEE International Workshop on Source Code Analysis and Manipulation, pp. 90–97. IEEE (2001)
5. Kamiya, T., Kusumoto, S., Inoue, K.: Ccfinder: a multilinguistic token-based code clone detection system for large scale source code. IEEE Transactions on Software Engineering 28(7), 654–670 (2002)
6. Kontogiannis, K.: Evaluation experiments on the detection of programming patterns using software metrics. In: Proceedings of the Fourth Working Conference on Reverse Engineering, pp. 44–54. IEEE (1997)
7. Li, Z., Lu, S., Myagmar, S., Zhou, Y.: Cp-miner: Finding copy-paste and related bugs in large-scale software code. IEEE Transactions on Software Engineering 32(3), 176–192 (2006)
8. Jiang, L., Misherghi, G., Su, Z., Glondu, S.: Deckard: Scalable and accurate tree-based detection of code clones. In: Proceedings of the 29th International Conference on Software Engineering, pp. 96–105. IEEE Computer Society (2007)
9. Lague, B., Proulx, D., Mayrand, J., Merlo, E.M., Hudepohl, J.: Assessing the benefits of incorporating function clone detection in a development process. In: Proceedings of the International Conference on Software Maintenance, pp. 314–321. IEEE (1997)
10. Baxter, I.D., Yahin, A., Moura, L., Sant'Anna, M., Bier, L.: Clone detection using abstract syntax trees. In: Proceedings of the International Conference on Software Maintenance, pp. 368–377. IEEE (1998)
11. Mayrand, J., Leblanc, C., Merlo, E.M.: Experiment on the automatic detection of function clones in a software system using metrics. In: Proceedings of the International Conference on Software Maintenance 1996, pp. 244–253. IEEE (1996)
12. Kapser, C.J., Godfrey, M.W.: Supporting the analysis of clones in software systems. Journal of Software Maintenance and Evolution: Research and Practice 18(2), 61–82 (2006)
13. Rysselberghe, F.V., Demeyer, S.: Evaluating clone detection techniques from a refactoring perspective. In: Proceedings of the 19th IEEE International Conference on Automated Software Engineering, pp. 336–339. IEEE Computer Society (2004)
14. Antoniol, G., Villano, U., Merlo, E., Di Penta, M.: Analyzing cloning evolution in the linux kernel. Information and Software Technology 44(13), 755–765 (2002)
15. Johnson, J.H.: Identifying redundancy in source code using fingerprints. In: Proceedings of the 1993 Conference of the Centre for Advanced Studies on Collaborative Research: Software Engineering, vol. 1, pp. 171–183. IBM Press (1993)

16. Grubb, P., Takang, A.A.: Software maintenance: concepts and practice. World Scientific (2003)
17. Baker, B.S.: A program for identifying duplicated code. Computing Science and Statistics, 49–49 (1993)
18. Ducasse, S., Nierstrasz, O., Rieger, M.: Lightweight detection of duplicated codea language-independent approach. Institute for Applied Mathematics and Computer Science, University of Berne, Switzerland (2004)
19. Ducasse, S., Rieger, M., Demeyer, S.: A language independent approach for detecting duplicated code. In: Proceedings of the IEEE International Conference on Software Maintenance (ICSM 1999), pp. 109–118. IEEE (1999)
20. Li, Z., Lu, S., Myagmar, S., Zhou, Y.: Cp-miner: A tool for nding copy-paste and related bugs in operating system code. In: OSDI, vol. 4, pp. 289–302 (2004)
21. Baker, B.S.: On finding duplication in strings and software. submitted for publication (1993)
22. Raza, A., Vogel, G., Plödereder, E.: Bauhaus – A tool suite for program analysis and reverse engineering. In: Pinho, L.M., González Harbour, M. (eds.) Ada-Europe 2006. LNCS, vol. 4006, pp. 71–82. Springer, Heidelberg (2006)
23. Komondoor, R., Horwitz, S.: Using slicing to identify duplication in source code. In: Cousot, P. (ed.) SAS 2001. LNCS, vol. 2126, pp. 40–56. Springer, Heidelberg (2001)
24. Komondoor, R.V.: Automated duplicated-code detection and procedure extraction. PhD thesis, UNIVERSITY OF WISCONSIN (2003)
25. Gallagher, K., Layman, L.: Are decomposition slices clones? In: 11th IEEE International Workshop on Program Comprehension, pp. 251–256. IEEE (2003)
26. Chen, W.K., Li, B., Gupta, R.: Code compaction of matching single-entry multipleexit regions. In: Cousot, R. (ed.) SAS 2003. LNCS, vol. 2694, pp. 401–417. Springer, Heidelberg (2003)
27. Patenaude, J.F., Merlo, E., Dagenais, M., Laguë, B.: Extending software quality assessment techniques to java systems. In: Proceedings of the Seventh International Workshop on Program Comprehension, pp. 49–56. IEEE (1999)
28. Raheja, K., Tekchandani, R.: An emerging approach towards code clone detection: metric based approach on byte code. International Journal of Advanced Research in Computer Science and Software Engineering 3(5) (2013)
29. Basit, H.A., Jarzabek, S.: Detecting higher-level similarity patterns in programs. ACM SIGSOFT Software Engineering Notes 30(5), 156–165 (2005)
30. De Lucia, A., Francese, R., Scanniello, G., Tortora, G.: Reengineering web applications based on cloned pattern analysis. In: Proceedings of the 12th IEEE International Workshop on Program Comprehension, pp. 132–141. IEEE (2004)
31. Marcus, A., Maletic, J.I.: Identification of high-level concept clones in source code. In: Proceedings of the 16th Annual International Conference on Automated Software Engineering (ASE 2001), pp. 107–114. IEEE (2001)
32. Gil, J.Y., Maman, I.: Micro patterns in java code. In: ACM SIGPLAN Notices, vol. 40, pp. 97–116. ACM (2005)
33. Shi, N., Olsson, R.A.: Reverse engineering of design patterns from java source code. In: 21st IEEE/ACM International Conference on Automated Software Engineering, ASE 2006, pp. 123–134. IEEE (2006)
34. Bakota, T., Ferenc, R., Gyimothy, T.: Clone smells in software evolution. In: IEEE International Conference on Software Maintenance, ICSM 2007, pp. 24–33. IEEE (2007)
35. Dangel, A., Pelisse, R.: Pmd is a source code analyzer (2014) (accessed March 10, 2014)
36. Kapdan, M.: Aktaş, M., Yiğit, M.: Yapısal kod klon analizinde metrik tabanlı teknikler. In: Ulusal Yazilim Muhendisligi Sempozyumu (UYMS), pp.1–19 (2013)

# An Empirical Study of Software Reuse and Quality in an Industrial Setting

Berkhan Deniz[1] and Semih Bilgen[2]

[1] Aselsan Electronics Inc.,
Defense System Technologies (SST) Group, Ankara, Turkey
[2] Middle East Technical University,
Electrical and Electronics Engineering Dept., Ankara, Turkey
berkhand@aselsan.com.tr,
semih-bilgen@metu.edu.tr

**Abstract.** Software reuse is known to be generally effective in reducing development and maintenance time and cost as well as increasing quality. In this paper, the effects of reuse on software quality in an industrial setting are empirically investigated within the framework of three different case studies. Throughout this study, we worked with Turkey's leading defense industry company Aselsan's software engineering department. We collected and calculated reuse and quality metrics as well as performance measures of individual embedded software modules and staff productivity rates. By analyzing these measurements, we developed suggestions to further benefit from reuse through systematic improvements to the reuse infrastructure and process.

**Keywords:** Software Reuse, Quality Metrics, Embedded Software, Fault-proneness, Industrial Study.

## 1 Introduction

Software reuse is generally accepted to reduce development and maintenance time and cost. Any software life cycle product can be reused, not only fragments of source code [1].

As software assets are reused, the accumulated defect fixes result in higher quality [2]. Therefore, a high degree of reuse correlates with a low defect density.

In this study, we investigated the reuse and quality relations in real-life software projects carried out by Turkey's leading defense industry company Aselsan. We designed three separate case studies in which we collected and calculated object-oriented quality metrics, reuse rates and performance of individual modules, fault-proneness of components, and productivity rates of the products. Then, by analyzing these metrics, we reached some useful conclusions: Based on these case studies, we developed suggestions to further benefit from reuse through systematic improvements to the reuse infrastructure and process.

The remaining sections of this paper are organized as follows: In Section 2, background information about reuse types, methods of measuring reuse and quality are

B. Murgante et al. (Eds.): ICCSA 2014, Part V, LNCS 8583, pp. 508–523, 2014.
© Springer International Publishing Switzerland 2014

presented. Also, previous studies on the effects of reuse on software quality are briefly reviewed. In Section 3, our research hypotheses and the designed case studies are presented. The collected reuse and quality metrics are introduced and the discussions of these measurements are provided for all case studies, respectively. Section 4 is about validity of the study. In Section 5, the collected data is analyzed and the research hypotheses are verified. Finally, in Section 6, suggestions to improve the reuse infrastructure are formulated based on the reported case studies. Section 7 concludes the paper; the work done and obtained results are summarized, achievements and difficulties of the study are reviewed; suggestions for future studies are offered.

## 2    Background

### 2.1    Reuse Types

Developers can resort to reuse via most software related entities such as requirements, system specifications, design reports, and processes such as domain engineering for product lines, as well as any other process artifact [1, 3]. There are two main reuse categories: Components-based and transformation-based [12]. In the first category, developers choose appropriate components, and reuse them, with modifications, if and where. Open source software components, commercial-off-the-shelf components, software architecture modules, and product-line components are some examples of reusable components [6]. In the second category, an automated engine produces outputs by transforming appropriate inputs. Most reuse-engines are examples of transformation-based reuse.

### 2.2    Measuring Reuse

Source line of code (SLOC) is the most often suggested metric for reuse size measurements in literature [1, 11]. Furthermore, researchers suggest object-oriented function points for measuring reuse, since this metric is more straightforward to implement [3]. Additionally, for component-based software, "size" is not an available metric as it is for standard systems. Therefore, "the number of use cases" is suggested as an alternate means of size measurements [4].

### 2.3    Measuring Quality

**ISO/IEC 25000 Quality Model.** The latest quality model released by ISO/IEC is Software product Quality Requirements and Evaluation (SQuaRE). This model covers software quality requirements with a systems perspective. The new standard includes a mechanical parts section including mechanics, hydraulics, electronics, and human processes. Hence, the new system-description investigates a wide range of applications [14]. ISO/IEC 25023 Measurement of the system and software product quality model of SQuaRE collects and replaces ISO/IEC 9126-2 and ISO/IEC 9126-3 revised [15].

**Code-Based Software Quality Metrics.** The above standard suggests both internal and external metrics for measuring quality, and various other metrics derived using this model. However, since the standard measures a software product via a large number of aspects; it does not provide specific code-based metrics; hence using this standard does not make sense for code-based measurements. Therefore, we decided to use another model for code-based metrics. The advantages of code-based metrics are summarized as:

- System level forecasting [8, 17]
- Prior identification of unsafe components [13]
- Development of safety design and programming instructions [8, 17]
- Identify the quality and structure of the software design and code [13]
- Prediction of fault-proneness [13]
- Prediction of development and testing efforts
- Validation of the software design quality [5]
- Improve software quality and productivity [5]
- Rapid response when a new technology is adopted [3]
- Reduction of implementation and maintenance cost [3]

The literature on CK metrics signifies that, in most of the studies, the measured metrics correlate with software quality, especially with fault-proneness [5]. Additionally, it is widely observed that OO concepts of coupling and complexity are strongly correlated with fault-proneness [16].

**Effects of Reuse on Software Quality.** Software products' quality usually increases with reuse; because as software artifacts are reused, the collection of the defect corrections in sequential versions brings about a higher quality [2]. Furthermore, the defects detected in the reused components are given higher priority. Additionally, components to be reused are designed and tested more carefully.

## 3    Research Hypotheses and Case Studies

In this study, we examined the relations between software reuse and quality in Aselsan. We formulated four hypotheses as shown below and in order to confirm these, we designed three different case studies.

- Hypothesis 1 – Code-based Quality: The quality of software products is improved as reuse rates of the products increase.
- Hypothesis 2 – Performance of Embedded Software: Performance of the embedded software products decays as reuse rate of the products increase.
- Hypothesis 3 – Fault-proneness: The number of defects detected in components decreases as these components are reused in various products.
- Hypothesis 4 – Productivity: The productivity rates of products increase as the reuse rates of these products increase.

Throughout this study, we worked with three different software teams. These teams were selected due to their accessibility and the possibility of communication with them arising from the author's employment.

The summary of metrics employed in each case study is displayed in Table 1.

**Table 1.** Case studies and the corresponding metrics employed

| Case Study / Metrics | Case Study 1 | Case Study 2 | Case Study 3 |
|---|---|---|---|
| Code-based Quality | + | | + |
| Performance | + | | + |
| Fault proneness | | + | |
| Productivity | | + | + |

## 3.1    Case Study 1

In this study, we worked with team 1. This team develops real-time embedded software for various air defense weapon systems produced in Aselsan and creates software in C++ language.

Three different software modules are investigated in this case study.

The first module is the User Command and Control Interface of an embedded system. This module opens up a TCP/IP socket interface and the external users of the system connect and control the system through this interface. This module is responsible for getting commands through socket, parsing them, and entering the commands into the rest of the system. This module also sends updates and requests to external users; it formats messages in bytes level and sends through the socket.

Since message taking and sending include parsing and formatting in bytes level, it is time-consuming to add a new message to the command interface. In order to simplify message management, a new reuse engine was developed and used by the developers in this team. This engine creates a middleware for all parsing and formatting parts of the socket interface. The user of this tool just decides on the interface functions, and the auto-created middleware is inserted into the main code.

This reuse engine is an example of transformation-based reuse since this engine reuses all the process needed for message sending, receiving, and parsing. The developed middleware can be used without any changes in many different systems. Additionally, this engine can also be used by the test engineers and reduces test efforts. Furthermore, the documentary outputs of this engine can be used for documentation purposes as an example for documentation reuse.

The second module does the same work as the first module, but is implemented using the above-mentioned reuse engine.

The third module is also a product of the reuse engine; however, its developers are different from the second one. It does work that is similar to the work in the second one, but it is a small module: It has fewer messages than the second one.

In addition to CK metrics at the class level, we employed additional complexity metrics in the experiment. In the OO environment, design concepts such as inheritance, coupling, and cohesion have been argued to embrace complexity [5].

512    B. Deniz and S. Bilgen

Also, complexity metrics have been shown to correlate with defect density in a number of case studies [13]. Hence, we selected additional complexity metrics from [7]: McCabe Cyclomatic Complexity (McCabeCC - the number of flows in a part of the code), Nested Block Depth (NBD - the extent of nested blocks of code), and Percent Branch Statements (% Branches – percentage of the statements that create a break in the sequential execution of statements).

In this study, the only available physical metric is the number of cycles for the same work to be done for each module. Therefore, we used this metric to investigate the change of performance metrics for embedded systems with changing reuse rates.

The reuse rates of the modules were calculated by using reused (auto-generated) non-comment line of codes, and total non-comment line of codes (Module 1: 0%, Module 2: 81%, Module 3: 52%).

**Table 2.** Extracted software quality metrics

| Metrics Type | Module 1 | Module 2 | Module 3 |
|---|---|---|---|
| CBO | 2,311111 | 1,944444 | 2,166667 |
| DIT | 0,333333 | 0,651166 | 0,066667 |
| NOC | 0,422222 | 0,686047 | 0,133333 |
| WMC | 4,777778 | 2,166667 | 2,9 |
| LCOM | 0,0820 | 0,0495 | 0,0823 |
| SLOC | 2819 | 4117 | 765 |
| McCabeCC | 2,49 | 1,30 | 1,59 |
| NBD | 1,71 | 0,84 | 1,10 |
| % Branches | 18,2 | 7,4 | 8,9 |
| NoCycles | 7914 | 9756 | 9648 |

The only performance metric calculated is the number of cycles (NoCycles) for the modules to receive the command from the system and send the corresponding data. (Table 2).

**Discussion of the Measurements.** We group the measured metrics according to their primary OO concepts i.e. Size, inheritance, coupling, and complexity, and compare these metrics with respect to increasing reuse rates. We have found a strong correlation between complexity metrics and reuse rate (Figure 1). Between coupling metrics and reuse rate, a positive relationship is observed (Figure 2). We did not observe a strong relation between the other quality metrics and reuse rate.

The usc of a reuse engine causes a reduction in complexity. In Figure 1, a reduction is noticed in all complexity metrics (WMC, McCabeCC, NBD, and % branches) with increasing reuse rate.

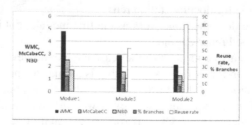

**Fig. 1.** Comparing complexity metrics and reuse rate

In Figure 2, an improvement is observed in terms of coupling as reuse rate increases. The change of the architecture for reuse and introducing interface classes in the system make the system less coupled.

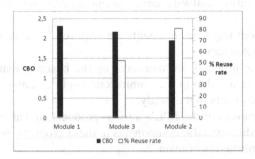

**Fig. 2.** Comparing coupling metric and reuse rate

Figure 3 shows the variation of performance metric (NoCycles) with different reuse rates. As expected, the number of cycles increases with increasing reuse. In the first module, after the command is received from the system, it is sent through related socket directly; however in second and third modules the system architecture changed in order to reuse the middleware and now between sending and receiving, a middleware is introduced which increases the number of cycles.

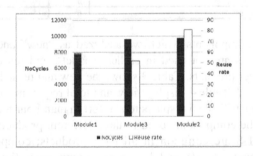

**Fig. 3.** Comparing performance metric and reuse rate

## 3.2 Case Study 2

In this study, we worked with team 2. This team develops command and control software for tactical fire support systems using a software product line. The team creates software in .NET environment. SPL of team 2 is composition-oriented. In it, there are two types of components: common platform components reused in various projects and product specific components developed for every single product. A product developed on this SPL consists of various numbers of common platform components and product specific components in it.

The measurements in case study 2 include defect counts of the components developed by this team, which are reused in different products and the productivity rates of these products.

Defect counts are obtained from the problem reporting system used in Aselsan. As a result of company politics, all defects detected during system integration and acceptance testing and also those reported by customers are kept in the problem reporting system i.e. Defects during software development process are not included in these measurements.

Measurements about requirement counts are acquired from the requirements management tool used in Aselsan.

Total efforts of the products are measured by the business management software used in Aselsan. However, due to the commercial confidentiality, we do not provide exact measures of the efforts in this study.

About the specifications of the SPL's various products, and about the properties of the common and product-specific components in these products, we worked with one of the configuration managers of team 2.

In this work, three different products are explored. All products have at least one product specific component, and other components are common platform components. The three products were developed sequentially with six months between the completions of each one.

**Table 3.** New and reused component counts in three different products

| Product No / Component count | New | Reused |
|---|---|---|
| Product 1 | 19 | 0 |
| Product 2 | 3 | 16 |
| Product 3 | 6 | 15 |

We classified the components which we analyzed as "new" and "reused" components. New components are not used in earlier products, and reused components are used previously in other products. Table 3 shows the new and reused component counts in the products analyzed. Table 4 displays new and total requirement counts in all components for each product. Table 5 displays total effort in man-hour for each product.

Defect counts of the components used in three different products are shown in Table 6. Components 1-8 are common in all three products; component 11 is partly common; and components 9, 10, and 12 are new components (i.e. They are used in corresponding products for the first time). All 12 components are common-platform components.

**Table 4.** New and total requirement counts in all components for each product

| Components / Product No | Product 1 | Product 2 | Product 3 |
|---|---|---|---|
| C1 | 291 / 291 | 4 / 295 | 0 / 295 |
| C2 | 383 / 383 | 10 / 293 | 0 / 293 |
| C3 | 261 / 216 | 0 / 216 | 1 / 217 |
| C4 | 167 / 167 | 0 / 167 | 0 / 167 |
| C5 | 301 / 301 | 4 / 305 | 0 / 305 |
| C6 | 304 / 304 | 11 / 315 | 0 / 315 |
| C7 | 126 / 126 | 21 / 147 | 1 / 148 |
| C8 | 220 / 220 | 32 / 252 | 27 / 279 |
| C9 | - | - | 211 / 211 |
| C10 | - | - | 177 / 177 |
| C11 | 275 / 275 | - | 14 / 289 |
| C12 | - | 141 / 141 | - |

**Table 5.** Total effort for each product

| Product No | Total Effort (man-hour) |
|------------|-------------------------|
| Product 1  | 2,75 * N                |
| Product 2  | 1,5 * N                 |
| Product 3  | N                       |

**Discussion of the Measurements.** According to Figure 4, the average defect count in the first product is more than 50, less than 3 in the following product and less than 1 in the third product. More than 95 % of the total defects are detected in the first product. There are various reasons for this improvement: the reused components are less modified than non-reused ones; therefore, they are more stable. Additionally, the reused components are designed more intensely; since defects in them affect different products. Furthermore, the employment of the common components in various products causes them to become faultless and finished components.

**Table 6.** Defect counts of the components in three different products

| Components / Product No | Product 1 | Product 2 | Product 3 |
|-------------------------|-----------|-----------|-----------|
| C1  | 87  | 2  | 0  |
| C2  | 35  | 5  | 0  |
| C3  | 54  | 1  | 1  |
| C4  | 20  | 0  | 0  |
| C5  | 63  | 0  | 0  |
| C6  | 100 | 6  | 0  |
| C7  | 24  | 0  | 1  |
| C8  | 48  | 3  | 0  |
| C9  | -   | -  | 24 |
| C10 | -   | -  | 34 |
| C11 | 55  | -  | 7  |
| C12 | -   | 27 | -  |

In Figure 5, defect counts of the product-specific components (i.e. Components 9, 10, and 12), and the average defect counts of common components are shown. For all three components, defect counts are more than 20. Since, these components are not reused, we observe a similar distribution as the defect counts of the common components in the first product they are used.

Defect count of component 11 is shown in Figure 6. This component is partially-common in products 1 and 3 (see Table 4 and Table 6). In product 1, more than 50 defects are detected, and in product 3 almost 10 defects are detected. The defect distribution of this component is similar to common components' distribution of products 1 and 2.

Table 7 shows productivity rates calculated by dividing number of new requirements (Table 4) by total efforts (Table 5).

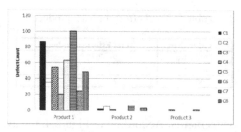

**Fig. 4.** Defect counts of the common components (C1-C8)

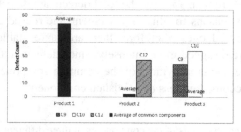

**Fig. 5.** Defect counts of the product-specific components (C9, C10, and C12)

Table 8 shows productivity rates calculated with the total number of requirements.

In Figure 7, productivities are compared. As components are reused in different products, productivity rates increase remarkably, which is not surprising.

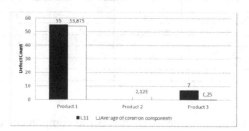

**Fig. 6.** Defect counts of the partially-common component (C11)

Comparison of the productivity rates with new requirements exhibits a sharp reduction between the first and second products; and a noteworthy expansion between the second and third products. We interpret the first case as a conclusion of the development by employing a product-line. Since the reused components are already developed in previous products, the productivity rates increase significantly.

**Table 7.** Productivity rates using new requirements

| Product No | New Requirements | Productivity (requirements / man-hour*N) |
|---|---|---|
| Product 1 | 2283 | 830,2 |
| Product 2 | 223 | 148,7 |
| Product 3 | 431 | 431 |

Furthermore, the first reduction in the second case is explained as an adaptation period of the development with the product line. Although there are developed-products, ready to be used, it is still time-consuming to gather up these components and integrate them with the recently developed components. Finally, the expansion between the second and third products in the second case is interpreted as an evidence of being trained in development with the product line. It is expected that, in future products, the productivity rates using the new requirements will exceed the productivity rate of the first product.

**Table 8.** Productivity rates using total requirements

| Product No | Total Requirements | Productivity (requirements / man-hour*N) |
|------------|--------------------|------------------------------------------|
| Product 1  | 2283               | 830,2                                    |
| Product 2  | 2231               | 1487,3                                   |
| Product 3  | 2796               | 2796                                     |

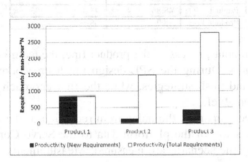

**Fig. 7.** Comparison of productivity rates of each product using new requirements and total requirements

## 3.3   Case Study 3

In this study, we worked with team 3. This team develops real time embedded software for fire control systems using a composition-oriented SPL in C++ language. The capabilities, which can be included in the product line or excluded from the product line, are modeled as separate components. In this SPL, there are various common-platform components reused in different products and also there are product-specific components.

Measurements of case study 3 include of changing productivity rates as reuse rates vary for different products developed sequentially using SPL of team 3. In addition, the performance measurements of a critical scenario in this domain will be measured and compared before and after the SPL is employed.

Reuse rates are calculated using reused non-comment line of codes and total non-comment line of codes. Productivity rates are calculated by dividing total non-comment line of codes by total effort to develop the so called product. Total efforts are measured by the business management software used in Aselsan. However, due to commercial confidentiality, we do not provide total efforts and total source line of code metrics here. For making these measurements, we worked with one of the configuration managers of the team.

In Table 9, reuse and productivity rates for the products developed using SPL of team 3 are given.

In order to compare the performance before and after employing the product line, one of the most critical scenarios in the system, the automatic video tracking scenario is investigated. The performance metrics used are time delay and CPU usage in this scenario. These metrics are measured using the provided embedded operating system functions.

**Table 9.** Reuse and productivity rates for products in SPL of team 3

| Product No | % Reuse Rate | Productivity (SLOC / man-hour) |
|---|---|---|
| Product 1 | 35,73 | 54,22 |
| Product 2 | 39,25 | 54,92 |
| Product 3 | 48,86 | 43,38 |
| Product 4 | 48,9 | 68,39 |

During the development process of the product line, the team introduced two more layers into the scenario. During the SPL design the team worked on two different approaches i.e. Pull and push strategies, while the first approach was more reusable with higher abstraction level.

Separately for three scenarios, the delay from reception of the track data from VT System to the transmission of the platform data to the Servo Controller System and the CPU usage during the scenario are measured. Measurements are shown below in Table 10.

**Table 10.** Measurements of the AVT scenario

| Scenarios / Measurements | Minimum Delay (ms) | Maximum Delay (ms) | Average Delay (ms) | % CPU Usage |
|---|---|---|---|---|
| Before SPL | 0,85 | 1,05 | 0,9 | 72,3 |
| Pull strategy | 2,12 | 35,0 | 20,0 | 81,5 |
| Push strategy | 2,12 | 5,5 | 3,2 | 79,9 |

**Discussion of the Measurements.** In Figure 8, the comparison of reuse and productivity rates of the products is displayed. Reuse rates increase from product 1 to product 4; however productivity change does not have the equivalent attitude. Productivity rates increase slightly between the first two products, and then productivity rate decreases from product 2 to product 3. However, between the last two products, productivity rate differs noticeably. When we discussed this situation with the team, we found out the following factors:

• There was a serious waste of time during the development of the non-reused (new) parts in product 3,
• Most of the developers of the product 3 were unfamiliar with software development by employing the product-line.

We can conclude that utilization of some normalizing factors i.e. Code complexity for the non-reused parts and experience of the developers, in measuring productivity rates can be useful. Furthermore, during the initial products, it is not surprising to observe productivity decays; since it is time consuming to get used to the product line in a software development team.

Previously, three different implementation methods of the AVT scenario are explained. The first method was before the team developed the SSRM SPL. In the second and third methods, there are two additional layers which are due to the product line employment and in order to increase the reuse of the scenario software.

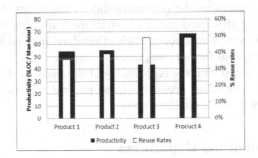

**Fig. 8.** Comparison of reuse and productivity for products in SPL of team 3

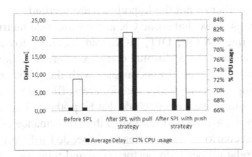

**Fig. 9.** Comparison of average delays and CPU usages of three different implementations of AVT scenario

The AVT scenario delay is measured for three cases (see Table 10), and comparison of the average delay and CPU usage are shown in Figure 9. According to these comparisons, we can conclude that while transforming the software into more reusable, and more abstract from the interfaces; we lose from the performance. Therefore, the developers should decide on the limit of this trade-off. Sometimes, the performance requirements allow these improvements; however sometimes performance requirements are too heavy. The second method was the most appropriate one in terms of reusability, however the developers had to perform and apply the third method. It was also more reusable compared to the first method, and it was acceptable when the performance requirements were considered.

# 4    Validity of the Case Studies

Four aspects of validity are summarized for empirical studies as "construct validity", "internal validity", "external validity", and "reliability" [10].

- Construct validity is about the compliance of the measurement and interpretation of the theoretical constructs. In our case studies, the metrics are either measured by the researcher or collected from the company databases.
- Internal validity is about whether the causal relations are studied. Since our case studies are not scientific experiments, we did not suffer for this validity.
- External validity focuses on the generalizability of the results. In our case studies, we aimed collecting all related metrics from all the teams; but for current states of the teams and projects, it was not possible. Therefore, our results and suggestions mainly are advantageous for our company.
- Reliability is concerned with the replicability of the study by different researchers. We claim that any researcher accessible to the internal metrics of a software company may come up with the similar results.

# 5    Verification of the Hypotheses

In this section, the results of the measurements in the case studies will be analyzed, and it will be stated whether or not the hypotheses are verified.

Case study 1 findings show that some CK metrics and size metrics do not correlate with changing reuse rate: SLOC, DIT, NOC, and LCOM. However, Coupling and Complexity CK metrics and the additional complexity metrics show a strong correlation with the changing reuse rate. In accordance with the relevant literature, the improvements in coupling and complexity metrics are sufficient to claim an increase in software quality. Therefore, we can conclude that Hypothesis 1 is verified.

In case study 1, we measured and compared the performance of a message receiving and transmitting scenario in three different embedded software modules. We find a strong negative correlation between performance and reuse rate; which is consistent with the related arguments in the literature.

In case study 3, we measured and compared the performance metrics of a critical scenario of an embedded software system before and after employing a product-line approach. We observed that, the case before the product line approach was the best regarding the performance, and as the software turned into more reusable and more abstract from other parts of the software, the performance of the software decayed.

Consequently, the measurements from two different case studies have verified Hypothesis 2.

In case study 2, measurements are taken from a product line which is used in subsequent products. When the components are not reused, we observe a large number of defects. Furthermore, we find that the decrease in the defect counts is independent of the product types. When the component is firstly used in product 3, again we detect a similar distribution as if the component is firstly used in product 2.

To conclude, as components are reused in several products, we observed that their defect counts decrease significantly and so their fault-proneness. Therefore, we can conclude that hypothesis 3 has been verified in this study.

In case study 2, productivity rates of the three products developed using the SPL approach are presented. Productivity rates are measured using the number of requirements using the requirements count and total effort. The results showed that, if the productivity is measured using the total number of requirements in the deployed product, the productivity rates improve significantly. Additionally, productivity is measured also by using the new requirements. In that case, we observed a reduction in productivity between the first and second products; and an increase in productivity between the second and third products. This situation is interpreted as an adaptation period of the product line approach.

In case study 3, we compared productivity rates of products implemented by another product line approach with increasing reuse rates. In this measurement, we also observed a positive correlation with reuse and productivity rates. However, the change of the members of the team during the development of product 3 caused a reduction in the productivity rate of this product. This situation is interpreted, similarly in case study 2, as an adaptation period of the product line.

Hence, we concluded that, if the effects of the adaptation period of the product line approach are ignored, the productivity rates improve significantly as the rate of reuse increases in a product line. Therefore, hypothesis 4 has been verified.

# 6      Suggestions for Further Benefit from This Study

In this section, suggestions regarding the reuse infrastructure and process to improve benefits of reuse are formulated.
- Use of Reference Metrics

Software developers should incorporate software quality metrics into their software development processes, and before and after serious decisions on design, technology or infrastructure; the change of these metrics should be investigated.

Therefore, in order to succeed in the employment of these metrics, the software developers should select reference metrics specific to their software domain and periodically monitor the changes of these metrics.
- Automated Detection of Architectural Effects

Real time embedded software developers should monitor the performance requirements after employing extensive architectural modifications; furthermore, they must update the modifications if the performance of the software eventually becomes unacceptable for the system.

Thereupon, the embedded software developers should develop methods in order to automate the process of detecting the architectural modifications which include the chance of worsening the software performance below system requirements.
- Recording Software Development Process Defects

In order to improve the management of defects, and investigate the defects intensely; the severity of the defects should also be provided after being corrected.

Additionally, the defects detected during the software development process should also be recorded; since the defects of the components during the development process is a key metric in order to improve the reuse infrastructure of a product line.

• Association of Defects with Design Concepts

In order to improve the management of defects, and investigate the defects intensely; the software developers should identify each defect with corresponding component, and the design concept.

• Recording Rework Efforts and Efforts Associated with Reuse

In order to be able to measure and analyze rework and reuse efforts, with changing reuse rates; these metrics should be recorded, and for this purpose the relevant infrastructure should be developed.

• Explicit Accounting for Code Reuse

During analysis of the productivity rates of the products, it was found that lack of the experience of the developer team was a significant factor of the declines in productivity; since, the employment of product lines requires an extra effort such as an adaptation period. Henceforth, during effort estimations of the products developed by a product line approach; the experiences of the developers, about the product line, should also be considered. Finally, the developers should also estimate and record efforts separately for reused and non-reused components, in order to analyze the impacts of reuse on the productivity rates deeply.

## 7     Conclusion

In this study, we worked with software engineering department of a defense industry company Aselsan. We examined their software projects and follow reuse and quality relations for these projects. For this purpose, we collected and compared some software measurements such as OO quality metrics, fault-proneness, performance, and productivity with changing reuse rates. Finally, we have formulated suggestions in order to improve the reuse infrastructure and process, after verifying reuse and quality relations in this setting. Throughout this study, we accomplished three different case studies and measured and compared different concepts in all three cases; however we were not able to obtain all these measures in all cases. Therefore, for future studies, it would be a significant improvement if all types of the measurements could be collected and compared with changing reuse rates, for all case studies separately.

## References

1. Frakes, W., Terry, C.: Software reuse: metrics and models. ACM Computing Surveys 28(2), 415–435 (1996)
2. Lim, W.C.: Effects of reuse on quality, productivity, and economics. IEEE Software 11(5), 23–30 (1994)
3. Jamali, S.M.: Object Oriented Metrics (A Survey Approach). Department of Computer Engineering Sharif University of Technology, Tehran, Iran (2006)

4. Sedigh-Ali, S., Ghafoor, A., Paul, R.A.: Metrics and models for cost and quality of component-based software. In: Proceedings of IEEE International Symposium on Object-Oriented Real-Time Distributed Computing, pp. 149–155 (2003)
5. Subramanyam, R., Krishnan, M.S.: Empirical analysis of CK metrics for object-oriented design complexity: implications for software defects. IEEE Transactions on Software Engineering 29(4), 297–310 (2003)
6. Mohagheghi, P., Conradi, R., Killi, O.M., Schwarz, H.: An Empirical Study of Software Reuse vs. Defect-Density and Stability. In: Proceedings of International Conference on Software Engineering, pp. 282–291 (2004)
7. Oliveira, M.F.S., Redin, R.M., Carro, L., da Cunha Lamb, L., Wagner, F.R.: Software Quality Metrics and their Impact on Embedded Software. In: 5th International Workshop on Model-based Methodologies for Pervasive and Embedded Software, MOMPES 2008, pp. 68–77 (2008)
8. El-Emam, K.: Object-oriented metrics: A review of theory and practice. In: Advances in Software Engineering, pp. 23–50. Springer-Verlag New York, Inc., New York (2002)
9. Chidamber, S.R., Kemerer, C.F.: A metrics suite for object oriented design. IEEE Transactions on Software Engineering 20(6), 476–493 (1994)
10. Runeson, P., Höst, M.: Guidelines for conducting and reporting case study research in software engineering. Empirical Software Engineering 14(2), 131–164 (2009)
11. Mohagheghi, P., Conradi, R.: Quality, productivity and economic benefits of software reuse: a review of industrial studies. Empirical Software Engineering 12(5), 471–516 (2007)
12. Dusink, L., van Katwijk, J.: Reuse Dimensions. In: SSR 1995 Proceedings of the 1995 Symposium on Software Reusability, pp. 137–149 (1995)
13. Nagappan, N., Ball, T., Zeller, A.: Mining metrics to predict component failures. In: ICSE 2006 Proceedings of the 28th International Conference on Software Engineering, Shanghai, China, pp. 452–461 (2006)
14. Boegh, J.: A New Standard for Quality Requirements. IEEE Software 25(2), 57–63 (2008)
15. ISO/IEC, Systems and software engineering – Systems and software Quality Requirements and Evaluation (SQuaRE) – Measurement of system and software product quality, ISO, ISO/IEC WD 25023 (2011)
16. Deniz, B.: Investigation of The Effects of Reuse on Software Quality in an Industrial Setting. M.S. thesis, Electrical and Electronics Engineering Dept., Middle East Technical University, Ankara, Turkey (2013)
17. Lincke, R., Lundberg, J., Löwe, W.: Comparing software metrics tools. In: ISSTA 2008 Proceedings of the 2008 International Symposium on Software Testing and Analysis, Seattle, Washington, USA, pp. 131–142 (2008)

# A Case Study of Refactoring Large-Scale Industrial Systems to Efficiently Improve Source Code Quality

Gábor Szőke, Csaba Nagy, Rudolf Ferenc, and Tibor Gyimóthy

Department of Software Engineering, University of Szeged, Hungary

**Abstract.** Refactoring source code has many benefits (e.g. improving maintainability, robustness and source code quality), but it takes time away from other implementation tasks, resulting in developers neglecting refactoring steps during the development process. But what happens when they know that the quality of their source code needs to be improved and they can get the extra time and money to refactor the code? What will they do? What will they consider the most important for improving source code quality? What sort of issues will they address first or last and how will they solve them? In our paper, we look for answers to these questions in a case study of refactoring large-scale industrial systems where developers participated in a project to improve the quality of their software systems. We collected empirical data of over a thousand refactoring patches for 5 systems with over 5 million lines of code in total, and we found that developers really optimized the refactoring process to significantly improve the quality of these systems.

**Keywords:** software engineering, refactoring, software quality.

## 1   Introduction

With short deadlines or lack of resources, developers tend to neglect refactoring steps during development and if they see a quick and easy way to get a test working and a ten-minute way to get it working with a simpler design, they will go for the quicker way, although the correct choice should be to spend ten minutes on refactoring. This usually results in the deterioration of the software. One way to combat this deterioration is to continuously re-engineer the code. Continuous reengineering is not only mentioned by popular development principles such as eXtreme programming [3], but the software engineering community realized that instead of spending money on maintenance tasks periodically it may be cheaper and more effective to continuously maintain the code and check its quality. For instance, Demeyer et al. say in [5] that "there is good evidence to support the notion that a culture of continuous reengineering is necessary to obtain healthy, maintainable software systems."

In our paper, we investigate how programmers re-engineer their code base if they have the time and extra money to improve the quality of their software systems. In a project we worked together with five companies where one of the goals was to improve the quality of some systems being developed by them. It was

B. Murgante et al. (Eds.): ICCSA 2014, Part V, LNCS 8583, pp. 524–540, 2014.

interesting to see how these companies optimized their efforts to achieve the best quality improvements at the end of the project. They are all profit-orientated companies, so they really tried to get the best ROI in terms of software quality. To achieve it, they had to make important decisions on what, where, when and how to re-engineer. We collected this information as experimental data and here we present our evaluation in the form of a case study. We found that developers really optimized the refactoring process to improve the quality of these systems; they usually went for the most critical but least risky types of refactorings. The results presented in this study could serve as a guideline for designing a re-engineering process.

The main contributions of this paper are:

- A case study on software refactorings with experimental data gathered from re-engineering large-scale proprietary software systems.
- Guidelines to re-engineer large-scale projects effectively.

The paper is organized as follows. In the next section we present related research work and then in Section 3 we introduce the motivational background of our case study. After, in Section 4 we present our results by answering our research questions. We discuss threats to validity and other results we got in Section 5 and finally we conclude the paper.

## 2   Related Work

Refactoring has been a hot topic since the appearance of Fowler's book [10] and Opdyke's PhD thesis [15]. There are many papers published in this area and it is not the aim here to systematically summarize these studies. In this section, we will give a general overview of software refactoring, and present some case studies which are similar to ours.

Mens et al. published a survey to provide an extensive overview of existing research in the area of software refactoring [12]. They identified six main refactoring activities. These are:

- Identifying precisely where the software should be refactored
- Determining which refactoring(s) should be applied to the places identified
- Ensuring that the refactoring applied preserves behaviour.
- Appling the refactoring
- Assessing the effect of the refactoring on quality characteristics of the software (e.g. complexity, understandability and maintainability) or the process (e.g. productivity and cost effort).
- Maintaining a consistency between the refactored program code and other software artifacts such as documentation, design documents, software requirements specifications and tests

Our study can be viewed as a piece of research work which attempts to support decisions on the first five activities.

Many papers have been published on where and how software code should be refactored – e.g. by applying automatic tools to identify bad smells [2,11], change smells [17] and by using static rule checkers such as CheckStyle[1], FindBugs[2] and PMD[3] for Java. Code clones may be regarded as a special type of bad smells and they are also typical targets of refactorings [4,20,19].

To determine which refactoring(s) should be applied, most of the studies investigate the effects of refactorings on metrics or quality attributes. Alshayeb et al. studied how refactoring improves external quality attributes such as adaptability and maintainability [1]. Stroggylos et al. analyzed source code version control system logs of popular open source software systems to detect changes marked as refactorings and examine how the software metrics are affected by this process [18]. DuBois et al. studied the impact of refactorings on cohesion and coupling metrics in [7] and found that benefits can occur, and described how and when the application of refactoring could improve selected quality characteristics [6]. Fontana et al. studied the impact of refactoring applied to reduce code smells on the quality evaluation of the system [9].

Murphy et al. studied four methods to collect empirical data on refactorings [14]: mining the commit log, analyzing code histories, observing programmers and logging refactoring tool use. In our study, we combine these methods. A similar study was conducted by Moser et al. [13] as they observed small teams working in similar, highly volatile domains and assessed the impact of refactoring in a close-to industrial environment [13]. Pinto et al. investigated what programmers said about refactoring on the popular Stack Overflow site [16].

# 3    Background

## 3.1    Project

The research work presented here formed part of an EU project. The main goal of the project was to develop a software refactoring framework, methodology and software products to support the 'continuous reengineering' methodology, hence provide support to identify critical code parts in a system and to restructure them to enhance maintainability. During the project, we developed an automatic/semi-automatic refactoring framework and tested this technology on the source code of industrial partners, having an in-vivo environment and live feedback on the tools. So partners not only participated in this project to develop the refactoring framework, but they also tested the tool set on the source code of their own product. This provided a good chance for them to refactor their own code and improve its quality.

In the initial step of the project we asked them to manually refactor their own code, and provide a detailed documentation of each refactoring, explaining what they did and why to improve the targeted code fragment. We gave them

---

[1] http://checkstyle.sourceforge.net/

[2] http://findbugs.sourceforge.net/

[3] http://pmd.sourceforge.net/

support by continuously monitoring their code base and automatically identifying problematic code parts using a static code analyzer based on the Columbus technology of the University of Szeged [8], namely the SourceMeter product of FrontEndART Ltd.[4] Companies had to fill in a survey with questions targeting the initial identification of steps; that is, evaluating the reports of SourceMeter looking for really problematic code fragments and explaining in the survey why that code part was actually a good target for refactoring. After identifying coding issues, they refactored each issue one-by-one and filled out another questionnaire for each refactoring, to summarize their experiences after improving the code fragment. There were around 40 developers involved in the project (5-10 on average from each company) who were asked to fill in the survey and carry out the refactorings.

## 3.2   Survey Questions

The survey consisted of two parts for each issue. The developers had to fill in the first part before they began refactoring the code, and the second part after the refactoring. In the first part, they asked the following questions:

- Which PMD rule violations helped you identify the issue?
- Which Bad Smells helped you to find the issue?
- Estimate how much it would take to refactor the problem.
- How big is the risk in carrying out the refactoring? (1-5)
- How do you think the refactoring will improve the quality of the whole system's code? (1-5)
- How do you think the refactoring will improve the quality of the current local code segment? (1-5)
- How much improvement do you think the refactoring will make to the current code segment? (1-5)
- How many files will the refactoring have an impact on?
- How many classes will the refactoring have an impact on?
- How many methods will the refactoring have an impact on?

We asked some questions after developers had finished the refactoring task. These were the following:

- Which PMD rule violations did the refactoring fix?
- Which Bad Smells did the refactoring fix?
- How much time did the refactoring task take?
- Did any automated solution help you to fix the problem?
- How much of the fix for this problem could be automated? (1-5)

For most of the questions, we provided some basic options. For the first question for example we provided a list of PMD rule violations with their names, to help the developers answer the questions quickly. In the questions on the classes and methods impacted, we provided different ranges, namely 1-5, 5-10, 10-25, 25-50, 50-100, 100+. Each question had a text field where the developers could explain their answers and they could also suggest possible improvements and add comments.

---

[4] http://frontendart.com

## 3.3  Systems under Investigation

In the study, we had chance to work together with five experienced companies in the ICT sector. These companies were founded in the last two decades and some of their projects were initiated before the millennium. The 5 given projects consisted of about 5 million lines of code altogether, written mostly in Java. The projects covered different ICT areas like ERPs, ICMS and online PDF Generation. More details can be found in Table 1.

**Table 1.** Systems that we examined

| Id | LOC | Domain |
|---|---|---|
| Company A | 200k | Specific Business Solutions |
| Company B | 4,300k | Enterprise Resource Planning (ERP) |
| Company C | 170k | Integrated Business Management |
| Company D | 128k | Integrated Collection Management Systems (ICMS) |
| Company E | 100k | Web-based PDF Generation |

Each project had Web/online modules and some of them could run as standalone applications too. Companies A and B commenced their projects with the first releases of Java Enterprise Edition. At that time there were no application frameworks (like SpringFramework) available, so they implemented their own versions. Therefore the core of their systems can be regarded as legacy Java systems, but still under active development.

# 4  Case Study

## 4.1  RQ1: What Kinds of Issues Did the Companies Find Most Reasonable to Refactor?

Our first research question focused on which issue types the companies considered the most important to refactor. We asked the companies which indicators helped them best in finding problematic code fragments in their systems. In our survey, companies could select Bad Code Smells and Rule Violations as indicators on how they found the issues.

In our evaluation, we distinguish a special kind of bad smell which suggests code clones in the system. In Figure 1, a distribution can be seen for the issues which helped the companies to identify the problematic code fragments in their code. The intersections in the figure came from the fact that developers could select more than one indicator per issue. The reason why bad smells and clones had no elements in their intersection was because a clone is a special kind of bad smell, as mentioned earlier. The same applies for the intersection of the former group and the rules group (an empty set cannot intersect anything).

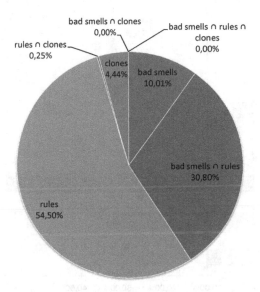

**Fig. 1.** Distribution of issue indicators

When we look at the results in Figure 1, we see that the companies found the majority of issues lay in the sets of rule violations and bad smells. It can also be seen that rule violations alone cover 85% of all the issues found. This also includes 75% of all the bad smells (because of the intersection). So the assumption is that rule violations are the best candidates for highlighting issues. However to confirm this, we also had to look at how many issues the companies fixed in order to choose the best indicator of refactorings.

Figure 2 shows the percentage of each fixed issue found from our survey. When we examine the ratio of fixed issues, we see that the bad smells are mostly refactored issues. However if we include the total number of issues, it is clear that rule violations gave the most advance.

Based on the fact that 85% of all issues were rule violations and developers mostly fixed these issues instead of the others, in future RQs we will focus on rule violations.

## 4.2  RQ2: What Are Those Attributes of Refactorings That Can Help in Selecting Them?

The rule violations in the survey were provided by the *PMD source code analyzer* tool. In our study, we categorized and aggregated these rules into groups. The groups we used were the Rulesets taken from the PMD website. The companies filled in the survey for 961 PMD refactorings altogether. These 961 refactorings produced 71 different rule violation types over 19 rulesets.

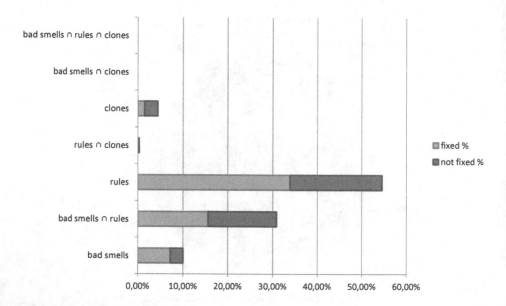

**Fig. 2.** Percentages of fixed issues for different problem types

Below, we will examine these rulesets based on different attributes. Based on our survey questions, we created the following attributes:

- **number of refactorings** indicates how many issues were fixed for a certain kind of PMD or ruleset.
- **average and total time required** tells us the total and average time that companies spent on a refactoring. (Values are in work hours.)
- **estimated time** shows how companies estimated the time that a refactoring operation would take. (Values were enumerated between 1 hour and 3-4 days.)
- **local improvement** indicates the subjective opinion of developers on how much the local code segment was improved by the refactoring (Values are between 1-5.)
- **global improvement** indicates the subjective opinion of developers on how much the code improved globally. (Values are between 1-5.)
- **risk** indicates the subjective opinion of developers on how risky the refactoring is. (Values are between 1-5.)
- **impact** is an aggregated number that tells us how many files/classes/methods a refactoring affected. (Values are enumerated between 1-100.)
- **priority** tells us how dangerous a rule violation is, and how important it is to fix it. The priority attribute did not come from the survey; we used the prioritisation of the underlying toolchain. (Values lie between 1-3.)

### 4.3    RQ3: Which Refactoring Operations were the Most Desirable Based on to the Attributes Defined Above?

The attributes above tell us how risky a refactoring operation is and how much time it will usually take to fix. By combining these attributes, we can discover which rules or rulesets are the most beneficial or riskiest; or by aggregating the first two attributes with time required, we can see which rules will best return the effort we invested in refactoring. Next, we investigate the number of refactorings, time required, improvement and risk.

**Number of Refactorings.** Now let us look into the most obvious attribute, namely the number of refactorings the companies performed. The results in Figure 3 indicate that the companies dealt with almost every kind of rule violation. The majority of refactored rule violations were found in the *Design* ruleset. This ruleset contains rules that flag suboptimal code implementations, so fixing these code fragments should significantly improve the software quality and perhaps even the performance. The *Design* ruleset is followed by the *Strict Exceptions*, *Unused Code* and *Braces* categories, which focuses on throwing and catching exceptions correctly, removing unused or ineffective code, and also the use and placement of braces. Some rule violations in the following categories were also fixed in large numbers under the *Basic, Migration, Optimization, String and StringBuffer* rulesets. The other rulesets scarcely came up (like *Empty Code*) or not at all (like *Android*). This is probably due to the fact that the projects did not contain these kinds of violations or contained only false positives.

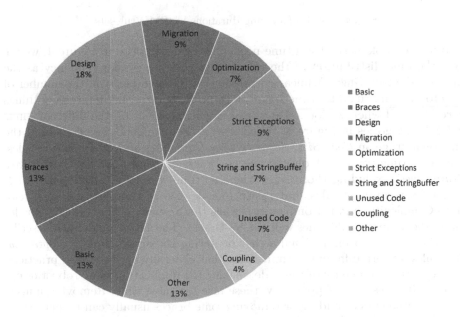

**Fig. 3.** Distribution of refactorings by PMD rulesets

**Average and Total Time Required.** After investigating how many refactorings the companies made, we will now examine how much time a refactoring operation took. (Here, we consider the time the developers spent on refactoring their source code, excluding the time they spent on testing and verifying the code.)

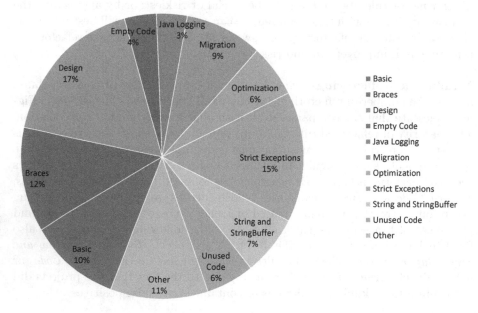

**Fig. 4.** Total refactoring durations by PMD rulesets

When we look at the total time needed for the categories in Figure 4, we see that the time distribution of the refactorings shows a similar tendency as the number of refactorings. A linear correlation can be seen between the number of refactorings and the total time spent on them. However, other interesting things were observed when we looked at the average time spent on the different kinds of PMD categories in Figure 5. It seems as if the companies spent most of the time on average on *Code Size*, *Security Code Guidelines* and *Optimization* rules. The least time was spent on average on Braces, *Import Statements* and *Java Beans* rules (excluding those rules where no time was spent at all). The *Code Size* ruleset contains rules that relate to code size or complexity (e.g. CyclomaticComplexity, NPathComplexity), while the *Security Code Guidelines* rules check the security guidelines defined by Oracle. The latter guidelines describe violations like exposing internal arrays or storing the arrays directly. *Optimization* rules concern different optimizations that generally apply to best practices. Reducing the complexity of the code, making the application more robust or optimizing it takes time. Apparently, these take the most time. Removing unused import statements or adding or removing some braces usually can be performed quickly, but to find which independent statements to extract so as to reduce the complexity is a hard task.

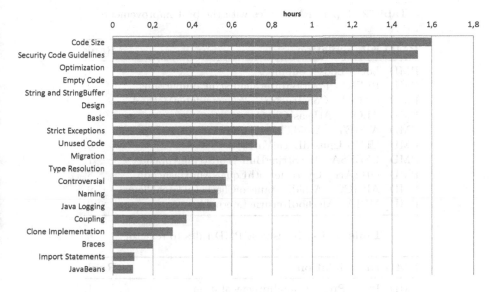

**Fig. 5.** Average refactoring durations by PMD Rulesets

**Global and Local Improvement.** To learn which PMD rule violations fit the attributes best, we summarized and averaged both the global and local improvement values got from the survey. We ranked both sets of values by their position in their data set. The average of the two former values gave us a list of the best improving PMD rulesets. From our results, the best improvements locally and globally are given by the *Strict Exceptions, Coupling and Basic* PMD rulesets. However, rulesets contain a lot of different rules, and hence the categories alone did not give us the proper information we sought. To get further information, a per-rule statistic was required.

For the per-rule statistics, we filtered the results with those cases where the companies did fewer than 4 refactorings of a single kind of PMD rule. This ensured that only relevant data was included in the statistics, and a single-refactored PMD rule could not harm the average values.

Table 2 shows a top list of the best improving PMD rule violations. The top list was made by taking the average of the local improvements and summing the average of global improvements, in descending order.

**Risk.** Table 3 shows the riskiest PMD rules used to refactor based on the replies by company experts. We can observe that in most cases the riskiest refactorings are for basic Java functionalities. The list includes rules concerning *java.lang.Object*'s *clone, hashCode* and *equals* method implementation, proper *catch* blocks and *throws* definitions, array copying and unused variables. All of the previous refactorings increased the quality of the software (by definition), but fixing these rule violations can have some unexpected consequences. These unexpected consequences are also caused by a previous improper implementation. Of course, if the software code had been written properly in the first place,

**Table 2.** Top 10 PMD rules with the best improvements

| PMD rule violation | Rank |
|---|---|
| PMD_LoC - LooseCoupling | 1. |
| PMD_PLFIC - PositionLiteralsFirstInComparisons | 2. |
| PMD_CCOM - ConstructorCallsOverridableMethod | 3. |
| PMD_ALOC - AtLeastOneConstructor | 4. |
| PMD_ATRET - AvoidThrowingRawExceptionTypes | 5. |
| PMD_ULV - UnusedLocalVariable | 6. |
| PMD_USBFSA - UseStringBufferForStringAppends | 7. |
| PMD_OBEAH - OverrideBothEqualsAndHashcode | 8. |
| PMD_AICICC - AvoidInstanceofChecksInCatchClause | 9. |
| PMD_MRIA - MethodReturnsInternalArray | 10. |

**Table 3.** Top 10 riskiest PMD rules to refactor

| PMD rule violation | Rank |
|---|---|
| PMD_PCI - ProperCloneImplementation | 1. |
| PMD_ALOC - AtLeastOneConstructor | 2. |
| PMD_SDTE - SignatureDeclareThrowsException | 3. |
| PMD_ACNPE - AvoidCatchingNPE | 4. |
| PMD_LoC - LooseCoupling | 5. |
| PMD_OBEAH - OverrideBothEqualsAndHashcode | 6. |
| PMD_AICICC - AvoidInstanceofChecksInCatchClause | 7. |
| PMD_ULV - UnusedLocalVariable | 8. |
| PMD_AISD - ArrayIsStoredDirectly | 9. |
| PMD_ATNPE - AvoidThrowingNullPointerException | 10. |

these unexpected results would have been appeared earlier, and could have been fixed during the development phase.

### 4.4    RQ4: Which Refactoring Operations Give the Best ROI?

In the above we saw the most beneficial and riskiest PMD rules, but which rule violations should we fix to improve the code the most with the least risk and as speedily as possible? To discover this, we defined an index value which indicates the 'return of investment' or ROI for short. To calculate this index, we ranked the averages of each attributes according to their percentage values with all averages in the same attribute using the *percentrank*[5] function.

$$ROI_{refactoring} = \text{percentrank}(\text{average}(improvement_{local}))$$
$$+ \text{percentrank}(\text{average}(improvement_{global}))$$
$$- \text{percentrank}(\text{average}(risk)) - \text{percentrank}(\text{average}(time))$$

---

[5] Here, **percentrank** returns the rank of a value in a data set as a percentage of the data set.

**Table 4.** Top 15 PMD rules to refactor with the best ROI values

| PMD rule violation | Rank |
|---|---|
| PMD_LoC - LooseCoupling | 1. |
| PMD_ATRET - AvoidThrowingRawExceptionTypes | 2. |
| PMD_USBFSA - UseStringBufferForStringAppends | 3. |
| PMD_OBEAH - OverrideBothEqualsAndHashcode | 4. |
| PMD_PLFIC - PositionLiteralsFirstInComparisons | 5. |
| PMD_MRIA - MethodReturnsInternalArray | 6. |
| PMD_LVCBF - LocalVariableCouldBeFinal | 7. |
| PMD_ALOC - AtLeastOneConstructor | 8. |
| PMD_SDTE - SignatureDeclareThrowsException | 9. |
| PMD_CCOM - ConstructorCallsOverridableMethod | 10. |
| PMD_PST - PreserveStackTrace | 11. |
| PMD_UPF - UnusedPrivateField | 12. |
| PMD_ULV - UnusedLocalVariable | 13. |
| PMD_AICICC - AvoidInstanceofChecksInCatchClause | 14. |
| PMD_CC - CyclomaticComplexity | 15. |

After we got the index number, we ordered the rule violations and took the first 15, which are listed in Table 4. Based on our findings the best ROI is indicated by mostly small, local refactorings of those possible errors that can cause big inconsistencies in future development or other parts of the software.

ROI statistics can tell developers which rule violations need to be fixed in order to get the most out of their refactoring efforts. They can improve the effectiveness of the software maintenance process, and can fix more issues; thus they can help to make the system more robust and also reduce the overall maintenance costs.

### 4.5   RQ5: How Can We Schedule Refactoring Operations Efficiently?

Now we will describe a way of scheduling refactoring operations. First, we will examine how the industrial partners scheduled their refactorings and then we will make recommendations based on these observations.

**How Did Companies Schedule Their Refactorings?** We asked the companies how they scheduled their refactoring operations when fixing rule violations. Each of the companies used the priority attribute that was given for each kind of rule violation, by using the toolchain that was used to extract the rule violations. Priorities were 1, 2, 3, indicating different levels of threat for each rule violation.

- **Priority 1** indicates dangerous programming flows.
- **Priority 2** indicates not so dangerous, but still risky or unoptimized code segments.
- **Priority 3** indicates violations to common programming and naming conventions.

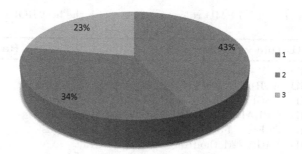

**Fig. 6.** Fix rate according to Priority

In Figure 6, we can see the percentage values of all the issues that were fixed for each priority level. They reveal that companies fixed Priority 1 issues the most and Priority 2 issues the second most. This means that companies here opted to fix the most threatening rule violations detected in the code.

Given these attributes, the most efficient way is to start refactoring those issues that had Priority 1 level rule violations. To find out how the companies actually scheduled their refactorings, we split the refactorings into two sets. The first set contains refactorings which were made in the first half of the project, and the other set contains refactorings made in the second half. The results of these experiments are represented in Figure 7. They tell us in percentage terms how much was fixed for each priority level in the first half and second half of the project. They indicate that the companies fixed most Priority 1 rule violations in the first half of the project and fixed most Priority 2 rules in the second half. This is consistent with what the companies told us and they provided good feedback on how they scheduled their refactoring process.

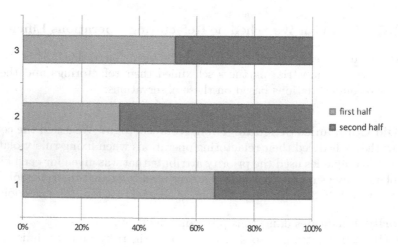

**Fig. 7.** Fixes in the first and second half according to Priority

**How Should the Refactoring Be Scheduled?** Learning from the experiences that the participating companies gained in the project, and from the results presented in Section 4.4, we suggest two kinds of scheduling. They are:

- Schedule refactorings by the priority-level of the issue, starting with the most threatening ones.
- Schedule refactorings by fixing issues with the best ROI score (see Table 4).

Choosing either of the above approaches should give an effective refactoring procedure. Scheduling by priority-level concentrates on fixing the most threatening issues, while concentrating on the ROI score should bring about the best improvement with the least effort and time. Moreover, combining the these former methods can also lead to a very efficient refactoring procedure, which offers the best of both approaches.

## 5 Discussion

Next, we will elaborate on potential threats to validity and some other interesting results that we obtained from our survey.

**Threats to Validity**

We identified some threats that can affect the construct, internal and external validity of our results.

The first one we encountered was the subjectivity of the survey. The answers to our survey questions were given by developers on a self-assessment basis. We did not measure the time needed or enhancement of refactorings with any automated solution; instead we let the developers answer the survey freely. Nevertheless, we carried out the survey with five industrial partners and therefore with many experts, which surely makes the results statistically relevant.

Another threat that we anticipated was that developers got 'unlimited' extra money and time to do the refactorings, so we could monitor how they refactored their system without any budget pressure. Although they got extra time and money in part of the project, there were still limits that might affect the results and the refactoring process.

Turning to external validity, the generalizability of our results depends on whether the selected programming language and rule violations are representative for general applications. The Java programming language was selected in the assessment together with the companies. These refactorings were made mostly on issues identified by PMD rule violations, hence they were Java specific. However, most of these rules could be generalized to abstract Object-Orientated rules, or they can be specifically defined for other programming languages.

Another threat is that whether fixing PMD rule violations can be viewed as refactoring or not. PMD refactorings are not like traditional refactoring operations that most studies examine (e.g. pull up, push down, move method, rename method, replace conditional with polymorphism). Despite this, Fowler [10] defined refactoring as "the process of changing a software system in such a way that it does not alter the external behavior of the code yet improves its internal structure." During the project we encountered several PMD rule violations and our general experience is that the refactoring of these violations does not alter external behavior, so they can by definition be treated as refactoring.

Overall, our methods were evaluated on large-scale industrial projects, with contributions from expert developers, on a big set of data, which is a rather unique case study in the refactoring research area.

### Other Results

In our case study (see Section 4) we summarized our results based on research questions addressed to experts working in five ICT companies. However we ran into several interesting cases which were worth mentioning, but could not be incorporated into our research questions.

One of the interesting cases we found was when we searched for the longest-lasting refactorings. We found that *Company A* carried out a *SignatureDeclareThrowsException* refactoring, which lasted 16 hours. The issue occurred in a method of a widely implemented interface, and the problem was that the method threw a simple *java.lang.Exception* Exception-type. This is not recommended because it hides information and it is harder to handle exceptions. The developer assigned to the issue estimated that the work took 1-2 days, and said that the risk was high because it impacted 10-25 files, but it was worth refactoring because the extra information they gained after the refactoring helped improve the maintainability of the source code.

Another intriguing example was with the same search as before. We found that *Company D* performed several *AvoidDuplicateLiterals* refactorings, which took them 7 hours on average to do; and each of the refactorings impacted on more than 100 classes. According to the comments in the survey, they used NetBeans IDE[6] to fix these kinds of issues. NetBeans IDE has a integrated refactoring suite that helps developers to refactor their source code. Here, they used this suite to extract duplicated literals to constant variables. The survey comments revealed that the refactoring suite really helped them in this refactoring task, and it would be great help if automated solutions could be devised and implemented to tackle other of issues as well.

## 6    Conclusions and Future Work

In our study, we evaluated five research questions on refactoring in Java programs. The main goal of our experiments was to learn how developers refactor in an industrial context when they have the required resources (both time and money) to do so.

---

[6] https://netbeans.org/

Our experiments were carried out on 5 large-scale industrial Java projects of different sizes and complexity. We studied refactorings on these systems, and learned which kinds of issues developers fixed the most, and which of these refactorings were best according to certain attributes defined in Section 4.2.

We also found that developers tried to optimize their refactoring process to improve the quality of these systems. We recommended two methods to schedule refactorings; one was based on priority and the other was based on a return in investments. Forming a refactoring process from either of these or simply combining them should lead to a very efficient refactoring process, making the system more robust, more maintainable, and most of all with lower costs.

In our experiments we gathered really big data on manual refactorings in an in-vivo industrial context. In this case study, we limited the context to the numerical evaluation of these results and investigated how to best select code fragments to effectively refactor our code base so as to improve software quality. In the future, with the data we obtained, we would like to investigate the effects of refactorings on source code quality and implement automatic techniques based on these results. We would also like to investigate the usage of these automatic algorithms as well.

**Acknowledgements.** This research was supported by the Hungarian national grant GOP-1.2.1-11-2011-0002. Here, we would like to thank all the participants of this project for their help and cooperation.

# References

1. Alshayeb, M.: Empirical investigation of refactoring effect on software quality. Inf. Softw. Technol. 51(9), 1319–1326 (2009)
2. Arcelli Fontana, F., Braione, P., Zanoni, M.: Automatic detection of bad smells in code: An experimental assessment. Journal of Object Technology 11(2), 1–38 (2012)
3. Beck, K.: Extreme Programming Explained: Embrace Change, 1st edn. Addison-Wesley Professional (1999)
4. Choi, E., Yoshida, N., Ishio, T., Inoue, K., Sano, T.: Extracting code clones for refactoring using combinations of clone metrics. In: Proceedings of the 5th International Workshop on Software Clones, IWSC 2011, pp. 7–13. ACM (2011)
5. Demeyer, S., Ducasse, S., Nierstrasz, O.: Object-Oriented Reengineering Patterns. Morgan Kaufmann (2002)
6. Du Bois, B.: A Study of Quality Improvements by Refactoring. Ph.D. thesis (2006)
7. Du Bois, B., Gorp, P.V., Amsel, A., Eetvelde, N.V., Stenten, H., Demeyer, S.: A discussion of refactoring in research and practice. Tech. rep. (2004)
8. Ferenc, R., Beszédes, Á., Tarkiainen, M., Gyimóthy, T.: Columbus – Reverse Engineering Tool and Schema for C++. In: Proceedings of the 18th International Conference on Software Maintenance (ICSM 2002), pp. 172–181. IEEE Computer Society (October 2002)
9. Fontana, F.A., Spinelli, S.: Impact of refactoring on quality code evaluation. In: Proceedings of the 4th Workshop on Refactoring Tools, WRT 2011, pp. 37–40. ACM (2011)

10. Fowler, M.: Refactoring: Improving the Design of Existing Code. Addison-Wesley Longman Publishing Co., Inc. (1999)
11. Maiga, A.: Impacts and Detection of Design Smells. Ph.D. thesis (2012)
12. Mens, T., Tourwé, T.: A survey of software refactoring. IEEE Trans. Softw. Eng. 30(2), 126–139 (2004)
13. Moser, R., Abrahamsson, P., Pedrycz, W., Sillitti, A., Succi, G.: A Case Study on the Impact of Refactoring on Quality and Productivity in an Agile Team. In: Meyer, B., Nawrocki, J.R., Walter, B. (eds.) CEE-SET 2007. LNCS, vol. 5082, pp. 252–266. Springer, Heidelberg (2008)
14. Murphy-Hill, E., Black, A.P., Dig, D., Parnin, C.: Gathering refactoring data: A comparison of four methods. In: Proceedings of the 2nd Workshop on Refactoring Tools, WRT 2008, pp. 7:1–7:5. ACM (2008)
15. Opdyke, W.F.: Refactoring Object-oriented Frameworks. Ph.D. thesis (1992)
16. Pinto, G.H., Kamei, F.: What programmers say about refactoring tools?: An empirical investigation of Stack Overflow. In: Proceedings of the 2013 ACM Workshop on Workshop on Refactoring Tools, WRT 2013, pp. 33–36. ACM (2013)
17. Ratzinger, J., Fischer, M., Gall, H.: Improving evolvability through refactoring. SIGSOFT Softw. Eng. Notes 30(4), 1–5 (2005)
18. Stroggylos, K., Spinellis, D.: Refactoring–does it improve software quality? In: Proceedings of the 5th International Workshop on Software Quality, WoSQ 2007, p. 10. IEEE Computer Society (2007)
19. Tairas, R.: Clone detection and refactoring. In: Companion to the 21st ACM SIGPLAN Symposium on Object-oriented Programming Systems, Languages, and Applications, OOPSLA 2006, pp. 780–781. ACM (2006)
20. Zibran, M.F., Roy, C.K.: Towards flexible code clone detection, management, and refactoring in IDE. In: Proceedings of the 5th International Workshop on Software Clones, IWSC 2011, pp. 75–76. ACM (2011)

# Modeling of Embedded System Using SysML and Its Parallel Verification Using DiVinE Tool

Muhammad Abdul Basit Ur Rahim[1], Fahim Arif[1], and Jamil Ahmad[2]

[1] Military College of Signals,
National University of Science and Technology,
Islamabad, Pakistan
{basit.phd9,fahim}@mcs.edu.pk
[2] Research Center for Modeling and Simulation,
National University of Science and Technology
Islamabad, Pakistan
jamil.ahmad@rcms.nust.edu.pk

**Abstract.** SysML is a modeling language that can be used for the modeling of embedded systems. It is rich enough to model critical and complex embedded systems. The available modeling tools have made the modeling of such large and complex systems much easier. They provide sufficient support for the specification of functional requirements in the elicitation phase as well as in the design phase by graphical modeling. These systems must be properly validated and verified before their manufacturing and deployment in order to increase their reliability and reduce their maintenance cost. In this paper, we have proposed a methodology for the modeling and verification of embedded systems in parallel and distributed environments. We demonstrate the suitability of the framework by applying it on the case study of embedded security system. The parallel model checking tool DiVinE has been used because the available sequential verification tools either fail or show poor performance. DiVinE supports Linear Temporal Logic (LTL) for defining nonfunctional requirements and DVE language for specifying models. First,the case study is modeled using SysML's state machine diagrams and then semantics are described to translate these state machine diagrams to DVE based model. The translated model is verified against specified LTL properties using DiVinE.

**keywords:** Model based validation, verification, security systems,SysML, DiVinE, Parallel verification.

# 1 Introduction

SysML reuses and extends the subset of UML 2.1 [1] and it is specifically designed for system engineering [2]. Both UML and SysML are currently being managed by Object Modeling Group (OMG). Few diagrams are common in both modeling languages like use case, activity, sequence and state diagrams. For modeling industrial applications, SysML provides the graphical notations for ports, parts, connectors, and

B. Murgante et al. (Eds.): ICCSA 2014, Part V, LNCS 8583, pp. 541–555, 2014.

clocks. SysML also has its own diagrams like requirement diagram, block definition diagram, internal block diagram and parametric diagram. Furthermore, it provides support for requirements specification and their traceability. Constraints can be specified for elicited requirements using parametric diagram. State of the art tools are available for SysML modeling language [3].

In SysML, functional requirements can be specified using requirement diagram. Before the availability of robust model checking tools, these requirements were hard to verify in earlier phases of the software development life cycle. Now, state of the art sequential and parallel robust model checking tools are available that can be helpful in validating and verifying the specified requirement.

The modeling of security system is really challenging due to its complex nature. A flaw in its model can cause a loss of money or even loss of precious lives. Thus, a model must be verified in the earlier phase of system development cycle in order to avoid such loses. SysML is a semi-formal modeling language and many authors have also proposed methodologies for the formal specification of requirements [20][21][22]. With the help of these methodologies, SysML has become more suitable for formal specification of user requirements. So, the functionality of an embedded system can be graphically modeled in SysML and its requirements can be specified in a formal language. Furthermore, it is desirable that the functionality of the system verifies these requirements.

Formal methods play an active role in developing a secure, reliable and deadlock free embedded system. The performance and accuracy of desired application can be evaluated before development. The model checking tools perform simulation as well as exhaustive verification of a model to evaluate the performance of application.

The quality attribute must be satisfied at design time especially in security systems. These systems must be safe, live and deadlock-free. All of these attributes can only be satisfied with formal specifications. This can be helpful in finding the unknown and undesired behaviors of the system in early design phase. If the system is formally defined then it can be verified against specific requirements. These requirements are called properties and can be formally specified in temporal logic. The functionality of the model is verified against these specified properties. These properties are also very helpful to verify the functional and non-functional requirements of the system. There are different types of temporal logic like Computational Tree Logic (CTL), Linear Temporal Logic (LTL) [4] to define these properties. In this paper, we use LTL to formally specify the requirements of security system. In LTL temporal logic, the operators $G$, $F$ and $W$ are used to define the properties. The operator $G$ states that the property will hold for all paths. The operator $F$ states that the property will hold for some parts of all paths.

The basic concept of verification and validation is to ensure that the right product is being developed in a right way. For this purpose different types of model checking tools are being used like sequential or parallel. PAT, SMV, UPPAAL are examples of sequential model checking tools, while DiVinE and Spin are examples of parallel model checking tools [5] [6].

The state machine diagram is used to describe the functionality of individual devices and in DiVinE, these state machine diagrams are considered as automata or process. To program these automata, DiVinE uses DVE language. In this paper, we have described the formal DVE semantics to translation the state machine diagram to DVE based model.

DiVinE is a parallel verification tool that supports LTL property specification and is more suitable tool for verification of large and complex systems. Its support of multi-core and many-core processors (CPUs and GPUs) for parallel verification makes it a better choice for computationally extensive model checking than sequential model checking. DiVinE is a discrete model checker therefore it does not support real-time modeling. Time can be handled using some third party tools like YAWL2DVE-t based on the concept of ticks [8].

The motivation of this paper is to graphically model a large scale embedded system and verify it against the formally specified requirements. The use of graphical language (SysML) is a major advantage to model a system with all aspect. Moreover, a parallel verification strategy is adopted in order to show its usefulness for large scale security system which may be helpful in exposing the earlier design errors which can lead to losses in terms of cost and human lives.

We adopt a parallel verification methodology for the embedded systems at a high level abstraction where SysML is considered more appropriate for the modeling of their specifications. This methodology is proved to be very useful both in terms of verification as well as in reducing the computation time. Our results, suggest the use of DiVinE for parallel verification for large scale applications. By increasing number of cores it verifies the model in very less time. Its simulation and verification modules not only informs about deadlocks and other errors but also provides counter example to make it error free. In this paper, we have applied this methodology on a case study of embedded security system.

The case study is first modeled with SysML and their non-functional properties are defined and verified using DiVinE tool. The remainder of paper contains following: section 2 shows related work, section 3 presents the proposed methodology, section 4 is about efficient use of resources, section 5 demonstrates the basic elements of DVE language, section 6 introduces case study and its verification, section 7 outlines the non-functional properties for security system, section 8 presents the analysis of results and section 9 is the short conclusion round out the paper.

## 2    Related Work

In [8], authors present a tool YAWL2DVE-t that translates graphical model of YAWL into DiVinE's own DVE language. In YAWL2DVE-t, a concept of tick is used for time constraints. A unified approach for V&V is presented that establishes synergy between formal verification, program analysis and software engineering techniques by combining data analysis and quality measuring [9]. In [10], SysML's activity diagram is extended by annotating with time constraints and SysML's activity diagram is mapped to discrete time Markov chain. Furthermore, PRISM model checking tool is used for evaluation of results. In [11], a methodology is proposed by using SystemC for temporal specifications and SysML for modeling SoC to determine and evaluate non-functional requirements. In [12], mathematical structure is used to specify the design and then the specifications are used in subset of SysML and MARTE diagrams. In [13], an extension of SysML is proposed in description of continuous data flow between blocks, time assignment to event-driven behavior and coupling of continuous time and event-driven simulation. The extension of SysML deals with continues time behavior and synchronize it with Simulink. In [14], formal methods are combined with SysML to verify safety properties of a case study. In [15], a methodology is presented for modeling and verification of a factory system. Author has used SysML and Petri Nets for modeling and specification and used TINA model checker for verification of properties. In [16], a case study is modeled using SysML and HyTech model checker is used for the analysis. In [17], a profile of SysML is proposed that is named as TEmporal Property Expression (TEPE) and verification is performed using UPPAAL model checker.

All the above SysML diagrams or their profiles are verified using sequential model checking tools. These model checking tools take much time for verification of large scale models. The parallel model checking tools verify model in distributed environment that could accelerate verification process. SPIN model checker is a parallel model checker that can be useful for the verification of SysML [19]. SPIN is a discrete model checker that cannot be useful for the verification of real-time systems. DiVinE is also a discrete model checker and it can be used with UPPAAL for the verification of real-time systems [19]. In [19], we have modeled a case study using state machine diagram of SysML and verified using UPPAAL model checking tool. The UPPAAL model checker is considered as a benchmark among model checkers and it's very useful for the verification of real-time properties. UPPAAL model can be further used for the verification of untimed properties in parallel distributed environment using DiVinE model checker.

## 3     Proposed Methodology

In this paper, we have proposed a methodology for verification of embedded system. The system is modeled to meet user requirements of functional and non-functional types. The functionality is specified using state machine diagram (SMD) of SysML. The DVE semantics have been described to translate SMD of SysML to DVE model where SMDs are considered as automata which are verified against some properties formally expressed in DiVinE supported LTL form.

The DiVinE provides a user friendly GUI, verifier and a simulator as well. It also generates a counter example in case a property fails. For example if the property deadlock-freeness fails then it shows the counter example which is a trajectory from the initial state to the deadlock state. With parallel model checking the verification process is accelerated since very large scale model contains a huge set of states. Hence, the parallel strategy is much more suitable for large scale system modeled with SysML. As the DiVinE does not provide execution time of a model therefore a utility named Execution Time Calculator (ETC) is developed to calculate the execution time of a model on multiple processors. In ETC, number of processors can be specified to verify the model and every time it returns the execution time for specified processors. Fig. 1 shows that how an embedded system can be modeled using SysML and transformed to DVE language.

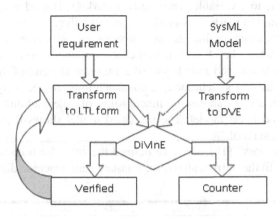

**Fig. 1.** Proposed Methodology

# 4     Efficient Use of Resources

With the use of DiVinE model checker, the resources can be used efficiently. Now, the resources can be utilized as per requirement. While in other model checker it is not possible to optimize the use of resources. As the DiVinE provides support for GPU based systems, so, the large scale applications can be verified in very less time by optimizing the use of resources. Using DiVinE model checker the number of processors or memory can be increased or decreased as per requirement case study. DiVinE provides algorithm for verification in concurrent distributed environment.

SysML is a graphical modeling language which is used in the earlier phase of system life cycle. To develop an error free application, it must be verified in analysis phase. With good understanding of DVE most of the SysML's diagrams can be translated into DVE language [7][18][19]. In this way, the graphical model of a system is translated into DVE and which further verified using parallel verification tool that ultimately saves verification cost.

# 5     DVE Semantics for State Machine Diagram

In this section, we describe the DVE semantics for translation of SysML's state machine diagram to DVE language for parallel verification of SMD.

The translated SMD is treated as automaton in DiVinE. The channels are used for communication among the automata and they are defined using *channel* keyword in DVE. The *process* keyword defines a process and each process of DiVinE presents the individual state machine diagram of SysML. The *state* keyword is used to define states of a SMD. The *Init* keyword is used to define an initial state of a SMD.

The *trans* keyword defines transaction among states of same process or states of different processes. The *sync* keyword is used to send or receive a value using defined channel. The *guard* keyword  is used to set a guard condition. The *effect* keyword is used to assign value to a variable. The question mark (*?*) is used with channel to send a value and exclamation mark (*!*) is used to receive a value.

In SysML, we cannot define the synchronous or asynchronous communication among the SMDs while the DiVinE model checker supports both types of communication. The *system async* and *sync* keyword are used at the end of the DVE model to define asynchronous and synchronous communications, respectively. Fig. [2] is the state machine diagram that shows the functionality of a keypad and Fig. 3 shows the respective translated DVE model. The translated model shows the transition among the state using (->) symbol.

In our previous work  [18], we have defined the rules for translation of sequence diagrams to DVE. In the current study, we translate state machine diagram to DVE.

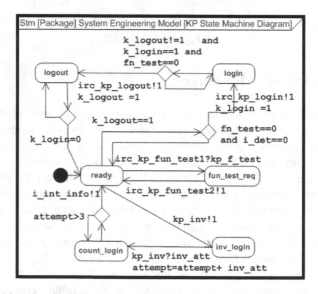

**Fig. 2.** State machine diagram of a keypad

# 6    Case Study

As the embedded security systems are critical systems, so, their safety measures should be satisfied. The case study of security system is comprised of input devices, input receiving unit, premises control unit, alarm server and output devices. The three input devices are: keypad, carbon monoxide detector and video camera. The three output devices are: alarm, lights and siren. Input devices communicate with Input Receiving Unit (IRU) which communicates with Premises Control Unit (PCU). PCU alerts the operator with the current status of the system. Then, the operator takes necessary actions accordingly. The PCU send instruction to Alarm Server (AS) and then AS communicates with output devices. Functionalities of all devices are defined as under:

**Keypad:** An administrator can login or logout to the system by using keypad. User logins to deactivate the system for a while and system activates again when user logouts. In case output devices switches on due to false alarm, the user can switch off the output by logging in the system. As the keypad is IP based, so, on receiving functionality test request from IRU, it sends back an acknowledgement to the PCU through IRU. The keypad can detects intruder on third invalid attempt and passes the message to IRU to switch on the lights and siren. Functionality of a keypad is shown in Fig. 2 along with sending and receiving channels.

**Smoke, Heat and Carbon Monoxide (SHC) Detector:** SHC detects smoke, heat and carbon monoxide and passes message to IRU. As the SHC is IP based, so, on receiving functionality test request from IRU, it sends back an acknowledgement to the PCU through IRU.

**Video Camera:** It provides live video stream to PCU through IRU and detects intruder by passing the message to IRU. As the video cam is IP based, so, on receiving functionality test request from IRU, it sends back an acknowledgement to the PCU through IRU.

**Input Receiving Unit (IRU):** IRU receives inputs from input devices on detection of intruders and receives functionality test request from PCU to pass messages to input devices. It also receives an acknowledgement from input devices and passes it to PCU. Moreover, it has the capability to detect intruder in video and inform the PCU by sending messages. On user's logout, it sends request to PCU to activate the system. It authorizes access to valid user on successful login to deactivate security system, except camera, for a certain location and sends messages to PCU.

**Premises Control Unit(PCU):** PCU checks the status of all individual I/O devices to verify their proper functionality by sending messages and receiving acknowledgements through IRU and AS. It can also check the status of IRU and AS. PCU receives inputs and alerts from IRU to maintain a log when security system is activated or deactivated and sends messages to AS to switch on alarm, lights or fire alarm.

It prompts the operator to call person-I. If person-I cannot attend the call then it prompts to call person-II. On confirmation about the entrance of an intruder, it prompts the operator to call police. In case of fire, it prompts the operator to call fire brigade and ambulance.

**Alarm Server (AS):** AS receives the signal from PCU to switch on and off the alarm, siren and lights. It receives functionality test request from PCU and acknowledges the request. In the same way, it receives functionality test request for output devices and forwards it to output devices. It also receives acknowledgements from output devices and forwards to PCU.

**Output Devices (Fire Alarm, Siren, Lights):** All output devices send acknowledgement to the alarm server on switching on and switching off. These devices also send acknowledgements to AS.

```
process KeyPad{
byte kp_f_test,inv_att;

state ready, login,logout, fun_test_req, inv_login, count_login;

init ready;
trans
ready->login{guard fn_test==0 and i_det==0; sync
irc_kp_login!1; effect k_login =1;},

login->logout{guard k_logout!=1 and k_login==1 and
fn_test==0; sync irc_kp_logout!1; effect k_logout =1;},

logout->ready{guard k_logout==1; effect k_login=0;},
ready->inv_login{sync kp_inv!1; },

inv_login->count_login{sync kp_inv?inv_att; effect
attempt=attempt+ inv_att;},

count_login->ready{guard attempt>3; sync i_int_info!1;},

ready->fun_test_req{sync irc_kp_fun_test1?kp_f_test;},

fun_test_req->ready{sync irc_kp_fun_test2!1;};
}
```

**Fig. 3.** Translated code from keypad's SMD

Fig. 8 is an activity diagram of the whole system that shows interaction among all devices. It shows how input devices send data to IRU and how it interacts with PCU. PCU shows status of the system to operator and passes instructions to AS for switching on or switching off the output devices. The case study of embedded security is modeled using SysML and the state machine diagrams are used to specify the functionality of individual devices (Fig. 2). Although SysML lacks in defining LTL properties, still it provides significant modeling support for embedded security system.

# 7    Properties

In DiVinE, user requirements (properties) can be defined in both ways, either in LTL form or its own programming format. To ensure the security of the system these properties must be satisfied. Fig. 4 shows DVE code for respective properties.

1. If intruder is detected then siren and lights should be switched on  (Fig. 4(a)).
2. If smoke, heat and carbon monoxide (SHC) detector detects fire then fire alarm should be switched on (Fig. 4 (b)).

```
process LTL_property {
state q1, q2;
init q1;
accept q2;
trans
q1 -> q1 {},
q1 -> q2 { guard (i_det==1 and
k_login==1 and (not
Premises_Control_Unit.switch_on_
lights_siren)); },
}
```
(a)

```
process LTL_property {
state q1, q2;
init q1;
accept q2;
trans
q1 -> q1 {},
q1 -> q2 { guard (f_det==1 and
not
Premises_Control_Unit.switch_on_
alarm); };
}
```
(b)

```
process LTL_property {
state q1, q2;
init q1;
accept q2;
trans
q1 -> q1 {},
q1 -> q2 { guard (f_det==1 and
k_login==1 and not
Premises_Control_Unit.switch_on_al
arm); };
}
```
(c)

```
process LTL_property {
state q1, q2;
init q1;
accept q2;
trans
q1 -> q1 {},
q1 -> q2 { guard ( (f_det==0 and not
Premises_Control_Unit.switch_on_al
arm ) and ( i_det==0 and not
Premises_Control_Unit.switch_on_lig
hts_siren )); };}
```
(d)

```
process LTL_property {
state q1, q2;
init q1;
accept q2;
trans
q1 -> q1 {},
q1 -> q2 { guard (f_det==1 and
k_login==1 and not
Premises_Control_Unit.switch_on_al
arm); };
}
```
(e)

**Fig. 4.** LTL property for: (a) Lights and Siren, (b) SHC and Alarm, (c) Invalid User and Fire Alarm, (d) No False Alarm and (e) Functionality Test

3. If intruder or fire is detected and user is log on then no alarming device should be switched on except the fire alarm (Fig. 4(c)).
4. There should be no false alarm (Fig. 4(d)).
5. If an operator sends request for the functionality test of devices, there should be an acknowledgement from the requested device (Fig. 4(e)).

# 8    Analysis of Results

As the embedded systems are becoming more larger and complex, so there is a dire need of model checkers with high computing performance. The difference of the proposed methodology and related work is that the accelerated computational speed of the verification of System Modeling Language will ultimately save modeling and verification cost and will also increase the reliability of a system. The proposed methodology shows significant performance in the field of software verification and suggests the use of SysML and DiVinE for the modeling and verification of large scale embedded systems.

**Table 1.** Execution time in seconds for LTL properties

| Property No. | Status | States | Number of Processors | | | |
|:---:|:---:|:---:|:---:|:---:|:---:|:---:|
| | | | 1 | 2 | 4 | 8 |
| 1 | True | 80463 | 567 | 327 | 156 | 121 |
| 2 | True | 71778 | 510 | 258 | 130 | 118 |
| 3 | True | 61291 | 407 | 308 | 109 | 89 |
| 4 | True | 46420 | 273 | 139 | 76 | 70 |
| 5 | True | 46420 | 154 | 137 | 93 | 74 |

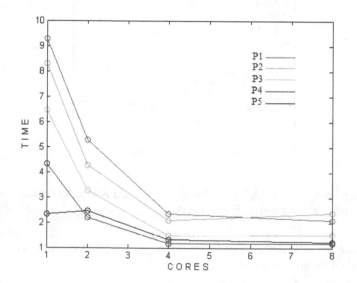

**Fig. 5.** Graph for execution time of execution time on multiple-core

All of the specified requirements (properties) are verified on different number of processors. Table 1 shows the states, status and execution time of individual property of the model of embedded security system. All the specified properties are satisfied and the performance is graphically shown in Fig. 6 where the execution time reduces with the increase in the number of processors. Number of states for individual property along with execution time are also shown in TABLE 1. DiVinE supports GPU which differentiate it from sequential model checking tools. By using DiVinE on GPU based systems, the execution speed and efficiency can be improved many times.

On the basis of execution time, the speedup and efficiency are calculated and also presented graphically. The DiVinE model checker has shown significant performance by increasing the number of processors. Tables (2 and 3) and graphs (Fig. 6 and Fig. 7) for speedup and efficiency show the significant difference in execution time on different number of processors. By using eight CPUs, more than four times increase in the speed of execution for property 1 can be seen. While the minimum increase in the speed of execution for all properties is also more than two times than the speed of execution on single processor (as speedup shown in table 2 for property 5). There is

**Table 2.** Speedup of model on number of processors

| Property No. | No. of Cores | | | |
|:---:|:---:|:---:|:---:|:---:|
| | 1 | 2 | 4 | 8 |
| 1 | 1 | 1.73 | 3.63 | 4.68 |
| 2 | 1 | 1.97 | 3.93 | 4.32 |
| 3 | 1 | 1.32 | 3.73 | 4.57 |
| 4 | 1 | 1.96 | 3.59 | 3.90 |
| 5 | 1 | 1.12 | 1.65 | 2.08 |

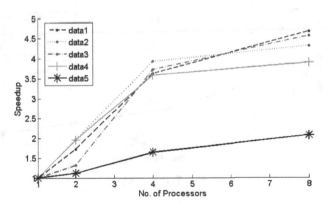

**Fig. 6.** Speedup

significant difference of efficiency for all individual properties on different number of processors, as shown in table 3. The results show the suitability of DiVinE for the verification of System Modeling Language (SysML). DiVinE is more suitable for very large size of model and shows more significant performance on GPU based systems where more processors can be added to verify an embedded system. In this way, the large size of graphical model of embedded system can be verified in very less time that saves the modeling and verification cost.

After modeling the case study, it is also found that the SysML lacks in formally defining the LTL properties. Functional requirements can be specified using requirement diagram but LTL properties cannot be specified. Some other possible specification techniques are required for the specification of the LTL properties.

The proposed methodology is more suitable for system or software engineering field especially for development of industrial control system and intelligent systems where the verification process takes much time. Now with less efforts, the software engineers can model and verify a very large complex system that runs in parallel and distributed environment. Likewise, system engineers can also verify the reliability, safety, fairness and liveness of the system.

**Table 3.** Efficiency of model on number of processors

| Property No. | No. of Cores | | | |
|:---:|:---:|:---:|:---:|:---:|
| | **1** | **2** | **4** | **8** |
| **1** | 1 | 0.86 | 0.90 | 0.58 |
| **2** | 1 | 0.98 | 0.98 | 0.540 |
| **3** | 1 | 0.66 | 0.93 | 0.57 |
| **4** | 1 | 0.98 | 0.89 | 0.48 |
| **5** | 1 | 0.56 | 0.41 | 0.26 |

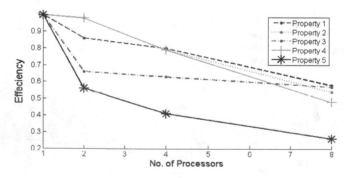

**Fig. 7.** Efficiency

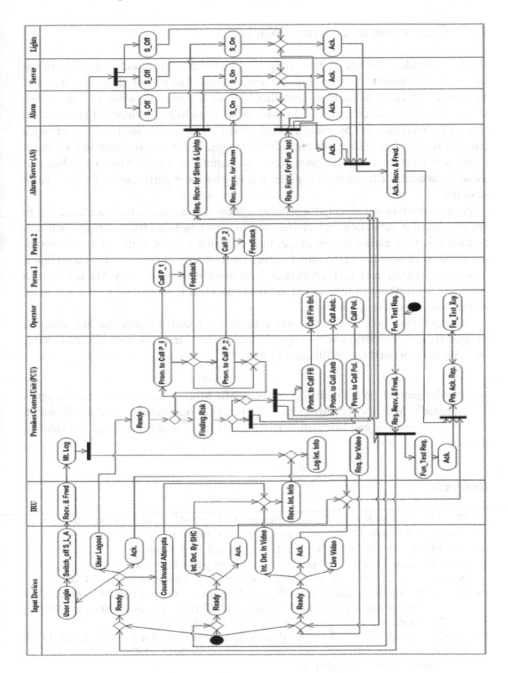

**Fig. 8.** Activity Diagram of Security System

# 9     Conclusion and Future Work

As the embedded systems are becoming more complex, so the modeling and verification technique need to be more reliable and efficient. The proposed methodology has significant effects in the field of modeling large scale embedded systems. The methodology shows an effective use of SysML in the modeling of embedded systems and also in their verification by using DiVinE model checker. The methodology presents parallel verification strategy for embedded systems. It shows significant improvement in the computational time as compare to sequential model checking tools. Because these sequential verification tools either fail or show poor performance for large scale systems.

As a case study we modeled a security system using SysML and its translation to DVE model for validation and verification in very less time. By modeling such embedded system in earlier design phase, time and huge amount of money can be saved. A limitation of SysML is also found during the modeling of the security system that it lacks in specification on LTL properties. Requirement Diagram of SysML is unable to show the interlinking of logical requirements. DiVinE is more easy to use and more suitable for large size applications.

In future, a tool will be developed to generate DVE code for state machine diagram of SysML. LTL properties will be graphically defined and LTL formula for DiVinE will also be generated for verification of model.

**Acknowledgement.** We would like to thank Mr. Mohsin Riaz for technical support and feedback. We also acknowledge the technical support provided by Research Centre for Modeling and Simulation (RCMS, NUST) for implementing and running the case study.

# References

1. O.M.G. OMG System Modeling Language (OMG SysML) specification. Specification document of System Modeling Language (SysML) (June 2012)
2. Vanderperren, Y., Dehaene, W.: SysML_SE_applied_to_SoC.pdf,
   http://www.omgsysml.org/SysML_SE_applied_to_SoC.pdf
3. SysML Tools, http://www.sysmltools.com/ (accessed September 24, 2013)
4. Bouyer, P.: Model-checking Timed Temporal Logics. In: Bouyer, P. (ed.) Proceedings of the 5th Workshop on Methods for Modalities (M4M5 2007). Electronic Notes in Theoretical Computer Science, vol. 231, pp. 323–341 (March 25, 2009)
5. Barnat, J., et al.: DiVinE 3.0 – An Explicit-State Model Checker for Multithreaded C & C++ Programs. In: Sharygina, N., Veith, H. (eds.) CAV 2013. LNCS, vol. 8044, pp. 863–868. Springer, Heidelberg (2013)
6. Siegel, S.F.: Verifying Parallel Programs with MPI-Spin 2007,
   http://pvmmpi07.lri.fr/tutorial2.html (accessed November 2, 2013)
7. Basit Ur Rahim, M.A., Arif, F., Ahmad, J.: Parallel verification of UML using DiVinE tool. In: 2013 5th International Conference on Computer Science and Information Technology (CSIT), Amman, Jordan, March 27-28 (2013)

8. Mashiyat, A.S., Rabbi, F., Wang, H., MacCaull, W.: An automated translator for model checking time constrained workflow systems. In: Kowalewski, S., Roveri, M. (eds.) FMICS 2010. LNCS, vol. 6371, pp. 99–114. Springer, Heidelberg (2010)
9. Jarraya, Y., Alawneh, L., Hassaine, F., Bebbabi, M.: A unified approach for verification and validation of systems and software. In: 13th Annual IEEE International Symposium and Workshop on Engineering of Computer Based Systems, ECBS 2006, Potsdam, March 27-30 (2006)
10. Debbabi, M., Jarraya, Y., Soeanu, A., Mourad, D.: Automatic verification and performance analysis of tme-constrained sysML activity diagrams. In: 14th Annual IEEE International Conference and Workshops on the Engineering of Computer-based Systems, ECBS 2007, Tucson, AZ, March 26-29 (2007)
11. Viehl, A., Sander, B., Bringmann, O., Rosenstie, W.: Integrated requirement evaluation of non-functional. In: Forum on Specification, Verification and Design Languages, FDL 2008, Stuttgart, September 23-25 (2008)
12. Marcello, M., Murillo, L.G., Prevostini, M.: Model-based design space exploration for RTES with SysML and MARTE. In: Forum on Specification, Verification and Design Languages, FDL 2008, Stuttgart, September 23-25 (2008)
13. Kawahara, R., Nakamura, H.: Verification of embedded system's specification using collaborative simulation of SysML and simulink models. In: International Conference on Model-based Systems Engineering, MBSE 2009, Haifa, March 2-5 (2009)
14. Petin, J.F., Evrot, D., Morel, G., Lamy, P.: Combining SysML and formal models for safety requirements verification. In: 22nd International Conference on Software & Systems Engineering and their Applications, Paris, November 1-5
15. Linhares, M., de Oliveira, R., Farines, J., Vernadat, F.: Introducing the modeling and verification process in SysML. In: ETFA 2007 - 12th IEEE Int. Conf. on Emerging Technologies and Factory Automation, Patras, September 25-28 (2007)
16. Mazzini, S., Puri, S., Mari, F., Melatti, I., Enrico, T.: Formal verification at system level. In: DAta Systems in Aerospace (DASIA), Org. EuroSpace, Canadian Space Agency, CNES, ESA, EUMETSAT, Instanbul, Turkey (2009)
17. Knorreck, D., Apvrille, L.: TEPE: A SysML language for time-sonstrained property modeling and formal verification. ACM SIGSOFT Software Engineering Notes 1 (January 2011)
18. Basit Ur Rahim, M.A., Arif, F., Ahmad, J.: Formal verification of sequence diagram using DiVinE. In: International Conference on Computer Software and Applications (ICCSA 2014), Hammamet, Tunisia (January 2014)
19. Basit Ur Rahim, M.A., Arif, F., Ahmad, J.: Modeling of real-time embedded system using SysML and its verification using UPPAAL and DiVinE. In: International Conference on Software Engineering and Service Science (ICSESS 2014), Beijing, Chine, June 2-29 (2014)
20. Petin, J.F., Evrot, D., Morel, G., Lamy, P.: Combining SysML and formal models for safety requirements verification. In: 22nd International Conference on Software & Systems Engineering and their Applications (ICSSEA 2010), Paris, France, December 7-9 (2010)
21. Jarraya, Y., Debbabi, M.: Formal Specification and Probabilistic Verification of SysML Activity Diagrams. In: Sixth International Symposium on Theoretical Aspects of Software Engineering, TASE 2012, Beijing, China, July 4-6 (2012)
22. COMPASS, COMPASS_WP_03.pdf, http://www.compass-research.eu/ (accessed April 14, 2014)

# Refactoring Smelly Spreadsheet Models

Pedro Martins and Rui Pereira

HASLab / INESC TEC & Universidade do Minho, Portugal
{prmartins,ruipereira}@di.uminho.pt

**Abstract.** Identifying bad design patterns in software is a successful and inspiring research trend. While these patterns do not necessarily correspond to software errors, the fact is that they raise potential problematic issues, often referred to as code smells, and that can for example compromise maintainability or evolution.

The identification of code smells in spreadsheets, which can be viewed as software development environments for non-professional programmers, has already been the subject of confluent researches by different groups.

While these research groups have focused on detecting smells on concrete spreadsheets, or spreadsheet instances, in this paper we propose a comprehensive set of smells for abstract representations of spreadsheets, or spreadsheet models. We also propose a set of refactorings suggesting how spreadsheet models can become simpler to understand, manipulate and evolve. Finally we present the integration of both smells and refactorings under the MDSheet framework.

**Keywords:** Spreadsheets, Smells, Model-Driven Engineering, ClassSheets.

## 1 Introduction

For many people, a spreadsheet system is the programming language and programming environment of choice. A spreadsheet programmer is often referred to as an *end user* and is a secretary, an accountant, a teacher, a student, an engineer, or any other person that is not a professional programmer and simply wants to solve a problem [31]. Although spreadsheets were introduced to help end users to perform simple mathematical operations, they quickly evolved into powerful software systems heavily used in industry by professional programmers to process complex and large data. For example, spreadsheets are often used in industry as a simple mechanism to adapt data produced by one system to the format required by a different one.

Regardless of their huge commercial success and wide acceptance, spreadsheets are also known for being highly error-prone! This is supported by both research studies [33,36,37,32] and errors reported in the media[1].

These facts suggest that spreadsheets are a particularly significant target for the application of software engineering principles. Surprisingly, only recently the

---

[1] A list of spreadsheets horror stories is maintained by the *European Spreadsheet Risks Interest Group* at http://www.eusprig.org/horror-stories.htm.

B. Murgante et al. (Eds.): ICCSA 2014, Part V, LNCS 8583, pp. 556–571, 2014.
© Springer International Publishing Switzerland 2014

research community started investigating the use of software engineering techniques in spreadsheets [24]. In software engineering, models were widely adopted as a suitable abstraction mechanism to specify/express complex software. Naturally, spreadsheets followed this trend, namely in the use of model-driven software development techniques [23,19,5,3,21], software refactoring and evolution techniques [20,10,22,14,6], and software metrics and quality models [7,16,8,15].

In this paper we present techniques to detect and eliminate *spreadsheet smells* in the context of model-driven spreadsheet development. A model-driven spreadsheet smell is not an error, but an indication of a poorly designed spreadsheet model. As a consequence spreadsheet smells can be used to improve a spreadsheet model's maintainability and quality. After presenting ClassSheets and a motivational example in Section 2 we present our contributions:

- We adapt a catalog of spreadsheet smells to work on model-driven spreadsheet. That is to say that for each smell defined at the spreadsheet data level we define an equivalent at the ClassSheet level. (Section 3);
- For each smell on ClassSheet models that we propose, we present a possible refactoring or set of refactorings to eliminate it (Section 4). The refactorings are expressed as model evolutions. Thus eliminating smells at the model level automatically eliminates smells at the data level;
- We automated smell detection with an implementation under the MDSheet framework (Section 5.1);
- We also implemented automated refactorings under MDSheet (Section 5.2);

We conclude this paper exposing related work in Section 6 and drawing some conclusions in Section 7.

## 2  ClassSheets– Models for Spreadsheets

Before presenting spreadsheet model smell detection, let us introduce as a running example model-driven spraedsheet. Figure 1 presents a spreadsheet used to grade students' scores. This spreadsheet contains information on various *Marks*, *Exams*, and *Students*. Each *Exam* has 3 parts, each having its own weight value.

| | A | B | C | D | E | F | G | H | I | J | K | L | M | N | O | P |
|---|---|---|---|---|---|---|---|---|---|---|---|---|---|---|---|---|
| 1 | Marks | Exam | First | | | Exam | Second | | | | Exam | Third | | ... | | |
| 2 | | Part 1 | Part 2 | Part 3 | Mark | Part 1 | Part 2 | Part 3 | Mark | Part 1 | Part 2 | Part 3 | Mark | ... | Final Mark | Grade |
| 3 | | 25 | 25 | 50 | 10 | 10 | 10 | 80 | 5 | 10 | 10 | 80 | 5 | ... | 100 {A,...,F} | |
| 4 | Student | | | | | | | | | | | | | ... | | |
| 5 | A | 50 | 70 | 20 | 4 | 35 | 10 | 80 | 3.425 | 40 | 35 | 70 | 3.175 | ... | 51.75 | C |
| 6 | B | 65 | 20 | 35 | 3.875 | 20 | 45 | 35 | 1.725 | 60 | 20 | 40 | 2 | ... | 39.375 | D |
| 7 | C | 90 | 100 | 85 | 9 | 75 | 95 | 80 | 4.05 | 80 | 85 | 90 | 4.425 | ... | 89.25 | A |
| 8 | D | 100 | 45 | 75 | 7.375 | 40 | 60 | 50 | 2.5 | 50 | 75 | 55 | 2.825 | ... | 65.125 | B |
| 9 | E | 35 | 60 | 80 | 6.375 | 60 | 50 | 55 | 2.75 | 40 | 20 | 70 | 3.1 | ... | 62.875 | B |
| 10 | F | 70 | 35 | 35 | 4.375 | 45 | 45 | 60 | 2.85 | 30 | 60 | 50 | 2.45 | ... | 46.375 | C |
| 11 | G | 95 | 80 | 80 | 8.375 | 70 | 80 | 80 | 3.95 | 70 | 90 | 85 | 4.2 | ... | 83.875 | A |
| 12 | H | 75 | 55 | 65 | 6.5 | 20 | 50 | 20 | 1.15 | 50 | 50 | 65 | 3.1 | ... | 63.5 | B |
| 13 | | ⋮ | ⋮ | ⋮ | ⋮ | ⋮ | ⋮ | ⋮ | ⋮ | ⋮ | ⋮ | ⋮ | ⋮ | ... | ⋮ | ⋮ |
| 14 | | | | | | | | | | | | | | ... | | |

Fig. 1. Instance of a grading spreadsheet

Every *Student* has a mark from each exam, calculated by a formula using the weights of each part, and the value the student obtained in the respective part. A student's final mark is the average mark across all exams.

This spreadsheet business logic can be abstracted and represented by a model. In this paper we consider ClassSheet model [23,12]: a high-level object-oriented abstract representation for spreadsheets. They can be compared to UML class diagrams [18], but for spreadsheets and with layout specification. Indeed they are formed by classes and attributes, which we will explain in the following paragraphs using the marks spreadsheet example. Figure 2 is the ClassSheet model to abstract and specify this grading spreadsheet. There are 3 classes,

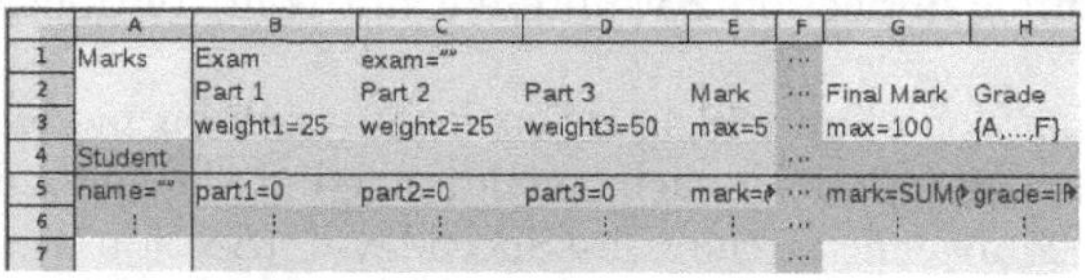

| | A | B | C | D | E | F | G | H |
|---|---|---|---|---|---|---|---|---|
| 1 | Marks | Exam | exam="" | | | | | |
| 2 | | Part 1 | Part 2 | Part 3 | Mark | ⋯ | Final Mark | Grade |
| 3 | | weight1=25 | weight2=25 | weight3=50 | max=5 | ⋯ | max=100 | {A,...,F} |
| 4 | Student | | | | | | | |
| 5 | name="" | part1=0 | part2=0 | part3=0 | mark=♦ | ⋯ | mark=SUM( | grade=IF |
| 6 | ⋮ | ⋮ | ⋮ | ⋮ | ⋮ | ⋯ | | |
| 7 | | | | | | ⋯ | | |

Fig. 2. ClassSheet model of a grading spreadsheet

**Marks**, **Student**, and **Exam**. Each **Student**, represented by row 5, has an attribute termed *name*, which is set in cell A6. Each *attribute* in a ClassSheet is defined by a string (e.g., name), the equal sign, and by a default value or a formula. Note that students expand vertically, that is, each student is added after the previous one in the next row. This is encoded in the model by the ellipsis shown in row 6. Each **Exam** has an attribute with the weight for each of the 3 parts, *weight1*, *weight2*, and *weight3* respectively, and expands horizontally. This is indicated by the ellipsis in column F. The relationship between a **Student** and an **Exam** gives us 4 attributes. The first 3, *part1*, *part2*, and *part3*, state the mark the student obtained in the specific part. The last attribute, *mark*, states the mark the student obtained on the exam. This *mark* formula is defined as:

```
mark=(weight1*part1 + weight2*part2 + part3*weight3) /
  ↪ SUM(weight1,weight2,weight3) * max /
  ↪ SUM(weight1,weight2,weight3)
```

The **Marks** class has 2 attributes. The first, *mark* (in cell G5, only partially), states the final mark a student obtained throughout the exams. The second, *value* (in cell H5, hidden by the *mark* formula), states the mark value given (A,B,C,D, or F). These two formulas are defined respectively as:

```
mark=SUM(Exam.mark*max/Exam.max)) / COUNT(Exam.mark)
grade=IF(mark<=(max/5),"F", IF(mark<=(max/5*2),"D",
  ↪ IF(mark<=(max/5*3),"C", IF(mark<=(max/5*4),"B","A"))))
```

Note that these formulas only differ from regular spreadsheet formulas by the use of attributes instead of cell references. This can be compared to the use of *cell names* in standard spreadsheet systems. Also note that some references can be directed to particular attributes as follows Exam.max.

# 3    Spreadsheet Model Smells

The concept of code smell [25] (or simply bad smell) was introduced as a concrete evidence that a piece of software may have a problem. Usually a smell is not an error in the program, but a characteristic that may cause problems understanding the software (for example, a long class in an object-oriented program) and updating and evolving the software. Although they were initially designed for objected-oriented programming, code smells have been adapted for other contexts, including spreadsheets [8,7,28,27]. In this section we survey the code smells defined for spreadsheet formulas introduced in [28]. Moreover, we adapt each of these five smells to the spreadsheet models described in [10,11,13,18,12].

## 3.1    Multiple Operations

The first smell we present is the *multiple operations* smell. This smell can be considered as present in a spreadsheet if a formula is created using too many operations. Clearly it is important to understand what "too many" exactly means in this scenario, and we will explain it in detail in Section 5.1. Note that this smell can be detected in individual cells.

Let us look at a formula extracted from our running example (from Figure 2), in this case the formula to evaluate the mark of a student for a specific exam:

```
mark=(part1*weight1 + part2*weight2 + part3*weight3) /
  ↪ SUM(weight1,weight2,weight3) * max /
  ↪ SUM(weight1,weight2,weight3)
```

As shown in this formula, ClassSheet formulas are quite similar to the ones written in regular spreadsheet systems. Indeed the operations (operators and formula names) are used in a similar way. Thus, we consider the model smell as it is considered in regular spreadsheets.

In this case we have 9 operations, and as we will see these are indeed too many to have in a formula and still be considered maintainable. Indeed having so many operations in the same cell has several disadvantages: for once, it becomes hard to understand what such a formula does; it also becomes difficult to update or change such a formula.

## 3.2    Multiple References

Let us consider the same formula shown in the previous smell. This formula has a second quality problem: it uses too many references. When defining ClassSheet models, instead of using regular spreadsheet references such as A1 or B2, the user writes the name of the attributes used formulas. In this case, part1 or weight1 are references to the attributes with the same name. This can be compared to the use of *cell names* in standard spreadsheet systems. Thus, in this case the smell adapts very well from plain spreadsheets to the ClassSheet models. In the mark formula shown before there are 12 references to other attributes which makes this formula hard to understand and maintain. Finally note that, as with the previous smell, this can be detected for each cell defined by a formula.

### 3.3   Condition Complexity

The third smell we consider is well known and applicable whenever conditional construction is possible in programming [29]. Thus this is also true for spreadsheets, and for spreadsheet models.

Let us consider the formula to present the *final mark* of the student:

```
value=IF(mark<=20,"F",IF(mark<=40,"D", IF(mark<=60,"C",
 ↪ IF(mark<=80,"B","A"))))
```

This formula is defined using 4 if conditions, which again makes it hard to manipulate. Indeed this is true in many other programming languages, but is especially true in spreadsheets as it must be defined in one line without any possible indentation as possible in most regular programming languages.

Since conditional programming is possible in these models, this smell also applies to their formulas. Once again, this smell can potentially occur in each and every cell of the model.

### 3.4   Long Calculation Chain

This smell appears in spreadsheets that require long paths of cell dereferencing to calculate the values of their formulas. For instance, if a formula has a reference to cell A1, and A1 has a reference to A2, and A2 has a reference to A3, then there is a long path that is necessary to run to compute the value of A1. This has a considerable impact on the formula's quality since it is necessary to go through several cells to see its arguments and understand it. Also, its result may change if one of its references changes its value, which may be difficult to notice it the path is too long or even comprises of different worksheets.

Let us again look at the formula example of the previous smell. Such a formula depends on the attribute mark, which depends on the attribute part1. Thus this smell also applies nicely to ClassSheet models. Indeed, instead of normal spreadsheet references we have references to attributes. This also makes it difficult to understand and evolve such formulas.

Finally note that in this case it is not possible to discover this smell only by looking at a cell. It is necessary to follow its references to other attributes recursively.

### 3.5   Duplicated Formulas

The last smell we adapt for ClassSheet models is *duplicated formulas*. This has similar problems to duplicated code in a regular programming language. In the case of spreadsheets, this occurs in a formula if part of its code is repeated.

Once again considering the mark formula, one can see that part of its definition, SUM(weight1,weight2,weight3), appears twice. Thus, this smell also applies to spreadsheet models.

Note that this smell can also occur when considering a range of cells. If part of a formula is repeated through a set of cells, then this repeated part of the formula is also considered duplication. Thus, this smell can occur in individual cells, but also in groups of cells.

### 3.6  MDSheet with Smell Detection

The smell detection presented in this paper was implemented within MDSheet. In MDSheet's toolbar, a new button is available to evaluate the smells on a model. After clicking the button, the model is annotated with comments that describe the smells present in the annotated cells.

In Figure 3, three cells with smelly attributes were annotated with comments, namely mark (in cell E5), mark (in cell G5), and grade (in cell H5).

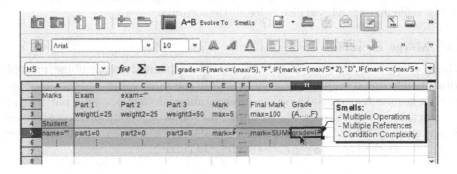

**Fig. 3.** Smell detection in MDSheet

## 4  Refactorings for Spreadsheet Models

In the previous section, we presented a catalog of smells for ClassSheets. The detection of such smells has been implemented in a model-driven spreadsheet environment. Before we present such environment, let us present techniques to automatically eliminate the smells. To this end we rely on *program refactorings* that change a particular piece of code, removing the smell, and not changing the semantics of the code. In this section we present a set of refactorings that can be used to eliminate the smells identified in the previous section. The following refactorings were defined by Hermans et al. [28] in order to remove the smells that they adapted to spreadsheets. We will explain each one and make the necessary adaptation to make them useful for ClassSheets.

### 4.1  Extract Subformula

This refactoring is intended to extract subformulas from exiting ones. Looking at the arguments at the root of a formula, we can extract new formulas from each argument at the root. This will create new formulas which will be referenced by the original formula.

Although this refactoring has been originally proposed for regular spreadsheet formulas, it can also be applied to models. In the case of models however this process requires the creation of a new attribute for each subformula extracted. These attributes will be created in the same class the subsformulas was extracted from.

## 4.2    Extract Common Subformula

This is an additional supporting refactoring which can be applied if a formula contains the same subformula multiple times. This refactoring follows the same procedure as *Extract Subformula*.

Once again, this refactoring can be applied to ClassSheets. For example, looking at the `mark` formula one can see `SUM(weight1,weight2,weight3)` repeated twice. Thus such subformula can be extracted, and a new attribute can be created using it: `total_weight=SUM(weight1,weight2,weight3)`. It in now possible to simply reference `total_weight` twice in the formula.

## 4.3    Merge Branches

When we have a complex `IF` condition, we can sometimes combine multiple branches into one if multiple branches result in the same value. This is a refactoring that is quite transversal to many programming languages, and in particular it is applicable to ClassSheets. For example, the following conditional construction `IF(mark>=3,"PASSED",IF(part3>=5,"PASSED","FAILED"))` can be rewritten as `IF(OR(mark>=3,part3>=5),"PASSED","FAILED")`.

## 4.4    Merge Formulas

One of the introduced smell is the long calculation chain. This refactoring tries to eliminate it. It can be applied when we detect that there are certain calculation steps in a chain that do not occur in other chains. When this is detected, those two calculation steps are merged into one formula without affecting the other computations as the merged calculation is not being used anywhere else.

As we explained, the long calculation chain is adaptable to the context of spreadsheet models. Moreover, this is also true for this refactoring. In the case of models this can be applied if an attribute is defined by a formula and it is only used in one place. In such cases, such attribute can be removed and the formula integrated into the other computation. In our model both `mark` formulas could be integrated in the formulas that use them. Note this would probably create more smells, and thus should not be applied in this particular situation.

## 4.5    Other Refactorings

In [28] two more refactorings are presented, namely *Group References* and *Move References*.

*Group References* can be applied when a spreadsheet cell references a series of adjacent cells, for example `SUM(A1;A2;A3;A4;A5)`. In this case one can restructure this into a lower number of ranges such as `SUM(A1:A5)`. This smell however cannot be applied to ClassSheet as no such range feature is present.

*Move References* is a refactoring which realocates the cells in a spreadsheet. If a formula is referencing multiple cells in different parts of the spreadsheet, for example `SUM(A1:A5;B5;B19;C4;C7)` one could move the values in B5, B19, C4, and C7 to A6 to A9. This allows to rewrite the formula as `SUM(A1:A9)`. Again, given that no ranges are possible in ClassSheets this refactorings does not apply.

## 4.6    Relationship between Smells and Refactorings

As shown in Table 1, several of the refactorings just presented can be applied to remove different ClassSheet smells. For instance, the extraction of subformula can be used to remove all smells except the long calculation chain. The way these refactorings are applied to remove the smells is detailed in Section 5.

**Table 1.** ClassSheet smells and refactorings that can be applied to reduce them

|  | Multiple Operations | Multiple References | Condition Complexity | Long Calculation Chain | Duplicated Formulas |
|---|---|---|---|---|---|
| Extract Subformula | ✓ | ✓ | ✓ |  | ✓ |
| Extract Common Sub. | ✓ |  | ✓ |  |  |
| Merge Branches |  |  | ✓ |  |  |
| Merge Formulas |  |  |  | ✓ |  |

# 5    Model-Driven Spreadsheet Framework

In the previous sections we have introduced the concept of *smell* and *refactoring* adapted to spreadsheet models, that is, to ClassSheets. In section 5.1 we will present how such smells can be detected under the MDSheet framework [13,11]. Moreover, in section 5.2 we will explain how the refactorings presented before can be incorporated under the same framework.

We choose this framework because it already integrates several model-driven spreadsheet features, including evolution [10], or data querying [34,9,4,17]. Moreover, given its modular design it allows for easy integration of new functionality, as the one we propose in this paper.

## 5.1    Detection of Smells in Spreadsheet Models

In this section we present in detail the detection of smells in ClassSheet models. As previously introduced, some of the smells depend on software metrics of certain artifacts that appear in formulas, e.g. the number of operations or the number of references. Following the approach presented in [28], each smell is classified as having *low*, *moderate*, or *high* risk, depending on these metrics. Obviously, the best case scenario is when no smell is present meaning there is no risk. This is the first step our implementation must handle. Thus next we present a Haskell [35] data type that represents these four options:

**data** $Risk = None \mid Low \mid Moderate \mid High$

In order to find a smell, and the associated risk, we defined a set of functions that operate on spreadsheet models. Their signatures are presented next:

$$smellMultipleOperations \quad :: Model \rightarrow [(Cell, Risk)]$$
$$smellMultipleReferences \quad :: Model \rightarrow [(Cell, Risk)]$$

$$smellConditionComplexity \ :: Model \rightarrow [(\ Cell, Risk)]$$
$$smellLongCalculationChain :: Model \rightarrow [(\ Cell, Risk)]$$
$$smellDuplicatedFormulas \quad :: Model \rightarrow [(\ Cell, Risk)]$$

These functions receive the model under consideration as an argument and return the list of cells of the models that contain attributes, and the respective risk.

The risk of a smell is obtained by evaluating the related metric on the model and then mapping it to the corresponding risk as set by the thresholds.

$$smellMultipleOperations \quad = map \ (id \times risk \ 4 \ 5 \ 9) \ \circ metricMultipleOperations$$
$$smellMultipleReferences \quad = map \ (id \times risk \ 3 \ 4 \ 6) \ \circ metricMultipleReferences$$
$$smellConditionComplexity = map \ (id \times risk \ 2 \ 3 \ 4) \ \circ metricConditionComplexity$$
$$smellLongCalculationChain = map \ (id \times risk \ 4 \ 5 \ 7) \circ metricLongCalculationChain$$
$$smellDuplicatedFormulas \quad = map \ (id \times risk \ 6 \ 9 \ 13) \circ metricDuplicatedFormulas$$

To help with the conversion of the value of the metric with its associated risk, an auxiliary function *risk* was created and, given the threshold values, converts the value of a metric to its corresponding risk. Note that we use the thresholds proposed and validated for regular spreadsheet formulas [28], since the formulas in ClassSheet models are defined in a similar way.

$$risk :: Int \rightarrow Int \rightarrow Int \rightarrow Int \rightarrow Risk$$
$$risk \ low \ moderate \ high \ n \ | \ n < low \qquad = None$$
$$| \ n < moderate = Low$$
$$| \ n < high \qquad = Moderate$$
$$| \ otherwise \qquad = High$$

Each of the smell functions use a metric function. These functions receive a model and evaluate the metric for each cell containing an attribute, returning the list of cells and the number of times the corresponding problem occurs in it. This can be represented with $Model \rightarrow [(\ Cell, Int)]$ as the type of the metric functions. For each result pair of cell and number of times the problem occurs, we apply the product of function *id* and *risk*. The function *id* will maintain the cell intact, and the risk will transform the number of the occurrence of the problem (for instance, number of operations) into the correct risk.

In the following sub-sections, we explain in more detail how we implement each of the metrics introduced before.

**Multiple Operations Metric.** To evaluate this metric, we iterate over all the items in the formula of each attribute and count all those representing operations, namely functions (*ExpFun*) and operators (*ExpUnoOp* and *ExpBinOp*):

$$metricMultipleOperations = countIf \ isOperation \circ modelAttributes$$
$$\textbf{where} \ isOperation \ (ExpFun \quad \_ \_) \quad = True$$
$$isOperation \ (ExpUnoOp \_ \_) \quad = True$$
$$isOperation \ (ExpBinOp \ \_ \_ \_) = True$$
$$isOperation \ \_ \qquad\qquad = False$$

Note that we show here a simplified version of this function as the implementation requires certain details not useful to understand our work.

**Multiple References Metric.** To evaluate the *multiple references* metric, we count the number of references present in the formula:

$$metricMultipleReferences = countIf\ isReference \circ modelAttributes$$
$$\textbf{where}\ isReference\ (ExpRef\ \_\ \_) = True$$
$$isReference\ \_ \qquad\qquad = False$$

**Condition Complexity Metric.** To evaluate the *condition complexity* metric, we count the number of *IF*s in the formula. Its implementation is very similar to the previous metrics and thus we do not show it here. Note that an *IF* is a function and counts as an operation for the *multiple operations* metric.

**Long Calculation Chain Metric.** To evaluate thi metric, a dependency tree is generated for each attribute and then its height is calculated:

$$metricLongCalculationChain\ m = map\ (\pi_2 \triangle chainLength)\ attrs$$
$$\textbf{where}\ chainLength = treeHeight \circ evalDeps\ [\ ]$$
$$attrs = map\ (((getCellClassName\ (classes\ m) \circ \pi_1) \triangle cellName \circ \pi_2) \triangle \pi_2)$$
$$(modelAttributes'\ m)$$
$$evalDeps\ l\ attr = Node\ attr\ deps$$
$$\textbf{where}\ deps = map\ (\lambda r \rightarrow evalDeps\ ((\pi_1\ attr) : l)\ (r, getAtt\ r)))\ refs$$
$$refs\ = filter\ (\lambda r \rightarrow \neg\ (r \in l))\ (references\ (\pi_2\ attr))$$
$$references\ c = [(cn, an)\ |\ ExpRef\ cn\ an \leftarrow universeBi\ c]$$
$$getAtt = fromJust \circ ('lookup'attrs)$$

**Duplicated Formulas.** To evaluate the *duplicated formula* metric, a list with the formula parts of each attribute is generated and compared to the lists of the other attributes. The number of attributes with a sub-formula match is returned.

$$metricDuplicatedFormulas\ m = map\ (\pi_1 \triangle countDuplicates)\ attrs$$
$$\textbf{where}\ countDuplicates\quad = length \circ filter\ id \circ findDuplicates$$
$$findDuplicates\ attr = [isDup\ attr\ attr'\ |\ attr' \leftarrow attrs, attr' \neq attr]$$
$$isDup\ attr\ attr'\quad = or\ (map\ (\lambda x \rightarrow x \in (\pi_2\ attr'))\ (drop\ 1\ (\pi_2\ attr)))$$
$$attrs\quad\qquad\qquad = map\ (id \triangle cellSubForm)\ (modelAttributes\ m)$$
$$cellSubForm\ (CellFormula\ (Formula\ \_\ e)) =$$
$$[e'\ |\ e' \leftarrow universeBi\ e, \neg\ (isTerm\ e')]$$
$$\textbf{where}\ isTerm\ (ExpVal\ \_)\quad = True$$
$$isTerm\ (ExpRef\ \_\ \_) = True$$
$$isTerm\ \_\qquad\qquad = False$$

The Haskell code used to specify the metrics makes use of generic programming using Uniplate [30], which allows for a concise way to traverse of data structures.

We have shown how the detection of ClassSheet smells presented in Section 3 are concisely implemented in Haskell. We use the Haskell programming language because is the implementation language of the model-driven spreadsheet framework MDSheet, where we will define the smell refactoring/elimination as ClassSheet evolution.

## 5.2  Refactoring of ClassSheet Formulas

The refactorings presented in Section 4 can involve the creation or removal of formulas from the cells, but all of them change the cell contents. The MDSheet environment does not take into account evolution of cell formula specifically. It focuses more one the layout of the spreadsheet and the correct referencing of cells using named attributes to provide a more user-friendly interface to end users. Nevertheless, our environment supports setting cell values, and changes to the layout and cell contents are needed in order to perform these refactorings.

The refactoring of formulas in MDSheet lies on top of its bidirectional transformation engine [6]. This engine is specified as a set of operations that can be performed on the spreadsheet models, another set that can be performed on the spreadsheet data, and a relationship between these two sets. The relation between operations on these two artifacts, model and data, describes the equivalent set of operations that is needed to be applied on the other artifact after the original artifact is evolved with the set of operations defined on it.

**Model Evolution.** In order to evolve spreadsheet models, a set of operations were defined on them, as expressed by the following data type:

$$
\begin{aligned}
\textbf{data } Op_M : Model &\rightarrow Model = \\
addColumn_M \quad & Where\ Index & \text{-- add a new column} \\
\mid delColumn_M \quad & Index & \text{-- delete a column} \\
\mid addRow_M \quad & Where\ Index & \text{-- add a new row} \\
\mid delRow_M \quad & Index & \text{-- delete a row} \\
\mid setLabel_M \quad & (Index, Index)\ Label & \text{-- set a label} \\
\mid setFormula_M \quad & (Index, Index)\ Formula & \text{-- set a formula} \\
\mid replicate_M \quad & ClassName\ Direction\ Int\ Int & \text{-- replicate a class} \\
\mid addClass_M \quad & ClassName\ (Index, Index)\ (Index, Index) & \\
& & \text{-- add a static class} \\
\mid addClassExp_M \quad & ClassName\ Direction\ (Index, Index)\ (Index, Index) & \\
& & \text{-- add an expandable class}
\end{aligned}
$$

This set of operations are the ones to be applied any time that a refactoring needs to be applied to evolve formulas with smells. Using these operations, we get the automatic coevolution of the instance using the bidirectional environment of MDSheet, that is, user will get the spreadsheet data automatically evolved after removing the smells from the model.

**Data Evolution.** The refactoring of model formulas presented in this paper not only affects spreadsheet models, but they are also applied to their instances with the following operations used by the bidirectional transformation engine:

$$
\begin{aligned}
\textbf{data } Op_D : Data &\rightarrow Data = \\
addColumn_D \quad & Where\ Index & \text{-- add a column}
\end{aligned}
$$

| | | |
|---|---|---|
| $\mid$ $delColumn_D$ | $Index$ | -- delete a column |
| $\mid$ $addRow_D$ | $Where\ Index$ | -- add a row |
| $\mid$ $delRow_D$ | $Index$ | -- delete a row |
| $\mid$ $AddColumn_D$ | $Where\ Index$ | -- add a column to all instances |
| $\mid$ $DelColumn_D$ | $Index$ | -- delete a column from all instances |
| $\mid$ $AddRow_D$ | $Where\ Index$ | -- add a row to all instances |
| $\mid$ $DelRow_D$ | $Index$ | -- delete a row from all instances |
| $\mid$ $replicate_D$ | $ClassName\ Direction\ Int\ Int$ | -- replicate a class |
| $\mid$ $addInstance_D$ | $ClassName\ Direction\ Model$ | -- add a class instance |
| $\mid$ $setLabel_D$ | $(Index, Index)\ Label$ | -- set a label |
| $\mid$ $setValue_D$ | $(Index, Index)\ Value$ | -- set a cell value |
| $\mid$ $SetLabel_D$ | $(Index, Index)\ Label$ | -- set a label in all instances |
| $\mid$ $SetValue_D$ | $(Index, Index)\ Value$ | -- set a cell value in all instances |

With this we guarantee that the improvements made to the model are also passed on to the their respective instances.

**Model and Data Coevolution.** One guarantee provided my MDSheet is the always conformance of the instances to their models. This is achieved by relating model operations to data ones, and vice versa. Whenever a model operation is performed, this operation is converted to a set of data ones that perform the necessary evolution steps in the instance so that the conformity to the model is restored. The inverse is also available, that is, changes to the instances also automatically coevolve the model so both the instance and the model are always synchronized.

**Refactoring by Evolution.** There are multiple evolution steps that can be performed to refactor the formula of an attribute. For all refactorings, one of the operations to perform is to refactor the formula which is done as defined before.

For the *extract subformula* refactoring, a new attribute is needed to store the extracted subformula. This implies the use of a $setFormula_M$ to create the new attribute and another $setFormula_M$ to update the old formula, after extracting the subformula. However, this can be impossible to realize if there is no place where to add the new attribute. Thus, a $addColumn_M$ or a $addRow_M$ may be necessary to allocate space for the new attribute.

For the *extract common subformula* refactoring, the evolution steps are the same as for the *extract subformula* refactoring. The difference is that multiple subformulas can be extracted with this refactoring.

For the *merge branches* refactoring, only a $setFormula_M$ is needed, updating the old formula with its refactored version.

For the *merge formulas* refactoring, a $setFormula_M$ has to be applied to set the formula with the other merged formulas. Moreover, for each of the merged formulas, a $setLabel_M$ or $setFormula_M$ can be applied to remove the formulas that were merged. If any row or column became empty, functions $delRow_M$ or $delColumn_M$ can be performed to remove them.

## 5.3    MDSheet with Smell Elimination

After detecting the spreadsheet smells, the system will then provide a set of
refactorings which can be applied to eliminate each smell. The user may choose
to apply such refactoring(s) to the model so it no longer contains smells. If we
look back at the smells detected in Figure 3, we can apply our refactorings and
end up with the table shown on the right in Figure 4.

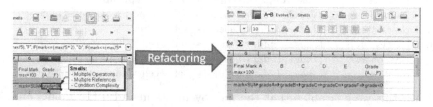

**Fig. 4.** Smell elimination in MDSheet

Note that the data will be automatically coevolved so that it always maintains
the instance's conformity to the model. This smell elimination on the model also
eliminates smells in the data.

# 6    Related Work

There are several works that have focused on techniques based on smells detection to help improve the overall quality of spreadsheets.

One example of such work is presented in [28]. Here, the authors point a
set of smells on formulas, define different threshold for these smells based on
how frequent they appear on typical spreadsheets and suggest refactorings that
improve the spreadsheet readability and maintainability. This is the work we
adapt in this paper. A similar approach is taken by the same authors in [27],
but this time they introduce smells that can point dangerous relations between
worksheets in the same spreadsheet.

In [2], the authors created a extension (*RefBook*) for Microsoft Excel that
detects and refactors a set of smells that find parts of the spreadsheet that
contain, for example, unnecessary complexity and duplicated expressions. In
this work, three empirical studies are performed to evaluate the capacity of this
tool in improving the quality of spreadsheets and improving their readability,
and the authors conclude that users prefer the improved quality of refactored
sheets. There has also been tools built with the specific purpose of pointing and
quantifying cells in a spreadsheet that can lead to potential problems. In [8], the
authors use a smells-based technique, together with strategies that detect chains
of inter-dependent cells and create new worksheets with color-based warnings.

In [26] the authors developed an extension for Excel, called *BumbleBee*, that
is capable of detecting repetition of formulas in different zones of a spreadsheet
and transforms them through a set of strategies. The authors conclude that more

than 70% of spreadsheets have a potential for the application of these transformations and that users are more capable of performing changes to spreadsheets using *BumbleBee*.

Another interesting work is presented in [7], where the authors suggest a new catalog of smells that can be used to further improve existing techniques such as the ones described in this section. Smells in Models is not an area as widely researched as smells in typical software programs. Of relevant reference however is [1]. Here, the authors propose a model-based quality assurance process that uses techniques that perform model quality analysis and model smells detection. Similarly to our approach, they also implement a tool that analysis and refactors models based on the Eclipse Modeling Framework.

## 7   Conclusion

The detection of code smells is a widely recognized software engineering technique that contributes to assessing the overall quality of a code repository. Indeed, identifying such bad design practices has now already been successfully explored in different contexts other than just source code, and namely in the context of spreadsheet engineering.

In the context of spreadsheets, however, smells have only been tackled at the level of concrete spreadsheets. This leaves out reasoning in the same way about spreadsheet abstract models.

This paper closes precisely this gap: we propose a catalog of smells for spreadsheet models, exploiting and adapting the catalog that has been proposed for spreadsheet instances. Finally, we follow the standard approach of associating refactorings with smells, in such a way that if these refactorings are adopted the identified smells disappear.

**Acknowledgments.** We would like to thank Jorge Mendes, Jácome Cunha, and João Saraiva for the help incorporating the ClassSheet smells in the MDSheet framework.

This work is part funded by ERDF - European Regional Development Fund through the COMPETE Programme (operational programme for competitiveness) and by National Funds through the FCT - Fundação para a Ciência e a Tecnologia within projects FCOMP-01-0124-FEDER-022701and Network Sensing for Critical Systems Monitoring (NORTE-01-0124-FEDER-000058), ref. BIM-2013_BestCase_RL3.2_UMINHO. The authors were funded by FCT grants BIM-2013_BestCase_RL3.2_UMINHO,    BI3-2013PTDC/EIA-CCO/116796/2010, respectively.

## References

1. Arendt, T., Taentzer, G.: Integration of smells and refactorings within the eclipse modeling framework. In: Proceedings of the Fifth Workshop on Refactoring Tools, WRT 2012, pp. 8–15. ACM, New York (2012)

2. Badame, S., Dig, D.: Refactoring meets spreadsheet formulas. In: Proceedings of the 2012 IEEE International Conference on Software Maintenance, ICSM 2012, pp. 399–409. IEEE Computer Society, Washington, DC (2012)
3. Beckwith, L., Cunha, J., Fernandes, J.P., Saraiva, J.: End-users productivity in model-based spreadsheets: An empirical study. In: Piccinno, A. (ed.) IS-EUD 2011. LNCS, vol. 6654, pp. 282–288. Springer, Heidelberg (2011)
4. Belo, O., Cunha, J., Fernandes, J.P., Mendes, J., Pereira, R., Saraiva, J.: Querysheet: A bidirectional query environment for model-driven spreadsheets. In: VL/HCC, pp. 199–200 (2013)
5. Cunha, J., Erwig, M., Saraiva, J.: Automatically inferring classsheet models from spreadsheets. In: Proceedings of the 2010 IEEE Symposium on Visual Languages and Human-Centric Computing, VLHCC 2010. IEEE Computer Society (2010)
6. Cunha, J., Fernandes, J.P., Mendes, J., Pacheco, H., Saraiva, J.: Bidirectional transformation of model-driven spreadsheets. In: Hu, Z., de Lara, J. (eds.) ICMT 2012. LNCS, vol. 7307, pp. 105–120. Springer, Heidelberg (2012)
7. Cunha, J., Fernandes, J.P., Ribeiro, H., Saraiva, J.: Towards a Catalog of Spreadsheet Smells. In: Murgante, B., Gervasi, O., Misra, S., Nedjah, N., Rocha, A.M.A.C., Taniar, D., Apduhan, B.O. (eds.) ICCSA 2012, Part IV. LNCS, vol. 7336, pp. 202–216. Springer, Heidelberg (2012)
8. Cunha, J., Fernandes, J.P., Mendes, J., Martins, P., Saraiva, J.: Smellsheet detective: A tool for detecting bad smells in spreadsheets. In: Proceedings of the 2012 IEEE Symposium on Visual Languages and Human-Centric Computing, VLHCC 2012, pp. 243–244. IEEE Computer Society, Washington, DC (2012)
9. Cunha, J., Fernandes, J.P., Mendes, J., Pereira, R., Saraiva, J.: Querying model-driven spreadsheets. In: 2013 IEEE Symposium on Visual Languages and Human-Centric Computing (VL/HCC), pp. 83–86 (2013)
10. Cunha, J., Fernandes, J.P., Mendes, J., Saraiva, J.: Embedding and evolution of spreadsheet models in spreadsheet systems. In: 2011 IEEE Symposium on Visual Languages and Human-Centric Computing, VLHCC 2011, pp. 186–201 (2011)
11. Cunha, J., Fernandes, J.P., Mendes, J., Saraiva, J.: A bidirectional model-driven spreadsheet environment. In: 34rd International Conference on Software Engineering, ICSE 2012, pp. 1443–1444 (June 2012)
12. Cunha, J., Fernandes, J.P., Mendes, J., Saraiva, J.: Extension and implementation of classsheet models. In: 2012 IEEE Symposium on Visual Languages and Human-Centric Computing, VLHCC 2012, pp. 19–22 (2012)
13. Cunha, J., Fernandes, J.P., Mendes, J., Saraiva, J.: MDSheet: A Framework for Model-driven Spreadsheet Engineering. In: Proceedings of the 34rd International Conference on Software Engineering, ICSE 2012, pp. 1412–1415. ACM (2012)
14. Cunha, J., Fernandes, J.P., Mendes, J., Saraiva, J.: Towards an evaluation of bidirectional model-driven spreadsheets. In: User Evaluation for Software Engineering Researchers, USER 2012, pp. 25–28. ACM Digital Library (2012)
15. Cunha, J., Fernandes, J.P., Mendes, J., Saraiva, J.: Complexity Metrics for Classsheet Models. In: Murgante, B., Misra, S., Carlini, M., Torre, C.M., Nguyen, H.-Q., Taniar, D., Apduhan, B.O., Gervasi, O. (eds.) ICCSA 2013, Part II. LNCS, vol. 7972, pp. 459–474. Springer, Heidelberg (2013)
16. Cunha, J., Fernandes, J.P., Peixoto, C., Saraiva, J.: A quality model for spreadsheets. In: 8th Int. Conf. on the Quality of Information and Communications Technology, Quality in ICT Evolution Track, QUATIC 2012, pp. 231–236 (2012)
17. Cunha, J., Fernandes, J.P., Pereira, R., Saraiva, J.: Graphical querying of model-driven spreadsheets. In: Yamamoto, S. (ed.) HCI 2014, Part I. LNCS, vol. 8521, pp. 419–430. Springer, Heidelberg (2014)

18. Cunha, J., Fernandes, J.P., Saraiva, J.: From Relational ClassSheets to UML+OCL. In: Proceedings of the Software Engineering Track at the 27th Annual ACM Symposium On Applied Computing, SAC 2012, pp. 1151–1158. ACM (2012)
19. Cunha, J., Saraiva, J., Visser, J.: Discovery-based edit assistance for spreadsheets. In: 2009 IEEE Symposium on Visual Languages and Human-Centric Computing (VL/HCC), pp. 233–237 (2009)
20. Cunha, J., Saraiva, J., Visser, J.: From spreadsheets to relational databases and back. In: Proceedings of the 2009 ACM SIGPLAN Workshop on Partial Evaluation and Program Manipulation, PEPM 2009, pp. 179–188. ACM (2009)
21. Cunha, J., Saraiva, J., Visser, J.: Model-based programming environments for spreadsheets. In: de Carvalho Junior, F.H., Barbosa, L.S. (eds.) SBLP 2012. LNCS, vol. 7554, pp. 117–133. Springer, Heidelberg (2012)
22. Cunha, J., Visser, J., Alves, T., Saraiva, J.: Type-safe evolution of spreadsheets. In: Giannakopoulou, D., Orejas, F. (eds.) FASE 2011. LNCS, vol. 6603, pp. 186–201. Springer, Heidelberg (2011)
23. Engels, G., Erwig, M.: ClassSheets: automatic generation of spreadsheet applications from object-oriented specifications. In: Proceedings of the 20th IEEE/ACM International Conference on Automated Software Engineering. ACM (2005)
24. Erwig, M.: Software Engineering for Spreadsheets. IEEE Software 29(5) (2009)
25. Fowler, M.: Refactoring: Improving the Design of Existing Code. Addison-Wesley (August 1999)
26. Hermans, F., Dig, D.: Bumblebee: A transformation environment for spreadsheet formulas. Tech. rep. (2013), http://dx.doi.org/10.6084/m9.figshare.813347
27. Hermans, F., Pinzger, M., van Deursen, A.: Detecting and visualizing inter-worksheet smells in spreadsheets. In: Glinz, M., Murphy, G.C., Pezzè, M. (eds.) ICSE, pp. 441–451. IEEE (2012)
28. Hermans, F., Pinzger, M., Deursen, A.: Detecting and refactoring code smells in spreadsheet formulas. Empirical Software Engineering, 1–27 (2014)
29. McCabe, T.J.: A complexity measure. IEEE Trans. Software Eng. 2(4) (1976)
30. Mitchell, N., Runciman, C.: Uniform boilerplate and list processing. In: ACM SIG-PLAN Workshop on Haskell Workshop, Haskell 2007, pp. 49–60. ACM (2007)
31. Nardi, B.A.: A Small Matter of Programming: Perspectives on End User Computing, 1st edn. MIT Press, Cambridge (1993)
32. Panko, R.: Facing the problem of spreadsheet errors. Decision Line 37(5) (2006)
33. Panko, R.: Spreadsheet errors: What we know. what we think we can do. In: Proceedings of the 2000 European Spreadsheet Risks Interest Group, EuSpRIG (2000)
34. Pereira, R.: Querying for Model-Driven Spreadsheets. Master's thesis, University of Minho (2013)
35. Peyton Jones, S.: Haskell 98: Language and libraries. Journal of Functional Programming 13(1), 1–255 (2003)
36. Powell, S.G., Baker, K.R., Lawson, B.: A critical review of the literature on spreadsheet errors. Decision Support Systems 46(1), 128–138 (2008)
37. Rajalingham, K., Chadwick, D.R., Knight, B.: Classification of spreadsheet errors. In: Proceedings of the 2001 European Spreadsheet Risks Interest Group (EuSpRIG), Amsterdam (2001)

# Towards Managing Understandability of Quality-Related Information in Software Development Processes

Vladimir A. Shekhovtsov and Heinrich C. Mayr

Institute for Applied Informatics, Alpen-Adria-Universität Klagenfurt, Austria
{Volodymyr.Shekhovtsov,Heinrich.Mayr}@aau.at

**Abstract.** Establishing common understanding between the parties in the software process is important for dealing with quality of the prospective software. This process is difficult to organize because the parties (especially, developers and business stakeholders) perceive quality based on different world views. To address this problem, we aim at a solution for managing understandability of quality-related information in the software process. This solution provides the set of understandability assessment activities (aimed at diagnosing problems with communicated terms not belonging to the view of the target party) and understandability improvement activities (aimed at resolving these problems by translating problematic terms between world views and providing necessary explanations). These activities are supported by a modular ontology incorporating available quality-related knowledge; particular configuration of the ontology modules describes the quality view of the involved party. The proposed solution is expected to reduce the time and effort for establishing a communication basis while discussing software quality, thus cutting costs and strengthening the mutual trust of the parties.

**Keywords:** understandability, software process, software quality, quality-related communicated information.

## 1 Introduction

To organize successful software development processes, it is necessary to involve the affected business stakeholders throughout the development lifecycle. Such involvement, however, cannot be organized without establishing common understanding between the parties in the process, such as software developers and business stakeholders. In particular, it is necessary to have such understanding while dealing with quality of the software under development at different lifecycle stages. If the parties fail to understand each other on this issue, they tend to postpone all quality-related communication activities until the later stages of the project (such as acceptance testing); this decision could significantly increase the related costs and effort.

Establishing such common understanding is problematic because the parties (in particular, software developers and business stakeholders) think in different conceptualizations of the real world and use different terminologies, especially when dealing with the quality of the software under development: agreeing on a common point of view is usually time-consuming and often fails.

B. Murgante et al. (Eds.): ICCSA 2014, Part V, LNCS 8583, pp. 572–585, 2014.

To address the above problem, we propose to elaborate a tool-supported framework aimed, in part, at supporting the process of establishing common understanding between the parties on the quality of the software under development; the important component of this framework is aimed at assessing and improving such quality characteristic as *understandability* of quality-related information to be communicated between parties. In this paper, we present the conceptual foundations and the implementation procedures for assessing and improving this characteristic; this research is being conducted as part of the ongoing QuASE project[1] established in cooperation with four local software development companies.

The paper is structured as follows. In Section 2, we establish a context for the presented research by defining the place of understandability management in the framework of the QuASE project. Section 3 provides the background information about understandability as quality characteristic and understandability conflicts; Section 4 introduces understandability management process, Section 5 introduces ontological support for understandability management, the procedures for understandability assessment and improvement are described in Section 6. Section 7 outlines possible usage scenarios for understandability management, Section 8 describes related work; it is followed by conclusions.

## 2    Background: QuASE Project

Research activities related to understandability management for quality-related information in the software process are performed in a course of the QuASE project. This project is devoted to the research and tool development aiming at improving the process of quality-related communication between parties in the software process.

In addition to understandability of information described in this paper, this process addresses *quality of decisions* based on communicated information, addressing this characteristic is supposed to be achieved, in particular: by issuing recommendations on the ways of conducting quality-related communications based on the analysis of the past experience of organizing such communications; it is also supposed to support these decisions by forecasting communication parameters and prospective outcomes. The goal of these recommending activities is to increase the parties' awareness of the communication context and the possible ways of action prior to and during the communication, lowering the effort necessary for coming to the right decisions w.r.t. implementing the communication.

Addressing understandability and quality of decisions is supposed to be implemented based on knowledge-oriented access interface to the communication-related data (representing communicated information) collected in industry software development projects which is supposed to be implemented as a part of QuASE software solution. This interface can be exemplified as follows:

---

[1] The QuASE Project is sponsored by the Austrian Research Promotion Agency (FFG) in the framework of the Bridge 1 program (http://www.ffg.at/bridge1); Project ID: 3215531.

1. Communication-related data in the industrial software development projects is kept in the project repositories, in particular, those controlled by issue management systems (IMS) such as JIRA [8]. To implement knowledge-oriented access interface it is proposed to provide an access to the data available in such databases through an ontology (*QuOntology*) capturing the concepts related to the domain of quality-related communications between parties in the software process; this ontology is supposed to contain not only the generic domain concepts, but also the concepts specific for the particular development contexts.

2. The provided interface includes the set of concepts and their attributes extending the structure of the data available in project repositories (referred to as *primary project information*) with *semantic annotations*, represented by additional concepts, concept attributes, and inter-concept relationships necessary for addressing the above quality characteristics. In particular, such *annotation information* can correspond to the factors influencing the attitude to software quality possessed by the parties in the software process, and, as a result, to the process of these parties' decision making in a course of communication.

# 3    Background: Information Understandability

## 3.1    Understandability: A Definition

In selecting understandability as the quality characteristic to be addressed in QuASE project we follow ISO/IEC 9126 quality model [6] (which defines understandability as a quality characteristic for software artifacts) by extending it to the case of quality-related communicated information in the software process: understandability is defined as *"the capability of the quality-related communicated information to enable the target party to understand its meaning and follow procedures defined therein exactly as intended by the originating party"*.

To elaborate on this, we follow the work of Adolph et al [1] who investigated the process of negotiating different perspectives in software development; in a course of this investigation, they defined the concept of *perspective mismatch*. According to this paper, "the inability to get everyone on the same page is a significant impediment to getting the job done; we referred to the source of these impediments, created by differing points of view to getting the job done, as a perspective mismatch".

Based on this concept, it is possible to state a draft operational definition for understandability which could be the source for defining the respective metrics: "*understandability of the quality-related communicated information is defined as a reciprocal of the perspective mismatch between the originated and the target party revealed from this information*." Such mismatch can be measured as the distance between the point of view expressed in the particular piece of quality-related information and the point of view of the target party. As we defined understandability as a reciprocal, the higher the information understandability, the shorter has to be this integrated distance.

## 3.2    Understandability Conflicts

In classifying the understandability conflicts, we also follow Adolph et al [1] who define the following categories of such conflicts:

1. *Translation-inducting conflicts* are related to the situations when only the terminology differs, but perspectives are aligned i.e. people are talking about the same things and share the common communication goals;
2. *Broadening-inducting conflicts* are related to the situations when it is necessary to broaden the understanding of the job by trying to understand the other's point of view; these conflicts involve aligning the perspectives (views on quality) as precautions. An example of such conflict could be the situation when the communicated information is explained from the business point of view but not only with different terminology but with different e.g. quality view; e.g. business consequences of the particular decisions are not explained;
3. *Scouting-inducting conflicts* refer to the cases when neither side can express the perspective to the other so it is necessary for both sides to acquire additional information to perform this task; the problem with such conflicts is that sometimes the sides are not able to do so which could lead to endless cycles in the software process;

These three categories of conflicts serve as a motivation for QuASE understandability research. We aim at the following activities:

1. addressing translation-inducting conflicts by directly supporting the terminology translation with a goal of reducing the effort necessary to perform this task;
2. reducing the impact of broadening-inducting conflicts by supplementing the information with the necessary explanations broadening the perspective (e.g. supplementing particular technical terms with the appropriate explanations related to their influence to business);
3. eliminating or reducing the number of scouting-inducting cases and translate such cases into the cases of broadening-inducting conflicts by supplying the sides with the necessary information helping to explain their perspective to the other side.


## 3.3    Research Goals

The problem of addressing the above conflicts for the specific case of understandability of quality-related communicated information in the software process leads to establishing the following research goals:

1. Adapt the existing systems of categories of understandability conflicts such as defined in [1] to the current case of the understandability of quality-related communicated information in the software process;
2. Establish and evaluate understandability quality subcharacteristics (assessment criteria) and relevant metrics for quality-related communicated information in the software process;

3. Define and implement the procedures for revealing the prospective understandability conflicts based on quality-related communicated information in the software
process and the appropriate criteria to assess the applicability of these procedures;
4. Define and implement the procedures for resolving the understandability conflicts
based on quality-related communicated information in the software process and the
appropriate criteria to assess the applicability of these procedures;

## 4     Understandability Management Process: An Overview

The above goals are planned to be addressed by the specific *understandability management process* comprised of the following activities:

1. *understandability assessment activities* aimed at revealing the prospective understandability conflicts; these activities evaluate the understandability for the particular pieces of quality-related communicated information and provide the relevant
recommendations;
2. *understandability improvement activities* aimed at resolving the understandability
conflicts; these activities aim at maximizing the understandability criteria related to
target parties for the particular pieces of quality-related communicated information.

Establishing these activities is based on the auxiliary activities addressing research
goals 1 and 2 as the former rely on the classification system of the understandability
conflicts and the appropriate set of quality subcharacteristics and quantitative metrics.

## 5     Ontological Support for Understandability Management

The support for information understandability management in QuASE solution is
based on implementing the access to the communicated information (whose understandability is being managed) through the modular ontology (QuOntology) providing
the capabilities of translating between world views. Initial research on QuOntology
has been published in [15], whereas [13] aims at presenting the current version of the
relevant conceptualizations.

The structure of QuOntology is depicted on Fig.1; it includes the following three
layers:

1. QuOntology core;
2. Domain ontology layer;
3. Context ontology layer.

*QuOntology core* represents a stable subset of the knowledge available as a result of
research and industrial practice; the knowledge represented in core does not depend
on the particular problem domain and the particular context. We use Unified Foundational Ontology (UFO) [4, 5] as a foundation for QuOntology core.

*Domain ontologies* represent the specifics of the particular problem domain which is addressed by the particular software under development (finance, banking, oil and gas etc.), the concepts from domain ontologies extends base concepts represented in QuOntology core [5]. Predefined domain ontology is *the ontology of quality harmonization* which includes the set of concepts specific for the domain of quality-related negotiations in software engineering; supplying additional domain ontologies is a separate task which should be completed while adapting QuASE solution to support development for the particular domain (it is planned to supplement the particular configuration of the QuASE tool with the set of domain ontologies).

*Context ontologies* represent the knowledge depending on particular components of the knowledge context; here we define the knowledge context as a set of all concepts connected to the particular communication process (organization, organization type, project, project type, stakeholder, stakeholder type, etc.), in particular, it is possible to distinguish:

1. Context ontology defined for the particular organization type;
2. Context ontology for the particular organization;
3. Context ontology for the particular project type;
4. Context ontology for the particular stakeholder category.

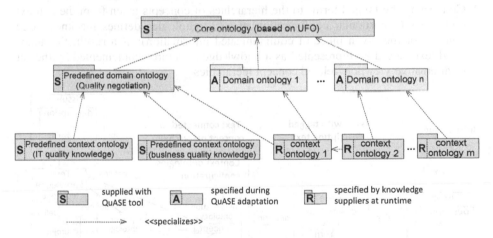

**Fig. 1.** QuASE ontology layers

All the concepts represented in context ontologies must extend the generic concepts represented in QuOntology core and the domain ontologies; in addition, it should be possible to extend the concepts represented other context ontologies. Two predefined context ontologies are planned to be implemented: *IT quality ontology*: the ontology representing common knowledge about quality possessed by IT people and *business quality ontology* representing common knowledge about quality possessed by business stakeholders; the additional context ontologies are supposed to be supplied by involving the experts (knowledge suppliers) at runtime.

## 6     Understandability Management Process: Activities

### 6.1     Text-Based Semantic Annotation

Both understandability assessment and understandability improvement activities rely on the common auxiliary activity aimed at *annotating primary information* (available in the project repositories) based on the additional knowledge provided by the QuOntology or explicitly specified by the user.

We refer to *semantic annotation* as to an activity of extending the set of attributes of quality-related communicated information or the set of its additional connected concepts with a set of attributes and concepts (*annotation information*) important for reaching particular goals of the QuASE solution.

In *text-based semantic annotation*, the values for the additional attributes of quality-related communicated information and the data for the additional connected concepts are extracted from natural language texts available in project repositories (in particular, such texts can represent issue descriptions).

Text-based semantic annotation is performed in two stages (Fig.2):

1. Analyzing natural language text with a purpose of recognizing and tagging the terms to be annotated;
2. Connecting the tagged terms to the hierarchies of concepts taken from the context ontologies (a particular configuration of such ontologies defines the knowledge context for the given piece of communicated information); as a result, the annotated text should be represented as a knowledge structure supplemented by the set of alternative viewport-related concept hierarchies.

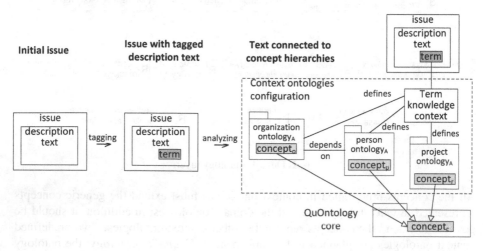

**Fig. 2.** Text-based semantic annotation (domain ontology layer is omitted, adapted from [14])

The results of text-based semantic annotation (annotation information) are used by information understandability management activities (both understandability assessment and understandability improvement); the specifics of this usage are outlined in the following sections.

## 6.2    Understandability Assessment

Understandability assessment activities involve assessing the particular piece of quality-related communicated information (e.g. represented by the annotated issue data available in the project database controlled by the issue management system) for the presence of understandability problems and the applicability of the particular understandability improvement techniques.

In particular, in a course of understandability assessment the following activities could be performed:

1. Calculating the distance between the current piece of communication information and the stored perspective information;
2. Reporting the distance and the categories of mismatch;
3. Forming the recommendations for performing further understandability improvement activities.

Particular recommendations could be related to the fact that in some situations (e.g. while dealing with a broadening-inducting problem) only the explanations are necessary to be provided and in other situation (i.e. while dealing with a translation-inducting problem) the terminology translation could be possible.

Additional research activities related to the understandability assessment could be, in particular, exemplified as follows:

1. Investigating the properties of the particular piece of communicated information which makes its translation feasible;
2. Investigating what makes a particular natural language text a good text from the point of view of its understandability for other parties and what makes a text a bad one;
3. For the case of IMS issues, investigating what makes a particular issue taken as a whole (as a set of attributes and natural language texts) a good issue from the point of view of its understandability for other parties;
4. Investigating the characteristics of the amount of effort to be applied to the particular text to reach the particular target related to its understandability.

## 6.3    Understandability Improvement

The goal of understandability improvement procedures is minimizing the distance between the current piece of quality-related communicated information and the perspective of the target party (the *understandability gap*). The following preliminary activities could be performed beforehand:

1. extracting the perspective information from the project repositories and storing the extracted perspectives in the knowledge base;
2. calculating the distance between the current piece of information and the stored perspective (by the means of understandability assessment)

After that, it is necessary to minimize the distance by applying the understandability improvement (perspective mismatch resolution) procedure.

The generic improvement approach could be specified as follows

1. Finding the source of misunderstanding and separating this source from the rest of the information; for terminology translation case this could be exemplified by singling out the set of problematic terms;
2. Bringing the source of misunderstanding closer to the perspective of the target party; for terminology translation case this can be exemplified by applying the translation procedures to the separated misunderstanding source; the appropriate termination criteria have to be defined for this process;
3. Merging the separated parts back to obtain the adapted textual description.

The following research questions have to be addressed prior to implementing understandability improvement procedures:

1. How to define the knowledge structure representing the point of view of the particular party in the software process?
2. How to define the distance between the particular piece of quality-related communicated information (e.g. a particular issue) and the point of view of the particular party? What are the characteristics of this distance?

**Translation-Based Understandability Improvement Procedure.** For the particular case of translation-inducting understandability conflict, it is proposed to apply the following translation-based understandability procedure aimed at transforming issue-related information between "world views" of the communicating parties. This procedure has to be applied to the natural language text, it is also possible to have a scenario of converting the structured information, but the main goal of the transformation subsystem is related to dealing with natural language specifications and stakeholder opinions.

To support understandability improvement through terminology translation it is necessary to implement switching between knowledge contexts related to recognized terms; this allows translating between world views of different communicating parties. The understandability improvement procedure for the case of translation-inducting conflict is performed in two stages (Fig.3):

1. Resolving (by applying the ontological reasoner) the corresponding concepts defined in context ontologies into the generic concepts defined in QuOntology core or domain ontologies;
2. Switching to the target context, looking up the target context concepts corresponding to the generic concepts; these target concepts form the translation results.

On the first stage, it should be possible to run the ontological reasoner to establish the connection between the recognized term and generic party-independent knowledge. In particular, if the description contains the term "Oracle RDBMS" (IT specific) the analysis process with a help of the reasoner should be able to relate it to more neutral term (e.g. "data storage"). By looking for all the generic concepts it would be possible to figure out all party-independent knowledge related to the particular issue description or opinion fragments; so the purpose of this improvement technique is to enable matching the independent concepts in QuOntology core to the concepts found in the particular description.

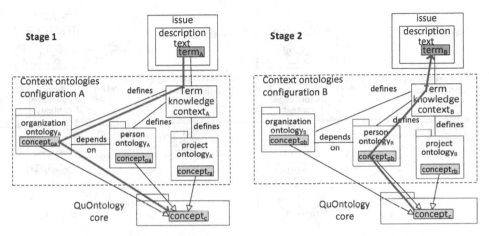

**Fig. 3.** Understandability improvement for the case of terminology conflicts

On the second stage, the different context ontology configurations should be used so the generic concepts have to be translated back to the context-specific concepts and then to corresponding target terms.

**Explanation-Based Understandability Improvement Procedure.** Besides the translation-inducting understandability improvement procedure, additional improvement techniques have to be specified to address the cases where the terminology translation is not sufficient. In particular, explanation-based improvement procedure has to be defined for the case of broadening-inducting conflicts; it has to supplement the problematic terms with extensive explanations aimed at support broadening of the particular point of view. To make able applying this procedure, the context ontology should contain explanations for concepts. A particular explanation could be defined:

1. in the same ontology as the concept to be explained: this reflects the situation when the target party directly knows about the particular concept, but needs additional information to understand it completely;
2. in a different related ontology at the same level (e.g. the ontology for other related context element): this reflects the situation when the target party can get the explanation for the particular term from the related context knowledge (e.g. the knowledge related to the relevant project or the involved organization);
3. in the ontology below the concept to be explained (Fig.4); this reflects the situation when the target party only knows about some generic definition of the concept, but needs additional target-specific information to understand it in necessary detail.

The procedure for explanation-based understandability improvement for the case when explanation is defined in the ontology below the concept to be explained is depicted on Fig.4. This procedure also involves switching between knowledge contexts and obtaining the necessary explanation from the target context.

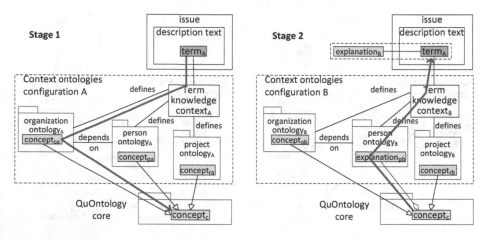

**Fig. 4.** Explanation-based understandability improvement

## 7    Usage Scenarios

In this section, we outline the usage scenarios for understandability assessment and understandability improvement for the case when communicated information is represented by issue descriptions, and the knowledge context is defined for particular business stakeholders.

Understandability assessment scenarios involve calculating understandability metrics for the source issues according for the given stakeholder's context, identifying the problems with the issue: showing the list of understandability problems for the given issue, and issuing recommendations for dealing with these problems.

For calculating understandability metrics the user has to:

1. select the source context (the original context for the issue e.g. its author) and the target context (the context to check the issue against e.g. the business stakeholder);
2. select the issue or the set of issues;
3. obtain the values characterizing the effort necessary to understand the issue in a target context (i.e. by the target stakeholder).

For diagnosing understandability problems for the particular issue, the user has to:

1. select source and target stakeholders and the set of issues as defined above;
2. obtain the set of problematic terms from the issue description text.

Besides understandability assessment scenarios, QuASE aims at supporting both translation-based and explanation-based understandability improvement scenarios. In translation-based scenario (Fig.5) the user has to:

1. select the issue or the set of issues or supplies the arbitrary text; Fig.5 reflects the situation when the arbitrary text has to be supplied;
2. select the source and target stakeholder;
3. obtain the version of the specified text with translated terms.

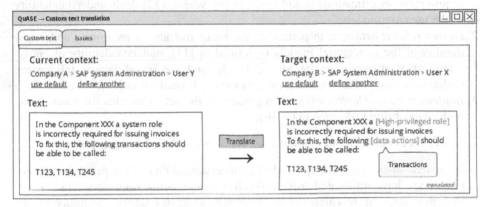

**Fig. 5.** Prototype UI for translation-based understandability improvement scenario

Explanation-based scenario entails performing the following activities (Fig.6):

1. selecting source and target stakeholders and the set of issues as defined above;
2. highlighting the problematic terms in the document and obtaining explanations for these terms.

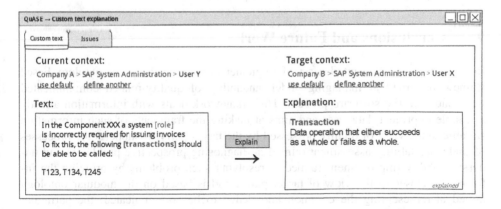

**Fig. 6.** Prototype UI for explanation-based understandability improvement scenario

## 8     Related Work

Current research addressing understandability of the information in the software process mainly deals with this quality characteristic defined for the following software process artifacts (we group these artifacts by the stage of the development process):

1. *Requirement engineering-related artifacts*: in particular, understandability of the requirement specifications is addressed in [9], whereas [2] deals understandability of the use case models;
2. *Design-related artifacts*: in particular, the set of metrics for measuring understandability of the conceptual models is defined in [11], understandability of entity-relationships diagrams is introduced in [3], and the set of factors influencing understandability of the business process models is outlined in [12];
3. *Implementation-related artifacts*: in particular, the set of metrics for source code understandability is defined in [7, 10];

The differences between our approach and the above techniques are as follows:

1. Most state-of-the-art techniques address understandability of the particular categories of development-related artifacts (such as requirements, source code, or conceptual models); our research, to the contrary, addresses understandability of the generic fragments of communicated information which could be contained in documents belonging to different categories; in this paper, these documents are exemplified by issues;
2. These techniques, as a rule, do not specifically address understandability of quality-related information;
3. They do not employ ontology-based approach for establishing common understanding between parties in the software process.

## 9     Conclusions and Future Work

In this paper, we presented a set of implementation procedures for ontology-based framework aimed at managing understandability of quality-related communicated information in the software process. This framework deals with information already available in project databases and aims at making the fragments of such information suitable to the view of quality possessed by the target party. It supports the processes of understandability assessment (aimed at diagnosing prospective problems) and understandability improvement (aimed at resolving such problems by bringing the information closer to the view of target party) and is based on the modular ontology aimed at representing the common knowledge to be communicated; the particular configuration of the ontology modules describes the knowledge on quality possessed by the particular understandability context such as the customer organization or the particular stakeholder; this ontology is applied to the raw issue data prior to communication. Establishing such framework allows us to address the problem of providing the parties in the software process an easy way of adapting the information to the point of view of other parties.

We plan to continue our research by defining detailed implementation procedures and the tool support for understandability management activities; in addition, the ongoing research aims at establishing the structure of knowledge base (QuIRepository) mapping the ontological knowledge into the project database data.

# References

1. Adolph, S., Kruchten, P., Hall, W.: Reconciling perspectives: A grounded theory of how people manage the process of software development. The Journal of Systems and Software 85, 1269–1286 (2012)
2. Anda, B., Jørgensen, M.: Quality and understandability of use case models. In: Lindskov Knudsen, J. (ed.) ECOOP 2001. LNCS, vol. 2072, pp. 402–428. Springer, Heidelberg (2001)
3. Genero, M., Poels, G., Piattini, M.: Defining and validating metrics for assessing the understandability of entity–relationship diagrams. Data & Knowledge Engineering 64, 534–557 (2008)
4. Guizzardi, G.: Ontological foundations for structural conceptual models. University of Twente (2005)
5. Guizzardi, G., Falbo, R., Guizzardi, R.S.: Grounding software domain ontologies in the unified foundational ontology (UFO): The case of the ODE software process ontology. In: Proceedings of the XI Iberoamerican Workshop on Requirements Engineering and Software Environments, pp. 244–251 (2008)
6. ISO/IEC 9126-1:2001: Software Engineering – Product Quality – Part 1: Quality Model. International Organization for Standardization, Geneva (2001)
7. Jin-Cherng, L., Kuo-Chiang, W.: A Model for Measuring Software Understandability. In: CIT 2006. IEEE (2006)
8. JIRA Issue Tracking System, http://www.atlassian.com/software/jira (accessed May 08, 2014)
9. Kamsties, E., von Knethen, A., Reussner, R.: A controlled experiment to evaluate how styles affect the understandability of requirements specifications. Information and Software Technology 45, 955–965 (2003)
10. Lin, J.-C., Wu, K.-C.: Evaluation of software understandability based on fuzzy matrix. Fuzzy Systems, 2008. In: FUZZ-IEEE 2008, pp. 887–892. IEEE (2008)
11. Mehmood, K., Cherfi, S.S.: Data quality through model quality: a quality model for measuring and improving the understandability of conceptual models. In: MDSEDQS 2009, pp. 29–32. ACM (2009)
12. Reijers, H.A., Mendling, J.: A study into the factors that influence the understandability of business process models. IEEE Transactions onSystems, Man and Cybernetics, Part A: Systems and Humans 41, 449–462 (2011)
13. Shekhovtsov, V.A., Mayr, H.C., Kop, C.: Harmonizing the Quality View of Stakeholders. In: Mistrik, I., Bahsoon, R., Eeles, R., Roshandel, R., Stal, M. (eds.) Relating System Quality and Software Architecture. Elsevier (in print, 2014)
14. Shekhovtsov, V.A., Mayr, H.C.: Managing Quality Related Information in Software Development Processes. In: CAiSE 2014 Forum. CEUR-WS.org (in print, 2014)
15. Shekhovtsov, V.A., Mayr, H.C., Kop, C.: Towards Conceptualizing Quality-Related Stakeholder Interactions in Software Development. In: Kop, C. (ed.) UNISON 2012. LNBIP, vol. 137, pp. 73–86. Springer, Heidelberg (2013)

# Web Accessibility in Social Networking Services

Janaína R. Loureiro, Maria Istela Cagnin, and Débora M.B. Paiva

College of Computing, Federal University of Mato Grosso do Sul (UFMS)
Campo Grande, MS, Brazil, PO Box 549, 79070-900
{janrloureiro,istela,dmbpaiva}@gmail.com

**Abstract.** Considering the rise of social networking services occurred in recent years and that about 15% of the world's population have some form of disability, it's important to estimate the software quality sub-characteristic of web accessibility offered by such services. The objective of this study is to evaluate web accessibility of a social networking sites sample, considering the most popular portals of this domain, by the perspective of one of the most common disabilities: visual impairment. These assessments were planned using GQM paradigm and were carried out considering two stages, including evaluations by WCAG 2.0 auto-mated tools and computer specialists with expertise in web accessibility. With results generated, it was possible to indicate which web accessibility issues need more attention from web developers in social networking context and to observe that none of the sample sites analyzed did not reach even the lowest conformance level of web accessibility described by WCAG 2.0.

**Keywords:** Web accessibility, social networking services, WCAG 2.0.

## 1 Introduction

In 2005, around 20% of world's population were using the Internet, number that has increased to 40% in 2014, as stated by International Telecommunication Union (ITU) [9]. In 2011, World Health Organization (WHO) reported in [27] that over a billion people in the wold have some kind of disability, representing 15% of global population, 5% more than they had previously estimated by the 1970s' reports.

The power of the Web is in its universality. Access by everyone regardless of disability is an essential aspect [2]. Thus, we may question why currently only 2% of the Web pages are accessible to disabled people [14]. After all, as stated by Radabaugh [15], "For people without disabilities, technology makes things easier. For people with disabilities, technology makes things possible."

Despite efforts to regulate web accessibility, the attention received in practice by this subject is still insuficient, being necessary to make it even more present among developers and researchers. In this context, this study approaches the issue of web accessibility, submitting the most popular social networking services in the world (Facebook [6], LinkedIn [12], ResearchGate [16] and Twitter [21])

B. Murgante et al. (Eds.): ICCSA 2014, Part V, LNCS 8583, pp. 586–601, 2014.

to the evaluation of WCAG 2.0 automated tools and experts. Social networking services were chosen because statistics from ITU shows that the number of people using social media services reached the one billion mark in 2012 [8].

The decision to address the visually impaired was based in the study of Ruth-Janneck [17], in which a group of users with different disabilities (visual, hearing, mental and motor) was interviewed. The study showed that users with total visual impairment are the ones that most consider web accessibility as important, crediting to it much of its social inclusion by enabling their access to the Internet. Additionally, it was observed that visual impairment it is the one that creates more barriers to web browsing in social networking services.

This paper is organized as follows: Section 2 reviews important concepts and definitions to the study, Section 3 defines the GQM plan adopted for evaluations and details the results obtained in each one, while Section 4 discusses about what was observed in evaluations and makes final remarks regarding the web accessibility in social networking services.

## 2    Concepts Review

### 2.1    Web Accessibility

Given the collaborative nature of Internet and its importance in facilitating communication, it is essential to consider, refine and expand technical issues and to highlight the need of considering accessibility from web developer and user perspective [5]. Accessibility is a goal that involves generation of documents that can be rendered by different devices. This includes access by blind people using screen readers with speech synthesizers [7].

### 2.2    WCAG 2.0

The World Wide Web Consortium (W3C), whose mission is to manage Internet to reach its full potential by developing protocols and guidelines that ensure its long-term growth [22], maintains a segment called Web Accessibility Initiative (WAI) that, in order to promote web accessibility, developed WCAG guidelines (Web Content Accessibility Guidelines). The latest version of WCAG is the 2.0, launched in 2008. It covers a wide range of recommendations for making web content more accessible [23] through 12 guidelines that are organized around four principles [24]: Perceivable, Operable, Understandable and Robust.

The concept of conformance level is related to the guidelines satisfaction [10]. If all success criteria (CS) of level A are met, the document reaches conformance level A. If all success criteria from levels A and AA are met, the document is in compliance with level AA. Finally, if all the success criteria from levels A, A and AAA are met, the document receives conformance level AAA classification [4].

### 2.3    WCAG 2.0 Automated Evaluation Tools

Automated evaluation tools consider the HTML code and make an analysis of their content by examining whether the site fits the principles and guidelines of the WCAG 2.0, significantly reducing the time and effort required for conducting evaluations of web accessibility [26]. Usage of automated tools was the first of two stages developed in the study. Three tools were chosen to be applied: AccessMonitor [1], Taw [19] and Total Validator [20], in order that one result could complete or validate the other. AccessMonitor was selected because the researcher already had experience with this tool, and the others because of their relevance, appearing in the list of tools specified by the Web Accessibility Initiative [25].

### 2.4    Social Networking Sites Sample

From the total of four sites within the domain of social networks in the sample, three were chosen according to their number of global users, according to ITU [8]. Therefore, the three most popular social networks were: Facebook (900 millions users), Twitter (200 millions users) and LinkedIn (150 millions users). The fourth and final element of the sample was ResearchGate, chosen for being described as a mash-up of Facebook, Twitter and LinkedIn [11], but focused to the academic society.

### 2.5    Goal Question Metric Paradigm (GQM)

The GQM approach is goal-oriented, and claims that a meaurement must always be justified by a purpose. Thus, the first stage is to specify the intended goals, link those goals to the data that will define them operationally, and finally provide a framework to interpret the data in relation to established goals. As a result of implementing the Goal Question Metric strategy, we have the specification of a system of measures to a particular set of issues, as well as the rules for the interpretation of the data obtained [3].

## 3    Evaluations

To measure web accessibility, that is a subcharacteristic of software quality attribute of Usability, of selected social networking services sample, two kinds of reviews were applied: automated evaluation tools based on WCAG 2.0 guidelines and computer experts with prior knowledge about WCAG 2.0. The decision of developing a study considering two different evaluations was made as an attempt to cover all perspectives about web accessibility of the sample: from the code syntax of pages through semantic evaluation of their content.

Reviews were written by eight computer experts from Faculty of Computing/Federal University of Mato Grosso do Sul, located at Campo Grande, MS, Brazil. Because it involves other people, the research project was previously submitted to the Ethics Committee for Research Involving Human Subjects of Platform Brazil CEP / CONEP, receiving its approval in December 2013, under the report number 497406 and CAAE 25054113.0.0000.0021. All participants of this study signed an informed consent form, voluntarily agreeing to make the assessments proposed by researchers and allowing them to use the data generated.

## 3.1  GQM Plan Definition

For the two types os evaluation developed (automated tools and computer experts), we used the GQM paradigm, in which the first step is to establish the study goal. The goal to be considered is as follows:

**Goal 1:** Evaluate which success criteria of WCAG 2.0 are not met by the sample of social networking sites being analyzed, and if any conformance level of web accessibility is satisfied.

Two questions must be answered to assess the web accessibility in order to meet the first goal:

- **Question 1:** Which success criteria of WCAG 2.0 are not satisfactorily met by each portal?
- **Question 2:** Which success criteria of WCAG 2.0 are not satisfactorily met by the sample of social networking sites considered?

**Table 1.** Subset of WCAG 2.0 success criteria from level A considered

| Principle | Guideline | Success Criterion |
|---|---|---|
| 1 Perceivable | 1.1 Text Alternatives | 1.1.1 Non-text Content: All non-text content that is presented to the user has a text alternative that serves the equivalent purpose. |
| | 1.2 Time-based Media | 1.2.1 Audio/Video-only (Prerecorded): An alternative for time-based media is provided that presents equivalent information for prerecorded audio/video content. |
| | | 1.2.2 Captions (Prerecorded): Captions are provided for all prerecorded audio content in synchronized media. |
| | | 1.2.3 Audio Description or Media Alternative (Prerecorded): An alternative for time-based media or audio description of the prerecorded video content is provided for synchronized media. |
| | 1.3 Adaptable | 1.3.1 Info and Relationships: Information, structure, and relationships conveyed through presentation can be programmatically determined or are available in text. |
| | | 1.3.2 Meaningful Sequence: When the sequence in which content is presented affects its meaning, a correct reading sequence can be programmatically determined. |
| | | 1.3.3 Sensory Characteristics: Instructions provided for understanding and operating content do not rely solely on sensory characteristics of components such as shape, size, visual location, orientation, or sound. |

**Table 1.** (*continued*)

| | 1.4 Distinguishable | 1.4.1 Use of Color: Color is not used as the only visual means of conveying information, indicating an action, prompting a response, or distinguishing a visual element. |
|---|---|---|
| | | 1.4.2 Audio Control: If any audio on a Web page plays automatically for more than 3 seconds, either a mechanism is available to pause or stop the audio, or a mechanism is available to control audio volume independently from the overall system volume level. |
| 2 Operable | 2.1 Keyboard Accessible | 2.1.1 Keyboard: All functionality of the content is operable through a keyboard interface without requiring specific timings for individual keystrokes. |
| | | 2.1.2 No Keyboard Trap: If keyboard focus can be moved to a component of the page using a keyboard interface, then focus can be moved away from that component using only a keyboard interface. |
| | 2.3 Seizures | 2.3.1 Three Flashes or Below Threshold: Web pages do not contain anything that flashes more than three times in any one second period. |
| | 2.4 Navigable | 2.4.1 Bypass Blocks: A mechanism is available to bypass blocks of content that are repeated on multiple Web pages. |
| | | 2.4.2 Page Titled: Web pages have titles that describe topic or purpose. |
| | | 2.4.3 Focus Order: If a Web page can be navigated sequentially and the navigation sequences affect meaning or operation, focusable components receive focus in an order that preserves meaning and operability. |
| | | 2.4.4 Link Purpose (In Context): The purpose of each link can be determined from the link text alone or from the link text together with its programmatically determined link context. |
| 3 Understandable | 3.3 Input Assistance | 3.3.1 Error Identification: If an input error is automatically detected, the item that is in error is identified and the error is described to the user in text. |
| | | 3.3.2 Labels or Instructions: Labels or instructions are provided when content requires user input. |
| 4 Robust | 4.1 Compatible | 4.1.1 Parsing: In content implemented using markup languages, elements have complete start and end tags, elements are nested according to their specifications, elements do not contain duplicate attributes, and any IDs are unique. |
| | | 4.1.2 Name, Role, Value: For all user interface components, the name and role can be programmatically determined and notification of changes to these items is available to user agents, including assistive technologies. |

For each proposed question a metric was defined, representing the third GQM element. The assessments are differentiated by index $a$: evaluation by automated tools ($a = 1$) and by experts ($a = 2$).

In order to keep the study scope well-defined, we chose a subset of WCAG 2.0 success criteria regarding only to level A. This decision was made because while conducting a preliminary evaluation of the sample sites with automated tools, trying to determine which conformance level the sites reached, we observed that none of the sample sites was successfull on level A. Therefore, as conformance levels are progressive, it was decided to focus only on success criteria level A.

Twenty level A success criteria were included in the analysis carried out in this study, as presented in Table 1. This subset was selected according to Ruth-Janneck's study [17], that reports these success criteria as essential to web navigation of visual impaired. The success criteria are presented keeping the hierarchical structure of WCAG 2.0, indicating principles and guidelines that each criterion belongs to.

Metrics defined in GQM plan are presented as follows:

**Metric 1.** This metric is related to the first question (Q1: Which success criteria of WCAG 2.0 are not satisfactorily met by each portal?). Consider that $P_{aecs}$ is a binary variable that assumes value 1 if, during the appraisal $a$, the evaluator $e$ (automated tool or expert), detected at least one error related to success criterion $c$ at site $s$, and 0 otherwise. Let $TP_{acs}$ be an integer variable that tells how many evaluators indicated errors related to the success criterion $c$ at site $s$. If $TP_{acs}$ assumes a value that is greater or equal to half of the universe of evaluators, thus indicating a consensus within the majority, then success criterion $c$ is said as not met by the site $s$. Below we describe formulas to calculation and analysis of $TP_{acs}$ on each evaluation, valid to $\forall c, 1 \leq c \leq 20, \forall s, 1 \leq s \leq 4$:

$$TP_{1cs} = \sum_{c=1}^{3} P_{1ecs}$$

if $TP_{1cs} \geq 2$, $c$ is not met

(a) Automated tools

$$TP_{2cs} = \sum_{e=1}^{2} P_{2ecs}$$

if $TP_{2cs} = 2$, $c$ is not met

(b) Experts

Evaluation made by experts deserves special attention. Since two specialists reviewed each site, it is possible that a draw happens in case they disagree. In these situations, it will be adopted as a first tiebreaker the results of this criterion on the evaluation made by automated tools. If the tie persists, meaning that the criterion was not presented as not met by the previous evaluation, the opinion from the current researcher will be considered. This way, a success criterion is said not met after evaluated by the experts if it receives exactly two negative evaluations.

**Metric 2.** This metric was defined to answer the second question (Q2: Which success criteria of WCAG 2.0 are not satisfactorily met by the sample of social networking sites considered?). Consider $P_{aecs}$ as described on Metric 1. Now being $TPD_{ac}$ an integer variable that tells, for each evaluation $a$, how many evaluators $e$ indicated errors related to the success criterion $c$ at the sample of social networking sites. Below we present formulas to calculation and analysis in each evaluation, valid for $\forall c, 1 \leq c \leq 20$:

$$TPD_{1c} = \sum_{s=1}^{4}\sum_{e=1}^{3} P_{1ecs}$$

if $TPD_{1c} \geq 6$, $c$ is not met

(a) Automated tools

$$TPD_{2c} = \sum_{s=1}^{4}\sum_{e=1}^{2} P_{2ecs}$$

if $TPD_{2c} \geq 4$, $c$ is not met

(b) Experts

## 3.2 Evaluation by Automated Tools

Evaluation by automated tools provides a first estimate of the web accessibility level of social networking sites that comprehend the sample analyzed, supporting the planning of further steps according to points that require greater attention.

In October 2013, all sample sites (Facebook, LinkedIn, ResearchGate and Twitter) were submitted to each of the three automated tools. The reports containing the results of the evaluations performed by them were saved for further analysis.

### Metric 1 - Answering Question 1

1. **Facebook.** Regarding the evaluation of automated tools, the application of Metric 1 for Facebook resulted in $TP_{1c1} \geq 2$ in five situations: for the success criteria 1.1.1, 1.3.1, 3.3.2, 4.1.1 and 4.1.2, hence considered as not met. Table 2 describes in details such success criteria and which were the problems related to them detected by automated tools.

**Table 2.** WCAG 2.0 success criteria not met by Facebook according to automated evaluation tools

| WCAG 2.0 SC | Detected Problems | $TP_{1c1}$ |
|---|---|---|
| 1.1.1 Non-text Content | Element input without tag label nor title. | 2 |
| 1.3.1 Info and Relationships also presented non-visually | Element input without tag label nor title. Tag label without textual content and association. Headings hierarchy violated. HTML element applied to control design. | 3 |
| 3.3.2 Labels or Instructions | Element input without tag label nor title. | 2 |
| 4.1.1 Code Parsing | HTML errors detected. Duplicated id. Tag label without textual content. | 3 |
| 4.1.2 Name, Role, Value | Element input without tag label nor title. Tag label without association. | 3 |

**Table 3.** WCAG 2.0 success criteria not met by LinkedIn according to automated evaluation tools

| WCAG 2.0 SC | Detected Problems | $TP_{1c2}$ |
|---|---|---|
| 2.4.1 Bypass Blocks | Inexistence of links to bypass content blocks. First page link doesn't lead to the main content area. | 2 |
| 2.4.4 Link Purpose (In Context) | Link with only a image as content, and without atribute alt. | 3 |
| 4.1.1 Code Parsing | HTML errors detected. | 2 |

**Table 4.** WCAG 2.0 success criteria not met by ResearchGate according to automated evaluation tools

| WCAG 2.0 SC | Detected Problems | $TP_{1c3}$ |
|---|---|---|
| 1.3.1 Info and Relationships also presented non-visually | HTML element applied to control design. Headings hierarchy violated. | 3 |
| 2.4.1 Bypass Blocks | Inexistence of links to bypass content blocks. First page link doesn't lead to the main content area. | 2 |

A great number of detected problems in Facebook is related to absence of description tags as label and title, especially in form elements. Errors in HTML code were also encountered, which reflects in how the page could be interpreted by an assistive technology.

2. **LinkedIn.** Computing results reported by each automated tool about LinkeIn, the variable $TP_{1c2}$ assumed a value greater or equal to 2 in three success criteria: 2.4.1, 2.4.4 and 4.1.1. The description of these criteria and problems related to them encountered by automated tools are presented in Table 3. The principal issue detected in LinkedIn concerns to its links, that not always have an alternative text to help users understand their purpose, and do not offer an option to skip blocks of contents in which the users aren't interested into.

3. **ResearchGate.** Considering ResearchGate, $TP_{1c3} \geq 2$ was observed in only two success criteria: 1.3.1 and 2.4.1, therefore labeled as not met by consensus of the automated tools. The criteria and their problems indicated by automated tools are shown in Table 4.

   In ResearchGate, HTML code was used for formatting visual content of the page, while the recommended is to make use of CSS code for that. This use of HTML code can result in a mistaken interpretation of the page by screen readers. Another problem detected that can influence screen readers performance is violation of headings hierarchy.

4. **Twitter.** The value of variable $TP_{1c4}$ establishes that criterion c should be classified as not met by Twitter according to automated evaluation tools whenever $TP_{1c4} \geq 2$. Regarding the subset of success criteria related to the visually impaired, from WCAG 2.0 level A, Twitter did not reach seven of

**Table 5.** WCAG 2.0 success criteria not met by Twitter according to automated evaluation tools

| WCAG 2.0 SC | Detected Problems | $TP_{1c4}$ |
|---|---|---|
| 1.1.1 Non-text Content | Element input without tag label nor title. Consecutive text and image links with the same target destination. | 3 |
| 1.3.1 Info and Relationships also presented non-visually | Element input without tag label nor title. Tag label without association. Headings hierarchy violated. HTML element applied to control design. Table without tag caption. Element fieldset without description. | 3 |
| 2.4.1 Bypass Blocks | Inexistence of links to bypass content blocks. First page link doesn't lead to the main content area. | 2 |
| 2.4.4 Link Purpose (In Context) | Link with only a image as content, and without atribute alt. | 2 |
| 3.3.2 Labels or Instructions | Element input without tag label nor title. Element fieldset without description. | 3 |
| 4.1.1 Code Parsing | HTML errors detected. Duplicated id. | 3 |
| 4.1.2 Name, Role, Value | Element input without tag label nor title. Tag label without association. Element frame and iframe without tag title. | 3 |

a total of 20: 1.1.1, 1.3.1, 2.4.1, 2.4.4, 3.3.2, 4.1.1 and 4.1.2. A description of the criteria and issues related to them by automated tools are presented in Table 5.

Regarding to all sites from the sample evaluated, Twitter presented the greatest amount of errors. Alternative text is missing in links, preventing that users be able to understand its purpose, as well as it is not possible to bypass all blocks of information. Elements like forms, fieldsets and frames are also lacking description tags.

## Metric 2 - Answering Question 2

To account the success criteria that presented recurrent problems in the sample sites, indicating critical points in the domain of social networking sevices, we calculated how often each success criterion was considered as not met by automated tools during the evaluation of the four sites, applying the formula from Metric 2 ($TPD_{1c} = \sum_{s=1}^{4} \sum_{e=1}^{3} P_{1ecs}$). Success criteria that showed problems in at least six of the twelve reviews of the sample ($TPD_{1c} \geq 6$) were told as not met by the sample of social networking services domain.

In at least one automated tool, 11 success criteria were considered as not met ($P_{1ecs} = 1$). After applying the Metric 2 formula to calculate $TPD_{1c}$, five success criteria had accumulated at least six errors occurrences associated with them by automated tools within the sample of social networking sites: 1.1.1 , 1.3.1 , 2.4.1, 4.1.1 and 4.1.2. Those are the success criteria of WCAG 2.0 level A considered according to Metric 2 as not met by the sample of the social networking sites domain.

In this context, results related to the Robust Principle represent an issue to worry about: both success criteria of this principle, 4.1.1 and 4.1.2, were labeled as not met. It is a worrying result once that principle assesses if the content is robust enough to be interpreted concisely by different user agents, including assistive technologies [24].

## 3.3    Evaluation by Experts

Despite automated validators used in the previous step be based on WCAG 2.0, they are not able to check guidelines completely, because there are some aspects that can not be fully assessed automatically and need a human judgment, as stated by Sierkowski [18]. This happens because an automated tool verifies the syntax of HTML, but not the semantics of the content. In addition to assessing the existence of accessible element, one must consider whether the alternative offered actually matches the content displayed on the page. For example, an automated tool can detect whether an alternative text is provided to an image, but can not distinguish if the meaning conveyed is the same in both cases - for sighted and visually impaired people [13]. This semantic analysis should then be carried out by specialists.

In January and February 2014, each expert reviewed a site from the social networking services sample, simulating navigation using a screen reader as assistive technology, analyzing and recording the behavior of the screen reader in response to the elements of the site, and also indicating details about the experience.

The aim was that experts could simulate the perspective of visually impaired user while navigating in social networkng sites, assessing then their conformance to guidelines.

The experts were instructed to consider success criteria listed in Table 1, with exception of 4.1.2, that would requires a detailed code evaluation, already held by automated tools. Success criterion 4.1.1 was always the first one evaluated, using the W3C HTML validator, so the HTML errors found could give the experts a first idea of potential accessibility problems they should look for.

## Metric 1 - Answering to Question 1

By applying the formula described in Metric 1 to calculate how many experts have detected problems related to the success criterion $c$ ($TP_{2cs} = \sum_{e=1}^{2} P_{2ecs}$), it was necessary that the two evaluators agreed in their answer to it. If there was disagreement, the criterion was considered not met if this was the decision of previous assessment with automated tools or of the researchers. Therefore, a success criterion was said to be not met whenever $TP_{2cs} = 2$.

1. **Facebook:** The variable $TP_{2c1}$ was equal to 2 in eight of the 20 success criteria evaluated, which were classified as not attended according to experts evaluation, representing barriers to the navigation of visually impaired: 1.1.1, 1.2.1, 1.2.2, 1.2.3, 1.3.1, 2.4.4, 3.3.2 and 4.1.1. The success criteria not met, with main considerations provided by experts who evaluated Facebook, are listed in Table 6.

**Table 6.** WCAG 2.0 success sriteria not met by Facebook according to experts

| SC | Experts Observations |
|---|---|
| 1.1.1 | Images processed just as a link, without meaningful alternative text. |
| 1.2.1 | Videos shared by users do not have transcripts. |
| 1.2.2 | Videos uploaded by facebook itself does not have a subtitle option. |
| 1.2.3 | The media does not present an alternative media or audio description. |
| 1.3.1 | Required fields are not stated. Error message without feedback to the user. |
| 2.4.4 | Links identification are not enough to indicate their purpose. |
| 3.3.2 | Not all forms have tag label to indicate its purpose. |
| 4.1.1 | Page is not well formed, 45 errors detected by HTML validator from W3C. |

**Table 7.** WCAG 2.0 Success Criteria not met by LinkedIn according to experts

| SC | Experts Observations |
|---|---|
| 1.1.1 | Images processed just as a link, without meaningful alternative text. |
| 1.3.1 | Error feedback highlighting field in red, without aditional information to the screen reader. |
| 1.3.3 | Right sided block of information is not accessible via keyboard while there are older posts to be displayed. |
| 1.4.1 | If the password is incorrect, error feedback requests to correct field highlighted, preventing reproduction by screen reader. |
| 2.1.1 | Right sided block of information is not accessible by keyboard. |
| 2.4.1 | Users can not skip blocks they are not interested into. |
| 2.4.4 | Links with null alt attribute, and without target description |
| 3.3.1 | Error feedback without information that can help user understands what went wrong. |
| 4.1.1 | Page is not well formed, 36 errors detected by HTML validator from W3C |

Experts that evaluated Facebook reported problems involving non-textual elements, as videos and images. They stated that videos shared by users do not have neither subtitles nor transcriptions, and that images alternative texts not always are meaningful. Similar happens to links and forms, which purpose is not clearly represented by their identification. All these problems may confuse users, once it's not possible for them to understand what are being displayed on screen without textual alternatives.

They also made observations about lack of errors feedback to indicate when something unexpected happens.

2. **LinkedIn:** In LinkedIn evaluation, the analysis of experts answers and tie-breakers application when neccessary resulted in classification of nine success criteria as not met ($TP_{2c2} = 2$): 1.1.1, 1.3.1, 1.3.3, 1.4.1, 2.1.1, 2.4.1, 2.4.4, 3.3.1 and 4.1.1. Table 7 brings the main observations of experts on these criteria.

LinkedIn do not offer support manipulation of videos, so success criteria 1.2.1, 1.2.2 and 1.2.3 were not evaluated in this site.

One of the most important observations made by experts about LinkedIn is that the right information block of the page is barely accessible via keyboard,

**Table 8.** WCAG 2.0 Success Criteria not met by ResearchGate according to experts

| SC | Experts Observations |
|---|---|
| 1.3.1 | Required fields are indicated with an *, but without explanation of its meaning. If these fields are not completed, feedback is given only by color. |
| 1.4.1 | Incorrectly filled fields are indicated only by color, without non-visual alternative. |
| 2.1.1 | Right sided block of information is not accessible via keyboard while there are older posts to be displayed. |
| 2.4.1 | Users can not skip blocks they are not interested into. |
| 2.4.4 | Links with null alt attribute, and without target description. |
| 3.3.1 | Error feedback is only visually presented, without alternative detectable by screen reader. |
| 3.3.2 | Description is not provided in text entry fields. |
| 4.1.1 | Page is not well formed, 2 errors detected by HTML validator from W3C. |

because the posts of the main block on the left are automatically recharged. This issue prevents visually impaired to access all page content.

Moreover, recurrent problems in other reviews were also detected. Alternative texts for images and links are not always enough to describe them as well as the feedback errors are not much informative. In some cases, an error is indicated only highlighting the field that needs action with a different color, and nothing else is informed to screen readers.

3. **ResearchGate:** When applying Metric 1 to experts evaluation of Research-Gate, $TP_{2c3}$ were equal to 2 in eight success criteria, considered then as not met: 1.3.1, 1.4.1, 2.1.1, 2.4.1, 2.4.4, 3.3.1, 3.3.2 and 4.1.1. Table 8 indicates major considerations of expert about each success criterion not met.

It's important to notice that success criteria 1.2.1, 1.2.2 and 1.2.3 were not applicable to ResearchGate, because the site do not support neither video upload nor sharing.

The ResearchGate expert evaluation was similar to LinkedIn, in which links were not well represented, access to information located on the right sided block of the page depends on the amount of posts on the main feed section and errors are displayed only using a different color for the field.

4. **Twitter:** Application of Metric 1 to Twitter's experts evaluation resulted in eight success criteria labeled as not met by the microblogging site: 1.1.1, 1.3.1, 2.1.1, 2.4.1, 2.4.4, 3.3.1, 3.3.2 and 4.1.1, showed in Table 9 with experts notes about each one.

In addition to the web accessibility barriers already mentioned in previous reviews, the main issue observed in Twitter by experts was layered windows and buttons that are not accessible via keyboard, making it impossible to visually impaired users to use all features offered by the site.

## Metric 2 - Answering Question 2

In order to determine which success criteria should be considered as not met according to expert evaluation by the social networking sites sample evaluated

**Table 9.** WCAG 2.0 Success Criteria not met by Twitter according to experts

| SC | Experts Observations |
|---|---|
| 1.1.1 | Images are not directly related to their description. Non-visible elements are read by screen reader. |
| 1.3.1 | Required fields are not stated. Error message without feedback to the user. Association between text and elements are not clear. |
| 2.1.1 | There are some buttons and layered windows that are not accessible via keyboard. |
| 2.4.1 | Users can not skip blocks they are not interested into. |
| 2.4.4 | Links with null alt attribute, and without target description. |
| 3.3.1 | Error feedback without information that can help user understands what went wrong. |
| 3.3.2 | Description is not provided in text entry fields. |
| 4.1.1 | Page is not well formed, 9 errors detected by HTML validator from W3C. |

in general, Metric 2 ($TPD_{2c} = \sum_{s=1}^{4} \sum_{e=1}^{2} P_{2ecs}$) was applied. $TPD_{2c}$ assumed a value different from 0 for 14 success criteria, but only in ten cases this value was greater or equal to 4: 1.1.1, 1.3.1, 1.3.3, 1.4.1, 2.1.1, 2.4.1, 2.4.4, 3.3.1, 3.3.2 and 4.1.1. Thus, those are the success criteria of WCAG 2.0 Level A considered as not met by the sample of the social networking sites domain under study.

These results shows that the Understandable Principle was the most problematic one, showing that sites from the sample had failures in what concers to providing input assistance to users [23]. All sites from the sample do not met succes criteria 1.3.1, 2.4.4 and 4.1.1, indicating that they displayed errors using only colors or other visual elements to highlight information; that their links descriptions were not informtive enough to indicate their purposes and that their pages were not well formed according to HTML recommendations.

## 4    Discussion and Final Remarks

After applying the two evaluations in the social networking sites sample studied, we have a review of code structure (syntax) of pages provided by automated evaluation tools and another review focused on the meaning of the site (semantics) carried out by computer experts. Therefore, it's possible to list success criteria that need to be inspected regarding the sample web accessibility.

The four social networking sites considered did not meet requirements to achieve the minimum conformance level of web accessibility according to WCAG 2.0. Thus, no conformance level can be attributed to the sample.

When considering the whole sample, all success criteria indicated as not met by automated validators also received this classification in the experts evaluation: 1.1.1, 1.3.1, 2.4.1 and 4.1.1. The exception falls at 4.1.2, which has not been evaluated by experts. Besides that, six other success criteria that required a subjective evaluation were labeled as not met by experts: 1.3.3, 1.4.1, 2.1.1, 2.4.4, 3.3.1 and 3.3.2.

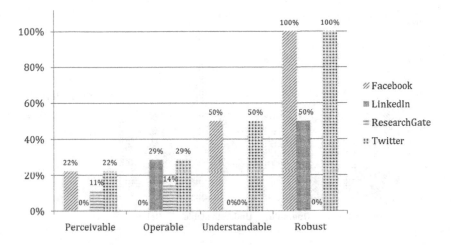

**Fig. 1.** Percentage of success criteria not met by site evaluated and WCAG 2.0 principle, according to evaluation by automated tools

The graph of Figure 1 shows the percentage of success criteria not met, according to the automated tools, by each site from social networking services sample organized by principles, taking into account only the 20 success criteria listed on Table 1.

It is possible to notice from graph in Figure 1 that Twitter was the one that had the highest number of criteria not met, distributed over the four WCAG 2.0 principles, while ResearchGate presents problems only in Perceivable and Operable Principles.

Figure 2 shows a graph ilustrating the percentage of success criteria not met, according to the experts, by each site from social networking services sample organized by principles, considering again only success criteria from Table 1.

Facebook was the site that presented more problems related to Perceivable Principle, because it was the only site in which experts indicated issues in success criteria involving videos. In the other hand, Twitter was the one that had less success criteria not met in the same principle. In other principles, sites performances were balanced, showing some consistency in problems presented.

From all that was observed, it's possible to conclude that the sample sites need to improve the way how their feedback errors are displayed for the visually impaired via screen readers. Also, more attention should be spend in alternative and/or descriptive texts for images, audios, videos and links in order to guarantee that everyone has full access to social networking services, regardless of their disabilities.

In relation to navigation, it's necessary to look for ways to optimize it, not being always neccessary to go through all the items and blocks of the page, because this makes a portal monotonous and inefficient for visually impaired to browse. It's also important to remember that HTML codes well formed are essential to

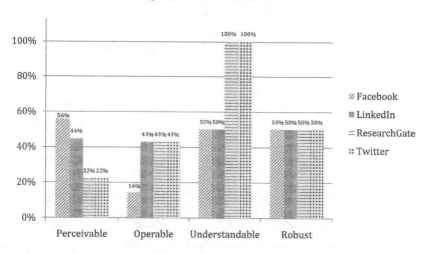

**Fig. 2.** Percentage of success criteria not met by site evaluated and WCAG 2.0 principle, according to evaluation by experts

assure adequate interpretation by screen readers, ensuring that information is transmitted correctly to visually impaired.

Finally, this study held a complete web accessibility evaluation of a sample of social networking sites, pointing out issues that ask for changes in order to meet WCAG 2.0 recommendations, therefore improving the software quality of these portals for visually impaired and providing a more enjoyable and effective navigation to them.

# References

1. Access Monitor (2014), http://www.acessibilidade.gov.pt/accessmonitor/
2. Berners-Lee, T.: World Wide Web Consortium Launches International Program Office for Web Accessibility Initiative (1997),
   http://www.w3.org/Press/IPO-announce
3. Basili, V.R.: Software Modeling and Measurement: The Goal Question Metric Paradigm. Computer Science Technical Report Series, CS-TR-2956 (UMIACS-TR-92-96), University of Maryland, College Park, Md. (1992)
4. Branco, R.G., de, F.A.P., Paiva, D.M.B.: Evaluation of Accessibility in Brazilian Municipal Sites. In: Annals of XXXI Brazilian Computer Society Congress, pp. 1265–1278 (2011) (in Portuguese)
5. Cusin, C.A., Vidotti, S.A.B.G.: Digital Inclusion via Web Accessibility. Liinc in Magazine 5(3), 45–65 (2009) (in Portuguese)
6. Facebook (2014), https://www.facebook.com/
7. Freire, A.P.: Accessibility in Web Systems Development: a study of Brazilian Scenario. Masters thesis, Institute of Math Sciences and Computing, University of São Paulo, São Carlos, Brazil (2008) (in Portuguese)
8. ITU, International Telecommunication Union. Trends in Telecommunication Reform 2012: Smart regulation for a broadband world (2012)

9. ITU, International Telecommunication Union. The World in 2014 - ICT Figures and Facts (2014), http://www.itu.int/en/ITU-D/Statistics/Documents/facts/ICTFactsFigures2014-e.pdf
10. de Lara, S.M.A.: Support Mechanisms for Usability and Accessibility in Older Adults Interaction on the Web. PhD thesis, Institute of Math Sciences and Computing, University of São Paulo, São Carlos, Brazil (2012) (in Portuguese)
11. Lin, T.: Cracking Open the Scientific Process. The New York Times (2012), http://www.nytimes.com/2012/01/17/science/open-science-challenges-journal-tradition-with-web-collaboration.html?_r=4&pagewanted=all
12. LinkedIn (2014), https://www.linkedin.com/
13. Martín, Y., Yelmo, J.: Guidance for the Development of Accessibility Evaluation Tools Following the Unified Software Development Process. Procedia Computer Science 27, 302–311 (2014)
14. Myers, K.: Trends on Web Accessibility and Mobile Payments. CIAB-Febraban (2012)
15. Radabaugh, M.P.: Director of IBM National Support Center for Persons with Disabilities. IBM Trainning Manual (1991)
16. ResearchGate (2014), http://www.researchgate.net/
17. Ruth-Janneck, D.: Experienced barriers in web applications and their comparison to the WCAG guidelines. In: Holzinger, A., Simonic, K.-M. (eds.) USAB 2011. LNCS, vol. 7058, pp. 283–300. Springer, Heidelberg (2011)
18. Sierkowski, B.: Achieving Web Accessibility. In: Proceedings of the 30th annual ACM SIGUCCS conference on User Services, Providence, Rhode Island, USA (2002)
19. TAW (2014), http://www.tawdis.net/
20. Total Validator (2014), http://www.totalvalidator.com/
21. Twitter (2014), https://twitter.com/
22. W3C, World Wide Web Consortium. About (2012), http://www.w3.org/Consortium/
23. W3C, World Wide Web Consortium. Web Content Accessibility Guidelines (WCAG) 2.0 (2008), http://www.w3.org/WAI/intro/wcag
24. WAI, Web Accessibility Initiative. Understanding the Four Principles of Accessibility (2013), http://www.w3.org/TR/UNDERSTANDING-WCAG20/intro.html
25. WAI, Web Accessibility Initiative. Complete List of Web Accessibility Evaluation Tools (2006), http://www.w3.org/WAI/ER/tools/complete
26. WAI, Web Accessibility Initiative. Selecting Web Accessibility Evaluation Tools (2005), http://www.w3.org/WAI/eval/selectingtools.html
27. WHO, World Health Organization. World Report on Disability (2011), http://whqlibdoc.who.int/publications/2011/9789240685215_eng.pdf?ua=1

# A Visual DSL for the Certification of Open Source Software[*]

Tiago Carção and Pedro Martins

HASLab / INESC TEC, Universidade do Minho, Portugal
{tiagocarcao,prmartins}@di.uminho.pt

**Abstract.** Quality assessment of open source software is becoming an important and active research area. One of the reasons for this recent interest is the consequence of Internet popularity. Nowadays, programming also involves looking for the large set of open source libraries and tools that may be reused when developing our software applications. In order to reuse such open source software artifacts, programmers not only need the guarantee that the reused artifact is certified, but also that independently developed artifacts can be easily combined into a coherent piece of software.

In this paper we improve over previous works and describe a visual language that allows programmers to graphically describe how software artifacts can be combined into powerful software certification processes. This paper introduces the visual language and describes how its elements are available to the user through an intuitive interface.

**Keywords:** Software Analysis, Software Certification, Open Source Software, Programming Languages, Process Management, Web.

## 1 Introduction

The advent and massive use of the Internet not only changed how people communicate, for example, via social networks, but also how software developers build their large and complex software systems. For software developers one of the main results of the Internet was the creation of large open source software repositories, such as *sourceforge*, where we may find libraries and tools for an immense variety of problems.

As a consequence, developing software nowadays not only involves reusing libraries provided by the underlying programming language, but also reusing libraries and tools that have been built by other software engineers and that are available as open source software.

[*] This Project is funded by ERDF - European Regional Development Fund through the COMPETE Programme (operational programme for competitiveness) and by National Funds through the FCT - Fundao para a Cincia e a Tecnologia (Portuguese Foundation for Science and Technology) within project Network Sensing for Critical Systems Monitoring (NORTE-01-0124-FEDER-000058), ref. BIM-2013_BestCase_RL3.2_UMINHO.

B. Murgante et al. (Eds.): ICCSA 2014, Part V, LNCS 8583, pp. 602–617, 2014.

In order to reuse such open source software, developers need to trust that software and, as a consequence, they often need to be certified that a reused software artifact does satisfy certain properties. For example, a developer may need to be sure that the copyrights involved in the reused software allow its usage. In our context, we refer to analyzing a property of piece of open source software as its certification.

In this paper we extend our work on developing a customizable web portal [1] for the certification of reusable, open source software packages. One of the main features of our web portal for the certification of open source software is the Domain Specific Language (DSL), introduced in [2]. This DSL allows the users of the web portal to define/combine new certifications as the combination of several different software tools. This DSL is implemented in the web portal as an Embedded DSL (EDSL): a DSL embedded in the Haskell programming language. Although this approach is powerful and did allow a quick development of the web portal, it contains two main issues: First, to define a new certification in that EDSL, the software developer needs to be an expert in the Haskell programming language. Secondly, and most important, the EDSL provides both poor syntax and error reporting since they are provided by the host language.

The purpose of this paper is two-fold:

- Firstly, we introduce a Visual Domain Specific Language (VDSL) where it is easy to specify the reuse and combination of (open source) software tools. Moreover, we build a proper compiler for such VDSL which reports errors in the web portal domain. This compiler translates the visual DSL to our former textual Haskell-based EDSL.
- Secondly, we present a case study where a tool to monitor energy consumption is specified in our visual DSL. Thus, we show how the widely used (open source) graph visualization tool *GraphViz* tool is instrumented in order to produce a report showing the energy consumption per its functions.

*This paper is organized as follows.* In Section 2 we provide an overview of the motivation and potential challenges this work faces. In Section 3 we introduce our visual language together with small examples of its usage. In Section 4 we present the process of creating a certification. The case study that we have used to demonstrate our VDSL is described in Section 5. In Section 6 we provide an overview of related works, and finally in Section 7 we conclude.

## 2   Motivation

The CROSS portal developed encloses the composition of tools to certify software. By a *Certification* we mean the execution of a software analysis tool that is capable of processing a source code file and of producing an information report.

These certifications can be composed out of various elements which can be arranged in a way that produces the wanted analysis, through a textual DSL presented in the portal. This DSL combines components, which are simple processes which together compose a certification.

In the same certification multiple analysis can be done, as in the DSL we have a notion of flow of information: from the program the user sequences components that transform inputs and produce new results which are themselves fueled to other components. In each certification, multiple flows can be implemented and then joined by a special type of component, an aggregator. Figure 1 shows a certification illustrating an analysis to a program with five different components.

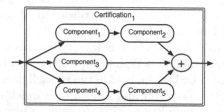

**Fig. 1.** An example of a certification composed by five components

Despite the powerful characteristics of the DSL, and the powerful analysis that can be implemented with it, the fact that this is a textual language embedded in hosting languages creates some disadvantages in this environment.

Firstly, having a textual representation forces the user to learn a new language. It is defined by constructs and primitives that user has to be aware of, as well as a set of syntactic and semantic rules mandatory to the creation of certifications. A visual language is much more intuitive, with drag and drop interfaces and real-time visualization of the flow of information, the learning curve is greatly diminished. Secondly, one great disadvantage of embedded DSLs is that error messages are related to the hosting language, not to the domain. This means that in the example of the web portal CROSS, whenever a user accidentally makes an error he is presented with error messages that have no relation whatsoever with the domain of software analysis or processes composition. This can make using the DSL hard for users not familiar with the hosting language.

A visual language make everything easier to handle: graphical representations of process relation are always visible, some errors can be avoided just by forbidding specific, invalid certifications on the graphical environment and errors are targeted to the specific domain of software analysis, making debug of certifications much easier and faster.

## 3    A Visual Language for Certifications

In this section we will introduce the visual language developed to allow easy implementation of processes in our portal. To desing such VDSL, we started by identifying all syntactic elements of the textual language; and by understanding how to recreate them visually. Their graphic representation would allow a more intuitive use of the language and reduce its associated learning curve.

In order to create a new certification, users always need to specify a flow that the certifying programs must follow. This flow starts in the inserted input and goes through a series of components to generate a report. From the same input, multiple customized flows of information can be used, but they must eventually be explicitly aggregated to generate a report. These elements are linked together by a connection that symbolizes the notion of the program flow between them.

The existing elements in the textual language which should be represented in the visual language are: *input, connection, components* and *aggregators*, which we shall present next.

The components of our visual language have been implemented in a system that is not restrictive on their organization and layout. This allows users to rearrange positions and connections with little effort. The fact that this is a flexible environment but with controlled actions, unobtrusively enforces the correct construction of certifications.

The elements which were identified in the textual language and implemented in the visual language are explained next in detail.

## 3.1 Input

The input represents the program to certify, and from here a certification can be composed. Each added component will indicate the action to be applied to the input. The textual representation can be seen in Listing 1.1.

**Listing 1.1.** Input specified in the textual language

```
Input
```

The graphic design of the input element, as is shown in Figure 2, is similar to the UML input for flow diagrams.

**Fig. 2.** The Input graphical representation

## 3.2 Component

A component is an element of the flow of information throughout a certification. The functionality of this element can be see as a single process to which information is fed and producing a desired result. All components have at least on input type and one input type, but some have various which have to be defined by the user.

Each component has a standard format: a parallelogram with unique and useful visual information. From the textual language, presented in Listing 1.2, we can see that each component has a name (which by definition is unique within all components), and the description of its input and output type.

**Listing 1.2.** A Component specified in the textual language

```
(readFile ,   "-j",   "-s")
```

Our visual language allows easy navigation and selection of the available components in the portal. For each component, relevant information is shown: the description of the component, and the input/output types it supports. To choose a component and add it to the certification creation process, it is necessary to select the component from the set of available components and to customize it by choosing its input and output types.

In the example of Figure 3, the user selected the component *readFile*, with *Java* as the input type and text as its output type.

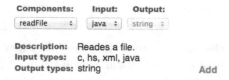

**Fig. 3.** Information presented to the user, associated with the available components

The graphical representation of a component, as can be seen in Figure 4, shows its name and the chosen languages for input and output separated by a symbol of transformation ->. Also indicated is a green circle representing the point of input where one can connect the flow either from the initial input, from other component or from an aggregator. The blue circle is the output point.

**Fig. 4.** A Component specified in the visual language

### 3.3   Aggregation

An aggregator is a specific type of component that aggregates various flows of information. This element can be seen as a special type of component and therefor the graphical representation of an aggregator is similar to the representation of a component.

In our textual representation, an aggregator would be represented as seen in Listing 1.3.

**Listing 1.3.** Textual version of an aggregator that takes the number of lines of multiple sources and produces a report

```
>|> (Aggregator, "-i", "-rep")
```

Adding an aggregator, in our visual environment is similar to adding a typical component: the user has to choose from a list of available aggregators. As other components they where they can see detailed information about it, such as its functionality and type.

The graphical representation of an aggregator, which can be seen in Figure 5, has a name that is unique within all aggregators and the input and output language separated by a symbol of transformation ->.

**Fig. 5.** Example of an aggregator, that takes the number of lines of multiple sources and produces a report

Since there are differences between how an aggregator and a typical component handles input information, with an aggregator being an element that will receive information from multiple sources, we changed the aspect of the input point to be represented as a green square meaning it can receive multiple flows. In component the input point, as shown in Subsection 3.2, is a green circle. Since the output works like a normal component, the graphical output representation is a blue circle.

### 3.4 Connection

The connection is the element linking two components or a component to an aggregator. Each connection has an arrow that showing of the flow and a label displaying the types that flow will handle, as can be seen in Figure 6 represented by the yellow arrow.

## 4  Creating Certifications

With the visual language elements, defined in the previous section, we shall now describe how different types of certifications with multiple components and layouts can be defined in our setting.

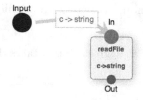

**Fig. 6.** Connecting two components

## 4.1 Sequential Flow

Our visual language allows a wide type of certifications, based on flows of information, to be implemented. One of these possibilities is the sequential flow.

**Listing 1.4.** Sequential flow of information in a textual form.

```
Input
    >- (readFile,    "-c",  "-s")
    >- (text2NLines, "-s",  "-i")
    >- (int2Report,  "-i",  "-rep")
```

This flow is a connection of tools in a sequential order, allowing the analysis of a software program all the way from the Input to the generation of a final report, chaining different components throughout the process.

One example of a certification is one where we want to analyze the number of lines in a C program. In the textual language we have to write the sequence of components linked by the combinator >- and make sure that the types and their representation is correct, as can be seen in Listing 1.4.

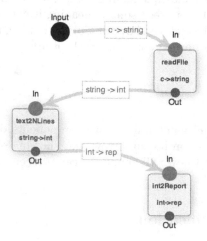

**Fig. 7.** Sequential flow of information

In the visual language, this process is faster and more intuitive. The user only needs to select the tools and its types, and link them with connections. Since our visual setting is restricted by invalid connections, the user does not have to worry about the correctness of the language syntax, since these are simply not allowed.

An example of the implementation of the certification to analyze the number of lines in a C program can be seen in Figure 7.

### 4.2   Parallel Flow

In addition to the sequential flow, our setting also supports parallel flows of information obtained when combining the components in two or more process chains. This flow, can have multiple paths, where each of these paths is sequential and joined by the element aggregator.

For instance, let us imagine that we want to analyze a tool, and know how many *ifs* conditions and *for* loops it has. This can be done by implementing a certification with two paths, where one counts the number of ifs conditions and the other counts the number of *for* loops. Afterwards, the results are joined by an aggregator producing a report.

**Listing 1.5.** Parallel flows of information described textually

```
Input
     >- (readFile ,  "-c",  "-s")
     >- (text2NFors,  "-s",  "-i")      >|
Input
     >- (readFile ,  "-c",  "-s")
     >- (text2NIfs ,  "-s",  "-i")      >|>
     (Aggregator,  "-i",  "-rep")
```

This certification can be expressed in the textual language by building two different paths. For each path, the user has to define the flow from the Input and include the combinator  >|  in the end to give an indication that the returned information will be aggregated. This can be seen in Listing 1.5.

In order to build this certification we need to add two paths. We need to start each of these paths with the readFile component. In order to count the number of *for* loops, in the left path, we add the text2NFors component. We do the same in the right path but this time to count the *if* loops, using the element text2NFors component. These two components produce integers in their analysis, which are funneled to an aggregator which generates a report with the relevant informations. The graphical representation of this process is in Figure 8.

### 4.3   Verification

As explained before, creating certifications in our settings has some rules, related to type correctness and to the structure of the computation chain that composes

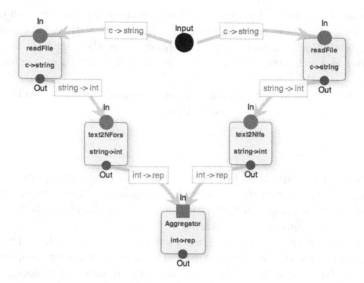

**Fig. 8.** Parallel flows of information

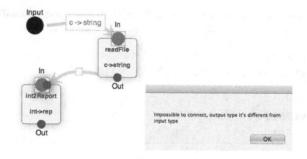

**Fig. 9.** Trying to establish a connection between two incompatible components

the certification. One important rule comes from the connection of a `component1` to a `component2`, where the first must have the same output type as the input type of the second.

These rules, in the textual language, can only be verified when the certification is fully constructed. Only after the user describes the certification can the the syntax and semantics of the language be analyzed.

In our visual language, this verification is done in real time. When one tries to link two components and their type do not match, an alert notification appears and the system simply forbids the creation of this connection. This is exemplified in Figure 9.

Another important rule when defining a certification is that one path must be continuous and sequential, i.e., all components in the certification must be linked together in a sequential manner. Elements of a certification are have to be fed

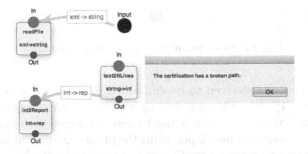

**Fig. 10.** Trying to create a certification with a broken path

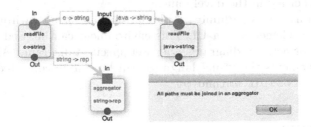

**Fig. 11.** Trying to create a certification with multiple paths not joined by an aggregator

an input to produce their result. This verification is automatically performed and all errors found are signaled and presented to the user, as we can see in Figure 10.

In Figure 11 we see yet another example of how our visual language aids in certification construction. In this example there are multiple paths which should be joined by an aggregator. This, as we saw in Figure 10, is also verified and any problems found are presented to the user..

Due to its interactive nature, a visual language improves over a previous textual approach as problems are easier to detect and are presented to the user in a setting that is easier to understand and to correct. Furthermore, some problems, typical of a textual setting, are simply inexistent here as the visual environment denies certain compositions of elements.

## 4.4   Technical Details

In order to facilitate familiarization with the language and to improve user-experience when dealing with it, we were influenced by existing visual elements of the Unified Modeling Language (UML) to represent some of its items. In Figure 12, for example, we see a graphical representation of the Input very similar to the one found in UML. Another example of similarities is in the flow presented in activity diagrams, with boxes containing important nodes in the flow of information and arrow relating these boxes.

●

**Fig. 12.** Start point of UML activity diagrams

This language was conceived to be embedded in an open source analysis web portal - CROSS [1][1]. Since we are dealing with a web environment, the chosen techniques relate to the ones widely used when developing web applications.

The visual language was developed using the programming language JavaScript. Alongside JavaScript, we used a well-known framework for the language, jQuery[2]. This framework simplifies the use of JavaScript and adds the possibility of using plugins to further aid in the development.

One plug-in used was jsPlumb[3], a plug-in that allows an environment where one can connect elements in a UI. This environment can be used to represent state machine or activity diagrams and user-specified diagrams. All the source code that implements the visual language can be can be consulted in the web portal, at www.cross.di.uminho.pt.

## 5  Case Study

The visual language for certifications presented in this work was designed to be used with different components and in different contexts, covering a wide range of possible analysis.

In this section we will present an example of how the functionality of the web postal can be used to analise a software program, written in the C programming language, and produce a report with information regarding energy consumption.

To do so, we use a work that has been developed in the context of power consumption in software [3]. In this particular work, techniques were developed to measure the impact of software design in energy consumption, a topic of high relevance with the current wide usage of smartphones and other mobile devices.

The analysis that we intend to make, regarding power consumption, goes through different phases. Initially it is necessary to instrument all the software modules so that each function can produce an output in order to know how much energy was spent on processing when the function was called. Once the software has been instrumented, it must be compiled to machine code.

One of the known methods we can use is the Spectrum-based Fault Localization (SFL) [4]. This method uses the information of the running spectrum of the program and, along with the information about the input and the expected output, can indicate what are the faults in the software.

Adapting an SFL-based algorithm to the energy analysis consumption creates an adapted SFL model for energy consumption, with which we can use the energy

---

[1] www.cross.di.uminho.pt

[2] http://jquery.com

[3] http://jsplumbtoolkit.com/

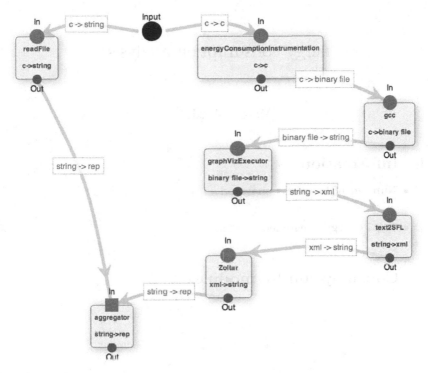

**Fig. 13.** Certification of energy consumption analysis for GraphViz

consumption information and build a matrix with the spectrum of the program consumption and, again, apply SFL techniques to obtain information from a matrix. All this can be seen in [3].

This analysis of the energy consumption can be made with a certification using a combination of various tool and components. Doing so, aids the user in implementing the analysis as he/she does not have to fiddle with the typical textual language constructs we have seen before.

As input for the study case we chose the open source tool GraphViz[4]. GraphViz is a software tool that allows textual representations to be presented in a visual format, such as graphs. This tool has different modes of presentation and is widely configurable.

In order to allow the representation of every phase of the energy consumption analysis in the certification creation process it is necessary to choose which tools must be available. The instrumentation of the software modules can be made by using a tool developed which uses clang to instrument the software.

The software compilation is made with gcc[5]. To execute Graphviz with different inputs we have used a tool that contains a textual graph sample and runs

---

[4] www.graphviz.org
[5] http://gcc.gnu.org

# Energy consumption analysis

March 14, 2014

## 1   Information

- **Number of functions analyzed - 753**

- **Total energy consumed - 3000mW**

## 2   Consumption by modules

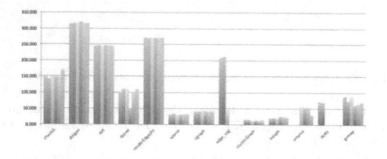

## 3   Consumption by functions

1

**Fig. 14.** The report generated in the certification for GraphViz

Graphviz with different flags, generating different visualizations. We have also used a tool that collects the outputs of power consumption of each function and builds the SFL matrix.

The tool to analyze the SFL matrix and to present the functions with problems was Zoltar [5] as. Thus, a tool that gathers the Zoltar output and generates a report describing software consumption using a graph bar which is explained in [3] was also used. This tool has the additional function of pretty printing the software with information about functions' energy consumption.

This certification implemented in our visual language can be seen in Figure 13. In this certification we can see that there are two paths, one of the paths represents the flow of code instrumentation, gathering results, and analyzing the SFL information. In the other path we have the flow that will allow the pretty printing. Both paths are joined in an aggregator that will assimilate and produce a report which is the conjunction of the results of both paths.

The certification produces a report with the GraphViz consumption values, as can be seen in Figure 14. Here we can see that in the first section we have a summary of the total functions analyzed and the total power consumed. In section two, we have two graph bars. In both graphs the $yy$ axis represents the power consumption. The $xx$ axis represents the modules and functions in graph one and two respectively. Each $xx$ value has 6 series - the 6 series represent the 6 different inputs to which GraphViz was tested.

In this section we presented a case study that analysis a source code of a regular programming language. Our VDSL, however, can define certifications for other software artifacts, by composing both simple tools, like for example HaLeX for reason about regular expressions[6], and complex software systems, like the GuiSurfer framework to reason about intercatice specfications [7, 8], or the LRC attribute grammar system [9]. By combining such tools we are able to define powerful software certifications.

# 6   Related Work

This works improves on previous ones, related to both the construction of a language for process management and the implementation of a web portal to certify software. In [2] we present a textual language for process management. This language is based on the functional language Haskell and uses zipper-based techniques, as the ones found in [10] to develops attribute grammar techniques to defines semantic analysis on the language [11]. Because we express our VDSL in an AG setting, we get for free well-kown techniques to analise and optimize our visual programs/specifications, namely the detection of circularities [12], the optimization of such circularities  [13–16], and the incremental execution of our programs [17]. In [1] and [18] we presented the portal and develop a framework that allows not only the certifications and analysis of open source software, but we also introduce novel language-independent techniques to further aid generating information about computer programs submitted to our portal.

There are works on languages for process management. Of relevant reference is [19], an implementation of the orchestration language Orc [20] is introduced as an embedded domain specific language in Haskell. In this work, Orc was realized as a combinator library using the lightweight threads and the communication and synchronization primitives of the Concurrent Haskell library [21].

# 7 Conclusions and Future Work

In this paper we presented a visual language that allows the creation of software certifications in the CROSS web portal. This language enables the user to create certifications to analyze software without needing to deal with the typical disadvantages of an embedded DSL.

For this visual language we created different visual elements and implemented the notion of sequential and parallel flows of information. We also worked on the validation associated with the construction of certifications. We also present a full example through a case study which analyzes energy consumption on a open source tool, GraphViz. We show how this certification can be implemented in our visual language and the type of results we can produce with our environment.

As a future work, we intend to extend the visual language to also allow reports generation and customization, by allowing specific results to be agglomerated into chapter, sections and subsections of the report.

# References

1. Martins, P., Fernandes, J.P., Saraiva, J.: A Web Portal for the Certification of Open Source Software. In: Cerone, A., Persico, D., Fernandes, S., Garcia-Perez, A., Katsaros, P., Ahmed Shaikh, S., Stamelos, I. (eds.) SEFM 2012 Satellite Events. LNCS, vol. 7991, pp. 245–261. Springer, Heidelberg (2014)
2. Martins, P., Fernandes, J.P., Saraiva, J.: A purely functional combinator language for software quality assessment. In: Symposium on Languages, Applications and Technologies (SLATE 2012), Schloss Dagstuhl. OASICS, vol. 21, pp. 51–69 (2012)
3. Carção, T.: Spectrum-based Energy Leak Localization. Master's thesis, University of Minho, Portugal (in preparation, 2014)
4. Abreu, R., Zoeteweij, P., van Gemund, A.: On the accuracy of spectrum-based fault localization. In: Proceedings of the Testing: Academic and Industrial Conference Practice and Research Techniques – Mutation (Mutation 2007), pp. 89–98 (2007)
5. Janssen, T., Abreu, R., van Gemund, A.J.: Zoltar: A spectrum-based fault localization tool. In: Proceedings of the 2009 ESEC/FSE Workshop on Software Integration and Evolution @ Runtime, SINTER 2009, pp. 23–30. ACM (2009)
6. Saraiva, J.: HaLeX: A Haskell Library to Model, Manipulate and Animate Regular Languages. In: ACM Workshop on Functional and Declarative Programming in Education. University of Kiel - TR 0210, pp. 133–140 (September 2002)
7. Silva, J.C., Saraiva, J., Campos, J.C.: A generic library for gui reasoning and testing. In: Proceedings of the 2009 ACM Symposium on Applied Computing, SAC 2009, pp. 121–128. ACM, New York (2009)
8. Silva, J.C., Silva, C., Gonçalo, R.D., Saraiva, J., Campos, J.C.: The guisurfer tool: Towards a language independent approach to reverse engineering gui code. In: Proceedings of the 2Nd ACM SIGCHI Symposium on Engineering Interactive Computing Systems, EICS 2010, pp. 181–186. ACM, New York (2010)

9. Kuiper, M., Saraiva, J.: Lrc - A Generator for Incremental Language-Oriented Tools. In: Koskimies, K. (ed.) CC 1998. LNCS, vol. 1383, pp. 298–301. Springer, Heidelberg (1998)
10. Martins, P., Fernandes, J.P., Saraiva, J.: Zipper-based attribute grammars and their extensions. In: Du Bois, A.R., Trinder, P. (eds.) SBLP 2013. LNCS, vol. 8129, pp. 135–149. Springer, Heidelberg (2013)
11. Swierstra, S.D., Alcocer, P.R.A., Saraiva, J.: Designing and implementing combinator languages. In: Swierstra, S.D., Oliveira, J.N. (eds.) AFP 1998. LNCS, vol. 1608, pp. 150–206. Springer, Heidelberg (1999)
12. Kastens, U.: Ordered attribute grammars. Acta Informatica 13, 229–256 (1980)
13. Saraiva, J., Swierstra, S.D.: Data Structure Free Compilation. In: Jähnichen, S. (ed.) CC 1999. LNCS, vol. 1575, pp. 1–17. Springer, Heidelberg (1999)
14. Fernandes, J.P., Saraiva, J.: Tools and Libraries to Model and Manipulate Circular Programs. In: PEPM 2007: Proceedings of the ACM SIGPLAN 2007 Symposium on Partial Evaluation and Program Manipulation, pp. 102–111. ACM Press (2007)
15. Pardo, A., Fernandes, J.P., Saraiva, J.: Shortcut fusion rules for the derivation of circular and higher-order programs. Higher-Order and Symbolic Computation, pp. 1–35. Springer (2011)
16. Fernandes, J.P., Pardo, A., Saraiva, J.: A shortcut fusion rule for circular program calculation. In: ACM SIGPLAN Haskell Workshop, Haskell 2007, pp. 95–106. ACM, New York (2007)
17. Saraiva, J., Swierstra, S.D., Kuiper, M.F.: Functional incremental attribute evaluation. In: Watt, D.A. (ed.) CC 2000. LNCS, vol. 1781, pp. 279–294. Springer, Heidelberg (2000)
18. Martins, P., Carvalho, N., Fernandes, J., Almeida, J., Saraiva, J.: A framework for modular and customizable software analysis. In: Murgante, B., Misra, S., Carlini, M., Torre, C.M., Nguyen, H.-Q., Taniar, D., Apduhan, B.O., Gervasi, O. (eds.) ICCSA 2013, Part II. LNCS, vol. 7972, pp. 443–458. Springer, Heidelberg (2013)
19. Campos, M., Barbosa, L.: Implementation of an orchestration language as a haskell domain specific language. Elect. Notes Theor. Comput. Sci. 255, 45–64 (2009)
20. Kitchin, D., Quark, A., Cook, W., Misra, J.: The orc programming language. In: Lee, D., Lopes, A., Poetzsch-Heffter, A. (eds.) FMOODS/FORTE 2009. LNCS, vol. 5522, pp. 1–25. Springer, Heidelberg (2009)
21. Peyton Jones, S., Gordon, A., Finne, S.: Concurrent haskell. In: 23rd Symposium on Principles of Programming Languages, POPL 1996, pp. 295–308. ACM (1996)

# Study on the Social Impact on Software Architecture through Metrics of Modularity

Braulio Siebra[1], Eudisley Anjos[1,2], and Gabriel Rolim[1]

[1] CI, Centre for Informatics,
Federal University of Paraiba, Joao Pessoa – Brasil
eudisley@ci.ufpb.br,
{braulio.siebra,gabrielsrolim}@gmail.com
[2] CISUC, Centre for Informatics and Systems,
University of Coimbra, Portugal
eusis@dei.uc.pt

**Abstract.** Software systems have constantly increased in size and complexity. At the same time, software architecture also grows and becomes difficult to maintain leading to failures or abandonment of systems. According to Mirroring Hypothesis (MH), the organizational structure of the development team is a mirror of software architecture. So, the importance in understanding what changes in social structure can impact in the software architecture is crucial to avoid architectural problems. This work compares modularity metrics, applied to open-source systems, with the structure of developers inside the organization. The results show the relationship between the architecture and organization and contribute to guide the evolution and maintenance of systems.

**Keywords:** Modularity, Social Metrics, Software Architecture.

## 1    Introduction

The distributed software development has taking the attention of researchers since the 1990s [1]. This occurs because organizations have changed socially, economically and geographically, so they can distribute the resources aiming to increase productivity, improve quality and reduce costs in software development [ 2 ] . But, the physical distribution of teams amplifies existing problems in the system development process. Cultural differences, language, and other factors increase the difficulty in communication and coordination during the development process [ 3 ] , [ 4 ] .

When you have a large system, it is normal that appear several difficulties understanding and consequently maintaining it. The definition of " divide and conquer" used for algorithms can be used during the design process and reinforces the thought that the use of the concept of modularity offers quick results. Dividing a large problem into smaller problems, it requires less effort to find solutions. Thus, modularity can be understood as a way to facilitate maintenance.

The focus on dividing the system into modules and allow concurrent development has led researchers to compare the organization of the developers with the structure of

B. Murgante et al. (Eds.): ICCSA 2014, Part V, LNCS 8583, pp. 618–632, 2014.

the system modules, the system architecture. In their research they found that the structure of the product architecture was equal to the structure of the organization and called it Conways Law or Mirrorring Hypothesis [ 5 ], [ 6 ].

Although the structure of the organization has been taken into consideration in academia, many aspects inherent to the developer and the influence that the social environment can have over the system have still to be researched. These aspects can be named as: number of people who collaborate with the project , geographic location , number of commits , etc. , tell us social information that can be related to system modularity . In this work we use modularity metrics to understand the system architecture and compare them to information about the development team of open-source systems. We expect to obtain results that identify whether there is any relationship between modularity of a system and social aspects of the contributors. The results contribute to understand what can be done to improve the software evolution and maintainability.

## 2      Conceptual Background

In this section, we present the theoretical basis relevant to understand this work. The main topics are: software architecture, modularity, open-source software and mirrorring hypothesis.

### 2.1      Software Architecture

Software systems can increase in size and complexity, and when these systems grow it is important that the project has a good architecture. The problem of building software also involves decisions about the structures that form the system, the overall structure of control, communication, synchronization, and data access protocols, assignment of functionality to elements of the system, or on physical distribution of system elements [7]. These demonstrates the importance of understand the system architecture. Software architecture can play an important role in at least seven aspects of software development: Understanding Reuse, Construction, Development, Analysis, Management and Communication [8].

### 2.2      Modularity

According to [9], the concept of modularity emerged in the 1960s. Developed countries were facing a crisis due to the mismatch in the growth of hardware on the evolution of the software, the hardware in this crisis was developing rapidly while the technical software development progressed slowly. Since then, this concept has been growing and becoming essential in the development of a system.

In [10], several researchers have tried to define modularity. One of the best definitions was conceived by *Michael Jackson,* where he described modularity as a property of structuring software into modules, and module as an artifact designed to be constructed and understood separately.

Already, after *Parnas*, a module is independent, ie, you can change one without affecting the other. *Parnas* also says that a module can be developed concurrently, that is, one can develop while the other team is responsible. In [5], he stressed the idea of independence of modules in the sense of both cohesion and coupling, and created a concept about the apportionment of information that has become one of the most cited general conceptions in software literature. This concept is the "information hiding" or protection of information. He says that each module should keep to themselves the information that only he matters. To be applied to systems, this concept generates a weak coupling, because of this concealment of information.

## 2.3    Open Source Software

The term open source or open code was created by OSI (Open Source Initiative) and refers to free software. OSI is an organization dedicated to promoting open source software or free software organization. It was founded as a form of incentive for organizations to employ that concept. Its function is to check which licenses qualify as free software licenses, and are disseminating this concept showing the technological and economic advantages.

The OSI determines that the open source program should ensure free distribution, i.e., the license must not restrict the marketing or distribution of the program. The program must include source code and must allow distribution in compiled form.

## 2.4    Conway's Law and Mirrorring Hypothesis

The concept of Conway's Law came in 1968 when the scientist and programmer Melvin Conway said any organization that makes system design will produce a design whose structure is a copy of the structure of the enterprise architecture [11]. That means organizations that want to design systems or generate products that copy the structure of how the company is organized, i.e. its architecture. It appears these days with a new terminology known as Mirroring Hypothesis.

The Mirroring Hypothesis allows the relationship between technological modularity and organizational structure. Fewer lines of code can result in a faster time with fewer defects. Therefore, the reuse is important not only to write less code, but also because it means find and fix problems quickly and at no charge.

There are other works approaching the relation between modularity and open-source systems. In [12], for example, was made a survey showing that the open-source systems were less coupled than proprietary systems. Besides prefer modular systems, open source communities also invest in tools and social practices based on openness and transparency technique.

# 3    Metrics of Modularity

According to [13], software metrics are designed to identify, measure and allow controlling the main parameters that affect software development. The software metrics used in this work are described below. It is important to mention that all metrics were selected to be applied to open-source object-oriented systems.

- **TLOC – Total Lines of Code:** The count of lines of code is generally used as a benchmark for other metrics. As the name implies, this metric will count the number of rows in the code.
- **NOPK - Number of Packages:** Number of packages is defined as the count of the packages in the selected scope for analysis. The metric includes all sub packages.
- **NBD - Nested Block Depth:** This metric represents the maximum number of blocks of code nested in a particular method of a class.
- **CC – Cyclomatic Complexity:** Developed by Thomas J. McCabe [14], the CC measures the level of complexity of a method or function by counting the paths in code with independent execution. This metric is largely used in academia and industry and one of the most important to us.
- **LCOM – Lack of Cohesion of Methods:** According to [15], cohesion is a qualitative indication of the degree to which a module focuses on just one thing. This metric measures how a class is not cohesive. I.e., the higher the number of LCOM the less cohesive is the class.
- **Ca - Afferent Coupling:** The measure proposed by afferent coupling in [16] is the number of different classes that relate to the current class by means of fields or parameters
- **Ce - Efferent Coupling:** Opposed to the afferent coupling, [16] also proposed the efferent coupling metric which is the number of different classes that the current class references through fields or parameters.
- **WMC - Weighted Methods per Class:** According to [17], this metric measures the individual complexity of a single class. The number of methods of a class and its complexities are indicators that time and effort are required to the development and maintenance of classes.

# 4     Methodology

In this section we describe: the open-source systems used as case study, the tools and methods used to collect the values of the metrics for software modularity and social analysis, and how the tests were performed.

## 4.1     Open Source Systems

The open-source systems provide their code openly and work with distributed development; therefore it is possible that contributor's social and cultural aspects influence the project modularity. For this reason and the fact that in open-source systems the source code is at our disposal, these systems have been our focus for testing.

It was necessary to compile the code for each project to obtain the numerical values of each metric. Eight systems written in Java were selected for case study:

FindBugs, HyperSQL Database Engine, JasperReports, jEdit, JMeter, Poi, TomCat and Vuze.

We point out that the systems choice mentioned above was also a consequence of their presence in GitHub repository software. Projects that were not hosted on GitHub were excluded from the scope of our research.

A version of each of the selected systems was chosen for testing. Each system and its selected version are shown in Table 1:

**Table 1.** Description and version of the systems used as study cases

| Open Source Systems | Description | Version |
|---|---|---|
| FindBugs | Uses static analysis to look for bugs in Java code. | 3.0.0 |
| HyperSQL Database Engine | It is a basic server written entirely in the Java language data. | 2.3.2 |
| JasperReports Library | It is a reporting tool that produces documents that can be viewed, printed or exported to various formats (PDF, HTML, XML, etc.) | 5.5.2 |
| JEdit | It is a text editor developed in Java. | 1.6 |
| JMeter | It is a tool used for load testing services offered by computer systems. | 2.12 |
| Poi | Provides pure Java libraries for reading and writing files in Microsoft Office formats such as Word, PowerPoint and Excel. | 3.11 |
| Tomcat | It is a Java web server. The Tomcat is a JEE application server, but it is not a server EJBs. | 8.0 |
| Vuze | It is a Java program that allows downloading files via the BitTorrent protocol | 5.3.0.0 |

## 4.2   Git / GitHub

Git is a version control system widely used by software development companies. Git helps large contributors teams to keep the systems organized and documented. Anything implemented in software or commented can be found anywhere at any time.

GitHub is a web hosting service designed for distributed projects that uses the Git version control. Thus, it is used as an online repository of source code for open source projects. Information about all commits (updates) of projects can be found there. It uses a social network that allows people to follow the project development; a feature to view graphs with the amount of updates per contributor, and other features. The eight systems used in the case study are hosted by GitHub.

## 4.3    Choice of Tools

After several readings and studies Eclipse Plugin 1.3.6 Metrics tool was chosen. This decision was based on the fact that all the chosen systems were written in Java and needed to be compiled to generate the metrics. This tool collected various metrics related to project modularity. Eight metrics that can influence systems modularity have been selected after reading some articles.

The research to find mining tools repository was done cautiously. The selected tool - GitStats- generates statistics of any project that is hosted on Git and outputs data such as: quantity of contributors, most active contributors, amount of commits, etc... The choice for GitStats happened for many reasons. Primarily because it accesses the Git server, where the eight selected systems are hosted. Furthermore, we assume the GitHub repositories as one of the best nowadays.

Both the Eclipse Metrics Plugin 1.3.6 and GitStats- are easily found tools used by businesses. The use of such tools allows us to perform a job that can be easily replicated in an inexpensively way by companies- not limiting this research to academic context.

### 4.3.1  Metrics Collected via Eclipse Metrics Plugin 1.3.6

In collecting data 23 metrics were found. Here, we present the eight metrics used in our research. There were two size metrics, three complexity metrics, two coupling metrics and one cohesion metric. The classification of each used metric is shown below in Table 2.

### 4.3.2  Metrics Collected by GitStats

The GitStats was used to generate some relevant social characteristics to come across the software metrics. This tool provided the quantity of contributors present in the GitHub repository of each project, the amount of commits by author and the total amount of commits for each system.

An essential feature that is not presented by GitStats is geographic location of each contributor. The location is very important to understand how modular the organization we are evaluating is. This idea was presented before by [12] and helped us to compare the organization structure with the source code structure. To obtain the contributors' location we executed solicitations using GitHub API to get the e-mail address and name. Through this information we could find some contributors' locations.

**Table 2.** Classification of metrics

| Metrics | Size | Complexity | Cohesion | Coupling |
|---------|------|------------|----------|----------|
| TLOC | X | | | |
| NPOK | X | | | |
| NBD | | X | | |
| CC | | X | | |
| WMC | | X | | |
| LCOM | | | X | |
| Ca | | | | X |
| Ce | | | | X |

We assumed that only the 20 most active contributors of each project were relevant. The amount of commits guided this assumption. We noticed that beyond 20, the  amount of commits became irrelevant for our purposes . It is important to mention here that it was not possible to find the location of all twenty contributors in each system. The difficulty was due to lack of information from the GitHub on the contributors. Sometimes just a nickname or a duplicated name was provided.

Thus, the geographical distribution is given as shown in (1) and the results of the related metrics are shown in the Table 5:

$$\text{Number of Countries / Number of Contributors} = D_G . \qquad (1)$$

## 5     Results

The first results were obtained using the tools mentioned in the section 4.3: Eclipse Metrics 1.3.6 and GitStats. Eclipse Metrics calculated metrics for each of the eight projects. We selected the most relevant metrics for this paper and the results are shown in Table 3 and Table 4.

The GitStats was used to generate some relevant social features to compare with the software metrics. Among the many metrics, this tool showed we selected the three most important ones : the quantity of contributors present in the GitHub repository of each project, the amount of commits of each author and the total amount of commits for each system.

**Table 3.** Size and complexity using *Metrics for Eclipse*

| Open-Source Systems | TLOC | NPOK | NBD | CC | WMC |
|---|---|---|---|---|---|
| FindBugs | 20.573 | 77 | 1,379 | 2,678 | 26.410 |
| HyperSQL | 69.191 | 32 | 1,781 | 3,569 | 36.824 |
| JasperReports | 235.798 | 116 | 1,460 | 2,045 | 37.726 |
| JEdit | 117.366 | 42 | 1,590 | 3,054 | 22.347 |
| JMeter | 32.764 | 43 | 1,485 | 1,879 | 5.882 |
| Poi | 86.762 | 48 | 1,317 | 1,863 | 17.148 |
| Tomcat | 150.834 | 102 | 1,553 | 2,611 | 33.013 |
| Vuze | 565.168 | 488 | 1,851 | 2,713 | 85.275 |

**Table 4.** Coupling and cohesion using *Metrics for Eclipse*

| Open Source Systems | LCOM | Ca | Ce |
|---|---|---|---|
| FindBugs | 0,267 | 25,494 | 11,325 |
| HyperSQL | 0,366 | 35,625 | 12,00 |
| JasperReports | 0,247 | 40,397 | 17,129 |
| JEdit | 0,222 | 18,929 | 8,429 |
| JMeter | 0,221 | 16,256 | 6,465 |
| Poi | 0,201 | 26,917 | 14,562 |
| Tomcat | 0,289 | 17,804 | 5,667 |
| Vuze | 0,318 | 22,795 | 4,814 |

## 5.1 Results Analysis

Here we present a comparison between modularity and social metrics. To evaluate the results we formulated five hypotheses according to [6] [17], [18]. We divided into subsections for more relevant comparisons and discuss each of the results.

**Table 5.** Result of social metrics

| Open Source Systems | Contributors | Commits | $D_G$ |
|---|---|---|---|
| FindBugs | 26 | 14.743 | 0,36 |
| HyperSQL | 12 | 5.361 | 0,44 |
| JasperReports | 20 | 6.961 | 0,37 |
| JEdit | 37 | 6.581 | 0,58 |
| JMeter | 29 | 10.332 | 0,55 |
| Poi | 33 | 5.483 | 0,57 |
| Tomcat | 24 | 12.067 | 0,60 |
| Vuze | 42 | 24.828 | 0,54 |

**Table 6.** More active author of each project

| Open Source Systems | Author/Commits (%) |
|---|---|
| FindBugs | Bill Pugh (EUA) / 50,04% |
| HyperSQL | Fredt (ING) / 46,32% |
| JasperReports | Teodord (ROM) / 48,36% |
| JEdit | Slava Pestov (RUS) / 33,33% |
| JMeter | Sebastian Bazley (ING) / 66,72% |
| Poi | Nick Burch (ING) / 29.45% |
| Tomcat | Mark Thomas (ING) / 66,81% |
| Vuze | Parg (EUA) / 27,96% |

### 5.1.1   Correlation between Coupling and Geographical Distribution

A análise entre a distribuição geográfica e o acoplamento foi direcionada a partir da hipótese que segue:

$H_1$: *to maintain system modularity an   increase in geographical distribution implies coupling reduction.*

This happens because once they have contributors in different countries, the system must be least coupled to facilitate development.

In order to see better what happens, we analyze in particular the result of metrics for two systems: Tomcat and JMeter. As shown in the previous tables, the most active contributor in each system was responsible for over 66% of the total commits . This is an indication that the code should be more coupled because a single person dominates and understands more the code. But when we analyze the coupling metric (Ca, Ce) we realized that Tomcat and JMeter are among the least coupled.

Although there is a contributor who dominates the code more than the others , those systems are among the most widely distributed geographically, as shown in Table 5. With greater geographical distribution, the code tends to become   least coupled . A weak coupling indicates less dependence between modules, i.e., there is a tendency for systems to be more modular. One of the issues raised here for future investigation is which attribute is more powerful than the other.

### 5.1.2   Correlation between Quantity of Contributors and Complexity

This analysis was based on the following hypothesis:

$H_2$: *to maintain system modularity the growth of contributors results in a decrease in system complexity.*

This happens because once there are many developers , the system needs to be less complex to facilitate development.

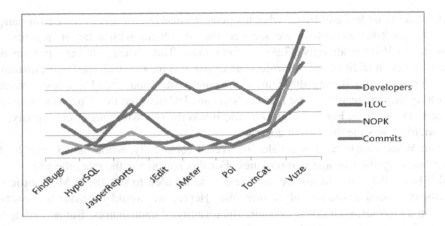

**Fig. 1.** Overview of each system

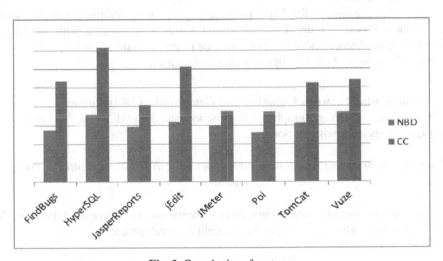

**Fig. 2.** Complexity of systems

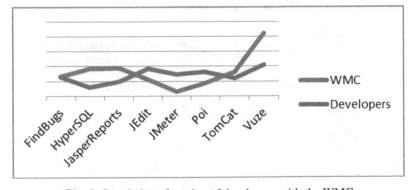

**Fig. 3.** Correlation of number of developers with the WMC

In order to understand better what happens, we analyze especially the performance of Vuze system. In Table 6 we see that the contributor with a larger quantity of commits in Vuze holds only 27.96% of commits . Thus, Vuze is better  split among contributors. In addition, of all projects analyzed, it is the one with the largest quantity of contributors. By having this feature, its complexity should tend to a lower value. But to collect the results of the metrics NBD and DC detected that Vuze is a complex system, as shown in Fig 2. In NBD metric it was the one with greater complexity, in DC metric it was the third most complex.

The WMC metric returns a value of overall complexity, while the CC and NBD calculate only the average complexities. For this reason we thought  the use of CC and NBD metrics would not be sufficient  and decided to use the WMC metric to assess the social evolution of architecture. Hence, we would evaluate the system version altogether, so it could consider the quantity of contributors. But according to Figure 3, we concluded that Vuze is still the most complex even when we analyze the system complete version.

A possible explanation for Vuze to be among the most complex systems can be seen when observing Figure 1, which shows Vuze as the system with the largest quantity of contributors, more lines of code, the largest amount of packages and larger quantity of commits of all the eight systems investigated.

### 5.1.3  Correlation between Cohesion and Geographical Distribution
In the analysis of the geographical distribution and LCOM metric we used as a starting point the following hypothesis:

$H_3$: *to maintain system modularity, high geographic distribution implies in greater cohesion.*

This happens because once they have contributors in different countries; the system must be more cohesive in order to facilitate development.

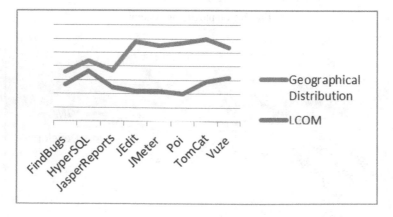

**Fig. 4.** Correlation of geographical distribution with cohesion

In Figure 4 we can see that jEdit, Poi and JMeter systems are among the systems which are geographically dispersed and are among the most cohesive. Moreover, Vuze and HyperSQL and FindBugs are among the least geographically dispersed and are those with low cohesion. However, the most surprising result was Tomcat, it was the system most widely dispersed and unexpectedly was among the least cohesive. The study of the unexpected behavior of this system is currently being developed by our researchers.

### 5.1.4 Correlation between the Quantity of Contributors and the Amount of Packages

This analysis was based on the following hypothesis:

*$H_4$: to maintain the system modularity the growth in t quantity of contributors implies in a growth in the amount of architectural components.*

This is because as the quantity of contributors increases, the system tends to increase the amount of packages.

Due to the data we collected from metrics, we considered the architectural components as packages. In order to see better what happens, we analyze in particular the behavior of Vuze and HyperSQL systems. Vuze is the one with more people contributing and thus more packages. HyperSQL is the one with fewer contributors and less packages. These two systems follow the trend we assumed in our hypothesis. However, as some of the systems do not follow this trend, we cannot conclude the hypothesis is met in all cases. To better understand this, we must know how close these people are in software design to understand cohesion, perhaps a social network analysis could give us more information on this comparison. Or even we could examine the architectural components considering another scope, larger or smaller than package.

### 5.1.5 Correlation between Geographical Distribution and System Complexity

The values obtained were quite motivating . Whereas a modular system has a low complexity, the hypothesis initially raised for this comparison was as follows:

*$H_5$: the best geographically distributed a system is, the least complex.*

This is because once they have contributors in different countries; the system needs to be less complex to facilitate development.

We can note from Figure 5 that when geographical distribution increases, there is in most cases, a reduction in system complexity. Some systems might not follow the trend, although this issue can be further tested by making a study of the evolution of the system and proving that for a particular system, when the geographical distribution increases, the complexity decreases. These tests are being conducted by our team for future work.

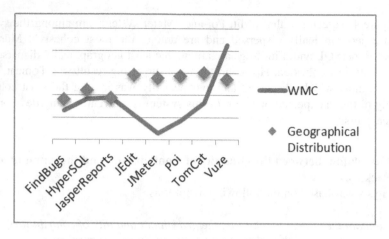

**Fig. 5.** Geographic distribution graph relating to Complexity

## 6     Conclusion and Future Works

Among the 8 open source systems investigated, there are few who escape completely the standard required to understand their behavior before the results of the metrics of modularity. They are:

- JasperReports: It is in 6th place in the complexity, i.e., it has low complexity and with only 20 people, it is the 2nd project with fewer contributors. Its geographical distribution is the 2nd smallest, which does not explain the low complexity. Therefore, we could not correlate the social aspects of modularity metrics in this system.

- JEdit: It is 2nd  largest system in quantity  of contributors and the 2nd largest in geographical distribution. The complexity of this system is the 2nd largest of all. To evaluate the system considering  its modularity, system complexity should have been lower.

- Poi: It has large number of contributors, but the system complexity is low. Several people developing the project in many different countries makes the code less complex since it is highly splitted.

- JMeter: It is among the projects with the largest number of contributors, there are 29. It is among the largest geographic distributions. Because of this, its complexity is the 2nd lowest.

These results imply that we need to research more to understand why these systems are exceptions. Maybe calculating more metrics could let us see new features.  The other results were more encouraging, proving that there is some relationship between social metrics and modularity in the development of a system.

The research confirms the initial idea that social parameters for system projects can be related to modularity metrics. However we still have difficulties on how to obtain social data for analysis. Although we have seen here and in other researches is that, according to the Mirroring Hypothesis, the organization mirrors the code, nevertheless working with open-source systems is a bit complicated. There are many peculiarities to consider with those systems, thus a great amount of metrics has to be analyzed to obtain a better result.

The results evidence the necessity of trade-off analysis among the metrics. This will underpin the peculiarities of each metric. Besides, to improve results it is necessary to increase the quantity of versions and the amount of metrics for each system. These tests are being conducted by our team.

We also intend to restrict the term "Social Factors", because it is far reaching. We will try to analyze other social factors associated to the contributors to get more responses. These   social factors would be: education, age, occupation of the contributors among others.

We will investigate the subcomponents of the system package structure. It is usually just a physical hierarchy, but logically there are other components, such as different subprojects, systems and libraries all of them maintained by different sub teams of contributors.

We will also check possible explanation on comparing and collecting every threat that affect the results. Every threat will be analyzed in order to ascertain its influence and its validity concerning the research.

# References

1. Carmel, E., Tija, P.: Offshoring Information Technology: Sourcing and Outsourcing to a Global Workforce. Cambridge University Press, New York (2005)
2. Audy, J., Prikladnicki, R.: Desenvolvimento Distribuído de Software:Desenvolvimento de software com equipes distribuídas. Elsevier, Rio de Janeiro (2008)
3. Lanubile, F., Damian, D., Oppenheimer, H.: Global Software Development: Technical, Organizational, and Social Challenges. ACM SIGSOFT Software Engineering Notes 28(6) (November 2003)
4. Pilatti, I., Audy, J.L.N., Prikladnicki, R..: Software configuration management over a global software development environment: lessons learned from a case study. In: Proceedings of the 2006 International Workshop on Global Software Development for the Practitioner (GSD 2006), pp. 45–50. ACM, New York (2006)
5. Morris, R., Parnas, D.L.: On the Criteria To Be Used in Decomposing Systems into Modules. Magazine Communications of the ACM (1972)
6. Baldwin, C.Y.: Modularity and Organizations, Harvard Business School Finance Working Paper No. 13-046 (2012), http://ssrn.com/abstract=2178640 or http://dx.doi.org/10.2139/ssrn.2178640
7. Garlan, D., Shaw, M.: An introduction to software architecture. Advances in software engineering and knowledge engineering 1, 1–40 (1993)
8. Garlan, D.: Software Architecture. School of Computer Science at Research Showcase (2001)
9. Rezende, D.A.: Engenharia de Software e sistemas de informação. 2nd ed. Brasport, Rio de Janeiro (2002)
10. Gabriel, R.P., Jackson, M.: Definitions of Modularity. Retrospective on Modularity. AOSD, Porto de Galinhas, Brazil (2011)

11. Conway, M.E.: How Do Committees Invent? Magazine Datamation (1968)
12. Maccormack, A., Rusnak, J., Baldwin, C.: Exploring the Duality between Product and Organizational Architectures: A Test of the "Mirroring" Hypothesis. Publication in Research Policy (2011)
13. Mills, E.E.: Software Metrics (CMU/SEI-88-CM-012). Software Engineering Institute, Carnegie Mellon University (1988)
14. McCabe, J.: A Complexity Measure. In: Proceedings of the 2nd international conference on Software engineering (ICSE 1976), p. 407. IEEE Computer Society Press, Los Alamitos (1976)
15. Pressman, R.S.: Engenharia de Software, 5th edn., p. 843. McGraw-Hill, Rio de Janeiro (2002)
16. Martin, R.C.: Prentice-Hall, Inc., Upper Saddle River (1995)
17. Rosemberg, L.H.: Applying and Interpreting Object Oriented Metrics. [S.l.]: NASA Software Assurance Technology Center, SACT (2007)
18. Cai, Y., Huynh, S.: Measuring Software Design Modularity, pp. 5–6 (2008)

# Towards a Formal Library for Precise and Accurate Measurements

Meryem Lamrani, Younès El Amrani, and Aziz Ettouhami

Laboratoire Conception et Systèmes, Department of Computer Science
University Mohamed V Agdal – Rabat, Morocco
{lamrani,elamrani,touhami}@fsr.ac.ma

**Abstract.** Several software quality metrics have been proposed during the last decades to reduce risks and faultiness during software development. Despite these many efforts, most defined measurements fail to reach a satisfactory level of performance due to the ambiguities detected in the way their definitions are expressed and later on interpreted. The paper introduces a formal extension of a well-known library of measures named FLAME. While this library is expressed using the semi-formal language OCL (Object Constraint Language) upon the UML meta-model, our extension make use of formal methods to provide a more precise and accurate measurements by proposing formal definitions of software metrics upon a formal expression of the UML metamodel. Unlike the original definition of the library, this extension supports formal proofs and guarantees a unique interpretation which allows the built of tools support.

## 1 Introduction

*"Accurate and minute measurement seems to the non-scientific imagination, a less lofty and dignified work than looking for something new. But nearly all the grandest discoveries of science have been but the rewards of accurate measurement and patient long-continued labour in the minute sifting of numerical results"* [38]. When dealing with measurements, it is essential to focus on both precision and accuracy to get useable data. The first one guaranties repeatability while the second one evaluates correctness. According to the ISO 5725 standard [9], precision is defined as the closeness of agreement between independent test or measurement results obtained under stipulated conditions, and accuracy is the closeness of agreement between a test or measurement result and the true value.

The scope of this paper covers measurements in the software engineering area. Quality in software engineering implies measurement of the desired properties such as reusability, maintainability, efficiency and reliability, known as quality indicators. For this reason, several software metrics were defined to measure these attributes. They are often classified into three categories:

B. Murgante et al. (Eds.): ICCSA 2014, Part V, LNCS 8583, pp. 633–648, 2014.
© Springer International Publishing Switzerland 2014

- **Product metrics:** describe the characteristics of the product such as size, complexity and performance [9].

- **Process metrics:** used to improve software development, testability and maintenance. Examples are the effectiveness of defect removal during development, the pattern of testing defect arrival and the response time of the fix process [9].

- **Project metrics:** informs about project characteristics and execution. Examples are number of software developers, the staffing pattern over the life cycle of the software, cost, schedule, and productivity [9].

Process metrics will be our main concern in the scope of this contribution. Because experiments have shown that software metrics calculated earlier in the software lifecycle are more likely to help faultiness detection and improve the control over time and cost estimation, we especially focus on software metrics defined at design phase, referred from now on, as software design metrics. Consequently, the control of software quality will be undeniably affected by the measurement quality at this phase. The quality of measurements depends on the accuracy and precision of the entire data acquisition system, which is represented by software metrics in this particular case. Unfortunately, most existing software metrics lack in precision and accuracy due to their ambiguous definitions.

In our previous contributions, we introduced a novel approach to model software measurement definitions system [1, 3]. Based on formal methods, it consists on building the basis of solid metrics definitions upon a formal expression of the UML meta-model [5]. Later on, we concentrated on the importance of having a formal semantics of OCL and described for the first time, the OCL predefined properties in a formal expression [2] complementing, in this way, the many researches done in the formalization of OCL expressions.

In the present paper, we extend this approach to cover further functions and definitions in order to present a consistent, precise and accurate version of FLAME, a library of measures [4].

## 1.1 Existing Approaches

Since our approach concerns different research area, we will proceed in the following to present the relevant related works in the formal representation of UML structures, the formalization of software metrics and finally, the formalization of OCL expressions.

### 1.1.1 The Formal Representation of UML Structures

UML continues to gain increased popularity making it the de-facto standard to model software design systems. Hence, it explains the many attempts in proposing a formalism of its structure. Among them, the use of a language based on higher-order logic called PVS specification language (PVS-SL) [10] where constituents of UML diagrams are represented as PVS theories. Others proposed the use of Description Logics (DLs) [11-12] where Object-oriented concepts are modeled in means of concepts. But almost all relevant efforts are based on the Z language such as Hall [13-14] and Hammond [15]. Malcolm Shroff and Robert B. France [16-17] based their approach on the Hall and Hammond's Z formalization approach of the class structures with the particularity of introducing inheritance relationship as an attribute in the inheriting class. In Lamrani et al. [3], we build our model according to the Laurent Henocque concept [18] who elaborated a formal specification of Object Oriented Constraint Programs. This approach was adopted to express formally the UML metamodel.

### 1.1.2 The Formalization of Software Quality Metrics

Several existing software metrics [19-21] have proved themselves to be useful in the process of measuring software quality. Often introduced as a set, we will relate the ones that are commonly recognized by the community for their relevance in bringing information on quality over certain software properties.

Among measurements defined at design level of software lifecycle, we find the MOOD and MOOD2 (Metrics for Object-Oriented Design) [22], MOOSE (Metrics for Object-Oriented Software Engineering) also known as the CK metrics [23], EMOOSE (Extended MOOSE) [24] and QMOOD (Quality Model for Object-Oriented Design) [25]. No standard definition exists to interpret these metrics in an unambiguous way especially when we know that they were initially defined using natural language for some or semi-formal languages for others. Thus, many researchers contribute in proposing formal definitions to these metrics such as El-Wakil et al. [26] who used XQuery [27] language. Harmer and Wilkie [28] present them as SQL queries over a relational schema. While most contributions based their formalization on the use of OCL language such as McQuillan et al. [29] whom extended the UML metamodel 2.0 to offer a framework for metric definitions and Goulao et al. [30] whom also used OCL language for defining component based metrics and used the UML 2.0 metamodel as a basis for their definitions. These latter efforts based their work on Baroni et al. approach which proposed a Formal Library for Aiding Metrics Extraction (FLAME) [4]. Although this library of measures is called Formal, it is essential to mention that its metric definition model lies on the use of OCL which remains till today a semi-formal language with no possibility for a further analysis and verification through theorem provers. Nevertheless, the use of OCL has its advantages since it remains a user friendly language and the most eligible for a potential spread usage of measurements in the industry area. Consequently, we propose an extension of FLAME since we agree that OCL should stay as foreground definitions when it comes to software measurements but with a significant importance of having an even more solid definitions based on formal methods that will remain in the background.

### 1.1.3 The Formalization of OCL Expressions

The OMG standard of OCL [6] contains an elaborated chapter on formalizing OCL semantics. This chapter corresponds to the work of Mark Ritchers et al. [31] and is considered as the most rigorous attempts to express OCL formally. Beside the fact that this approach requires a strong mathematical background, it does, unfortunately, not propose a formal alternative to the predefined properties which constitute an important part of OCL expressions. Flake et al. [32, 33] extended this formal semantics by giving descriptions of ordered sets, global OCL variables definitions, UML statechart states and OCL messages. These different notions do not constitute a part of the need for the present contribution. Other interesting works were proposed by both Kyas et al. [34] who used the language of the theorem prover PVS to translate the UML and OCL constraints, and Brucker et al. [35] who introduce transformations rules of OCL constraints into B formal expressions. Whereas, a new formalism were established by Gergly et al. [36] on the Abstract State Machines technique called OCLASM. A recent contribution elaborated by Lamrani et al. [2] introduces for the first time a formal definition of OCL predefined properties by building a solid OCL expression based on a formal definition of OCL types metamodel. This latter approach will be used in the present paper to deal with OCL expressions in general and OCL predefined properties in particular.

## 1.2  Paper Objectives

The main objective of this paper is to propose a standardized library to measure software design quality. Software measurement is fully recognized as an important sub-discipline of software engineering but it is almost ignored by many industrial stakeholders. The major reason remains the lack of tools support. By extending a library of measures such as FLAME, we aim to provide a rigorous definition of its functions to formally express any set of software design metrics defined upon the UML metamodel with possibility to elaborate proofs through this formal specification.

FLAME is defined as a formal library for aiding metrics extraction, but even though it is called formal, the use of OCL, considered as a semi-formal language, results in weakly defined functions. OCL is based on mathematical logic but its semantics remains incomplete, furthermore, its syntax is given by a grammar description and no metamodel is available unlike the metamodel of UML which means that it suffers from an absence of well-formedness rules. As a significant example, the "iterate" expression which is known to be potentially non-deterministic since there is no precision on order evaluation leading to different possible results [31]. Also, the usage of many keyworks such as self, asSet and some predefined functions such as OclAsType, OclIsKindOf, without providing any precise and rigorous definition, constitute significant limitations to consider OCL as completely formal.

The choice of the Z language [7-8] is essentially justified by its maturity and simplicity. Its grouping concept illustrated by schemas offers a clear semantics, meanwhile, the ability to be type-checked and verified through formal proofs responds to the objectives settled for a formal specification language. The whole approach including formal definitions of FLAME methods is fully type checked using Z/EVES tool [37] which is a proof tool based on EVES and ZF set theory that supports the Z notation.

The first step is to express UML metamodel in a formal language without any reflexive reference to UML, it results in more clarity. The second step is to express the OCL semantics in a more rigorous manner based on a formal definition of the OCL types metamodel and finally express FLAME functions and then metrics in a precise and accurate definition that enables to check certain system properties involving metrics. This could not be achieved with previous definitions using OCL.

The paper is organized as follows: Section 1 presents an overview of FLAME. Section 2 briefly introduces essential aspects of Z-based model for metrics formalization including the UML metamodel formalization and the formal expression of OCL semantics. Section 3 proposes the extension of FLAME methods based on the combined approach described in previous sections. Finally, Section 4 draws conclusion and perspectives.

## 2 FLAME Overview

Formal Library for Aiding Metrics Extraction called FLAME is a library of measures build with the Object constraint Language (OCL) upon the UML metamodel. Initially based on the idea of MoodLib, a library of functions used to calculate MOOD and MOOD2 (Metrics for Object-Oriented Design) [22].

The choice of FLAME is justified by its multiple advantages:
- Its focus on metrics defined earlier in the software lifecycle (design metrics).
- Based on UML metamodel and built with OCL. Both are familiar OMG standard which have increasing popularity among software designers.
- Based on object oriented paradigms and able to describe any set of metrics built upon the core of UML metamodel.

FLAME presents its functions in a grouping context according to the UML metamodel. The functions are classified as general, set, percentage or counting functions. General are those that return Booleans.

**Table 1.** Function At Attribute Context

| Acronym | Name | Return Type |
|---------|------|-------------|
| FCV | Feature to Classifier Visibility | Boolean |

Set return set of elements which can have the type of any meta-class in the UML metamodel.

**Table 2.** Function At Classifier Context

| Acronym | Name | Return Type |
|---------|------|-------------|
| - | Coupled Classes | Set(Classifier) |

Percentage return a value representing a percentage and finally counter functions return integers.

**Table 3.** Function At Operation Context

| Acronym | Name | Return Type |
|---------|------|-------------|
| OUN | Operation Use Number | Integer |

As illustrated before, FLAME library [4] presents first tables of acronyms of the functions used in FLAME grouped by their context in the UML metamodel. Then, it describes the OCL definitions of these functions according to the same logical grouping.

**Table 4.** Classifier Set Function

| Name | Feature2AttributeSet |
|------|----------------------|
| Informal Definition | Subset of Attributes (from one set of Features) belonging to the current Classifier. |
| FLAME definition | **Classifier :: feature2AttributeSet( s: Set( Feature ) ): Set( Attribute )** = s -> select( f: Feature \| f.oclIsKindOf( Attribute ) ) -> collect( f \| f.oclAsType( Attribute ) ) -> asSet |

Since we propose in this contribution an extension of FLAME library, we will proceed to present FLAME functions gradually as we introduce the formal version.

## 3   Model Based Approach

This section describes the formal approach adopted to express the metrics definitions rigorously. This approach [3] is based on a leveled formalization. At first, it gives a formal specification of the UML metamodel part on which the second level, consisting on metrics formalization, is defined.

### 3.1 Methodology

The methodology is an adaptation of the Laurent Henocque contribution [18] about the formal specification of object-oriented constraint programs. The following definitions are basics notion of this approach. Each definition is presented with its Z [7-8] representation:

- **ObjectReference:** a set of object references as an uninterpreted data type.

  *Z representation:* [ObjectReference]

- **ReferenceSet:** a finite set of object references used to model object types.

  *Z representation:* **ReferenceSet** $==$ $\mathbb{F}$ *ObjectReference*

- **CLASSNAME:** class names defined using free type syntax of Z.

  *Z representation:* **CLASSNAME** ::= ClassElement | ClassNamedElement | ...

- **ObjectDef:** a predefined super class for all future classes.

  *Z representation:*

  $$\begin{array}{|l}\hline \_\_ObjectDef_____ \\ ref: ObjectReference \\ class:CLASSNAME \\ \hline \end{array}$$

- **Instances:** a function mapping class names to the set of instances of that class

  Z representation: | **instances:** *CLASSNAME* → *ReferenceSet*

- **NIL:** Undefined Object

  Z representation: | *NIL: ObjectDef*

- **Class:** implemented via two constructs:

  **a) A class definition:** a schema in which we find, in its invariant part, both the class attributes and the inheritance relationships and in its predicate part, specification of class invariants.

  *Z representation:*

  $$\begin{array}{|l}\hline \_\_ClassDefElement_____ \\ name: seq\ CHAR \\ \hline \end{array}$$

  **b) A class specification:** a combination of a class definition extended with the ObjectDef and class references.

  *Z representation:*
  **ClassSpecElement** ≅ *ClassDefElement* ∧ *[ObjectDef| class =ClassElement]*

The adaptation of Laurent Henocque approach [18] consists on using and extending the above notions to formalize UML class structures including inheritance, aggregation, relationships and visibility to obtain a formal expression of the UML metamodel.

### 3.2 UML Metamodel Formalization

The following schema is extracted and combined according to the UML metamodel specifications [5]. It consists on the core package.

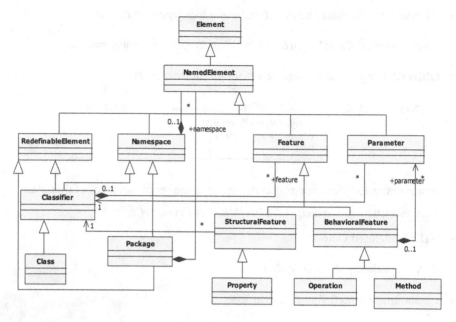

**Fig. 1.** A partial representation of the UML metamodel core package

Since we refer to a more recent version of the UML metamodel (version 2.3) than the one used in FLAME library, we proceed in the scope of this contribution to an update through our formalization approach and extension.

**Inheritance**

Simple inheritance and multiple inheritance relationships are built simply by importing the schema definition of inherited superclasses into the class that inherit from them.

*Z representation (simple inheritance):*

```
__ClassDefNamedElement_____
 ClassDefElement
|_____
```

*Z representation (multiple inheritance):*

```
__ClassDefClassifier_____
 ClassDefNamespace
 ClassDefRedefinableElement
|_____
```

**Relationship with Multiplicities**

General relations are free of constraints, which mean that every tuple can be accepted. The multiplicity is naturally stated in the predicate part as the cardinal of related target objects for each source object.

*Z representation:*

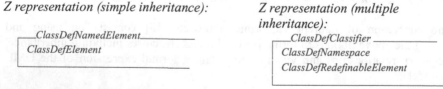

pc: Parameter ↔ Classifier
_____
∀ c: Classifier · # (pc ( {c} )) ⩽ 1

## Aggregation

The aggregate relation is more constrained than a general one, thus we have to change the type of relation to make a distinction between both. In the different aggregate relations given in our UML metamodel fragment, the multiplicity is of 0..1 which means that each component occurs in at most one composite. Consequently, its relational inverse is an injective partial function.

*Z representation:*

| hasNamedElement: Namespace $\leftrightarrow$ NamedElement |
| --- |
| hasNamedElement $\tilde{\ } \in$ NamedElement $\rightarrowtail$ Namespace |

The $\rightarrowtail$ symbol represents the partial function and the $\tilde{\ }$ stands for the relational inverse.

## Class Types

Class types are defined using an axiomatic definition. The declaration part contains type sets that correspond to the existing classes of our given model. Each type is defined as a finite set of object references.

*Z representation:*

| Element, NamedElement, Namespace, .... : ReferenceSet |
| --- |
| Predicate Part |

The predicate part describes the properties of these sets:

- A type is equal to the union of the corresponding class instances and the type of all its subclasses.

  *Z representation:*
  Element = instances ClassElement $\cup$ NamedElement
  NamedElement = instances ClassNamedElement $\cup$ Namespace $\cup$ RedefinableElement $\cup$ Feature
  ...
  instances ClassElement = { o: ClassSpecElement | o.class = ClassElement • o.ref }
  instances ClassNamedElement = { o: ClassSpecNamedElement | o.class = ClassNamedElement • o.ref }
  ...

- Each object reference is used at most once for an object which means that no two distinct object bindings share the same object reference.

  *Z representation:*
  $\forall$ i: instances ClassElement • $\exists$ x: ClassSpecElement • x.ref = i
  $\forall$ i: instances ClassNamedElement • $\exists$ x: ClassSpecNamedElement • x.ref = i
  ...

## 3.3 Formal OCL Expressions

OCL is a typed language since each expression has a type. All types used in the expression must follow the rules of type conformance, illustrated in the OCL Standard Library [6]:

**Table 5.** Type Conformance Rules

| Type | Conforms to/ Is a subtype of | Condition |
|---|---|---|
| Set(T1) | Collection(T2) | If T1 conforms to T2 |
| Sequence(T1) | Collection(T2) | If T1 conforms to T2 |
| Bag(T1) | Collection(T2) | If T1 conforms to T2 |
| OrderedSet(T1) | Collection(T2) | If T1 conforms to T2 |
| Integer | Real | |
| InlimiledNatural | Integer | *is an *invalid* Integer |

The conformance of different types is determined by a type hierarchy represented in the OCL metamodel types.

### 3.4 OCL Types Metamodel Formalization

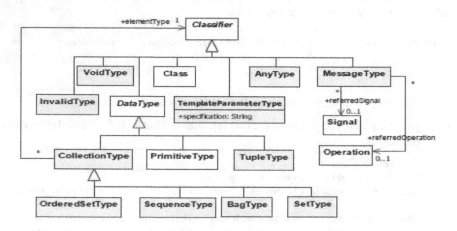

**Fig. 2.** Abstract syntax kernel metaodel for OCL types

The conformance of types in OCL is guaranteed by the presence of a metamodel for OCL types. The same formalization approach adopted for the UML metamodel is applied to formalize the OCL types metamodel.

The detailed representation of each class type is given through an axiomatic definition where the declaration part contains the type sets corresponding to all the classes in the OCL metamodel types while several axioms constitute the properties:

*Classifier, Class, VoidType,...: ReferenceSet*

*Classifier = instances ClassClassifier ∪ VoidType ∪ Class ∪ AnyType ∪ MessageType ∪ InvalidType ∪ DataType ∪ TemplateParameterType*

*VoidType = instances Class VoidType*
*Class = instances ClassClass*
...

Each type is the union of all the corresponding class instances, and of the types of its subclasses (if existing).

*instances ClassClassifier ={o: ClassSpecClassifier | o.class = ClassClassifier · o.i}*
*instances ClassClass = {o: ClassSpecClass | o.class = ClassClass · o.i}*
...

The set "instances" holds the schema bindings. These sets are pairwise disjoint by construction.

$\forall\ i : instances(ClassClassifier) \bullet\ (\exists\ x : ClassSpecClassifier \bullet\ x.i = i)^*$
$\forall\ i : instances(ClassClass) \bullet\ (\exists\ x : ClassSpecClass \bullet\ x.i = i)$
...

The same object reference cannot be used for two distinct objects in the same class.

# 4    FLAME Extended to Formal Methods

## 4.1 Classifier Set Functions

| Name | feature2AttributeSet |
|---|---|
| *Informal Definition* | Subset of Attributes (from one set of Features) belonging to the current Classifier. |
| *FLAME Definition* | **Classifier:: feature2AttributeSet( s: Set( Feature ) ): Set( Attribute )**<br>= *s -> select( f: Feature | f.oclIsKindOf( Attribute ) ) -> collect( f | f.oclAsType( Attribute ) ) -> asset* |
| *FLAME Definition Ext.* | *feature2AttributeSet: ObjectDef × $\mathbb{P}$ Feature → $\mathbb{P}$ Property*<br>$\forall$ *o: ObjectDef; S: $\mathbb{P}$ Feature  \| instances o.class = Feature $\wedge$ S = { f: Feature \| oclIsKindOf (o, Property) = TRUE } · feature2AttributeSet (o, S) = { f: S\| oclAsType (o, Property) = o }* |

| Name | newAttributes |
|---|---|
| *Informal Definition* | Set of Attributes declared in the current Classifier. |
| *FLAME Definition* | **Classifier:: newAttributes( ): Set( Attribute )**<br>= *definedAttributes( ) - allInheritedAttributes( )* |
| *FLAME Definition Ext.* | *newAttributes: ObjectDef × Classifier → $\mathbb{P}$ Property*<br>$\forall$ *o: ObjectDef; C: Classifier; A1,A2: $\mathbb{P}$ Property \| A1 = definedAttributes (o, C) $\wedge$ A2 = allInheritedAttributes (o, C) · newAttributes (o, C) = A1 - A2* |

| Name | definedAttributes |
|---|---|
| *Informal Definition* | Set of Attributes declared in the Classifier, including overridden Attributes. |
| *FLAME Definition* | **Classifier:: definedAttributes( ): Set( Attribute )**<br>= *feature2AttributeSet( self.definedFeatures( ) )* |
| *FLAME Definition Ext.* | **definedAttributes: ObjectDef × Classifier → P Property**<br>∀ *o: ObjectDef; C: Classifier; A:* P *Property* \| *A = feature2AttributeSet (o,*<br>*(definedFeatures (o, Classifier))) • definedAttributes (o, C) = A* |

| Name | allInheritedAttributes |
|---|---|
| *Informal Definition* | Set of Attributes declared in the Classifier, including overridden Attributes. |
| *FLAME Definition* | **Classifier:: allInheritedAttributes( ): Set( Attribute )**<br>= *feature2AttributeSet( self.allInheritedFeatures( ) )* |
| *FLAME Definition Ext.* | **allInheritedAttributes: ObjectDef × Classifier → P Property**<br>∀ *o: ObjectDef; C: Classifier; A:* P *Property* \| *A = feature2AttributeSet (o,*<br>*(allInheritedFeatures (o, C))) • allInheritedAttributes (o, C) = A* |

| Name | allAttributes |
|---|---|
| *Informal Definition* | Set containing all Attributes of the Classifier itself and all its inherited Attributes (both directly and indirectly). |
| *FLAME Definition* | **Classifier:: allAttributes( ): Set( Attribute )**<br>= *feature2AttributeSet( self.allFeatures( ) )* |
| *FLAME Definition Ext.* | **allInheritedAttributes: ObjectDef × Classifier → P Property**<br>∀ *o: ObjectDef; c: Classifier; S:* P *Property* \| *S = feature2AttributeSet (o,*<br>*(allFeatures (o, c))) • allAttributes (o, c) = S* |

## 4.2  Classifier Counting Functions

| Name | DAN – Defined Attributes Number |
|---|---|
| *Informal Definition* | Number of defined Attributes in the Classifier. |
| *FLAME Definition* | **Classifier:: DAN ( ): Integer**<br>= *definedAttributes( ) -> size( )* |
| *FLAME Definition Ext.* | **DAN: ObjectDef × Classifier → ℕ**<br>∀ *o: ObjectDef; C: Classifier; A:* P *Property* \| *A = definedAttributes (o,*<br>*C) • DAN (o, C) = # A* |

| Name | IAN – Inherited Attributes Number |
|---|---|
| *Informal Definition* | Number of inherited Attributes in the Classifier. |
| *FLAME Definition* | **Classifier:: IAN ( ): Integer**<br>= *allInheritedAttributes( ) -> size( )* |
| *FLAME Definition Ext.* | **IAN: ObjectDef × Classifier → ℕ**<br>∀ *o: ObjectDef; C: Classifier; A:* P *Property* \|*A = allInheritedAttributes (o, C)*<br>*• IAN (o, C) = # A* |

| Name | AAN – Available Attributes Number |
|---|---|
| *Informal Definition* | Number of Attributes in the Classifier. |
| *FLAME Definition* | **Classifier:: AAN ( ): Integer**<br>= *allAttributes( ) -> size( )* |
| *FLAME Definition Ext.* | **AAN: ObjectDef × Classifier → ℕ**<br>∀ *o: ObjectDef; C: Classifier; A:* ℙ *Property \| A = allAttributes (o, C) · AAN (o, C) = # A* |

### 4.3 GeneralizableElement Set Functions

| Name | CHIN – Children Number |
|---|---|
| *Informal Definition* | Number of directly derived Classes. |
| *FLAME Definition* | **GeneralizableElement:: CHIN ( ): Integer**<br>= *children( ) -> size( )* |
| *FLAME Definition Ext.* | **CHIN: ObjectDef × RedefinableElement → ℕ**<br>∀ *o: ObjectDef; r: RedefinableElement; S:* ℙ *RedefinableElement*<br>∀ *\| S = children (o, r) · CHIN (o, r) = # S* |

### 4.4  Package Counting Functions

| Name | CN – Classes Number |
|---|---|
| *Informal Definition* | Number of Classes in the Package. |
| *FLAME Definition* | **Package:: CN ( ): Integer**<br>= *allClasses( ) -> size( )* |
| *FLAME Definition Ext.* | **CN: ObjectDef × Package → ℕ**<br>∀ *o: ObjectDef; P: Package; C:* ℙ *Class; n:* ℕ *\| C = allClasses (o, P) ∧ \| n = # C · CN (o, P) = n* |

| Name | PIAN – Package Inherited Attributes Number |
|---|---|
| *Informal Definition* | Number of Attributes inherited in the Package. |
| *FLAME Definition* | **Package:: PIAN ( ): Integer**<br>= *allClasses( ) -> iterate( elem: Class; acc: Integer = 0 \| acc + elem.IAN( ) )* |
| *FLAME Definition Ext.* | **PIAN: ObjectDef × Package → ℕ**<br>∀ *o: ObjectDef; P: Package; C:* ℙ *Class; n:* ℕ<br>*\| C = allClasses (o, P) ∧ (∀ c: C · n = n + IAN (o, c)) · PIAN (o, P) = n* |

| Name | PAAN – Package Available Attributes Number |
|---|---|
| *Informal Definition* | Number of available Attributes inherited in the Package. |
| *FLAME Definition* | **Package:: PAAN ( ): Integer**<br>= *allClasses( ) -> iterate( elem: Class; acc: Integer = 0 \| acc + elem.AAN( ) )* |
| *FLAME Definition Ext.* | **PAAN: ObjectDef × Package → ℕ**<br>∀ *o: ObjectDef; P: Package; C:* ℙ *Class; n:* ℕ<br>*\| C = allClasses (o, P) ∧ (∀ c: C · n = n + AAN (o, c)) · PAAN (o, P) = n* |

## 5   Application of FLAME Extension to Metrics

This section presents an application of FLAME extended functions to software design metrics belonging to different set of well-known metrics such as MOOSE (Metrics for Object-Oriented Software Engineering) [23] and EMOOSE (Extended MOOSE) [24] which act at the classifier level, and MOOD (Metrics for Object-Oriented Design) [22] and QMOOD (Quality Model for Object-Oriented Design) [25] that measure the whole system

| Name | *NOC – Number of Children (from the MOOSE set)* |
|---|---|
| *Informal Definition* | The number of classes that inherit directly from the current Class. |
| *FLAME Definition* | **Classifier:: NOOC( ): Integer**<br>$= self.CHIN()$ |
| *FLAME Definition Ext.* | *NOC: ObjectDef $\times$ Classifier $\rightarrow \mathbb{N}$*<br>$\forall\ o: ObjectDef;\ c: Classifier;\ n: \mathbb{N}\ \mid\ n = CHIN(o,\ c) \cdot NOC(o,\ c) = n$ |

| Name | *SIZE 2 (from the EMOOSE set)* |
|---|---|
| *Informal Definition* | Number of local Attributes and Operations defined in the Class. |
| *FLAME Definition* | **Classifier:: SIZE 2( ): Integer**<br>$= self.DON() + self.DAN()$ |
| *FLAME Definition Ext.* | *SIZE2: ObjectDef $\times$ Classifier $\rightarrow \mathbb{N}$*<br>$\forall\ o: ObjectDef;\ c: Classifier;\ n1, n2: \mathbb{N}\ \mid\ n1 = DON(o,\ c) \wedge n2 = DAN(o,\ c) \cdot SIZE2(o,\ c) = n1 + n2$ |
| *Comments* | *DON( ) is the equivalent of DAN( ), but for operations.* |

| Name | *DSC – Design Size in Classes (from the QMOOD set)* |
|---|---|
| *Informal Definition* | Number of Classes in the Package. |
| *FLAME Definition* | **Package:: DSC( ): Integer**<br>$= self.CN()$ |
| *FLAME Definition Ext.* | *DSC: ObjectDef $\times$ Package $\rightarrow \mathbb{N}$*<br>$\forall\ o: ObjectDef;\ P: Package;\ n: \mathbb{N}\ \mid\ n = CN(o,\ P) \cdot DSC(o,\ P) = n$ |

| Name | *AIF – Attributes Inheritance Factor (from the MOOD set)* |
|---|---|
| *Informal Definition* | Quotient between the number of inherited Attributes in all Classes of the Package and the number of available Attributes (locally defined plus inherited) for all Classes of the current Package. |
| *FLAME Definition* | **Package:: AIF( ): Percentage**<br>$= self.PIAN() / self.PAAN()$<br>*pre: self.PAAN() $> 0$* |
| *FLAME Definition Ext.* | *AIF: ObjectDef $\times$ Package $\rightarrow \mathbb{R}$*<br>$\forall\ o: ObjectDef;\ P: Package \mid PAAN(o, P) > 0$<br>$\cdot\ AIF(o, P) = PIAN(o, P)\ \text{div}\ PAAN(o, P)$ |

# 6 Conclusion and Perspectives

The present contribution synthesizes and extends previous efforts done towards the constitution of a formal model for software design measurements. It first presents a formal specification of UML metamodel upon which measurements are defined, then, it expresses formally the OCL types metamodel for a rigorous definition of the OCL predefined properties and thus OCL expressions. The whole model represents the basic foundation for an accurate and precise definition of FLAME functions that will help providing standardized expression to various set of metrics commonly recognized as useful and efficient to bring information over software quality.

As future work, we plan to integrate a large part of the UML metamodel which will allow to cover other set of metrics such as code metrics since the experiments have shown that an hybrid use of design metrics with code metrics are more likely to detect faultiness during software lifecycle.

# References

1. Lamrani, M., El Amrani, Y., Ettouhami, A.: A Formal Definition of Metrics for Object Oriented Design: MOOD Metrics. Journal of Theoretical and Applied Information Technology 49, 1–10 (2013)
2. Lamrani, M., El Amrani, Y., Ettouhami, A.: On Formalizing Predefined OCL Properties. Journal of WASET 73, 327–331 (2013)
3. Lamrani, M., El Amrani, Y., Ettouhami, A.: Formal Specification of Software Design Metrics. In: The Sixth International Conference on Computer and Software Engineering Advances, pp. 348–355 (2011)
4. Baroni, A.L., Abreu, F.B.: A Formal Library for Aiding Metrics Extraction. In: International Workshop on Object-Oriented Re-Engineering at ECOOP (2003)
5. The Object Management Group, UML 2.3 superstructure specification (2010)
6. The Object Management Group, Object Constraint Language (February 2010)
7. Spivey, M.: The Z Notation. Prentice-Hall (1992)
8. Woodcock, J., Davies, J.: Using Z: Specification, Proof and Refinement. In: Prentice Hall International Series in Computer Science (1996)
9. Kan, S.H.: Metrics and Models in Software Quality Engineering, 2nd edn., pp. 85–126. Addison-Wesley Professional (2002)
10. Aredo, D.B., Traore, I., Stølen, K.: Towards a formalization of UML Class Structure in PVS. Research Report no. 272, University of Oslo (1999)
11. Calì, A., Calvanese, D., De Giacomo, G., Lenzerini, M.: Reasoning on UML Class Diagrams in Description Logics. In: Proceedings of IJCAR Workshop on Precise Modelling and Deduction for Object-oriented Software Development (PMD) (2001)
12. Efrizoni, L., Wan-Kadir, W.M.N., Mohamad, R.: Formalization of UML Class using Description Logics. In: The International Symposium in Information Technology (2010)
13. Hall, J.A.: Specifying and Interpreting Class Hierarchies in Z, pp. 120–138. Bowen and Hall
14. Bowen, J.P., Hall, J.A. (eds.): Z User Workshop, Cambridge, Workshops in Computing. Springer, New York (1994)
15. Wilkie, F., Harmer, T.: Tool support for measuring complexity in heterogeneous object-oriented software. In: Proceedings of IEEE International Conference on Software Maintenance, Montreal, Canada (2002)

16. France, R.B., Bruel, J.M., Larrondo-Petrie, M.M., Shroff, M.: Exploring the Semantics of UML Type Structures with Z. In: Proceedings of the Formal Methods for Open Object-based Distributed Systems, pp. 247–257. Springer (1997)

17. Shroff, M., France, R.B.: Towards a Formalization of UML Class Structures in Z. In: Proceedings of the 21st Computer Software and Application Conference, pp. 646–651. IEEE Press (1997)

18. Henocque, L.: Z specification of Object Oriented Constraint Programs. RACSAM (2004)

19. Chidamber, S.R., Kemerer, C.F.: A metric suite for Object Oriented Design. Journal IEEE Transactions on Software Engineering 20(6), 476–493 (1994)

20. Lorenz, M., Kidd, J.: Object-Oriented Software Metrics. Prentice Hall Object-Oriented Series (1994)

21. Fenton, N.E., Lawrence, S.P.: Software Metrics: A Rigorous and Practical Approach. International Thompson Computer Press (1996)

22. Abreu, F.B., Carapuça, R.: Object-Oriented Software Engineering: Measuring and Controlling the Development Process. In: 4th International Conference on Software Quality. McLean (1994)

23. Chidamber, S.R., Kemerer, C.F.: A metric suite for Object Oriented Design. Journal IEEE Transactions on Software Engineering 20(6), 476–493 (1994)

24. Li, W., Henry, S., Kafura, D., Schulman, R.: Measuring object-oriented design. Journal of Object-Oriented programming 8(4), 48–55 (1995)

25. Bansiya, J., Davids, C.: Automated metrics and object- oriented development. Dr. Dobbs Journal, 42–48 (1997)

26. El-Wakil, M.M., El-Bastawisi, A., Riad, M.B., Fahmy, A.: A novel approach to formalize Object- Oriented Design. In: 9th International Conference on Empirical Assessment in Software Engineering (EASE 2005) (2005)

27. XQuery 1.0 Standard by W3C XML Query Working Group

28. Wilkie, F., Harmer, T.: Tool support for measuring complexity in heterogeneous object-oriented software. In: Proceedings of IEEE International Conference on Software Maintenance, Montreal, Canada (2002)

29. McQuillan, J.A., Power, J.F.: Towards re-usable metric definitions at the meta-level. In: PhD Workshop of the 20th European Conference on Object- Oriented Programming (ECOOP 2006) (2006)

30. Goulao, M., Abreu, F.B.: Formalizing metrics for COTS. In: Proceedings of the ICSE Workshop on Models and Processes for the Evaluation of COTS Components, Edinburgh, Scotland (2004)

31. Richters, M., Gogolla, M.: On Formalizing the UML Object Constraint Language OCL. Wirtschaftsinformatik 50, 449–464 (1998)

32. Flake, S., Mueller, W.: Formal Semantics of OCL Messages. Electronic Notes in Theoretical Computer Science 102, 77–97 (2003)

33. Flake, S.: Towards the Completion of the Formal Semantics of OCL 2. 0. In: Reproduction, 73-82 (2003)

34. Kyas, M., et al.: Formalizing UML Models and OCL Constraints in PVS1. Electronic Notes in Theoretical Computer Science 115, 39–47 (2005)

35. Marcano, R., Levy, N.: Transformation rules of OCL constraints into B formal expressions. In: Workshop on critical systems development with UML 5th International Conference on the Unified Modeling Language (2002)

36. Mezei, G., Levendovszky, T., Charaf, H.: Formalizing the Evaluation of OCL Constraints. Acta Polytechnica Hungarica 4, 89–110 (2007)

37. Saaltink, M.: Z and EVES. Z User Workshop York, 223-242 (1991)

38. Thomson, W. (Baron Kelvin): Report of the Forty-First Meeting of the British Association for the Advancement of Science (1872) at Presidential inaugural address, to the General Meeting of the British Association, Edinburgh (August 1871)

# Framework for Maintainability Measurement of Web Application for Efficient Knowledge-Sharing on Campus Intranet

Sanjay Misra[1] and Fidel Egoeze[2]

[1] Covenant University, Ota, Nigeria
[2] Federal University of Technology, Minna, Nigeria
sanjay.misra@covenantuniversity.edu.ng,
egoeze@st.futminna.edu.ng

**Abstract.** Web Application is placed and accessed over a network, which could be an intranet, extranet or the internet. An intranet is identified as an ideal platform for knowledge- sharing and collaboration in organizations, or institutions. But it is at times hampered by maintainability issue which indeed is a key quality attribute of web applications. This paper presents an explicit description of a process for prediction of maintainability of web application based on design metrics and statistical approach. The work investigates whether a set of measures identified for UML class diagram structural properties (size, complexity, coupling, cohesion) could be good predictors of class diagram maintainability based on the sub-characteristics; understandability, analyzability, and modifiability. Results indicate that useful prediction models can be built from the measures and identified the strongest predictors from the proposed metrics. This framework can be applied to construct maintainability prediction models to control the maintenance tasks of the system and promote efficient collaboration in university campus.

**Keywords:** web application, maintainability, Intranet, metrics, correlation, regression, analysis.

## 1 Introduction

Web applications are software systems placed and accessed over a network. This could be an intranet providing private communication within an organization, an extranet for semi-private internet work communication or the internet providing world-wide interaction. Web application has grown from simple websites presenting hyperlink text documents to large – scale complex applications, supporting e-commerce and collaborative activities. Tramontana [14] describes web application as an extension of web site. Conallen [2] categorized web applications into two types as;

- Presentation-oriented: A simple document-sharing website consisting of hyper linked text document providing information to users.
- Service-oriented: A complicated web application that provides some services to users, for example, e-commerce applications.

B. Murgante et al. (Eds.): ICCSA 2014, Part V, LNCS 8583, pp. 649–662, 2014.
© Springer International Publishing Switzerland 2014

Web – based systems provide varied functionalities to diverse users, governments and organizations, and enterprises now recourse to the web for their activities. With this increased reliance on the web as a convenient platform for delivering complex applications, there is the need to ensure that web applications are reliable and of high quality.

A major attribute of quality software is maintainability which is about the ease with which to affect maintenance in a system.  According to IEEE 1990 [6], maintainability is the ease with which a software system or component can be modified to correct faults, improve performance or other characteristics, or adapt to a changed environment. Saini et al [13] describes maintainability as the ease with which a product can be maintained in order to correct defects, meet new requirements, make future maintenance easier or cope with a changed environment. A maintainable system undergoes changes with a degree of ease.

ISO 9126 [7] identified maintainability with such sub-characteristics as; Analyzability, changeability, stability, and testability described as follows;

- Analyzability: How easy or difficult it is to diagnose the system for deficiencies or to identify the parts that need to be modified
- Changeability: How easy or difficult it is to make adaptations to the system.
- Stability: How easy or difficult it is to keep the system in a consistent state during modification.
- Testability: How easy or difficult it is to test the system after modification.

Maintainability can be quantified with the use of metrics which can be employed to develop models for predicting maintenance costs and effort. Metrics refers to analytical measurement applied to quantify the state of a system.  It is a quantitative measure of the degree to which a system, component, or process possesses a given attribute [3].

There is no specific agreement on how to measure maintainability. Welker and Oman [15] suggested measuring maintainability in terms of Cyclomatic complexity, lines of code (LOC) and lines of comments. Polo et al [12] used number of modification requests, mean effort per modification request and type of correction to examine maintainability. Muthannaet[10] developed maintainability model using polynomial linear regressions. Kiewkanya et al [8] measured maintainability by measuring both modifiability and understandability. Ghosheh et al [5] used UML web design metrics to assess if the maintainability of the system can be improved by comparing and correlating the results with different measures of maintainability. Relationship between UML metrics and maintainability measures including Understandability time, and Modifiability time were studied. Mendes et al [9] propose an effort prediction for design and authoring effort usingsource code metrics. Size measured in terms of length, functionality, and complexity was the independent variable.

Maintainability can be measured at the later or early stage of the development process. At the later stage maintainability measurement can be used to evaluate the system for improvements. The measurement at the early stage is important for early identification of possible risks, and efficient allocation of resources. For early evaluation of maintainability, indicators based on properties of early artifacts [1] such as the structural properties of class diagrams are necessary. Class diagram is a product of unified modeling language (UML). Generoet al[4] noted that UML class diagrams are the key outcome of those early phases and the foundation for all later design and implementation. By considering the importance of maintainability by using class diagrams this paper presents a framework for maintainability measurement of the web application based on design metrics applied to UML class diagrams.

The paper is organized as follows; Section 2 presents scope and motivation of the work. Selection of metrics and the data selection and analysis are in section 3 and section 4 respectively. The concluding remarks are in section 5.

## 2    Scope of the Work

Maintainability can be evaluated by measuring some of its sub- characteristics. A number of sub characteristics can be found but this work limits to evaluate maintainability by measuring the following sub-characteristics; understandability, analyzability, and modifiability of class diagrams, defined as;

- Understandability: Ease with which to understand the system. Everyone involved in the development should be able to understand what the code does.
- Analyzability: Ease with which to diagnose the system for deficiencies or to identify the parts that need to be modified
- Modifiability: Ease with which to make adaptations to the system

The motivation of this work is to establish that the web application maintainability depends on measurable (design) metrics. It is to investigate through statistical approach whether a set of measures identified for UML class diagram structural properties (size, complexity, coupling, cohesion) could be good predictors of class diagram maintainability based on the sub- characteristics mentioned above. To determine this, the following steps are taken: - Metrics selection, Data collection, Data analysis and model development. Data analysis involves Correlation, and Regression analysis. All these steps are explained in coming sections.

## 3    Metrics Selection

In this section, we identify the metrics that affect the dependent variables (maintainability). Several design metrics have been defined for software maintainability. For this work, design metrics based on UML extension for web application as proposed by Conallen [2] are used. Design metrics provide early identification of potential risks at the development stage. There are the four major structural properties commonly used to represent the quality of any software design irrespective of the development paradigm in use [11]. The design attributes being measured are size, complexity, coupling and cohesion. These attributes have been identified as the most popular software design predictors. These design attributes are summarized as follows:-

- Size: -Size quantifies the structural elements of a system. This can be measured by counting lines of code. The number of lines of code (LOC) is the common way of measuring size. For this work, size of the UML diagram is determined by counting the number of components in the class diagram.
- Complexity: - Measures the complexity of program control flow by calculating the number of loops and branches in a component. It indicates the degree of difficulty in understanding the internal and external structure of class and its relationship. Complexity is measured in this work by the number of associations and relations in the class diagram
- Coupling: - An indication of the degree of interdependency between modules. It is the degree of interaction between two components. Low coupling results in high maintainability. A class is coupled to another, if methods of one class use methods or attributes of another class. For this work, coupling is measured from the relationship and components in the UML class diagram.
- Cohesion: - An indication of the degree to which the methods and attributes of a class bind together. It defines the internal consistency within a class.A class is cohesive when its parts are highly correlated. This work measures cohesion using the relationship and components in the UML class diagram.

The metrics proposed for the study are defined in Gosheh et al [5], though some of them were first defined in Genero et al [3]. The metrics are summarized in Table 1.

## 4    Data Collection and Analysis

This stage describes the collection of data for the study. This study seeks to establish if any correlation exists between the metric values and the subjective ratings on maintainability sub characteristics. To this end, values of both variables need to be collected. First, the metric values are collected automatically using metric tool. For the subjective ratings on the maintainability sub-characteristics, data were collected from web developers and professionals through controlled experiment. This is for empirical validation of the proposed metrics. Key specifications of the experiment include;

    - Goal definition: The experiment is to:-

(i)    Determine if any relationship exists between the metrics and the maintainability sub characteristics: understandability, analyzability, and modifiability
(ii)   Determine the metrics that are strong candidates to establish prediction models for maintainability of the web application.

    - Participants (subjects): The participants were experienced industrial practitioners in the field. A total of fifteen participants were involved. All the participants had good background knowledge of the tasks.

**Table 1.** Considered Metrics

| Type | Metric Name | Description |
|------|-------------|-------------|
| Size | Server Page (NServP) | Number of server Page |
| | Client Page (NClienP) | Number of client pages |
| | Webpage (NWebP) | Number of web pages = server Pages + client pages |
| | Form Pages (NFrmP) | Number of Form Pages |
| | Form Elements (NFrmE) | Number of Form Elements |
| | Client Scripts Components (NCsrC) | Number of Client scripts components |
| | ServerScriptComponents (NSscrC) | Number of server scripts components |
| | Class (NC) | Number of classes |
| | Attributes (NA) | Number of attributes |
| | Methods | Number of methods |
| Complexity | Associations (NAss) | Number of associations |
| | Aggregations (NAgg) | Number of aggregation relationships |
| | Link (NLnk) | Number of link relationship |
| | Submit (NSbmt) | Number of Submit relationship $\times$ Form Elements |
| | Build (NBlds) | Number of builds relationship $\times$ (Server Script Component + Client Scripts Components) |
| | Forward (NFwrd) | Number of forward relationships |
| | Include (NIncld) | Number of include relationships |
| Coupling | WebDataCoupling (Wdata) | Number of data exchangedover number of server pages (Form Elements/Server Page) |
| | WebControlCoupling (WContr) | Number of relationships over number of web Pages (Link + submit + build + Forward + include) / web Pages |
| | EntropyCoupling (EntCoup) | $1/n \times (-\log 1/(1 + m))$ Where n = number of elements m = number of relationships |
| Cohesion | EntropyCohesion (EntCoh) | Total entropy coupling/ entropy coupling of one class diagram. |

- Experiment materials and tasks: This work is on intranet web application for promotion of knowledge- sharing in academic institutions. The materials used for the study consist of 21 class diagrams of an Academic Management Information System. As indicated earlier, the subjects are to rate each maintainability sub-characteristics on a five- point scale as depicted below;

| Very Easy | Easy | Medium | Difficult | Very Difficult |
|-----------|------|--------|-----------|----------------|
| 5 | 4 | 3 | 2 | 1 |

The metric values of the UML class diagrams are shown in Appendix (Table 6), and the subjects' ratings on the maintainability sub characteristics are in Appendix (Table 7). The subjects' ratings presented are the mean values of the rating.

-Variable definition: As indicated above, values of two variables are involved in the maintainability measurement. The variables are defined as follows;

- The independent variables are the metric value of the UML class diagram size, complexity, coupling, and cohesion.
- The dependent variables are the maintainability sub-characteristics; understandability, analyzability, and modifiability measured using the subjects' ratings

## 4.1   Data Analysis

As noted earlier, this work is to determine the correlation between the metric values and the subjective ratings on maintainability. It is also to identify from the correlation, the strongest predictors for model development. These tasks are accomplished using statistical analysis including correlation, and regression analysis.

### 4.1.1  Correlation Analysis

This is used to establish the correlation between the dependent variables and the metrics proposed (independent variables) and identify the metrics that are the strongest predictors. This can be accomplished by examining the correlation between each metric and the subjective ratings provided by the software developers [10]. The higher the value of the coefficient of determination, the stronger the relationship between maintainability and the metric considered. Thus, metrics with high correlation coefficients are candidates in the maintainability prediction. The metrics are ranked in order of the correlation coefficients. Spearman rank correlation was used to determine the correlation of the data collected in the study. Each of the metrics was correlated separately to the subjective ratings. To calculate Spearman correlation coefficient (R), data are arranged in ranks and the difference in rank (d) computed for each pair. Spearman correlation equation is expressed as follows;

$$R = 1 - \frac{6\,(\Sigma d^2)}{n\,(n^2 - 1)} \qquad (1)$$

Where

d = difference in ranks for each pair of items

n = number of observations.

Now, as a way of illustration computation of the Spearman Rank correlation for the first predictor, NServP, with the sub characteristics; Understandability, Analysability, and Modifiability are demonstrated in Table 2, Table 3 and Table 4. Spearman Correlation Summary for the metrics and the dependent variables are summarized in Table 5.

**Table 2.** Metrics Correlation with Understandability: NServpVs Understandability

| NServP | NservP Rank | Underst. | Underst. Rank | d | $d^2$ |
|--------|-------------|----------|---------------|------|-------|
| 4 | 4.5 | 3 | 8 | -3.5 | 25 |
| 4 | 4.5 | 4 | 15 | -10.5 | 10.25 |
| 3 | 1.5 | 4 | 15 | -13.5 | 182.25 |
| 5 | 8.5 | 2 | 2.5 | 6 | 36 |
| 3 | 1.5 | 4 | 15 | -13.5 | 182.25 |
| 6 | 11.5 | 4 | 15 | -3.5 | 12.25 |
| 4 | 4.5 | 5 | 20 | -15.5 | 240.25 |
| 5 | 8.5 | 2 | 2.5 | 6 | 36 |
| 6 | 11.5 | 3 | 8 | 3.5 | 12.25 |
| 7 | 13 | 3 | 8 | 5 | 25 |
| 9 | 14.5 | 3 | 8 | 6.5 | 42.25 |
| 5 | 8.5 | 4 | 15 | -6.5 | 42.25 |
| 11 | 17 | 3 | 8 | 9 | 81 |
| 14 | 19.5 | 5 | 20 | -0.5 | 0.25 |
| 10 | 16 | 4 | 15 | 1 | 1 |
| 4 | 4.5 | 5 | 20 | -15.5 | 240.25 |
| 12 | 18 | 3 | 8 | 10 | 100 |
| 15 | 21 | 2 | 2.5 | 18.5 | 342.25 |
| 9 | 14.5 | 3 | 8 | 6.5 | 42.25 |
| 5 | 8.5 | 4 | 15 | -6.5 | 42.25 |
| 14 | 19.5 | 2 | 2.5 | 17 | 289 |

$$\sum d^2 = \underline{2071.5}$$

R = 1 − 6x2071.5

        21x440

= 1 − 1.345    = 0.345

**Table 3.** Metrics Correlation with Analyzability: NservPVs Analyzability

| NservP | NservPRank | Analyz ab. | Analyza. Rank | d D | $d^2$ |
|---|---|---|---|---|---|
| 4 | 4.5 | 2 | 4 | 0.5 | 0.25 |
| 4 | 4.5 | 2 | 4 | 0.5 | 0.25 |
| 3 | 1.5 | 3 | 11 | -9.5 | 90.25 |
| 5 | 8.5 | 3 | 11 | -2.5 | 6.25 |
| 3 | 1.5 | 3 | 11 | -9.5 | 90.25 |
| 6 | 11.5 | 2 | 4 | 7.5 | 56.25 |
| 4 | 4.5 | 2 | 4 | 0.5 | 0.25 |
| 5 | 8.5 | 4 | 16.5 | -8 | 64 |
| 6 | 11.5 | 2 | 4 | 7.5 | 56.25 |
| 7 | 13 | 2 | 4 | 9 | 81 |
| 9 | 14.5 | 3 | 11 | 3.5 | 12.25 |
| 5 | 8.5 | 2 | 4 | 4.5 | 20.25 |
| 11 | 17 | 3 | 11 | 6 | 36 |
| 14 | 19.5 | 3 | 11 | 8.5 | 72.25 |
| 10 | 16 | 3 | 11 | 5 | 25 |
| 4 | 4.5 | 5 | 20 | -15.5 | 240.25 |
| 12 | 18 | 4 | 16.5 | 2.5 | 6.25 |
| 15 | 21 | 4 | 16.5 | 4.5 | 20.25 |
| 9 | 14.5 | 5 | 20 | -5.5 | 30.25 |
| 5 | 8.5 | 5 | 20 | -11.5 | 132.25 |
| 14 | 19.5 | 4 | 16.5 | 3 | 9 |
| | | | | | 1,040 |

$$R = 1 - \frac{6 \times 1040}{21 \times 440}$$
$$= 1 - 0.675 \quad = 0.325$$

**Table 4.** Metrics Correlation with Modifiability:NservpVs Modifiability

| NServP | NServP Rank | Modifiability | Modifiability Rank | D | $d^2$ |
|---|---|---|---|---|---|
| 4 | 4.5 | 1 | 1.5 | 3 | 9 |
| 4 | 4.5 | 2 | 5.5 | 1 | 1 |
| 3 | 1.5 | 2 | 5.5 | 4 | 16 |
| 5 | 8.5 | 2 | 5.5 | -3 | 9 |
| 3 | 1.5 | 1 | 1.5 | 0 | 0 |
| 6 | 11.5 | 3 | 12 | -0.5 | 0.25 |
| 4 | 4.5 | 2 | 5.5 | 1 | 1 |
| 5 | 8.5 | 2 | 5.5 | -3 | 9 |

**Table 4.** (*continued*)

| | | | | | |
|---|---|---|---|---|---|
| 6 | 11.5 | 3 | 12 | -0.5 | 0.25 |
| 7 | 13 | 3 | 12 | -1 | 1 |
| 9 | 14.5 | 3 | 12 | -2.5 | 6.25 |
| 5 | 8.5 | 3 | 12 | 3.5 | 12.25 |
| 11 | 17 | 5 | 20 | 3 | 9 |
| 14 | 19.5 | 4 | 17 | -2.5 | 6.25 |
| 10 | 16 | 4 | 17 | 1 | 1 |
| 4 | 4.5 | 2 | 5.5 | 1 | 1 |
| 12 | 18 | 5 | 20 | 2 | 4 |
| 15 | 21 | 5 | 20 | -1 | 1 |
| 9 | 14.5 | 3 | 12 | -2.5 | 6.25 |
| 5 | 8.5 | 3 | 12 | 3.5 | 12.25 |
| 14 | 19.5 | 4 | 17 | 2.5 | 6.25 |
| | | | | | 112 |

$$R = 1 - \frac{6 \times 112}{21 \times 440}$$
$$= 1 - 0.073 \ = 0.927$$

The entire correlation results are summarized as follows:

**Table 5.** Spearman Correlation Summary for the metrics and the dependent variables

| Metrics | Understandability | Analyzability | Modifiability |
|---|---|---|---|
| NServP | -0.345 | 0.325 | **0.927** |
| NClienP | **0.942** | -0.131 | -0.083 |
| NWbP | 0.196 | 0.332 | **0.855** |
| NFrmP | -0.124 | 0.162 | 0.415 |
| NFrmE | 0.065 | 0.527 | 0.319 |
| NCSrC | -0.018 | 0.411 | 0.480 |
| NAssR | 0.045 | **0.904** | 0.259 |
| NAggR | **0.882** | 0.066 | 0.098 |
| NLnkR | 0.166 | 0.195 | 0.518 |
| NSbmtR | 0.236 | -0.202 | 0.499 |
| NBldsR | 0.016 | 0.113 | -0.239 |
| FwdR | 0.302 | 0.266 | 0.265 |
| NIncdR | 0.017 | 0.50 | 0.475 |
| Wdata | 0.407 | -0.191 | -0.739 |
| Wcontr | 0.017 | -0.178 | -0.798 |
| Entcoh | -0.139 | 0.566 | 0.430 |

From the Table, it is observed that the strongest predictors for the Maintainability sub-characteristics are:

(i)     Understandability – NclienP (0.942), and NAggR (0.882)
(ii)    Analyzability – NAss (0.904)
(iii)   Modifiability – NservP (0.927), and NwebP (0.855)

So, these are the variables to be used to build the prediction models.

## 4.1.2  Regression Analysis

Regression analysis is a statistical method for predicting the value of a dependent variable from one or more predictors (independent) variables. Researchers usually rely on (multiple) regression to predict dependent (criterion) variable from independent (predictor) variables. A model of relationship is hypothesized and estimates of the parameter values are used to develop an estimation regression equation [16]. Various tests are then used to evaluate the model. The general linear regression equation applied to develop the model is given as:

$$Y = a + b_1X_1 + b_2X_2 + \cdots + b_kX_k$$

where

$Y$ = dependent variable
$X_1, X_2, \cdots X_k$ = independent variables
$b_1, b_2, \cdots b_k$ = coefficients of regression
$a$ = Intercept.

For multiple regression involving two independent variables,

$$b_1 = \frac{SDy}{SDx_1} \times \beta_1 \quad (SD = \text{standard deviation}) \tag{2}$$

$$b_2 = \frac{SDy}{SDx_2} \times \beta_2 \tag{3}$$

$$a = \overline{Y} - b_1\overline{X} - b_2\overline{X_2} \tag{4}$$

where

$$\beta_1 = \frac{r_{y1} - r_{12}\, r_{y2}}{1 - r^2_{12}} \tag{5}$$

$$\beta_2 = \frac{r_{y2} - r_{12}\, r_{y1}}{1 - r^2_{12}} \quad (r = \text{Pearson correlation coefficient}) \tag{6}$$

$$r_{y1} = \frac{Sx_1y}{\sqrt{Sx^2_1 Sy^2}} \tag{7}$$

$$r_{12} = \frac{Sx_1x_2}{\sqrt{Sx^2_1 Sx^2_2}} \tag{8}$$

$$ry_2 = \frac{Sx_2y}{\sqrt{Sx^2_2 Sy^2}} \tag{9}$$

Where

$$Sx_1^2 = \Sigma X_1^2 - \frac{(\Sigma X_1)^2}{n} \quad \text{(n = number of observations)} \tag{10}$$

$$Sx_2^2 = \Sigma X_2^2 - \frac{(\Sigma X_2)^2}{n} \tag{11}$$

$$Sy^2 = \Sigma y^2 - \frac{(\Sigma y)^2}{n} \tag{12}$$

$$Sx_1 x_2 = \Sigma X_1 X_2 - \frac{(\Sigma X_1)(\Sigma X_2)}{n} \tag{13}$$

$$Sx_1y = \Sigma X_1Y - \frac{(\Sigma X_1)(\Sigma Y)}{n} \tag{14}$$

$$Sx_2y = \frac{\Sigma X_2y - (\Sigma X_2)(\Sigma Y)}{n} \tag{15}$$

It should be noted that for a case of one independent variable, simple regression equation is applied. This is expressed as;

$$Y = a+bx \tag{16}$$

Where

$$b = \frac{Sxy}{Sx^2} \tag{17}$$

$$a = \overline{Y} - b\overline{X} \tag{18}$$

$$Sxy = \Sigma XY - \frac{\Sigma X \Sigma Y}{n} \tag{19}$$

and

$$Sx^2 = \Sigma X^2 - \frac{(\Sigma X)^2}{n} \tag{20}$$

These are the underlying computations in the development of regression models involving two independent variables, and one independent variable respectively. However, computations using statistical software packages prove to be easier and more dependable, especially for cases involving more than two independent variables.

## 5    Conclusion

This paper has presented a maintainability quantification framework that could be employed to develop prediction models to control the maintenance tasks of the intranet web application based on design metrics applied to UML class diagram, and statistical approach. The process involved metrics identification, data collection, correlation analysis and regression analysis and model development. Design attributes measured were size, complexity, coupling and cohesion and subjective ratings were obtained on maintainability sub-characteristics through controlled experiment. Data collected were analyzed using Spearman rank correlation analysis and from the analysis the strongest predictors that could be applied to build maintainability prediction model were identified. Computations underlying the regression models were presented. The framework presented in this paper has provided detailed steps to reliably measure maintainability. This can be used to effectively control the maintenance of intranet application for efficient knowledge- sharing on a university campus.

**Future Work:** In this paper we have presented regression model which explain that how to compute the regression coefficients (for developing the model). The application of the model is the task of the future work. We are collecting the separate data for the examining maintainability sub- characteristics.

## References

1. Briand, L., Arisholm, S., Counsell, F., Houdek, F., Thevenod-Fosse, P.: Empiricalstudies of object- oriented artefacts, method and processes; State of the art and future directives. Empirical software engineering 4(4), 357–404 (2000)
2. Conallen, J.: Building web applications with UML. Addison-Wesley Publishing Company, Reading (2003)
3. Genero, M., Piattini, M., Calero, C.: Empirical validation of class diagram metrics. In: Proceedings of the International Symposium in Empirical Software Engineering, pp. 195–203. IEEE Computer Society Press (2002)
4. Genero, M., Manso, E., Cantone, G.: Building UML class diagram maintainability prediction models based on early metrics. In: Proceedings of the 9th International Symposium on Software Metrics 2003, pp. 263–275 (2003)
5. Ghosheh, E., Black, S., Qaddour, J.: Design Metrics for Web Application maintainability Measurement. In: Proceeding of 6th IEEE/ACS International Conference Computer Systems and applications, pp. 778–784. IEEE, Doha (2008) ISBN 97814244196
6. IEE: Institute of Electronics Engineers. IEE standard computer Dictionary. A Compilation of IEE Standard Computer Dictionary. New York (1990)

7. ISO 9126-1 Software Engineering –Product quality- Part 1: Quality model (2001)
8. Kiewkanya, M., Jindasawat, N., Muenchaisri, P.: A methodology for constructing maintainability model of object- oriented design. In: Proceedings of 4th International Conference on Quality Software, pp. 206–213. IEEE Computer Society (2004)
9. Mendes, E., Mosley, S.: Council: Web Metrics – Estimating design and authoring effort. IEEE Multimedia 8(1), 50–57 (2001)
10. Muthanna, S., Kontogiannis, K., Ponnainbalam, K., Stacey, B.: A maintainability Model for Industrial Software Systems using Design Level metrics. In: Proceeding of the 7th Working Conference on Reverse Engineering, Brisbane, Australia, pp. 248–256 (2000) ISSN: 1095-1350
11. Perepletchikov, M.: Software design metrics for predicting maintainability of service – oriented software; Phd thesis, School of Computer Science and Information technology. RMIT University Melbourne, Austalia (2009)
12. Polo, M., Piattini, M., Ruiz, F.: Using code metrics to predict maintenance of legacy programs: A case study. In: Proceedings of the Internal Conference on Software Maintenance (ICSM), pp. 202–208. IEEE Computer Society, Florence (2001)
13. Saini, R., Dubey, S.K., Rana, A.: Analytical study of Maintainability models for quality evaluation. Indian Journal of Computer Science and Engineering 2(3) (2011) ISSN 0976-5106
14. Tramontana, P.: Reverse engineering of web Application: Universita Degli Studidi Napoli Federico II. PhD thesis (2005)
15. Welker, K.D., Oman, P.W.: Software maintainability metrics models in practice. Journal of Defense software engineering 8(6), 19–23 (1995), http://Crosstalkonline.org
16. Zhao, L., Hayes, J.: Maintainability Prediction: a regression analysis of measures of evolving systems. In: Proceeding of the 21st IEEE International Conference on Software Maintenance, September 25-29, pp. 601–604 (2005)

# Appendix

**Table 6.** Measure Values of UML Class Diagrams

| Diagram | ServP | NclP | NwbP | FrmP | FrmE | CSC | ASS | Aggr | Lnk | Sbmt | Blds | Fwd | Incd | Wdata | Wcontr | Entcoh |
|---|---|---|---|---|---|---|---|---|---|---|---|---|---|---|---|---|
| D1 | 4 | 3 | 7 | 1 | 4 | 0 | 1 | 0 | 2 | 4 | 3 | 1 | 2 | 1 | 1.7 | 1 |
| D2 | 4 | 4 | 8 | 1 | 4 | 0 | 1 | 1 | 3 | 3 | 2 | 2 | 2 | 1 | 1.5 | 1 |
| D3 | 3 | 4 | 9 | 2 | 3 | 1 | 2 | 2 | 2 | 3 | 3 | 1 | 1 | 1 | 1.1 | 1 |
| D4 | 5 | 1 | 5 | 2 | 4 | 0 | 2 | 2 | 2 | 4 | 3 | 1 | 2 | 0.8 | 2.4 | 2 |
| D5 | 3 | 3 | 6 | 2 | 4 | 2 | 2 | 1 | 2 | 4 | 2 | 2 | 2 | 1.3 | 2 | 2 |
| D6 | 6 | 4 | 10 | 3 | 3 | 2 | 1 | 0 | 3 | 4 | 3 | 1 | 2 | 0.5 | 1.3 | 1 |
| D7 | 4 | 6 | 10 | 2 | 4 | 3 | 1 | 1 | 2 | 4 | 3 | 1 | 2 | 1 | 1.2 | 3 |
| D8 | 5 | 1 | 6 | 3 | 4 | 3 | 3 | 0 | 2 | 3 | 4 | 2 | 1 | 0.8 | 2 | 3 |
| D9 | 6 | 3 | 9 | 1 | 3 | 2 | 1 | 2 | 3 | 3 | 3 | 2 | 2 | 0.5 | 1.6 | 2 |
| D10 | 7 | 3 | 10 | 3 | 4 | 4 | 1 | 3 | 4 | 3 | 1 | 2 | 2 | 0.6 | 1.2 | 4 |
| D11 | 9 | 3 | 12 | 5 | 4 | 4 | 2 | 0 | 5 | 4 | 2 | 1 | 3 | 0.4 | 1.3 | 4 |
| D12 | 5 | 4 | 9 | 4 | 4 | 3 | 2 | 2 | 4 | 3 | 2 | 3 | 1 | 0.3 | 1.4 | 3 |
| D13 | 11 | 2 | 13 | 1 | 2 | 1 | 1 | 0 | 5 | 3 | 2 | 2 | 3 | 0.2 | 1.2 | 1 |
| D14 | 14 | 6 | 20 | 2 | 4 | 2 | 2 | 3 | 3 | 2 | 1 | 3 | 2 | 0.3 | 0.6 | 2 |
| D15 | 10 | 4 | 14 | 3 | 4 | 4 | 2 | 3 | 1 | 3 | 2 | 1 | 2 | 0.4 | 0.6 | 4 |
| D16 | 4 | 6 | 10 | 1 | 4 | 6 | 3 | 3 | 4 | 4 | 2 | 4 | 3 | 1 | 1.7 | 6 |
| D17 | 12 | 3 | 15 | 4 | 6 | 6 | 3 | 2 | 3 | 1 | 2 | 4 | 3 | 0.5 | 0.9 | 6 |
| D18 | 15 | 2 | 17 | 6 | 5 | 6 | 4 | 0 | 5 | 2 | 3 | 1 | 2 | 0.3 | 0.8 | 6 |
| D19 | 9 | 3 | 12 | 3 | 4 | 2 | 10 | 6 | 4 | 3 | 2 | 1 | 3 | 0.4 | 0.9 | 2 |
| D20 | 5 | 4 | 9 | 1 | 4 | 2 | 12 | 7 | 2 | 2 | 4 | 2 | 2 | 0.3 | 1.3 | 5 |
| D21 | 14 | 1 | 15 | 2 | 6 | 4 | 2 | 7 | 4 | 3 | 1 | 2 | 3 | 0.4 | 0.9 | 4 |

**Table 7.** Subjective ratings on maintainability

| Diagram | Understandability | Analyzability | Modifiability |
|---|---|---|---|
| D1 | 3 | 2 | 1 |
| D2 | 4 | 2 | 2 |
| D3 | 4 | 3 | 2 |
| D4 | 2 | 3 | 2 |
| D5 | 4 | 3 | 1 |
| D6 | 4 | 2 | 3 |
| D7 | 5 | 2 | 2 |
| D8 | 2 | 4 | 2 |
| D9 | 3 | 2 | 3 |
| D10 | 3 | 2 | 3 |
| D11 | 3 | 3 | 3 |
| D12 | 4 | 2 | 3 |
| D13 | 3 | 3 | 5 |
| D14 | 5 | 3 | 4 |
| D15 | 4 | 3 | 4 |
| D16 | 5 | 5 | 2 |
| D17 | 3 | 4 | 5 |
| D18 | 2 | 4 | 5 |
| D19 | 3 | 5 | 3 |
| D20 | 4 | 5 | 3 |
| D21 | 2 | 4 | 4 |

# Use of Graph Databases in Tourist Navigation Application

Anahid Basiri[1], Pouria Amirian[2], and Adam Winstanley[2]

[1] Nottingham Geospatial Institute, the University of Nottingham, UK
Anahid.basiri@nottingham.ac.uk
[2] Department of Computer Science, National University of Ireland, Maynooth, Ireland
pouria.amirian@nuim.ie

**Abstract.** Navigation services, such as car navigation services, are widely used nowadays. However current car navigation systems are not fully suitable for the navigational needs of tourists. In contrast with drivers, tourists are not constrained by road networks and can walk in places where vehicles are not allowed to move. As current turn-by-turn navigational instructions to be given to vehicle's derivers are mostly based on street network-based algorithms, this way of navigating is not fully suitable for tourists as they do not only move on streets. In addition, Tourists want to see important feature of the area, no matter they take longer path rather than shortest. They want to get navigated through the most touristic path. In order to provide such tourist-specific navigation services, a landmark-based solution was considered. it calculates a route passing more landmarks. This may help user to visit attractive part of a place. It is possible to provide users with the navigational instructions landmark-by-landmark rather than turn-by-turn. In this application, a graph database is used because of having highly connected data and also need to remove the mapping layer between physical storage layer and application logic layer to have more availability and responsiveness.

**Keywords:** Landmark-based navigation, Graph databases, Tourist navigation, Location Based Services (LBS).

# 1 Introduction

Nowadays, car navigation has become one of the most widely used examples Location-Based Services (LBSs). But current car navigation systems are not fully suitable for the navigational needs of all categories of users such as tourists and visitors, pedestrians, users with special needs such as handicaps. There is a need to develop different navigation application for different categories of users to consider their specific needs, constraints and preferences. Tourists and visitors can easily go into a building or underground to get to their destination where GPS's signals are unavailable. Seamless Indoor and outdoor navigation is one of the most important features which should be handled in their navigation application and is still topic of many research projects (Karimi, 2011), (Li et al., 2013), (Hansen et al., 2009).

B. Murgante et al. (Eds.): ICCSA 2014, Part V, LNCS 8583, pp. 663–677, 2014.
© Springer International Publishing Switzerland 2014

In addition to seamless positioning solutions, another aspect of tourist navigation application which has not been covered in car navigation is non-turn-by-turn navigational instruction delivery. Tourists and visitors would like to get some information about each Point of Interest (POI) while traversing. Instead of having two different applications; one for navigation and the other one for tourist guidance, that would be better to combine them into one application and navigate users through most touristic path (rather than shortest path) and giving navigational instructions landmark-by-landmark (rather than junction-by-junction).

In both mentioned challenges of pedestrian navigation systems; seamless indoor/outdoor positioning and non-turn-by-turn navigational instruction giving, landmarks can help. It is possible to calculate position of a user relative to landmarks' positions. In Landmark-based positioning, positions of user are sensed or calculated respect to landmarks. Relative position of users' devices can be sensed using ultrasound (Holm, 2009), dead-reckoning (Etienne and Séguinot, 1993), collaborative positioning techniques (Lee et al., 2012) or inertial sensors (Vepa, 2011) mostly. Since landmarks can be detected and labeled indoors and outdoors, both, (see figure 1) it is possible to have positions of users seamlessly (Millonig and Schechtner, 2005).

**Fig. 1.** Indoor landmarks

In addition, it is possible to provide users with the navigational instructions landmark-by-landmark rather than turn-by-turn. This way of navigating is called landmark-based navigation (May et al., 2005). Land mark based navigation is a kind of navigation service in which users are provided with navigational instructions, such as turn right, go straight, turn left, etc. whenever they approach each landmarks (Fang et al., 2011). One of the most important advantages of landmark-based navigation is making user sure that they are on the right way which is one of the tourists' preferences since usually they are not very familiar with the area they are visiting. They may not get lost, since they are seeing the very landmark which was used as a part of navigational instruction. From another point of view this approach is more suitable for tourists and visitors since it is possible to add some descriptive information or image of landmarks while navigating, so they will see more while visiting an area and also being navigated too at the same time (Basiri et al, 2013), (Basiri et al, 2014).

In order to implement such application, a graph database is used. By nature LBS applications and specially navigation applications need to be highly responsive in real time or near real time. In addition, users of such applications usually include huge number of users who receive information with high volume, such as video or image of landmarks so LBS application must provide appropriate scalability and performance measures. In this case, many LBS applications resort to proprietary network processing technologies. Most conventional and some modern storage systems such as Relational Database Management Systems (RDBMS) usually build a network representation on top of the physical storage model and keep the network structure in the main memory. In addition, some of the network processing technologies process widely used network analysis and store the results beside the network data itself. However when additional network elements (such as new land marks or new roads) have to be added to the data, all the representation and built network must be rebuilt and recompiled. This recompilation process would be serious processing task especially when the newly added elements have lots of connection to the existing elements.

In order to prevent facing with such issues, it is better to store connected data in their natural representation and this is where the graph database comes into play (storing data elements as well as their relationships as graphs in the graph database). In summary, graph databases don't need a mapping layer between physical storage layer and application logic layer. The mentioned integration of storage and application layer also results in flexibility for handling consistency of the data when the size of data is very large and as a result cannot be stored in main memory of single server.

Section 2 describes tourist navigation service and related requirements. Section 3 explains graph databases and their use in tourist navigation services. And in the section 4 implementation of the application is shown.

# 2    Landmark-Based Pedestrian Navigation

A pedestrian has several possible navigational strategies to find a desired goal (Redish, 1999); the individual has no information and is forced to search randomly (random navigation), the individual moves towards a visible cue which leads to the arrival point (taxon navigation), The individual follows a fixed motor program (praxic navigation, e.g. "turn left after 200 meters, then turn right after 150 meters"), the individual associates directions with visual cues (route navigation, e.g. "turn left at the church") and when the individual forms a mental representation of the surroundings and is able to plan routes between any locations within the area (locale navigation).

Landmarks can have an important role in random navigation, taxon navigation, route navigation and local navigation. Several researchers in the field of spatial cognition assert that navigating humans rely on three forms of spatial knowledge: landmark, route and survey knowledge (Siegel and White, 1975), (Werner et al., 1997). Exploring an unfamiliar environment, pedestrians first notice outstanding objects or structures at fixed locations. These unique objects or places are easy to recognize and can be kept in memory without difficulty (Schechtner, 2005).

This shows the importance of the meaning of landmarks for human navigational tasks. Landmarks are stationary, distinct and salient objects or places, which serve as cues for structuring and building a mental representation of the surrounding area. Any object can be perceived as a landmark, if it is unique enough in comparison to the adjacent items. The importance of landmarks for pedestrian navigation and wayfinding instructions is proved by many researches (Hung,. 2012), (Michon and Denis, 2001), (Denis, 2003), (Basiri et al, 2012) and (Raubal and winter, 2002).

This section landmark and landmark-based navigation concepts and related definitions are explained. Since in landmark-based navigation firstly landmarks should be stored in a database and then it becomes possible to localize users or provide them with landmark-based navigational instruction, firstly landmark definition and attributes are discussed in more details. Then landmark-based positioning and landmark-based path finding which are two important components of a landmark-based navigation service are explained.

## 2.1     What Is a Landmark?

A landmark can be defined as anything which is easily recognizable, such as a monument or a building. Landmarks are one of the interests of tourists probably due to notable physical features or historical significance. Landmarks are often used for casual navigation by ordinary people, such as giving directions.

In urban studies as well as in geography, a landmark is furthermore defined as an external point of reference that helps orienting in a familiar or unfamiliar environment (Lynch, 1960), (Schoier, 2012). Landmarks are also used in verbal route instructions. These two properties of landmarks; being used as references to orient objects and being used in verbal route instructions, are very important and potentially helpful in navigation systems and services.

Landmarks can be geometric shapes, and they may include additional information (e.g., in the form of bar-codes and geo-tags). In general, landmarks have a fixed and known position, relative to which users can localize themselves. Landmarks' data must be stored in a database to be used in landmark-based positioning.

Landmarks should be carefully chosen to be easy to identify; for example, a large building has got priority to a small one. A feature which has got sufficient contrast to the background is a good option to be considered as landmark since its image would be recognizable to users. Such objects have to possess a certain saliency, which makes them remarkable and distinctive. So the surrounding area determines the characteristics a point must have to be perceived as a landmark (e.g. a shopping center may not be very outstanding in urban areas, but becomes a salient landmark when being situated in a rural village). In the third section, the process of landmark extraction is explained in more detail.

After extracting important features as landmarks, best path between current location of user and selected destination should be found. This make the application enable to provide users with navigational instructions on the calculated route. Next subsection is focused on landmark-based localization. Then landmark-based path finding algorithm is explained.

## 2.2    Landmark-Based Positioning

Positioning is one of the most important components of any navigation system. Tourists and in general Pedestrians need seamless indoor and outdoor wayfinding assistance, so the system needs reliable and accurate positioning techniques in the situation where Globe Navigation Satellite Systems (GNSS) signals are not available. Unlike cars, tourists can easily enter buildings or even their destination is a roofed area such as gallery or museum; so an ideal navigation system should work seamlessly in and out of doors. This paper discusses landmark-based positioning technique which can be implemented as a seamless indoor/outdoor positioning solution since landmarks are available both indoors and outdoors. One of the advantages of using landmarks in tourist navigation services is being sure about availability of landmarks around users since the purpose of their visit is seeing touristic features which usually can be considered as a landmark because of their uniqueness.

In general, two main categories of landmark-based positioning techniques can be imagined; image-based and non-image-based techniques. Some examples of image based landmark positioning are QRCode-based positioning (Basiri et al, 2014) and photo-based positioning. Examples of non-image-based positioning are Radio Frequency Identification (RFID) and Bluetooth network positioning.

This paper focuses on image-based landmark positioning techniques since most mobile phones are equipped with cameras so less hardware requirements are needed. RFID tags and Bluetooth networks are not available ubiquitously. The hardware needs to be installed both on users' handheld devices (e.g. RFID readers) and also on the landmarks (e.g. RFID tags). This means extra cost for both service provider and users. In addition, in most image-based positioning approaches the computation and processing phase is done on the server side so such approaches have got less power consumption in comparison with Bluetooth positioning and similar techniques where users need to keep their mobile phones' Bluetooth on all the time. In addition, many landmarks are historically registered features, or absolutely huge objects. In such cases, it is almost impossible to install or affix a tag or any signal transmitter for positioning purposes. Being much cheaper, having less power consumption and ability of being used using available mobile devices may make image-based positioning more accepted by users.

In Image-based or camera-based positioning user can be viewed, identified and tracked by a network of cameras (such as CCTVs). Another approach which is categorized in image-based positioning techniques is user takes a photo of a registered landmark and then send/upload it for further image processing, feature matching and finally to find his/her location. Since second approach needs less hardware infrastructure, it has been implemented in this project. In this part image-based positioning using mobile devices' cameras is explained in more detailed.

In image-based positioning using mobile devices' cameras, user can take a photo of a registered landmark and then send/upload it for further image processing and feature extraction to find relative location respect to the landmark. Usually this process is handled on server-side machines. Based on feature extraction and image matching techniques, it is possible to find landmark of which the photo has taken. As landmarks are usually unique and distinctive objects, the feature matching process is usually accurate enough for positioning purposes. Then scale of the photo and angel of view (rotation) can be easily calculated since the absolute sizes of different façades of the landmark are stored in a database. Based on scale and angel of view, relative location of user respect to the landmark can be calculated.

Positioning of a user based on camera positioning system has two main steps generally; image processing to identify the landmark (feature detection) and finding scale and rotation (localization). In the first step, i.e. landmark detection, uniqueness of landmark make it much easier to find matching image in the database. Then using actual size and shape of the corresponding landmark which has been stored in a database, it is possible to calculate scale and rotation of photo taken by user. This piece of information is used to calculate relative position of user respect to the landmark. Since absolute positions of landmarks are available in the database, absolute position of user can easily be calculated and used in path finding and navigation service. In next subsection, landmark-based path finding using calculated location of user and specified destination is explained.

## 2.3 Landmark-Based Path Finding

In order to find the shortest path between two points, path finding algorithms are looking for minimum distance traversed, or to find the fastest path, the route with minimum time is preferable. In the tourist navigation the most touristic path should be calculated whose output is a path with maximum landmark features on the way to the destination. Users of a tourist navigation application want to see monuments and landmarks which may need deviation from the shortest path. Landmark-based path finding algorithm is providing more attractive and at the same time more reliable path. Since users are seeing more landmarks on the way, they are surer that they are taking the right way.

In landmark-based path finding we look for a path which traversing less distance to get the destination passing more landmarks on the way. So landmark-based path finding algorithms are trying to maximize the result of:

$$\text{Number of landmarks of each edge / length of that edge.} \qquad (1)$$

The same shortest path algorithm can be implemented but the cost or distance, which is usually called weight and supposed to be optimized, will be replaced by the value of *number of landmark/length*. Based on landmark-based path finding algorithm, the route is calculated then it is possible to navigate user providing image, informatics text of landmarks can be viewed from this route (Basiri et al., 2014).

As it explained in the section 2.2, the application uses image-based positioning technique to localise users. Also the tourist navigation application sends some descriptive information as well as image of landmarks to be seen on the way.

Exchange of such high volume data is an issue which should be solved to maintain responsiveness of the application. Next section explains how graph database can handle such issue.

# 3   Graph Databases

Highly connected network data are at the heart of most LBS applications. Elements in Location-based networks, road networks, junctions, landmarks, moving users and in general any producer and consumer network constitute highly connected data. Storage and processing of such highly connected and interrelated data can be problematic for most conventional and modern storage systems such as Relational Database Management Systems (RDBMS) and most types of NoSQL databases. The mentioned issue gets worse if the LBS applications need to provide responsive interactivity to the end users.

By nature LBS applications need to be highly responsive in real time or near real time. In addition, since users of tourist navigation applications include huge number of users who send and receive high volume data such as images, such application must provide appropriate scalability and performance measures. In this case, many navigation applications resort to proprietary network processing technologies. The mentioned technologies usually build a network representation on top of the physical storage model and keep the network structure in the main memory. In addition, some of the network processing technologies process widely used network analysis and store the results beside the network data itself. However when additional network elements (such as new landmarks or new roads) have to be added to the data, all the representation and built network must be rebuilt and recompiled. This recompilation process would be serious processing task especially when the newly added elements have lots of connection to the existing elements. In order to prevent facing with such issues, it is good idea to store connected data in their natural representation and this is where the graph database comes into play (storing data elements as well as their relationships as graphs in the graph database). In summary, graph databases don't need a mapping layer between physical storage layer and application logic layer. The mentioned integration of storage and application layer also results in flexibility for handling consistency of the data when the size of data is very large and as a result cannot be stored in main memory of single server.

Graph theory was pioneered by Euler in the 18th century, and has been actively researched and improved by mathematicians, sociologists, anthropologists, and others ever since. However, it is only in the past few years that graph theory and graph thinking have been applied to information management. The graph databases can be categorized as one of several models of modern NoSQL databases. All the other models of modern NoSQL databases lack capabilities to handle huge volume of highly connected data (Amirian et al. 2013). At the other hand, the conventional RDBMS systems can handle relationship, but they are not designed with highly connected data in mind.

This section tries to explain the use of graph databases for handling highly connected data (landmark networks) in comparison with relational (SQL) and NoSQL databases. In this paper relational DBMS are called SQL databases.

## 3.1   SQL Database and Handling Highly Connected Data

It has been more than thirty years since relational DBMSs (SQL DBMSs) became the major solution for storing all kinds of data in many types of applications. They manage data using relational algebra and relational calculus as their theoretical foundation. They use tables, relationships, keys and Structured Query Language (SQL) to perform all sorts of functions with data. One of the important advantages of SQL systems is the normalization process which ensure about storage of data in separate tables and only once in whole database. They usually are the best solutions when the schema of data is fixed and predefined. In other words, RDBMSs are ideal solutions to managing structured data. The SQL systems can be effectively used in many common Geospatial-related workflows. Since they support transaction and locking features, they provide robust consistency and backend for enterprise GIS systems. Usually geospatial data have a fixed schema and in most cases they are not used in isolation. That is a join of two or more datasets and connecting data through spatial operations is needed in most GIS workflows. For this reason managing fixed schema geospatial data with limited connectivity and using them in GIS workflows can usually be done effectively through SQL systems.

The SQL databases handle connected data using relationship and they retrieve connected data using joins. However joins are one of the most computationally expensive processes for SQL databases (Amirian et al., 2010). In most cases, joins are the bottleneck of SQL databases. In order to avoid many joins (which is needed in handling highly connected data in SQL databases), denormalization process can be used to store data items several times. But there are several issues associated with denormalization process especially with providing consistency in large datasets (Amirian et al., 2013). In addition to issues related to handling highly connected data, some other problems arise with SQL systems when scalability is needed by adding more servers and technologies to bind them together. With more loads on a SQL system, vertical partitioning, denormalization and removal of the relational constraints comes into play. In summary, to achieve high scalability in SQL systems the normalized relational model of data storage has to be compromised and deviated from relational model.

## 3.2   NoSQL Databases and Handling Highly Connected Data

The NoSQL (Not only SQL) DBMSs are a broad class of DBMSs identified by non-adherence to the SQL (relational) model. There are different types of NoSQL databases, each with distinct set of characteristics but they all can deal with large amounts of (semi-structured and unstructured) data and are able to support a large set of read and write operations and they are designed with scalability and distribution of data in mind. For this reason NoSQL and relational models are not in contrast with each other rather they complement each other. The most widely accepted taxonomy of NoSQL databases are: key-value, document, columnar and graph (Tiwari, 2011).

The key-value database is the simplest type of NoSQL databases. As the name implies, this type of database stores schema-less data using keys. The key is usually a string and the stored values can be any valid type such as a primitive programming data type (string, integer, etc.) or a BLOB (Binary Large Object) without any predefined schema. It provides a simple API to access stored data (Fowler and Sadalage, 2012). In most cases, this type of NoSQL database solution provides very little functionality beyond key-value storage. There is no support for relationships (Xiang et al. 2010). In terms of concurrency, they usually provide eventual consistency but don't provide optimistic concurrency especially in highly scalable environments. Queries are just limited to accessing values using keys but since there is one request to access the value, the queries are executed quickly. Transactions are limited to a single key. The database contains no semantic model. In other words, the client is in charge of interpreting and understanding values. Key-value databases can be utilized to store geospatial data but their complexity hinders spatial searches especially for polylines and polygons. For this reason, it needs to be spatially indexed for fast data retrieval which, in most cases, gives lower performance than a relational database. Many researchers and developers recommend using indexing techniques such as grids or tiles, quadkeys, space filling curves and similar approaches. In summary, key-value data stores are ideal for inserting, deleting and searching huge amount of data items using their unique identifiers (keys). In the case of highly connected data or spatial searches they are not good solutions.

A document database in its simplest form is a key-value database in which the database understands its values (Hecht and Jablonski, 2011). In other words, values inside the database are based on predefined formats such as XML, JSON or BSON (Binary JSON). This feature of document databases provides many advantages over key-value databases. Instead of putting too much logic in the application layer, many operations can be done by the database itself. Queries in this type of NoSQL databases are quite flexible and some document databases even have their own query languages. Similar to key-value databases, there is no need to adhere to a predefined schema to insert data. There is only limited support for relationships and joins as each document is stand alone. However, more concurrency options, such as optimistic concurrency and eventual consistency are available (Chang, 2006). Transaction integrity is supported for one document or document fragment .

The document databases can be used for managing geospatial data more effectively than key-value databases. Since geospatial data inside the document database can be retrieved using flexible queries, they can be used for storing and managing geospatial data in multiple use cases. In fact many document databases support geospatial data natively or through extensions. Some applications of LBS such as proximity queries can be efficiently implemented using these document databases. As mentioned before, relationships and joins are not supported the way they are supported in relational databases. Often in common GIS workflows relationships and joins have to be used. However, the document-oriented nature of the system has some major effects on the way that data can be retrieved. For example if the application needs data items from the same documents it would be very fast. However whenever the data items are part of different types of documents there is no efficient approach to reduce the number of index-lookups. In summary indexing is just based on documents and there is no notion of relationships in document databases.

The columnar (or column family) databases store data in set of columns and distribute data based on columns (rather than rows in SQL databases). The column is the smallest unit of data and it is a triplet that contains a key, value and timestamp (Hecht and Jablonski, 2011). Columnar databases store all values beside the name of the columns and stores null values simply by ignoring the column. Usually, related columns compose a column-family. All the data in a single column family will be stored on the same physical set of files (Xiang et al., 2010). This feature provides higher performance for search, data retrieval and replication operations. A super column is a column that contains other columns but it cannot contain other super columns ((Hecht and Jablonski, 2011). Most columnar databases use a distributed file-system to store data to disk and so provide a horizontally scalable system. In fact columnar databases are designed to run on a large number of machines. Queries in this type of NoSQL databases are limited to keys and in most cases they don't provide a way to query by column or value. By limiting queries to just keys, columnar databases ensure that procedure to find the machine containing actual data is quite fast. There is no join capability and, as in other types of NoSQL databases, there is limited support for transactions.

Columnar databases are ideal for storing huge amounts of data when high availability is needed. Similar to document databases, there are many columnar databases which support the management and simple analysis of geospatial data. Any Geospatial Information System (GIS) related application which needs heavy data insertion and fast data retrieval with simple queries can efficiently make use of columnar databases. As an example, an Automatic Vehicle Location (AVL) application which needs to track the location of many vehicles simultaneously can store the incoming data from vehicles in a columnar database and respond to queries efficient. In summary this type of databases doesn't support relationships and in order to handle highly connected data, there is a need for mapping layer to create network structure (which is not efficient).

As the name implies graph databases are based on graph theory and employ nodes, properties and edges as their building blocks. The nodes and edges can have properties. In the graph databases various nodes might have different properties. The graph databases are well suited for data which can be modeled as networks such as road networks. Their main feature is the fact that each node contains a direct pointer to its adjacent node, so no index lookups are necessary for traversing connected data (which is really valuable). As a result they can manage huge amount of highly connected data since there is no need for expensive join operations. Some of graph databases support transactions in the way that relational databases support them. In other words the graph database allows the update of a section of the graph in an isolated environment, hiding changes from other processes until the transaction is committed. Geospatial data can be modeled as graphs. Since graph databases support topology natively, topological relationship (especially connectivity) between geospatial data can be easily

managed by this type of NoSQL databases. In most GIS workflows, topological relationships play a major role. In addition, since graph databases are ideal for managing
data with evolving schema, they can be effectively used in Volunteered Geographic
Information (VGI) and crowd sourcing applications. Also, graph databases are the
best choice for managing huge linear networks (such as roads) and for routing and
navigation applications (Amirian et al., 2013). In addition since each edge in graph
database can have different set of properties, they provide flexibility in traversal of
network based on various properties. For example it is possible to combine time; distance, number of points of interest and user preferences in finding best path and the
mentioned path would be unique for each user. This allows us to optimize number of
landmarks/length of path.

## 4    Implementation

The landmark-based navigation system is intended to provide navigation services to
tourist in Maynooth, a small town where National University of Ireland is located.
From general point of view, there are mainly four steps in our landmark-based navigation system design. Firstly, Landmarks must be defined, extracted and stored in a
database. In this step both geometrical and non-geometrical characters of each landmark is stored in the graph database.

Another component of this system is positioning component, see figure 2. The positioning component is responsible for calculating users' positions using image-based
positioning using photo taken by users' cameras where GPS signals are not available
wherever GPS signals are available, then GPS gives the location of users. In both
situations; availability and unavailability of GPS signals, it is possible to use image-
based positioning service however it is not recommended to use it where GPS signals
are received. This is recommended to users to make requests as minimum as possible
to reduce firstly user's devices' battery consumption and also network data exchange
and secondly prevent any potential problem to positioning component.

Then based on a routing algorithm, the most touristic path is calculated. Finally,
navigational instructions which help users to get their destination are provided using
landmarks' information and photos on the way.

Using landmarks stored in the database, it is possible to find the best path and navigate users to get their destination. The architecture used to provide user with landmark-based navigational instructions is illustrated in figure 2. The landmark-based
navigation system implemented in National University of Ireland is consists of four
main components; positioning component, service and data database, navigation service calculation engine and users and their mobile devices.

Positioning component is responsible of calculation position of the user using GPS
or image-based positioning techniques which explained previously and also tracking
them. It delivers its output, position of the user, to the navigation service calculation
engine.

The navigation service calculation engine uses user's position as input of two other
services; first it calculates the best path using user's position and selected destination

and secondly, based on user's position, it can calculate visible landmarks on the way. In order to do both of these tasks, the navigation service calculation engine needs to have access to the spatial database where landmarks' information, such as location, size, etc is stored. In this project, a graph data base is used. Elements in landmark-based navigation services are changing with quite high degree of frequency, since landmark which has got one of the essential role can be visible from one user but not another one. Processing of such highly connected and interrelated data can be problematic for most conventional and modern storage systems such as Relational Database Management Systems (RDBMS). Also when additional network elements (such as new land marks or new roads) have to be added to the data, all the representation and built network should not be rebuilt and recompiled over and over. In order to prevent facing with such issues, it is better to store connected data in their natural representation using a graph database.

The navigation service calculation engine uses landmarks, edges and nodes data to find more reliable path. This is calculated based on landmark-based path finding algorithm explained in previous section.

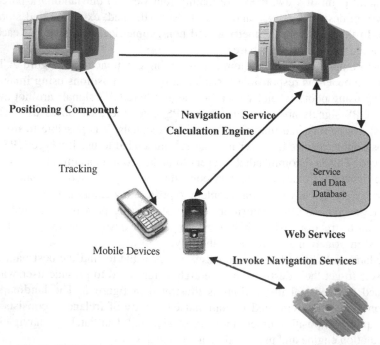

**Fig. 2.** NUIM Campus Landmark-based Navigation System Arcitecture

After route calculation, user should be provided with navigational instructions to follow calculated route. In this step, the navigation service calculation engine uses positional information provided by positioning engine and also data stored in the database such as information of landmarks to calculate the landmarks to be seen from user's location. Whenever user's location changes or new image-based positioning is requested by user, this process will be reseated and a new set of navigational instruction is provided.

1. Figure 3 shows, the web application interface of the system which provides image of landmarks as a part of navigation service. As it is shown in figure 3, three modes of travel; pedestrians, cars and wheelchair can use this service.

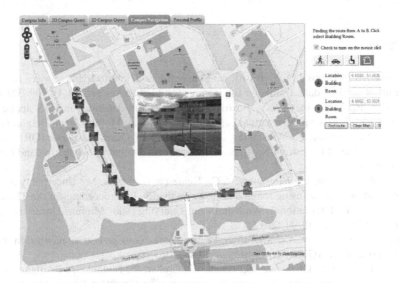

**Fig. 3.** Navigation services in eCampus web application

## 5    Conclusion

Landmarks are often used for casual navigation, such as giving directions, by ordinary people. Also they might be interests of tourists and visitors because of their uniqueness. So landmarks can be applied in tourist navigation services to navigate tourists and visitors to get their destinations and at the same time give additional information about landmarks passing by. Landmark-based navigation is a kind of navigation service in which positions of users are calculated based on the nearest landmark. Then based on current position of user which can be calculated using image of nearest landmark and selected destination, a landmark-based path finding algorithm, which calculates a route with maximum landmarks on the way and minimum distance passed, calculates most reliable path. Then users are provided with some information about interesting or important features, landmarks, around themselves to make sure they are on the correct way. In order to implement this application a graph database has been used to store landmark data and network information to avoid unavailability of the system due to high volume data exchange for positioning purpose. This paper explains these three steps in detail and implemented a landmark-based navigation application.

**Acknowledgement.** Research presented in this paper was funded by a Strategic Research Cluster grant (07/SRC/I1168) by Science Foundation Ireland under the National Development Plan. The authors gratefully acknowledge this support.

This work was financially supported by EU FP7 Marie Curie Initial Training Network MULTI-POS (Multi-technology Positioning Professionals) under grant nr. 316528.

# References

2. Amirian, P., Alesheikh, A.A., Basiri, A.: Standard-based, interoperable services for accessing urban services data. Computer Environment and Urban Systems 34(4), 309–321 (2010)
3. Amirian, P., Winstanley, A.C., Basiri, A.: Using Graph databases in LBS applications: Storing and Processing Navigational and Tracking data. In: Mobile Gehnt, Belgium (2013)
4. Basiri, A., Amirian, P., Winstanley, A.C.: The Use of Quick Response (QR) Codes in Landmark-Based Pedestrian Navigation. International Journal of Navigation and Observation (2014)
5. Basiri, A., Amirian, P., Winstanley, A.C., Kuntzsch, C., Sester, M.: Uncertainty han-dling in navigation services using rough and fuzzy set theory. In: Proceedings of the Third ACM SIGSPATIAL International Workshop on Querying and Mining Uncertain Spatio-Temporal Data, pp. 38–41 (2012)
6. Basiri, A., Winstanley, A.C., Amirian, P.: Landmark-based pedestrian navigation. In: 21st GIS Research UK (GISRUK) Conference, UK (2013)
7. Chang, F., Dean, J., Ghemawat, S., Hsieh, W., Gruber, R.: Bigtable: A distributed stor-age system for structured data. In: Seventh Symposium on Operating System Design and Implementation (2006)
8. Tom, A., Denis, M.: Referring to Landmark or Street Information in Route Directions: What Difference Does It Make? In: Kuhn, W., Worboys, M.F., Timpf, S. (eds.) COSIT 2003. LNCS, vol. 2825, pp. 362–374. Springer, Heidelberg (2003)
9. Elias, B.: Determination of Landmarks and Reliability Criteria for Landmarks. Technical Paper, ICA Commission on Map Generalization, 5th Workshop on Progress in Automated Map Generalization. IGN, Paris, France (2003)
10. Etienne, S., Séguinot, V.: Navigation by Dead Reckoning and Local Cues. Journal of Navigation 46, 364–370 (1993), doi:10.1017/S0373463300011802.
11. Fang, Z., Li, Q., Zhang, X., Shaw, S.L.: A GIS data model for landmark-based pe-destrian navigation. International Journal of Geographical Information Science (2011), doi:10.1080/13658816.2011.615749
12. Fontaine, S., Denis, M.: The Production of Route Instructions in Underground and Urban Environments. In: Freksa, C., Mark, D.M. (eds.) COSIT 1999. LNCS, vol. 1661, pp. 83–94. Springer, Heidelberg (1999)
13. Fowler, M., Sadalage, P.: NoSQL distilled: a brief guide to the emerging world of polyglot persistence. Addison-Wesley Publication (2012)
14. Gaisbauer, C., Frank, A.U.: Wayfinding Model for Pedestrian Navigation. In: The AGILE International Conference on Geographic Information Science, pp. 1–9 (2008)
15. Hecht, R., Jablonski, S.: NoSQL Evaluation A Use Case Oriented Survey. In: International Conference on Cloud and Service Computing, pp. 336–341 (2011)
16. Hansen, R., Wind, R., Jensen, C.S., Thomsen, B.: Seamless Indoor/Outdoor Positioning Handover for Location-Based Services in Streamspin. In: Tenth International Conference on Mobile Data Management: Systems, Services and Middleware, pp. 267–272 (2009)

17. Holm, S.: Hybrid ultrasound-RFID indoor positioning: Combining the best of both worlds. In: IEEE Int. Conf. RFID, Orlando, FL, pp. 155–162 (2009)
18. Hung, J.C.: The smart-travel system: utilising cloud services to aid traveller with personalised requirement. IJWGS 8(3), 279–303 (2012)
19. Karimi, H.: Universal Navigation on Smartphones. Springer (2011) ISBN-10: 1441977406
20. Lee, J.K., Grejner-Brzezinska, D.A., Toth, C.: Network-based Collaborative Navigation in GPS-Denied Environment. Journal of Navigation 65, 445–457 (2012), doi:10.1017/S0373463312000069.
21. Li, X., Wang, J., Li, T.: Seamless Positioning and Navigation by Using Geo-Referenced Images and Multi- Sensor Data. Journal of Sensors 13(7), 9047–9069 (2013)
22. Lynch, K.: The image of the city, p. 48. MIT Press (1960)
23. May, A.J., Ross, T., Bayer, S.H.: Incorporating Landmarks in Driver Navigation System Design: An Overview of Results from the REGIONAL Project. Journal of Navigation 58, 47–65 (2005), doi:10.1017/S0373463304003054.
24. Michon, P.-E., Denis, M.: When and Why Are Visual Landmarks Used in Giving Directions? In: Montello, D.R. (ed.) COSIT 2001. LNCS, vol. 2205, pp. 292–305. Springer, Heidelberg (2001)
25. Millonig, A., Schechtner, K.: Developing Landmark-based Pedestrian Navigation Systems. In: Proceedings of the 8th International IEEE Conference on Intelligent Transportation Systems, pp. 196–202 (2005) 0-7803-9215-9/05
26. Pielot, M., Boll, S.: "In Fifty Metres Turn Left": Why Turn-by-turn Instructions Fail Pedestrians. In: Haptic, Audio and Visual Interfaces for Maps and Location Based Services (2010)
27. Raubal, M., Winter, S.: Enriching Wayfinding Instructions with Local Landmarks. In: Egenhofer, M., Mark, D.M. (eds.) GIScience 2002. LNCS, vol. 2478, pp. 243–259. Springer, Heidelberg (2002)
28. Redish, D.: Beyond the cognitive map: from place cells to episodic memory. MIT, Cambridge (1999)
29. Schechtner, M.K.: Developing Landmark-based Pedestrian Navigation Systems. In: Proceedings of the 8th International IEEE Conference on Intelligent Transportation Systems, Vienna (2005)
30. Siegel, W., White, S.H.: The Development of Spatial Representations of Large-scale Environments. In: Reese, H.W. (ed.) Advances in Child Development and Behaviour, vol. 10, pp. 9–55. Academic Press, New York (1975)
31. Schoier, G., Borruso, G.: Spatial Data Mining for Highlighting Hotspots in Personal Navigation Routes. IJDWM 8(3), 45–61 (2012)
32. Tiwari, S.: Professional NoSQL. Wrox Publication (2011)
33. Vepa, R.: Ambulatory Position Tracking of Prosthetic Limbs Using Multiple Satellite Aided Inertial Sensors and Adaptive Mixing. Journal of Navigation 64, 295–310 (2011), doi:10.1017/S0373463310000494
34. Werner, S., Krieg-Brückner, B., Mallot, H., Schweizer, K., Freksa, C.: Spatial Cogni-tion: The Role of Landmark, Route and Survey Knowledge in Human and Robot Navigation. In: Jarke, M., Pasedach, K., Pohl, K. (eds.) Informatik aktuell, pp. 41–50. Springer, Berlin (1997)
35. Xiang, P., Hou, R., Zhou, Z.: Cache and consistency in NoSQL. In: 3rd IEEE International Conference on Computer Science and Information Technology, pp. 117–120 (2010)

# Evaluation of Data Management Systems
# for Geospatial Big Data

Pouria Amirian[1], Anahid Basiri[2], and Adam Winstanley[1]

[1] Department of Computer Science, National University of Ireland Maynooth, Ireland
[2] Nottingham Geospatial Institute, The University of Nottingham, UK
amirian@cs.nuim.ie, anahid.basiri@nottingham.ac.uk
adam.winstanley@nuim.ie

**Abstract.** Big Data encompasses collection, management, processing and analysis of the huge amount of data that varies in types and changes with high frequency. Often data component of Big Data has a positional component as an important part of it in various forms, such as postal address, Internet Protocol (IP) address and geographical location. If the positional components in Big Data extensively used in storage, retrieval, analysis, processing, visualization and knowledge discovery (geospatial Big Data) the Big Data systems need certain type of techniques and algorithms for management, analytics and sharing.

This paper describes the concept of geospatial Big Data management with focus on using typical and modern database management systems. Then the typical and modern types of databases for management of geospatial Big Data are evaluated based on model for storage, query languages, handling connected data, distribution models and schema evolution. As the results of the evaluations and benchmarks of this paper illustrate there is no single solution for efficient management of geospatial Big Data and in order to utilize unique characteristics of geospatial Big Data (such as topological, directional and distance relationship) a polyglot geospatial data persistence system is needed.

**Keywords:** geospatial Big Data, graph database, XML document database, column-family database, spatial database, geospatial Big Data Management, polyglot geospatial data persistence.

## 1 High Level Introduction to Geospatial Big Data

Often data component in Big Data has a geospatial component as an important part of it in various forms, such as postal address, Internet Protocol (IP) address and geographical location (geospatial Big Data). As it mentioned in many research papers, management and analysis of geospatial data is complex and requires specific storage, processing, analysis and publication mechanisms [1, 2, 3, 4, 5 and 6]. In fact management and analysis of geospatial data have been always revealed the limitations of information systems and computational frameworks. In a nutshell, unique characteristics of geospatial data such as high volume, various type of relationships between geospatial objects (e.g. distance, directional and topological relationships), need for long transactions, computationally intensive algorithms of processing and inclusion of

B. Murgante et al. (Eds.): ICCSA 2014, Part V, LNCS 8583, pp. 678–690, 2014.
© Springer International Publishing Switzerland 2014

time component, makes the management and analysis of geospatial Big Data even more complicated. Some researchers agreed that geospatial data may represent the biggest Big Data challenge of all [7]. If the positional components in Big Data extensively used in storage, retrieval, analysis, processing, visualization and knowledge discovery (geospatial Big Data) the Big Data systems need certain type of technologies, techniques and algorithms for management, analytics and sharing. [8]. Using geospatial Big Data provides unprecedented opportunities for providing improved, more adaptive, more intelligent and cost-effective services in government, private and science and research sectors [9]. In summary management of geospatial data has several challenges in the storage, processing, analysis, visualization and publication areas. This paper focuses on management of geospatial Big Data for standard online sharing and publication.

## 2 Standard Publication of Geospatial Big Data Using Web Services

Publication of geospatial Big Data in standard manner provides opportunities for executing distributed and collaborative data preparation, data mining and knowledge discovery tasks. Also the ever-increasing access to geospatial data on the Web results in enhanced system efficiency through cost and time reduction in data collection, data preparation and information retrieval. Moreover, such access helps decision-makers to manage their assets better, enables faster responses for time-sensitive decisions, and improves the communication process across diverse agencies. In this regard, geospatial data should be shared and accessed using standard services which are openly published over the Web [10]. In this context, there are generally two approaches for publishing geospatial data in standard manner. The first approach is to use the specifications published and managed by Open Geospatial Consortium (OGC). The mentioned specifications (geospatial services) consist of defined set of request/responses to access geospatial resources [11, 12]. The second approach is to use web services technologies and use the standard messaging and standard interface definition mechanisms. The mentioned standard messaging and standard interface mechanisms are inherent to web services technologies and there is no need to predefine set of request/response in order to exposing geospatial resource over the web [10].

These two approaches are just standard approaches for exposing geospatial resources over the web. There are also other approaches [13] that utilize proprietary and platform-dependant solutions for exposing geospatial resources. This paper focuses on standard approaches. The first approach is standard and well supported in Geospatial Information (GI) community. The second approach belongs to the broader and more dynamic Information Technology (IT) community. As it mentioned before the second approach utilizes web services technologies that consist of several technologies such as XML, XSD, WSDL, SOAP as core technologies. These technologies can be used over the web (HTTP) or any other protocol. At the other hand the first approach (using geospatial services) limited to the web [10]. This is a serious issue in publishing geospatial resources to the users. Although some OGC specification can be defined using core web service technologies, but using core web service technologies for some OGC services is not possible (in standard manner) [14]. As an example Web

Map Service (WMS) specification is the most implemented geospatial service. It creates image of geospatial data (map in OGC terminology) as response to GetMap request over the HTTP. So the WMS service provides sharing of geospatial data at image-level. Since supported data types of web services defined by XSD and there is no native support for binary data in XSD, creating wrapper web service for WMS results in non-standard web service [10, 12, 13, 14 , 15]. The main geospatial service for sharing geospatial data at object level is Web Feature Service (WFS). WFS provides access to geospatial data using GML format which contains both geometrical as well as attribute properties of each geospatial data items [17]. Since GML is a XML grammar, there is no serious difference in using both kinds of services for publication and sharing of geospatial Big Data over the web at the object-level.

Often the huge volume of geospatial data is the reason for the complexity of publication of geospatial Big Data issue. In addition to huge volume of geospatial data, sometimes the velocity and variety components in geospatial Big Data are major reasons for the issue. For example in disaster management and when real-time or near-real time decision making is critical, it is necessary to access various voluminous geospatial data from different sources (satellite data, surveillance systems and social network data) with high frequency of change. In these situations the velocity of change of data and variety of data sources are as important as the volume of data.

Traditionally Relational Database Management Systems (RDBMS or SQL databases) with spatial extensions (Spatial Databases on top of relational or object-relational systems) were used as backend system for geospatial services [18]. Nowadays these systems still can be used in many geospatial data-related tasks but the mentioned systems are not efficient enough to handle geospatial Big Data, especially when the volume, velocity and variety of datasets are far beyond the capacity of a single server and the datasets need to be handled in distributed manner. As it illustrated in Figure 1, in a typical system (using the SQL database) there are four layers in the system for publication of geospatial data. In this case geospatial services (such as WFS) just provide access to geospatial data through a service layer for various kinds of clients. All the request and response processing is done in business logic layer. If geospatial datasets were stored in SQL or spatial databases in data layer, the business logic layer must contains a mapping layer. In the case of WFS and if the client requests to get data in GML format (application data model), the business logic layer must retrieve data from databases and then create a GML document.

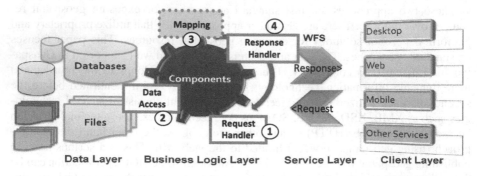

**Fig. 1.** A typical system for publication of geospatial data with relational or spatial databases

The mapping layer has negative effects on scalability, availability and performance of the system. In this case NoSQL databases can provide the required quality of services for standard publication of geospatial Big Data. In a nutshell, since the storage data model and application data model can be the same in NoSQL databases, the mapping layer is not necessary in the system. In addition, NoSQL databases are designed with the idea of distribution of data and processes in cluster of machines which is a major advantage in comparison with relational and object-relational databases [19, 20]. Following sections first explain the major types of NoSQL databases and then provide an evaluation of using a relational (SQL database), a spatial and a NoSQL for handling geospatial Big Data with focus on standard publication.

## 3    SQL and NoSQL Databases

Relational DBMSs (SQL databases) use tables, columns, keys and Structured Query Language (SQL) to perform all sorts of tasks with data. One of the important facets of SQL databases is the normalization process which ensure about storage of data items in separate tables and only once in whole database. The SQL databases usually are the best solutions when the schema of data is fixed and strong consistency is most needed feature. In other words, SQL databases are ideal solutions to managing structured data such that all users can access to the same set of data in same state at all times (strong consistency). The SQL databases can be effectively used in many common geospatial-related workflows. Since they support transaction and locking features, they provide efficient backend for enterprise GIS systems. Also geospatial data have a fixed schema and in most cases they are not used in isolation. As a result join of two or a few more datasets and connecting data through spatial operations is needed in most GIS workflows. For this reason managing fixed schema geospatial data with limited connectivity and using them in GIS workflows can usually be done effectively through SQL databases.

The SQL databases handle connected data using relationship and they retrieve connected data using joins. So connections between related data tables are stored using primary and foreign keys and join between connected tables are needed when retrieving data. So the related data are stored in SQL databases separately and they can be related using joins. However joins are one of the most computationally expensive processes for SQL databases. In most cases, joins are the bottleneck of SQL databases. In order to avoid many joins (which is needed in handling highly connected data in SQL databases) denormalization process can be used to store data items several times in single large tables. But there are several issues associated with denormalization process especially with providing consistency in large datasets. In addition to issues related to handling highly connected data, some other problems arise when SQL databases need to handle high volume of data and when scalability is needed by adding more servers and technologies to bind them together. When distribution is needed in SQL databases and with more loads on a SQL database, vertical partitioning and denormalization process are needed which results in complexity in providing strong consistency. In summary, to achieve high scalability in SQL databases the normalized relational model of data storage has to be compromised and deviated from relational model.

The NoSQL (Not only SQL) DBMSs are a broad class of DBMSs identified by non-adherence to the SQL (relational) model. There are different types of NoSQL databases, each with distinct set of characteristics but they all can deal with large amounts of (semi-structured and unstructured) data and are able to support a large set of read and write operations and they are designed with scalability and distribution of data in mind. Since they are designed with distribution in mind, there are other alternatives for providing consistency (most notably eventual consistency model). For this reason NoSQL and relational models are not in contrast with each other rather they complement each other. The most widely accepted taxonomy of NoSQL databases are: key-value, document, column-family and graph [19].

The key-value database is the simplest type of NoSQL databases. As the name implies, this type of database stores schema-less data using keys. The key is usually a string and the stored values can be any valid type such as a primitive programming data type (string, integer etc) or a BLOB (Binary Large Object) without any predefined schema. It provides a simple API to access stored data. In most cases, this type of NoSQL database solution provides very little functionality beyond key-value storage. There is no support for relationships [20]. Queries are just limited to accessing values using keys but since there is one request to access the value, the queries are executed very quickly. Transactions are limited to a single key. The database contains no semantic model. Key-value databases can be utilized to store geospatial data but the complexity of geospatial data hinders spatial searches especially for polyline and polygon objects. For this reason, it needs to be spatially indexed for fast data retrieval which, in most cases, gives lower performance than a SQL or spatial database. In summary, key-value data stores are ideal for inserting, deleting and searching huge amount of simple data items using their unique identifiers (keys).

A document database in its simplest form is a key-value database in which the database understands its values [21]. In other words, values inside the database are based on predefined formats such as XML, JSON or BSON. This feature of document databases provides many advantages over key-value databases. Queries in this type of NoSQL databases are quite flexible. Similar to key-value databases, there is no need to adhere to a predefined schema to insert data. There is only limited support for relationships and joins as each document is stand alone. The document databases can be used for managing geospatial data more effectively than key-value databases. Since geospatial data inside the document database can be retrieved using flexible queries, they can be used for storing and managing geospatial data in multiple use cases. In fact many document databases support geospatial data natively or through extensions. Some of them can store geospatial data using GeoJSON format. Some queries such as proximity queries can be efficiently implemented using these document databases [22]. As mentioned before, relationships and joins are not supported the way they are supported in relational databases. The document-oriented nature of this NoSQL database has some major effects on the way that data can be retrieved. For example if the application needs data items from the same collection (documents with same schema) it would be very fast. However whenever the data items are part of different types of documents there is no efficient approach to reduce the number of index-lookups. Special kinds of document database can partially handle relationships efficiently. XML document databases store documents as XML documents. This kind of document databases can be configured to enforce adherence to set of predefined XML schemas. In addition to all the advantages of document databases, XML document databases are able to utilize many XML technologies to provide further functionality. For example, they can use XQuery and XPath to perform various queries and create flexible result sets, they can make use of XPointer to reference other documents thus modeling a relationship and they can use XSD and RelaxNG to enforce schema validation [22]. Geography Markup Language (GML), as a standard mechanism for storing, modeling and exchanging geospatial data [17], is an XML-based grammar and so XML document databases are an ideal choice for managing geospatial data in GML format. Since the storage data model and application data model is XML document in XML document databases, they can be efficiently utilized when the standard publication of geospatial Big Data is needed.

The column-family databases store data in set of columns and distribute data based on columns. The column is the smallest unit of data and it is a triplet that contains a key, value and timestamp [19]. Column-family databases store all values beside the name of the columns and stores null values simply by ignoring the column. Usually, related columns compose a column-family. All the data in a single column family will be stored on the same physical set of files [20]. This feature provides higher performance for search, data retrieval and replication operations. A super column is a column that contains other columns but it cannot contain other super columns [21]. Most column-family databases use a distributed file-system to store data to disk and so provide a horizontally scalable system. In fact column-family databases are designed to run on a large number of machines. Queries in this type of NoSQL databases are limited to keys and in most cases they don't provide a way to query by column or value. By limiting queries to just keys, column-family databases ensure that procedure to find the machine containing actual data is quite fast. There is no join capability and, as in other types of NoSQL databases, there is limited support for transactions. Column-family databases are ideal for storing huge amounts of data when high availability is needed. Similar to document databases, there are many column-family databases which support the management and simple analysis of geospatial data. Any GIS related application which needs heavy data insertion and fast data retrieval with simple queries can efficiently make use of column-family databases. In summary this type of databases doesn't support relationships and in order to handle highly connected data, there is a need for mapping layer to create network structure (which is not efficient).

As the name implies graph databases are based on graph theory and employ nodes, properties and edges as their building blocks. The nodes and edges can have properties. In the graph databases various nodes might have different properties. The graph databases are well suited for data which can be modeled as networks such as road networks, social networks, biological networks and semantic webs. Their main feature is the fact that each node contains a direct pointer to its adjacent node, so no index lookups are necessary for traversing connected data. As a result they can manage huge amount of highly connected data since there is no need for expensive join operations. Some of graph databases support transactions in the way that relational databases support them. In other words the graph database allows the update of a section of the graph in an isolated environment, hiding changes from other processes until the transaction is committed. Geospatial data can be modeled as graphs. Since graph databases support topology natively, topological relationship (especially connectivity) between geospatial data can be easily managed by this type of NoSQL databases [8]. In most GIS workflows, topological relationships play a major role. In addition since each edge in graph database can have different set of properties, they provide flexibility in traversal of network based on various properties. For example it is possible to combine time, distance, number of points of interest and user preferences in finding best path and the mentioned path would be unique for each user. In summary the storage model of graph databases is a graph and there is a need for mapping layer whenever other data structure is needed in application layer.

## 4    Implementation and Benchmarks

In order to find the best database model for standard publication of Big Data three systems were implemented based on the architecture illustrated in figure 1. Three different models of databases are used in the mentioned systems: relational (SQL), spatial and XML document. Since the evaluation is for database models rather than specific product, for consistency of the benchmark in this research Microsoft SQL Server 2012 (MS SQL Server) is used in three different models. The MS SQL Server is a relational DBMS (SQL database) with built-in support for spatial data using geography and geometry data types. In addition to spatial data types it has several advanced features that make MS SQL Server a capable spatial database (such as support for various spatial reference systems, diverse spatial indexing, implementation of OGC simple features specification and spatial topology handling based on 9 intersection model. Also the MS SQL Server has a native XML support through XML data type, XML indexing, support for XQuery and XPath and query optimization for XML queries. In other words, MS SQL Server is a spatial and XML document database on top of relational engine.

For the evaluation purpose, four geospatial datasets containing polygon features are created. The mentioned datasets contain hundred thousand (100k), one million (1m), ten million (10m) and a hundred million (100 m) polygon features. Each polygon feature has at least three points (three vertices) and at most 2000 vertices. In addition each polygon has at least one part and at most five parts (multi-part polygon). The EPSG 4326 was used as the spatial reference system to store coordinates of vertices.

Microsoft C# programming language was used for implementing business and service layer as well as client layer. The client application was a simple application for calling the WFS service to retrieve geospatial data based on several predefined queries. The mentioned application also utilized several profilers to record the metrics for the performance and scalability benchmarks.

## 4.1    Storage

Storage of polygon features (and multi-part polygon features) requires at least four tables in relational model. Figure 2, illustrates the conceptual model for the mentioned four tables using a ER diagram.

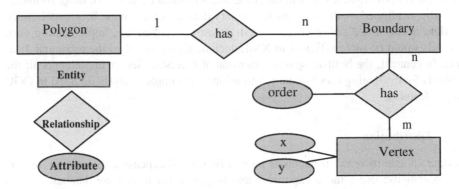

**Fig. 2.** Conceptual model of SQL Database for storing polygon features (in ERD)

A polygon in spatial database is defined using at least one closed ring. So there is one table for a polygon dataset. For XML document database, there is one GML document fragment per polygon (even multipart polygon), so there is one collection of documents for each polygon dataset.

## 4.2    Query Language and Retrieval

For relational and spatial databases the SQL language is utilized as the query language. As it mentioned before there are at least four tables for polygon features. As a result in order to retrieve data of a polygon three joins are needed. Since spatial database of this research implements the OGC Simple Features specification, all the spatial operators and methods are available as extension to standard SQL. There is no need for joins in spatial database since the polygon features are stored in a single table. The XML document database uses a XML-specific query language (XQuery) in order to retrieve geospatial data. In order to retrieve polygon features, there is no need for joins since the polygon features are stored as documents in a single collection.

## 4.3    Handling Attribute Relationships

Attribute relationships are non-spatial relationships (for example ownership relationship between an owner and a land). In relational and spatial databases this kind of relationship can be defined using primary and foreign keys. In order to retrieve related data in the mentioned databases, joins are needed. In XML document databases the attribute relationship can be defined by XPointer and for traversing between different document fragments XLink can be used. A link to one or more related documents can be found in the origin document fragment in XML document databases.

## 4.4    Topological Relationships

There is no support for evaluating topological relationships between geospatial data in SQL and XML document databases in MS SQL Server. Implementing methods for evaluating topological relationships or methods for simple analysis of geospatial data is similar in both XML document database and relational database. In order to implement the mentioned methods some in-memory data structures must be created. Since the data about single polygon are stored in just one collection, implementing such methods would be more efficient in XML document database than the relational database. In contrast, the built-in spatial extension of MS SQL Server implements all the methods for evaluating topological relationships and simple analysis outlined in OGC simple features.

## 4.5    Distribution

As data volume increase, it becomes more difficult and expensive to scale up (or vertical scalability; use a more expensive and bigger server to run the database on). A more efficient approach is scale out (or horizontal scalability; use of a cluster of servers to run a database). In general NoSQL databases are designed and implemented with focus on horizontal scalability and high volume of data. Depending on the distribution model, database can handle larger quantities of data and process a greater read and write traffic or more availability in the face of network slowdowns or breakages [19]. Usually there are two approaches for data distribution; replication and sharding. With replication copies of same data are stored on multiple nodes. So each bit of data can be found in multiple places [23]. In contrast, sharding puts different data on different nodes so each server acts as the single source for subset of data [20].

In the case of geospatial Big Data in most cases a combination of sharding and replication provides the highest availability, scalability and performance. Sharding in NoSQL database are easier since the natural unit of distribution is often the same as the unit of storage. In XML document database, distribution can be done based on the XML fragments. In other words, different XML fragments (storage unit) can be distributed on different nodes. In contrast sharding in SQL databases are not as straightforward as for the NoSQL database. The natural unit of distribution for geospatial Big Data is a geospatial feature. But data of a geospatial feature in SQL databases is spread over multiple tables which makes the distribution complex. For spatial database, the sharding is much easier than the SQL database but still is complicated in comparison with NoSQL database. It is possible to use replication for three models. But as it mentioned before, using different tables for single geospatial feature makes replication hard and complex for SQL database.

## 4.6    Schema Definition and Evolution

Schema definition for SQL and Spatial database is done using SQL language commands for creating tables, columns and indexes. For XML document databases the

XSD can be used for defining the schema of XML document. Both SQL and XSD support various data types. The schema change or evolution during the life cycle of application development is common and complex practice for databases. It is true that all NoSQL databases are schemaless (or schema-free) but in order to use the data inside NoSQL databases there is an implicit schema in the applications that use NoSQL databases. Table 1 summarizes the comparison between SQL, Spatial and NoSQL databases for handling geospatial Big Data.

**Table 1.** Various characteristics of three different approaches for geospatial Big Data storage

| Item | SQL database | Spatial database | XML document NoSQL database |
|---|---|---|---|
| Logical Storage unit | Row | Geospatial feature | GML fragment |
| Logical Storage of a geospatial dataset | Multiple tables | Single table | Single collection |
| Query Language and Retrieval | SQL language | Extended SQL language with OGC Simple Features Specification | XQuery |
| Attribute Relationships | Primary and Foreign key and Joins | Primary and Foreign key and Joins | XLink and XPointer |
| Topological Relationships | No Native Support /Requires mapping layer | Extended SQL with OGC Simple Features Specification | No Native Support /Requires mapping layer |
| Distribution | Hard Replication Hard Sharding | Easier than SQL database Easier than SQL database | Easy Replication Easy Sharding |
| Schema Definition and Evolution | SQL language | SQL language | XSD language |

## 4.7    Performance and Scalability Evaluations

In order to perform performance and scalability benchmarks, the mentioned three databases were filled with polygon datasets which include vast amount of features from 100,000 to 100,000,000 multi-part polygons. For performance tests, the response time was used as the metric. Figure 3 and 4, illustrate the results of the performance tests for single feature retrieval and feature retrieval using range queries respectively.

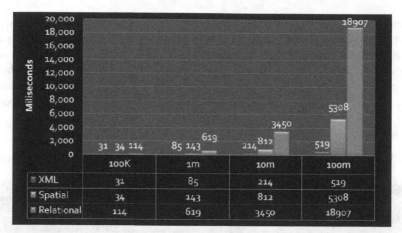

**Fig. 3.** Performance test for retrieval of single geospatial feature (lowest is the best)

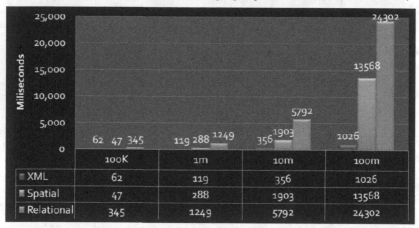

**Fig. 4.** Performance test for retrieval of group of geospatial features using various range queries (lowest is the best)

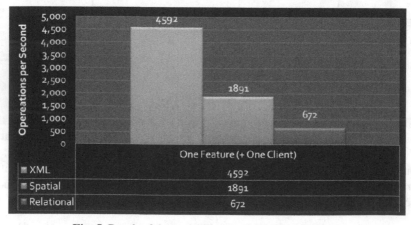

**Fig. 5.** Result of the scalability test (highest is the best)

In order to perform scalability benchmark, operation per second was utilized as metric. Figure 5 illustrates the result of the scalability test.

The results of tests proved that XML document database (NoSQL) provides better performance and scalability for standard publication of geospatial Big Data for some specific queries.

# 5    Conclusion

Geospatial data have specific characteristics that often reveal the limitation of computing systems. The storage and analysis of geospatial Big Data (high volume of high frequency of change geospatial data from various data sources) is challenging and complex and needs horizontal scalability and various models for consistency, data access and distribution. As the evaluations and benchmarks of this paper illustrate, the NoSQL databases provide several qualities needed for efficient analysis and management of geospatial Big Data. But this doesn't mean relational (SQL) or spatial databases don't have any place in geospatial Big Data landscape. The authors of this paper believe that polyglot geospatial data persistence approach is efficient model for geospatial Big Data handling. In other words using various database models for different tasks in a single system (polyglot data persistence).

# References

1. Taniar, D., Rahayu, W.: A taxonomy for nearest neighbour queries in spatial databases. Journal of Computer and System Sciences 79(7), 1017–1039 (2013)
2. Amirian, P., Basiri, A., Alesheikh, A.: Interoperable exchange and share of urban services data through geospatial services and XML database. In: Complex, Intelligent and Software Intensive Systems, CISIS (2010)
3. Mahboubi, H., Bimonte, S., Deffuant, G., Chanet, J., Pinet, F.: Semi-Automatic Design of Spatial Data Cubes from Simulation Model Results. IJDWM 9(1), 70–95 (2013)
4. Basiri, A., Amirian, P., Winstanley, A.: The USE of Quick Response (QR) Code in Landmark-Based Pedestrian Navigation. International Journal of Navigation and Observation (2014)
5. Yildizli, C., Pedersen, T., Saygin, Y., Savas, E., Levi, A.: Distributed Privacy Preserving Clustering via Homomorphic Secret Sharing and Its Application to (Vertically) Partitioned Spatio-Temporal Data. IJDWM 7(1), 46–66 (2011)
6. Safar, M., Ebrahimi, D., Taniar, D.: Voronoi-based reverse nearest neighbor query processing on spatial networks. Multimedia Systems 15(5), 295–308 (2009)
7. Minelli, M., Chambers, M., Dhiraj, A.: Big Data, Big Analytics: Emerging Business Intelligence and Analytic Trends for Today's Businesses. Wiley (2013)
8. Amirian, P., Basiri, A., Winstanley, A.: Efficient Online Sharing of Geospatial Big Data Using NoSQL XML Databases. In Proceedings of IEEE Fourth International Conference on Computing for Geospatial Research and Application (COM. Geo) (2013)
9. Amirian, P., Basiri, A., Winstanley, A.: Implementing Geospatial Web Services Using Service Oriented Architecture and NoSQL Solutions. In: The Third International Conference on Digital Information and Communication Technology and its Applications (2013)

10. Amirian, P., Basiri, A., Alesheikh, A.: Standards-based, interoperable services for accessing urban services data for the city of Tehran. Computers, Environment and Urban Systems (2010)

11. Foerster, T., Schäffer, B.: A client for distributed geo-processing on the web. In: Ware, J.M., Taylor, G.E. (eds.) W2GIS 2007. LNCS, vol. 4857, pp. 252–263. Springer, Heidelberg (2007)

12. Sample, J., Shaw, K., Tu, S., Abdelguerfi, M.: Geospatial services and applications for the Internet. Springer (2008)

13. Stollberg, B., Zipf, A.: OGC Web Processing Service Interface for Web Service Orchestration Aggregating Geo-processing Services in a Bomb Threat Scenario. In: Ware, J.M., Taylor, G.E. (eds.) W2GIS 2007. LNCS, vol. 4857, pp. 239–251. Springer, Heidelberg (2007)

14. Schäffer, B., Baranski, B., Foerster, T., Brauner, J.: A Service-Oriented Framework for Real-time and Distributed Geoprocessing. In: Geospatial Free and Open Source Software in the 21st Century. Lecture Notes in Geoinformation and Cartography. Springer (2010)

15. Foerster, T., Schaeffer, B., Brauner, J., Baranski, B.: Geospatial Web Services for Distributed Processing - Applications and Scenarios. In: Geospatial Web Services: Advances in Information Interoperability, pp. 245–286. Information Science Reference (2011)

16. Vretanos, A.: OpenGIS Web Feature Service 2.0 Interface Standard, OGC 09-025r1 and ISO/DIS 19142 (2010)

17. Lake, R.: The application of geography markup language (GML) to the geological sciences. Computers & Geosciences 31(9), 1081–1094 (2005)

18. Oosterom, P.: Research and development in geo-information generalisation and multiple representation. Computers, Environment and Urban Systems (2010)

19. Fowler, M., Sadalage, P.: NoSQL Distilled. Addison Wesely (2013)

20. Celko, J.: Complete Guide to NoSQL, What Every SQL Professional Needs to Know about Non-Relational Databases. Morgan Kaufman (2014)

21. Chang, F., Dean, J., Ghemawat, S., Hsieh, W., Gruber, R.: Bigtable: A distributed storage system for structured data. In: Seventh Symposium on Operating System Design and Implementation (2006)

22. Amirian, P., Alesheikh, A.: Publishing Geospatial Data through geospatial web service and xml database system. American Journal of Applied Science 5(10) (2008)

23. Hecht, R., Jablonski, S.: NoSQL Evaluation A Use Case Oriented Survey. In: International Conference on Cloud and Service Computing, pp. 336–341 (2011)

24. McCreary, D., Kelly, A.: Making Sense of NoSQL. Manning (2013)

# Introducing iSCSI Protocol on Online Based MapReduce Mechanism[*]

Shaikh Muhammad Allayear[1], Md. Salahuddin[1],
Fayshal Ahmed[1], and Sung Soon Park[2]

[1] Department of Computer Science and Engineering
East West University, Bangladesh
[2] Anyang University, Korea
allayear@ewubd.edu, {2010-2-60-001,2010-2-60-015}@ewu.edu.bd,
sspark@anyang.ac.kr

**Abstract.** In large internet enterprise and data management system, Hadoop MapReduce is a popular framework. For data intensive batch jobs, MapReduce provided its impact. To build consistent, hi-availability (HA) and scalable data management system to serve peta bytes of data for the massive users, are the main focused objects. MapReduce is a programming model that enables easy development of scalable parallel applications to process vast amount of data on large cluster. Through a simple interface with two functions map and reduce, this model facilities parallel implementation of real world tasks such as data processing for search engine and machine learning. Earlier version of Hadoop MapReduce has several performance problems like connection between Map to Reduce task, data overload and time consumption. In this paper, we proposed a modified MapReduce architecture MRA (MapReduce Agent) which is a fusion of iSCSI protocol and the downloaded reference code of Hadoop*. Our developed MRA can reduce completion time, improve system utilization and give better performance.

## 1 Introduction

Nowadays, dealing with datasets in the order of terabytes or even petabytes is a reality. Therefore, processing such big datasets in an efficient way is a clear need for many users. In this context, Hadoop MapReduce is a big data processing framework that has rapidly become the important factor in both industry and academia. Google proposed MapReduce. The MapReduce framework simplifies the development of large-scale distributed applications on clusters of commodity machines. It has become widely popular, e.g., Google uses it internally to process more than 20 PB per day. Yahoo!, Facebook and others use Hadoop, an open-source implementation of MapReduce. MapReduce has emerged as a popular way to harness the power of large clusters of

---

[*] This research (Grants NO. 2013-140-10047118) was supported by the 2013 Industrial Technology Innovation Project Funded by Ministry Of Science, ICT and Future Planning.
The source code for HOP can be downloaded from http://code.google.com/p/hop

B. Murgante et al. (Eds.): ICCSA 2014, Part V, LNCS 8583, pp. 691–706, 2014.

computers. The programmer only needs to write the logic of a Map function and Reduce function. This eliminates the need to implement fault-tolerance and low-level memory management in the program. A key benefit of MapReduce is that it automatically handles failures, hiding the complexity of fault-tolerance from the programmer. If a node crashes, MapReduce reruns its tasks on a different machine. MapReduce is typically applied to large batch-oriented computations that are concerned primarily with time to job completion. The Google MapReduce framework [1] and open-source Hadoop system reinforce this usage model through a batch-processing implementation strategy: the entire output of each map and reduce task is materialized to a local file before it can be consumed by the next stage. Materialization allows for a simple and elegant checkpoint/restart fault tolerance mechanism that is critical in large deployments, which have a high probability of slowdowns or failures at worker nodes.

To solve the above discussed problem we propose a modified MapReduce architecture that is MapReduce Agent (MRA).Our developed MRA provides several important advantages to a MapReduce framework. We highlight the potential benefits first:

- In map reduce framework, data transmit from map to reduce stage. So there may be connection problem. To solve this problem MRA creates iSCSI[2] Multi-Connection and Error Recovery Method [3] to avoid drastic reduction of transmission rate from TCP congestion control mechanism and guarantee fast retransmission of corruptive packet without TCP re-establishment.
- For fault tolerance and workload, MRA creates Q-chained cluster. Q-chained cluster [3] is able to balance the workload fully among data connections in the event of packet losses due to bad channel characteristics.

Basically Hadoop performs its I/O operation by dividing into blocks as well as iSCSI protocol and that motivates us iSCSI may provide better performance in Hadoop architecture.

### 1.1   Structure of the Paper

The rest of this paper is organized as follows. Overview of the iSCSI protocol, Hadoop MapReduce architecture and pipelining mechanism [5] described in section 2. At section 3 we described our research motivations. We describe our proposed model of Map Reduce Agent (MRA) in brief in section 4. We evaluated the performance and result in section 5. Finally, at section 6 we provided the conclusion of this paper.

## 2     Background

In this section, besides the iSCSI protocol we review the MapReduce programming model and describe the salient features of Hadoop, a popular open-source implementation of MapReduce.

## 2.1 iSCSI Protocol

iSCSI [2](Internet Small Computer System Interface) is a transport protocol that works on top of TCP. iSCSI transports SCSI packets over TCP/IP. iSCSI client-server model describes clients as iSCSI initiator and data transfer direction is defined with regard to the initiator. Outbound or outgoing transfers are transfer from initiator to the target.

iSCSI read/write operation parameters[6] values are determined during login phase and full feature phase by mutual understanding of iSCSI initiator and target data transfer capability. During login phase iSCSI target authenticate iSCSI initiator and then allow entering into full feature phase. iSCSI commands and data are exchanged in this phase. According to iSCSI operations there are two classes of parameters need to negotiate during login phase. While iSCSI write operation occurred, MaxBrustLength and MaxRecvDataSegmentLength (MRDSL) parameters are negotiated between iSCSI initiator and target [7]. According to target capability the values are set. At iSCSI read operation, iSCSI initiator requests target to provide desired data based on initiator capability read operations parameters are set Number of sector per command, MaxRecvDataSegmentLength and Phase Collapse.

## 2.2 Programming Model

To use MapReduce, the programmer [4] expresses their desired computation as a series of jobs. The input to a job is an input specification that will yield key-value pairs. Each job consists of two stages: first, a user-defined map function is applied to each input record to produce a list of intermediate key-value pairs. Second, a user-defined reduce function is called once for each distinct key in the map output and passed the list of intermediate values associated with that key. The MapReduce framework automatically parallelizes the execution of these functions and ensures fault tolerance.

Optionally, the user can supply a combiner function [1]. Combiners are similar to reduce functions, except that they are not passed all the values for a given key: instead, a combiner emits an output value that summarizes the Input values it was passed. Combiners are typically used to perform map-side "pre-aggregation," which reduces the amount of network traffic required between the map and reduce steps.

```
public interface Mapper < K1, V1, K2, V2>{

    void map(K1 key, V1 value, OutputCollector<K2, V2> output);

    void close ();

}
```

**Fig. 1.** Map function interface

## 2.3    Hadoop Architecture

Hadoop [4] is composed of Hadoop MapReduce, an implementation of MapReduce designed for large cluster, and the Hadoop Distributed File System (HDFS), a file system optimized for batch-oriented workloads such as MapReduce. In most Hadoop jobs, HDFS is used to store both the input to the map step and the output of the reduce step. Note that HDFS is not used to store intermediate results (e.g., the output of the map step): these are kept on each node's local file system.

A Hadoop installation consists of a single master node and many worker nodes. The master, called the Job-Tracker, is responsible for accepting jobs from clients, dividing those jobs into tasks, and assigning those tasks to be executed by worker nodes. Each worker runs a Task-Tracker process that manages the execution of the tasks currently assigned to that node. Each TaskTracker has a fixed number of slots for executing tasks (two maps and two reduces by default).

## 2.4    Map Task Execution

Each map task is assigned a portion of the input file called a split. By default, a split contains a single HDFS block (64 MB by default)[4], so the total number of file blocks determines the number of map tasks. The execution of a map task is divided into two phases.

- The map phase reads the task's split from HDFS, parses it into records (key/value pairs), and applies the map function to each record.
- After the map function has been applied to each input record, the commit phase registers the final output with the TaskTracker, which then informs the JobTracker that the task has finished executing.

Figure 1 contains the interface that must be implemented by user-defined map function. After the map function has been applied to each record in the split the close method is invoked.

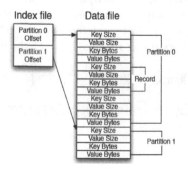

**Fig. 2.** Map task index and data file format (2 partition/reduce case)

The third argument to the map method specifies an OutputCollector instance, which accumulates the output records produced by the map function. The output of the map step is consumed by the reduce step, so the OutputCollector stores map output in a format that is easy for reduce tasks to consume. Intermediate keys are assigned to reducers by applying a partitioning function, so the OutputCollector applies that function to each key produced by the map function, and stores each record and partition number in an in-memory buffer. The OutputCollector spills this buffer to disk when it reaches capacity.

A spill of the in-memory buffer involves first sorting the records in the buffer by partition number and then by key. The buffer content is written to the local file systems an index file and a data file (figure 2). The index file points to the offset of each partition in the data file. The data file contains only the records, which are sorted by the key within each partition segment.

During the commit phase, the final output of the map task is generated by merging all the spill files produced by this task a single pair of data and index files. These files are registered with the Task Tracker before the task completes. The Task Tracker will read these files when servicing requests from reduce tasks.

## 2.5     Reduce Task Execution

The execution of a reduce task is divided into three phases.

- The shuffle phase fetches the reduce task's input data. Each reduce task is assigned a partition of the key range produced by the map step, so the reduce task must fetch the content of this partition from every map task's output.
- The sort phase groups records with the same key together.

```
public interface Reducer<K2, V2, K3, V3>{

    void reducer (K2 key,Iterator<V2> values, OutputCollector<K3, V3> output);

    void close();
}
```

**Fig. 3.** Reduce function interface

- The reduce phase applies the user-defined reduce function to each key and corresponding list of values.

In the shuffle phase, a reduce task fetches data from each map task by issuing HTTP requests to a configurable number of Task Trackers at once(5 by default). The Job Tracker relays the location of every Task Tracker that hosts map output to every Task Tracker that is executing a reduce task. Note that a reduce task cannot fetch the output of a map task until the map has finished executing and committed its final output to disk.

After receiving its partition from all map outputs, the reduce task enters the sort phase. The map output for each partition is already sorted by the reduce key. The reduce task merges these runs together to produce a single run that is sorted by key. The task then enters the reduce phase, in which it invokes the user-defined reduce function for each distinct key in sorted order, passing it the associated list of values. The output of the reduce function is written to a temporary location on HDFS. After the reduce function has been applied to each key in the reduce task's partition, the task's HDFS output file is atomically renamed from its temporary location to its final location.

In this design, the output of both map and reduce tasks is written to disk before it can be consumed. This is particularly expensive for reduce tasks, because their output is written to HDFS. Output materialization simplifies fault tolerance, because it reduces the amount of state that must be restored to consistency after a node failure. If any task (either maps or reduces) fails, the Task Tracker simply schedules a new task to perform the same work as the failed task. Since a task never exports any data other than its final answer, no further recovery steps are needed.

### 2.6    Pipelining Mechanism

In pipelining version [5] of Hadoop they developed the Hadoop online prototype (HOP) that can be used to support continuous queries: MapReduce jobs that run continuously. They also proposed a technique known as online aggregation which can provide initial estimates of results several orders of magnitude faster than the final results. Finally the pipelining can reduce job completion time by up to 25% in some scenarios.

## 3    Motivations

In pipelining mechanism [5] they used naïve implementation to send data directly from map to reduce tasks using TCP. When a client submits a new job to Hadoop, the JobTracker assigns the map and reduce tasks associated with the job to the available TaskTracker slots. They modified Hadoop so that each reduce  task contacts every map task upon initiation of the job and opens a TCP socket which will be used to send the output of the map function. But there may be some drawbacks occurred in TCP connection and TCP congestion during data transmission. For that reason, TCP connection being disconnected and after that data can be retransmitted which takes long time. So we proposed MRA that can send data without retransmission using iSCSI multi-connection and also manage load balancing of data because iSCSI protocol works over TCP. Another motivation is that iSCSI protocol is block I/O based and Hadoop's map task also assigns HDFS block for input process.

## 4    Proposed Model: MapReduce Agent (MRA)

Traditional MapReduce implementation provides a poor interface for interactive data analysis, because they do not emit any output until the map task has been executed to completion. After producing output of map function, our proposed MRA creates multi-connection with reducer rapidly. If one connection falls or data overload problem occurs then the rest of job will distribute to other connections. Our Q-Chained cluster [3] load balancer maintains this job. So that the reducer can continue its mechanism and that reduce job completion time.

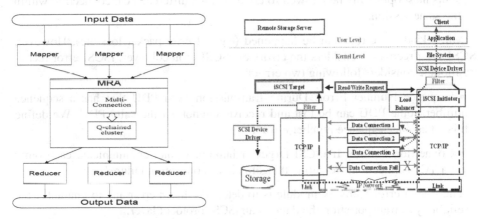

**Fig. 4.** Map Reduce Agent Architecture (MRA)

**Fig. 5.** Overview of Multi-connection and Error Recovery Method of iSCSI [3]

## 4.1  Multi-connection and Error Recovery Method of iSCSI

In order to alleviate the degradation of iSCSI-based remote transfer service caused by TCP congestion control, we propose MRA Multi-Connection and Error Recovery method for one session which uses multiple connections for each session. As mentioned in [8], in a single TCP network connection when congestion occurs by a time-out or the reception of duplicate ACKs (Acknowledgement) then one half of the current window size is saved in *sstresh* (slow start window). Additionally, if the congestion is indicated by a timeout, *cwnd* (congestion window) is set to one segment. This may cause a significant degradation in online MapReduce performance. On the other hand in Multi-Connection case, if TCP congestion occurs within connection, the takeover mechanism selects another TCP connection.

The general overview of the proposed Multi-Connection and Error Recovery based on iSCSI protocol scheme which has been designed for iSCSI based transfer system. When the mapper (worker) is in active mode or connected mode for reduce job that time session is started. This session is indicated to be a collection of multiple TCP connection. If packet losses occur due to bad channel characteristics in any connection, our proposed scheme will pick out Q-Chained Cluster's balanced redistribute data by the other active connections.

### 4.1.1  Error Recovery Procedure in iSCSI Protocol

Error recovery is strongly required for iSCSI protocol. The following two considerations prompted the design of much of the error recovery functionality in iSCSI [9].

- PDU may fail the digest check and be dropped, despite being received by the TCP layer. The iSCSI layer must optionally be allowed to recover such dropped PDUs.

- A TCP connection may fail at any time during the data transfer. All the active tasks must optionally be allowed to continue on a different TCP connection within the same session.

Many kinds of errors can be happened (e.g. bit error,packet loss etc). However, iSCSI error recovery considers the errors on iSCSI protocol layer. iSCSI error recovery module considers following two errors:

- Sequence Number Error: During transmission Data PDU that has a sequence number, some PDU can be lost and receiver cannot get the valid PDU. We define this situation, "sequence number error".

- Connection Failure: If iSCSI target or initiator cannot communicate each other via a TCP connection, we define this situation, "connection failure".

The role of error recovery module is to detect the listed error and to guarantee the reliability of transportation the data on an iSCSI protocol layer.

*Error Recovery Procedure.* iSCSI protocol with error recovery checks the sequence number of every iSCSI PDU. If iSCSI target or initiator receives an iSCSI PDU with an out of order sequence number, then it requests an expected sequence number PDU again. In Connection failure case, when a connection has no data communication during the engaged time, iSCSI protocol with error recovery checks the connection status by the nop-command [9]. We assume the multiple connections.

*Sequence Number Error.* When an initiator receives an iSCSI status PDU with an out of order or a SCSI response PDU with an expected data sequence number (ExpDataSN) that implies missing data PDU(s), it means that the initiator detected a header or payload digest error one or more earliest ready to transmission (R2T) PDUs or data PDUs. When a target receives a data PDU with an out of order data sequence number (DataSN), it means that the target must have hit a header or payload digest error on at least one of the earlier data PDUs. The target must discard the PDU and request retransmission with recovery R2T.

The following cases lend themselves to connection recovery:

- TCP connection failure: The initiator must close the connection. It then must either implicitly or explicitly logout the failed connection with the reason code "remove the connection for recovery" and reassign connection allegiance for all commands still in progress associated with the failed connection on one or more connections. For an initiator, a command is in progress as long as it has not received a response or a Data-in PDU including status.

- Receiving an Asynchronous Message [9] that indicates one or all connections in a session has been dropped. The initiator must handle it as a TCP connection failure for the connection(s) referred to in the Message. At an iSCSI target, the following cases lend themselves to connection recovery

• TCP connection failure: The target must close the connection and, if more than one connection is available, the target should send an Asynchronous Message that indicates it has dropped the connection. Then, the target will wait for the initiator to continue recovery

## 4.2    Q-Chained Cluster Load Balancer

Q-chained cluster is able to balance the workload fully among data connections in the event of packet losses due to bad channel characteristics. When congestion occurs in a data connection, this module can do a better job of balancing the workload which is originated by congestion connection, will be distributed among N-1 connections instead of a single data connection. However, when congestion occurs in a specific data connection, balancing the workload among the remaining connections can become difficult, as one connection must pick up the workload of the component where it takes place. In particular, unless the data placement scheme used allows the workload, which is originated by congestion connection to be distributed among the remaining operational connections.

Figure 6 illustrates how the workload is balanced in the event of congestion occurrence in a data connection (data connection 1 in this example) with Q-chained cluster. For example, with the congestion occurrence of data connection 1, primary data Q1 is no longer transmitted in congestion connection for the TCP input rate to be throttled and thus its recovery data q1 of data connection 1 is passed to data connection 2 for conveying storage data. However, instead of requiring data connection 2 to process all data both Q2 and q1, Q-chained cluster offloads 4/5ths of the transmission of Q2 by redirecting them to q2 in data connection 3. In turn, 3/5ths of the transmission of Q3 in data connection 3 are sent to q3. This dynamic reassignment of the workload results in an increase of 1/5th in the workload of each remaining data connection.

| Data connection | 0 | 1 | 2 | 3 | 4 | 5 |
|---|---|---|---|---|---|---|
| Primary Data | $Q_0$ | F | $1/5Q_2$ | $2/5Q_3$ | $3/5Q_4$ | $4/5Q_5$ |
| Recovery Data | $1/5q_5$ | F | $q_1$ | $4/5q_2$ | $3/5q_3$ | $2/5q_2$ |

**Fig. 6.** Q-Chained Load Balancer

## 4.3    MRA between Jobs

Although MapReduce was originally designed as a batch oriented system [5], it is often used for interactive data analysis. A user submits a job to extract information from a data set. Traditional MapReduce implementation provides a poor interface for interactive data analysis, because they do not emit any output until the map task has been executed to completion.But in MRA, the data records produced by map tasks are sent to reduce tasks shortly after each record is generated [see the figure 8: the flow-chart of MRA]. As a result we can produce output more quickly.

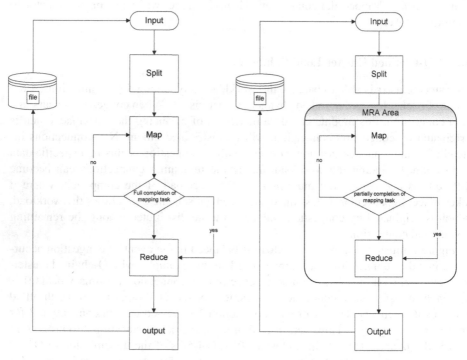

**Fig. 7.** Flow Chart of Hadoop          **Fig. 8.** Flow Chart of MRA

## 4.4    Continuous Map Reduce Jobs

A bare-bones implementation of continuous MapReduce jobs is easy to implement using MRA. No changes are needed to implement continuous map tasks: map output is already delivered to the appropriate reduce task shortly after it was generated. Our implemented MRA that allows map functions to force their current output to reduce tasks. When a reduce task is unable to accept such data, the mapper framework stores it locally and sends it after few time. With proper scheduling of reducers, this MRA allows a map task to ensure that an output record is promptly sent to the appropriate reducer. To support continuous reduce tasks, the user-defined reduce function must be periodically invoked on the map output available at that reducer. Applications will have different requirements for how frequently the reduce function should be in-voked; possible choices include periods based on wall-clock time, logical time (e.g., the value of a field in the map task output), and the number of input rows delivered to the reducer. The output of the reduce function can be written to HDFS.

In our current implementation, the number of map and reduce tasks is fixed, and must be configured by the user. To maintain workload in remote transfer MRA creates Q-chained cluster

## 4.5    Fault Tolerance

Our MRA Hadoop implementation is robust to the failure of both map and reduces tasks. To recover from map task failures, we added bookkeeping to the reduce task to record which map task produced each MRA spill file. To simplify fault tolerance, the reducer treats the output of a MRA map task as "tentative" until the JobTracker informs the reducer that the map task has committed successfully. The reducer can merge together spill files generated by the same uncommitted mapper but will not combine those spill files with the output of other map tasks until it has been notified that the map task has committed. Thus, if a map task fails, each reduce task can ignore any tentative spill files produced by the failed map attempt. The JobTracker will take care of scheduling a new map task attempt, as in stock Hadoop. If a reduce task fails and a new copy of the task is started, the new reduce instance must be sent all the input data that was sent to the failed reduce attempt. To reduce transmission failure we used iSCSI multi-connection so that output of mapper data can transmit within a moment and avoid load balance Q-chained cluster provide better performance.

# 5    Performance Evaluation

As per as [5] we also evaluate the effectiveness of online aggregation, we performed two experiments on Amazon EC2 using different data sets and query workloads. In their first experiment [5], they wrote a "Top-K" query using two MapReduce jobs: the first job counts the frequency of each word and the second job selects the K most frequent words. We ran this workload on 5.5GB of Wikipedia article text stored in HDFS, using a 128MB block size. We used a 60-node EC2 cluster; each node was a "high-CPU medium" EC2 instance with 1.7GB of RAM and 2 virtual cores. A virtual core is the equivalent of a 2007-era 2.5Ghz Intel Xeon processor. A single EC2 node executed the Hadoop Job- Tracker and the HDFS NameNode, while the remaining nodes served as slaves for running the TaskTrackers and HDFS DataNodes.

A thorough performance comparison between pipelining, blocking and MRA is beyond the scope of this paper. In this section, we instead demonstrate that MRA can reduce job completion times in some configurations. We report performance using both large (512MB) and small (32MB) HDFS block sizes using a single workload (a word count job over randomly-generated text). Since the words were generated using a uniform distribution, map-side combiners were ineffective for this workload. We performed all experiments using relatively small clusters of Amazon EC2 nodes. We also did not consider performance in an environment where multiple concurrent jobs are executing simultaneously.

## 5.1    Performance Results of iSCSI Protocol

Our proposed scheme throughputs in different RTTs are measured for each number of connections in Figure 9. We see the slowness of the rising rate of throughput between 8 connections and 9 connections. This shows that reconstructing the data in turn influences throughputs and the packet drop rates are increased when the number of TCP connections is 9 as the maximum use of concurrent connections between initiator and target.

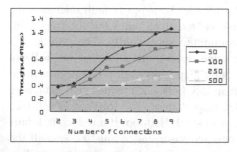 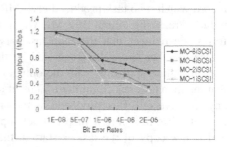

**Fig. 9.** Throughput of Multi-Connection iSCSI System. Y axis is containing Throughput easurement With Mbps & X axis is for number of connections. 50,100,250 and 500 RTT are measured by ms.

**Fig. 10.** Throughput of Multi-Connection iSCSI vs iSCSI At different error rates. Y axis is Throughput & X axis is for Bit error rate.

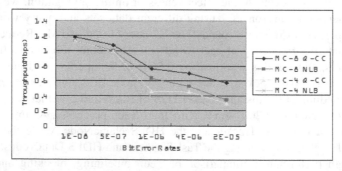

**Fig. 11.** Q-Chained Cluster Load Balancer vs No Load Balancer. MC: Multi Connection, Q-CC: Q-Chained Cluster NLB: No Load Balancer.

Therefore, 8 is the maximum optimal number of connections from a performance point of view. Multi-Connection iSCSI mechanism also works effectively because the data transfer throughputs increase linearly when the round trip time is larger than 250ms.

In Figure 10, the performance comparison of Multi-Connection iSCSI and iSCSI at different bit-error rates is shown. We see that for bit-error rates of over $5.0 \times 10^{-7}$ the Multi-Connection iSCSI (2 connections) performs significantly better than the iSCSI

(1 connection), achieving a throughput improvement about 24 % in SCSI read. More-over, as bit-error rates go up, the figure shows that the rising rate of throughput is getting higher at 33% in 1.0×10-6, 39.3% in 3.9×10-6and 44% in 1.5×10-5. Actually, Multi-Connection iSCSI can avoid the forceful reduction of transmission rate effi-ciently from TCP congestion control using another TCP connection opened during a service session, while iSCSI does not make any progress. Under statuses of low bit error rates (< 5.0×10-7), we see little difference between Multi-Connection iSCSI and iSCSI. At such low bit errors iSCSI is quite robust at handling these.

In Figure 11, Multi-Connection iSCSI(8 connections) with Q-Chained cluster shows the better average performance about 11.5%. It can distribute the workload among all remaining connections when packet losses occur in any connection. To recall an example given earlier, with M = 6, when congestion occurs in a specific connection, the workload of each connection increases by only 1/5. However, if Mul-ti-Connection iSCSI (proposed Scheme) establishes a performance baseline without load balancing, any connection, which is randomly selected from takeover mechan-ism, is overwhelmed.

## 5.2    Performance Results and Comparison on MapReduce

In the Hadoop map reduce architecture [4, 5]; their first task is to generate output which is done by map task consume the output by reduce task. The whole thing makes the process lengthy because reduce task have to wait for the output of the map task. In pipelining mechanism [5], they send output of map task immediately after generation of per output to the reduce task so it takes less time than Hadoop MapRe-duce[4]. During the transmission (TCP) if any problem occurred then they retransmit again which took more time and drastically reduce the performance of MapReduce mechanism.

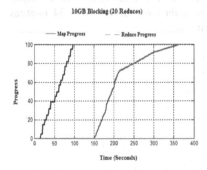

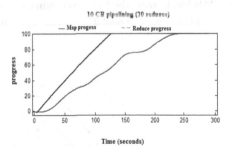

**Fig. 12.** CDF of map and reduce task comple-tion times for a 10GB wordcount job using 20 map tasks and 20 reduce tasks (512MB block size). The total job runtimes were 361 seconds for blocking.

**Fig. 13.** CDF of map and reduce task com-pletion times for a 10GB wordcount job using 20 map tasks and 20 reduce tasks (512MB block size). The total job runtimes were 290 seconds for pipelining.

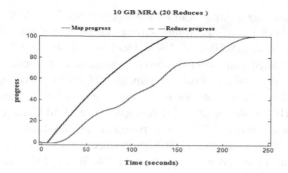

**Fig. 14.** CDF of map and reduce task completion times for a 10GB wordcount job using 20 map tasks and 20 reduce tasks (512MB block size). The total job runtimes were 240 seconds for MRA.

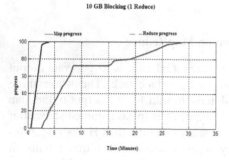

**Fig. 15.** CDF of map and reduce task completion times for a 10GB wordcount job using 20 map tasks and 1 reduce tasks (512MB block size). The total job runtimes were 29 minutes for blocking.

**Fig. 16.** CDF of map and reduce task completion times for a 10GB wordcount job using 20 map tasks and 1 reduce tasks (512MB block size). The total job runtimes were 34 minutes for pipelining.

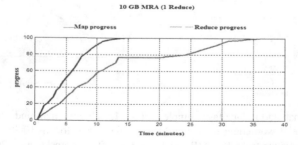

**Fig. 17.** CDF of map and reduce task completion times for a 10GB wordcount job using 20 map tasks and 1 reduce tasks (512MB block size). The total job runtimes were 36 minutes for MRA.

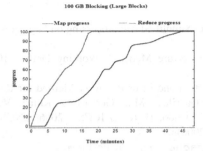

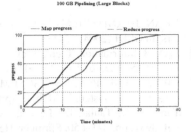

**Fig. 18.** CDF of map and reduce task completion times for a 100GB wordcount job using 240 map tasks and 60 reduce tasks (512MB block size). The total job runtimes were 48 minutes for blocking.

**Fig. 19.** CDF of map and reduce task completion times for a 100GB wordcount job using 240 map tasks and 60 reduce tasks (512MB block size). The total job runtimes were 36 minutes for pipelining.

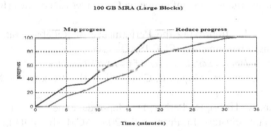

**Fig. 20.** CDF of map and reduce task completion times for a 100GB wordcount job using 240 map tasks and 60 reduce tasks (512MB block size). The total job runtimes were 32 minutes for MRA.

On the other hand our proposed mechanism (MRA) recovers the drawback by using multi-connection and Q-chained load balancer method. In these circumstances MRA may prove its better time of completion.

# 6    Conclusion

MapReduce has added new dimension for large scale parallel programming. Our paper demonstrated that MapReduce can be more useful if we use MRA. We attribute this success to several reasons. First, the model is easy to use, even for programmers without experience with parallel and distributed systems, since it hides the details of parallelization, fault-tolerance, locality optimization, and load balancing. Second, we have developed an implementation of MapReduce that scales to large clusters of machines comprising thousands of machines. The implementation makes efficient use of these machine resources and therefore is suitable for use on many of the large computational problems. Third, MRA can reduce the time to job completion.

# References

[1]  DEAN, J., AND GHEMAWAT, S. MapReduce: Simplified dataprocessing on large clusters. In OSDI (2004).

[2]  SAM-3 Information Technology – SCSI Architecture Model 3, Working Draft, T10 Project 1561-D, Revision7 (2003)

[3]  Allayear, S.M., Park, S.S.: iSCSI Multi-connection and Error Recovery Method for Remote Storage System in Mobile Appliance. In: Gavrilova, M.L., Gervasi, O., Kumar, V., Tan, C.J.K., Taniar, D., Laganá, A., Mun, Y., Choo, H. (eds.) ICCSA 2006. LNCS, vol. 3981, pp. 641–650. Springer, Heidelberg (2006)

[4]  Hadoop, HYPERLINK, http://hadoop.apache.org/mapreduce/

[5]  Condie, T., Conway, N., Alvaro, P., Hellerstein, J.M.: UC Berkeley: MapReduce Online. Khaled Elmeleegy, Russell Sears (Yahoo! Research)

[6]  Allayear, S.M., Park, S.S., No, J.: iSCSI Protocol Adaptation with 2-way TCP Hand Shake Mechanism for an Embedded Multi-Agent Based Health Care Service. In: Proceedings of the 10th WSEAS International Conference on Mathematical Methods, Computational Techniques and Intelligent Systems, Corfu, Greece (2008)

[7]  Allayear, S.M., Park, S.S.: iSCSI Protocol Adaptation With NAS System Via Wireless Environment. In: International Conference on Consumer Electronics (ICCE), Las Vegus, USA (2008)

[8]  Caceres, R., Iftode, L.: Improving the Performance of Reliable Transport Protocols in Mobile Computing Environments. IEEE JSAC

[9]  RFC 3270, http://www.ietf.org/rfc/rfc3720.txt

[10]  Verma, A., Zea, N., Cho, B., Gupta, I., Campbell, R.H.: Breaking the MapReduce Stage Barrier*

[11]  Yang, H., Dasdan, A., Hsiao, R., Parker, D.: Map-reduce-merge: simplified relational data processing on large clusters. In: Proc. of the 2007 ACM SIGMOD International Conference on Management of Data (January 2007)

[12]  Hellerstein, J.M., Haas, P.J., Wang, H.J.: Online aggregation. In: SIGMOD (1997)

[13]  Shah, M.A., Hellerstein, J.M., Brewer, E.A.: Highly-available, fault-tolerant, parallel dataflows. In: SIGMOD (2004)

[14]  Thusoo, A., Sarma, J.S., Jain, N., Shao, Z., Chakka, P., Anthony, S., Liu, H., Wyckoff, P., Murthy, R.: Hive—a warehousing solution over a Map-Reduce framework. In: VLDB (2009)

[15]  Wu, S., Jiang, S., Ooi, B.C., Tan, K.-L.: Distributed online aggregation. In: VLDB (2009)

[16]  Yang, C., Yen, C., Tan, C., Madden, S.: Osprey: Implementing MapReduce-style fault tolerance in a shared-nothing distributed database. In: ICDE (2010)

[17]  Chan, J.O.: An Architecture for Big Data Analytics

[18]  Daneshyar, S., Razmjoo, M.: Large-Scale Data Processing Using Mapreduce in Cloud Computing Environment

[19]  Ji, C., Li, Y., Qiu, W., Awada, U., Li, K.: Big Data Processing in Cloud Computing Environments

[20]  Padhy, R.P.: Big Data Processing with Hadoop-MapReduce in Cloud Systems

[21]  Stokely, M.: Histogram tools for distributions of large data sets

# Big Data Clustering: A Review

Ali Seyed Shirkhorshidi[1], Saeed Aghabozorgi[1],
Teh Ying Wah[1], and Tutut Herawan[1,2]

[1] Department of Information Systems
Faculty of Computer Science and Information Technology
University of Malaya
50603 Pantai Valley, Kuala Lumpur, Malaysia
[2] AMCS Research Center, Yogyakarta, Indonesia
shirkhorshidi_ali@siswa.um.edu.my,
{saeed,tehyw,tutut}@um.edu.my

**Abstract.** Clustering is an essential data mining and tool for analyzing big data. There are difficulties for applying clustering techniques to big data duo to new challenges that are raised with big data. As Big Data is referring to terabytes and petabytes of data and clustering algorithms are come with high computational costs, the question is how to cope with this problem and how to deploy clustering techniques to big data and get the results in a reasonable time. This study is aimed to review the trend and progress of clustering algorithms to cope with big data challenges from very first proposed algorithms until today's novel solutions. The algorithms and the targeted challenges for producing improved clustering algorithms are introduced and analyzed, and afterward the possible future path for more advanced algorithms is illuminated based on today's available technologies and frameworks.

**Keywords:** Big Data, Clustering, MapReduce, Parallel Clustering.

## 1   Introduction

After an era of dealing with data collection challenges, nowadays the problem is changed into the question of how to process these huge amounts of data. Scientists and researchers believe that today one of the most important topics in computing science is Big Data. Social networking websites such as Facebook and Twitter have billions of users and they produce hundreds of gigabytes of contents per minute, retail stores continuously collect their customers' data, You Tube has 1 billion unique users which are producing 100 hours of video each an hour and its content ID service scans over 400 years of video every day [1], [2]. To deal with this avalanche of data, it is necessary to use powerful tools for knowledge discovery. Data mining techniques are well-known knowledge discovery tools for this purpose [3]–[9]. Clustering is one of them that is defined as a method in which data are divided into groups in a way that objects in each group share more similarity than with other objects in other groups [1]. Data clustering is a well-known technique in various areas of computer science and related domains. Although data mining can be considered as the main origin of

B. Murgante et al. (Eds.): ICCSA 2014, Part V, LNCS 8583, pp. 707–720, 2014.

clustering, but it is vastly used in other fields of study such as bio informatics, energy studies, machine learning, networking, pattern recognition and therefore a lot of research works has been done in this area [10]–[13]. From the very beginning researchers were dealing with clustering algorithms in order to handle their complexity and computational cost and consequently increase scalability and speed. Emersion of big data in recent years added more challenges to this topic which urges more research for clustering algorithms improvement. Before focusing on clustering big data the question which needs to be clarified is how big the big data is. To address this question Bezdek and Hathaway represented a categorization of data sizes which is represented in table 1 [14].

**Table 1.** Bezdek and Hathaway categorization for big data

| | | | Big data | | |
|---|---|---|---|---|---|
| Bytes | $10^6$ | $10^8$ | $10^{10}$ | $10^{12}$ | $10^{>12}$ |
| "Size" | Medium | Large | Huge | Monster | Very large |

Challenges of big data have root in its five important characteristics [15]:

- **Volume:** The first one is Volume and an example is the unstructured data streaming in form of social media and it rises question such as how to determine the relevance within large data volumes and how to analyze the relevant data to produce valuable information.
- **Velocity:** Data is flooding at very high speed and it has to be dealt with in reasonable time. Responding quickly to data velocity is one of the challenges in big data.
- **Variety:** Another challenging issue is to manage, merge and govern data that comes from different sources with different specifications such as: email, audio, unstructured data, social data, video and etc.
- **Variability:** Inconsistency in data flow is another challenge. For example in social media it could be daily or seasonal peak data loads which makes it harder to deal and manage the data specially when the data is unstructured.
- **Complexity:** Data is coming from different sources and have different structures; consequently it is necessary to connect and correlate relationships and data linkages or you find your data to be out of control quickly.

Traditional clustering techniques cannot cope with this huge amount of data because of their high complexity and computational cost. As an instance, the traditional K-means clustering is NP-hard, even when the number of clusters is k=2. Consequently, scalability is the main challenge for clustering big data.

The main target is to scale up and speed up clustering algorithms with minimum sacrifice to the clustering quality. Although scalability and speed of clustering algorithms were always a target for researchers in this domain, but big data challenges underline these shortcomings and demand more attention and research on this topic. Reviewing the literature of clustering techniques shows that the advancement of these techniques could be classified in five stages as shown in Figure 1.

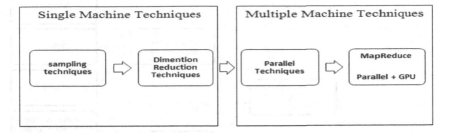

**Fig. 1.** Progress of developments in clustering algorithms to deal with big data

In rest of this study advantages and drawbacks of algorithms in each stage will be discussed as they appeared in the figure respectively. In conclusion and future works we will represent an additional stage which could be the next stage for big data clustering algorithms based on recent and novel methods.

Techniques that are used to empower clustering algorithms to work with bigger datasets trough improving their scalability and speed can be classified into two main categories:

- Single-machine clustering techniques
- Multiple-machine clustering techniques

Single machine clustering algorithms run in one machine and can use resources of just one single machine while the multiple-machine clustering techniques can run in several machines and has access to more resources. In the following section algorithms in each of these categories will be reviewed.

## 2    Big Data Clustering

In general, big data clustering techniques can be classified into two major categories: single-machine clustering techniques and multiple-machine clustering techniques. Recently multiple machine clustering techniques has attracted more attention because they are more flexible in scalability and offer faster response time to the users. As it is demonstrated in Fig. 2 single-machine and multiple-machine clustering techniques include different techniques:

- Single-machine clustering
  - Sample based techniques
  - Dimension reduction techniques
- Multiple-machine clustering
  - Parallel clustering
  - MapReduce based clustering

In this section advancements of clustering algorithms for big data analysis in categories that are mentioned above will be reviewed.

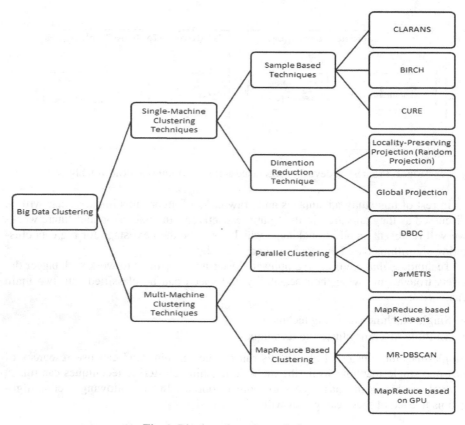

**Fig. 2.** Big data clustering techniques

## 2.1    Single-Machine Clustering Techniques

### 2.1.1  Sampling Based Techniques

These algorithms were the very first attempts to improve speed and scalability of them and their target was dealing with exponential search space. These algorithms are called as sampling based algorithms, because instead of performing clustering on the whole dataset, it performs clustering algorithms on a sample of the datasets and then generalize it to whole dataset. This will speed up the algorithm because computation needs to take place for smaller number of instances and consequently complexity and memory space needed for the process decreases.

*Clustering Large Applications based on Randomized Sampling (CLARANS)*
Before introducing CLARANS [16] let us take a look at its predecessor Clustering Large Applications (CLARA) [17]. Comparing to partitioning around medoids (PAM) [17], CLARA can deal with larger datasets. CLARA decreases the overall quadratic complexity and time requirements into linear in total number of objects. PAM calculates the entire pair-wise dissimilarity matrix between objects and store it in central memory, consequently it consume $O(n^2)$ memory space, thus PAM cannot

be used for large number of n. To cope with this problem, CLARA does not calculate the entire dissimilarity matrix at a time. PAM and CLARA can be regarded conceptually as graph-searching problems in which each node is a possible clustering solution and the criteria for two nodes for being linked is that they should be differ in 1 out of k medoids.

PAM starts with one of the randomly chosen nodes and greedily moves to one of its neighbors until it cannot find a better neighbor. CLARA reduce the search space by searching only a sub-graph which is prepared by the sampled O (k) data points.

CLARANS is proposed in order to improve efficiency in comparison to CLARA. like PAM , CLARANS, aims to find a local optimal solution by searching the entire graph, but the difference is that in each iteration, it checks only a sample of the neighbors of the current node in the graph. Obviously, CLARA and CLARANS both are using sampling technique to reduce search space, but their difference is in the way that they perform the sampling. Sampling in CLARA is done at the beginning stage and it restricts the whole search process to a particular sub-graph while in CLARANS; the sampling is conduct dynamically for each iteration of the search procedure. Observation results shows that dynamic sampling used in CLARANS is more efficient than the method used in CLARA [18].

*BIRCH*
If the data size is larger than the memory size, then the I/O cost dominates the computational time. BIRCH [19] is offering a solution to this problem which is not addressed by other previously mentioned algorithms. BIRCH uses its own data structure called clustering feature (CF) and also CF-tree. CF is a concise summary of each cluster. It takes this fact into the consideration that every data point is not equally important for clustering and all the data points cannot be accommodated in main memory.

CF is a triple <N,LS,SS> which contains the number of the data points in the cluster, the linear sum of the data points in the cluster and the square sum of the data points in the cluster, the linear sum of the data points in the cluster and the square sum of the data points in the cluster. Checking if CF satisfies the additive property is easy, if two existing clusters need to be merged, the CF for the merged cluster is the sum of the CFs of the two original clusters. The importance of this feature is for the reason that it allows merging two existing clusters simply without accessing the original dataset.

There are two key phase for BIRCH algorithm. First, it scans the data points and build an in memory tree and the second applies clustering algorithm to cluster the leaf nodes. Experiments conducted in [20] reveal that in terms of time and space, BIRCH performs better than CLARANS and it is also more tolerable in terms of handling outliers. Fig. 3 represents a flowchart which demonstrates steps in BIRCH algorithm.

*CURE*
Single data point is used to represent a cluster in all previously mentioned algorithms which means that these algorithms are working well if clusters have spherical shape, while in the real applications clusters could be from different complex shapes. To deal with this challenge, clustering by using representatives (CURE) [21] uses a set of

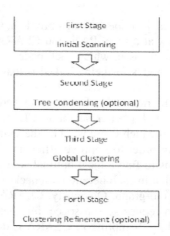

**Fig. 3.** BIRCH algorithm flowchart

well-scattered data points to represent a cluster. In fact CURE is a hierarchical algorithm. Basically it considers each data point as a single cluster and continually merges two existing clusters until it reaches to precisely k clusters remaining. The process of selecting two clusters for merging them in each stage is based on calculating minimum distance between all possible pairs of representative points from the two clusters. Two main data structures empower CURE for efficient search. Heap is the first, which is used to track the distance for each existing cluster to its closest cluster and the other is k-d tree which is used to store all the representative points for each cluster.

CURE also uses sampling technique to speed up the computation. It draws a sample of the input dataset and run mentioned procedure on the sample data. To clarify the necessary sample size Chernoff bound is used in the original study. If the dataset is very large, even after sampling, the data size is still big and consequently the process will be time consuming. To solve this issue, CURE uses partitions to accelerate the algorithm. If we consider $n$ as the original dataset and the $n'$ as the sampled data, then CURE will partition the $n'$ into $p$ partitions and within each partition it runs a partial hierarchical clustering until either a predefined number of clusters is reached or the distance between the two clusters to be merged exceeds some threshold. Then another clustering runs and passes on all partial clusters from all the $p$ partitions. At the final stage all non-sampled data points will assign to the nearest clusters. Results [21] represent that in comparison to BIRCH, execution time for CURE is lower while it maintain the privilege of robustness in handling outliers by shrinking the representative points to the centroid of the cluster with a constant factor. Fig. 4 demonstrate the flowchart of CURE algorithm.

### 2.1.2 Dimension Reduction Techniques

Although the complexity and speed of clustering algorithms is related to the number of instances in the dataset, but at the other hand dimensionality of the dataset is other influential aspect. In fact the more dimensions data have, the more is complexity and it means the longer execution time. Sampling techniques reduce the dataset size but they do not offer a solution for high dimensional datasets.

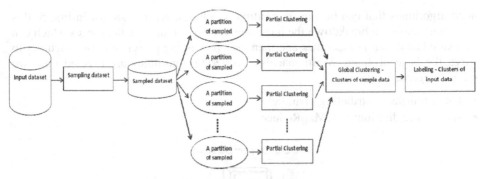

**Fig. 4.** CURE flowchart

*Locality-Preserving Projection (Random Projection)*
In this method, after projecting the dataset from d-dimensional to a lower dimensional space, it is desired that the pairwise distance to be roughly preserved. Many clustering algorithms are distance based, so it is expected that the clustering result in the projected space provide an acceptable approximation of the same clustering in the original space. Random projection can be accomplished by a linear transformation of the data matrix $A$, which is the original data matrix. If $R$ be a $d \times t$ and rotation matrix ($t << d$) and its elements $R(i, j)$ be independent random variables, then $A' = A.R$ is the projection of matrix $A$ in $t$ dimensional space. It means each row in $A'$ has $t$ dimensions. Construction of rotation matrix is different in different random projection algorithms. Preliminary methods propose Normal random variables with mean 0 and variance 1 for $R(i, j)$ although there are studies which represent different methods for allocating values to $R(i, j)$ [22]. After generating the projected matrix into lower dimensional space, clustering can be performed. Samples of algorithm implementation is illustrated in [23], [24] and recently a method is proposed to speed up the computation [25].

*Global Projection*
In global projection the objective is for each data point it is desirable that the projected data point to be as close as possible to original data point while in local preserving projection the objective was to maintain pairwise distances roughly the same between two pairs in projected space. In fact if the main dataset matrix is considered to be $A$ and $A'$ be the approximation of it, in global projection aim is to ize $||A' - A||$. Different approaches are available to create the approximation matrix such as SVD (singular value decomposition) [26], CX/CUR [27], CMD [28] and Colibri [29].

## 2.2   Multi-machine Clustering Techniques

Although sampling and dimension reduction methods used in single-machine clustering algorithms represented in previous section improves the scalability and speed of the algorithms, but nowadays the growth of data size is way much faster than memory and processor advancements, consequently one machine with a single processor and a memory cannot handle terabytes and petabytes of data and it underlines the need algorithms that can be run on multiple machines. As it is shown in Fig. 5, this technique allows to breakdown the huge amount of data into smaller pieces which can be loaded on different machines and then uses processing power of these machines to solve the huge problem. Multi machine clustering algorithms are divided into two main categories:

- Un-automated distributing– parallel
- Automated distributing– MapReduce

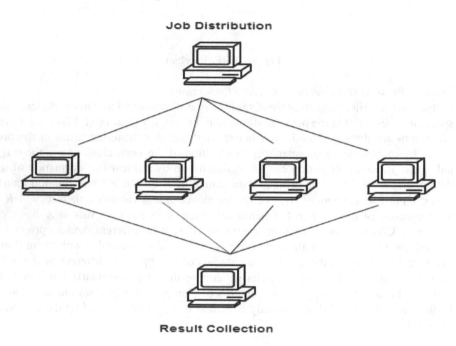

**Fig. 5.** General concept of multi-machine clustering techniques

In parallel clustering, developers are involved with not just parallel clustering challenges, but also with details in data distribution process between different machines available in the network as well, which makes it very complicated and time consuming. Difference between parallel algorithms and the MapReduce framework is in the comfortless that MapReduce provides for programmers and reveals them form unnecessary networking problems and concepts such as load balancing, data distribution, fault tolerance and etc. by handling them automatically. This feature allows huge parallelism and easier and faster scalability of the parallel system. Parallel and distributed clustering algorithms follows a general cycle as represented below:

In the first stage, data is going to be divided into partitions and they distribute over machines. Afterward, each machine performs clustering individually on the assigned partition of data. Two main challenges for parallel and distributed clustering are minimizing data traffic and its lower accuracy in comparison with its serial equivalent.

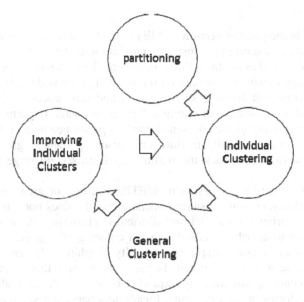

**Fig. 6.** General cycle for multi-machine clustering algorithms

Lower accuracy in distributed algorithms could be caused by two main reasons, first, it is possible that different clustering algorithms deploy in different machines and secondly even if the same clustering algorithm is used in all machines, in some cases the divided data might change the final result of clustering. In the rest of this study, parallel algorithms and MapReduce algorithms will be discussed subsequently, then more advanced algorithms proposed recently for big data will be covered.

## 2.3 Parallel Clustering

Although parallel algorithms add difficulty of distribution for programmers, but it is worthfull because of the major improvements in scaling and speed of clustering algorithms. At the following parts some of them will be reviewed.

*DBDC*
DBDC [30], [31] is a distributed and density based clustering algorithm. Discovery of clusters of arbitrary shapes is the main objective of density based clustering. The density of points in each cluster is much higher than outside of the cluster, while the density of the regions of noise is lower than the density in any of the clusters. DBDC [32] is an algorithm which obeys the cycle mentioned in Figure 2 . At the individual clustering stage, it uses a defined algorithm for clustering and then for general clustering, a single machine density algorithm called DBSCAN is used for finalizing the results. The results show that although DBDC maintain the same clustering quality in comparison to its serial interpretation, but it runs 30 times faster than that.

*ParMETIS*

ParMETIS [33] is the parallel version of METIS [34] and is a multilevel partitioning algorithm. Graph partitioning is a clustering problem with the goal of finding the good cluster of vertices. METIS contains three main steps. First step is called as coarsening phase. In this stage maximal matching on the original graph is done and then the vertices which are matched create a smaller graph and this process is iterated till the number of vertices become small enough. The second stage is partitioning stage in which k-way partitioning of the coarsened graph is performed using multilevel recursive bisection algorithm. Finally in third un-coarsening stage, a greedy refinement algorithm is used to project back the partitioning from second stage to the original graph.

ParMETIS is a distributed version of METIS. Because of graph based nature of ParMETIS it is different from clustering operations and it does not follow the general cycle mentioned earlier for parallel and distributed clustering. An equal number_of vertices are going to distribute initially, then a coloring of a graph will compute in machines. Afterward, a global graph incrementally matching only vertices of the same color one at a time will be computed. In partitioning stage this graph broadcast to machines and recursive bisection by exploring only a single path of the recursive bisection tree performs in each machine. Finally un-coarsening stage is consisting of moving vertices of edge-cut. Experiments represent that ParMETIS was 14 to 35 times faster than serial algorithm while maintaining the quality close to the serial algorithm.

*GPU based parallel clustering*

A new issue is opened recently in parallel computing to use processing power of GPU instead of CPU to speed up the computation. G-DBSCAN [35] is a GPU accelerated parallel algorithm for density-based clustering algorithm, DBSCAN. It is one of the recently proposed algorithms in this category. Authors distinguished their method by using a graph based data indexing to add flexibility to their algorithm to allow more parallelization opportunities. G-DBSCAN is a two-step algorithm and both of these steps have been parallelized. The first step constructs a graph. Each object represents a node and an edge is created between two objects if their distance is lower than or equal to a predefined threshold. When this graph is ready, the second step is to identify the clusters. It uses breath first search (BFS) to traverse the graph created in the first step. Results show that in comparison to its serial implementation, G-DBSCAN is 112 times faster.

## 2.4    MapReduce

Although parallel clustering algorithms improved the scalability and speed of clustering algorithms still the complexity of dealing with memory and processor distribution was a quiet important challenge. MapReduce is a framework which is illustrated in Fig. 7 initially represented by Google and Hadoop is an open source version of it [36]. In this section algorithms which are implemented based on this framework are reviewed and their improvements are discussed in terms of three features:

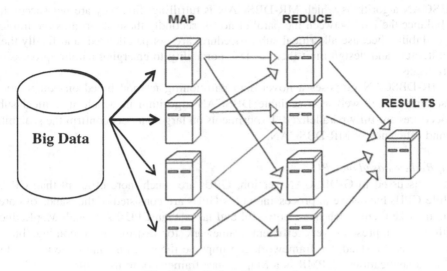

**Fig. 7.** MapReduce Framework

- **Speed up:** means the ratio of running time while the dataset remains constant and the number of machines in the system is increased.
- **Scale up:** measures if x time larger system can perform x time larger job with the same run time
- **Size up:** keeping the number of machines unchanged, running time grows linearly with the data size

*MapReduce based K-means (PK-means)*
PKMeans [37] is distributed version of well-known clustering algorithm K-means [38], [39]. The aim of k-means algorithm is to cluster the desire dataset into k clusters in the way that instances in one cluster share more similarity than the instances of other clusters. K-means clustering randomly choose k instance of the dataset in initial step and performs two phases repeatedly: first it assigns each instance to the nearest cluster and after finishing the assignment for all of the instances in the second phase it updates the centers for each cluster with the mean of the instances.
PKMeans distributes the computation between multiple machines using MapReduce framework to speed up and scale up the process. Individual clustering which contain the first phase happens in the mapper and then general clustering perform second phase in the reducer.

PKMeans has almost linear speed up and also a linear size up. It also has a good scale up. For 4 machines it represented a scale up of 0.75. At the other hand, PKMeans is an exact algorithm, it means that it offer the same clustering quality as its serial counterpart k-means.

*MR-DBSCAN*
A very recent proposed algorithm is MR-DBSCAN [40] which is a scalable MapReduce-based DBSCAN algorithm. Three major draw backs are existed in parallel

DBSCAN algorithms which MR-DBSCAN is fulfilling: first they are not successful to balance the load between the parallel nodes, secondly these algorithms are limited in scalability because all critical sub procedures are not parallelized, and finally their architecture and design limit them to less portability to emerging parallel processing paradigms.

MR-DBSCAN proposes a novel data partitioning method based on computation cost emission as well as a scalable DBSCAN algorithm in which all critical sub-procedures are fully parallelized. Experiments on large datasets confirm the scalability and efficiency of MR-DBSCAN.

*MapReduce based on GPU*

As it discussed in G-DBSCAN section, GPUs are much more efficient than CPUs. While CPUs have several processing cores GPUs are consisted of thousands of cores which make them much more powerful and faster than CPUs. Although MapReduce with CPUs represents very efficient framework for distributed computing, but if GPUs is used instead, the framework can improve the speed up and scale up for distributed applications. GPMR is a MapReduce framework to use multiple GPUs. Although clustering applications are not still implemented in this framework, but the growth of data size urge researcher to represent faster and more scalable algorithms so maybe this framework could be the appropriate solution to fulfill those needs.

## 3     Conclusion and Future Works

Clustering is one of the essential tasks in data mining and they need improvement nowadays more than before to assist data analysts to extract knowledge from terabytes and petabytes of data. In this study the improvement trend of data clustering algorithms were discussed. to sum up, while traditional sampling and dimension reduction algorithms still are useful, but they don't have enough power to deal with huge amount of data because even after sampling a petabyte of data, it is still very big and it cannot be clustered by clustering algorithms, consequently the future of clustering is tied with distributed computing. Although parallel clustering is potentially very useful for clustering, but the complexity of implementing such algorithms is a challenge. At the other hand, MapReduce framework provides a very satisfying base for implementing clustering algorithms. As results shows, MapReduce based algorithms offer impressive scalability and speed in comparison to serial counterparts while they are maintaining same quality. Regarding to the fact that GPUs are much powerful than CPUs as a future work, it is considerable to deploy clustering algorithms on GPU based MapReduce frameworks in order to achieve even better scalability and speed.

**Acknowledgments.** This work is supported by University of Malaya High Impact Research Grant no vote UM.C/625/HIR/MOHE/SC/13/2 from Ministry of Education Malaysia.

# References

1. Havens, T.C., Bezdek, J.C., Palaniswami, M.: Scalable single linkage hierarchical clustering for big data. In: 2013 IEEE Eighth International Conference on Intelligent Sensors, Sensor Networks and Information Processing, pp. 396–401. IEEE (2013)
2. YouTube Statistic (2014),
   http://www.youtube.com/yt/press/statistics.html
3. Williams, P., Soares, C., Gilbert, J.E.: A Clustering Rule Based Approach for Classification Problems. Int. J. Data Warehous. Min. 8(1), 1–23 (2012)
4. Priya, R.V., Vadivel, A.: User Behaviour Pattern Mining from Weblog. Int. J. Data Warehous. Min. 8(2), 1–22 (2012)
5. Kwok, T., Smith, K.A., Lozano, S., Taniar, D.: Parallel Fuzzy c-Means Clustering for Large Data Sets. In: Monien, B., Feldmann, R.L. (eds.) Euro-Par 2002. LNCS, vol. 2400, pp. 365–374. Springer, Heidelberg (2002)
6. Kalia, H., Dehuri, S., Ghosh, A.: A Survey on Fuzzy Association Rule Mining. Int. J. Data Warehous. Min. 9(1), 1–27 (2013)
7. Daly, O., Taniar, D.: Exception Rules Mining Based on Negative Association Rules. In: Laganá, A., Gavrilova, M.L., Kumar, V., Mun, Y., Tan, C.J.K., Gervasi, O. (eds.) ICCSA 2004. LNCS, vol. 3046, pp. 543–552. Springer, Heidelberg (2004)
8. Ashrafi, M.Z., Taniar, D., Smith, K.A.: Redundant association rules reduction techniques. Int. J. Bus. Intell. Data Min. 2(1), 29–63 (2007)
9. Taniar, D., Rahayu, W., Lee, V.C.S., Daly, O.: Exception rules in association rule mining. Appl. Math. Comput. 205(2), 735–750 (2008)
10. Meyer, F.G., Chinrungrueng, J.: Spatiotemporal clustering of fMRI time series in the spectral domain. Med. Image Anal. 9(1), 51–68 (2004)
11. Ernst, J., Nau, G.J., Bar-Joseph, Z.: Clustering short time series gene expression data. Bioinforma. 21(suppl. 1), i159–i168 (2005)
12. Iglesias, F., Kastner, W.: Analysis of Similarity Measures in Times Series Clustering for the Discovery of Building Energy Patterns. Energies 6(2), 579–597 (2013)
13. Zhao, Y., Karypis, G.: Empirical and theoretical comparisons of selected criterion functions for document clustering. Mach. Learn. 55(3), 311–331 (2004)
14. Hathaway, R., Bezdek, J.: Extending fuzzy and probabilistic clustering to very large data sets. Comput. Stat. Data Anal. 51(1), 215–234 (2006)
15. Big Data, What is it and why it is important,
    http://www.sas.com/en_us/insights/big-data/
    what-is-big-data.html
16. Ng, R.T., Han, J.: CLARANS: A method for clustering objects for spatial data mining. IEEE Trans. Knowl. Data Eng. 14(5), 1003–1016 (2002)
17. Kaufman, L., Rousseeuw, P.J.: Finding Groups in Data: An Introduction on Cluster Analysis. John Wiley and Sons (1990)
18. Ng, R.T., Han, J.: CLARANS: A method for clustering objects for spatial data mining. IEEE Trans. Knowl. Data Eng. 14(5), 1003–1016 (2002)
19. Zhang, T., Ramakrishnan, R., Livny, M.: BIRCH: An efficient data clustering method for very large database. In: SIGMOD Conference, pp. 103–114 (1996)
20. Zhang, T., Ramakrishnan, R., Livny, M.: BIRCH: An efficient data clustering method for very large database. In: SIGMOD Conference, pp. 103–114 (1996)
21. Guha, S., Rastogi, R.: CURE: An efficient clustering algorithm for large database. Inf. Syst. 26(1), 35–58 (2001)

22. Achlioptas, D., McSherry, F.: Fast computation of low rank matrix approximations. J. ACM 54(2), 9 (2007)
23. Fern, X.Z., Brodley, C.E.: Random projection for high dimensional data clustering: A cluster ensemble approach. In: ICML, pp. 186–193 (2003)
24. Dasgupta, S.: Experiments with random projection. In: UAI, pp. 143–151 (2000)
25. Boutsidis, C., Chekuri, C., Feder, T., Motwani, R.: Random projections for k-means clustering. In: NIPS, pp. 298–306 (2010)
26. Golub, G.H., Van-Loan, C.F.: Matrix computations, 2nd edn. The Johns Hopkins University Press (1989)
27. Drineas, P., Kannan, R., Mahony, M.W.: Fast Monte Carlo algorithms for matrices III: Computing a compressed approximate matrix decomposition. SIAM J. Comput. 36(1), 132–157 (2006)
28. Sun, J., Xie, Y., Zhang, H., Faloutsos, C.: Less is More: Compact Matrix Decomposition for Large Sparse Graphs. In: SDM (2007)
29. Tong, H., Papadimitriou, S., Sun, J., Yu, P.S., Faloutsos, C.: Colibri: Fast mining of large static and dynamic graphs. In: Proceedings of the 14th ACM SIGKDD International Conference on Knowledge Discovery and Data Mining, pp. 686–694 (2008)
30. Januzaj, E., Kriegel, H.-P., Pfeifle, M.: DBDC: Density based distributed clustering. In: Bertino, E., Christodoulakis, S., Plexousakis, D., Christophides, V., Koubarakis, M., Böhm, K. (eds.) EDBT 2004. LNCS, vol. 2992, pp. 88–105. Springer, Heidelberg (2004)
31. Aggarwal, C.C., Reddy, C.K. (eds.): Data Clustering: Algorithms and Applications (2013)
32. Ester, M., Kriegel, H.P., Sander, J., Xui, X.: A density-based algorithm for discovering clusters in large spatial database with noise. In: KDD, pp. 226–231 (1996)
33. Karypis, G., Kumar, V.: Parallel multilevel k-way partitioning for irregular graphs. SIAM Rev. 41(2), 278–300 (1999)
34. Karypis, G., Kumar, V.: Multilevel k-way partititing scheme for irregular graphs. J. Parallel Disteributed Comput. 48(1), 96–129 (1998)
35. Andrade, G., Ramos, G., Madeira, D., Sachetto, R., Ferreira, R., Rocha, L.: G-DBSCAN: A GPU Accelerated Algorithm for Density-based Clustering. Procedia Comput. Sci. 18, 369–378 (2013)
36. Anchalia, P.P., Koundinya, A.K., Srinath, N.: MapReduce Design of K-Means Clustering Algorithm. In: 2013 International Conference on Information Science and Applications (ICISA), pp. 1–5 (2013)
37. Zhao, W., Ma, H., He, Q.: Parallel k-means clustering based on MapReduce. In: Cloud Computing, pp. 674–679 (2009)
38. Han, J., Kamber, M., Pei, J.: Data mining: concepts and techniques. Morgan Kaufmann (2006)
39. Mirkin, B.: Clustering for data mining a data recovery approach. CRC Press (2012)
40. He, Y., Tan, H., Luo, W., Feng, S., Fan, J.: MR-DBSCAN: a scalable MapReduce-based DBSCAN algorithm for heavily skewed data. Front. Comput. Sci. 8(1), 83–99 (2014)

# An Approachable Analytical Study
# on Big Educational Data Mining

Saeed Aghabozorgi[1], Hamidreza Mahroeian[2], Ashish Dutt[1],
Teh Ying Wah[1], and Tutut Herawan[1,3]

[1] Department of Information System
University of Malaya
50603 Pantai Valley, Kuala Lumpur, Malaysia
[2] University of Otago
New Zealand
[3] AMCS Research Center, Yogyakarta, Indonesia
{saeed,teh,tutut}@um.edu.my,
hamidreza.mahroeian@postgrad.otago.ac.nz,
ashish_dutt@siswa.um.edu.my

**Abstract.** The persistent growth of data in education continues. More institutes now store terabytes and even petabytes of educational data. Data complexity in education is increasing as people store both structured data in relational format and unstructured data such as Word or PDF files, images, videos and geo-spatial data. Indeed learning developers, universities, and other educational sectors confirm that tremendous amount of data captured is in unstructured or semi-structured format. Educators, students, instructors, tutors, research developers and people who deal with educational data are also challenged by the velocity of different data types, organizations as well as institutes that process streaming data such as click streams from web sites, need to update data in real time to serve the right advert or present the right offers to their customers. This analytical study is oriented to the challenges and analysis with big educational data involved with uncovering or extracting knowledge from large data sets by using different educational data mining approaches and techniques.

**Keywords:** Big Data, Educational Data, Educational Data Mining, Data Mining, Analytical Study.

# 1 Introduction

Big data can be considered as the theory of looking at voluminous didactic amounts of data be it in physical or digital format being stored in diverse repositories ranging from tangible account bookkeeping records of an educational institution to class test or examination records to alumni records [1]. These records continue to grow in size and variety. We learn from our mistakes as the old adage goes, in a similar fashion, today the businesses are being operated based on the decisions over the data that was collected by the business. Predictions, associations, clustering and many other

B. Murgante et al. (Eds.): ICCSA 2014, Part V, LNCS 8583, pp. 721–737, 2014.
© Springer International Publishing Switzerland 2014

commonly occurring business decisions are taken each day by corporate to enhance productivity and mutual growth [2]. And these significant decisions are dependent exclusively on the data collected during business operations and human judgments. This concept of big data has now been applied to various sectors like governments, businesses, hospital management to name a few but there has been little research work been done in its application in the educational sector. This is what we aim to find, through this research work. Tomes have been written on the efficacy of Big Data, the technologies that can be used to harness the sheer strength it exudes. But there has been very little to negligible research work on the application of big data in educational sector. Utilizing different data for making decisions is not new concept; corporations use complicated calculation on data generated by different customers for business intelligence or analytics. Various techniques used in Business intelligence can distinguish historical trends and customer patterns from data and can generate different models that can result in prediction of future patterns and trends [3]. Consist of proven methodologies from computer science, mathematics and statistics used for deriving non-redundant information from large scaled datasets (big data) [4].

One of the clear examples of exploiting useful data to discover online mode behavior is *Web analytics* with different methods that sign and report visits of Web page, specific region or particular domains and the different links that clicked through. To understand how people use the Web, Web analytics are applied, but corporations have utilized more complicated approaches and tools to track more sophisticated user interactions with their websites [5, 6]. Example of web analytics include analyzing the purchasing habits of the consumer, the application of recommendation algorithms in commercial websites search engines such that they are able to recommend the most likely product a consumer would like, notable examples are Netflix, Amazon. The same concept is now being applied to various e-learning systems for example Edmodo is a free open source LMS that is able to predict similar books or resources based on the learner's web activity on the e-LMS [7].

New approaches and methods are considered imperative for extraction and analysis of the aforementioned tasks so as to seamlessly integrate with the unstructured data that these information systems generate. Big data is voluminous and would be futile to bind it within a specific number boundary. One of the means by which it can be defined could be its net usability worth. According to Manyika *et al.* [10] a data set whose computational size exceeds the processing limit of software can be categorized as big data. Several studies have been conducted in the past that have provided detailed insights into the application of traditional data mining algorithms like clustering, prediction, association to tame the sheer voluminous power of big data.

Recent advances in machine learning field has provided with unique approaches to foresee knowledge discovery in datasets. These algorithms have been successful in finding correlations between unstructured data and one of their applications has been into predictive modeling. Such models can be treated as virtual prototypes of a real working model. When injected with real datasets in such models can help ascertain any debacles that can then be promptly addressed to thus mitigating operational costs of both man and machine labor.

Two specific fields that are significant to the exploitation of big data in education are *educational data mining* and *learning analytics*. Although there is no hard and fast distinction between these two areas, they have had different research histories to some extent and are developing as discrete research areas. In general, *educational data mining* tries to uncover new patterns in captured data, building new algorithms or new models, whilst *learning analytics* looks for identified predictive models in educational systems[1, 4]. As it can be seen in the figure 1, educational data such as log files, user (learner) interaction data, and social network data types are expected to grow in the near future. This research study is oriented to the challenges and analysis with big educational data involved with uncovering or extracting knowledge from large data sets by using different educational data mining approaches and techniques. It is arranged in the following ways: in the next section, background of study including importance of education and educational data, the nature of big data, the basic understanding of data mining or knowledge discovery techniques will be described. In section 3, from big educational data mining perspective, the concept of educational data mining, big educational data, as well as big data mining is further discussed. Section 4 details the major challenges concerned with big educational data mining, and finally related discussion and conclusion is outlined.

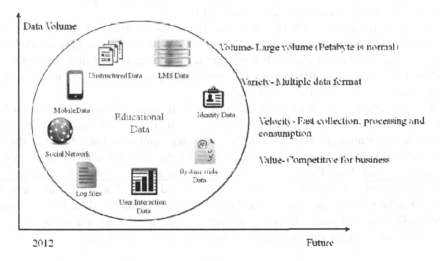

**Fig. 1.** Growth of different Educational Data

# 2    Rudimentary

## 2.1    Education

Learning providers, institutes, universities, schools and colleges always had the ability to generate huge amounts of educational data [8]. Even a small kindergarten school that only supply to a play group of children aged between 4-6 years can produce enormous quantities of data which is ranged from their academics to their peer activities, classroom activities and so forth. After the detonation of the buzzword, "Big data" in different industrial sectors, researchers and industry workers are collating towards vista's that could presumably be affected by this surge [9].

Recent advances in technology has made it now possible to explore any previously unknown information that lay buried in deep caveats of heaps of data sets [10]. However, the most basic question that needs to be answered first is that, "Is there really any big data in education?" or are we simply looking at an impasse.

## 2.2 Big Data

There are a number of similar definitions of big data. Perhaps the most well-known and popular version is derived from IBM,2 which proposed that big data could be differentiated by any or all of three V words to examine situations, events, and so on: volume, variety, and velocity [1, 9, 11].

*Volume* is attributed to larger quantities of data being produced from a various range of resources. For instance, big data can comprise data captured from the Internet of Things (IoT). As initially pictured, IoT is associated to the data collected from a range of different devices and sensors networked together, over the Internet [12]. Big data can also be cited to the explosion of information accessible on common social media such as Facebook and Twitter [13].

*Variety* is referred to utilizing numerous types of data to investigate a situation or event. On the IoT, millions of devices generating a steady flow of data results in not only a large volume of data but different kinds of data features of different situations. Furthermore, people on the Internet produce a highly various set of structured, semi-structured as well as unstructured data [9].

*Velocity* of data which is attributed to a rapid increase in data over time for both kind of structured and unstructured data, and more frequent decision making about that data is essential [1]. As the world becomes more global and developed, and as the IoT generates, there is a growing frequency of data capture and decision making procedure about those things as they progress throughout the world. Additionally, the velocity of social media use is in its obvious upward trend. The clear example would be 250 million tweets per day. As decisions are made using big data, those decisions eventually can have a substantial impact on the next data that's captured and analyzed, counting another dimension to velocity of big data [1, 10].

## 2.3 Data Mining

In databases, Data mining or knowledge discovery popularly known as KDD is the automatic mining of implied and appealing patterns from vast amounts of data [14]. Data mining is recognized as a field which is multidisciplinary in which a number of computing paradigm congregated such as decision tree construction, rule induction, artificial neural networks, instance-based learning, Bayesian learning, logic programming. In addition, some of the functional data mining techniques and methods are listed like statistics, visualization, clustering, classification and association rule mining [15,16]. These techniques discover new, implicit and practical knowledge based on students' usage data.

Data mining has been broadly applied in different kinds of educational systems. On one hand, there are common traditional classroom environments such as special education [17] and higher education [18]. On the other hand, there is education which is computer-based as well as web-based education like well-known and learning management systems known as LMS Systems [19], web-based adaptive hypermedia systems [20] and intelligent tutoring systems(ITS) [21]. The major difference between one and the other is the data accessible in each system. Traditional classrooms only have obtainable information about attendance of student, basic course syllabus; course objectives and learners plan data. However, web and computer-based education has much more readily information because these education systems can track all the data pertained to specific students' actions and interactions onto log files and databases, (e.g. generating log files data) [22].

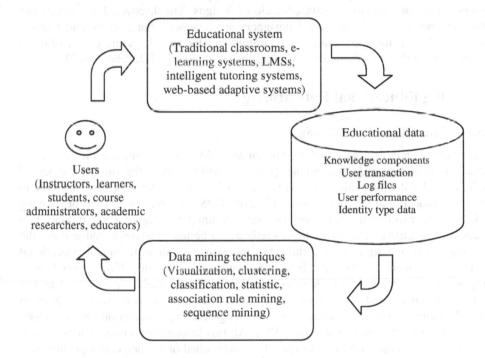

**Fig. 2.** Applying data mining to the design of educational systems

In order to improve learning effectiveness, the application of data mining approaches and techniques to educational systems, can be observed as a formative evaluation technique which is the evaluation of an educational program while it is still in development phase, and with the purpose of continually enhancing the program. Auditing the way students use the educational system is perhaps one common way to assess instructional design in a this manner would help learning developers to have the improved instructional materials which is going to result in having different data types such as log files, performance, transaction [23]. Data mining techniques should be applied to collect information that can be used to assist instructional designers/developers to build an educational foundation for judgments when designing or improving an environment's instructive approach. The application of data mining to the design of educational systems is an iterative cycle of hypothesis formation, testing, and refinement (see Figure 2). Extracted knowledge should go through the loop towards guiding, facilitating, and enhancing learning as a whole. In this process, the aim is not just to turn data into knowledge, but also to filter mined knowledge for decision making [16,24].

As it is represented in Figure 2, educators and educational designers (whether in school districts, curriculum companies, or universities) design, plan, create, and maintain educational systems. Students use those educational systems to learn. Building off of the available information about courses, students, usage, and interaction, data mining techniques can be applied in order to discover useful knowledge that helps to improve educational designs. The discovered knowledge can be used not only by educational designers and teachers, but also by end users—students. Hence, the application of data mining in educational systems can be oriented to supporting the specific needs of each of these categories of stakeholders[23].

# 3     Big Educational Data Mining

## 3.1     Educational Data Mining

Educational Data Mining popularly known as EDM is a field that exploits statistical, machine-learning, and data-mining (DM) algorithms over the different types of educational data. Its major objective is to analyze these types of data in order to resolve educational research issues [25,26]. EDM is concerned with developing methods to explore the relationships between unique types of data, produced in educational settings and, using these methods, to better understand students and the settings in which they learn. While, the increase in both instrumental educational software as well as state databases of student's information have created large repositories of data reflecting how students learn [26]. Whereas, the use of Internet in education has created a new context known as e-learning or web-based education in which large amounts of data about teaching–learning interaction are endlessly generated and ubiquitously available [23]. All this information provides a gold mine of educational data. EDM seeks to tap these untouched or maiden data repositories to better discern learners and learning abilities, and to develop computational approaches that combine data and theory to transform practice to benefit learners. EDM has emerged as a prolific research area in recent years for researchers all over the world from different and related research areas [7].

Education Data Mining can be extremely helpful in deducing inferences, make predictions and more to establish students behavior and attitude as well as concentration to its educational goals. The results deciphered by utilizing the traditional data mining algorithms to educational context  can help enhance the educational system as all stakeholders can look into the trends found once analytic reasoning is applied on the data of student related parameters [25]. Usually we use regression techniques to analyze data, When we unitize the data into statistical numbers for analytic reasons, usually the produced results can be plotted on a graphs and trends can be found in terms of lines or combination of several data points as a concentration of some student behavior to learning or researching or any such related activity [25]. EDM is involved with various groups of users such as learning developers, instructors, educators, researchers. Different groups consider educational data from different angles, based on their mission, vision, and major purpose for using data mining as it is depicted in Table 1.

**Table 1.** EDM Users/Stakeholders

| User/Actors | Objectives for using data mining |
| --- | --- |
| Learners/Students/pupils | To personalize e-learning, to recommend activities to learners resources and learning tasks that could further improve their learning, to suggest interesting learning experiences to the students[27] |
| Educators/Instructors/Teachers/Tutors | To get objective feedback about instructions, to analyze students' learning and behavior, to detect which student need support, to predict student performance, to classify learners into groups[28] |
| Course Developers/Educational Researchers | To evaluate and maintain courseware, to improve student learning, to evaluate structure of course content and its effectiveness in learning process[29] |
| Organizations/Learning Providers/Universities/Private Training Companies | To enhance the decision processes in higher learning institutions to streamline the efficiency in the decision making process, to achieve specific objectives[30] |
| Administrators/School District Administrators/Network Administrators/System Administrators | To develop the best way to organize institutional resources and their educational offer, to utilize available resources more effectively, to enhance educational program offers and determine the effectiveness of distance learning approach[31] |

Today, there exists a wide variety of educational data sets that can be downloaded for free from the Internet. Some widely acclaimed and used repositories are PSLC DataShop (The world's largest repository of learning interaction data), Data.gov (official website of United States Government on Educational data sets), NSES Data sets (is the primary federal entity for collecting and analyzing data related to education in United States) [26,32], Barro-Lee data set (the data set provided by researchers Barrow and lee whose contribution has been discussed in section 1), UNISTATS Dataset (website provides comparable sets of information about full or part time undergraduate courses and is designed to meet the information needs of prospective students), SABINS (The School Attendance Boundary Information System) provides free of charge, aggregate census data and GIS-compatible boundary files for school attendance areas, or school catchment areas, for selected areas in the United States for the 2009-10, 2010-11 and 2011-12 school years. UIS (is an UNESCO initiative), EdStats (A World Bank Initiative), Education Human

Development Network (A World Bank Initiative) and IPEDS Data Center (the primary source for data on colleges, universities, and technical and vocational postsecondary institutions in the United States), TLRP [33] .

### 3.2    Analysis of Current Tools Being Used for Educational Data Sets

At present statistical tools are predominantly being used to quantify and assess the educational data sets. Prominent ones are RapidMiner, SAS, IBM SPSS, KEEL [34] (is a knowledge extraction tool based on evolutionary learning). Programming language like R is mostly used for statistical analysis and plays a pivotal role in programming custom tests that may not be available in commercial software packages. There are some online web based data exploration tools typical java based that gives the user the freedom to choose from the varied dataset types and see a graphical representation of them. One of these is Education Data Explorer being provided by Oregon Department of Education, United States. Another one is Educational Data Analysis Tool (EDAT), it allows you to download NCES survey datasets to your computer. EDAT guides you through selecting a survey, population, and variables relevant to your analysis [30].

### 3.3    Educational Data Set Problem and Possible Solutions

Does the problem really exist or are we running behind a chimera? The "Education for All", a global monitoring report prepared by United Nations is the prime instrument to assess global progress towards its goals. It seems that there is a flurry of activity around big data and how it's touching and transforming every aspect of our life. Analysis of these large scale datasets can help improve the robustness and generalizability of educational research. The problem with most large scale secondary data-sets used in higher education research is that they are constructed using complex sample designs that often cluster lower level units (students), within higher level units (colleges) to achieve efficiencies in the sampling process [35]. As it is clearly shown in Figure 3, the term "Big Educational Data Mining" known as BEDM can be proposed for the extraction of useful big educational data from vast quantities of different large data sets.

### 3.4    Big Educational Data

Education has always had the capacity to produce a tremendous amount of data, compared to any other industry. First, academic study requires many hours of schoolwork and homework for several numbers of years. These extended interactions with materials produce a huge quantity of data. Second, education content is tailor-made for big data, generating cascading effects of insights thanks to the high correlation between concepts [31]. Recent advancement in technology and data science has made it possible to unlock/explore these large data sets [15]. The benefits range from more effective self-paced learning to tools that enable instructors to

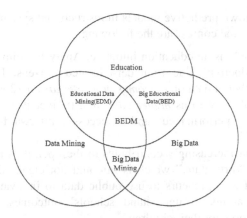

**Fig. 3.** Big Educational Data Mining (BEDM), Extracting new knowledge from Big Data Sets

**Table 2.** Educational Data Type classes

| No. | Data Type | Description |
|-----|-----------|-------------|
| 1 | Identity Data | Personal Information, Authority, Domain Rights, Geographical Information |
| 2 | User Interaction Data | engagement metrics, click rate, page views, bounce rate, etc |
| 3 | Inferred Content Data | How well does a piece of content perform across a group, or for any one subgroup, of students? What measurable student proficiency gains result when a certain type of student interacts with a certain piece of content? |
| 4 | System-Wide Data | Rosters, grades, disciplinary records, and attendance information are all examples of system-wide data. |
| 5 | Inferred Student Data | Exactly what concepts does a student know, at exactly what percentile of proficiency? What is the probability that a student will pass next week's quiz, and what can she do right this moment to increase it? |

pinpoint interventions, create productive peer groups, and free up class time for creativity and problem solving. For instance, as it is represented in Table 2, educational data can be categorized to five different categories: one pertaining to student identity and on boarding, and four student activity-based data sets that have the potential to improve learning outcomes. They are listed below in order to see how complicated they are to attain:

Two areas that are specific to the use of big data in education are *educational data mining* and *learning analytics*. Although there is no hard and fast distinction between these two fields, they have had somewhat different research histories and are developing as distinct research areas. Generally, *educational data mining* looks for new patterns in data and develops new algorithms and/or new models, while *learning analytics* applies known predictive models in instructional systems. Big Data practical examples in Educational context are the following:

- The clear example is an education initiative. Analysts estimate that £16 billion is wasted in productivity due to under-educated citizens. In response, the UK government gathered data on *outcomes of Kindergarten-12 education* (elementary and high school) as well as *higher education* (university). The data pertained to student school performance and "success" afterwards as measured by employment[36].

- The government increasingly contributes to the open data movement; it is okay with releasing "dirty data," which is raw and not cleansed. Open data enables individuals and entrepreneurs to use public data to innovate. Data visualization tools enable parents to understand schools' outcomes, so they can select appropriate schools for their children.

- Universities can use data in exciting ways; they analyze students' social media sharing, patterns in checking out library materials, what courses they take (and outcomes they achieve). This data helps them steer students to courses that are aligned with their goals. It helps with student retention.

- Big data enables interesting insights and correlations such as students that have high library fines tend to perform worse on tests. Universities also correlate performance data with socioeconomic and email data, so they can learn what student characteristics predict the best performance at their schools, and they use this to guide their recruitment. They are also starting to be able to predict which students will drop out before graduating, which helps them give additional support [9].

- Cost drivers [of education] are keys in big data adoption in the UK, which has developed the most comprehensive database of pupils (schoolchildren) in the world. It traces 600,000 pupils' performance from 3,000 elementary schools through career. It has ten years of data on pupils' exams, tests, socioeconomic status, geography, transport, free meals, behavior issues and many others. It is a rich dataset from which the government can learn and improve schools. It can answer political questions. The government is also combining its data with health, crime and welfare datasets. It studies what students' lives are like outside school, to try to develop a fuller picture of factors that affect performance. This can help challenge conventional thinking and guide policy.

- This initiative is teaching us many things. Socioeconomic status is not as important as we thought; school performance and responsiveness is very important. Schools can use data to change. For example, science, technology, engineering and math courses are far more important than we thought, even when students don't intend to pursue STEM careers.

- Privacy is an issue with these databases, but the government believes that the advantages outweigh the pupils' compromised privacy [37] .

- Another traditional belief is that poor pupils do poorly and that schools need more money to increase performance. The data are showing that how the money is invested is more important than how much money is in the school's budget. We are starting to be able to measure return on outcomes.

- The UK example is more complex, but it effectively illustrates how internal and external data can be mashed up to address complex problems such as school performance. It's an excellent example of big data.

## 3.5    Big Data Mining

In typical data mining systems, the mining procedures require computational intensive computing units for data analysis and comparisons. A computing platform is, therefore, needed to have efficient access to, at least, two types of resources: data and computing processors. For small scale data mining tasks, a single desktop computer, which contains hard disk and CPU processors, is sufficient to fulfill the data mining goals. Indeed, many data mining algorithm are designed for this type of problem settings. For medium scale data mining tasks, data are typically large (and possibly distributed) and cannot be fit into the main memory. Common solutions are to rely on parallel computing [43], [33] or collective mining [12] to sample and aggregate data from different sources and then use parallel computing programming (such as the Message Passing Interface) to carry out the mining process. For Big Data mining, because data scale is far beyond the capacity that a single personal computer (PC) can handle, a typical Big Data processing framework will rely on cluster computers with a high-performance computing platform, with a data mining task being deployed by running some parallel programming tools, such as MapReduce or Enterprise Control Language (ECL), on a large number of computing nodes (i.e., clusters). The role of the software component is to make sure that a single data mining task, such as finding the best match of a query from a database with billions of records, is split into many small tasks each of which is running on one or multiple computing nodes. For example, as of this writing, the world most powerful super computer Titan, which is deployed at Oak Ridge National Laboratory in Tennessee, contains 18,688 nodes each with a 16-core CPU. Such a Big Data system, which blends both hardware and software components, is hardly available without key industrial stockholders' support. In fact, for decades, companies have been making business decisions based on transactional data stored in relational databases [10]. Big Data mining offers opportunities to go beyond traditional relational databases to rely on less structured data: weblogs, social media, e-mail, sensors, and photographs that can be mined for useful information [1]. Major business intelligence companies, such IBM, Oracle, Teradata, and so on, have all featured their own products to help customers acquire and organize these diverse data sources and coordinate with customers' existing data to find new insights and capitalize on hidden relationships.

# 4    Major Challenges in Big Educational Data Mining

## 4.1    Is Education Data Big Enough to Call It Big Data?

Startups like Knewton [38] and Desire2Learn [10] have been founded on the concept of Big Data. We had seen similar e-commerce startup during the early nineties when the e-commerce boom was there but history is a mute audience to some of that startup's fate. Few of them have perished by now. However, the business startup's founded on big data in educational context would not face the similar fate because its foundation rests on the principle of didactic unstructured data that is already present in Informational systems. Perhaps one of the reasons for some of the ill-fated e-commerce startups to fail was that their business model did not rest on the availability of a constant flow of data from which information could be minded. But this is not the case here. All we need is specialized algorithms that are designed to work with educational datasets because we already have the data with us. Now companies like Yahoo, Google, Dell, HP to name a few have ventured into open-source development of big data software's like Apache foundation Hadoop to facilitate collective learning by using contests like hackadays or hackathons [25,9]. We also need to understand that there lies a gap between the application of big data in commerce and that in education sector. While the former has seen various advances in it but for the latter we are still dependent on traditional data mining algorithms. And the problem of using such algorithm is that they may not fit the dataset and that can cause a loss of valuable predictions that otherwise could have been ascertained by using the data mining algorithms that would fit the educational dataset. Educational experts have posed various deployment and implementation barriers to harness the power of big data in education and learning analytics that most importantly includes technical lacunae, institutional velocity, legal and quite often ethical issues by applying general data mining algorithms. For big data to be meaningful it will require the seamless integration of specifically tailored algorithms that could the power of this raging beast to tame it into knowledge that will be useful to both the learner and the educator [39].

### 4.2    BDM: Challenges in Applying DM Approaches on Big Data (from the Educational Perspective)

A conceptual view of the Big Data processing framework can be depicted in the figure 4, which includes three tiers from inside out with considerations on data accessing and computing (Tier I), data privacy and domain knowledge (Tier II), and Big Data mining algorithms (Tier III). The challenges at Tier I focus on data accessing and actual computing procedures.

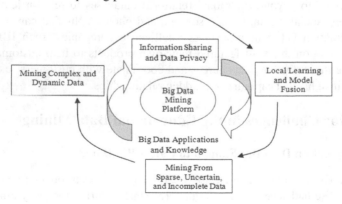

**Fig. 4.** A conceptual view of the Big Data processing framework [4]

Because Big Data are often stored at different locations and data volumes may continuously grow, an effective computing platform will have to take distributed large-scale data storage into consideration for computing [4,11]. For example, while typical data mining algorithms require all data to be loaded into the main memory, this is becoming a clear technical barrier for Big Data because moving data across different locations is expensive (e.g., subject to intensive network communication and other IO costs), even if we do have a super large main memory to hold all data for computing. The challenges at Tier II center on semantics and domain knowledge for different Big Data applications [40]. Such information can provide additional benefits to the mining process, as well as add technical barriers to the Big Data access (Tier I) and mining algorithms (Tier III). For example, depending on different domain applications, the data privacy and information sharing mechanisms between data producers and data consumers can be significantly different. Sharing sensor network data for applications like water quality monitoring may not be discouraged, whereas releasing and sharing mobile users' location information is clearly not acceptable for majority, if not all, applications [41]. In addition to the above privacy issues, the application domains can also provide additional information to benefit or guide Big Data mining algorithm designs. For example, in market basket transactions data, each transaction is considered independent and the discovered knowledge is typically represented by finding highly correlated items, possibly with respect to different temporal and/or spatial restrictions. In a social network, on the other hand, users are linked and share dependency structures. The knowledge is then represented by user communities, leaders in each group, and social influence modeling etc. Therefore, understanding semantics and application knowledge is important for both low-level data access and for high level mining algorithm designs [16]. At Tier III, the data mining challenges concentrate on *algorithm designs in tackling the difficulties raised by the Big Data volumes, distributed data distributions, and by complex and dynamic data characteristics.* The circle at Tier III contains three stages [4].

- Primarily, sparse, heterogeneous, uncertain, incomplete, and multi-source data are preprocessed by *data fusion techniques.*
- Secondarily, complex and dynamic data are mined after pre-processing.
- Tertiary, the global knowledge that is obtained by local learning and model fusion is tested and relevant information is fed back to the pre-processing stage.

Then the model and parameters are adjusted according to the feedback. In the whole process, information sharing is not only a promise of smooth development of each stage, but also a purpose of Big Data processing [30].

## 4.3    EDM: Challenges in Applying DM Approaches on Education Data

The recent advances in information technology have seen the proliferation of software's that can code a completely functional website replete with a backend database system in less than an hour. So this has led to a rampant growth of e-learning systems mostly cloud technology based. Most of these have incorporated recommendation features as used by their business oriented counterparts. And both of them are generating voluminous amounts of data. While online learning systems have proffered the educator, developer and researcher opportunities to create personalized learning systems but do note these personalization are using the traditional data mining algorithms [7]. So what's the problem then? One would ask. Well, one of the problem is which most of the e-learning systems are not able to ascertain from an educational point of view is that these systems are used by learners who have their individual learning styles. When a learner interacts with an LMS it leaves behind a trail of breadcrumbs or log text files for example its interactions within the LMS forum with either other students or with the course facilitator [33]. So it logically follows that if we have to mine this data then it becomes imperative to figure out the correct dataset to use so as to derive logical conclusions from it. Till now, there have been fewer instances where data mining methods [44-48] have been introduced within the e-learning systems to facilitate learner progress. The other problem from a developer's point of view would be to determine how to classify individual learning style of a learner so as to provide it with a truly personalized learning environment. While another challenge will be as we have repeatedly mentioned it in previous sections too on how to develop specific data mining algorithms [49-52] that can cater to the learning analytical domain. So essentially what really matters at this point is to find out methodologies that can help clean educational dataset so that it could further be processed [23, 42].

## 5    Discussion and Conclusion

In this analytical study, on the whole, the background of study regarding to importance of education and its educational data growth as big data, big data mining tools and techniques to mine these vast amounts of data has been discussed. Moreover, the challenges involved with big educational data mining and extraction of big educational data has been addressed from different educational data mining perspectives. Working with big data using data mining and analytics is rapidly becoming common in the commercial sector. Tools and techniques once confined to research laboratories are being adopted by forward-looking industries, most notably those serving end users through online systems [4,43]. Higher education institutions are applying learning analytics to improve the services they provide and to improve visible and measurable targets such as grades and retention. K–12 schools and school districts are starting to adopt such institution-level analyses for detecting areas for improvement, setting policies, and measuring results. Now, with advances in adaptive learning systems, possibilities exist to harness the power of feedback loops at the level of individual teachers and students [40]. Measuring and making visible students' learning and assessment activities open up the possibility for learner's to develop skills in monitoring their own learning and to see directly how their effort builds onto their success. Teachers gain views into students' performance that help them adapt their teaching or initiate interventions in the form of tutoring, tailored assignments, and the like. Personalized adaptive learning systems enable educators to quickly see the effectiveness of their adaptations and interventions, providing feedback for continuous improvement. The practical applications of open source data mining tools in an educational setting can augment both the researcher and developer to compare distinct prototypes bearing the same design functionalities. The results thus obtained could then be used to integrate within the existing in-house educational framework as used by institutions so as to keep pace with the rapid adoption of blended learning environment. Open source tools for adaptive learning systems, commercial offerings, and increased understanding of what data reveal are leading to fundamental shifts in teaching and learning systems. As content moves online and mobile devices for interacting with content enable teaching to be always on, educational data mining and learning analytics will enable learning to be always assessed. Educators at all levels will benefit from understanding the possibilities of the developments described in the use of big data herein. Besides challenges of this new field which is introduced as big educational data mining concerned with big identified educational data, the importance of analyzing big educational data captured, extracted from large scaled data sets using multiple approaches of big data and data mining analysis has to be considered in further studies.

**Acknowledgments.** This work is supported by University of Malaya High Impact Research Grant no vote UM.C/625/HIR/MOHE/SC/13/2 from Ministry of Education Malaysia.

# References

1. Sagiroglu, S., Sinanc, D.: Big data: A review. In: 2013 International Conference on Collaboration Technologies and Systems (CTS), pp. 42–47 (2013)
2. Pena-Ayala, A.: Educational data mining: A survey and a data mining-based analysis of recent works. Expert Systems with Applications (2013)
3. Siemens, G., Long, P.: Penetrating the fog: Analytics in learning and education. Educause Review 46, 30–32 (2011)
4. Wu, X., Zhu, X., Wu, G., Ding, W.: Data mining with big data (2012)
5. Bizer, C., Boncz, P., Brodie, M.L., Erling, O.: The meaningful use of big data: four perspectives–four challenges. ACM SIGMOD Record 40, 56–60 (2012)
6. Abraham, A.: Business intelligence from web usage mining. Journal of Information & Knowledge Management 2, 375–390 (2003)
7. Romero, C., Ventura, S.: Educational data mining: A survey from 1995 to 2005. Expert Systems with Applications 33, 135–146 (2007)
8. Gobert, J., Sao Pedro, M., Baker, R., Toto, E., Montalvo, O.: Leveraging educational data mining for real time performance assessment of scientific inquiry skills within microworlds. Journal of Educational Data Mining (accepted, 2012)
9. LaValle, S., Lesser, E., Shockley, R., Hopkins, M.S., Kruschwitz, N.: Big data, analytics and the path from insights to value. MIT Sloan Management Review 52, 21–31 (2011)
10. Manyika, J., Chui, M., Brown, B., Bughin, J., Dobbs, R., Roxburgh, C., et al.: Big data: The next frontier for innovation, competition, and productivity (2011)
11. Trelles, O., Prins, P., Snir, M., Jansen, R.C.: Big data, but are we ready? Nature Reviews Genetics 12, 224–224 (2011)
12. Bathuriya, G., Sai Nandhinee, M.: Implementation of Big Data for Future Education Development Data Mining Data Analytics
13. Centola, D.: The spread of behavior in an online social network experiment. Science 329, 1194–1197 (2010)

14. Kurgan, L.A., Musilek, P.: A survey of Knowledge Discovery and Data Mining process models. Knowledge Engineering Review 21, 1–24 (2006)
15. Romero, C., Ventura, S., García, E.: Data mining in course management systems: Moodle case study and tutorial. Computers & Education 51, 368–384 (2008)
16. Liao, S.-H., Chu, P.-H., Hsiao, P.-Y.: Data mining techniques and applications–A decade review from 2000 to 2011. Expert Systems with Applications 39, 11303–11311 (2012)
17. Tsantis, L., Castellani, J.: Enhancing learning environments through solution-based knowledge discovery tools: Forecasting for self-perpetuating systemic reform. Journal of Special Education Technology 16, 39–52 (2001)
18. Romero, C., Ventura, S., Zafra, A., de Bra, P.: Applying Web usage mining for personalizing hyperlinks in Web-based adaptive educational systems. Computers & Education 53, 828–840 (2009)
19. Romero, C., Ventura, S., De Bra, P.: Knowledge discovery with genetic programming for providing feedback to courseware authors. User Modeling and User-Adapted Interaction 14, 425–464 (2004)
20. Wang, Y.: Web mining and knowledge discovery of usage patterns. CS 748T Project (2000)
21. Cetintas, S., Si, L., Xin, Y.P., Hord, C.: Automatic detection of off-task behaviors in intelligent tutoring systems with machine learning techniques. IEEE Transactions on Learning Technologies 3, 228–236 (2010)
22. Witten, I.H., Frank, E.: Data Mining: Practical machine learning tools and techniques. Morgan Kaufmann (2005)
23. Romero, C., Ventura, S., Pechenizkiy, M., Baker, R.S.: Handbook of educational data mining. Taylor & Francis, US (2011)
24. Han, J., Kamber, M., Pei, J.: Data mining: concepts and techniques. Morgan kaufmann (2006)
25. Romero, C., Ventura, S.: Educational data mining: a review of the state of the art. IEEE Transactions on Systems, Man, and Cybernetics, Part C: Applications and Reviews 40, 601–618 (2010)
26. Baker, R., Yacef, K.: The state of educational data mining in 2009: A review and future visions. Journal of Educational Data Mining 1, 3–17 (2009)
27. Sael, N., Marzak, A., Behja, H.: Multilevel clustering and association rule mining for learners' profiles analysis (2013)
28. Anozie, N., Junker, B.W.: Predicting end-of-year accountability assessment scores from monthly student records in an online tutoring system. In: Proceedings of the American Association for Artificial Intelligence Workshop on Educational Data Mining (AAAI 2006), Boston, MA, July 17, pp. 1–6 (2006)
29. Razzaq, L., et al.: A web-based authoring tool for intelligent tutors: blending assessment and instructional assistance. In: Nedjah, N., de Macedo Mourelle, L., Borges, M.N., de Almeida, N.N. (eds.) Intelligent Educational Machines. SCI, vol. 44, pp. 23–49. Springer, Heidelberg (2007)
30. Peña-Ayala, A., Cárdenas, L.: How Educational Data Mining Empowers State Policies to Reform Education: The Mexican Case Study. In: Peña-Ayala, A. (ed.) Educational Data Mining. SCI, vol. 524, pp. 65–101. Springer, Heidelberg (2014)
31. Lara, J.A., Lizcano, D., Martínez, M.A., Pazos, J., Riera, T.: A System for Knowledge Discovery in E-Learning Environments within the European Higher Education Area- Application to student data from Open University of Madrid, UDIMA. Computers & Education (2013)
32. Berry, M.J., Linoff, G.: Data mining techniques: For marketing, sales, and customer support. John Wiley & Sons, Inc. (1997)

33. Chen, M.-S., Park, J.S., Yu, P.S.: Data mining for path traversal patterns in a web environment. In: Proceedings of the 16th International Conference on Distributed Computing Systems, pp. 385–392 (1996)
34. Alcalá-Fdez, J., Sánchez, L., García, S., del Jesús, M.J., Ventura, S., Garrell, J., et al.: KEEL: a software tool to assess evolutionary algorithms for data mining problems. Soft Computing 13, 307–318 (2009)
35. Fayyad, U., Piatetsky-Shapiro, G., Smyth, P.: The KDD process for extracting useful knowledge from volumes of data. Communications of the ACM 39, 27–34 (1996)
36. Siemens, G., de Baker, R.S.: Learning analytics and educational data mining: Towards communication and collaboration. In: Proceedings of the 2nd International Conference on Learning Analytics and Knowledge, pp. 252–254 (2012)
37. Evfimievski, A., Srikant, R., Agrawal, R., Gehrke, J.: Privacy preserving mining of association rules. Information Systems 29, 343–364 (2004)
38. Agrawal, D., Das, S., El Abbadi, A.: Big data and cloud computing: current state and future opportunities. In: Proceedings of the 14th International Conference on Extending Database Technology, pp. 530–533 (2011)
39. Chen, H., Chiang, R.H., Storey, V.C.: Business Intelligence and Analytics: From Big Data to Big Impact. MIS Quarterly 36, 1165–1188 (2012)
40. Zikopoulos, P., Eaton, C., DeRoos, D., Deutsch, T., Lapis, G.: Understanding big data. McGraw-Hill, New York (2012)
41. Bienkowski, M., Feng, M., Means, B.: Enhancing teaching and learning through educational data mining and learning analytics: An issue brief. SRI International, Washington, DC (2012)
42. Nisbet, R., Elder IV, J., Miner, G.: Handbook of statistical analysis and data mining applications. Access Online via Elsevier (2009)
43. Guide, P.: Getting Started with Big Data (2013)
44. Kalia, H., Dehuri, S., Ghosh, A.: A Survey on Fuzzy Association Rule Mining. International Journal of Data Warehousing and Mining 9(1), 1–27 (2013)
45. Waas, F., Wrembel, R., Freudenreich, T., Thiele, M., Koncilia, C., Furtado, P.: On-Demand ELT Architecture for Right-Time BI: Extending the Vision. International Journal of Data Warehousing and Mining 9(2), 21–38 (2013)
46. Abelló, A., Darmont, J., Etcheverry, L., Golfarelli, M., Mazon, J.-N., Naumann, F., Bach Pedersen, T., Rizzi, S., Trujillo, J., Vassiliadis, P., Vossen, G.: Fusion Cubes: Towards Self-Service Business Intelligence. International Journal of Data Warehousing and Mining 9(2), 66–88 (2013)
47. Williams, P., Soares, C., Gilbert, J.E.: A Clustering Rule Based Approach for Classification Problems. International Journal of Data Warehousing and Mining 8(1), 1–23 (2012)
48. Priya, R.V., Vadivel, A.: User Behaviour Pattern Mining from Weblog. International Journal of Data Warehousing and Mining 8(2), 1–22 (2012)
49. Kwok, T., Smith, K.A., Lozano, S., Taniar, D.: Parallel Fuzzy c-Means Clustering for Large Data Sets. In: Monien, B., Feldmann, R.L. (eds.) Euro-Par 2002. LNCS, vol. 2400, pp. 365–374. Springer, Heidelberg (2002)
50. Daly, O., Taniar, D.: Exception Rules Mining Based on Negative Association Rules. In: Laganá, A., Gavrilova, M.L., Kumar, V., Mun, Y., Tan, C.J.K., Gervasi, O. (eds.) ICCSA 2004. LNCS, vol. 3046, pp. 543–552. Springer, Heidelberg (2004)
51. Taniar, D., Rahayu, W., Lee, V.C.S., Daly, O.: Än Exception rules in association rule mining. Applied Mathematics and Computation 205(2), 735–750 (2008)
52. Ashrafi, M.Z., Taniar, D., Smith, K.A.: Redundant association rules reduction techniques. International Journal of Business Intelligence and Data Mining 2(1), 29–63 (2007)

# Genetic and Backtracking Search Optimization Algorithms Applied to Localization Problems

Alan Oliveira de Sá[1], Nadia Nedjah[2], and Luiza de Macedo Mourelle[3]

[1] Center of Electronics, Communications and Information Technology
Admiral Wandenkolk Instruction Center, Brazilian Navy, Rio de Janeiro, Brazil
[2] Department of Electronics Engineering and Telecommunication,
Engineering Faculty, State University of Rio de Janeiro, Brazil
[3] Department of System Engineering and Computation,
Engineering Faculty, State University of Rio de Janeiro, Brazil
alan.oliveira.sa@gmail.com, {nadia,ldmm}@eng.uerj.br

**Abstract.** The localization problem arises from the need of the elements of a swarm of robots, or of a Wireless Sensor Network (WSN), to determine its position without the use of external references, such as the Global Positioning System (GPS), for example. In this problem, the location is based on calculations that use distance measurements to anchor nodes, that have known positions. In the search for efficient algorithms to calculate the location, some algorithms inspired by nature, such as Genetic Algorithm (GA) and Particle Swarm Optimization Algorithm(PSO), have been used. Accordingly, in order to obtain better solutions to the localization problem, this paper presents the results obtained with the Backtracking Search Optimization Algorithm (BSA) and compares them with those obtained with the GA.

## 1 Introduction

Several applications of swarm robotics require an individual to be able to know its position. It may be absolute, according to an universal reference system, or relative, taken over other individuals, based on a local coordinate system. In the same way, the Wireless Sensor Network (WSN), whose prospects of application are broad and have attracted great attention from industry, in most cases are useless when it is not possible to know the position of its sensors [1].

In both cases, the devices, whether they are robots or sensors, have common characteristics such as the small size, limited power source and low cost. Thus, equipping each element with a Global Positioning System (GPS), is often not feasible.

The localization problem consists of calculating the position of a set of robots or sensors in a situation where it is not possible to use an external reference, such as GPS. Many localization algorithms rely on the ability of the nodes to measure their distances to anchors, whose positions are already known. Some of the most common techniques for distance measurement are based either on the received signal strength, the propagation time of the signal, or comparing the propagation time of two signals with different propagation speeds [2].

B. Murgante et al. (Eds.): ICCSA 2014, Part V, LNCS 8583, pp. 738–746, 2014.

Since the measurement techniques rely on signals propagation, one should consider a distance limit for such measurements. In the simple case, where all the anchors are within the distance measurement limit, the measures are straight and done using one-hop. However, in cases where one or more anchors are outside of the distance limit, measurements are done indirectly, using multiples hops, through algorithms such as Sum-dist or DV-hop [3]. Depending on the topology of the network/swarm, both multi and one-hop cases can coexist.

In [3], the authors propose three phases to address the multi-hop problem:

1. Determine the distances from each node to anchors.
2. Compute the position of each node based on the distances measured in phase 1.
3. Refine the position of each node using the position and distance informations of the neighbor nodes.

Bio-inspired optimization techniques have been applied to the localization problem, both in one-hop and multi-hop cases [1] [2]. In this paper, we compare the performance and accuracy of the Genetic Algorithm (GA) [4], with the Backtracking Search Optimization Algorithm (BSA) [5], in solving the localization problem with one-hop distance measurements.

This paper is organized as follows: First, in Section 2, we present some related works. Thereafter, in Section 3, we specify the localization problem. In Sections 4 and 5, we briefly describe the GA and BSA, respectively. Then, in Section 6, we present and compare the obtained results. Finally, in Section 7, we present the conclusions and some possibilities for future work.

## 2   Related Works

The hardware and energy limitations typical of the elements of a WSN, or swarm of robots, have motivated the search for more efficient localization algorithms.

In [1], the authors report the use of GA for solving the localization problem in a WSN without obstacles and without noise.

Another approach to the localization problem is given in [8], where the authors propose the use of mobile anchors and an algorithm based on GA to establish the location of static unknown nodes. However, this method may have disadvantages in networks with large coverage area, due to high expense of energy when anchors move trough the network.

In [2], the author presents a Swarm-Intelligent Localization (SIL) algorithm, based on the Particle Swarm Optimization (PSO), to solve the localization problem on a static sensor network, with two and three dimensions. In [9], the authors demonstrate the ability of SIL in solving the localization problem in mobile sensor networks.

In this context, this work aims to contribute to the search for more efficient algorithms for the localization problem by evaluating the performance of BSA.

# 3   Problem Specifications

The localization problem considered in this work presents two dimensions. As a premise, we consider the problem as one-hop, *i.e.*, all unknown nodes are able to measure their distance to an anchor directly. In this scenario no errors are introduced in the distance measurements.

To establish the fitness function, we must consider that, for a given unknown node, the square of the distance error for each anchor is given by (1), where, $d_a$ and $pos_a$ are the measured distance and the position of the anchor node $a$, respectively, and $pos_i$ is the estimated position for the unknown node $i$.

$$f_a(pos_i) = (d_a - ||pos_a - pos_i||)^2 \tag{1}$$

Therefore, the fitness function for the position of an unknown node $i$ is obtained by the sum of the square of the distance errors for each anchor node, as shown in (2), where $A$ is the number of anchors.

$$f(pos_i) = \sum_{a=1}^{A}(d_a - ||pos_a - pos_i||)^2 \tag{2}$$

Thus, to find the node position, we should minimize (2). In this scenario, where there are no errors in distance measurements, the minimum value of $f(pos_i)$ is 0.

# 4   The Genetic Algorithm

The GA is based on the Darwinian principle of evolution of species, where the fittest individuals have more probability to be selected to form the next generation. In this analogy, each individual represents a possible solution and the population represents a set of possible solutions. Each generation is an iterative cycle of the algorithm and the chromosome corresponds to the structure in which the solutions of a problem are encoded. At each generation, selection, crossover and mutation operators are applied, in order to drive the adaptive search to optimize the solution of a problem. The main steps of GA are shown in Algorithm 1 [6].

The selection operator selects individuals from the previous generation to initiate the composition of the current population. In this process, the individual's fitness, calculated by the fitness function, determines probability of that individual to be selected. The crossover operator groups together the selected individuals, forming couples that will exchange genetic material. The mutation operator, which represents the exploratory behavior of the algorithm makes alterations in some generated individuals, allowing the exploration of new possible solutions in the search space. This reduces the chance of the algorithm to converge to a local optimum.

To execute the GA, we used the implementation GAOT (Genetic Algorithms for Optimization Toolbox) [6], for MATLAB. Since GAOT is an implementation

---

**Algorithm 1.** Genetic Algorithm

---
Initializes the start population $P_0$ with $N$ individuals;
Evaluate($P_0$)
$i \leftarrow 1$
**repeat**
    $P'_i \leftarrow$ Selection($P_{i-1}$)
    $P_i \leftarrow$ Crossover-and-Mutation($P'_i$)
    Evaluate($P_i$)
    $i \leftarrow i + 1$
**until** Stopping Criterion
**return** Best Solution

---

that seeks the maximum fitness value, the fitness function was modified as in (3).

$$f(pos_i) = -\sum_{a=1}^{A}(d_a - ||pos_a - pos_i||)^2 \tag{3}$$

The selection, crossover and mutation operators were implemented by the GAOT functions *normGeomSelect*, *simpleXover* and *binaryMutation* [6], respectively. The individual is represented by 54 bits (27 bits for each dimension).

## 5 The Backtracking Search Optimization Algorithm

The BSA is a new evolutionary algorithm. It uses information obtained from past generations to search for better fitness solutions. The bio-inspired philosophy of BSA is analogous to a social group of living creatures that, at random moments, return to hunting areas that were previously considered good for finding food. The general structure of BSA is shown in Algorithm 2 [5].

---

**Algorithm 2.** Backtracking Search Optimization Algorithm

---
Initialization
**repeat**
    Selection-I
    **Generate new population**
        Mutation
        Crossover
    **end**
    Selection-II
**until** Stopping conditions are met

---

During the initialization phase, the algorithm generates and evaluates the initial population $P_0$ and starts a historical population $P_{old}$. The historical population constitutes the memory of BSA.

During the Selection-I, the algorithm randomly determines whether the current population $P$ is recorded as the historical population $P_{old}$. After that, the individuals of $P_{old}$ are shuffled.

The mutation operator creates $P_{modified}$, which constitutes an initial version of the new population $P_{new}$, according to (4). Therefore, $P_{modified}$ is the result of the movement of individuals of $P$ in the directions set by $(P_{old} - P)$. $F$ defines the motion amplitude, and is given by (5).

$$P_{modified} = P + F(P_{old} - P) \tag{4}$$

$$F = k \cdot rndn \tag{5}$$

The value of $k$ is adjusted empirically during prior simulations, whose best result is presented in Section 6. Moreover, $rndn \sim N(0, 1)$, where $N$ is the standard normal distribution. To create the final version of $P_{new}$, the crossover operator randomly crosses over elements from $P_{modified}$ with elements of $P$.

During Selection-II, the algorithm selects elements from $P_{new}$ (individuals after mutation and crossover) that have better fitness than elements of $P$ (individuals before mutation and crossover) and replaces them in $P$. Thus, $P$ only receives new evolved individuals. After meeting the stopping conditions, the algorithm returns the best solution found.

In this paper, the simulations using BSA are performed using an implementation in MATLAB and available in [7]. Unlike GAOT, this implementation of BSA seeks the minimum value of the fitness function, so, in this case, we used the function defined in (2).

# 6   Results

The simulations were performed in a search space of 100 × 100 measurement units. In this search space, we randomly distributed 1000 nodes, referred to as unknown nodes, as their actual position are actually unknown. Still, in this search space, were randomly allocated 3 anchor nodes. This is the minimum number of anchors for a problem with two dimensions [2]. The same simulations were repeated with 4 and 5 anchor nodes.

Considering the stochastic nature of the GA and BSA, and the fact that each unknown node represents a localization problem, we adopt the mentioned number of unknown nodes in this first simulations to provide reliability to the results.

Note that although randomly generated, the same scenarios were precisely reproduced for both GA and BSA, in order to have a fair comparison. Populations of 100 individuals were applied in both GA and BSA.

The average error of the positions ($AEP$) of all unknown nodes is computed, for each generation, as described in (6), wherein $g$ represents the generation number, $i$ represents the unknown node, $U$ is the total number of unknown nodes, $pos_{real}$ is the exact position of the node and $pos_{estimated}$ is the position computed by the optimization algorithm.

$$AEP_g = \sum_{i=1}^{U} \frac{||pos_{real}(i) - pos_{estimated}(i)||}{U} \tag{6}$$

Using BSA, simulations for different values of $k$ were carried out, in a range from 0.5 to 3. The best results were obtained with $k = 1$. In the same way, for the GA, we performed simulations for different values of mutation rate, crossover rate and probability of selecting the best individual. The surveyed ranges of the GA parameters were from 0.1 to 0.3 for the mutation rate, from 0.5 to 0.7 for the crossover rate and from 0.05 to 0.3 for the probability of selecting the best individual. The best results were achieved with the parameters settings of Table 1.

**Table 1.** Parameters used in the GA

| Parameter | Value |
|---|---|
| Mutation rate | 0.20 |
| Crossover rate | 0.60 |
| Probability of selecting the best individual | 0.10 |

The results of the $AEP$ per generation, obtained with the two algorithms for the mentioned settings, considering 3, 4 and 5 anchors, are presented in Figure 1. We can verify that the BSA has, in general, smaller errors than GA at each generation. We can also note a significant improvement in the results, for both algorithms, when we increase the number of anchors from 3 to 4 nodes. However, when 5 anchors were used, no considerable improvement was achieved. With 5 anchor nodes, after 100 generations the BSA presents an $AEP$ of the order of $10^{-5}$, while the GA presents an $AEP$ of the order of $10^{-2}$.

Additionally, we performed another set of simulations with 100 unknown nodes, where both algorithms ran for 1000 generations. The results showed that, 80% of the unknown nodes achieved the optimum result, with zero error, with the BSA. With the GA, no exact location of the unknown nodes was discovered and the node closer to the actual position presented an error of the order of $10^{-4}$.

In Figures 2 and 3 we present two maps: one for GA and another for BSA, containing the locations of 3 anchor nodes and of a sample of 100 unknown nodes. In these figures, it is possible to compare the final estimated position with the exact position of each node. In both maps, the errors are almost imperceptible considering the area of $100 \times 100$ measurement units, as, for 3 anchor nodes, the final $AEP$ is of the order of $10^{-3}$ with the BSA and $10^{-1}$ with the GA. In Figure 2 we highlight a node with the highest error, visually perceptible. This exemplifies the lower effectiveness of GA.

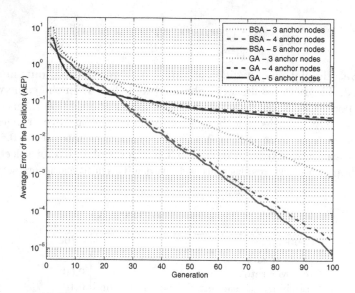

**Fig. 1.** Plots of AEP per generation, for GA and BSA

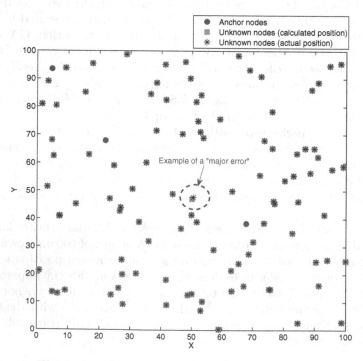

**Fig. 2.** Actual positions and positions calculated by GA

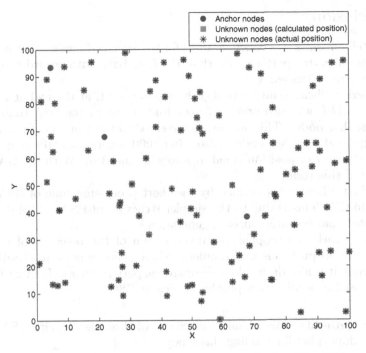

**Fig. 3.** Actual positions and positions calculated by BSA

Another advantage of the BSA is the average processing time, which is significantly less than that of the GA. Figure 4 presents a comparison between the processing times for both algorithms, normalized to the highest processing time.

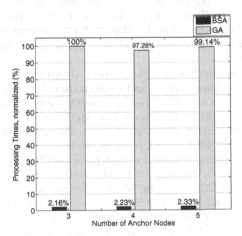

**Fig. 4.** Comparison of normalized processing times

# 7    Conclusion

Based on these results, we conclude that, for the specified localization problem, the BSA has a better performance than the GA, both with regard to efficacy and with regard to efficiency.

The superior efficacy can be verified by the fact that, at the end of 100 generations, the $AEP$ was two orders of magnitude lower in the BSA than in the GA, for 3 anchor nodes. This advantage increased to three orders of magnitude, when using 4 and 5 anchor nodes. Also, after 1000 generations 80% of unknown nodes had an error of position equal to zero with the BSA. With the GA, none node achieved this result.

The highest efficiency is verified by the short processing time of BSA when compared to GA. This is due to the simpler structure of BSA, which demands less computational resources at each generation.

For future work, we propose the investigation of the behavior of GA and BSA in multi-hop problems and problems where there is noise in the distance measurements. It is also desirable to compare the performance of BSA with other algorithms used in localization problems, such as PSO.

**Acknowledgement.**    We thank the State of Rio de Janeiro (FAPERJ, http://www.faperj.br) for funding this study.

# References

1. Sun, W., Su, X.: Wireless sensor network node localization based on genetic algorithm. In: 3rd Int. Conf. on Communication Software and Networks, pp. 316–319. IEEE (2011)
2. Ekberg, P.: Swarm-Intelligent Localization. Thesis, Uppsala Universitet, Uppsala, Sweden (2009)
3. Langendoen, K., Reijers, N.: Distributed Localization Algorithms. In: Zurawski, R. (ed.) Embedded Systems Handbook, pp. 36.1–36.23. CRC Press (2005)
4. Holland, J.H.: Adaptation in natural and artificial systems: An introductory analysis with applications to biology, control, and artificial intelligence. MIT Press, Cambridge (1975)
5. Civicioglu, P.: Backtracking Search Optimization Algorithm for numerical optimization problems. Applied Mathematics and Computation 219, 8121–8144 (2013)
6. Houck, C.R., Joines, J., Kay, M.: A genetic algorithm for function optimization: A Matlab implementation. ACM Transactions on Mathmatical Software (1996)
7. Civicioglu, P.: Backtracking Search Optimization Algorithm (BSA) For Numerical Optimization Problems - BSA code for Matlab2013a,
   http://www.pinarcivicioglu.com/bsa.html (accessed December 11, 2013)
8. Huanxiang, J., Yong, W., Xiaoling, T.: Localization Algorithm for Mobile Anchor Node Based on Genetic Algorithm in Wireless Sensor Network. In: Int. Conf. on Intelligent Computing and Integrated Systems, pp. 40–44. IEEE (2010)
9. Ekberg, P., Ngai, E.C.: A Distributed Swarm-Intelligent Localization for Sensor Networks with Mobile Nodes. In: 7th Int. Wireless Communications and Mobile Computing Conference, pp. 83–88 (2011)

# Optimization of the Multiechelon System for Repairable Spare Parts Using Swarm Intelligence Combined with a Local Search Strategy*

Orlando Duran[1] and Luis Perez[2]

[1] Pontificia Universidad Catolica de Valparaiso
Valparaiso, Chile
orlando.duran@ucv.cl
[2] Universidad Tecnica Federico Santa Maria
Valparaiso, Chile
luis.perez@usm.cl

**Abstract.** Repairable spare parts are referred to critical and expensive components which could have failures; this type of spare parts are common in mining, military and other industries with high value physical assets. Repairable components, after a failure, can be restored to operation by a repair procedure which does not constitute an entire substitution. Many industries deploy their operations and their inventories in a geographically distributed structure. That distribution is composed by a central installation or depot and a set of bases where processes normally, take place (i.e. mining in northern Chile). This type of arrangement is called multi-echelon systems. The main concern in this type of systems is: in what number and how to distribute those expensive and critical repairable spare parts. The decision is restricted by a limited budget and the necessity of not affecting the normal operations of the system. Multi-echelon, multi-item optimization problems are known for their hardness in solving them to optimality, and therefore heuristics methods are approached to near-optimally solve such problems. The most prominent model is the Multi-Echelon Technique for Recoverable Item Control (METRIC), presented by Sherbrooke in 1968 [10]. That model has been extensively used in the military world and in the last years in others industries such aviation and mining. Through this model availability values are obtained from the performance characterization of backorders at the bases. METRIC allocates spare parts in the system on a global basis, since the METRIC model considers all locations simultaneously in the performance analysis. This work proposes the use of Particle Swarm Optimization with local search procedures to solve the multi-echelon of repairable spare parts optimization problem. The major difference between our proposal and previous works lies in that we will combine population based methods with specific local search methods The use of hybridization of non-traditional techniques to attain better

---

* The author(s) declare(s) that there is no conflict of interests regarding the publication of this article.

B. Murgante et al. (Eds.): ICCSA 2014, Part V, LNCS 8583, pp. 747–761, 2014.
© Springer International Publishing Switzerland 2014

optimization performance, is the main challenge of this work. No previous works have already been devoted to the use of hybridization of such techniques in such a type of problems.

**Keywords:** Multiechelon systems, Particle Swarm Optimization, Repairable Spare parts.

# 1    Introduction

Repairable spare parts are referred to critical and expensive components which have infrequent failures; this type of spare parts are common in mining, military and in a variety of industries with high valued physical assets. Aircraft and warship engines, transportation equipment, and high cost electronics are typical examples of repairable items.

A component is said to be repairable if, after a failure, it can be restored to operation by a repair procedure which does not constitute an entire substitution.

Repairable spare parts inventories have particular importance for different industries because they are characterized by heavily utilized, relatively expensive, equipment. Those inventories constitute a high fraction of the total of all assets in a typical company. Because of that, in recent years increased attention has been given to the problems of management of this type of inventories.

The repairable spare parts inventory management is mainly focused in the decision of the ordering strategy and the stock levels to obtain economic efficiency and system availability. That type of decisions is a difficult one, not only because the probabilistic nature of the repair and resupply times, but also because it involves multi-item, multi-echelon, and some other factors.

Currently available solutions to the multi-echelon inventory problem of repairable spare parts are mainly based on the METRIC (Multi-Echelon Technique for Recoverable Item Control) model proposed by [10].

Sherbrooke [10] proved that METRIC is useful in single echelon cases by proving convexity of the objective function. However, and as [2] Diaz pointed out, stock allocation using METRIC is based on marginal analysis. Thus far, the literature has noted the inefficiencies of the marginal approach and some alternatives has been proposed, such as mathematical programming procedures based on Lagrangean multipliers. Others strategies such as closed queuing network models have been reported in literature but they have limited practical appeal because of prohibitive computational complexity, [14] ,[16],[15] ,[12], [19], [17], [18].

In literature we found two works that explore the use of metaheuristics in such a problem. [6] Zhao et al. (2010) proposed a genetic algorithm (GA) for the multi-echelon inventory problem of repairable spare parts. To improve the performance of the GA, a local search procedure, including two heuristic methods, was implemented and integrated into the algorithm. [9] proposed a new model for a two-echelon inventory system for non-repairable items with continuous reviews control and subject to constraints such as the average annual order frequency, the expected number of backorder, and budgets. The authors also explored the use of a GA to solve this problem efficiently.

To explore the use of other resolution methods, this work presents the results obtained in applying a swarm based optimization algorithm to the Repairable spare parts Multi-Echelon optimization problem. Also a local search procedure was applied in order to obtain better results. Comparisons with the results obtained by a Genetic Algorithm are also presented.

The remainder of the paper is organized as follows. In Section 2, the problem is defined precisely. Section 3 is dedicated to a comprehensive explanation of the methodology proposed to solve the model. In Section 4, numerical examples are given. In section 5 the results are presented and discussed. Finally, conclusions are provided and future research directions are proposed in Section 6.

## 2 Statement of the Problem

The system considered in this research project has several bases and a central depot with several types of repairable items (figure 1). At each base, there is a set of spare parts. When a components fails, it should be replaced by a spare part in good condition. If the base has at least one spare part in stock the component is immediately replaced; if no part is available, a backorder is generated at the base, and the base requests a unit of that spare part from the depot. If the base has the repair capacity, the failed part is repaired at the base; otherwise, it will be sent to the depot, and a resupply demand from the depot is placed. The problem considers four assumptions:

- The decision as to whether a base repairs an item does not depend on stock levels or workload;
- The base is resupplied by the depot, not by lateral supply from another base;
- continuous review;
- one for one replenishment ordering policy.

If we assume that we have a budget $B$ that is available to buy spares parts of different types. The problem can be expressed as: define the spare parts allocation and levels that contribute to the minimization of the sum of backorders at all bases, with the total investment (cost) of the spares (all types at all bases and the depot) not exceeding a given budget $B$.

The following notation is used for mathematical formulation of the problem:

- $m_{ij}$: the average annual demand of part $i$ at base $j$
- $s_{ij}$: the stock level of part $i$ at base $j$
- $T_{ij}$: the average repair time of part $i$ at base $j$
- $\mu_{ij}$: the corresponding average pipeline of part $i$ at base $j$
- $r_{ij}$: probability of repair of part $i$ at base $j$
- $O_j$: average order and ship time from depot to base $j$

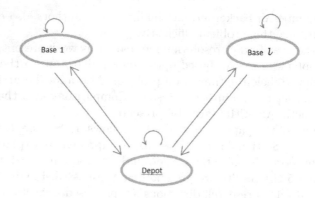

**Fig. 1.** Modelo generico de sistema multiescalonado de repuestos reparables

Note that, for the aforementioned variables, a 0 subscript refers to the central depot. Then, the average demand on the depot, $m_{i0}$, can be calculated as follows:

$$m_{i0} = \sum_{j=1}^{n} m_{ij}(1 - r_{ij}) \tag{1}$$

with:

$(i = 1, 2, ., l; j = 1, 2, , n)$

Where:

$n$: the total number of bases.
$l$: the total of different items of spare parts.
   The average pipeline for each demand at base $j$ can be computed as:

$$\mu_{ij} = m_{ij}\left[r_{ij}T_{ij} + (1 - r_{ij}) * \left(O_j + \frac{EBO[s_{i0} \mid m_{i0}T_{i0}]}{m_{i0}}\right)\right] \tag{2}$$

with:

$(i = 1, 2, ., l; j = 1, 2, , n)$

Where $EBO[s_{i0} \mid m_{i0}T_{i0}]$ is the expected backorder of part $i$ at depot when the average depot pipeline is $m_{i0}T_{i0}$, and the stock level of part $i$ at depot is $s_{i0}$. In addition, the expected backorders are:

$$EBO[s_{ij} \mid m_{ij}T_{ij}] = \sum_{x=s_{ij}+1}^{\infty} (x - s)P\{DI = x|\mu_{ij}T_{ij}\} \tag{3}$$

where the P{} terms are the steady-state probabilities for the number of units of stock due in, and $DI$ represents the number of units of stock due in from repair and resupply. The availability at base $j$ is:

$$A_j = 100 \prod_{i=1}^{l} \left( 1 - \frac{EBO[s_{ij} \mid m_{ij}T_{ij}]}{N_j Z_i} \right)^{Z_i} \tag{4}$$

with:

$$(i = 1, 2, ., l; j = 1, 2, , n)$$

Where:

$N_j$ : the number of equipment at base $j$,
$Z_i$: the number of occurrences of spare part $i$ on the equipment.

Using the presented formulation and the previously listed variables, the mathematical model of the multiechelon inventory problem is as follows:

$$max: \quad A = \frac{\sum_{j=1}^{n} A_j N_j}{\sum_{j=1}^{n} N_j} \tag{5}$$

s.t.

$$\sum_{i=1}^{l} \sum_{j=0}^{n} c_i s_{ij} \leq B \tag{6}$$

Where $c_i$ denotes the price of part $i$. The equation 5 corresponds to the objective function that maximizes the overall system availability. The maximization is constrained by the total investment of equipment support or budget $(B)$. The decision variables corresponds to the stock levels of part $i$ at base $j$ (integer values).

Multi-echelon, multi-item optimization problems are known for their hardness in solving them to optimality, and therefore heuristics methods are approached to near-optimally solve such problems.

## 3 Particle Swarm Optimization

The underlying PSOs concept is to generate and initialize a starting population with random solutions (integer values) called particles ([5]). Through a series of iterations, the initial population evolves to find optimal solutions. In PSO, each particle in population has a velocity, which enables them to travel through the solution space instead of mutating or perishing. Therefore, each particle i at the iteration k can be represented by a position $(x_i^k)$ and its respective velocity$(v_i^k)$. The modification of the particle position is performed by using its previous position information and its current velocity. Each particle keeps record of its best position (personal best, *pbest*) so far and the best position achieved in the group (global best, *gbest*) among all personal bests. These principles can be formulated as:

$$v_i^{k+1} = wv_i^k + c_1 r_1((pbest)_i^k - x_i^k) + c_2 r_2(gbest^k - x_i^k) \tag{7}$$

$$x_i^{k+1} = x_i^k + v_i^{n+1} \tag{8}$$

where:

- $w$ is inertia weight;
- $c_1$ and $c_2$ are two positive constants, called cognitive and social parameters, respectively;
- $i = 1,2,\ldots,S$, and $S$ is the size of the swarm;
- $r_1$ and $r_2$ are random numbers uniformly distributed in $[0, 1]$;
- $k = 1,2,\ldots; I$, denotes the iteration number;
- $I$ is the maximum allowable iteration number.

The first term on the right hand side of Eq. (7) is the previous velocity of the particle, which enables the particle to fly in the search space. The second and third terms are used to change the velocity of the agent according to *pbest* and *gbest*. The proposed PSO algorithm woks according to the scheme that is described as follows:

---
**Algorithm 1.** Proposed PSO algorithm
---
1: Set generation = 0
2: Create initial population of solutions randomly
3: Repeat the following until a termination condition is satisfied
4: Evaluate
5: Compare
6: Imitate
7: Set generation = generation +1
8: Set new particles positions
9: End repeat

---

The evaluation procedure measures how well each particle solves the problem at hand. The comparison phase identifies the best particles using the fitness function. The imitation phase produces new particle positions based on some of the best particle positions previously found. The objective is to find the particle that best solves the optimization problem.

## 3.1   Modified PSO Algorithm

The main difference between the traditional PSO technique and the one used in this work is that the latter does not use the velocity vector in a traditional manner. Here, a mechanism adapted from [1], the proportional likelihood has

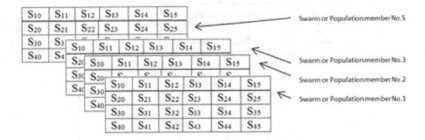

**Fig. 2.** General particle representation

| 1 | 2 | 1 | 2 | 3 | 3 |
|---|---|---|---|---|---|
| 1 | 1 | 2 | 1 | 2 | 3 |
| 1 | 2 | 3 | 2 | 1 | 3 |
| 1 | 2 | 2 | 3 | 1 | 3 |

**Fig. 3.** A sample particle representation

been especially designed to be used with discrete values in the PSO. The modified PSO algorithm has been used in previous works [3],[4] .

As previously mentioned, this algorithm deals with discrete variables, and its population of candidate solutions contains particles of a given size. Potential sets of solutions are represented by a swarm of particles. Each particle takes integer values and represents the stock level of part $i$ at base $j$. The swarm is compound of $S$ particles. Each particle keeps record of the best position it has ever attained. This information is stored in a separate particle, called $B(i)$. The algorithm also keeps record of the global best position attained by any particle in the swarm along the iterations. This information is also stored in a special particle called $G$. An integer matrix of dimension $(n + 1)l$ is used to represent each particle, where n is the total number of bases, $l$ is the total item of spare parts. Fig. 2a gives an example with 5 bases, and 4 spare parts. In the example shown in Fig. 2b, the stock level of part 1 at depot is 1, the stock level of part 2 at depot is 1, and the stock level of part 3 at base 3 is 2.

## 4    Proposed Strategy

The following sections are devoted to the series of steps that are included in the resolution strategy. Figure 4 depicts the phases that are included.

### 4.1    The Generation of Initial Population

The initial population generation procedure is performed as follows: for each particle, the algorithm randomly selects a part i and a location j, and adds

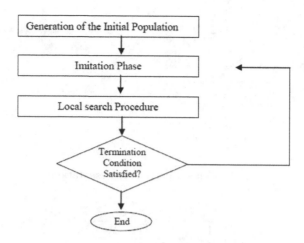

**Fig. 4.** Global Optimization Procedure

a random number between 1 and 10, considered as the maximum quantity or capacity of any stock location (corresponding stock level $s_{ij}$). This procedure will be repeated until the total investment cannot afford to increase more spare parts to the stock.

The procedure mentioned above is iterated until the size of population is generated.

After the initial generation of the population, the estimation of the fitness function is done (in this case, the estimation of the availability with each of the solutions suggested by the members of the swarm).

---

**Algorithm 2.** Initial population generation

---
1: Let $totalcost=0$
2: Select a random part $i$ and a random location $j$;
3: If $(totalcost + c_i) < B$, then let $s_{ij} = s_{ij} + 1$, $totalcost = totalcost + c_i$, return to step 2; otherwise, go to step 4;
4: Randomly select an inventory location $j$, and a part $i$ from parts with lower cost $(c_i)$ than the previous selected part i;
5: If $(totalcost + c_i) < B$, then let $s_{ij} = s_{ij} + 1$, $totalcost = totalcost + c_i$, return to step 4; otherwise, go to step 6;
6: If part $i$ is the part with the lowest price, the initial solution is completed; otherwise, return to step 4.

---

### 4.2   Imitation Phase

The imitation phase is based on the algorithm suggested by [3],[4]. The particle position modification is based on the difference between the actual position $(X_i^k)$ and the best position of the particle $(B_i)$, named $\Delta$ XB $= X_i^k - B_i$, and the difference between the actual position and the global best position in the swarm,

$$X_i^k = \begin{bmatrix} 1 & 2 & 2 & 3 \\ 1 & 3 & 1 & 1 \\ 2 & 3 & 1 & 1 \\ 1 & 2 & 3 & 1 \end{bmatrix} \qquad B_i = \begin{bmatrix} 2 & 1 & 2 & 3 \\ 1 & 3 & 1 & 2 \\ 2 & 1 & 2 & 2 \\ 1 & 1 & 2 & 1 \end{bmatrix}$$

$$X_i^k - B_i = \begin{bmatrix} -1 & 1 & 0 & 0 \\ 0 & 0 & 0 & -1 \\ 0 & 2 & -1 & -1 \\ 0 & 1 & 1 & 0 \end{bmatrix}$$

**Fig. 5.** Comparison between G.A. and PSO

$$X_j' = \begin{bmatrix} 1 & 1 & 2 & 3 \\ 1 & 3 & 1 & 1 \\ 2 & 1 & 2 & 1 \\ 1 & 2 & 2 & 1 \end{bmatrix}$$

**Fig. 6.** Comparison between G.A. and PSO

named $\Delta \, XG = X_i^k - G$. These differences represent the movements needed to change the actual position of a particle to the position given by the second term of each expression (See figure 5). The difference between $X_i^k$ and $B_i$, is a $(n+1)l$ matrix called $\Delta$ XB that represents the changes that are needed to move the particle $X_i^k$ to the best position of the swarm $(B_i)$. Therefore, $\Delta$XB could take values equal to and different to zero. If $\Delta XB \neq 0$, it means that there is an improvement potential through changing the position of the particle to the best global position. Let $\theta$ be the number of elements different from 0 in $\Delta XB$. If the difference between a given element of $X_i^k$ and $B_i$ is not null, it means that the corresponding particle is susceptible to change through the imitation process described in the following paragraphs. In the example shown in figure 5, the value of $\theta$ is 8.

Once the $\Delta XB$ matrix is obtained and the $\theta$ scalar is computed a new vector $\varphi$ is constructed. The vector $\varphi$ keeps record of the positions where the elements of $X_i^k$ and $B_i$ are not equal. In our example, the vector has 8 members:

$$\varphi = [(1,0);(1,1);(2,3);(3,1);(3,2);(3,3);(4,1);(4,2)]$$

After that, $\kappa$, a random number, is generated. The scalar $\kappa$ whose values are in the interval $(0,\theta)$ corresponds to the quantity of changes that will be made to $X_i^k$, based on the difference between $X_i^k$ and $B_i$. Suppose that, in our example, $\kappa$ is 4.

Then, a set $\tau$ of $\kappa$ randomly generated binary numbers is defined. If the binary number is 1, the change is made; on other hand, if the number is 0, the change is not performed.

Suppose that $\tau = [0\ 1\ 0\ 1\ 1\ 0]$

This means that the third, fifth and sixth elements of $\varphi$ have the positions of the elements in $X_i^k$ that will be changed by the correspondent elements in $B_i$. This operation is shown in figure 6.

To update the particle position in accordance to the best global position $(G)$ a set of similar operations are performed. In the case that several of the applied movements involve the same position $s_{ij}$, the change caused by the global best position, the second operation in our algorithm, has the priority. This process is repeated until the last particle member of the swarm is modified.

## 4.3    The Local Search Procedure

At the end of each iteration, a local search procedure is applied to the best individual to accelerate the convergence of the algorithm. The local search procedure tries to increase additional spare parts units to the solution aiming at making full use of the budget $(B)$. The procedure checks all spare parts that can be feasibly added to the solution, and add 1 to the stock level at the best location. This process is repeated until the residual budget cannot afford even an additional unit of the cheapest part.

---

**Algorithm 3.** Local Search Procedure

---
1: Select gbest
2: Let *totalcostbest* the total cost associated to the best solution
3: Select a random part $i$ and a random location $j$;
4: If $(totalcostbest + c_i) < B$, then let $s_{ij} = s_{ij} + 1$, $totalcostbest = totalcostbest + c_i$, return to step 3; otherwise, go to step 5;
5: Randomly select another inventory location $j$, and a part $i$ from parts with lower cost $(c_i)$ than the previous selected part i;;
6: If $(totalcostbest + c_i) < B$, then let $s_{ij} = s_{ij} + 1$, $totalcostbest = totalcostbest + c_i$, return to step 3; otherwise, go to step 6;
7: If part $i$ is the part with the lowest price, the initial solution is completed; otherwise, return to step 4.

---

# 5    Computational Experiments

## 5.1    Algorithm Implementation and Tuning

The purpose of this section is to show, through a numerical example, how the proposed formulation can be used to solve a multi echelon, multi item repairable spare parts problem. To test the performance of the improved discrete PSO algorithm, we solved the problem instance given in [6] Zhao et al. (2010). In that instance, there are one depot, ten bases, and fifteen different spare parts. The parameter values for the instance were set as follows:

The number of equipment at the bases:
$(N_j)$x10=(10,10,10,10,10,10,10,10,10,10);

The average annual demands of spare parts for each base:
$(m_{ij})$x15x10=(12.1,6.6,11.2,5.9,13.2,13.6,10.8,7.2,12.5,12.3,5.1,8.9,3.3,2.0,9.5)x10;

The price of parts:
$(c_i)$x10=(120,100,155,85,167,221,77,122,189,356,151,106,118,172,162);

The number of occurrences of part on the equipment:
$(Zi)$x10=(1,2,1,3,2,1,1,1,2,1,1,3,2,4,2).

The average repair time for spare parts at bases is 0.01 (year).
The average repair time for spare parts at the depot is 0.02 (year).
The repair probability at bases is 30%.
The average order and ship time from depot to bases is 0.01 (year).

In addition, the instance is solved under different budget $(B=\$7500, \$10000,$ ..., \$50000)$.

The PSO based algorithm was implemented in MATLAB environment.

For selecting the algorithms parameters nine test problems were designed. We tested three different swarm sizes, i.e. 20, 30 and 50 particles. Additionally the number of iterations was defined as 30, 60, 80 and 100 respectively. As can be seen in figure 7, there are no important differences among the different configurations. Therefore we selected the following configuration: 30 particles and 100 iterations as the ending condition (Figure 7). Those tests were executed using 1 unit as the maximum level of $s_{ij}$ (no information is available from Sun et al. (2010) about the upper limit used to test the GA). In addition, it can be observed that with lower budgets PSO overcomes GA. On the contrary, with higher budgets GA overcomes PSO. However, as will be seen later in this work, with higher capacity limits PSO shows better performances.

# 6    Experimental Results

A set of ten test problems was used to evaluate the proposed implementation of the PSO algorithm. Each problem was solved 100 times with different values of the maximum stock levels($s_{ij}$), from 1 unit to 10 units. The parameter setting obtained previously was used by the proposed PSO algorithm to run all the test instances. As it was expected, solving the test problems several times can increase the percentage of success. The comparison of computational results

**Table 1.** Comparison of the maximum availabilities between those obtained by the GA and the proposed PSO in %

| Budget $ | GA result | 3 | 4 | 5 | max $s_{ij}$ 6 | 7 | 8 | 9 | 10 |
|---|---|---|---|---|---|---|---|---|---|
| 7500 | 87.39 | 91.50 | 91.26 | 92.03 | 91.52 | 91.62 | 90.86 | 91.15 | 90.83 |
| 10000 | 89.18 | 92.11 | 92.03 | 93.06 | 92.84 | 92.58 | 92.46 | 91.68 | 91.46 |
| 12500 | 90.74 | 92.85 | 92.19 | 94.60 | 92.84 | 93.52 | 93.18 | 92.54 | 91.94 |
| 15000 | 92.63 | 95.84 | 93.74 | 94.60 | 93.22 | 95.46 | 93.58 | 93.61 | 93.76 |
| 17500 | 94.71 | 95.84 | 94.24 | 94.60 | 94.37 | 95.46 | 93.58 | 94.68 | 93.76 |
| 20000 | 95.97 | 97.42 | 95.58 | 95.74 | 94.63 | 95.46 | 93.58 | 94.68 | 93.76 |
| 22500 | 97.54 | 97.42 | 95.58 | 95.74 | 98.45 | 96.48 | 94.36 | 96.20 | 96.01 |
| 25000 | 98.44 | 99.40 | 99.40 | 95.74 | 98.45 | 96.48 | 94.62 | 96.20 | 96.01 |
| 27500 | 98.99 | 99.46 | 99.46 | 99.47 | 98.45 | 96.48 | 95.38 | 96.20 | 96.34 |
| 30000 | 99.24 | 99.46 | 99.46 | 99.47 | 98.45 | 99.54 | 96.88 | 96.20 | 96.34 |
| 32500 | 99.40 | 99.46 | 99.46 | 99.47 | 98.45 | 99.57 | 97.43 | 96.60 | 98.31 |
| 35000 | 99.52 | 99.46 | 99.46 | 99.47 | 99.62 | 99.57 | 97.43 | 97.89 | 98.77 |
| 37500 | 99.62 | 99.66 | 99.46 | 99.47 | 99.62 | 99.57 | 99.67 | 97.89 | 98.77 |
| 40000 | 99.71 | 99.70 | 99.46 | 99.47 | 99.62 | 99.57 | 99.67 | 97.89 | 98.77 |
| 42500 | 99.83 | 99.72 | 99.74 | 99.47 | 99.62 | 99.57 | 99.67 | 98.71 | 98.81 |
| 45000 | 99.89 | 99.75 | 99.74 | 99.87 | 99.86 | 99.57 | 99.67 | 98.71 | 98.81 |
| 47500 | 99.91 | 99.75 | 99.74 | 99.93 | 99.86 | 99.57 | 99.94 | 98.71 | 98.81 |
| 50000 | 99.96 | 99.81 | 99.74 | 99.93 | 99.86 | 99.57 | 99.94 | 98.71 | 99.52 |

**Table 2.** Performance metrics for B=$50000

| stock levels($s_{ij}$) | Best | Average | % | 1 % Opt | 5% Opt |
|---|---|---|---|---|---|
| 1 | 99.72 | 99.70 | 0.05 | 100.00 | 100.00 |
| 2 | 99.75 | 99.60 | 0.05 | 100.00 | 100.00 |
| 3 | 99.81 | 99.40 | 0.21 | 60.00 | 100.00 |
| 4 | 99.74 | 99.85 | 0.34 | 100.00 | 100.00 |
| 5 | 99.93 | 99.81 | 0.08 | 100.00 | 100.00 |
| 6 | 99.86 | 99.58 | 0.05 | 50.00 | 100.00 |
| 7 | 99.57 | 99.75 | 0.99 | 100.00 | 100.00 |
| 8 | 99.94 | 99.94 | 0.19 | 70.00 | 100.00 |
| 9 | 98.71 | 98.00 | 0.72 | 20.00 | 100.00 |
| 10 | 99.52 | 99.10 | 0.42 | 60.00 | 100.00 |

are given in Table 1. The first column presents the budgets available in each case. The second column presents the availability values obtained by the GA in [11], under different budget, and the last eight columns give the results of the improved PSO with different maximum stock levels. Figure 8 shows the best results obtained by the proposed PSO besides those results obtained by the GA.

For each one of the planned budgets and each one of the different values of the maximum stock levels($s_{ij}$), other performance measurements were computed. (For space economy reasons we have added In Table 2 only the results for B=$50000). Column "Best" shows the Availability (Best solution) found by

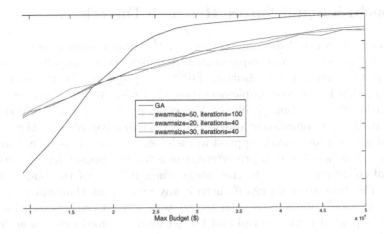

**Fig. 7.** Parameter definition

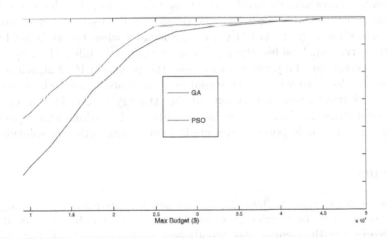

**Fig. 8.** Comparison between G.A. and PSO

the PSO in 100 runs. The second Columns shows the average value obtained by the 100 runs. Column "Average" shows the results from (% [Best-Ave]/Best), that is the deviation between the best solution and the averaged solution over the 100 runs. Column "1 % Opt." shows the percentage of solutions that differed by at most 1 % from the best solution found. Column "5 % Opt." shows the percentage of solutions that differed by at most 5% from the best solution found. For each problem, 65% on average of the solutions were within 1% of optimally. Finally, almost 100% of the solutions found by the PSO Algorithm were within the 5% of optimally.

# 7   Conclusion and Future Research Directions

In this research, a novel solution strategy for the two-echelon inventory system problem was developed and implemented. A hybrid search algorithm based on discrete particle swarm optimization (DPSO) is proposed. To the best of our knowledge, this is the first implementation of a discrete PSO in a repairable Multi-echelon optimization problems. Moreover, until now, any hybrid strategy has been used in combination of swarm intelligence to solve this kind of problems. The local search procedure is applied in each iteration to the best individual in the swarm. The local search procedure search for the possibility of increasing additional spare parts units to the stocks when full use of the budget $(B)$ is possible. The procedure checks if there is any spare part that can be feasibly added to the stock, and add one unit to the stock level at the best location. This process is repeated until the residual budget cannot afford even an additional unit of the cheapest part.

In order to demonstrate the effectiveness of the DPSO, we have carried out an experimental study and also explored the use of a GA to solve the problem. The main conclusion of this work is that the developed PSO strategy overcomes the performance shown by the GA (Figure 8). Specially when the available budgets are more restricted. Besides, the results show that the higher the upper limits of $s_{ij}$, the better are the performances using the proposed PSO algorithm. This technique is closer to real world scenarios and hence more feasible to apply in practice. Future research may extend the strategy by considering variations in PSO operators. Initialization strategies, fitness definitions and replacement strategies are obviously possible and might lead to more efficient solutions.

# References

1. Correa, E.S., Freitas, A., Johnson, C.G.: A new discrete particle swarm algorithm applied to attribute selection in a bioinformatics data set. In: M.K., et al. (eds.) Proceedings of the Genetic and Evolutionary Computation Conference, GECCO 2006, pp. 35–42. ACM Press, Seattle (2006)
2. Diaz. A, Multi echelon inventory models for repairable items. Doctor thesis. University of Maryland (1995)
3. Duran, O., Rodriguez, N., Consalter, L.A.: Collaborative Particle Swarm Optimization with a Data Mining Technique for Manufacturing Cell Design. Expert Systems with Applications 37(2), 1563–1567 (2010)
4. Duran, O., Perez, L., Batocchio, A.: Optimization of modular structures using Particle Swarm Optimization. Expert Systems with Applications 39(3), 3507–3515 (2012)
5. Eberhart, R.C., Kennedy, J.: Particle Swarm Optimization. In: Proceeding of the IEEE International Conference on Neural Networks. IEEE Service Center 1213, Perth (1995)
6. Zhao, F., Sun, J., Zhang, L., Ma, Z.: Genetic algorithm for the multi-echelon inventory problem of weapon equipment repairable spare parts. In: 2010 3rd IEEE International Conference on Computer Science and Information Technology (ICCSIT), July 9-11, vol. 2, pp. 619–622 (2010)

7. Goyal, Giri, B.C.: Recent trends in modelling of deteriorating inventory. European Journal of Operational Research 134(1), 1–16 (2001)
8. Nowicki, D.R., Randall, W.S., Ramirez-Marquez, J.E.: Improving the computational efficiency of metric-based spares algorithms. European Journal of Operational Research 219(2), 324–334 (2012)
9. Pasandideh, S.H.R., Niaki, S.T.A., Tokhmehchi, N.: A parameter-tuned genetic algorithm to optimize two-echelon continuous review inventory systems. Expert Systems with Applications 38(9), 11708–11714 (2012)
10. Sherbrooke, C.: METRIC: A multi-echelon technique for recoverable item control. Operations Research 12, 122–141 (1968)
11. Sun, J., Zhao, F., Zhang, L.: An improved genetic algorithm for the multi-echelon inventory problem of repairable spare parts. In: 2010 IEEE International Conference on Intelligent Computing and Intelligent Systems (ICIS), October 29-31, vol. 1, pp. 440–444 (2010)
12. Gross, D., Gu, B., Soland, R.M.: Iterative solution methods for obtaining steady state probability distributions of Markovian multi-echelon repairable items inventory systems. Computer and Operations Research 20, 817–828 (1993)
13. Gross, D., Hams, C.M.: Fundamentals of Queueing Analysis, 2nd edn. John Wiley and Sons, New York (1985)
14. Gross, D., Ince, J.F.: A closed queueing network model for multi-echelon repairable items provisioning. AIIE Transactions 10, 307–314 (1978)
15. Gross, D., Kioussin, L.C., Miller, D.R.: A network decomposition approach for approximate steady state behavior of Markovian multi-echelon repairable item inventory systems. Management Science 33, 1453–1468 (1987)
16. Gross, D., Miller, D.R., Soland, R.M.: A closed queueing network model for multi-echelon repairable item provisioning. IIE Transactions 15, 344–352 (1983)
17. Albright, S.C.: An approximation to the stationary distribution of a multiechelon repairable-item inventory system with finite sources and repair channels. Naval Research Logistics 36, 179–195 (1989)
18. Albright, S.C., Gupta, A.: Steady-state approximation of a multiechelon multi-indentured repairable-item inventory system with a single repair facility. Naval Research Logistics 40, 479–493 (1993)
19. Albright, S.C., Soni, A.: Markovian multiechelon repairable inventory system. Naval Research Logistics 35, 49–61 (1988)

# A Goal-Oriented Meta-Model for Scientific Research

Jérôme Dantan[1,2], Yann Pollet[2], and Salima Taibi[1]

[1] Esitpa, Agri'terr, Mont-Saint-Aignan, France
{jdantan,staibi}@esitpa.fr
[2] CNAM, CEDRIC, Paris, France
yann.pollet@cnam.fr

**Abstract.** In many research domains, studies need to be addressed in a multi-disciplinary manner: each expert deals with a particular aspect of the problem. It may be useful for experts to take into account new data, share partial results, and update their own indicators and models, in order to take advantage of new measures and updated indexes in real time. For this, an easily-understandable knowledge model for any raw data source, statistical operator, indicator or business process to be available for experts, is needed. In this paper, we propose a goal-oriented meta-model, to index and reuse treatments and an extension of existing semantic Web standards to index goal-oriented services and assist and/or automate their selection. These features enable capabilities for interoperability and information exchange between three layers of knowledge: goal, domain and data layers. An application with an existing ontology of the agriculture domain and farm durability indicators will be proposed.

**Keywords:** declarative language, knowledge, meta-model, ontology, semantic indexation.

## 1 Introduction

In many research domains, scientists have to cope with large amounts of data issued from multiple data sources, looking for regular patterns in raw data that may suggest models. Extraction of knowledge from data is widely used in fields such as biology, economics, etc. For this, experts develop models in their respective area of research. The efficiency of the experts depends on their ability to cooperate and to share pertinent information with the others. Indeed, each domain is both consumer of treatments provided by another team, and provider for the benefit of other areas. Each team of experts has to make available and publish business processes for the benefit of other teams, which allows them to have the most up to date treatments. As a result, we propose a meta-model system where heterogeneous data and models are held together by a core ontology that describes the whole domain of agriculture (AGROVOC, [7], [12]), so that the experts will be able to operate and cooperate. The rest of the paper is organized as follows: section 2 presents an analysis of the domain of agriculture, and then section 3 presents a state of the art and previous work. Section 4 introduces our approach. Section 5 illustrates the architecture of the system. Finally, section 6 presents our conclusions and perspectives for future work.

B. Murgante et al. (Eds.): ICCSA 2014, Part V, LNCS 8583, pp. 762–774, 2014.

## 2 Context: The Agriculture Domain

In this paper, we consider fields that belong to the agriculture domain such as agronomy, ecology, rural economy, durability model development, etc. Assessing the sustainability of human activities is becoming a worldwide major concern. Researchers in the agriculture domain do not only seek to measure and improve performance of farms, but also their sustainability (e.g. agroecological sustainability, economic sustainability, socio-territorial sustainability), through indicators to define. The development and computation of such indicators require being expert in multiple areas and requiring data relating to many domains. Indeed, various aspects about farms need to to be treated, such as: soil biology, crop diversity, waste management, economic viability, sales operations, etc.

Many models and indicators, which measure the sustainability of farms, already exist. They are used as tools for decision support for farmers, chambers of agriculture and agricultural researchers (agronomy, land use, chemistry ...). The agricultural experts have therefore to compute:

- Many data (from samples, surveys ...) stored in heterogeneous formats (Excel, Access databases, etc).
- Many indicators and models to evaluate the sustainability or durability of farms from such data. For example: the "IDEA" method [16], the "Indigo" Method, the "Farre" Self-diagnosis, the "Diaphyt" model (Waste Management Plant Protection Products), the "Dialogue" model (Farm agri-environmental diagnosis), the "Diage" model (global farma diagnosis), etc.
- Many computer programs (wrappers) have been developed to compute data and indicators.

Each indicator, model or wrapper is adapted to a specific context. For example, a French researcher does not have the same concerns as a Chinese researcher in terms of sustainability: input levels tolerated more or less, different requirements in terms of economic profitability and productivity ... In some countries, such models and indicators do not exist. That is why a meta-model is required. It is designed as a collaborative system between experts: everyone can share data, algorithms, models that have been indexed as a service via the Internet.

Let us briefly recall the examples presented in [5]: suppose that a researcher has developed a soil quality index, and that he wants to modify the IDEA model (which is a farm durability indicator) by inserting such a criterion in the sustainability calculation methodology and then applying the modified IDEA model on his/her own data. Our system may allow automatically assessing the effects of the new model, applying it on new data and sharing his /her findings with other experts around the world. For its part, another researcher will be able to test sustainability indicators (e.g. IDEA) indexed on the system on his/her own data and invent new diagnostic tools based on existing indicators. For example he/she may change the weight of some sub-criteria such as biodiversity, environmental protection, heavy metal rate, etc and create his/her own indicator by reusing existing indicators and the computer treatment that are already indexed.

# 3     Background

## 3.1     Knowledge Representation with Ontologies

Let us recall that "an ontology is a formal specification of a shared conceptualization" [2]. We will use Web Ontological Language (OWL) related technologies that provide model flexibility, explicit representation of semantics, out-of-the-box reasoners and proliferation of freely available background knowledge [4]. We have chosen the AGROVOC OWL ontology [7] [12] as core ontology, which formalizes the agriculture domain. OWL [17] distinguishes between two main categories of properties that an ontology builder may want to define (1) Datatype properties for which the value is a data literal and (2) object properties for which the value is an individual. A state of the art on both goal-oriented and semantic Web approaches is provided in [6].

The ontological database model that [11] propose includes the contribution of a query declarative language, OntoQL. The goal-oriented approach is more and more used in software engineering and looks promising [8]. However, the scale at which we work is different, because, in this case, data are spread across many heterogeneous databases and distributed over the Web. Indeed, we distinguish two parts in our model:

- The structure (i.e. the intention) which is in the ontology (schema, metadata). It consists very roughly of the ideas, properties, or corresponding signs that are implied or suggested by the concept in question.
- The data (i.e. the extension) which are spread across multiple heterogeneous databases (relational DB, Excel files, NoSQL DB) and distributed over the Web. They consist of things to which the model applies.

As a result, we have expressed data sources and goals under a common formalism understandable by domain experts, using a core ontology. Moreover, we have defined a knowledge model with three different layers and two types of derived properties which are not directly stored in the databases but are derived when the attribute is referenced within an OCL-style (Object Constraint Language) query language expression, as an extension of OWL [5]:

- The derived properties, which result in a reference path, that are indexed thanks to data-oriented services (data layer). For example, the "famer_age" derived property of the "farm" class is deduced from the "age" datatype property of the "farmer" class thanks to the "farmer.farmer.age" transitive link. The relevant data and metadata are provided by Data-services (D-services) that are expressed thanks to the elements of the domain ontology.
- The factor properties, which match to hierarchically organized goals and which are indexed by the knowledge-oriented services (goals layer). A K-service is characterized by the goal it instantiates, added to an OCL-style (AGRO-JD Object Constraint Language) pre-condition (if any). For example, the "SoilQualityIndex" factor property is defined by the rates of heavy metals and respiration to microbial biomass ratio ($qCO2$). The output is a "soil quality index" concept belonging to the

ontology of goals. The following code example provides the expression of a soil quality index, and of some heavy metal rates, that are formulated thanks to the AGRO-JD language:

```
SoilQualityIndex = SUM(Farm.qCO2Rate.MEDIAN,
Farm.HeavyMetalRate.MEDIAN)
Farm.HeavyMetalRate.MEDIAN = SUM(Farm.CadmiumRate.MEDIAN,
Farm.PlumbRate.MEDIAN, Farm.ArsenicRate.MEDIAN)
Farm.CadmiumRate =
SUM(NAVIGATION(Farm.Parcels).CadmiumRate)
Farm.PlumbRate = SUM(NAVIGATION(Farm.Parcels).PlumbRate)
Farm.ArsenicRate =
SUM(NAVIGATION(Farm.Parcels).ArsenicRate)
```

New elements of the goals ontology should be added to enrich sustainability models such as the IDEA model. The services that provide the goals to reach are called Knowledge Services (K-services). As a result, we provided a way to index services designed by experts thanks to the ontology of goals; an ontological formulation that formalizes the data, treatments, and goals developed by computer scientists, and ad hoc aggregated operators and factor properties that extend OWL through a common formalism. The declarative query language that we have defined is mapped to a standard query language of the Semantic Web.

### 3.2    SPARQL: an Ontology Query Language

Many query languages exist in the Semantic Web. According to [1] and [11], these languages are classified into seven categories (family of SPARQL, RQL family, language inspired by XPath, XSLT or XQuery languages English controlled language, languages with reactive rules, deductive and other languages). [11] propose the evaluation of one representative of two categories: SPARQL category and RQL category.

As we use an OWL format as core ontology (AGROVOC), we naturally have selected to a query language belonging to the SPARQL category, which considers all the information, either ontologies or data, as triples. In this category, there are languages such as SPARQL [14], RDQL [15], SquishQL [13] and TriQL [3].

SPARQL 1.0 has become a W3C standard [14]; it is the reference language for querying OWL ontologies. Moreover, it is often mentioned in literature and many studies in the field of Semantic Web. Finally, the 1.1 version of SPARQL is being finalized [9]. We have therefore chosen to analyze this language on the basis of requirements that we have detected. Some of these requirements were proposed by [11] about the OntoDB query language.

In this paper, we more accurately present the indexation of the goals, by proposing to index them thanks to an "entity-goal-relation" model and thanks to a declarative language that extends the SPARQL 1.1 semantic Web query language.

# 4    Approach

## 4.1    The Entity-Goal-Relationship Model

As already mentioned, the goals are hierarchically organised. Consequently, they are evaluated on the basis of their sub goals. For instance, we can only evaluate the over-all sustainability or durability of a farm by assessing its sub-goals i.e. the sustainabili-ty or the durability of the elements that compose the farm. Furthermore, the global sustainability is composed of the agro-ecological sustainability, the economic sustai-nability, etc. Therefore, we have simply expressed the fact that the sustainability goal is based on the evaluation of sub goals, including the objects which are entities related to the "farm" objects (parcels, etc.) for the other components of sustainability (eco-nomic, human, etc...).

For this, we have tagged the link between sub-goals and goals with the label "con-tribute_to". This relationship has been expressed in a high level constraint declarative language [5] using the concepts of the core OWL ontology "AGROVOC". This label can also be expressed in SPARQL. The meta-model we have created contains three types of elements: entities, the goals and the relationships. We have included our me-ta-model in the three-layered model data-goal-domain we defined in [5]: the domain layer (core ontology), the data layer (data extraction and metadata information) and the goals layer (hierarchically organised elements that have to be evaluated). The Entity-Goal-Relationship model is illustrated in Fig. 1.

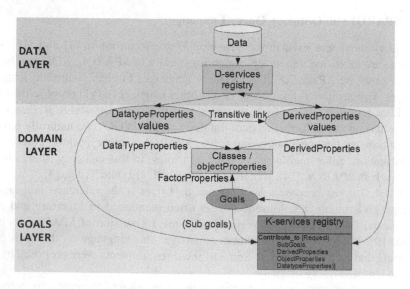

**Fig. 1.** The Entity-goal-relationship model

An example which illustrates the agriculture domain is provided in Fig. 2.

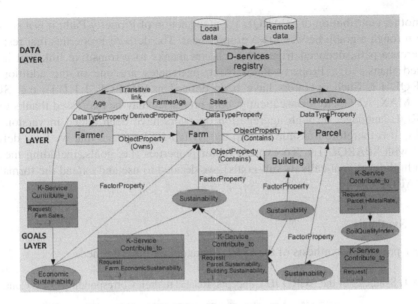

**Fig. 2.** Illustration of the agriculture domain

The entities match with the classes of the core domain ontology. The goals match with the hierarchy of factor properties (ontology of goals). The relationships are the labels linking the various elements of the core ontologies and goals. As already mentioned, goals and sub-goals are hierarchically linked thanks to a relationship which is labelled "contributes_to".

In the system we have designed, the operation of matching compares the abstract services that are defined by the user with the existing concrete services, to ensure their strict equivalence, isolate their common elements or enumerate their differences.

## 4.2    Exploiting SPARQL Version 1.1

We have explored the requirements on the power of expression as well as the implementation of the SPARQL 1.1 language. SPARQL 1.1 allows, like SPARQL 1.0: querying ontologies, querying both ontologies and data and defining a formal semantics.

Ontologies and data must be manipulated in order to map the AGRO-JD high-level language to the standard language of the Semantic Web SPARQL 1.1, which is the case: indeed, one of the most significant additions of SPARQL 1.1 is the ability to edit graphs. Until now, in SPARQL version 1.0, data were only accessible for reading and could not be updated; it required going through external languages. Two types of operations have been defined: operations on graphs (graph management) and operations on data in graphs (graph update). The available operations on graphs are CREATE and DROP to create and remove graphs; LOAD and CLEAR to add and delete all triples in the specified graph(s). The available operations on data and triples are INSERT DATA, DELETE DATA to insert/remove data given in the request, INSERT/DELETE to insert/remove triples in the specified graph(s).

Another contribution of SPARQL 1.1 is the notion of Property Path: it is a possible route through a graph between two graph nodes. The derived Properties that result in a reference path, deduced from other classes thanks to a transitive link that is expressed thanks to a Property Path. Next, aggregates are one of the additions in SPARQL 1.1. Such operators have been defined in SPARQL 1.1 [11], e.g. SUM, MIN, MAX, AVG, etc. Consequently, data properties may be expressed thanks to the SPARQL language, through aggregation operators have been added in version 1.1. Finally, it is therefore possible to map the AGRO-JD language that we have defined in [5] with SPARQL 1.1. To express factor properties (i.e. goals, including the fact that a high-level goal calls its sub-goals), we decided to use and extend the formalism of SPARQL 1.1.

### 4.3    Synthesis

Table 1 provides a synthesis of the OWL operators we have defined.

**Table 1.** Synthesis of the AGRO JD Concepts, Related to their Corresponding OWL Concepts

| AGRO JD concept | OWL concept | Description |
|---|---|---|
| Native property | DatatypeProperty and ObjectProperty | Basic OWL properties. |
| Derived property | Yes (Property Path) | Obtained thanks to a transitive link; indexed by D-services. |
| Factor property | No | Index the hierarchically organised goals; indexed by D-services (goal layer). |
| Generalized properties | No | Native properties UNION Derived property of a class |
| Description | EquivalentClass | Link between two class descriptions. Defines a new class C', which properties are a subset of the generalized properties of the C class. |
| Contribute_to | No | Link between sub-goals and goals, in the goals ontology. |

According to [10], there is no standard way to express "that the fillers of a data property can be derived from others by means of a formula" in OWL. It "seems yet unlikely to appear in the forthcoming OWL 2.0. Furthermore, there is no way to express the factor properties we have defined in our meta-model.

## 4.4 AGRO-JD-QL – SPARQL 1.1 Mapping

Table 2 illustrates the relation between the AGRO-JD-QL operators we have defined to the SPARQL basic operators.

**Table 2.** Synthesis of the AGRO-JD-QL Operators, Related to their Corresponding SPARQL Operators

| AGRO JD-QL operator | SPARQL1.1 operator mapping | Description |
|---|---|---|
| SELECTION | SELECT | Selection |
| WHERE | WHERE | Extracts only the records that fulfil a specified criterion |
| Aggregates (SUM, AVG...) | SPARQL aggregate operators | Aggregate operators |
| NAVIGATION([Class1].[Class2]) | Basic SPARQL query | Selection of an object property. |
| RESTRICTION ([Property], [Class]) | Basic SPARQL query | Restriction |
| VALUE | Basic SPARQL query | Property value |
| Concatenation of the whole operators provided above | Property path | May be chained thanks to the basic OWL operators such as navigation, restriction, and value, added to an additional aggregation operator. |
| CREATE GOAL UPDATE GOAL DELETE GOAL | No | Create, update or delete a goal. |

The instantiation, modification and deletion of goals are not implemented in the basic SPARQL 1.1 language. The parameters belonging to the core ontology can be expressed by using a label that can be expressed as a SPARQL query (associated with the "contribute_to" relationship).

## 4.5 Examples

The following code attaches a FactorProperty called "sustainability" to the "Farm" class, thanks to the AGRO JD-QL language. This FactorProperty is actually a sub-goal calculated from all that is contained in the farm (parcels, buildings, etc.).

```
CREATE GOAL {
  sustainability(SELECTION ?x WHERE Farm.Contains ?x)
contribute_to Farm.Sustainability}
```

The following code illustrates the selection request of all the objects contained in the farm called "The Green Farm", formulated in SPARQL.

```
<!-- Selection request of all the objects contained in the farm called
"The Green Fam". -->
PREFIX ag: <http://www.fao.org/aos/agrovoc#>
SELECT ?p
WHERE {
  ag:Farm ag:contains ?p .
  ag:Farm ag:name "The Green Fam" .
}
```

For instance, the sustainability of a farm is formulated thanks to each kind of durability that compose the overall durability of a given farm, e.g. parcels, buildings, etc. The following code illustrates the AGRO JD-QL level, for both Farm and Parcel classes:

```
<!-- Declaration of the "economicSustainability" and
"ecologicalSustainability" FactorProperties as farm sus-
tainability sub goals. -->
(Farm.economicSustainability AND
Farm.ecologicalSustainability) contribute_to
Farm.sustainability (Parcel.economicSustainability AND
Parcel.ecologicalSustainability) contribute_to Par-
cel.sustainability
```

### 4.6   A SPARQL Extension Proposal

The CREATE GOAL, UPDATE GOAL, SELECT GOAL and DELETE GOAL operators may be added to SPARQL 1.1. In the code below the "sustainability" goal is added to the "Farm" class thanks to the "CREATE GOAL" operator. Consequently, the "Farm.sustainability" Factor property can be expressed in the AGRO-JD language and computed from, on the one hand, the "sustainability" property of the objects contained in the Farm, and on the other, from "Farm.economicSustainability" and "Farm.ecologicalSustainability" Factor properties.

```
<!-- Declaration of the "sustainability" goal thanks to
new SPARQL "CREATE GOAL" operator. -->
CREATE GOAL {
    ag:farm owl:goal sustainability
    BEGIN
        SELECT SUM (?s
            WHERE {
                ?p jd:sustainability ?s
```

```
            ag:Farm ag:contains ?p .
      }
  )
  END
  BEGIN
      SELECT SUM (
          ag:farm owl:goal ecologicalSustainability,
          ag:farm owl:goal economicSustainability
      )
  END }
```

## 4.7   An OWL Extension Proposal

Once declared thanks to SPARQL, goals have to be defined thanks to an OWL formulation. The code below provides an example of OWL code which formulates the
Farm.sustainability goal thanks to:

- the "sustainability" property of the objects contained in the Farm,
- the "Farm.economicSustainability" and "Farm.ecologicalSustainability" Factor-
  Properties.

For this, we propose to add, as an extension of OWL, the following elements to the
XML formulation of ontologies:

- the "jd:FactorProperty" tag (declares a FactorProperty),
- the "jd:subGoalOf" attribute (declares that a goal is the sub goal of another goal),
- the jd:agregationMethod attribute (declares the method to aggregate sub goals in
  goals which belong to an upper level).

```
<!-- Example which illustrates the proposed OWL extension. -->
<!-- LINK BETWEEN FARM AND PARCEL -->
<owl:ObjectProperty rdf:ID="contains">
    <rdfs:domain rdf:resource="#Farm" />
    <rdfs:range rdf:resource="#Parcel" />
</owl:ObjectProperty>
<!-- PARCEL CLASS -->
<owl:Class rdf:ID="Parcel">
    <!-- "sustainability" is a FactorProperty of the
"Parcel" class -->
    <owl:DatatypeProperty rdf:ID="sustainability">
        <rdfs:domain rdf:resource="#Parcel" />
        <rdfs:range rdf:resource="##FactorProperty"/>
    </owl:DatatypeProperty>
</owl:Class>
<!-- FARM CLASS -->
<owl:Class rdf:ID="Farm">
```

```
<!-- "sustainability" is a FactorProperty of the
"Farm" class, deduced from the Sustainability property of
the classes "contained" in the "Farm" -->
    <jd:FactorProperty rdf:ID="sustainability">
        <rdfs:domain rdf:resource="#Farm" />
        <rdfs:range rdf:resource="SELECTION ?x WHERE
Farm.Contains ?x" jd:agregationMethod="SUM" />
    </jd:FactorProperty>
<!-- The "Farm.economicSustainability" and the
"Farm.ecologicalSustainability" factor properties contri-
bute to the "Farm.sustainability" property -->
    <jd:FactorProperty rdf:ID="economicSustainability">
        <jd:subGoalOf rdf:resource="#sustainability" /
    </jd:FactorProperty>
    <jd:FactorProperty rdf:ID="ecologicalSustainability">
        <jd:subGoalOf rdf:resource="#sustainability" />
    </jd:FactorProperty>
</owl:Class>
```

## 5   Architecture

Fig. 3 illustrates the components of the future collaborative system.

- The core ontology layer provides the ontology which formalizes the agriculture domain (e.g. AGROVOC).
- The services orchestration module provides the orchestration of the services.
- The DOS (Data-Oriented-Services or wrappers) may be developed and installed on the system by computing developers, assisted to mathematics researchers (algorithms, computations).
- The KOS (Knowledge-Oriented-Services) may be developed by the experts in agriculture, thanks to the AGRO-JD-QL declarative language.

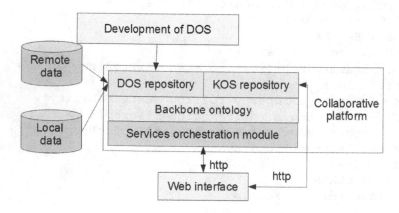

**Fig. 3.** Architecture of the system

# 6    Conclusion

As a conclusion, we provide a goal-oriented meta-model system, a declarative language that formulates the goals, its mapping to the SPARQL language and extensions of Semantic Web standards, to formulate the goals.

Much work remains to be done, including the development of a goal-oriented OWL extension, the orchestration of the services, and the implementation of the D-services and K-services into a SAAS cloud platform. A Column-oriented DBMS like Google Apps Datastore might store the ontologies of the domain and goals, linked to the future D-services and K-services that may be developed in the cloud.

# References

1. Bailey, J., Bry, F., Furche, T., Schaffert, S.: Web and Semantic Web Query Languages: A Survey. In: Eisinger, N., Małuszyński, J. (eds.) Reasoning Web. LNCS, vol. 3564, pp. 35–133. Springer, Heidelberg (2005)
2. Borst, W.N.: Construction of Engineering Ontologies. PhD thesis, University of Twente, Enschede, Holland (1997)
3. Carroll, J.J., Bizer, C., Hayes, P., Stickler, P.: Named Graphs, Provenance and Trust. In: Proceedings of the 14th International Conference on World Wide Web (WWW 2005), pp. 613–622. ACM Press, New York (2005)
4. Castellani, S., et al.: A knowledge-based system to support legal case construction. In: Proceedings of Knowledge Engineering and Ontology Development Conference (KEOD 2011), Paris, France, pp. 15–27 (2011)
5. Dantan, J., Pollet, Y., Taibi, S.: Semantic indexation of Web services for collaborative expert activities. In: Proceedings of IADIS International Conference Information Systems, Berlin, Germany, March 10-12, pp. 57–64 (2012)
6. Dantan, J., Pollet, Y., Taibi, S.: The GOAL Approach - A Goal-Oriented Algebraic Language. In: proceedings of ENASE International Conference on Evaluation of Novel Approaches to Software Engineering, Angers, France, July 4-6, pp. 173–180 (2013)
7. Food and Agriculture Organization of the United Nations (FAO), "AGROVOC" (2012), http://aims.fao.org/standards/agrovoc/about
8. Guzelian, G., Cauvet, C., Ramadour, P.: Conception et réutilisation de composants: une approche par les buts. In: Proceeding of INFORSID 2004, Biarritz, France, May 24-28, pp. 179–174 (2004)
9. Harris, S., Seaborne, A.: SPARQL 1.1 Query Language. W3C Working Draft 05 January 2012 (2012), http://www.w3.org/TR/sparql11-query/
10. Iannone, L., Rector, A.: Calculations in OWL. In: Proceedings of the Fifth OWLED Workshop on OWL: Experiences and Directions, collocated with the 7th International Semantic Web Conference (ISWC 2008), Karlsruhe, Germany, October 26-27 (2008)
11. Jean, S.: OntoQL, un langage d'exploitation des bases de données à base ontologique. PhD thesis, University of Poitiers, France (2007)
12. Lauser, B., Sini, M.: From AGROVOC to the agricultural ontology service/concept server: an OWL model for creating ontologies in the agricultural domain. In: Proceedings of DCMI 2006, International Conference on Dublin Core and Metadata Applications: Metadata for Knowledge and Learning, pp. 76–88 (2006)

13. Miller, L., Seaborne, A., Reggiori, A.: Three Implementations of SquishQL, a Simple RDF Query Language. In: Horrocks, I., Hendler, J. (eds.) ISWC 2002. LNCS, vol. 2342, pp. 423–435. Springer, Heidelberg (2002)
14. Prud'hommeaux, E., Seaborne, A.: SPARQL Query Language for RDF. W3C Recommendation 15 January 2008 (2008),
    http://www.w3.org/TR/rdf-sparql-query/
15. Seaborne, A.: RDQL - A Query Language for RDF. W3C Member Submission 9 January 2004 (2004), http://www.w3.org/Submission/RDQL/
16. Vilain, L., et al.: La méthode IDEA Indicateur de durabilité des exploitations agricoles. Educagri, France (2001)
17. World Wide Web Consortium (W3C). OWL Web Ontology Language reference (2004),
    http://www.w3.org/TR/owl-ref/

# A Decision Support System for Efficient Crop Production Supply Chain Management

Valeria Borodin[1], Jean Bourtembourg[2], Faicel Hnaien[1], and Nacima Labadie[1]

[1] Laboratory of Industrial Systems Optimization (LOSI),
Charles Delaunay Institute (ICD), University of Technology of Troyes (UTT),
ICD-LOSI, 12 rue Marie Curie - CS 42060, 10004 Troyes, France
{valeria.borodin,faicel.hnaien,nacima.labadie}@utt.fr
[2] Agricultural Cooperative Society in the Region of Arcis-sur-Aube,
Industrial Zone of Villette, 10700 Villette-sur-Aube, France
j.bourtembourg@scara.fr

**Abstract.** This paper presents a decision support system for efficient and responsive crop production supply chain management, which captures harvesting, transportation and storage activities, through taking a holistic perspective as a whole. Crop production represents one of the most significant stage in agricultural sector for both agricultural cooperatives and individual farmers, due to its high cost and considerable impact on crop quality and yield. Whilst considering the dynamic behaviour and the intricacy of the studied agricultural system, combined discrete event simulation and optimization approaches, scenario analysis and performance measurement tools are implemented for supporting decision making for cooperatives and individual farmers, respectively. Likewise, it enables to examine and evaluate alternative system re-configurations and strategies for an eventual supply chain rethinking or redesign. The decision support system proves to be particularly responsive and effective when applied to a real life agricultural case study.

**Keywords:** decision support system, agricultural supply chain, combined simulation optimization, scenario analysis, stochastic programming.

## 1 Introduction

In order to overcome the new challenges facing agricultural sector, the crop production supply chains must particularly be very reactive, flexible, with a high yield and at low cost. Its improving and eventual re-configuration can lead to an upgrade in efficiency, responsiveness, business integration and make it able to confront the market competitiveness. Among the main aims of agricultural supply chains management are to reduce the overall involved cost and to manage efficiently the activities of harvesting, transporting and storage of the raw production, from the growing fields to the long-time period storage facilities. The context is made even more complex due to the highly dynamic behaviour and stochastic environment of crop production supply chains that integrates variability in processing time at each operation and uncertainties in reception demand, queuing networks, customer satisfaction, crop quality degradation, etc.

B. Murgante et al. (Eds.): ICCSA 2014, Part V, LNCS 8583, pp. 775–790, 2014.

Due to the straightforward dependence of meteorological conditions, weather and climate variability must be considered within the framework of applications dedicated for decision supporting crop production management. This is particularly indispensable in terms of requirements imposed on the crop quality and sustainability. Premature or delayed harvest is critical since for product preservation, crop production must be maintained and adequately stored.

In the light of the above observations and in order to defy the challenges facing agricultural sector, operation research tools and techniques combined with simulation modelling are proving to be very amenable and efficient for supply chains management. Moreover, scenario analysis and system performance measurement enable to investigate and evaluate agricultural supply chain for an eventual enhancement or redesign.

Elicited by a real life case study encountered at a typical French agricultural cooperative, the purpose of the present paper is to illustrate an efficient and flexible crop production decision support system, which embeds harvesting, transportation, drying and storage activities. Based on combined discrete event simulation modelling and optimization approaches, scenario analysis and performance measurement, the proposed decision support system addresses operational, tactical and strategical issues of the crop production supply chain by providing decisions for both individual farmers and agricultural cooperatives.

The remainder of this paper is structured as follows: in the next Section, a brief review of the state of the art related to agricultural supply chain management and modelling approaches dealing with uncertain data is provided. In Section 3, the problem statement and crop production activities are presented in details. After that, in Section 4, the agricultural decision support system is exposed, whilst illustrating each used approach in details: the simulation and optimization modules, as well as the scenario analysis and performance measurement. In Section 5, several computational experiments are reported and discussed. Finally, in Section 6, some conclusions are drawn and topics for future research are outlined.

## 2    Research Background

With the growing complexity of the agricultural managerial decisions, many simulation models have been performed for tackling different activities in the wider context of the total supply chain network. There is extensive literature focusing on many decision-making problems arising from complex agricultural processes and operations that are described by using simulation modelling and improved via the sensibility or scenario-based analysis [1,15,21,22,23].

As far as in-field operations are concerned, various decision support system and advanced simulation applications have been introduced. [9] developed a decision support system for route planning optimization in terms of minimizing the risk of soil compaction for agricultural vehicles carrying time-depended loads. [7] proposed a discrete event simulation model of the harvesting and transportation systems of a sugar-cane plantation. [12] developed a simulation model for capacity planning in sugar-cane transport as part of a whole-of-system modelling framework. Another decision support system dedicated to advice farmers about in-field operational decision is presented in [11].

Regarding the harvest activity, scheduling and planning problems have drawn considerable attention from the research community. For example, [19] proposed a decision support system capable to provide computerised support to decision-makers charged with the task of scheduling sugar-cane harvesting operations in South Africa. Another study provided by [16] proposed the use of real-time scheduling methods based on dispatching rules for production system in processed canned fruit industry as a case study. Thereupon, a number of optimization approaches have been developed. [6] illustrated an operational model for generating short term planning decisions for fresh produce industry. [10] presented a practical tool for optimally scheduling wine grape harvesting operations taking into account both operational costs and grape quality. [20] provided a planning methodology to determine the farm areas and the seeding times for annual plants that survive for only a growing season while maximizing the total profit.

Expensive literature in the field of the distribution planning for agricultural supply chains is reviewed in [5]. The authors proposed a classification of models based on the most relevant features, such as: the used optimization approaches, the type of considered crops and the research aims. It is worthy of note that only a few studies handle uncertainties in agricultural managerial problems, while most consider only deterministic aspects. Related to this topic, [18] presented the current background and suggested some theoretical directions for future research by taking a broad view of supply chain uncertainty. The authors have argued that there are many sources of uncertainties and management strategies that still consider future research.

Nevertheless, from the previously mentioned research efforts, we can conclude that, although the agricultural supply chain is a sector which attracts a growing interest, there is a paucity in managerial applications or tools, dedicated to efficiently supervise and redesign the multi-product agricultural supply chain. This lack of decision support tools is especially discernible in case of advanced applications within the highly dynamic and uncertain environment.

## 3 Problem Statement

Consider an agricultural cooperative specialized in multi-seed production and commercialization. Let the cooperative be constituted of an union of several hundred of contractual farmers (named *adherents*) for whom it provides consulting, drying, storage, transportation and other customer services. Likewise, the cooperative offers its services for non-contractual farmers, called *customers*.

At harvest time, farmers proceed to the gathering activities. In order to be more appropriately preserved, the grains are carefully forwarded towards storage facilities (also named *silos*), especially designed for this purpose. In order to reduce their handling and transportation cost, some farmers entrust their harvest operations to competent agricultural contractors, other farmers share their equipment and machinery amongst each other.

The quantities to be received at each silo are unknown. The cooperative has only information of the global quantity to be received from their contractual farmers during the whole harvest season and only a probabilistical expectation about customers quantities to be received. The latters are not contractually obliged and are free to decide the silo and/or the quantity to deliver in particular, and the cooperative to be provided, in general.

The weather variability represents the deciding factor for seed production, that control and affect the cereal growth, development, ripeness and its quality degradation. Cereal harvest is possible only on days without rain. Moreover, premature or delayed harvest increases the risk of yield and quality degradation. Thus, crop harvesting must be achieved as soon as possible once seeds have reached their physiological maturity for an appropriate cereal storage.

While considering the highly uncertain conjuncture of a cereal crop production supply chain system, the aim is to provide decision making support at:

- **strategical level:** investigate alternative configurations of silos implantation for an eventual supply chain redesign;
- **tactical level:** provide the most appropriate subset of storage facilities to open each harvest season, in terms of the cost and level of cooperative reception area service;
- **operational level:** propose plausible harvest scheduling solution(s) by minimizing the crop quality degradation risk; manage harvesting, transportation and storage activities; operate drying process; manage loss queuing network.

and, also, to evaluate the whole supply chain, from growing fields to storage facilities, with a view of improving its performances and reducing the overall cost.

## 4    Decision Support System

The development of an efficient agricultural supply chain must consider heterogeneous activities and processes, in the context of a dynamic and highly uncertain environment. With the purpose to satisfy cooperative and farmers' economical profitability and to meet requirements imposed on crop quality and sustainability, this paper proposes a decision support system based on combined simulation optimization approach, scenario analysis and performance measurement, as illustrated in figure 1.

Simulation optimization problem is recognized as a hard problem [14]. Related to this subject, several research studies have been conducted and theoretically exposed and improved [8,3,4]. Combined simulation optimization interest is twofold: on the one hand, optimization addresses large scale decision making problems, and, on the other hand, simulation is able to tackle the stochastic nature of the considered system and to check up on the convergence of the optimization solution(s) to its practical objective function.

Applied to enhance the examined crop production supply chain performance, the optimization module provides solution(s) for the quality risk management

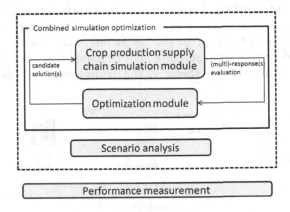

**Fig. 1.** Decision support system

problem in order to verify its feasibility and convergence by the simulation module, realistically models the activities performed along the whole supply chain system, for a desired number of simulation replications. It is worth to notice that there is a complete separation between the simulation modelling that represents entirely the studied system and the module that solves optimization problems.

The rest of this section presents the main assumptions of the considered decision support system and a detailed explanation of the handled crop production inherent activities.

### 4.1   Simulation Module

Without loss of generality, the simulation modelling designed below is based on the following assumptions: **a.** each parcel is considered to be single variety sown at the same time; **b.** a single post-maturity quality degradation indicator $r_a$ used hereafter, can express a combination of several degradation indicators, such as Hagberg falling number, over-maturity indicator, etc.; **c.** the time scale of harvest availability is based on daily meteorological conditions.

*Harvesting reception and storage.* The cooperative must ensure the reception availability of each silo and the appropriate preservation of harvested crop. For these purposes, many cooperatives dispose of two types of silos:

- **expedition silos:** are used for a long time period storage, dedicated to ensure the intrinsic seed quality. Typically, this kind of silos has a non-constraint storage capacity.
- **satellite silos:** serve as proximity facilities at time of harvest. Due to their limiting storage capacities, an inventory control level and crop transfer to expedition silos must be organized during the whole harvest season, in order to maintain satellite silos empty and, also, to ensure a high reception service and an adequately grain storage.

*Transportation and logistics.* Logistics and transportation activities represent an important component of the cooperative system. In most cases, farmers deliver harvested crop to the nearest storage facility. From its share, cooperative forwards received cereals from satellite silos to expedition ones, for the sake of satellite storage availability, as shown in the figure 2 depicted below.

**Fig. 2.** Harvest transportation: "—" farmer possible trips and "- - -": cooperative transfer

So as not to perturb farmer reception, in the current supply chain configuration, the cooperative empties satellite silos preponderantly during the night, when no farmer harvest is received. This practice has an important drawback: the night-time periods are not sufficient to empty satellite silos, such as to be able to receive whole farmers' harvesting crop, during daytime hours of the next day. Hence, the cooperative performs crop transfer also during the daytime period, that affects storage service availability to farmer disposal.

*Drying process.* Seed moisture content is one of the most important post-harvest factor influencing seed quality and storability, especially in case of oilseeds or seeds with a non-negligible moisture content (such as maize, rice, etc.). After being collected in so-called pre-drying storage, seeds are subsequently forwarded to available dryers $l$ ($l = \overline{1, L}$) with a limited capacity $T_l$, which must be preheated to a given temperature before use. Seeds should be dried to approximately 15% equilibrium relative humidity before long-term storage. Moreover, in order to avoid the seed collage, the crop contained in pre-drying storage must be blend at least once in each 24 hours. The drying process is explained in greater detail in the figure 3. Drying is a post-harvest process that not only affects product quality and storage life, but also consumes a considerable amount of energy. In order to reduce the involved cost and ensure an accurate storage, it is possible to investigate different setting and scenarios via the proposed drying handling.

*Queuing-loss system.* In the actual configuration of the supply chain, where farmers can improve a continuous process of gathering operation, the queues of trucks affect the synchronization between in-field activities and reception area service due to the limited flow rates of receiving pits. The congestion of vehicles nearby the silos' area may cause the idleness of machinery in the growing fields. By taking account of these impediments in harvesting accomplishment, the simulation model must also integrate queuing systems on silos' reception area. In this

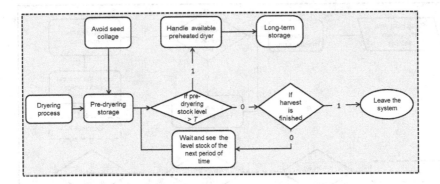

**Fig. 3.** Drying process

sense, the authors of [1] have been proposed and discussed a multi-echelon silo loss queuing system, that can serve three classes of arrival trucks: transfer, adherent and customer trucks. More precisely, in [1] a M/D/W/W ⊢ B/PR+FIFO queuing-loss system is proposed, where: $M$ represents a maximum number of expected customers to be served; the system service rate is considered deterministic (D); $W$ denotes the number of reception pits; $B$ indicates the buffer level, which once exceeded, the customers arriving later do not wait in the queue and leave the system; the queue discipline between farmers and transfer trucks is supposed with pre-emption for farmers reception (PR); the queue rule for farmers reception is considered first in first out (FIFO).

*Crop production supply chain modelling.* The supply chain operations related to harvesting activity can be modelled by the following main types of data:

- parcels, theirs seed varieties, physiological maturity dates and quality risk degradation indicators;
- equipment and machinery fleet composition for each farmer (or farmers' association);
- rates of Poisson process $\lambda_1$ and $\lambda_2$ that describe inter-arrivals time for both classes of clients: adherents and customers, respectively;
- distances between parcels and storage facilities;
- characteristics of each silo: localisation, number of receiving pits and their flow rates, number of dryers and their associated capacity and flow rates;
- cooperative heterogeneous truck fleet;
- schedule of opening hours for farmers reception of each storage facility;
- daily harvest availability discrete distributions.

The model is composed of several sub-models, each of which simulates activities related to harvesting, transportation and storage stages. In this respect, figures 4 and 5, illustrate schematically the basic operations of the logistics reception system encountered at the studied crop production cooperative.

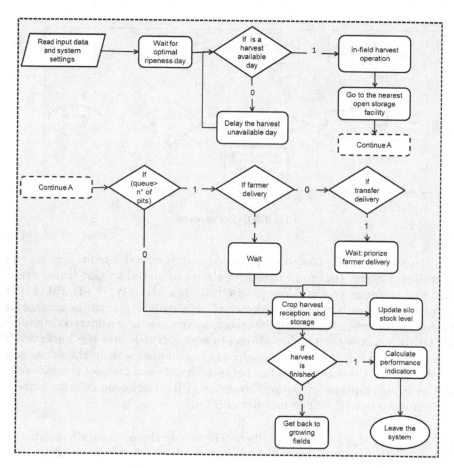

**Fig. 4.** Simulation modelling flowchart: *satellite silo inventory control and transfer operation to expedition storage facility*

More specifically, figure 4 shows the harvesting and crop transportation from growing fields to the nearest storage facility by using an individual or shared agricultural equipment fleet. All of these operations are made once the cereals have reached their physiological maturity and when the weather conditions are favourable for harvesting. As aforementioned above, adherent farmers are obliged to deliver their parcel gathered crop to cooperative storage facilities, thus, they must wait for harvest delivery, even if there is a queue nearby the reception area. That is not the case of contractual independent farmers, who leave the queuing system if there are more than $B$ farmers waiting for reception service.

The inventory control level and crop transfer operation of each satellite silo, during whole harvest season, are described in the figure 5. If there are some cereal stocks and available transportation resources, crop transfer is organized from satellite silos to expedition ones, in order to keep satellite silos empty and, also, to guarantee a high reception service and appropriate cereal storage. The sub-model depicted in the figure 5 is associated to each satellite storage facility.

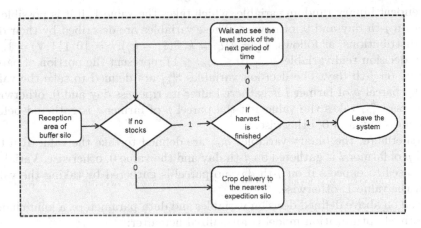

**Fig. 5.** Simulation modelling flowchart: *harvesting and crop delivery from growing fields to storage facility*

## 4.2   Optimization Module

As mentioned previously, gathering the harvest represents an important stage for both farmers and agricultural cooperatives that requires the risk control of crop quality degradation. In the context of gathering activity, weather conditions represent the key factor that affects the harvesting progress. On the other hand, crop harvesting must be completed as soon as possible after the cereal ripeness. Delayed harvest increases both the risk of yield and quality degradation. Consequently, harvest delaying becomes critical since the viability of the crop production must be maintained and adequately stored.

With a view to minimizing the risk of crop quality degradation under climate uncertainty, let us consider the typical activity of product harvesting, when farmers carry out to gather crops once the cereals have reached their maturity.

Hence, consider a crop production cooperative that integrates $I$ different adherent farmers. Each farmer $i, i = \overline{1, I}$ has $P_i$ parcels. The harvest planning time horizon is constituted of $J$ days. The quantity to be harvested $q_{ip}$ at each parcel $p$ of farmer $i$ is supposed to be known and it depends on sown acreage and crop yield. Moreover, it has been assumed that the ripeness dates $d_{ip}^*$ for all parcels $p$ of each farmer $i$, are known. No farmer $i$ can gather more than $c_i$ per day.

Cereals' ripeness dates can be determined via agronomic techniques before the harvest begins. The parameters $r_a$ and $r_b$ denote the degradation indicators when the crops are gathered after and before their physiological maturity. Without loss of generality, assume these indicators been identical for all parcels.

Based on the weather station data related to the volume of rainfall during the harvest season, it is possible to dispose information about daily harvest availability. Thus, let $\Omega$ be the problem weather scenarios set, where $\omega = (\xi_1^\omega, \xi_2^\omega, \cdots, \xi_J^\omega)$ with their associated joint discrete probability $\mathbb{P}(\omega) = \prod_{j=1}^J \mathbb{P}(\xi_j^\omega), \forall \omega \in \Omega$, where $\mathbb{P}$ denotes probability. Then, let the daily harvest availability $\xi_j^\omega$ be an independent binary random variable, which takes the value 1, if it is possible to gather on $j$-th day and 0, otherwise. These variables are described by their discrete distributions, as follows: $D_j = \{(\xi_j = k, \mathbb{P}(\xi_j = k)), k = \{0, 1\}\}, \forall j = \overline{1, J}$.

The decision real variables $x_{ipj}$ ($0 \le x_{ipj} \le 1$) represent the portion of parcel gathered on $j$-th day. The decision variables $z_{ipj}^a$ are defined to take the value 1, if the parcel $p$ of farmer $i$ is gathered after its ripeness day and 0, otherwise. Vice-versa, $z_{ipj}^b$ takes the value 1 if the parcel $p$ of farmer $i$ is gathered before its ripeness day and 0, otherwise.

Furthermore, the binary variables $g_{ipj}$ are defined to take the value 1, if the parcel $p$ of farmer $i$ is gathered on $j$-th day and the value 0, otherwise. Variables $o_j$ are used to express if on $j$-th day no parcel is gathered by taking the value 0, and the value 1, otherwise.

Based on above defined decision variables and data parameters, a joint chance constrained optimization model is formalized hereafter:

$$min \sum_{j=1}^{J} \sum_{i=1}^{I} \sum_{p=1}^{P_i} \{r_b \cdot z_{ipj}^b \cdot (j - d_{ip}^*) - r_a \cdot z_{ipj}^a \cdot (j - d_{ip}^*)\} \tag{1}$$

$$\sum_{j=1}^{J} x_{ipj} = 1, \qquad\qquad \forall i, \forall p \tag{2}$$

$$\sum_{p=1}^{P_i} x_{ipj} \cdot q_{ip} \le c_i, \qquad\qquad \forall i, \forall j \tag{3}$$

$$\mathbb{P}(x_{ipj} \le \xi_j^\omega, \forall i, \forall j, \forall p, \forall \omega) \ge 1 - \alpha, \tag{4}$$

$$x_{ipj} \le g_{ipj}, \qquad\qquad \forall i, \forall p, \forall j \tag{5}$$

$$M \cdot x_{ipj} \ge g_{ipj}, \qquad\qquad \forall i, \forall p, \forall j \tag{6}$$

$$o_j \le \sum_{i=1}^{I} \sum_{p=1}^{P_i} g_{ipj}, \qquad\qquad \forall j \tag{7}$$

$$M \cdot o_j \ge \sum_{i=1}^{I} \sum_{p=1}^{P_i} g_{ipj}, \qquad\qquad \forall j \tag{8}$$

$$g_{ipj} \ge g_{ip(j-1)} - \sum_{j'=1}^{j-1} x_{ipj'} + o_j - 1, \qquad\qquad \forall i, \forall p, \forall j > 2 \tag{9}$$

$$z_{ipj}^a \cdot (j - d_{ip}) \ge 0, \qquad\qquad \forall i, \forall p, \forall j \tag{10}$$

$$z_{ipj}^b \cdot (j - d_{ip}) \le 0, \qquad\qquad \forall i, \forall p, \forall j \tag{11}$$

$$z^a_{ipj} + z^b_{ipj} \geq x_{ipj}, \qquad\qquad \forall i, \forall p, \forall j \qquad (12)$$

$$0 \leq x_{ipj} \leq 1, \qquad\qquad \forall i, \forall p, \forall j \qquad (13)$$

$$z^a_{ipj}, z^b_{ipj} \in \{0,1\}, \qquad\qquad \forall i, \forall p, \forall j \qquad (14)$$

$$r_j, g_{ipj} \in \{0,1\}, \qquad\qquad \forall i, \forall p, \forall j \qquad (15)$$

Constraints (2) are prescribed, since each parcel must entirely be harvested. As all daily harvested quantities must not exceed the capacity of farmers $c_i$, constraints (3) are required. Non-linear constraints (4) are affected by uncertain harvest availability on only right-hand side and can be violated at least $\alpha$ times. Constraints (5) and (6) force the binary variables $g_{ipj}$ to take the value 1, if the parcel $p$ of farmer $i$ is gathered on $j$-th day and the value 0, otherwise. Note that $M$ is a constant user-defined big value (e.g. greater than $\sum_{i=1}^{I} P_i$).

Similarly, constraints (7) and (8) impose the binary variable $o_j$ to value 1 if at least one parcel is gathered on $j$-th day and 0, otherwise. Constraints (10), (11), and (12) oblige the binary variables $z^a_{ipj}$ to take the value 1, if the parcel $p$ of farmer $i$ is gathered after its ripeness day, and $z^b_{ipj}$ take the value 1, if the parcel $p$ of farmer $i$ is gathered before its ripeness day, respectively.

For economical reasons (combine harvester break is very onerous expenditure), constraints (9) force to continue the harvest process on parcel $p$ of farmer $i$ once it is started. It is authorised to change the parcel after a rain, which in any case provoke a combine harvester break.

Notwithstanding its non-linearity, in a recent study [2], a basic version of the quality risk management problem (1)-(15) have been discussed and solved via an equivalent linear mixed integer reformulation jointly with scenario based approaches. Also, the authors have been introduced a new $(1 - \alpha)$–scenario pertinence in order to handle efficiently the probabilistic constraints (4) and to reduce the computational time.

## 4.3   Scenario Analysis and Performance Measurement

As a complex large scale real-life system, agricultural supply chain captures many characteristics of a multi-responses optimization problem. In that vein, by setting up various performance indicators, it is possible to evaluate and investigate the supply chain for different system configurations. Hence, let us define the following relevant performance responses (also called *attributes*, *indicators* or *metrics*) of the studied crop production system:

- harvest season length;
- maximum distance to the nearest storage implantation for each parcel;
- risk of crop quality degradation;
- total operating cost of silos exploitation;

- total quantity distance of harvested crop, forwarded from parcels to expedition silos;
- etc.

Because the total operating cost is rather significant compared to the cost involved by transfer and logistics operations between satellite and expedition silos, the impact evaluation of closing/opening of one or more satellite silo(s) contributes to supply chain improvement and redesign. Also, it offers an analytical view of the system re-configuration(s) and its changing over time.

## 5    Computational Results

The decision support system presented in this paper, was conceived with a view to supervising and improving the crop production supply chain, for a cereal and oilseed agricultural cooperative society, situated in the region of Arcis-sur-Aube (France). More specifically, this is a grain and oilseed agricultural cooperative, for which efficient crop production management is crucial since: on the one hand, it permits to preserve the grain yield and quality in order to sell profitably cereal, and on the other hand, it is able to propose a high level reception area service to their adherents and to attract more customers.

Typically, the harvest lasts about one month. During this period, the climate forecasting data is derived from the weather stations, with an acceptable reliability level. The information concerning variety ripeness dates is determined by the technically competent organisations in agronomy and in particular, in crop agriculture.

The simulation module presented in section 3, have been implemented by using the commercial simulation language SIMAN in the Arena Rockwell Automation environment (version 13.0) for instances with up to 4000 parcels, 3 expedition and 11 satellite silos. The computational experiments have been performed on an Intel(R) Core(TM) i7-2720QM CPU 2.20GHz workstation. The chance constrained optimization problem (1)-(9) tackled via its equivalent linear reformulation [2] has been solved by using the OptQuest package for Arena, which is based on a combination of three different types of meta-heuristics including scatter search, tabu search and neural networks.

Without loss of generality, let us examine the scenarios analysis, tested for one geographical zone of the case study cooperative, which entails: 1 expedition and 3 satellite silos and more than 1100 parcels. For each experimental scenario, there were $10^5$ simulation replications to ensure the convergence of the optimization solution(s) to its practical objective function. The confidence level of $(1 - \alpha)$ is defined to take the value 0.95. Beforehand optimized, the right hand side of figure 7 illustrates the performance of 5 scenarios corresponding to two satellite silos closing in terms of the total operating cost of silos exploitation, quality risk degradation and total amount of harvested cereal crop, forwarded from parcels to expedition silos.

In order to ensure an accurate maize and sunflower drying, the figures 6 shows the start-ups of two cooperative dryers, performed in the framework of

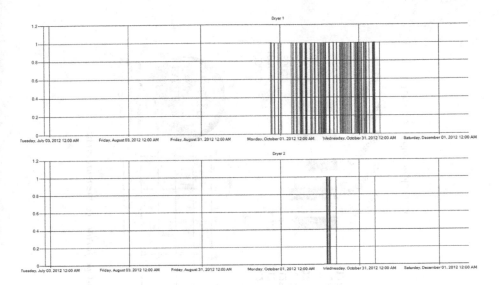

**Fig. 6.** Computational results: *dryers' start-ups*

the actual oilseed harvesting configuration. Other drying scenarios (with a fixed number of start-ups) must be tested for reducing the amount of drying energy consummation and the involved costs, respectively.

In what follows, let us examine the left-hand side of figure 7, which provides different system characteristics at operational level. The first left-hand side image reports the output results related to the client reception area service of the case study agricultural cooperative. As the pie depicted in the figure 7 highlights, about 18 % of customers leave the system, because of unsatisfactory reception area service. Hence, in order to improve the level of reception area service, other scenarios of storage facilities organization must be conceived and evaluated.

Next, let us discuss the second left-hand side plot shown in the figure 7, that present queuing length of one satellite silo, endowed with a single receiving pit. Such reports offer statistical results about queuing system of each storage facility. On the one hand, the longer waiting delays of adherents at the storage facilities, lead to the delayed returns of farmer trucks to the growing fields. Consequently, this fact affects the availability of machinery transporting fleet, as well as causes resource and worker idleness in the fields. On the other hand, it provokes customer dissatisfaction, which leads them to leave the system. Efficient queuing management can substantially ameliorate supply chain performance.

Finally, the third left-hand side plot illustrated in the figure 7, expresses the daily quality degradation risk of several adherent parcels. For the purpose of sustainable crop production supply chain management, it is primordial to consider crop quality degradation risk.

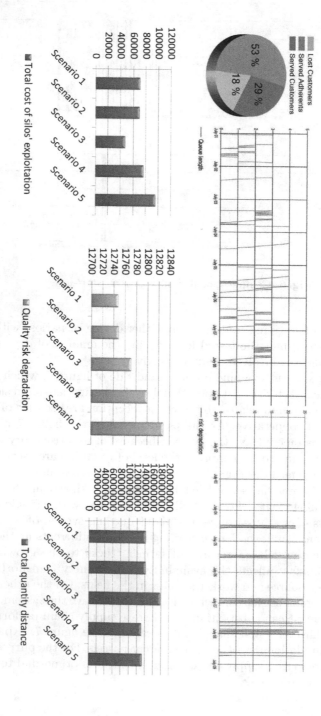

**Fig. 7.** Computational results: *scenario analysis and operational performance measurement*

As experimental results show, the scenarios analyse offers a spectrum of miscellaneous storage facility re-configurations towards with various agricultural performance indicators. Computational results provide managerial insights and measure the significance of the relevant agricultural indicators while making the supply chain more responsive and cost-effective. Moreover, this experimental implementation emphasizes the importance and the effectiveness of decision support tools based on combined simulation optimization approaches, scenario analysis and performance measurement.

# 6  Conclusions and Future Research

This paper presents a decision support system designed to provide decisions for both individual farmers and cooperatives, by taking a holistic perspective of the multi-seed crop production supply chain. Detailed simulation experiments were conducted in order to attest the practical efficiency of the developed decision support system.

Although the paper focused on a case study in multi-seed supply chain, the proposed polymorphic decision support system can be also applied in other agricultural based industries (such as sugar, wine grape, etc.), which capture the same main activities and are also affected by various uncertainties (weather, loss queuing network or customer delivery policies).

Future work would be dedicated to widen the agricultural decision support system at the tactical level by implementing and evaluating miscellaneous storage procedures (e.g. direct transfer from growing fields to expedition storage facilities, storage at the headlands, etc.), and also, various farmers' contractual delivery policies. At operational level, the next stage of the present study would be addressed to developing dynamic re-optimization approaches in order to obtain more pertinent harvest scheduling solutions.

# References

1. Borodin, V., Bourtembourg, J., Hnaien, F., Labadie, N.: A discrete event simulation model for harvest operations under stochastic conditions. In: 10th IEEE International Conference on Networking, Sensing and Control (ICNSC), pp. 708–713 (2013)
2. Borodin, V., Bourtembourg, J., Hnaien, F., Labadie, N.: A quality risk management problem: case of annual crop harvest scheduling. International Journal of Production Research 52(9), 2682–2695 (2014)
3. Fu, M.C.: Optimization for simulation: Theory vs. practice. Informs Journal of Computing, 192–215 (2002)
4. Yang, T., Pohung, C.: Solving a multiresponse simulation-optimization problem with discrete variables using a multiple attribute decision making method. Mathematics and Computers in Simulation, 9–21 (2005)
5. Ahumada, O., Villalobos, J.R.: Application of planning models in the agri-food supply chain: A review. European Journal of Operational Research, 1–20 (2008)

6. Ahumada, O., Villalobos, J.R.: Operational model for planning the harvest and distribution of perishable agricultural products. International Journal of Production Economics 133, 677–687 (2011)
7. Arjona, E., Bueno, G., Salazar, L.: An activity simulation model for the analysis of the harvesting and transportation systems of a sugarcane plantation. Computers and Electronics in Agriculture 32, 247–264 (2001)
8. Bachelet, B., Yon, L.: Model enhancement: Improving theoretical optimization with simulation. Simulation Modelling Practice and Theory, 703–715 (2007)
9. Bochtis, D.D., Sørensen, C.G., Green, O.: A DSS for planning of soil-sensitive field operations. Decision Support Systems 53, 66–75 (2012)
10. Ferrer, J., MacCawley, A., Maturana, S., Toloza, S., Vera, J.: An optimization approach for scheduling wine grape harvest operations. International Journal of Production Economics 112, 985–999 (2008)
11. Hameed, I.A., Bochtis, D.D., Sørensen, C.G., Vougioukas, S.: An object-oriented model for simulating agricultural in-field machinery activities. Computer and Electronics in Agriculture 81, 24–32 (2012)
12. Higgins, A., Davies, I.: A simulation model for capacity planning in sugarcane transport. Computers and Electronics in Agriculture 47, 85–102 (2005)
13. Kall, P., Wallace, S.N.: Stochastic Programming. John Wiley and Sons (1994)
14. Kleijnen, J.P., van Beer, W., van Nieuwenhuyse, I.: Constrained optimization in expensive simulation: Novel approach. European Journal of Operations Research 202, 164–174 (2010)
15. Le Gal, P.-Y., Le Masson, J., Bezuidenhout, C.N., Lagrange, L.: Coupled modelling of sugarcane supply planning and logistics as a management tool. Computer and Electronics in Agriculture 68, 168–177 (2009)
16. Parthanadee, P., Buddhakulsomsiri, J.: Simulation modeling and analysis for production scheduling using real-time dispatching rules: A case study in canned fruit industry. Computers and Electronics in Agriculture 70, 245–255 (2010)
17. Shapiro, A., Dentcheva, D., Ruszczyński, A.: Lectures on stochastic programming: Modelling and Theory. Society for Industrial and Applied Mathematics, Philadelphia (2009)
18. Simangunsong, E., Hendry, L., Stevenson, M.: Supply-chain uncertainty: a review and theoretical foundation for future research. International Journal of Production Research, 4493–4523 (2012)
19. Stray, B.J., van Vuuren, J.H., Bezuidenhout, C.N.: An optimisation-based seasonal sugarcane harvest scheduling decision support system for commercial growers in south Africa. Computers and Electronics in Agriculture 83, 21–31 (2012)
20. Tan, B., Cömden, N.: Agricultural planning of annual plants under demand, maturation, harvest and yield risk. European Journal of Operational Research, 539–549 (2012)
21. Teimoury, E., Nedaei, H., Ansari, S., Sabbaghi, M.: A multi-objective analysis for import quota policy making in a perishable fruit and vegetable supply chain: A system dynamics approach. Computer and Electronics in Agriculture 93, 37–45 (2013)
22. van der Vorst, J.G., Tromp, S.O., van der Zee, D.J.: Simulation modelling for food supply chain redesign; integrated decision making on product quality, sustainability and logistics. International Journal of Production Economics 47, 6611–6631 (2009)
23. Zhnag, F., Johnson, D.M., Johnson, M.A.: Development of a simulation model of biomass supply chain for biofuel production. Renewable Energy 44, 380–391 (2012)

# Conceptualizing Crop Life Cycle Events to Create a User Centered Ontology for Farmers

Anusha Indika Walisadeera[1,2], Athula Ginige[1], and Gihan N. Wikramanayake[2]

[1] School of Computing, Engineering & Mathematics,
University of Western Sydney, Penrith, NSW 2751, Australia
[2] University of Colombo School of Computing, Colombo 07, Sri Lanka
waindika@cc.ruh.ac.lk, a.ginige@uws.edu.au, gnw@ucsc.cmb.ac.lk

**Abstract.** People need contextualized information and knowledge to make better decisions. In case of farmers, the information that they require is available through agricultural websites, agriculture department leaflets and mass media. However, available information and knowledge are general, incomplete, heterogeneous, and unstructured. Since the farmers need the information and knowledge within their own context and need to represent information in complete and structured manner we developed a farmer centered ontology in the domain of agriculture. Because of the data complexity of the relationships among various concepts, to attenuate the incompleteness of the data, and also to add semantics and background knowledge about the domain we have selected a logic based ontological approach to create our knowledge repository. In this study, we have investigated how to model the actual representation of the domain and its challenges. The internal evaluation has been done to test the usefulness of the ontology during the design process. We have developed the online knowledge base that can be queried based on the farmer context.

**Keywords:** agricultural information/knowledge, contextualized information, knowledge representation, ontology modeling, knowledge base development, ontology evaluation.

## 1 Introduction

Farmers need agricultural information and knowledge to make informed decisions at various stages of the farming life cycle such as seasonal weather, best varieties or cultivars, seeds, fertilizers and pesticides, information on pest and diseases, control methods, harvesting and post harvesting methods, accurate market prices, current supply and demand, and information on farming machinery and practices [1] [2]. Farmers can get some of this information from multiple sources such as agricultural websites, agriculture department leaflets and mass media. However this information is general, incomplete, heterogeneous, and unstructured. Farmers require information within the context of their specific needs in a structured manner. Such information could make a greater impact on their decision-making process [3].

B. Murgante et al. (Eds.): ICCSA 2014, Part V, LNCS 8583, pp. 791–806, 2014.

Not having an agricultural knowledge repository that is consistent, well-defined, and provide a representation of the agricultural information and knowledge needed by the farmers within their own context is a major problem.

Social Life Networks for the Middle of the Pyramid (www.sln4mop.org) is an International Collaborative research project aiming to develop mobile based information system to support livelihood activities of people in developing countries [4]. The research work presented in this paper is part of the Social Life Network project, aiming to provide information and knowledge to farmers based on their own context in Sri Lanka using a mobile based information system.

To represent the information in context-specific manner, firstly, we have identified the farmers' context (i.e. farmers' context model) specific to the farmers in Sri Lanka such as *farm environment, types of farmers, farmers' preferences,* and *farming stages* [5]. We also identified the six farming stages related to our application such as Crop Selection, Pre-Sowing, Growing, Harvesting, Post-Harvesting, and Selling [5].

Next we have identified an optimum way to organize the information and knowledge in context using ontologies. An Ontology provides a structured view of domain knowledge and act as a repository of concepts in the domain [6]. The most quoted definition of ontology was proposed by Thomas Gruber as *"an ontology is an explicit specification of a conceptualization"* [7]. Mainly due to the complex nature of the relationships among various concepts, attenuate the incompleteness of the data, and also add semantics and background knowledge about the domain we have selected a first-order logic (FOL) based ontological approach to create our knowledge repository. In this project we have developed a general framework for ontology design that was used to design the ontology for farmers to represent the necessary agricultural information and knowledge within the farmers' context [8]. The ontology was implemented using protégé editor (based on OWL 2-DL).

In this paper, we describe the way we modeled the events associated with the farming life cycle and the associated challenges. When modeling complex real world domains within the user context the representation of the entities and their relationships at different time intervals and in different locations is a challenge. This paper also describes the internal evaluation procedure that we adopted. The remainder of the paper is organized as follows. Section 2 describes the domain modeling especially for second and third stages of the farming life cycle. The way we modeled the first stage of farming life cycle has been published earlier [8]. Section 3 provides a summary of ontology evaluation techniques used to test the usefulness of the ontology. Finally, section 4 concludes the paper and describes the future directions.

## 2     Domain Modeling

The second stage of the farming life cycle includes the information on quality agricultural inputs such as seed rate, plant nutrients and fertilizing, irrigation facilities and new techniques for field preparation. The growing stage (i.e. third stage) includes information related to managing the crop through its growing stage. Information on planting methods, good agricultural practices, common growing problems and their management is required in this stage.

Based on our ontology design approach (see Fig. 1), we first identify a set of questions that reflect various motivation scenarios. Next, we create a model to represent user context. Then derive the contextualized information incorporating user context and task modeling with generic knowledge module. We referred to these contextualized information as the competency questions to design our ontology.

In second stage of the life cycle, we will describe how we modeled the event "application of fertilizer" to provide fertilizer information and knowledge to farmers in their specific context. Farmers need to find the answers to *"What are the suitable fertilizers for selected crops and in what quantities?"* (i.e. motivation scenario question) based on the farmers context. Next we formulated the contextualized questions related to this question. Some examples are given in Table 1. The way of modeling the contextualized information related to this application is outside the scope of this paper and it is explained in [5] and [8].

By analyzing the gathered fertilizer information reviewed in [9] [10] [11] and Crop Knowledge Database created by the IT staff of the Agriculture Department in Sri Lanka, we have found that; when applying a fertilizer to a specific crop (or soil) it involves a fertilizer quantity. A fertilizer quantity depends on the many factors. For example, it depends on the fertilizer types (e.g. Chemical, Organic, or Biological fertilizers) and its specific sources (e.g. Nitrogen, Phosphorus, Potassium, etc.) and

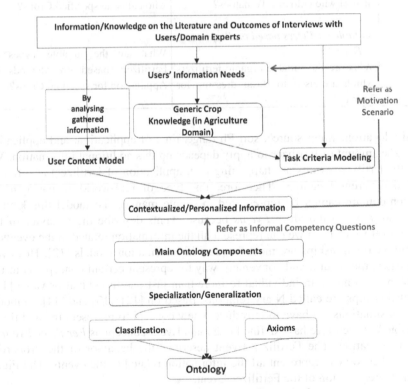

**Fig. 1.** Our Ontology Design Framework

**Table 1.** Farmers' Information Needs in Context for Second Stage of the Farming Life Cycle

| Farmers' Information Needs (i.e. Motivation Scenario) | Farmers' Information Needs in Context (i.e. Competency Questions) | Generalizing Contextualized Information |
|---|---|---|
| What are the suitable fertilizers for selected crops and in what quantities? | *Suitable fertilizers based on the Environment:* What are the suitable fertilizers and in what quantities for Banana which are grown in Dry Zone? | What are the suitable fertilizers and in what quantities for the Crops which are grown in specified Location (e.g. Climatic Zone)? |
| | Which fertilizers are the most appropriate for Chilli under rain-fed condition? | Which fertilizers are the most appropriate for Crops under specified conditions? |
| | What are the suitable fertilizers and in what quantities for farmers in Badulla district who cultivate Tomatoes? | What are the suitable fertilizers and in what quantities for farmers in specified Location (e.g. Districts) who cultivate specified Crops? |
| | *Suitable fertilizers based on Preferences of Farmers:* What are the suitable organic fertilizers which are used to Basal dressing for Tomato? | What are the suitable Types of Fertilizers based on Methods of Application for specified Crops? |

their ratio, location, water source, soil Ph range, time of application, and application method. The amount of fertilizer to apply depends on this additional information. We can see the fertilizer events are happening (e.g. application of fertilizer) at different times and different locations. Therefore, this type of real-world scenario in the application domain cannot be represented in simple manner. To model this kind of situations we have to introduce *Events* because events describe the behavior of the properties over time. Then we can represent all the information related to the events.

The binary relationships are most frequent in information models [12]. However, in some cases, the natural and convenient way to represent certain concepts is to use relationships to connect an individual to more than just one individual or value [13]. These relationships are called N-ary relations. Based on [12], [13] and [14], to model this kind of situations we have to introduce a new concept to represent the additional information. We therefore have defined one main Event concept as *Fertilizer Event* to handle this situation. The Fertilizer Event describes the behavior of the properties over time. Then we can represent all the information related to the events. The Fig. 2 shows the representation of the Fertilizer Event.

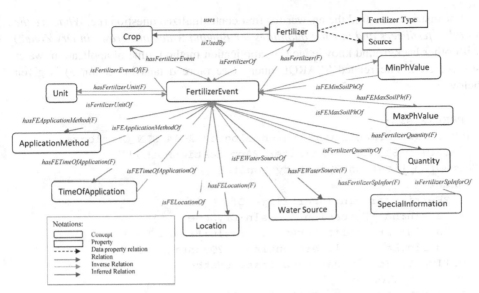

**Fig. 2.** Representation of the FertilizerEvent Concept

First we have identified the concepts, for example Crop, Fertilizer, TimeOfApplication, Location, ApplicationMethod, WaterSource, Unit, etc. as shown in Fig. 2. Next we have explicitly specified associative relationships and their inverse (when available) among concepts by maintaining the semantics (e.g. hasFertilizerUnit, isFertilizerUnitOf). The individual FertilizerEvent_1 (of FertilizerEvent concept) represents a single object. It contains all the information held in the above relations. We defined the cardinality restriction on the relationships by specifying that each instance of FertilizerEvent has exactly one value for Fertilizer, Unit, ApplicationMethod, TimeOfApplication and so on. It is specified as $F$ in Fig. 2 ($F$ refers the functional property in OWL). A crop has many FertilizerEvent (i.e. non-functional) but a FertilizerEvent has exactly one value for Crop (i.e. functional). Also we have identified the data properties that are related to the concepts, for example Fertilizer Type and Source are the data properties related to the Fertilizer concept.

We have inferred the knowledge such as Crop *uses* Fertilizer and Fertilizer *isUsedBy* Crop (i.e. inverse relationship of *uses*). This has been implemented by using object property chain in OWL 2 [15]. The relations *uses* and *isUsedBy* specify as a combination of existing object properties as depicted below. Note that, here o refers the property chain axiom notation in Protégé 4.2.

hasFertilizerEvent o hasFertilizer ⟶ uses
i.e. if **Crop** *hasFertilizerEventOf* **FertilizerEvent** and **FertilizerEvent** *hasFertilizer* **Fertilizer** then **Crop** *uses* **Fertilizer.**

isFertilizerOf o isFertilizerEventOf ⟶ isUsedBy
i.e. if **Fertilizer** *isFertilizerOf* **FertilizerEvent** and **FertilizerEvent** *isFertilizerEventOf* **Crop** then **Fertilizer** *isUsedBy* **Crop**

For instance, we find the answer for first contextualized question (i.e. *What are the suitable fertilizers and in what quantities for Banana which are grown in Dry Zone?*) with other background knowledge (i.e. application method, time of application, water source, etc.). The related SPARQL query (see more details in section 3) is given below:

```
PREFIX sln:
<http://www.semanticweb.org/ontologies/2014/SLN_Ontology#>
   SELECT DISTINCT ?Fertilizer ?ApplicationMethod
?TimeOfApplication ?Quantity ?Unit
   WHERE {
   { ?Fertilizer sln:isFertilizerOf    ?s   . }
   { ?s sln:hasFELocation       sln:DryZone  . }
   { ?s sln:isFertilizerEventOf     sln:Banana   . }
   { ?s sln:hasFertilizerQuantity   ?Quantity  . }
   OPTIONAL{ ?s sln:hasFEApplicationMethod
?ApplicationMethod . }
   OPTIONAL{ ?s sln:hasFETimeOfApplication
?TimeOfApplication . }
   OPTIONAL{ ?s sln:hasFEWaterSource     ?WaterSource . }
   OPTIONAL{ ?s sln:hasFertilizerSpInfor     ?SpecialInformation
. }
   OPTIONAL{ ?s sln:hasMaxSoilPh     ?MaxSoilPh . }
   OPTIONAL{ ?s sln:hasMinSoilPh     ?MinSoilPh . }
   OPTIONAL{ ?s sln:hasFertilizerUnit     ?Unit . }
   }
   LIMIT 250
```

The answer of the above query is as follows:

| Fertilizer | ApplicationMethod | TimeOfApplication | Quantity | Unit |
|---|---|---|---|---|
| http://www.semanticweb.org/ontologies/2014/SLN_Ontology#Urea | http://www.semanticweb.org/ontologies/2014/SLN_Ontology#Top_dressing_1 | http://www.semanticweb.org/ontologies/2014/SLN_Ontology#Apply_after_2_months_planting | http://www.semanticweb.org/ontologies/2014/SLN_Ontology#120 | http://www.semanticweb.org/ontologies/2014/SLN_Ontology#Kilograms_per_hectare |
| http://www.semanticweb.org/ontologies/2014/SLN_Ontology#Triple_Super_Phosphate(TSP) | http://www.semanticweb.org/ontologies/2014/SLN_Ontology#Top_dressing_1 | http://www.semanticweb.org/ontologies/2014/SLN_Ontology#Apply_after_2_months_planting | http://www.semanticweb.org/ontologies/2014/SLN_Ontology#80 | http://www.semanticweb.org/ontologies/2014/SLN_Ontology#Kilograms_per_hectare |
| http://www.semanticweb.org/ontologies/2014/SLN_Ontology#Muriate_of_Potash(MOP) | http://www.semanticweb.org/ontologies/2014/SLN_Ontology#Top_dressing_1 | http://www.semanticweb.org/ontologies/2014/SLN_Ontology#Apply_after_2_months_planting | http://www.semanticweb.org/ontologies/2014/SLN_Ontology#250 | http://www.semanticweb.org/ontologies/2014/SLN_Ontology#Kilograms_per_hectare |

The output explains that there are three fertilizers Urea, Triple Super Phosphate (TSP), and Muriate of Potash (MOP) and related quantities are 120, 80, and 250 respectively. It gives additional information such as this is used for Top dressing 1 and need to apply after two months of planting and the unit of the fertilizer quantity is Kilograms per hectare, etc.

Next we have extended the ontology structure for the growing problems and their control methods. By analyzing collected information from reliable sources [9] [10] [16] [17], we have identified that there are different types of growing problems especially for the vegetable cultivation. Mostly, crop diseases are caused by microorganisms such as fungi, bacteria, viruses, and nematodes. The prevention and termination methods are depending on the crops which farmers grow, the pests or diseases they are susceptible to as they affect crops differently, and environment and their location. There are different types of control methods available to reduce the amount of disease and pest to an acceptable level or eliminate. Disease control strategies need to be combined with methods for weed, insect and other production concerns (e.g. control weeds can use as a disease control method). Diseases can be prevented by utilizing proper cultural practices such as disease resistant cultivars or variety selections, irrigation and humidity management, plant and soil nutrition, pruning, row spacing, field sanitation and crop rotation, selecting suitable planting dates and rates, and burying certified seeds. Otherwise farmers can apply appropriate chemical pesticides such as fungicides (to control fungi), insecticides (to control pest) and weedicides (to control weed) which are available in the market to control pest and diseases. Farmers can also apply a suitable biological control method to prevent or control the attack, for an example Cytobagus salviniae to control Salvinia (Salvinia molesta). Normally, several methods are adapted at the same time in same place. Thus there are multiple factors to consider when selecting a suitable control method.

Based on the above findings we have defined a control method selection criterion (see Table 2) to deliver the information and knowledge related to third stage of the farming life cycle.

**Table 2.** Criteria for Control Method Selection

| Criteria (Factors) |
| --- |
| Environment:<br>• Soil<br>• Location<br>• Water |
| Farming Stage:<br>• Application Stage<br>  o Before Infestation (Avoid and Prevention)<br>  o After Infestation (Control) |
| Farmer Preferences:<br>• Control Method Types |

Based on the criteria we defined the contextualized or personalized information list given in Table 3.

**Table 3.** Farmers' Information Needs in Context for Third Stage of the Farming Life Cycle

| Farmers' Information Needs (i.e. Motivation Scenario) | Farmers' Information Needs in Context (i.e. Competency Questions) | Generalizing Contextualized Information |
|---|---|---|
| Which are the most suitable control methods to a particular disease? | *Suitable control methods based on the Environment:* What are the suitable control methods to control weed for Radish which is grown in Up Country? | What are the suitable control methods for different types of growing problems to specified Crop which are grown in specified Location? |
| | *Suitable control methods based on Preferences of Farmers:* What are the suitable cultural control methods to control Bacterial wilt for Brinjal? What are the suitable chemical control methods and in what quantities to control Damping-off for Tomato? | What are the different types of control methods to specified growing problem of a Crop? |
| | *Suitable control methods based on the Farming Stages:* What are the suitable control methods to control Bacterial wilt for Brinjal before infestation of the disease? | What is the suitable control method based on the specified farming stages to specified growing problem of a Crop? |

According to the data analysis, we have seen that for example Tomato and Brinjal face same diseases such as Bacterial Wilt and Damping-off. However, their symptoms and recommended control methods are different based on the crop (i.e. same diseases but different symptoms and control methods, sometimes it may be the same). Therefore a disease control method depends on the disease related to the specific crop. In addition to that we identified that crops have same disease problem (e.g. Damping-off) but has different causal agents (e.g. Pythium spp, Phytopthora spp, Rhizoctonia spp), different symptoms, and also it occurs in different stages (e.g. Nursery stage, Early stage, Mature Stage, or Any stage). Based on the causal agent, different control methods have been recommended. We have to define clearly the Growing Problem of a crop based on symptom, causal agent, and problem stage. Thus we have to define an event as *Growing Problem Event* to model the real world representation of the growing problem of a crop (see Fig. 3).

Next we need to identify the suitable control methods with respective to the events (i.e. Growing Problem Events). When selecting the control methods, it depends on many factors specially types of the control methods (e.g. cultural, chemical, or bio), soil type, location, time of application, and application stage. If the control method is chemical it involves the quantity (i.e. the amount of the fungicide, pesticide, or insecticide) then we need to give exact amount of it. We also need to provide its unit, application method, special information, etc. This is also complex real situation and need to represent all the information by describing their relationships. We therefore defined an event as *Control Method Event* to handle this complex situation. The Fig. 3 represents these events.

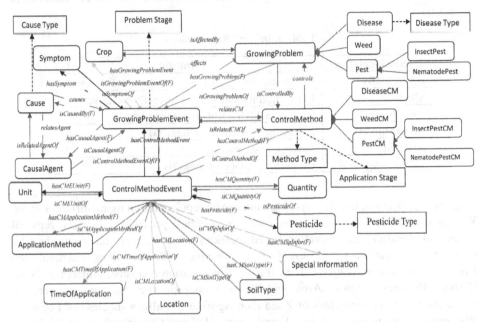

**Fig. 3.** Representation of the GrowingProblemEvent and the ControlMethodEvent

We defined the concepts such as GrowingProblem, Symptom, Cause, CausalAgent, ControlMethod, SoilType, Quantity, and Pesticide. Other concepts shown in Fig. 3 are same concepts defined in FertilizerEvent (Fig. 2) such as Unit, ApplicationMethod, TimeOfApplication, Location, etc. The GrowingProblem concept can be further categorized as Disease, Weed, InsectPest, and NematodePest. ControMethod can also be further categorized. We defined the data properties (if available), for example Disease has DiseaseType (e.g. Parasitic or Non-Parasitic), ControlMethod has ApplicationStage (e.g. After Infestation or Before Infestation). Then we have explicitly specified associative relationships and their inverse (if available) by maintaining the semantics (e.g. hasSymptom, hasControlMethod, hasCausalAgent, etc.). The individuals of GrowingProblemEvent or ControlMethodEvent represent a single object encapsulating all the information related to specified event. Here we also defined the cardinality restriction on the relationships. It is specified as *F* (i.e. functional property) in Fig. 3. A GrowingProblemEvent has many ControlMethodEvent (i.e. non-functional) but a ControlMethodEvent has exactly one value for the GrowingProblemEvent (i.e. functional).

Based on the existing information we have inferred the additional knowledge. For example, relation *isAffectedBy* defines as Crop *isAffectedBy* GrowingProblem and its inverse as GrowingProblem *affects* Crop. This has been implemented by using object property chain in OWL 2. These relations specify as a combination of existing object properties. The Table 4 shows few inferred relationships including its inverse. Some examples related to the growing problems and control methods are included in Annex 01.

Table 4. Inferred Relationships

| Inferred Relation | Definition |
|---|---|
| isAffectedBy | hasGrowingProblemEvent o hasGrowingProblem |
| affects | isGrowingProblemOf o isGrowingProblemEventOf |
| controls | isRelatedCMOf o hasGrowingProblem |
| isControlledBy | isGrowingProblemOf o relatesCM |
| relatesAgent | causes o hasCausalAgent |
| isRelatedAgentOf | isCausalAgentOf o isCausedBy |

In our application, the farmers' location can be represented in many different ways. We have identified the location as zones, agro zones, provinces, districts, regional areas and elevation (i.e. elevation based locations). We therefore have introduced a concept as Location to represent Zone, Province, District, RegionalArea, and Elevation (i.e. generalization). Then the concept Location is a super concept of Zone, District, Province, RegionalArea, and Elevation concepts (i.e. taxonomic hierarchy). Since AgroZone is a subclass of Zone then AgroZone is also a subclass of Location. We analyzed the maps for Agro-climatic zones [18] and Climatic zones in Sri Lanka [18]. We also have collected the information about regional areas, districts, provinces, and elevation based locations. Based on the data we identified that there is a conceptual overlap between concepts. For example Southern province covers the Galle, Matara, and Hambantota districts; Matara, Monaragala, and Kurunegala districts cover the Low Country Intermediate Zone. This concept needs to handle semantically. We therefore have modeled this by defining semantic relationships among the concepts. We have defined a relation as *belongsTo* and its inverse as *isBelonged*, and then specify the transitive property (see Fig. 4). This has been implemented by using transitive property in OWL 2.

The RegionalArea belong to the Districts; the Districts belong to the Provinces as well as the Zones (it also includes the AgroZone); Zones belong to Elevation. Then we can infer (using inference engine) the following information and knowledge (see Table 5). As an example now we can find (i.e. infer) the regional areas where each crop will grow.

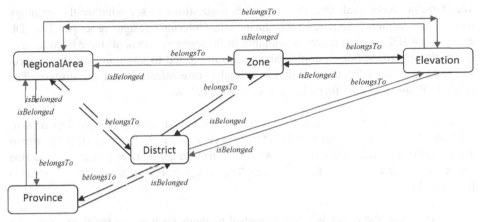

**Fig. 4.** Illustration of the Overlapping Concept

**Table 5.** Inferred Knowledge

| Transitive Property | Inferred Knowledge |
|---|---|
| **RegionalArea** belongsTo **District** belongsTo **Zone** | **RegionalArea** belongsTo **Zone** |
| **RegionalArea** belongsTo **District** belongsTo **Province** | **RegionalArea** belongsTo **Province** |
| **RegionalArea** belongsTo **Zone** belongsTo **Elevation** | **RegionalArea** belongsTo **Elevation** |
| **District** belongsTo **Zone** belongsTo **Elevation** | **District** belongsTo **Elevation** |

## 3    Ontology Evaluation

By doing comparative study of ontology languages and tools, we have selected protégé [19] as ontology development environment and Web Ontology Language (OWL) [20] as ontology language to implement our ontology. We also compared the operators and features of the conjunctive queries, Description Logics (DL), and FOL. We have identified that DL is more expressive than conjunctive queries but FOL offers a good expressive power in comparison of DL. However, DL is fully decidable fragment of the FOL [21] and also reduces the complexity when compared with FOL. In our application, decidability is very important as we need to retrieve agricultural information in context. We therefore selected the DL based (OWL 2-DL) approach to implement our ontology.

After implementing the ontology, we examine the deficiencies of the ontology in use. Through the internal evaluation we test the usefulness as well as consistency of the ontology during the design process.

To evaluate the usefulness, we need to check the ontological commitments (e.g. it is an agreement by multiple parties (e.g. persons and software systems) to adopt a particular ontology when communicating about the domain of interest, even though they do not have same experiences, theories or prescriptions about that domain [22]).

We therefore have evaluated the competency questions to see whether the ontology meets the farmers' requirements during the internal design process. The DL expressions (DL query facility is available in Protégé environment) have been used to query the ontology. For example, farmers are interested to know the suitable control methods to control Bacterial Wilt for Brinjal before infestation of the disease. The related DL query can be formulated as described below.

```
ControlMethod      and      hasControlMethodApplicationStage
value "Before Infestation"^^string and isRelatedCMOf some
((GrowingProblemEvent      and      hasGrowingProblem      value
Bacterial_Wilt)      and      (isGrowingProblemEventOf      value
Brinjal))
```

The output of the query is a list of control methods such as use resistance varieties for Bacterial wilt, deep drain to facilitate drainage, etc. by satisfying farmers' constraints (see more examples in Annex 01). By evaluating outputs of the queries we refine the ontology.

We also have tested the consistency and inferences using reasoners (we used FaCT++ and Hermit 1.3.8 reasoners plug-in with Protégé 4.2). The implicit knowledge is derived from the ontology through inference and reasoning procedures attached to the ontology.

We have developed the online knowledge base (SPARQL Endpoint: http://webe2.scem.uws.edu.au/arc2/select.php) using the Semantic Web technologies (we used ARC2-appmosphere RDF classes toolkit [23] as it is a good Semantic Web framework for PHP language which allows to easily manage structured data in RDF) to share and reuse the agricultural information and knowledge via the Web. We can retrieve and query the contextualized information and knowledge on the Web using the SPARQL queries [24]. We have evaluated the SPARQL queries as well as the inferences equivalent to competency questions through the ARC2 application. Some examples are given in Annex 01.

These two internal evaluation procedures (using protégé implementation and SPARQL Endpoint) are used to improve and refine our ontology further.

## 4     Conclusions

In this paper we have presented a way to conceptualize crop life cycle events to create a user centered ontology for farmers. Using this model we have developed an agricultural ontology for Sri Lankan farmers to address the problem of not having agricultural knowledge repositories that can be easily accessed by farmers within their own context.

In this paper, we have modeled the knowledge representation of the real-world complex scenarios by defining N-ary relations. We have handled the conceptual overlapping by using the transitive property. The online knowledge base with a SPARQL end-point was created to share and reuse the domain knowledge that can be queried based on farmer context. We also have used this SPARQL end-point as well as implemented ontology using protégé to internal ontology evaluation against the competency questions. Using this approach we have improved and refined the ontology further. Knowledge organized in this manner can better assist the decision making process.

Based on the techniques that we discovered, we developed a generalized framework for ontology design that can be used to create knowledge repositories which are capable of providing information according to user context. Using this approach we can add more concepts, relationships, and constraints with different scenarios to the ontology to provide even richer knowledge base.

We have described inclusion of crop life cycle related events related to the first three stages of the farming life cycle and our overall objective of this research project is to design and develop the ontology to cover all stages of farming life cycle and to provide agricultural information and knowledge to farmers in their context.

In future, we hope to develop an end to end ontology management system where the community can participate in keeping the Ontology current. We have already designed the architecture and hope to develop and deploy this system soon.

**Acknowledgment.** We acknowledge the financial assistance provided to carry out this research work by the HRD Program of the HETC project of the Ministry of Higher Education, Sri Lanka and the valuable assistance from other researchers working on the Social Life Network project. Assistance from the National Science Foundation to carry out the field visits is also acknowledged.

# References

1. De Silva, L.N.C., Goonetillake, J.S., Wikramanayake, G.N., Ginige, A.: Towards using ICT to Enhance Flow of Information to aid Farmer Sustainability in Sri Lanka. In: 23rd Australasian Conference on Information Systems (ACIS), Geelong, Australia, pp. 1–10 (2012)
2. Lokanathan, S., Kapugama, N.: Smallholders and Micro-enterprises in Agriculture: Information Needs and Communication Patterns. LIRNEasia, Colombo, Sri Lanka (2012)
3. Glendenning, C.J., Babu, S., Asenso-Okyere, K.: Review of Agriculture Extension in India Are Farmers' Information Needs Being Met? International Food Policy Research Institute (2010)
4. Ginige, A.: Social Life Networks for the Middle of the Pyramid, http://www.sln4mop.org//index.php/sln/articles/index/1/3
5. Walisadeera, A.I., Wikramanayake, G.N., Ginige, A.: An Ontological Approach to Meet Information Needs of Farmers in Sri Lanka. In: Murgante, B., Misra, S., Carlini, M., Torre, C.M., Nguyen, H.-Q., Taniar, D., Apduhan, B.O., Gervasi, O. (eds.) ICCSA 2013, Part I. LNCS, vol. 7971, pp. 228–240. Springer, Heidelberg (2013)
6. Gruber, T.R.: Toward Principles for the Design of Ontologies Used for Knowledge Sharing. International Journal of Human-Computer Studies 43, 907–928 (1995)
7. Gruber, T.R.: A Translation Approach to Portable Ontology Specifications. Technical Report KSL 92-71, Knowledge System Laboratory, Stanford University, California (1993)

8. Walisadeera, A.I., Wikramanayake, G.N., Ginige, A.: Designing a Farmer Centred Ontology for Social Life Network. In: 2nd International Conference on Data Management Technologies and Applications (DATA 2013), Reykjavík, Iceland (2013)
9. Department of Agriculture, Sri Lanka, http://www.agridept.gov.lk
10. NAVAGOVIYA, CIC (Private Sector), http://www.navagoviya.org
11. Fertilizer use by crop in India. Published by Food and Agricultural Organization of United Nations (FAO), Rome (2005)
12. Dahchour, M., Pirotte, A.: The Semantics of Reifying n-ary Relationships as Classes. In: 4th International Conference on Enterprise Information System (ICEIS 2002), Ciudad Real, Spain, pp. 580–586 (2002)
13. Noy, N., Rector, A.: Defining N-ary Relations on the Semantic Web, http://www.w3.org/TR/swbp-n-aryRelations
14. N-Ary Relations, http://www.neon-project.org
15. OWL Working Group (W3C): OWL 2 Web Ontology Language Primer, 2nd edn., http://www.w3.org/TR/owl2-primer/
16. Organic Crop Production: Disease Management, http://www.agriculture.gov.sk.ca/ Default.aspx?DN=e12b17d3-8563-4446-afe5-3cc44505789f
17. Products for Disease Control - Crops, Gardens & Plants, http://www.arbico-organics.com/category/natural-organic-plant-disease-control
18. Punyawardena, B.V.R.: Technical report on the characterization of the agro-ecological context in which Farm Animal Genetic Resources (FAnGR) are found: Sri Lanka, http://www.fangrasia.org/admin/admin_content/files/60202.pdf
19. Knublauch, H., Fergerson, R.W., Noy, N.F., Musen, M.A.: The Protégé OWL Plugin: An Open Development Environment for Semantic Web Applications. In: McIlraith, S.A., Plexousakis, D., van Harmelen, F. (eds.) ISWC 2004. LNCS, vol. 3298, pp. 229–243. Springer, Heidelberg (2004)
20. Patel-Schneider, P.F., Hayes, P., Horrocks, I.: OWL Web Ontology Language Semantics and Abstract Syntax. W3C Recommendation, http://www.w3.org/TR/owl-semantics
21. Baader, F., Calvanese, D., McGuinness, D.L., Nardi, D., Patel-Schneider, P.F. (eds.): The Description Logics Handbook – Theory and Applications, 2nd edn. Cambridge University Press (2008)
22. Holsapple, C.W., Joshi, K.D.: A Collaborative Approach to Ontology Design. Communications of the ACM – Ontology 44(2), 42–47 (2002)
23. Oldakowski, R., Bizer, C., Westphal, D.: RAP: RDF API for PHP. In: Proceedings of the 1st Workshop on Scripting for the Semantic Web, 2nd European Semantic Web Conference, ESWC 2005 (2005)
24. Prud'hommeaux, E., Seaborne, A.: SPARQL Query Language for RDF. W3C Recommendation, http://www.w3.org/TR/rdf-sparql-query

# Annex 01: Internal Evaluation Procedure

| Competency Questions | Tested DL Query | Answer of the DL Query | Equivalent SPARQL Query | Answer of the SPARQL Query |
|---|---|---|---|---|
| What are the suitable cultural control methods to control Bacterial Wilt for Tomato? | ControlMeth od and hasControlMeth odType value "Cultural"^^stri ng and isRelatedCMOf some ((GrowingProbl emEvent and hasGrowingPro blem value Bacterial_Wilt) and (isGrowingProb lemEventOf value Tomato)) | Instances (2) ❖ Use resistance_ varieties_for Bacterial_wilt ❖ Crop_rotation_with_non_solanace os_crops | PREFIX sln: <http://www.semanticweb.org/o ntologies/2014/SLN_Ontology#> SELECT DISTINCT ?ControlMethods WHERE { { ?ControlMethods sln:hasControlMethodType "Cultural"^^xsd:string .} { ?ControlMethods sln:isRelatedCMOf ?o .} { ?o sln:hasGrowingProblem sln:Bacterial_Wilt . } { ?o sln:isGrowingProblemEventOf sln:Tomato . } } LIMIT 250 | ControlMethods http://www.semanticweb.org/ontologies/2014/SLN_Ontolo gy#Crop_rotation_with_non_solanaceos_crops http://www.semanticweb.org/ontologies/2014/SLN_Ontolo gy#Use_resistance_varieties_for_Bacterial_wilt |
| What are the suitable fertilizers used by Banana? | Fertilizer and isUsedBy value Banana | Instances (5) ❖ Urea ❖ 'Muriate_of_Potash(MOP)' ❖ 'Triple_Super_Phosphate(TSP)' ❖ Rock_phosphate ❖ Compost | PREFIX sln: <http://www.semanticweb.org/o ntologies/2014/SLN_Ontology#> SELECT ?FertilizerName WHERE { { ?FertilizerName sln:isUsedBy sln:Banana . } } LIMIT 250 | FertilizerName http://www.semanticweb.org/ontologies/2014/SLN_Ontol ogy#Compost http://www.semanticweb.org/ontologies/2014/SLN_Ontol ogy#Muriate_of_Potash(MOP) http://www.semanticweb.org/ontologies/2014/SLN_Ontol ogy#Rock_phosphate http://www.semanticweb.org/ontologies/2014/SLN_Ontol ogy#Triple_Super_Phosphate(TSP) http://www.semanticweb.org/ontologies/2014/SLN_Ontol ogy#Urea |
| What are the organic fertilizers for Basal dressing for Banana? | Fertilizer and hasFertilizerTy pe value "Organic"^^stri ng and isFertilizerOf some (FertilizerEvent and isFertilizerEven tOf value Banana and | Instances (1) ❖ Compost | PREFIX sln: <http://www.semanticweb.org/o ntologies/2014/SLN_Ontology#> SELECT DISTINCT ?FertilizerName WHERE { { ?FertilizerName sln:hasFertilizerType "Organic"^^xsd:string . } { ?FertilizerName sln:isFertilizerO ?s . } { ?s | FertilizerName http://www.semanticweb.org/ontologies/2014/S LN_Ontology#Compost |

| Question | Description | Instances | SPARQL Query | Problems / Symptoms |
|---|---|---|---|---|
| | hasFEApplicati onMethod value Basal_dressing) | | sln:hasFEApplicationMethod sln:Basal_dressing .} { ?s sln:isFertilizerEventOf sln:Banana . } LIMIT 250 | **Problems** |
| What are the growing problems of Brinjal? | GrowingPro blem and affects value Brinjal | Instance (8) ❖ 'Shoor_and_Fruit_Borer_(SFB)' ❖ Mites ❖ Anthacnose ❖ Hoppers ❖ Foor_Rot ❖ Thrips ❖ Bacterial_Wilt ❖ Damping-off | PREFIX sln: <http://www.semanticweb.org/o ntologies/2014/SLN_Ontology# > SELECT ?Problems WHERE { { ?Problems sln:affects sln:Brinjal . } } LIMIT 250 | http://www.semanticweb.org/ontologies/2014/SLN_Ontolog y#Anthacnose<br>http://www.semanticweb.org/ontologies/2014/SLN_Ontolog y#Bacterial_Wilt<br>http://www.semanticweb.org/ontologies/2014/SLN_Ontolog y#Damping-off<br>http://www.semanticweb.org/ontologies/2014/SLN_Ontolog y#Foot_Rot<br>http://www.semanticweb.org/ontologies/2014/SLN_Ontolog y#Hoppers<br>http://www.semanticweb.org/ontologies/2014/SLN_Ontolog y#Mites<br>http://www.semanticweb.org/ontologies/2014/SLN_Ontolog y#Shoot_and_Fruit_Borer_(SFB)<br>http://www.semanticweb.org/ontologies/2014/SLN_Ontolog y#Thrips |
| | | | | **Symptoms** |
| What are the symptoms of Damping-off for Tomato? | Symptom and hasGPOfSympt om value Damping-off and isSymptomOf some (GrowingProble mEvent and isGrowingProbl emEventOf value Tomato) | Instances (6) ❖ Causes_root_rot ❖ Poor_germination_may_occur_du e_to_pre-emergence_damping_left ❖ Affected_seedlings_collapse_at_th e_base_of_stem_and_death_of_se edling ❖ Cause_rot_of_soft_tissue ❖ Causes_blackening_of_the_stem base_and_on_root_resulting_in_dr y_root ❖ Cause_brownish_black_discolorat ion_at_the_base_of_the_stem_and _death_at_seeding | PREFIX sln: <http://www.semanticweb.org/o ntologies/2014/SLN_Ontology# > SELECT ?Symptoms WHERE { { ?Symptoms sln:hasGPOfSymptom sln:Damping-off . } { ?Symptoms sln:CropOfSymptom sln:Tomato . } } LIMIT 250 | http://www.semanticweb.org/ontologies/2014/SLN_Ontology# Affected_seedlings_collapse_at_the_base_of_stem_and_death_of _seedling<br>http://www.semanticweb.org/ontologies/2014/SLN_Ontology# Cause_brownish_black_discoloration_at_the_base_of_the_stem_a nd_death_at_seeding<br>http://www.semanticweb.org/ontologies/2014/SLN_Ontology# Cause_rot_of_soft_tissue<br>http://www.semanticweb.org/ontologies/2014/SLN_Ontology# Causes_blackening_of_the_stem_base_and_on_root_resulting_in_ dry_root<br>http://www.semanticweb.org/ontologies/2014/SLN_Ontology# Causes_root_rot<br>http://www.semanticweb.org/ontologies/2014/SLN_Ontology# Poor_germination_may_occur_due_to_pre-emergence_damping_left |

# Erratum: Explicit Untainting to Reduce Shadow Memory Usage in Dynamic Taint Analysis

Young Hyun Choi[1], Min-Woo Park[1],
Jung-Ho Eom[2], and Tai-Myoung Chung[1]

[1] Dept. of Electrical and Computer Engineering, Sungkyunkwan University
300 Cheoncheon-dong, Jangan-gu, Suwon-si, Gyeonggi-do, 440-746, Republic of Korea
`{yhchoi,mwpark,tmchung}@imtl.skku.ac.kr`
[2] Military Studies, Daejeon University
62 Daehakro, Dong-gu, Daejeon, Republic of Korea
`eomhun@gmail.com`

B. Murgante et al. (Eds.): ICCSA 2014, Part V, LNCS 8583, pp. 175–184, 2014.
© Springer International Publishing Switzerland 2014

**DOI 10.1007/978-3-319-09156-3_56**

The paper "Explicit Untainting to Reduce Shadow Memory Usage in Dynamic Taint Analysis" authored by Young Hyun Choi, Min-Woo Park, Junh-Ho Eom, and Tai-Myoung Chung, DOI 10.1007/978-3-319-09156-3_13, appearing on pages 175-184 of this publication has been retracted by mutual consent of the volume editors and the authors. It contains a considerable amount of incorrect data.

The original online version for this chapter can be found at
http://dx.doi.org/10.1007/978-3-319-09156-3_13

# Author Index

Printed in the United States
By Bookmasters